Student's Solutions Manual
Part I

to accompany

Thomas' Calculus
Tenth Edition

Student's Solutions Manual
Part I

John L. Scharf
Carroll College

Maurice D. Weir
Naval Postgraduate School

to accompany

Thomas' Calculus

Tenth Edition

Based on the original work by

George B. Thomas, Jr.

Massachusetts Institute of Technology

As revised by

Ross L. Finney
Maurice D. Weir
and
Frank R. Giordano

Boston San Francisco New York
London Toronto Sydney Tokyo Singapore Madrid
Mexico City Munich Paris Cape Town Hong Kong Montreal

Reproduced by Addison Wesley Longman from camera-ready copy supplied by the authors.

ISBN 0-201-50381-6

1 2 3 4 5 6 7 8 9 10 VG 03 02 01 00

PREFACE TO THE STUDENT

This Student's Solutions Manual contains the solutions to all of the odd-numbered exercises in the 10th Edition of Thomas' CALCULUS as reviewed by Ross L. Finney, Maurice D. Weir and Frank R. Giordano, excluding the Computer Algebra System (CAS) exercises. We have worked each solution to ensure that it

- conforms exactly to the methods, procedures and steps presented in the text
- is mathematically correct
- includes all of the steps necessary so you can follow the logical argument and algebra
- includes a graph or figure whenever called for by the exercise, or if needed to help with the explanation
- is formatted in an appropriate style to aid you in its understanding

How to use a solution's manual

- solve the assigned problem yourself
- if you get stuck along the way, refer to the solution in the manual as an aid but continue to solve the problem on your own
- if you cannot continue, reread the textbook section, or work through that section in the Student Study Guide, or consult your instructor
- after solving the problem (if odd-numbered), carefully compare your solution procedure to the one in the manual
- if your answer is correct but your solution procedure seems to differ from the one in the manual, and you are unsure your method is correct, consult your instructor
- if your answer is incorrect and you cannot find your error, consult your instructor

TABLE OF CONTENTS

PRELIMINARY CHAPTER

P.1 LINES

1. (a) $\Delta x = -1 - 1 = -2$
$\Delta y = -1 - 2 = -3$

(b) $\Delta x = -1 - (-3) = 2$
$\Delta y = -2 - 2 = -4$

3. (a)

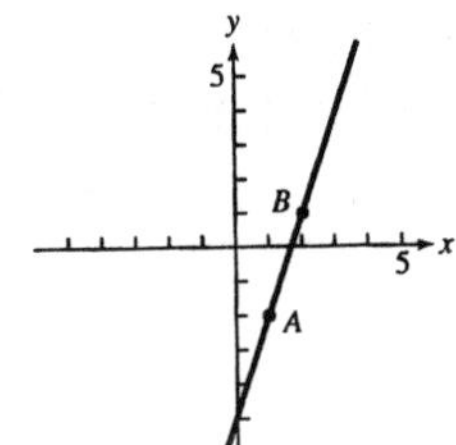

$$m = \frac{1-(-2)}{2-1} = \frac{3}{1} = 3$$

(b)

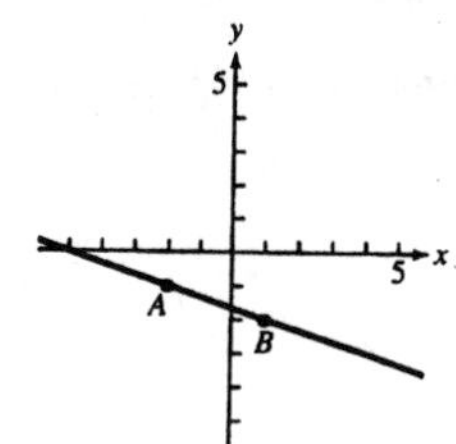

$$m = \frac{-2-(-1)}{1-(-2)} = \frac{-1}{3} = -\frac{1}{3}$$

5. (a) $x = 2$, $y = 3$

(b) $x = -1$, $y = \frac{4}{3}$

7. (a) $y = 1(x-1) + 1$

(b) $y = -1[x-(-1)] + 1 = -1(x+1) + 1$

9. (a) $m = \frac{3-0}{2-0} = \frac{3}{2}$

$y = \frac{3}{2}(x-0) + 0$

$y = \frac{3}{2}x$

$2y = 3x$

$3x - 2y = 0$

(b) $m = \frac{1-1}{2-1} = \frac{0}{1} = 0$

$y = 0(x-1) + 1$

$y = 1$

11. (a) $y = 3x - 2$

(b) $y = -1x + 2$ or $y = -x + 2$

13. The line contains $(0,0)$ and $(10,25)$.

$m = \frac{25-0}{10-0} = \frac{25}{10} = \frac{5}{2}$

$y = \frac{5}{2}x$

15. (a) $3x + 4y = 12$

$4y = -3x + 12$

$y = -\frac{3}{4}x + 3$

i) Slope: $-\frac{3}{4}$

ii) y-intercept: 3

(b) $x + y = 2$

$y = -x + 2$

i) Slope: -1

ii) y-intercept: 2

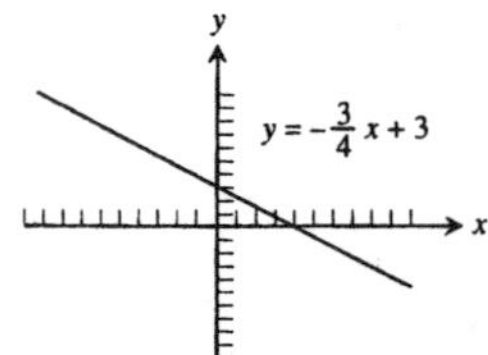

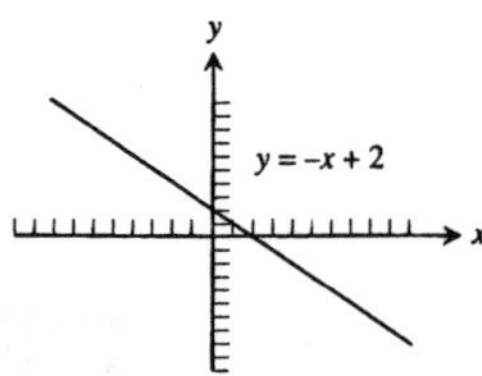

17. (a) i) The desired line has slope -1 and passes through $(0,0)$: $y = -1(x-0)+0$ or $y = -x$.

ii) The desired line has slope $\frac{-1}{-1} = 1$ and passes through $(0,0)$: $y = 1(x-0)+0$ or $y = x$.

(b) i) The given equation is equivalent to $y = -2x+4$. The desired line has slope -2 and passes through $(-2,2)$: $y = -2(x+2)+2$ or $y = -2x-2$.

ii) The desired line has slope $\frac{-1}{-2} = \frac{1}{2}$ and passes through $(-2,2)$: $y = \frac{1}{2}(x+2)+2$ or $y = \frac{1}{2}x+3$.

19. $m = \frac{9-2}{3-1} = \frac{7}{2}$

$f(x) = \frac{7}{2}(x-1)+2 = \frac{7}{2}x - \frac{3}{2}$

Check: $f(5) = \frac{7}{2}(5) - \frac{3}{2} = 16$, as expected.

Since $f(x) = \frac{7}{2}x - \frac{3}{2}$, we have $m = \frac{7}{2}$ and $b = -\frac{3}{2}$.

21.
$$-\frac{2}{3} = \frac{y-3}{4-(-2)}$$
$$-\frac{2}{3}(6) = y-3$$
$$-4 = y-3$$
$$-1 = y$$

23. $y = 1\cdot(x-3)+4$
$y = x-3+4$
$y = x+1$
This is the same as the equation obtained in Example 5.

25. (a) The given equations are equivalent to $y = -\frac{2}{k}x + \frac{3}{k}$ and $y = -x+1$, respectively, so the slopes are $-\frac{2}{k}$ and -1. The lines are parallel when $-\frac{2}{k} = -1$, so $k = 2$.

(b) The lines are perpendicular when $-\frac{2}{k} = \frac{-1}{-1}$, so $k = -2$.

27. Slope: $k = \frac{\Delta p}{\Delta d} = \frac{10.94-1}{100-0} = \frac{9.94}{100} = 0.0994$ atmospheres per meter
At 50 meters, the pressure is $p = 0.0994(50)+1 = 5.97$ atmospheres.

29. (a) Suppose x°F is the same as x°C.

$$x = \frac{9}{5}x + 32$$

$$\left(1 - \frac{9}{5}\right)x = 32$$

$$-\frac{4}{5}x = 32$$

$$x = -40$$

Yes, $-40°$F is the same as $-40°$C.

(b)

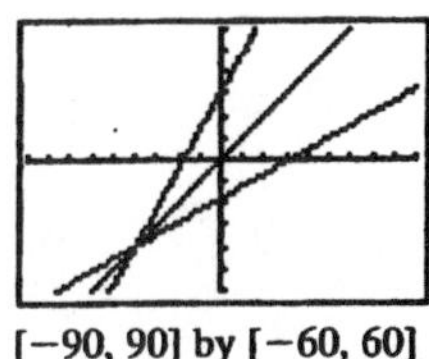

[−90, 90] by [−60, 60]

It is related because all three lines pass through the point $(-40, -40)$ where the Fahrenheit and Celsius temperatures are the same.

31.

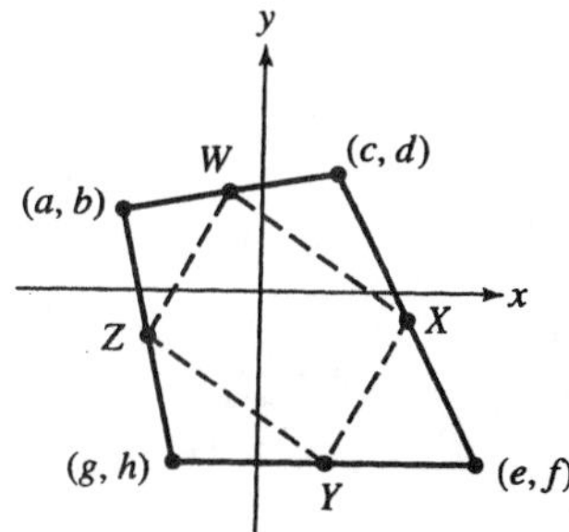

Suppose that the vertices of the given quadrilateral are (a, b), (c, d), (e, f), and (g, h). Then the midpoints of the consecutive sides are $W\left(\frac{a+c}{2}, \frac{b+d}{2}\right)$, $X\left(\frac{c+e}{2}, \frac{d+f}{2}\right)$, $Y\left(\frac{e+g}{2}, \frac{f+h}{2}\right)$, and $Z\left(\frac{g+a}{2}, \frac{h+b}{2}\right)$. When these four points are connected, the slopes of the sides of the resulting figure are:

$$\text{WX:} \quad \frac{\frac{d+f}{2} - \frac{b+d}{2}}{\frac{c+e}{2} - \frac{a+c}{2}} = \frac{f-b}{e-a}$$

$$\text{XY:} \quad \frac{\frac{f+h}{2} - \frac{d+f}{2}}{\frac{e+g}{2} - \frac{c+e}{2}} = \frac{h-d}{g-c}$$

$$\text{ZY:} \quad \frac{\frac{f+h}{2} - \frac{h+b}{2}}{\frac{e+g}{2} - \frac{g+a}{2}} = \frac{f-b}{e-a}$$

$$\text{WZ:} \quad \frac{\frac{h+b}{2} - \frac{b+d}{2}}{\frac{g+a}{2} - \frac{a+c}{2}} = \frac{h-d}{g-c}$$

Opposite sides have the same slope and are parallel.

33. (a) The equation for line L can be written as

$y = -\frac{A}{B}x + \frac{C}{B}$, so its slope is $-\frac{A}{B}$. The perpendicular line has slope $\frac{-1}{-A/B} = \frac{B}{A}$ and passes through (a, b),

so its equation is $y = \frac{B}{A}(x - a) + b$.

(b) Substituting $\frac{B}{A}(x - a) + b$ for y in the equation for line L gives:

$$Ax + B\left[\frac{B}{A}(x-a) + b\right] = C$$

$$A^2x + B^2(x - a) + ABb = AC$$

$$(A^2 + B^2)x = B^2a + AC - ABb$$

$$x = \frac{B^2a + AC - ABb}{A^2 + B^2}$$

Substituting the expression for x in the equation for line L gives:

$$A\left(\frac{B^2a + AC - ABb}{A^2 + B^2}\right) + By = C$$

$$By = \frac{-A(B^2a + AC - ABb)}{A^2 + B^2} + \frac{C(A^2 + B^2)}{A^2 + B^2}$$

$$By = \frac{-AB^2a - A^2C + A^2Bb + A^2C + B^2C}{A^2 + B^2}$$

$$By = \frac{A^2Bb + B^2C - AB^2a}{A^2 + B^2}$$

$$y = \frac{A^2b + BC - ABa}{A^2 + B^2}$$

The coordinates of Q are $\left(\frac{B^2a + AC - ABb}{A^2 + B^2}, \frac{A^2b + BC - ABa}{A^2 + B^2}\right)$.

(c) Distance $= \sqrt{(x-a)^2 + (y-b)^2}$

$$= \sqrt{\left(\frac{B^2a + AC - ABb}{A^2 + B^2} - a\right)^2 + \left(\frac{A^2b + BC - ABa}{A^2 + B^2} - b\right)^2}$$

$$= \sqrt{\left(\frac{B^2a + AC - ABb - a(A^2 + B^2)}{A^2 + B^2}\right)^2 + \left(\frac{A^2b + BC - ABa - b(A^2 + B^2)}{A^2 + B^2}\right)^2}$$

$$= \sqrt{\left(\frac{AC - ABb - A^2a}{A^2 + B^2}\right)^2 + \left(\frac{BC - ABa - B^2b}{A^2 + B^2}\right)^2}$$

$$= \sqrt{\left(\frac{A(C - Bb - Aa)}{A^2 + B^2}\right)^2 + \left(\frac{B(C - Aa - Bb)}{A^2 + B^2}\right)^2}$$

$$= \sqrt{\frac{A^2(C - Aa - Bb)^2}{(A^2 + B^2)^2} + \frac{B^2(C - Aa - Bb)^2}{(A^2 + B^2)^2}}$$

$$= \sqrt{\frac{(A^2+B^2)(C-Aa-Bb)^2}{(A^2+B^2)^2}}$$

$$= \sqrt{\frac{(C-Aa-Bb)^2}{A^2+B^2}}$$

$$= \frac{|C-Aa-Bb|}{\sqrt{A^2+B^2}}$$

$$= \frac{|Aa+Bb-C|}{\sqrt{A^2+B^2}}$$

35. $m = \frac{37.1}{100} = \frac{14}{\Delta x} \Rightarrow \Delta x = \frac{14}{.371}$. Therefore, distance between first and last rows is $\sqrt{(14)^2 + \left(\frac{14}{.371}\right)^2} \approx 40.25$ ft.

37. (a) $y = 0.680x + 9.013$
    (b) The slope is 0.68. It represents the approximate average weight gain in pounds per month.
    (c)

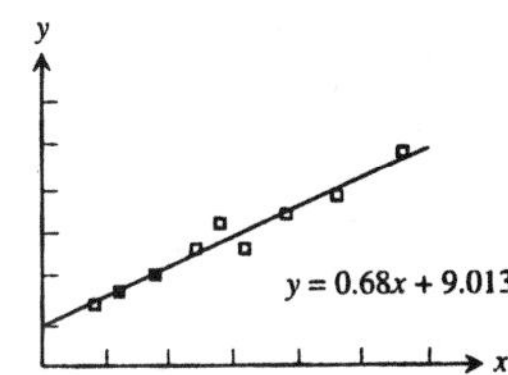

    (d) When $x = 30$, $y \approx 0.680(30) + 9.013 = 29.413$.
    She weighs about 29 pounds.

39. (a) $y = 5632x - 11{,}080{,}280$
    (b) The rate at which the median price is increasing in dollars per year
    (c) $y = 2732x - 5{,}362{,}360$
    (d) The median price is increasing at a rate of about \$5632 per year in the Northeast, and about \$2732 per year in the Midwest. It is increasing more rapidly in the Northeast.

P.2 FUNCTIONS AND GRAPHS

1. base $= x$; $(\text{height})^2 + \left(\frac{x}{2}\right)^2 = x^2 \Rightarrow \text{height} = \frac{\sqrt{3}}{2}x$; area is $a(x) = \frac{1}{2}(\text{base})(\text{height}) = \frac{1}{2}(x)\left(\frac{\sqrt{3}x}{2}\right) = \frac{\sqrt{3}}{4}x^2$; perimeter is $p(x) = x + x + x = 3x$.

3. Let D = diagonal of a face of the cube and ℓ = the length of an edge. Then $\ell^2 + D^2 = d^2$ and (by Exercise 2) $D^2 = 2\ell^2 \Rightarrow 3\ell^2 = d^2 \Rightarrow \ell = \frac{d}{\sqrt{3}}$. The surface area is $6\ell^2 = \frac{6d^2}{3} = 2d^2$ and the volume is $\ell^3 = \left(\frac{d}{\sqrt{3}}\right)^{3/2}$ $= \frac{d^3}{3\sqrt{3}}$.

5. (a) Not the graph of a function of x since it fails the vertical line test.
   (b) Is the graph of a function of x since any vertical line intersects the graph at most once.

7. (a) domain $= (-\infty, \infty)$; range $= [1, \infty)$
   (b) domain $= [0, \infty)$; range $= (-\infty, 1]$

9. $4 - z^2 = (2 - z)(2 + z) \geq 0 \Leftrightarrow z \in [-2, 2] =$ domain. Largest value is $g(0) = \sqrt{4} = 2$ and smallest value is $g(-2) = g(2) = \sqrt{0} = 0 \Rightarrow$ range $= [0, 2]$.

11. (a) Symmetric about the origin

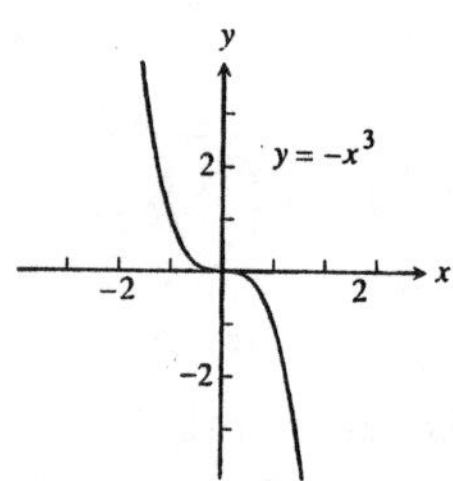

(b) Symmetric about the y-axis

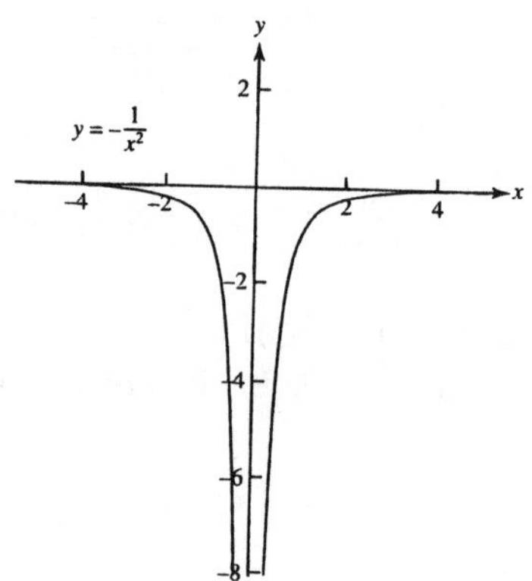

13. Neither graph passes the vertical line test
   (a)

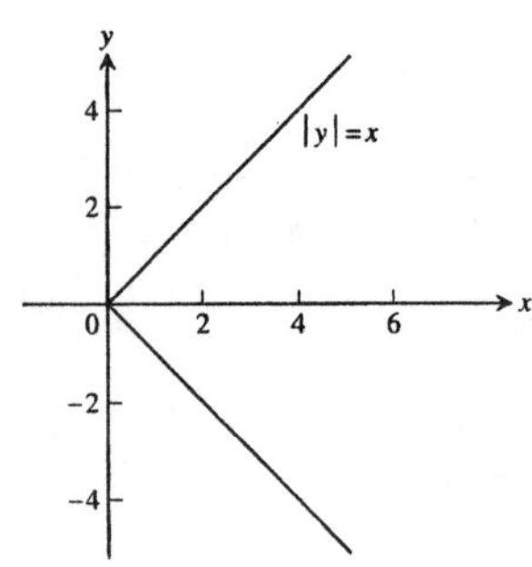

(b)

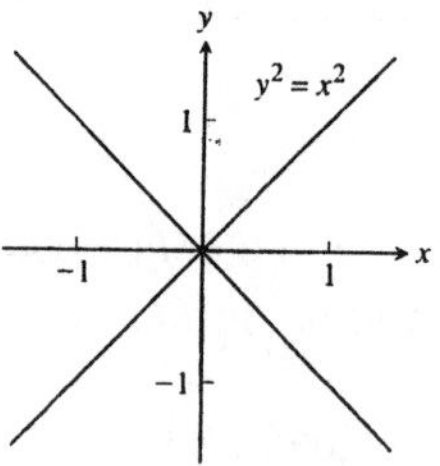

15. (a) even
   (b) odd

17. (a) odd
   (b) even

19. (a) neither
   (b) even

21. (a)

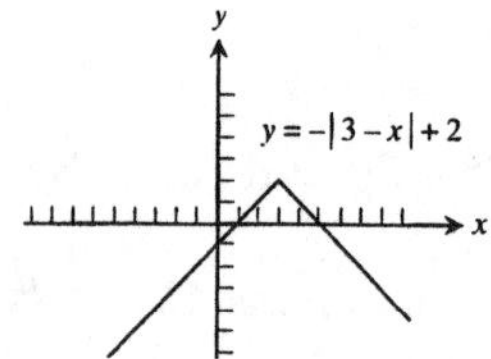

(b) The graph of f(x) is the graph of the absolute value function stretched vertically by a factor of 2 and then shifted 4 units to the left and 3 units downward

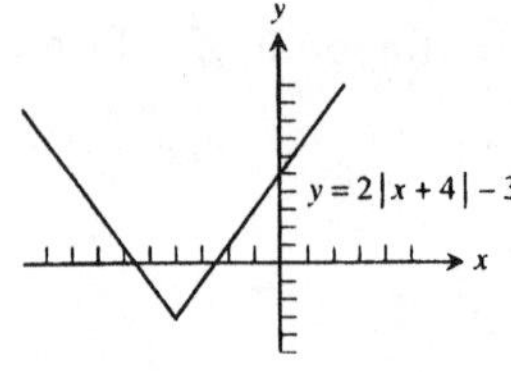

Note that $f(x) = -|x-3|+2$, so its graph is the graph of the absolute value function reflected across the x-axis and then shifted 3 units right and 2 units upward.

$(-\infty, \infty)$
$(-\infty, 2]$

$(-\infty, \infty)$ or all real numbers
$[-3, \infty)$

23. (a)

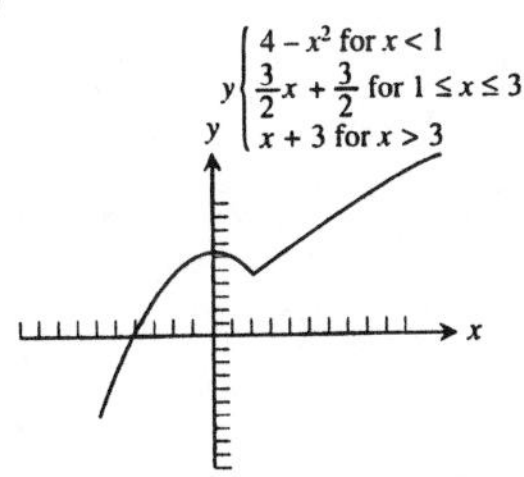

(b) $(-\infty, \infty)$ or all real numbers
(c) $(-\infty, \infty)$ or all real numbers

25. Because if the vertical line test holds, then for each x-coordinate, there is at most one y-coordinate giving a point on the curve. This y-coordinate corresponds to the value assigned to the x-coordinate. Since there is only one y-coordinate, the assignment is unique.

27. (a) Line through $(0,0)$ and $(1,1)$: $y = x$
Line through $(1,1)$ and $(2,0)$: $y = -x+2$

$$f(x) = \begin{cases} x, & 0 \le x \le 1 \\ -x+2, & 1 < x \le 2 \end{cases}$$

(b) $$f(x) = \begin{cases} 2, & 0 \le x < 1 \\ 0, & 1 \le x < 2 \\ 2, & 2 \le x < 3 \\ 0, & 3 \le x \le 4 \end{cases}$$

(c) Line through $(0,2)$ and $(2,0)$: $y = -x+2$
Line through $(2,1)$ and $(5,0)$: $m = \frac{0-1}{5-2} = \frac{-1}{3} = -\frac{1}{3}$, so $y = -\frac{1}{3}(x-2)+1 = -\frac{1}{3}x + \frac{5}{3}$

$$f(x) = \begin{cases} -x+2, & 0 < x \le 2 \\ -\frac{1}{3}x + \frac{5}{3}, & 2 < x \le 5 \end{cases}$$

(d) Line through $(-1,0)$ and $(0,-3)$: $m = \frac{-3-0}{0-(-1)} = -3$, so $y = -3x-3$

Line through $(0,3)$ and $(2,-1)$: $m = \frac{-1-3}{2-0} = \frac{-4}{2} = -2$, so $y = -2x+3$

$$f(x) = \begin{cases} -3x-3, & -1 < x \le 0 \\ -2x+3, & 0 < x \le 2 \end{cases}$$

29. (a) Position 4 (b) Position 1 (c) Position 2 (d) Position 3

31.

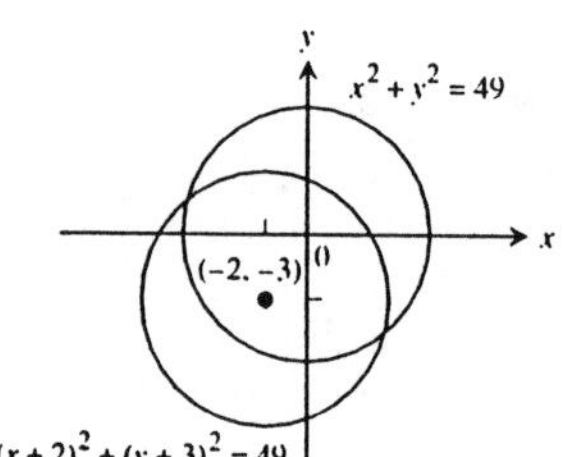

33.

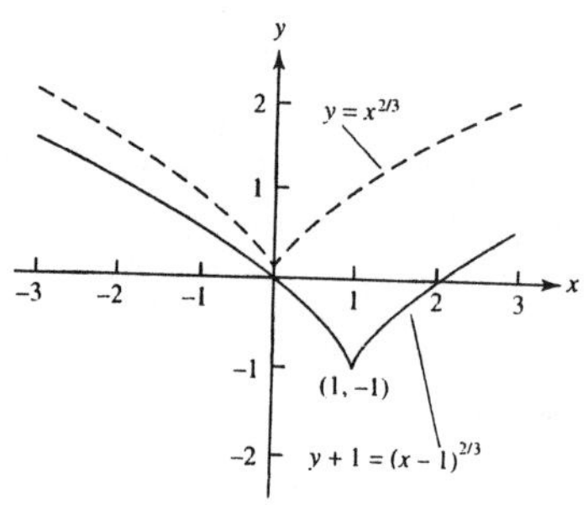

35.

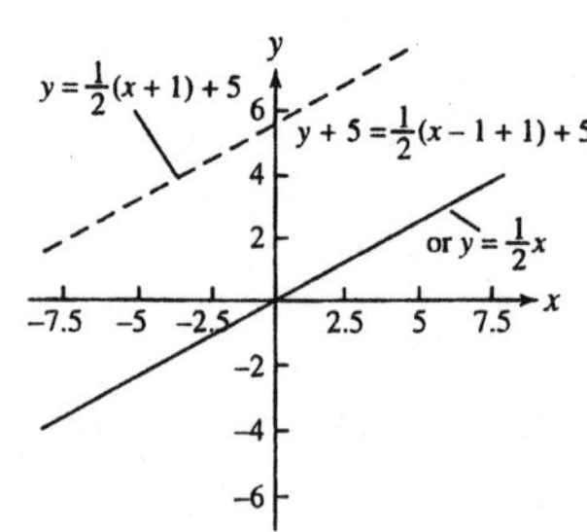

37. (a) $f(g(0)) = f(-3) = 2$

(b) $g(f(0)) = g(5) = 22$

(c) $f(g(x)) = f(x^2 - 3) = x^2 - 3 + 5 = x^2 + 2$

(d) $g(f(x)) = g(x + 5) = (x + 5)^2 - 3 = x^2 + 10x + 22$

(e) $f(f(-5)) = f(0) = 5$

(f) $g(g(2)) = g(1) = -2$

(g) $f(f(x)) = f(x + 5) = (x + 5) + 5 = x + 10$

(h) $g(g(x)) = g(x^2 - 3) = (x^2 - 3)^2 - 3 = x^4 - 6x^2 + 6$

39. (a) $u(v(f(x))) = u\left(v\left(\frac{1}{x}\right)\right) = u\left(\frac{1}{x^2}\right) = 4\left(\frac{1}{x}\right)^2 - 5 = \frac{4}{x^2} - 5$

(b) $u(f(v(x))) = u\left(f\left(x^2\right)\right) = u\left(\frac{1}{x^2}\right) = 4\left(\frac{1}{x^2}\right) - 5 = \frac{4}{x^2} - 5$

(c) $v(u(f(x))) = v\left(u\left(\frac{1}{x}\right)\right) = v\left(4\left(\frac{1}{x}\right) - 5\right) = \left(\frac{4}{x} - 5\right)^2$

(d) $v(f(u(x))) = v(f(4x - 5)) = v\left(\frac{1}{4x - 5}\right) = \left(\frac{1}{4x - 5}\right)^2$

(e) $f(u(v(x))) = f\left(u\left(x^2\right)\right) = f\left(4\left(x^2\right) - 5\right) = \frac{1}{4x^2 - 5}$

(f) $f(v(u(x))) = f(v(4x-5)) = f\left((4x-5)^2\right) = \frac{1}{(4x-5)^2}$

41. (a) $y = g(f(x))$ (b) $y = j(g(x))$
(c) $y = g(g(x))$ (d) $y = j(j(x))$
(e) $y = g(h(f(x)))$ (f) $y = h(j(f(x)))$

43. (a) Since $(f \circ g)(x) = \sqrt{g(x) - 5} = \sqrt{x^2 - 5}$, $g(x) = x^2$.

(b) Since $(f \circ g)(x) = 1 + \frac{1}{g(x)} = x$, we know that $\frac{1}{g(x)} = x - 1$, so $g(x) = \frac{1}{x-1}$.

(c) Since $(f \circ g)(x) = f\left(\frac{1}{x}\right) = x$, $f(x) = \frac{1}{x}$.

(d) Since $(f \circ g)(x) = f\left(\sqrt{x}\right) = |x|$, $f(x) = x^2$.
The completed table is shown. Note that the absolute value sign in part (d) is optional.

$g(x)$	$f(x)$	$(f \circ g)(x)$
x^2	$\sqrt{x-5}$	$\sqrt{x^2-5}$
$\frac{1}{x-1}$	$1+\frac{1}{x}$	$x, x \neq -1$
$\frac{1}{x}$	$\frac{1}{x}$	$x, x \neq 0$
$\sqrt{x}$	x^2	$\lvert x \rvert, x \geq 0$

45. (a) domain: $[0, 2]$; range: $[2, 3]$ (b) domain: $[0, 2]$; range: $[-1, 0]$

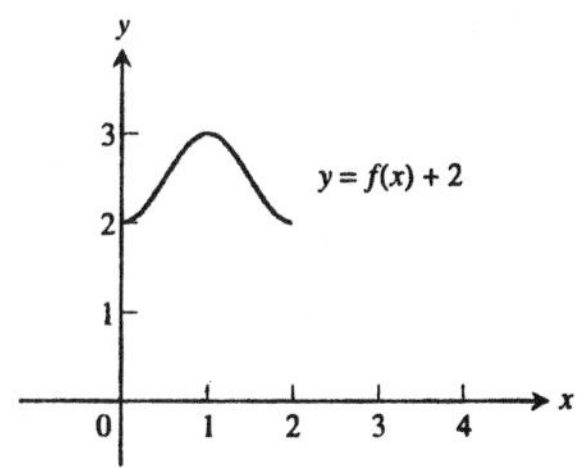

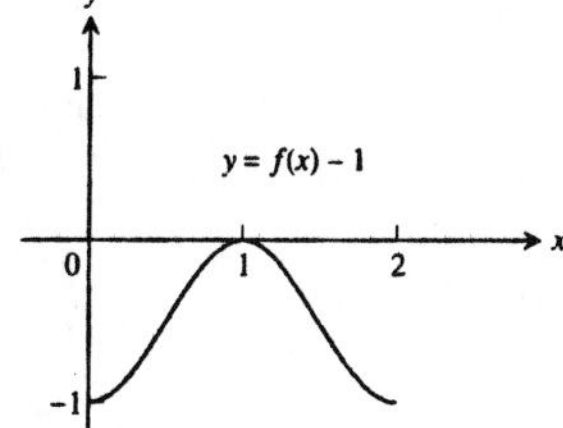

(c) domain: $[0,2]$; range: $[0,2]$

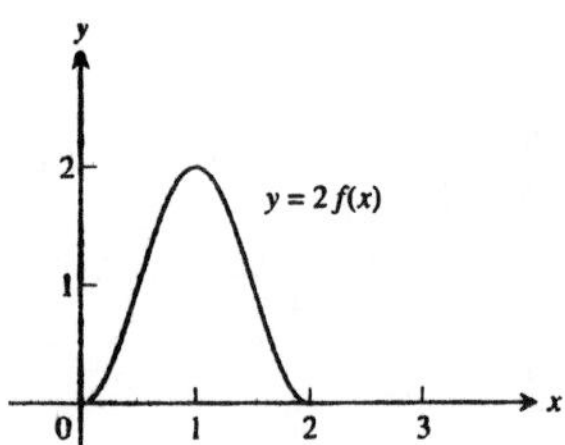

(d) domain: $[0,2]$; range: $[-1,0]$

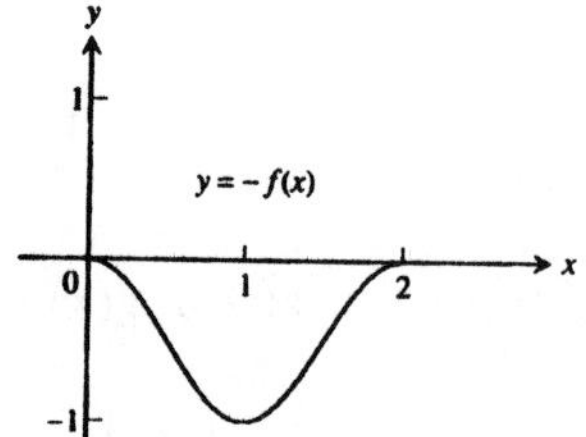

(e) domain: $[-2,0]$; range: $[0,1]$

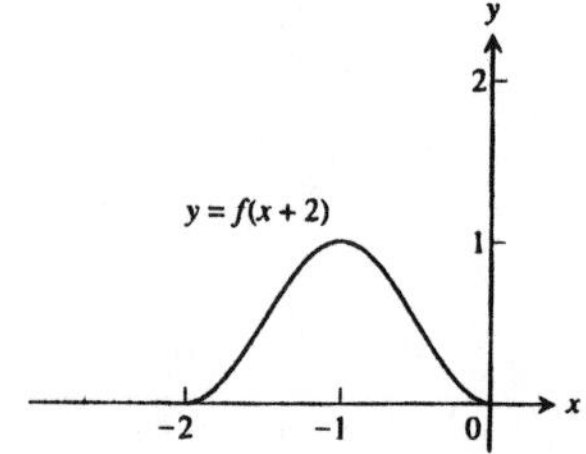

(f) domain: $[1,3]$; range: $[0,1]$

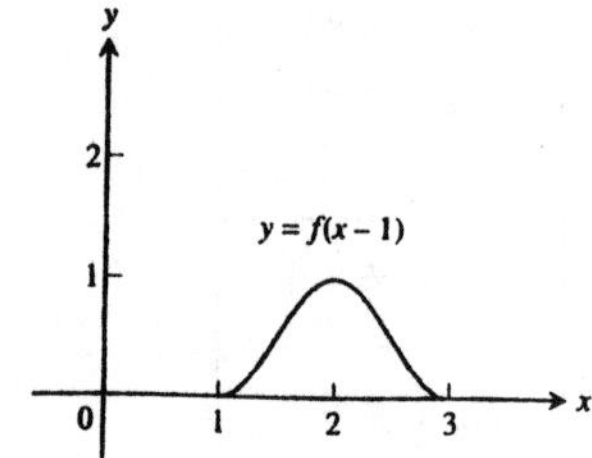

(g) domain: $[-2,0]$; range: $[0,1]$

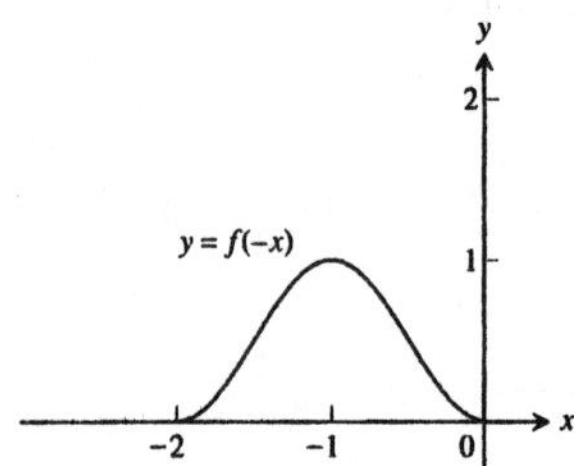

(h) domain: $[-1,1]$; range: $[0,1]$

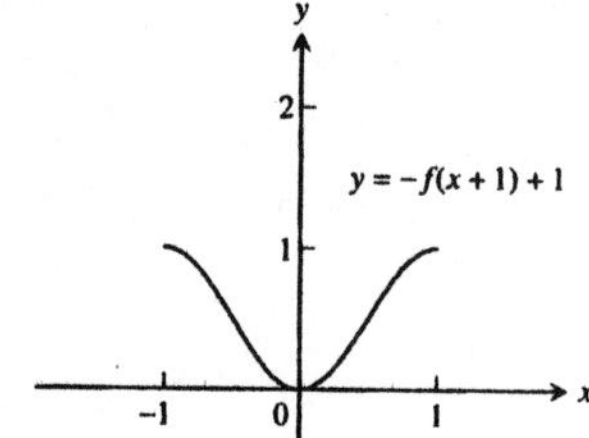

47. (a) Because the circumference of the original circle was 8π and a piece of length x was removed.

(b) $r = \frac{8\pi - x}{2\pi} = 4 - \frac{x}{2\pi}$

(c) $h = \sqrt{16 - r^2} = \sqrt{16 - \left(4 - \frac{x}{2\pi}\right)} = \sqrt{16 - \left(16 - \frac{4x}{\pi} + \frac{x^2}{4\pi^2}\right)} = \sqrt{\frac{4x}{\pi} - \frac{x^2}{4\pi^2}} = \sqrt{\frac{16\pi x}{4\pi^2} - \frac{x^2}{4\pi^2}} = \frac{\sqrt{16\pi x - x^2}}{2\pi}$

(d) $V = \frac{1}{3}\pi r^2 h = \frac{1}{3}\pi\left(\frac{8\pi - x}{2\pi}\right)^2 \cdot \frac{\sqrt{16\pi x - x^2}}{2\pi} = \frac{(8\pi - x)^2\sqrt{16\pi x - x^2}}{24\pi^2}$

49. (a) Yes. Since $(f \cdot g)(-x) = f(-x) \cdot g(-x) = f(x) \cdot g(x) = (f \cdot g)(x)$, the function $(f \cdot g)(x)$ will also be even.

(b) The product will be even, since

$$\begin{aligned}(f\cdot g)(-x) &= f(-x)\cdot g(-x)\\ &= (-f(x))\cdot(-g(x))\\ &= f(x)\cdot g(x)\\ &= (f\cdot g)(x).\end{aligned}$$

(c) Yes, $f(x) = 0$ is both even and odd since $f(-x) = -f(x) = f(x)$

51. (a)

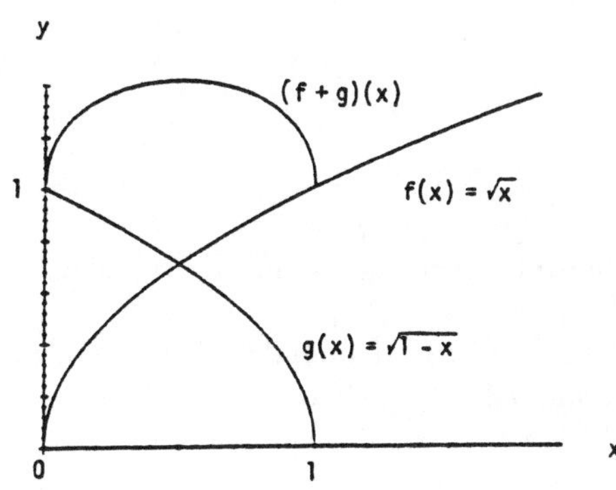

(b)

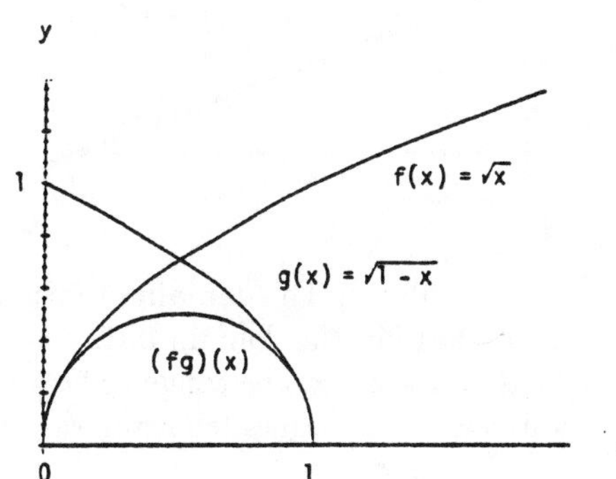

(c)

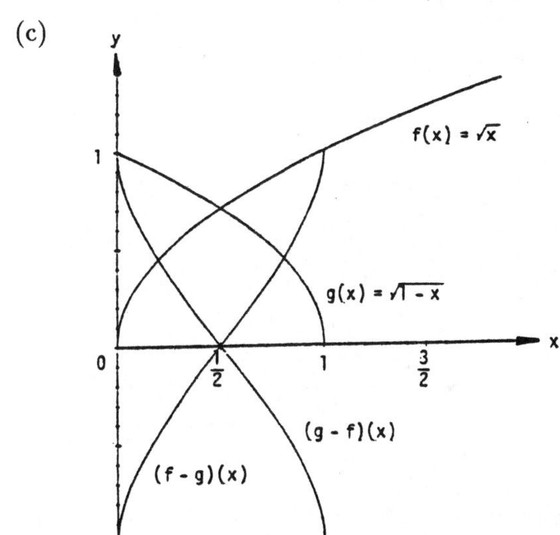

(d)

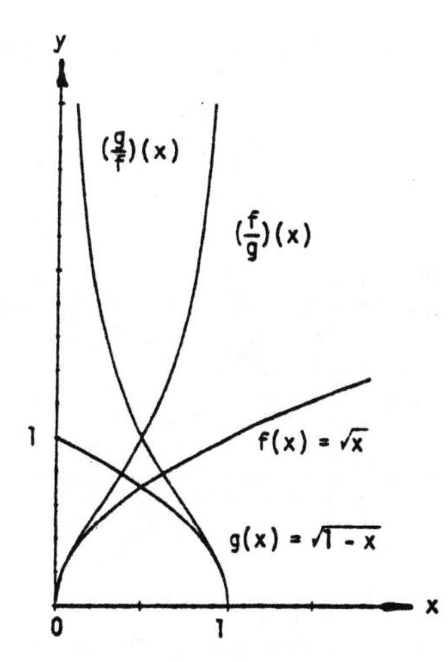

53. (a) $y_4 = (f\circ g)(x)$; $y_3 = (g\circ f)(x)$

(b)

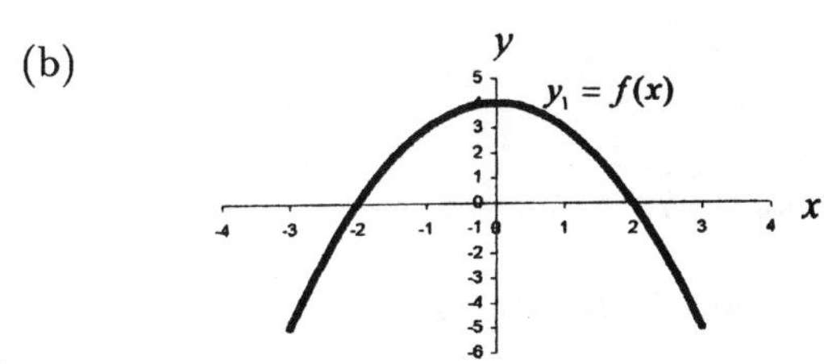

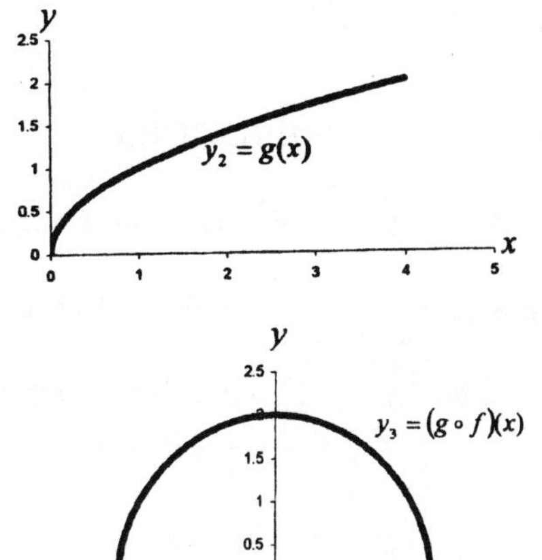

$D(g \circ f) = [-2, 2]$; The domain of $g \circ f$ is the set of all values of x in the domain of f for which the values $y_1 = f(x)$ are in the domain of g.
$R(g \circ f) = [0, 2]$; The range of $g \circ f$ is the subset of the range of g that includes all the values of g(x) evaluated at the values from the range of f where g(x) is defined.

(c) The graphs of $y_1 = f(x)$ and $y_2 = g(x)$ are shown in part (a).

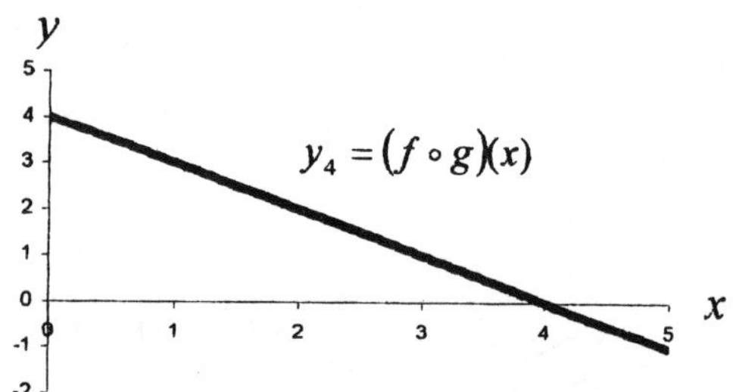

$D(f \circ g) = [0, \infty)$; The domain of $f \circ g$ is the set of all values of x in the domain of g for which the values $y_2 = g(x)$ are in the domain of f.
$R(f \circ g) = (-\infty, 4]$; The range of $f \circ g$ is the subset of the range of f that includes all the values of f(x) evaluated at the values from the range of g where f(x) is defined.

(d) $(g \circ f)(x) = \sqrt{4 - x^2}$; $D(g \circ f) = [-2, 2]$; $R(g \circ f) = [0, 2]$

$(f \circ g)(x) = 4 - (\sqrt{x})^2 = 4 - x$ for $x \geq 0$; $D(f \circ g) = [0, \infty)$; $R(f \circ g) = (-\infty, 4]$

55. (a) The power regression function on the TI-92 Plus calculator gives $y = 4.44647x^{0.511414}$
(b)

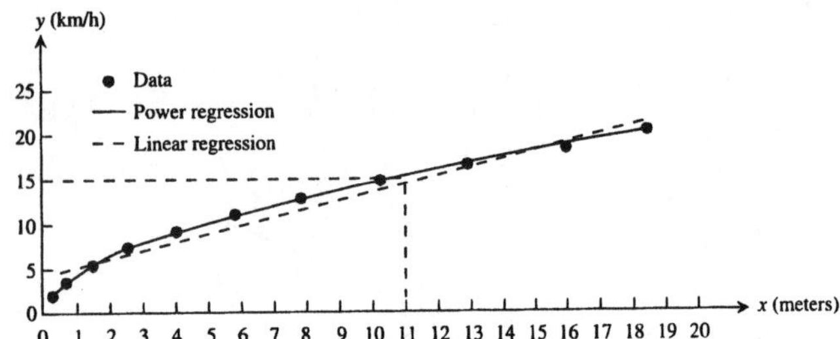

(c) 15.2 km/h
(d) The linear regression function on the TI-92 Plus calculator gives $y = 0.913675x + 4.189976$ and it is shown on the graph in part (b). The linear regression function gives a speed of 14.2 km/h when $y = 11$m. The power regression curve in part (a) better fits the data.

P.3 EXPONENTIAL FUNCTIONS

1. The graph of $y = 2^x$ is increasing from left to right and has the negative x-axis as an asymptote. (a)

3. The graph of $y = -3^{-x}$ is the reflection about the x-axis of the graph in Exercise 2. (e)

5. The graph of $y = 2^{-x} - 2$ is decreasing from left to right and has the line $y = -2$ as an asymptote. (b)

7.

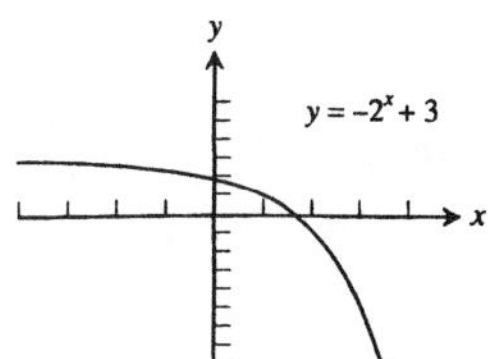

Domain: $(-\infty, \infty)$
Range: $(-\infty, 3)$
x-intercept: ≈ 1.585
y-intercept: 2

9.

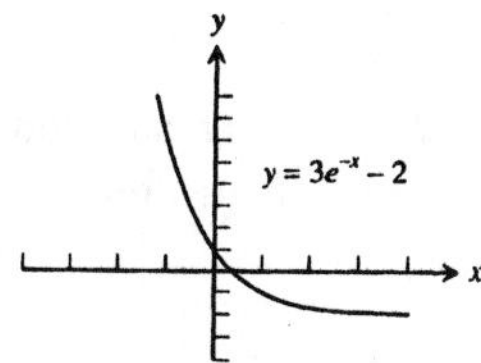

Domain: $(-\infty, \infty)$
Range: $(-2, \infty)$
x-intercept: ≈ 0.405
y-intercept: 1

11. $9^{2x} = (3^2)^{2x} = 3^{4x}$

13. $\left(\frac{1}{8}\right)^{2x} = (2^{-3})^{2x} = 2^{-6x}$

15.

x	y	Δy
1	−1	
		2
2	1	
		2
3	3	
		2
4	5	

17.

x	y	Δy
1	1	
		3
2	4	
		5
3	9	
		7
4	16	

19. The slope of a straight line is $m = \frac{\Delta y}{\Delta x} \rightarrow \Delta y = m(\Delta x)$. In Exercise 15, each $\Delta x = 1$ and $m = 2 \rightarrow$ each $\Delta y = 2$, and in problem 16, each $\Delta x = 1$ and $m = -3 \rightarrow$ each $\Delta y = -3$. If the changes in x are constant for a linear function, say $\Delta x = c$, then the changes in y are also constant, specifically, $\Delta y = mc$.

21. Since $f(1) = 4.5$ we have $ka = 4.5$, and since $f(-1) = 0.5$ we have $ka^{-1} = 0.5$.
Dividing, we have

$$\frac{ka}{ka^{-1}} = \frac{4.5}{0.5}$$

$$a^2 = 9$$

$$a = \pm 3$$

Since $f(x) = k \cdot a^x$ is an exponential function, we require $a > 0$, so $a = 3$. Then $ka = 4.5$ gives $3k = 4.5$, so $k = 1.5$. The values are $a = 3$ and $k = 1.5$.

23.

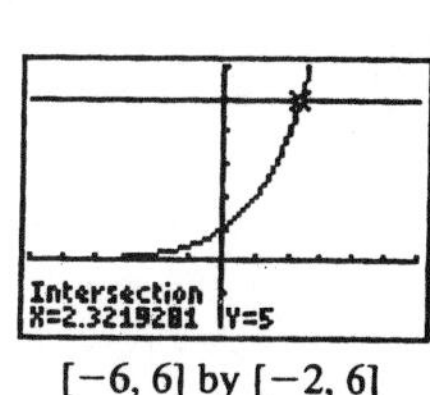

[−6, 6] by [−2, 6]

25.

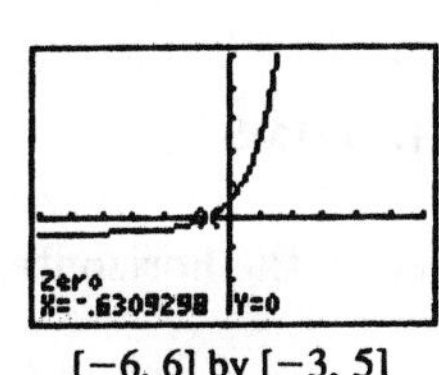

[−6, 6] by [−3, 5]

$x \approx 2.3219$ $\qquad$ $x \approx -0.6309$

27. $5422(1.018)^{19} \approx 7609.7$ million

29. Let t be the number of years. Solving $500{,}000(1.0375)^t = 1{,}000{,}000$ graphically, we find that $t \approx 18.828$. The population will reach 1 million in about 19 years.

31. (a) $A(t) = 6.6\left(\frac{1}{2}\right)^{t/14}$

    (b) Solving $A(t) = 1$ graphically, we find that $t \approx 38$. There will be 1 gram remaining after about 38.1145 days.

33. Let A be the amount of the initial investment, and let t be the number of years. We wish to solve $A(1.0625)^t = 2A$, which is equivalent to $1.0625^t = 2$. Solving graphically, we find that $t \approx 11.433$. It will take about 11.433 years. (If the interest is credited at the end of each year, it will take 12 years.)

35. Let A be the amount of the initial investment, and let t be the number of years. We wish to solve $Ae^{0.0625t} = 2A$, which is equivalent to $e^{0.0625t} = 2$. Solving graphically, we find that $t \approx 11.090$. It will take about 11.090 years.

37. Let A be the amount of the initial investment, and let t be the number of years. We wish to solve $A\left(1 + \frac{0.0575}{365}\right)^{365t} = 3A$, which is equivalent to $\left(1 + \frac{0.0575}{365}\right)^{365t} = 3$. Solving graphically, we find that $t \approx 19.108$. It will take about 19.108 years.

39. After t hours, the population is $P(t) = 2^{t/0.5}$ or, equivalently, $P(t) = 2^{2t}$. After 24 hours, the population is $P(24) = 2^{48} \approx 2.815 \times 10^{14}$ bacteria.

41. (a) Let $x = 0$ represent 1900, $x = 1$ represent 1901, and so on. The regression equation is $P(x) = 6.033(1.030)^x$.

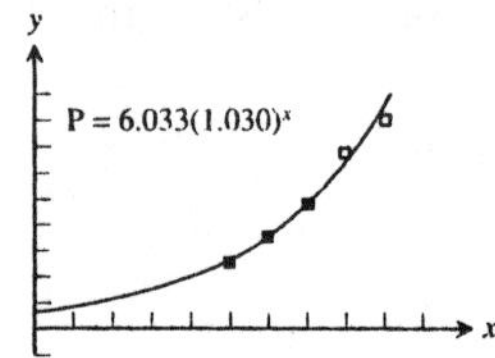

    (b) The regression equation gives an estimate of $P(0) \approx 6.03$ million, which is not very close to the actual population.

    (c) Since the equation is of the form $P(x) = P(0) \cdot 1.030^x$, the annual rate of growth is about 3%.

P.4 FUNCTIONS AND LOGARITHMS

1. Yes one-to-one, the graph passes the horizontal test.

3. Not one-to-one since (for example) the horizontal line $y = 2$ intersects the graph twice.

5. Yes one-to-one, the graph passes the horizontal test

7. Domain: $0 < x \leq 1$, Range: $y \geq 0$

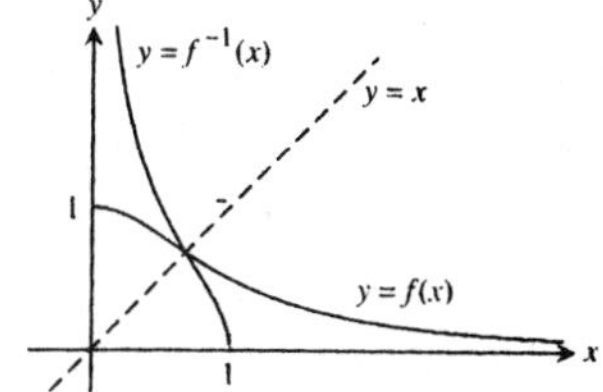

9.

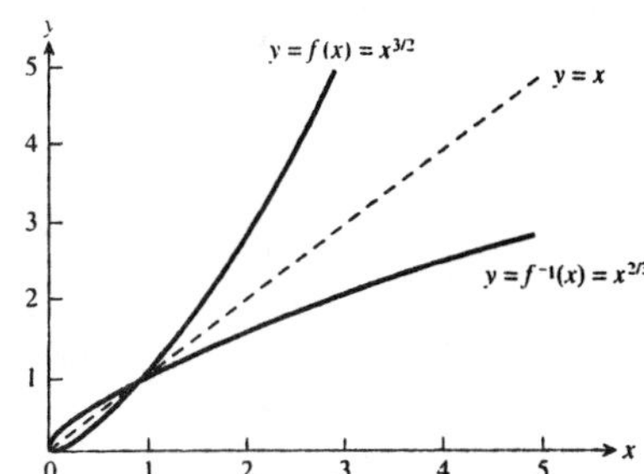

Domain: $x \geq 0$, Range: $y \geq 0$

11. Step 1: $y = x^2 + 1 \Rightarrow x^2 = y - 1 \Rightarrow x = \sqrt{y-1}$
Step 2: $y = \sqrt{x-1} = f^{-1}(x)$

13. Step 1: $y = x^3 - 1 \Rightarrow x^3 = y + 1 \Rightarrow x = (y+1)^{1/3}$
Step 2: $y = \sqrt[3]{x+1} = f^{-1}(x)$

15. Step 1: $y = (x+1)^2 \Rightarrow \sqrt{y} = x + 1 \Rightarrow x = \sqrt{y} - 1$
Step 2: $y = \sqrt{x} - 1 = f^{-1}(x)$

17. $y = 2x + 3 \rightarrow y - 3 = 2x \rightarrow \frac{y-3}{2} = x$. Interchange x and y: $\frac{x-3}{2} = y \rightarrow f^{-1}(x) = \frac{x-3}{2}$
Verify.
$(f \circ f^{-1})(x) = f\left(\frac{x-3}{2}\right) = 2\left(\frac{x-3}{2}\right) + 3 = (x-3) + 3 = x$

$(f^{-1} \circ f)(x) = f^{-1}(2x+3) = \frac{(2x+3)-3}{2} = \frac{2x}{2} = x$

19. $y = x^3 - 1 \rightarrow y + 1 = x^3 \rightarrow (y+1)^{1/3} = x$. Interchange x and y: $(x+1)^{1/3} = y$
$\rightarrow f^{-1}(x) = (x+1)^{1/3}$ or $\sqrt[3]{x+1}$
Verify.

$(f \circ f^{-1})(x) = f(\sqrt[3]{x+1}) = (\sqrt[3]{x+1})^3 - 1 = (x+1) - 1 = x$

$(f^{-1} \circ f)(x) = f^{-1}(x^3 - 1) = \sqrt[3]{(x^3-1)+1} = \sqrt[3]{x^3} = x$

21. $y = x^2,\ x \le 0 \to x = -\sqrt{y}$. Interchange x and y: $y = -\sqrt{x} \to f^{-1}(x) = -\sqrt{x}$ or $-x^{1/2}$

Verify.

For $x \ge 0$ (the domain of f^{-1}), $(f \circ f^{-1})(x) = f(-\sqrt{x}) = (-\sqrt{x})^2 = x$

For $x \le 0$ (the domain of f), $(f^{-1} \circ f)(x) = f^{-1}(x^2) = -\sqrt{x^2} = -|x| = x$

23. $y = -(x-2)^2,\ x \le 2 \to (x-2)^2 = -y,\ x \le 2 \to x - 2 = -\sqrt{-y} \to x = 2 - \sqrt{-y}$.

Interchange x and y: $y = 2 - \sqrt{-x} \to f^{-1}(x) = 2 - \sqrt{-x}$ or $2 - (-x)^{1/2}$

Verify.

For $x \le 0$ (the domain of f^{-1})

$(f \circ f^{-1})(x) = f(2 - \sqrt{-x}) = -[(2 - \sqrt{-x}) - 2]^2 = -(-\sqrt{-x})^2 = -|x| = x$

For $x \le 2$ (the domain of f),

$(f^{-1} \circ f)(x) = f^{-1}(-(x-2)^2) = 2 - \sqrt{(x-2)^2} = 2 - |x-2| = 2 + (x-2) = x$

25. $y = \frac{1}{x^2},\ x > 0 \to x^2 = \frac{1}{y},\ x > 0 \to x = \sqrt{\frac{1}{y}} = \frac{1}{\sqrt{y}}$.

Interchange x and y: $y = \frac{1}{\sqrt{x}} \to f^{-1}(x) = \frac{1}{\sqrt{x}}$ or $\frac{1}{x^{1/2}}$

Verify.

For $x > 0$ (the domain of f^{-1}), $(f \circ f^{-1})(x) = f\left(\frac{1}{\sqrt{x}}\right) = \frac{1}{(1/\sqrt{x})^2} = x$

For $x > 0$ (the domain of f), $(f^{-1} \circ f)(x) = f^{-1}\left(\frac{1}{x^2}\right) = \frac{1}{\sqrt{1/x^2}} = \sqrt{x^2} = |x| = x$

27. $y = \frac{2x+1}{x+3} \to xy + 3y = 2x + 1 \to xy - 2x = 1 - 3y \to (y-2)x = 1 - 3y \to x = \frac{1-3y}{y-2}$.

Interchange x and y: $y = \frac{1-3x}{x-2} \to f^{-1}(x) = \frac{1-3x}{x-2}$

Verify.

$$(f \circ f^{-1})(x) = f\left(\frac{1-3x}{x-2}\right) = \frac{2\left(\frac{1-3x}{x-2}\right)+1}{\frac{1-3x}{x-2}+3} = \frac{2(1-3x)+(x-2)}{(1-3x)+3(x-2)} = \frac{-5x}{-5} = x$$

$$(f^{-1} \circ f)(x) = f^{-1}\left(\frac{2x+1}{x+3}\right) = \frac{1-3\left(\frac{2x+1}{x+3}\right)}{\frac{2x+1}{x+3}-2} = \frac{(x+3)-3(2x+1)}{(2x+1)-2(x+3)} = \frac{-5x}{-5} = x$$

29. $y = (e^a)^x - 1 \to e^a = 3 \to a = \ln 3 \to y = e^{x \ln 3} - 1$

(a) $D = (-\infty, \infty)$

(b) $R = (-1, \infty)$

31. $y = 1 - (\ln 3)\log_3 x = 1 - (\ln 3)\,\frac{\ln x}{\ln 3} = 1 - \ln x$

(a) $D = (0, \infty)$

(b) $R = (-\infty, \infty)$

(c)

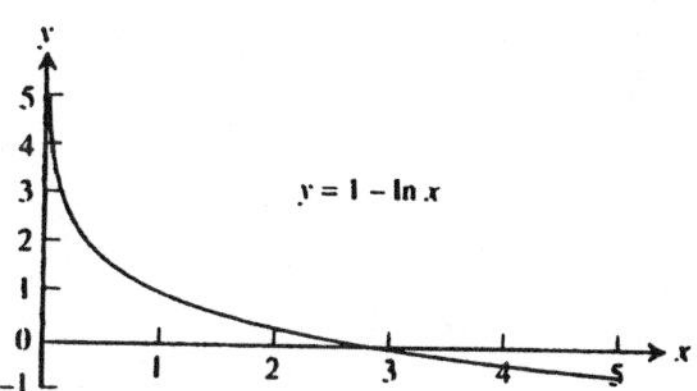

33. $(1.045)^t = 2$

$\ln (1.045)^t = \ln 2$

$t \ln 1.04 = \ln 2$

$t = \frac{\ln 2}{\ln 1.045} \approx 15.75$

Graphical support:

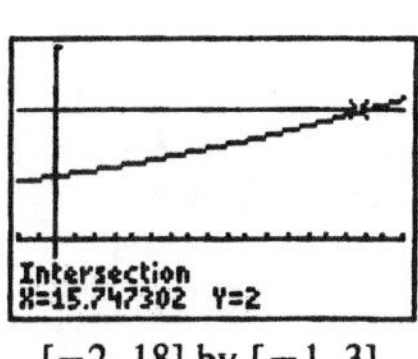

[−2, 18] by [−1, 3]

35. $e^x + e^{-x} = 3$

$e^x - 3 + e^{-x} = 0$

$e^x\left(e^x - 3 + e^{-x}\right) = e^x(0)$

$\left(e^x\right)^2 - 3e^x + 1 = 0$

$e^x = \frac{3 \pm \sqrt{(-3)^2 - 4(1)(1)}}{2(1)}$

$e^x = \frac{3 \pm \sqrt{5}}{2}$

$x = \ln\left(\frac{3 \pm \sqrt{5}}{2}\right) \approx -0.96 \text{ or } 0.96$

Graphical support:

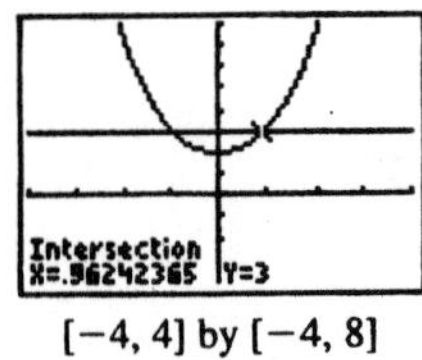

[−4, 4] by [−4, 8]

37. $\ln y = 2t + 4 \rightarrow e^{\ln y} = e^{2t+4} \rightarrow y = e^{2t+4}$

39. (a) $y = \frac{100}{1+2^{-x}} \rightarrow 1 + 2^{-x} = \frac{100}{y} \rightarrow 2^{-x} = \frac{100}{y} - 1 \rightarrow \log_2(2^{-x}) = \log_2\left(\frac{100}{y} - 1\right) \rightarrow -x = \log_2\left(\frac{100}{y} - 1\right)$

$\rightarrow x = -\log_2\left(\frac{100}{y} - 1\right) = -\log_2\left(\frac{100-y}{y}\right) = \log_2\left(\frac{y}{100-y}\right).$

Interchange x and y: $y = \log_2\left(\frac{x}{100-x}\right) \rightarrow f^{-1}(x) = \log_2\left(\frac{x}{100-x}\right)$

Verify.

$$(f \circ f^{-1})(x) = f\left(\log_2\left(\frac{x}{100-x}\right)\right) = \frac{100}{1 + 2^{-\log_2\left(\frac{x}{100-x}\right)}} = \frac{100}{1 + 2^{\log_2\left(\frac{100-x}{x}\right)}} = \frac{100}{1 + \frac{100-x}{x}}$$

$$= \frac{100x}{x + (100-x)} = \frac{100x}{100} = x$$

$$(f^{-1} \circ f)(x) = f^{-1}\left(\frac{100}{1+2^{-x}}\right) = \log_2\left(\frac{\frac{100}{1+2^{-x}}}{100 - \frac{100}{1+2^{-x}}}\right) = \log_2\left(\frac{100}{100(1+2^{-x}) - 100}\right)$$

$$= \log_2\left(\frac{1}{2^{-x}}\right) = \log_2(2^x) = x$$

(b) $y = \frac{50}{1+1.1^{-x}} \rightarrow 1 + 1.1^{-x} = \frac{50}{y} \rightarrow 1.1^{-x} = \frac{50}{y} - 1 \rightarrow \log_{1.1}(1.1^{-x}) = \log_{1.1}\left(\frac{50}{y} - 1\right) \rightarrow -x = \log_{1.1}\left(\frac{50}{y} - 1\right)$

$\rightarrow x = -\log_{1.1}\left(\frac{50}{y} - 1\right) = -\log_{1.1}\left(\frac{50-y}{y}\right) = \log_{1.1}\left(\frac{y}{50-y}\right).$

Interchange x and y: $y = \log_{1.1}\left(\frac{x}{50-x}\right) \rightarrow f^{-1}(x) = \log_{1.1}\left(\frac{x}{50-x}\right)$

Verify.

$$(f \circ f^{-1})(x) = f\left(\log_{1.1}\left(\frac{x}{50-x}\right)\right) = \frac{50}{1 + 1.1^{-\log_{1.1}\left(\frac{x}{50-x}\right)}} = \frac{50}{1 + 1.1^{\log_{1.1}\left(\frac{50-x}{x}\right)}} = \frac{50}{1 + \frac{50-x}{x}}$$

$$= \frac{50x}{x + (50-x)} = \frac{50x}{50} = x$$

$$(f^{-1} \circ f)(x) = f^{-1}\left(\frac{50}{1+1.1^{-x}}\right) = \log_{1.1}\left(\frac{\frac{50}{1+1.1^{-x}}}{50 - \frac{50}{1+1.1^{-x}}}\right) = \log_{1.1}\left(\frac{50}{50(1+1.1^{-x}) - 50}\right)$$

$= \log_{1.1}\left(\frac{1}{1.1^{-x}}\right) = \log_{1.1}(1.1^x) = x$

41. (a) Amount $= 8\left(\frac{1}{2}\right)^{t/12}$

(b) $8\left(\frac{1}{2}\right)^{t/12} = 1 \rightarrow \left(\frac{1}{2}\right)^{t/12} = \frac{1}{8} \rightarrow \left(\frac{1}{2}\right)^{t/12} = \left(\frac{1}{2}\right)^3 \rightarrow \frac{t}{12} = 3 \rightarrow t = 36$
There will be 1 gram remaining after 36 hours.

43. $375{,}000(1.0225)^t = 1{,}000{,}000 \rightarrow 1.0225^t = \frac{8}{3} \rightarrow \ln(1.0225^t) = \ln\left(\frac{8}{3}\right) \rightarrow t \ln 1.0225 = \ln\left(\frac{8}{3}\right)$
$\rightarrow t = \frac{\ln(8/3)}{\ln 1.0225} \approx 44.081$
It will take about 44.081 years.

45. Sound level with intensity = 10I is $10 \log_{10}(10I \times 10^{12}) = 10\left[\log_{10} 10 + \log_{10}(I \times 10^{12})\right]$
$= 10 + 10 \log_{10}(I \times 10^{12})$ = original sound level + 10 $\Rightarrow$ an increase of 10 db

47.

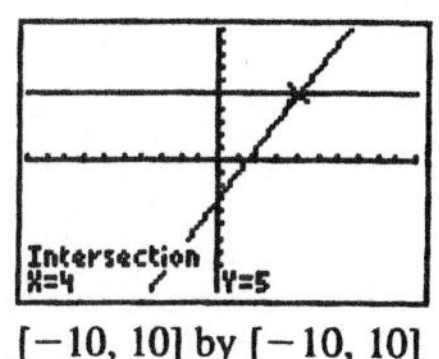

[−10, 10] by [−10, 10]

(4, 5)

49. (a)

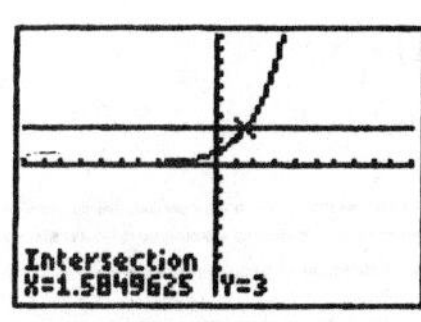

[−10, 10] by [−10, 10]

(1.58, 3)

(b) No points of intersection, since $2^x > 0$ for all values of x.

51. (a)

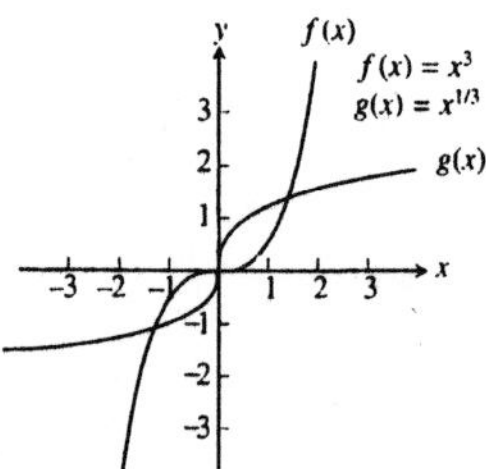

(b) and (c)

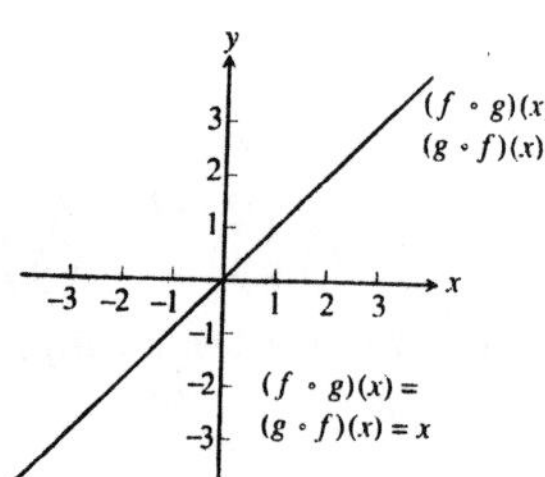

We conclude that f and g are inverses of each other because $(f \circ g)(x) = (g \circ f)(x) = x$, the identity function.

53. (a)

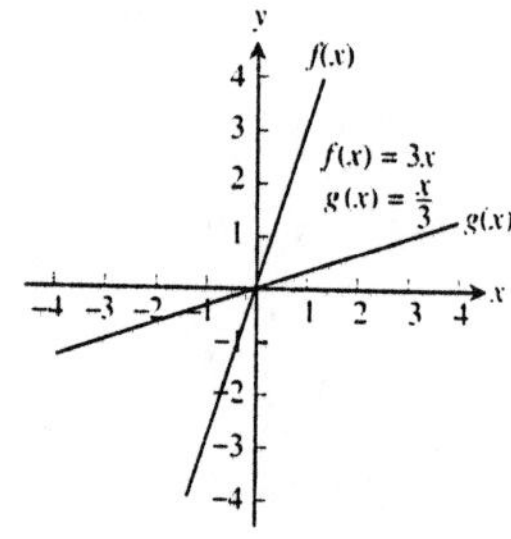

(b) and (c)

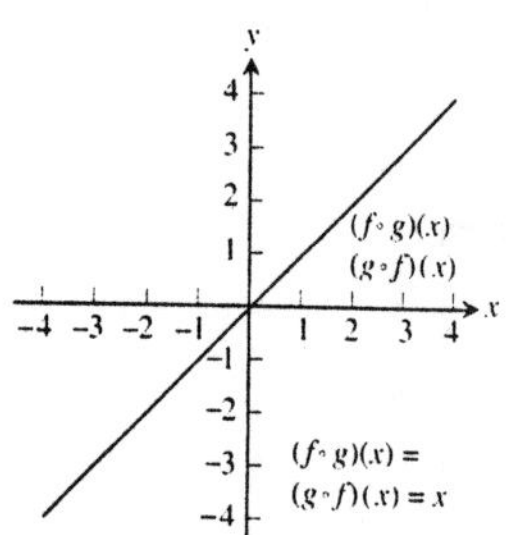

We conclude that f and g are inverses of each other because $(f \circ g)(x) = (g \circ f)(x) = x$, the identity function.

55. (a)

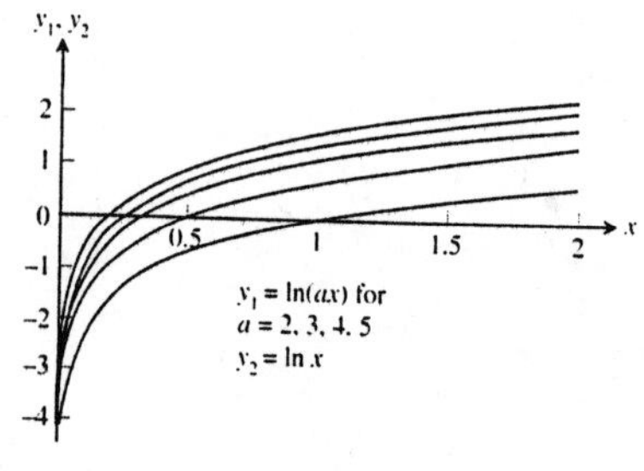

The graphs of y_1 appear to be vertical translates of y_2

(b)

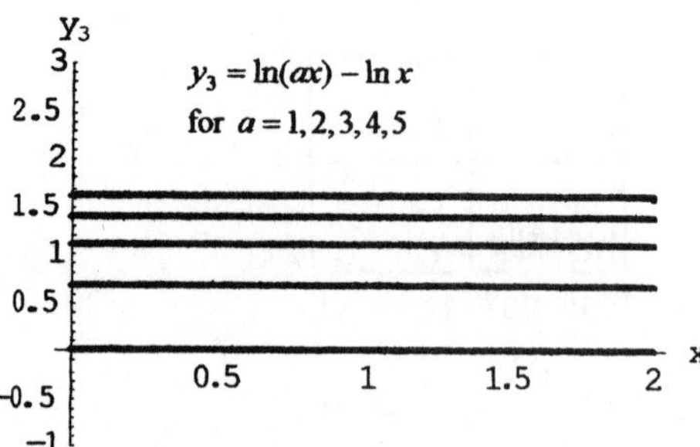

The graphs of $y_1 - y_2$ support the finding in part (a).

(c) $y_3 = y_1 - y_2 = \ln ax - \ln x = (\ln a + \ln x) - \ln x = \ln a$, a constant.

57. From zooming in on the graph at the right, we estimate the third root to be $x \approx -0.76666$

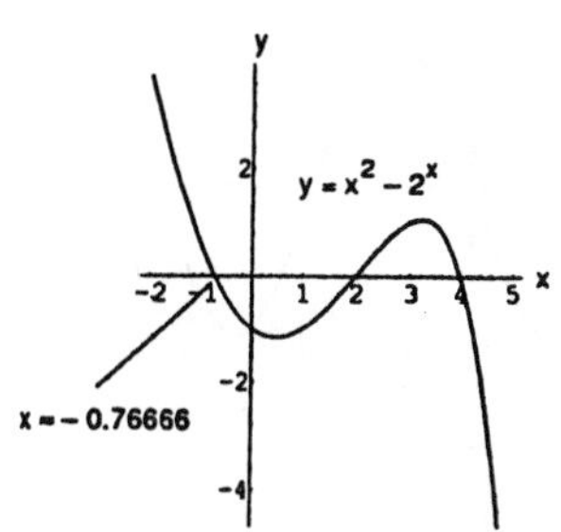

59. (a) The LnReg command on the TI-92 Plus calculator gives $y(x) = -474.31 + 121.13 \ln x$ $\Rightarrow y(82) = -474.31 + 121.13 \ln(82) = 59.48$ million metric tons produced in 1982 and $y(100) = -474.31 + 121.13 \ln(100) = 83.51$ million metric tons produced in 2000.

(b)

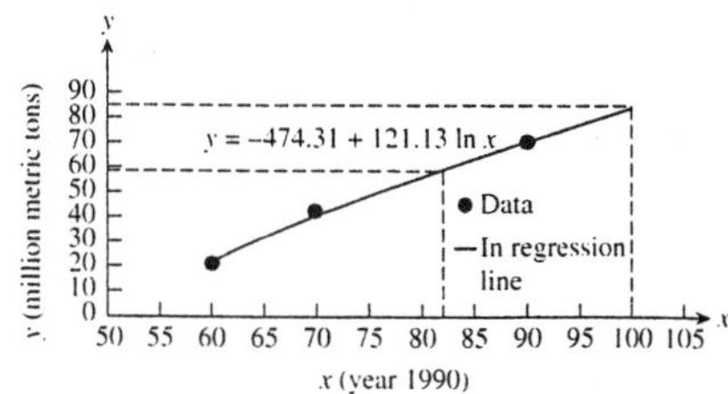

(c) From the graph in part (b), $y(82) \approx 59$ and $y(100) \approx 84$.

P.5 TRIGONOMETRIC FUNCTIONS AND THEIR INVERSES

1. (a) $s = r\theta = (10)\left(\frac{4\pi}{5}\right) = 8\pi$ m (b) $s = r\theta = (10)(110^\circ)\left(\frac{\pi}{180^\circ}\right) = \frac{110\pi}{18} = \frac{55\pi}{9}$ m

3.

θ	$-\pi$	$-\frac{2\pi}{3}$	0	$\frac{\pi}{2}$	$\frac{3\pi}{4}$
$\sin\theta$	0	$-\frac{\sqrt{3}}{2}$	0	1	$\frac{1}{\sqrt{2}}$
$\cos\theta$	-1	$-\frac{1}{2}$	1	0	$-\frac{1}{\sqrt{2}}$
$\tan\theta$	0	$\sqrt{3}$	0	und.	-1
$\cot\theta$	und.	$\frac{1}{\sqrt{3}}$	und.	0	-1
$\sec\theta$	-1	-2	1	und.	$-\sqrt{2}$
$\csc\theta$	und.	$-\frac{2}{\sqrt{3}}$	und.	1	$\sqrt{2}$

5. (a) $\cos x = -\frac{4}{5}$, $\tan x = -\frac{3}{4}$ (b) $\sin x = -\frac{2\sqrt{2}}{3}$, $\tan x = -2\sqrt{2}$

7. (a) 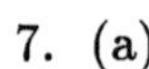

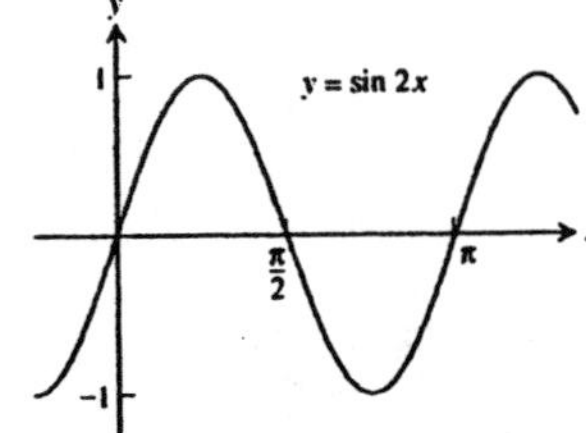

period $= \pi$

(b)

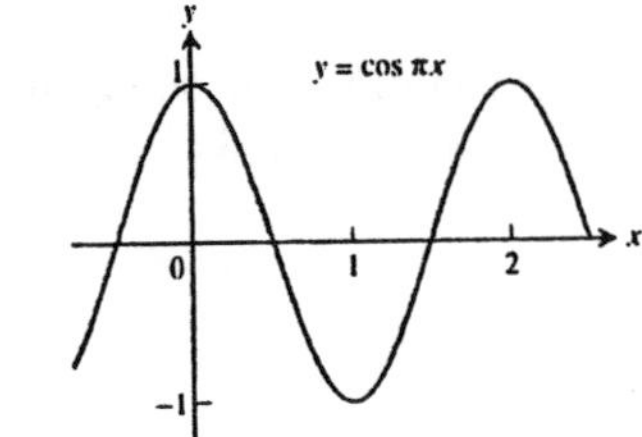

period $= 2$

9. (a)

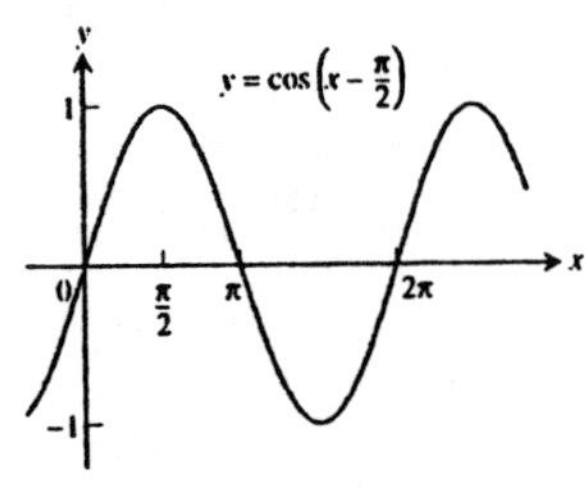

period $= 2\pi$

(b)

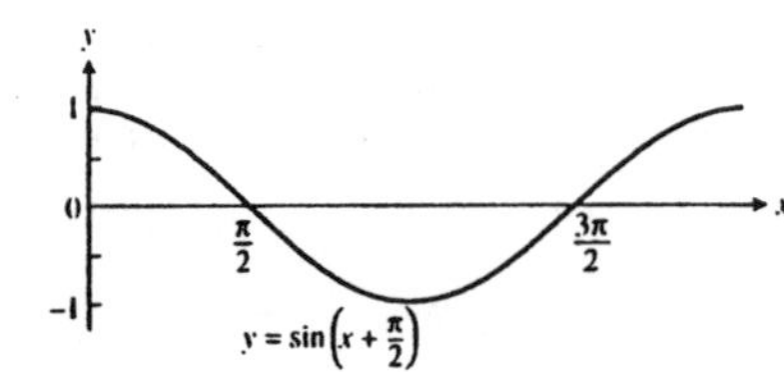

period $= 2\pi$

11. period $= \frac{\pi}{2}$, symmetric about the origin

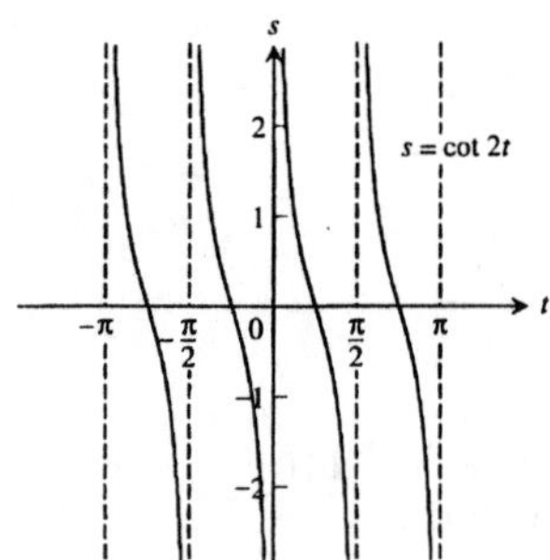

13. (a) $\cos(\pi + x) = \cos\pi\cos x - \sin\pi\sin x = (-1)(\cos x) - (0)(\sin x) = -\cos x$
    (b) $\sin(2\pi - x) = \sin 2\pi\cos(-x) + \cos(2\pi)\sin(-x) = (0)(\cos(-x)) + (1)(\sin(-x)) = -\sin x$

15. (a) $\cos\left(x - \frac{\pi}{2}\right) = \cos x\cos\left(-\frac{\pi}{2}\right) - \sin x\sin\left(-\frac{\pi}{2}\right) = (\cos x)(0) - (\sin x)(-1) = \sin x$

$$\cos(A - B) = \cos(A + (-B)) = \cos A\cos(-B) - \sin A\sin(-B) = \cos A\cos B - \sin A(-\sin B)$$
$$= \cos A\cos B + \sin A\sin B$$

17. If $B = A$, $A - B = 0 \Rightarrow \cos(A - B) = \cos 0 = 1$. Also $\cos(A - B) = \cos(A - A) = \cos A\cos A + \sin A\sin A$ $= \cos^2 A + \sin^2 A$. Therefore, $\cos^2 A + \sin^2 A = 1$.

19. (a) $A = 2$, $B = 2\pi$, $C = -\pi$, $D = -1$

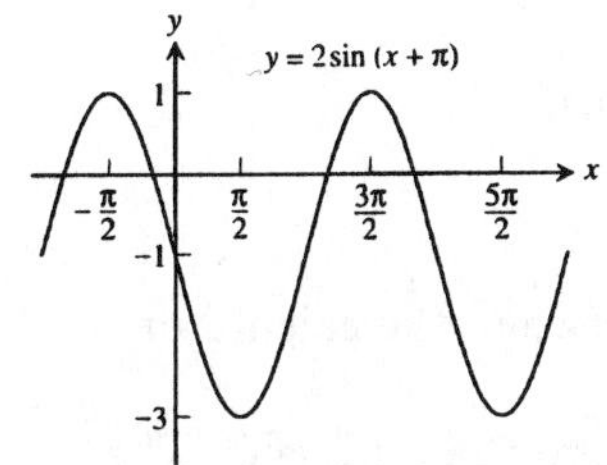

(b) $A = \frac{1}{2}$, $B = 2$, $C = 1$, $D = \frac{1}{2}$

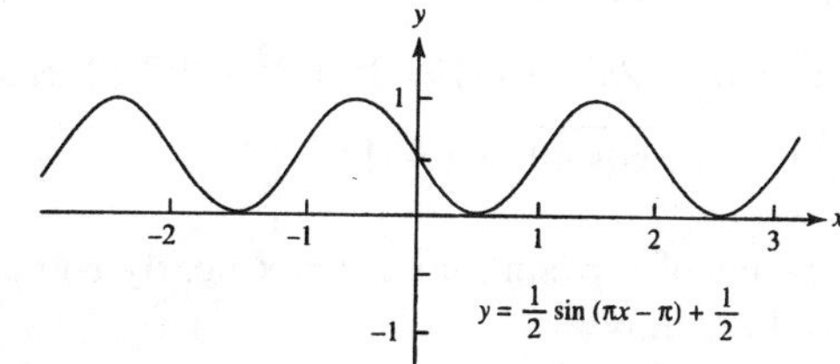

21. (a) amplitude $= |A| = 37$ (b) period $= |B| = 365$
(c) right horizontal shift $= C = 101$ (d) upward vertical shift $= D = 25$

23. (a) $\frac{\pi}{4}$ (b) $-\frac{\pi}{3}$ (c) $\frac{\pi}{6}$

25. (a) $\frac{\pi}{3}$ (b) $\frac{3\pi}{4}$ (c) $\frac{\pi}{6}$

27. The angle α is the large angle between the wall and the right end of the blackboard minus the small angle between the left end of the blackboard and the wall $\Rightarrow \alpha = \cot^{-1}\left(\frac{x}{15}\right) - \cot^{-1}\left(\frac{x}{3}\right)$.

29. According to the figure in the text, we have the following: By the law of cosines, $c^2 = a^2 + b^2 - 2ab\cos\theta = 1^2 + 1^2 - 2\cos(A - B) = 2 - 2\cos(A - B)$. By distance formula, $c^2 = (\cos A - \cos B)^2 + (\sin A - \sin B)^2 = \cos^2 A - 2\cos A\cos B + \cos^2 B + \sin^2 A - 2\sin A\sin B + \sin^2 B = 2 - 2(\cos A\cos B + \sin A\sin B)$. Thus $c^2 = 2 - 2\cos(A - B) = 2 - 2(\cos A\cos B + \sin A\sin B) \Rightarrow \cos(A - B) = \cos A\cos B + \sin A\sin B$.

31. Take each square as a unit square. From the diagram we have the following: the smallest angle α has a tangent of $1 \Rightarrow \alpha = \tan^{-1} 1$; the middle angle β has a tangent of $2 \Rightarrow \beta = \tan^{-1} 2$; and the largest angle γ has a tangent of $3 \Rightarrow \gamma = \tan^{-1} 3$. The sum of these three angles is $\pi \Rightarrow \alpha + \beta + \gamma = \pi$
$\Rightarrow \tan^{-1} 1 + \tan^{-1} 2 + \tan^{-1} 3 = \pi$.

33. $\sin^{-1}(1) + \cos^{-1}(1) = \frac{\pi}{2} + 0 = \frac{\pi}{2}$; $\sin^{-1}(0) + \cos^{-1}(0) = 0 + \frac{\pi}{2} = \frac{\pi}{2}$; and $\sin^{-1}(-1) + \cos^{-1}(-1) = -\frac{\pi}{2} + \pi = \frac{\pi}{2}$.
If $x \in (-1, 0)$ and $x = -a$, then $\sin^{-1}(x) + \cos^{-1}(x) = \sin^{-1}(-a) + \cos^{-1}(-a) = -\sin^{-1} a + \left(\pi - \cos^{-1} a\right)$
$= \pi - \left(\sin^{-1} a + \cos^{-1} a\right) = \pi - \frac{\pi}{2} = \frac{\pi}{2}$ from Equations (7) and (9) in the text.

35. From the figures in the text, we see that $\sin B = \frac{h}{c}$. If C is an acute angle, then $\sin C = \frac{h}{b}$. On the other hand, if C is obtuse (as in the figure on the right), then $\sin C = \sin(\pi - C) = \frac{h}{b}$. Thus, in either case, $h = b\sin C = c\sin B \Rightarrow ah = ab\sin C = ac\sin B$.

By the law of cosines, $\cos C = \frac{a^2 + b^2 - c^2}{2ab}$ and $\cos B = \frac{a^2 + c^2 - b^2}{2ac}$. Moreover, since the sum of the interior angles of a triangle is π, we have $\sin A = \sin(\pi - (B + C)) = \sin(B + C) = \sin B\cos C + \cos B\sin C$
$= \left(\frac{h}{c}\right)\left[\frac{a^2 + b^2 - c^2}{2ab}\right] + \left[\frac{a^2 + c^2 - b^2}{2ac}\right]\left(\frac{h}{b}\right) = \left(\frac{h}{2abc}\right)(2a^2 + b^2 - c^2 + c^2 - b^2) = \frac{ah}{bc} \Rightarrow ah = bc\sin A$.
Combining our results we have $ah = ab\sin C$, $ah = ac\sin B$, and $ah = bc\sin A$. Dividing by abc gives
$$\frac{h}{bc} = \underbrace{\frac{\sin A}{a} = \frac{\sin C}{c} = \frac{\sin B}{b}}_{\text{law of sines}}.$$

37. (a) $c^2 = a^2 + b^2 - 2ab\cos C = 2^2 + 3^2 - 2(2)(3)\cos(60^\circ) = 4 + 9 - 12\cos(60^\circ) = 13 - 12\left(\frac{1}{2}\right) = 7$.
Thus, $c = \sqrt{7} \approx 2.65$.
(b) $c^2 = a^2 + b^2 - 2ab\cos C = 2^2 + 3^2 - 2(2)(3)\cos(40^\circ) = 13 - 12\cos(40^\circ)$. Thus,
$c = \sqrt{13 - 12\cos 40^\circ} \approx 1.951$.

39. (a) The graphs of $y = \sin x$ and $y = x$ nearly coincide when x is near the origin (when the calculator is in radians mode).
(b) In degree mode, when x is near zero degrees the sine of x is much closer to zero than x itself. The curves look like intersecting straight lines near the origin when the calculator is in degree mode.

41. (a) Domain: all real numbers except those having the form $\frac{\pi}{2} + k\pi$ where k is an integer.
Range: $-\frac{\pi}{2} < y < \frac{\pi}{2}$

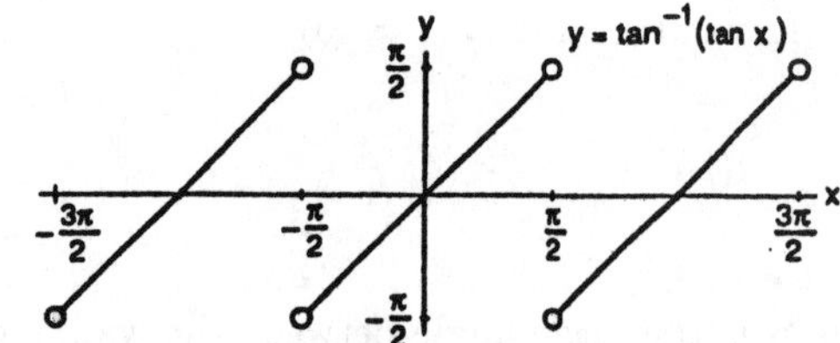

(b) Domain: $-\infty < x < \infty$; Range: $-\infty < y < \infty$
The graph of $y = \tan^{-1}(\tan x)$ is periodic, the graph of $y = \tan\left(\tan^{-1} x\right) = x$ for $-\infty \le x < \infty$.

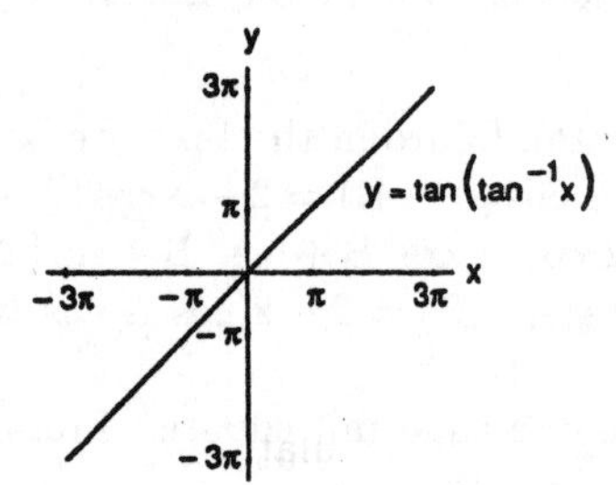

43. The angle $\tan^{-1}(2.5) \approx 1.190$ is the solution to this equation in the interval $-\frac{\pi}{2} < x < \frac{\pi}{2}$. Another solution in $0 \le x < 2\pi$ is $\tan^{-1}(2.5) + \pi \approx 4.332$. The solutions are $x \approx 1.190$ and $x \approx 4.332$.

45. This equation is equivalent to $\cos x = -\frac{1}{3}$, so the solution in the interval $0 \le x \le \pi$ is $y = \cos^{-1}\left(-\frac{1}{3}\right) \approx 1.911$. Since the cosine function is even, the solutions in the interval $-\pi \le x < \pi$ are $x \approx -1.911$ and $x \approx 1.911$.

47. (a)

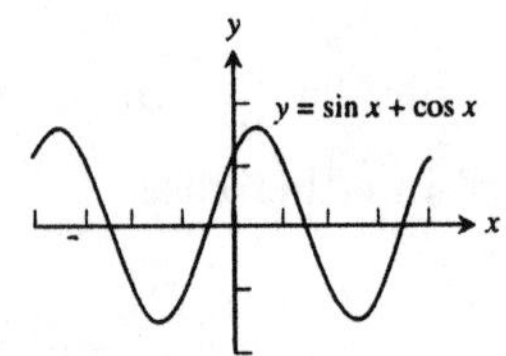

The graph is a sine/cosine type graph, but it is shifted and has an amplitude greater than 1.
(b) Amplitude ≈ 1.414 (that is, $\sqrt{2}$)
Period $= 2\pi$
Horizontal shift ≈ -0.785 $\left(\text{that is, } -\frac{\pi}{4}\right)$ or 5.498 $\left(\text{that is, } \frac{7\pi}{4}\right)$ relative to $\sin x$.

Vertical shift: 0

(c) $\sin\left(x+\frac{\pi}{4}\right) = (\sin x)\left(\cos\frac{\pi}{4}\right) + (\cos x)\left(\sin\frac{\pi}{4}\right)$

$= (\sin x)\left(\frac{1}{\sqrt{2}}\right) + (\cos x)\left(\frac{1}{\sqrt{2}}\right)$

$= \frac{1}{\sqrt{2}}(\sin x + \cos x)$

Therefore, $\sin x + \cos x = \sqrt{2}\,\sin\left(x+\frac{\pi}{4}\right)$

49. (a) The sinusoidal regression on the TI-92 Plus calculator gives $p = 0.599\sin(2479t - 2.801) + 0.265$

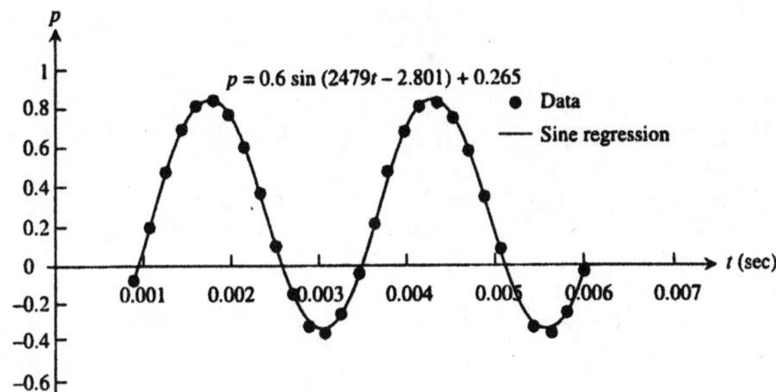

(b) The period is approximately $\frac{2\pi}{2479}$ seconds, so the frequency is approximately $\frac{2479}{2\pi} \approx 395$ Hz

51. (a) Using a graphing calculator with the sinusoidal regression feature, the equation is $y = 3.0014\sin(0.9996x + 2.0012) + 2.9999$.

(b) $y = 3\sin(x+2) + 3$

P.6 PARAMETRIC EQUATIONS

1. $x = \cos t,\ y = \sin t,\ 0 \le t \le \pi$
$\Rightarrow \cos^2 t + \sin^2 t = 1 \Rightarrow x^2 + y^2 = 1$

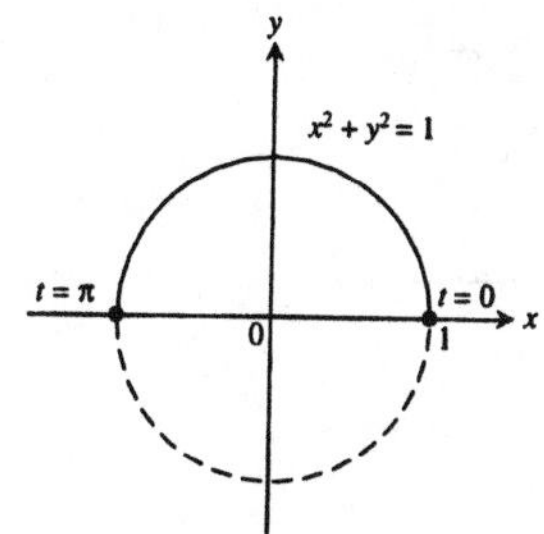

3. $x = \sin(2\pi t),\ y = \cos(2\pi t),\ 0 \le t \le 1$
$\sin^2(2\pi t) + \cos^2(2\pi t) = 1 \Rightarrow x^2 + y^2 = 1$

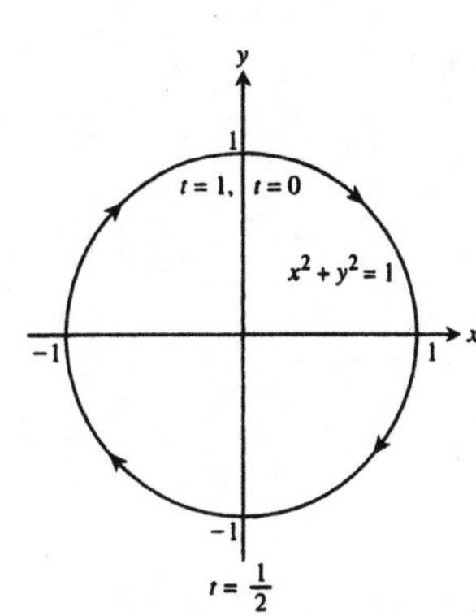

5. $x = 4\cos t,\ y = 2\sin t,\ 0 \le t \le 2\pi$

$\Rightarrow \Rightarrow \frac{16\cos^2 t}{16} + \frac{4\sin^2 t}{4} = 1 \Rightarrow \frac{x^2}{16} + \frac{y^2}{4} = 1$

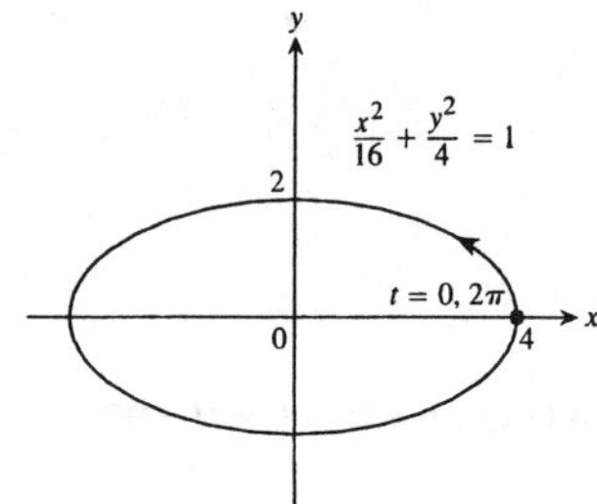

7. $x = 3t,\ y = 9t^2,\ -\infty < t < \infty \Rightarrow y = x^2$

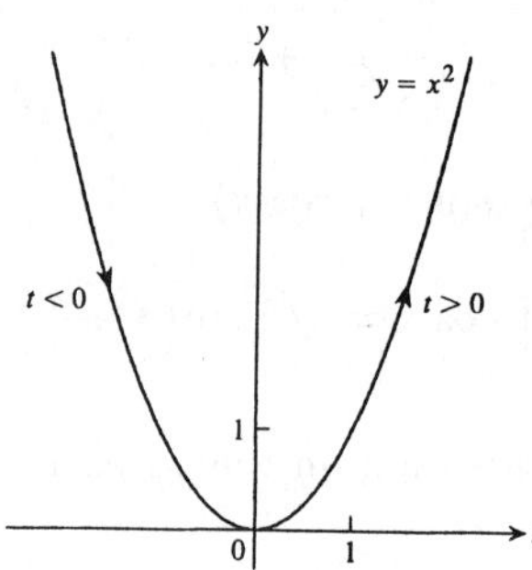

9. $x = t,\ y = \sqrt{t},\ t \ge 0 \Rightarrow y = \sqrt{x}$

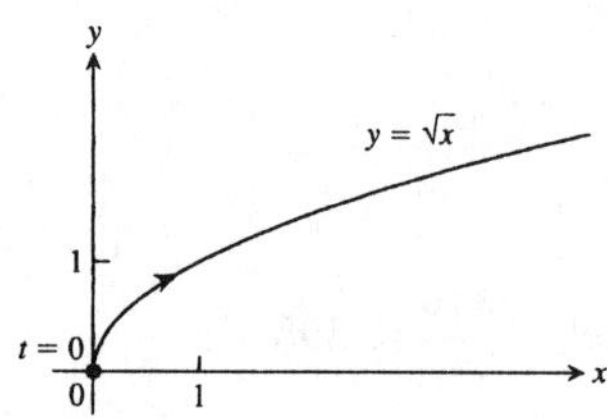

11. $x = -\sec t,\ y = \tan t,\ -\frac{\pi}{2} < t < \frac{\pi}{2}$

$\Rightarrow \sec^2 t - \tan^2 t = 1 \Rightarrow x^2 - y^2 = 1$

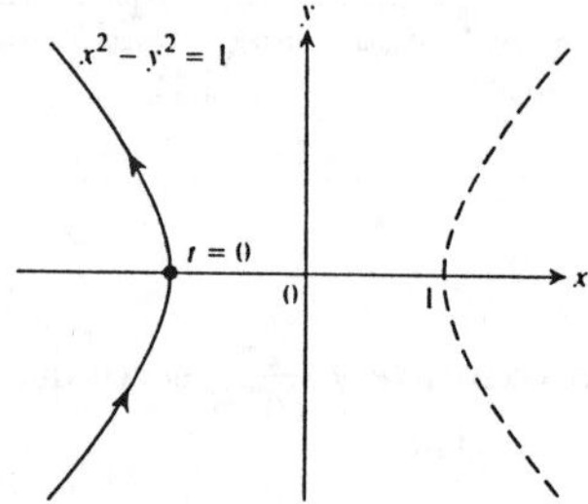

13. $x = x = 1 - t,\ y = 1 + t,\ -\infty < t < \infty$

$\Rightarrow 1 - x = t \Rightarrow y = 1 + (1 - x)$

$\Rightarrow y = -x + 2$

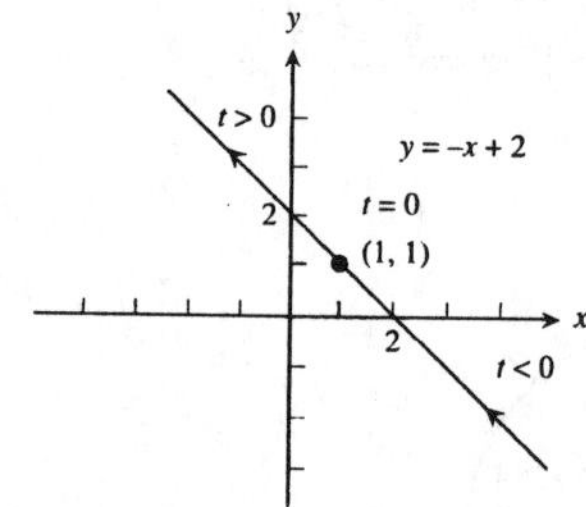

15. $x = t,\ y = \sqrt{1 - t^2},\ -1 \le t \le 0$

$\Rightarrow y = \sqrt{1 - x^2}$

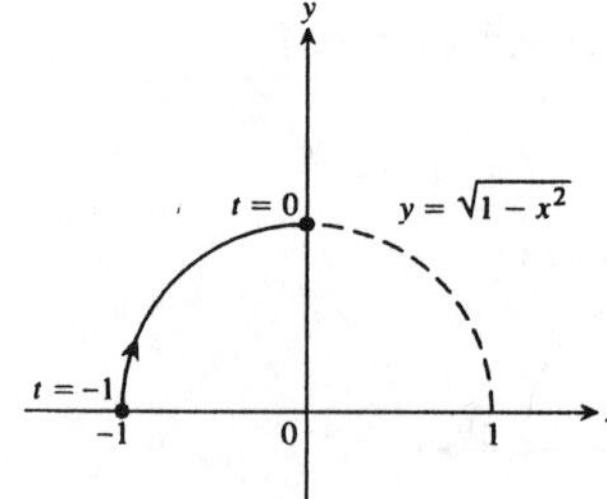

17. $x = e^t + e^{-t}$, $y = e^t - e^{-t}$, $-\infty < t < \infty$

$$\left(e^t + e^{-t}\right)^2 - \left(e^t - e^{-t}\right)^2 = \left(e^{2t} + 2 + e^{-2t}\right)$$
$$- \left(e^{2t} - 2 + e^{-2t}\right) = 4 \Rightarrow x^2 - y^2 = 4$$

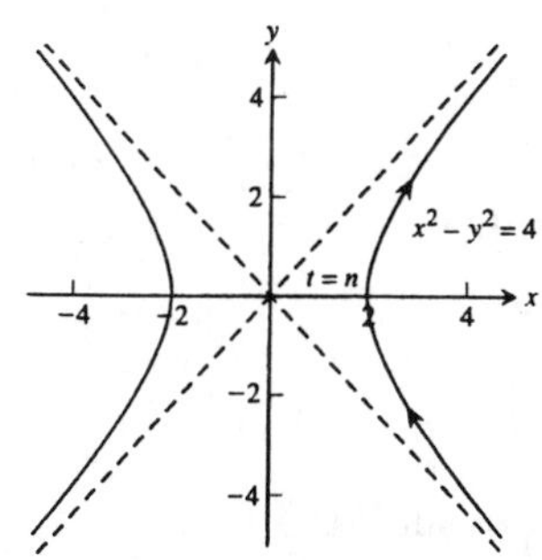

19. (a) $x = a \cos t$, $y = -a \sin t$, $0 \le t \le 2\pi$

(b) $x = a \cos t$, $y = a \sin t$, $0 \le t \le 2\pi$

(c) $x = a \cos t$, $y = -a \sin t$, $0 \le t \le 4\pi$

(d) $x = a \cos t$, $y = a \sin t$, $0 \le t \le 4\pi$

21. Using $(-1, -3)$ we create the parametric equations $x = -1 + at$ and $y = -3 + bt$, representing a line which goes through $(-1, -3)$ at $t = 0$. We determine a and b so that the line goes through $(4, 1)$ when $t = 1$.
Since $4 = -1 + a$, $a = 5$.
Since $1 = -3 + b$, $b = 4$.
Therefore, one possible parametrization is $x = -1 + 5t$, $y = -3 + 4t$, $0 \le t \le 1$.

23. The lower half of the parabola is given by $x = y^2 + 1$ for $y \le 0$. Substituting t for y, we obtain one possible parametrization $x = t^2 + 1$, $y = t$, $t \le 0$.

25. For simplicity, we assume that x and y are linear functions of t and that the point (x, y) starts at $(2, 3)$ for $t = 0$ and passes through $(-1, -1)$ at $t = 1$. Then $x = f(t)$, where $f(0) = 2$ and $f(1) = -1$.
Since slope $= \dfrac{\Delta x}{\Delta t} = \dfrac{-1 - 2}{1 - 0} = -3$, $x = f(t) = -3t + 2 = 2 - 3t$. Also, $y = g(t)$, where $g(t) = 3$ and $g(1) = -1$.
Since slope $= \dfrac{\Delta y}{\Delta t} = \dfrac{-1 - 3}{1 - 0} = -4$, $y = g(t) = -4t + 3 = 3 - 4t$.
One possible parametrization is: $x = 2 - 3t$, $y = 3 - 4t$, $t \ge 0$.

27. Graph (c). Window: $[-4, 4]$ by $[-3, 3]$, $0 \le t \le 2\pi$

29. Graph (d). Window: $[-10, 10]$ by $[-10, 10]$, $0 \le t \le 2\pi$

31. Graph of f: $x_1 = t$, $y_1 = e^t$

Graph of f^{-1}: $x_2 = e^t$, $y_2 = t$

Graph of $y = x$: $x_3 = t$, $y_3 = t$

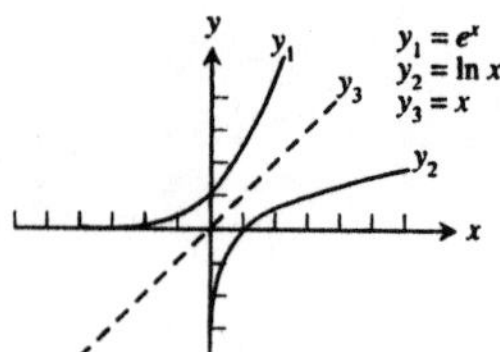

33. Graph of f: $x_1 = t$, $y_1 = 2^{-t}$

Graph of f^{-1}: $x_2 = 2^{-1t}$, $y_2 = t$

Graph of $y = x$: $x_3 = t$, $y_3 = t$

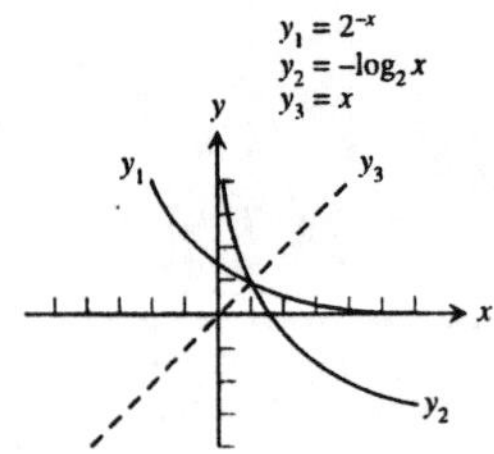

35. Graph of f: $x_1 = t$, $y_1 = \ln t$

Graph of f^{-1}: $x_2 = \ln t$, $y_2 = t$

Graph of $y = x$: $x_3 = t$, $y_3 = t$

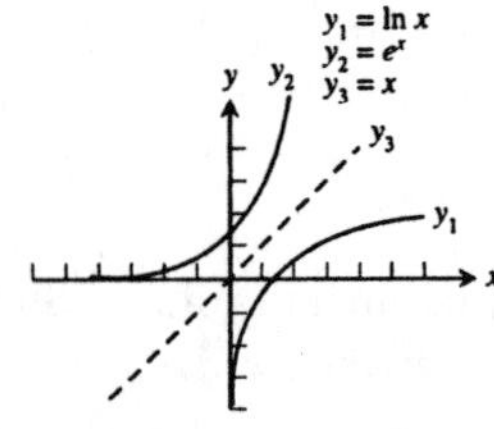

37. Graph of f: $x_1 = t$, $y_1 = \sin^{-1} t$

Graph of f^{-1}: $x_2 = \sin^{-1} t$, $y_2 = t$

Graph of $y = x$: $x_3 = t$, $y_3 = t$

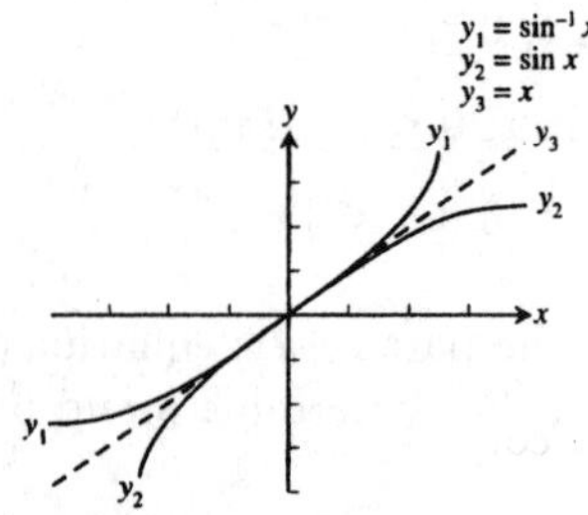

39. The graph is in Quadrant I when $0 < y < 2$, which corresponds to $1 < t < 3$. To confirm, note that $x(1) = 2$ and $x(3) = 0$.

41. The graph is in Quadrant III when $-6 \leq y < -4$, which corresponds to $-5 \leq t < -3$. To confirm, note that $x(-5) = -2$ and $x(-3) = 0$.

43. (a)

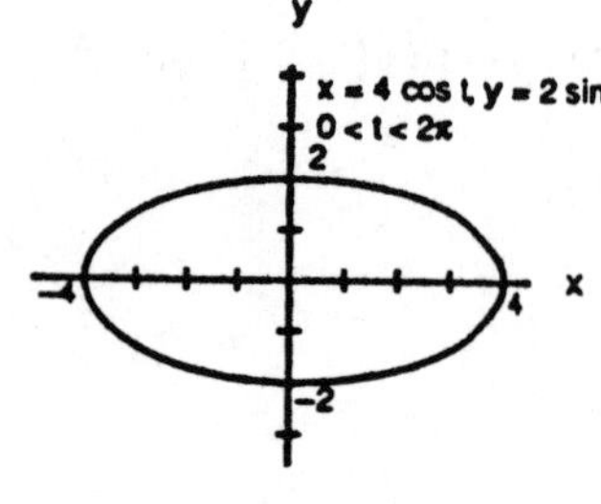

(b)

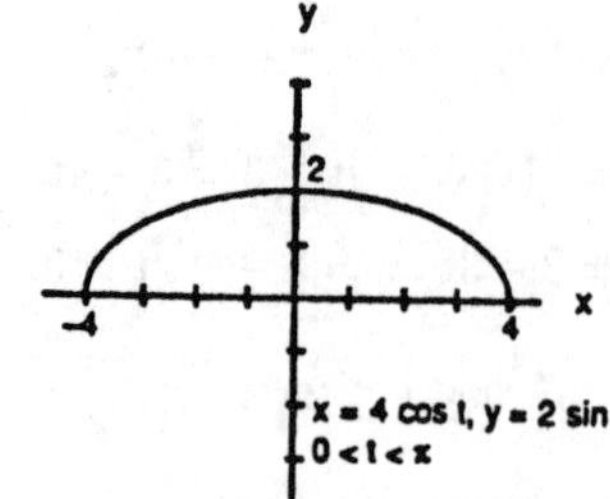

(c)

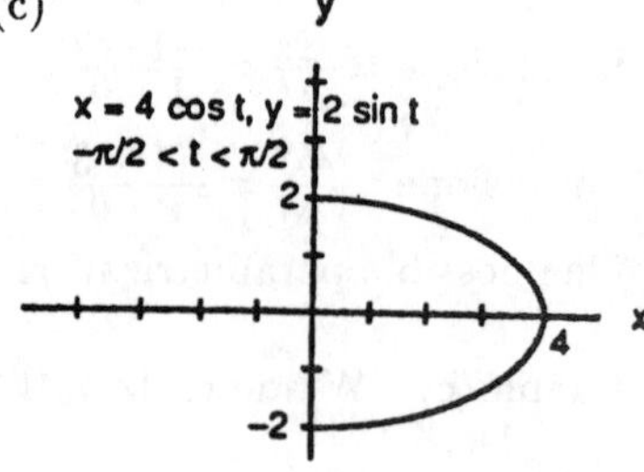

45.

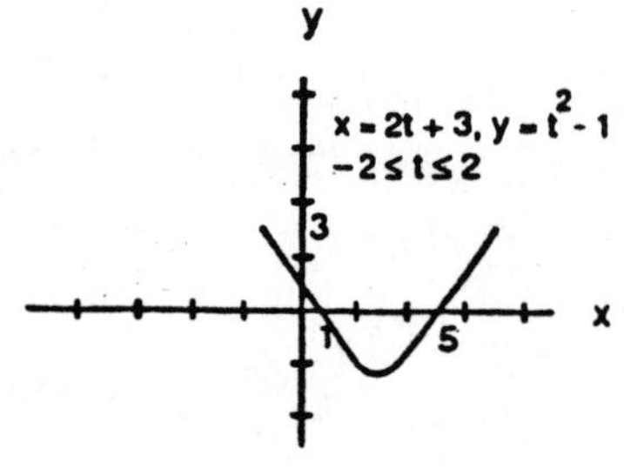

47. (a)

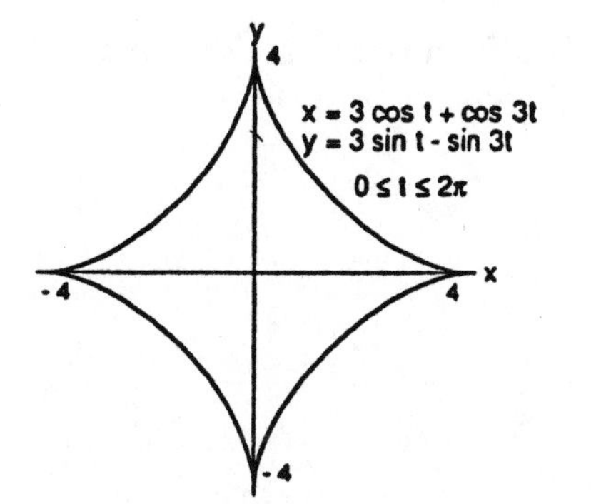

(b)

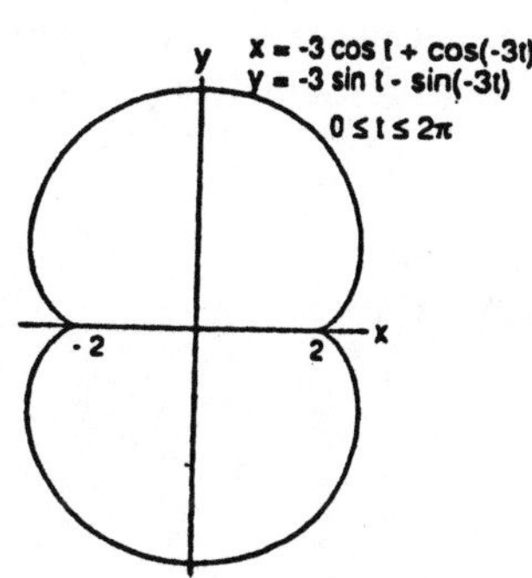

49. Extend the vertical line through A to the x-axis and let C be the point of intersection. Then $\mathrm{OC} = \mathrm{AQ} = x$ and $\tan t = \frac{2}{\mathrm{OC}} = \frac{2}{x} \Rightarrow x = \frac{2}{\tan t} = 2 \cot t$; $\sin t = \frac{2}{\mathrm{OA}}$ $\Rightarrow \mathrm{OA} = \frac{2}{\sin t}$; and $(\mathrm{AB})(\mathrm{OA}) = (\mathrm{AQ})^2 \Rightarrow \mathrm{AB}\left(\frac{2}{\sin t}\right) = x^2$ $\Rightarrow \mathrm{AB}\left(\frac{2}{\sin t}\right) = \left(\frac{2}{\tan t}\right)^2 \Rightarrow \mathrm{AB} = \frac{2 \sin t}{\tan^2 t}$. Next $y = 2 - \mathrm{AB} \sin t \Rightarrow y = 2 - \left(\frac{2 \sin t}{\tan^2 t}\right) \sin t =$ $2 - \frac{2 \sin^2 t}{\tan^2 t} = 2 - 2\cos^2 t = 2 \sin^2 t$. Therefore let $x = 2 \cot t$ and $y = 2 \sin^2 t$, $0 < t < \pi$.

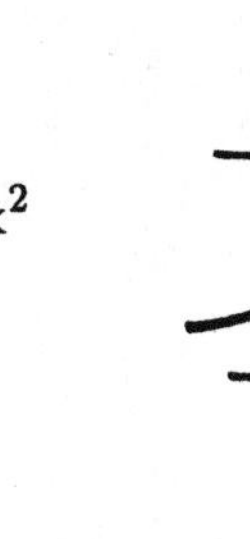

51. (a) $-5 \le x \le 5 \Rightarrow -5 \le 2 \cot t \le 5 \Rightarrow -\frac{5}{2} \le \cot t \le \frac{5}{2}$

The graph of cot t shows where to look for the limits on t.

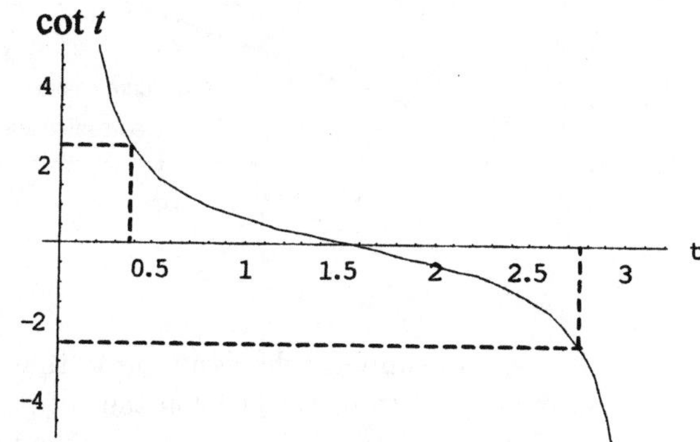

$\tan^{-1}\left(\frac{2}{5}\right) \le t \le \pi + \tan^{-1}\left(-\frac{2}{5}\right) \Rightarrow 0.381 \le t \le 2.761$

The curve is traced from right to left and extends infinitely in both directions from the origin.

(b) For $-\frac{\pi}{2} < t < \frac{\pi}{2}$, the curve is the same as that which is given. It first traces from the vertex at $(0, 2)$ to the left extreme point in the window, and then from the right extreme point in the window to the vertex point. For $0 < t < \frac{\pi}{2}$, only the right half of the curve appears, and it traces from the right extreme of the window to the vertex at $(0, 2)$ and terminates there. For $\frac{\pi}{2} < t < \pi$, only the left half of the curve appears, and it traces from the vertex to the left extreme of the window.

(c) For $x = -2 \cot t$, the curve traces from left to right rather than from right to left. For $x = 2 \cot(\pi - t)$, the curve traces from right to left, as it does with the original parametrization.

P.7 MODELING CHANGE

1. (a)

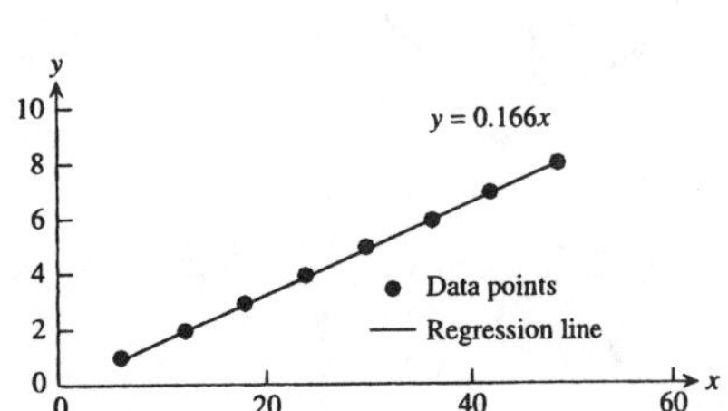

The graph supports the assumption that y is proportional to x. The constant of proportionality is estimated from the slope of the regression line, which is 0.166.

(b)

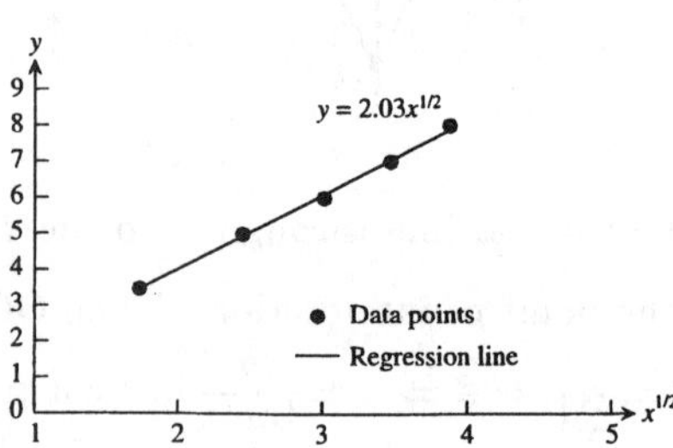

The graph supports the assumption that y is proportional to $x^{1/2}$. The constant of proportionality is estimated from the slope of the regression line, which is 2.03.

(c) Because of the wide range of values of the data, two graphs are needed to observe all of the points in relation to the regression line.

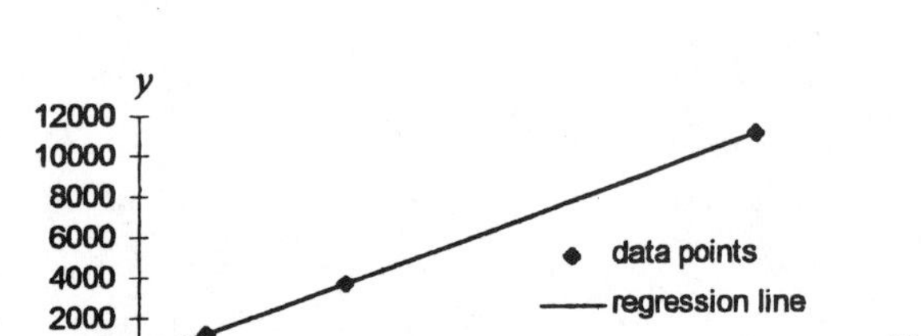

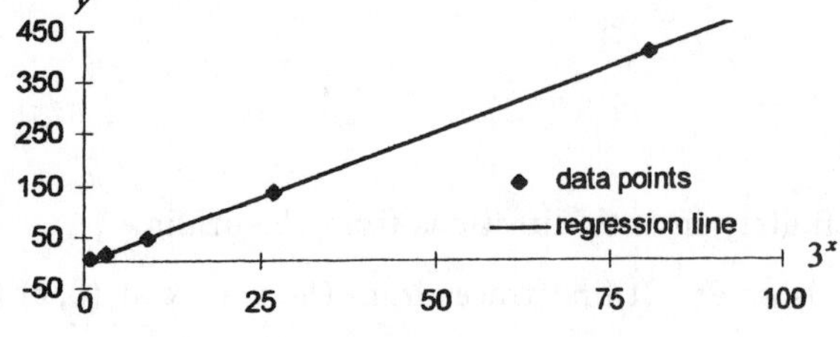

The graphs support the assumption that y is proportional to 3^x. The constant of proportionality is estimated from the slope of the regression line, which is 5.00.

(d)

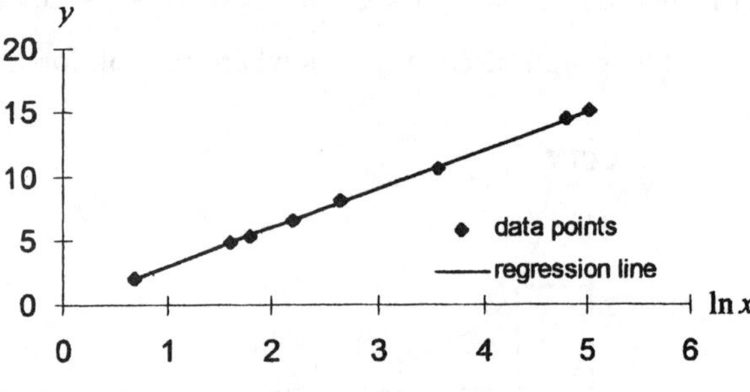

The graph supports the assumption that y is proportional to ln x. The constant of proportionality is estimated from the slope of the regression line, which is 2.99.

3. First, plot the braking distance versus the speed.

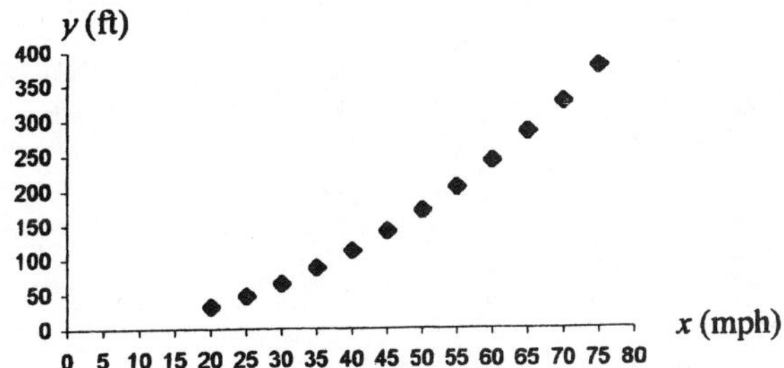

The shape of the graph suggests either a power function or an exponential function to describe the relationship. First, try to fit a quadratic function. Using quadratic regression on the TI-92 Plus calculator gives $y = 0.064555x^2 + 0.078422x + 4.88961$.

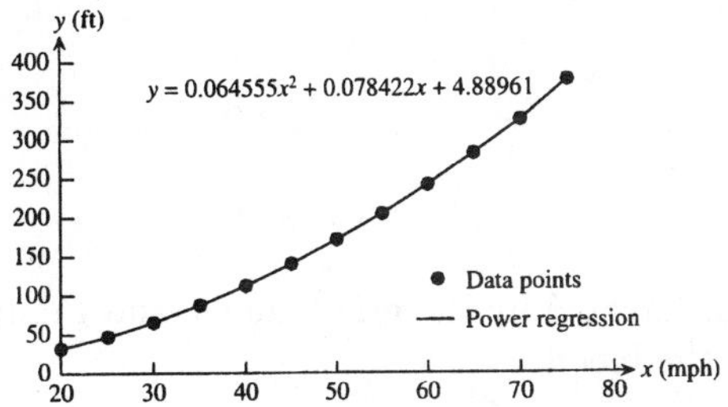

The quadratic regression fits the data well as seen by the following plot of the relative errors versus the actual stopping distance.

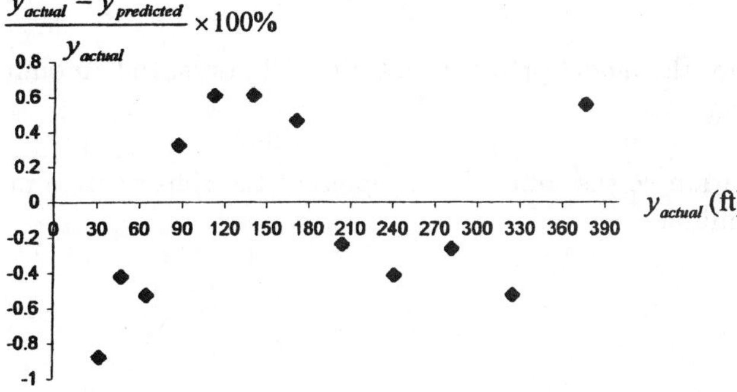

The largest relative error is less than 1%.

5. (a) First plot the amount of digoxin in the blood versus time.

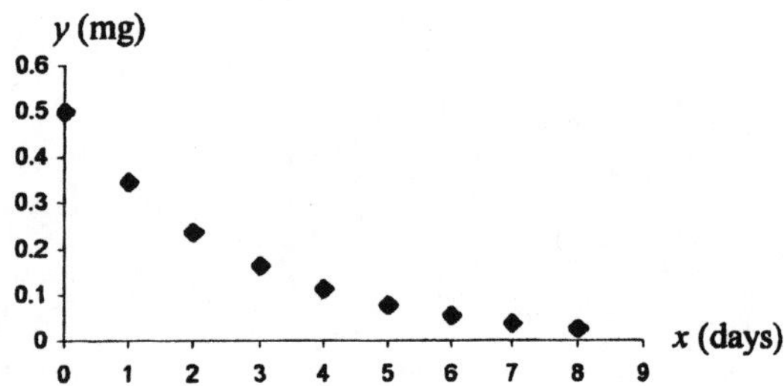

The graph suggests that the amount decays exponentially with time. The exponential regression on the TI-92 Plus calculator gives $y = 0.5(0.69^x) = 0.5e^{-0.371x}$.

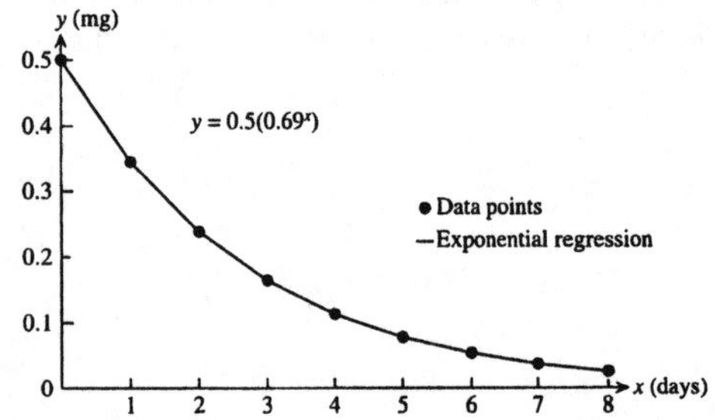

(b) The exponential function fits the data very well as demonstrated by the graph above and the following is a plot of the relative error versus the actual amount in the blood.

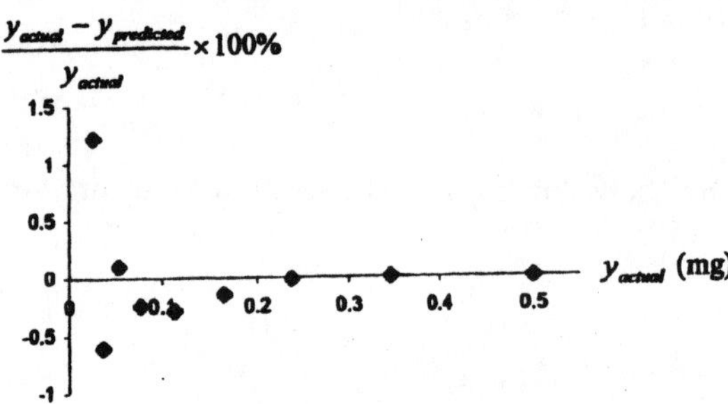

(c) $y(12) = 0.5e^{-0.371(12)} = 0.00583$, therefore, the model predicts that after 12 hours, the amount of digoxin in the blood will be less than 0.006 mg.

7. (a) First, plot a graph of the blood concentration versus time. Let t represent the elapsed time in days and C the blood concentration in parts per million.

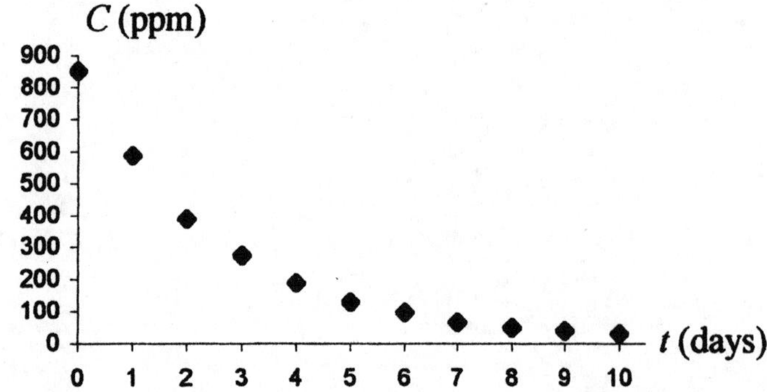

The graph suggests that the amount decays exponentially with time. The exponential regression function on the TI-92 Plus calculator gives $C = 770\left(0.7146^t\right) = 770e^{-0.336t}$.

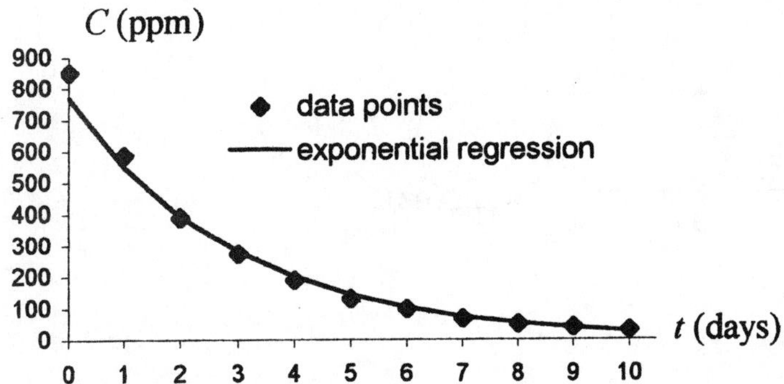

(b) The exponential function appears to capture a trend for this data. The following graph shows the relative errors in the model estimates.

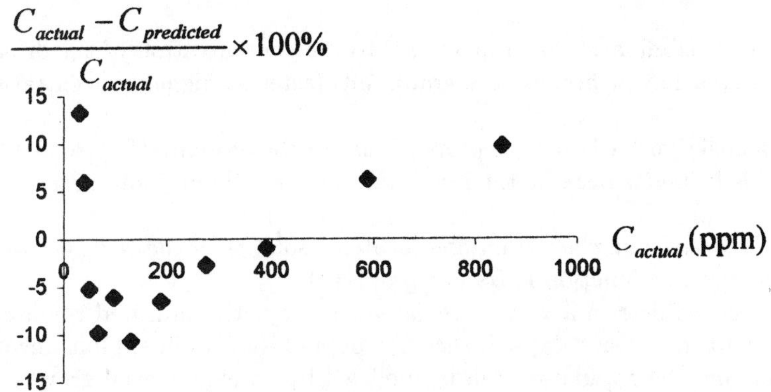

The relative errors in the predicted values are as large as 13.2% and the errors are large for small as well as large blood concentrations. The pattern of the residual errors does not suggest an obvious improvement of the model.

(c) $10 = 770e^{-0.336t} \Rightarrow t = -\frac{1}{0.336}\ln\left(\frac{10}{770}\right) = 12.93$ days. Therefore, the model predicts that the blood concentration will fall below 10 ppm after 12 days and 22 hours.

9.

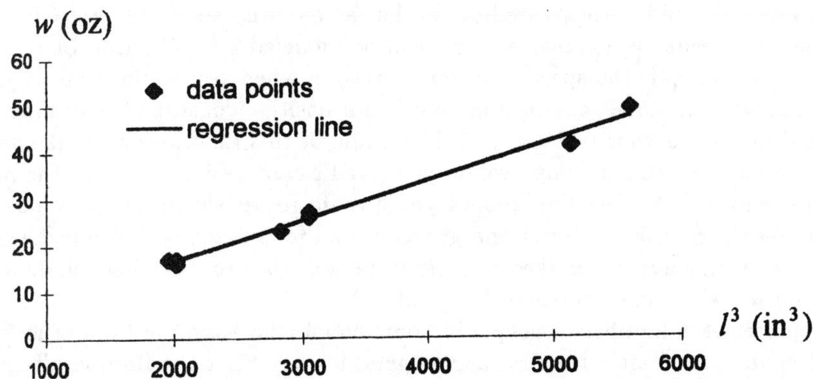

The slope of the regression line is 0.008435, so the model that estimates the weight as a function of L is $w = 0.008435L^3$. The model fits the data reasonably well as demonstrated by the following plot of the relative errors in the weight estimates by the model.

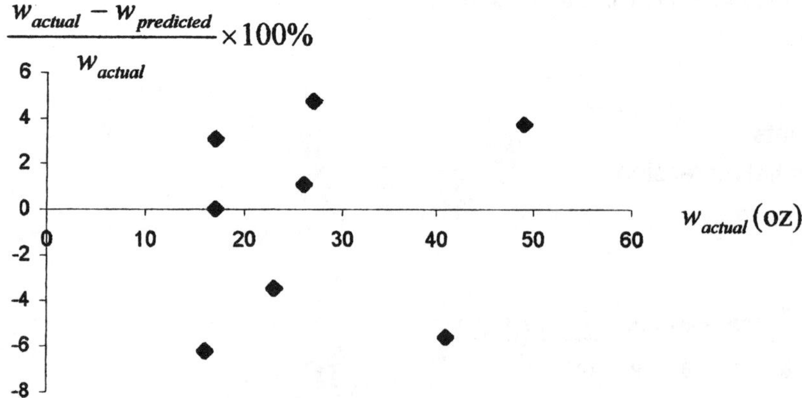

The relative error in the estimated values is always less than 7%.

11. Graph (c). For some drugs, the rate of elimination is proportional to the concentration of the drug in the blood-stream. Graph (c) matches this behavior because the graph falls faster at higher concentrations.

13. Graph (c). The rate of decay of radioactive Carbon-14 is proportional to the amount of Carbon-14 present in the artwork. Graph (c) matches this behavior because the graph falls faster at higher amounts.

15. (a) One possibility: If an item sells for $p and x is the number of items sold, then the revenue from sales will be $y = px$, and the graph of the revenue function looks like graph (a).
    (b) One possibility: If y is the number of deer in a very large game reserve with unlimited resources to support the deer and x represents the number of years elapsed, then the population would exhibit unconstrained growth over time. In this situation, the population can be modeled by an exponential growth function like $y = y_0 e^{kx}$, where y_0 is the initial deer population, k is a constant, and the growth of the function looks like graph (b).
    (c) One possibility: If y represents the selling price per unit that can be realized for a certain commodity, say grape jelly for example, and x represents the availability of the commodity, then the unit selling price for the commodity is often times inversely proportional to its availability. This relationship can be modeled with a function of the form $y = \frac{y_0}{(x+1)^{\alpha}}$, where y_0 is the unit selling price when no product is available, α is a positive constant, and the graph of the function looks like graph (c).
    (d) One possibility: Let y represent the speed of your car and x represent the amount of time after you punch the accelerator. At first you will rapidly accelerate but, as the car picks up speed, the rate of acceleration (i.e., the rate at which the car speeds up) decreases. This can be modeled by a function like $y = y_{new} + (y_0 - y_{new})e^{-kx}$, where y_0 is the speed you were traveling when you stepped on it, y_{new} is the new speed you achieve when you are done accelerating, and k is a positive constant (determined in part by the size of your engine and how good your traction is). The graph of this function looks like graph(d).
    (e) One possibility: Let y represent the amount you owe on your credit card and x represent the number of monthly payments you have made. At first the amount you owe decreases slowly because most of your payment goes toward paying the monthly interest charge. But, as the amount you owe decreases, the interest charge decreases and your payment makes a bigger difference toward reducing the debt. This can be modeled with a function like the one represented by graph (c).
    (f) One possibility: Let y represent the number of people in your school who have the flu and let x represent the number of days that have elapsed after the first person gets sick. At first the flu doesn't spread very quickly because there are only a few sick people to pass it on. But, as more people get sick the disease spreads more rapidly. The most volatile mixture is when half the people are sick, because then there are a lot of sick people to spread the disease and a lot of uninfected people who can still catch it. As time continues and more people get sick, there are fewer and fewer people available to catch the flu and the

spread of the disease begins to slow down. This behavior can be modeled with a function like the one represented by graph (f).

17.

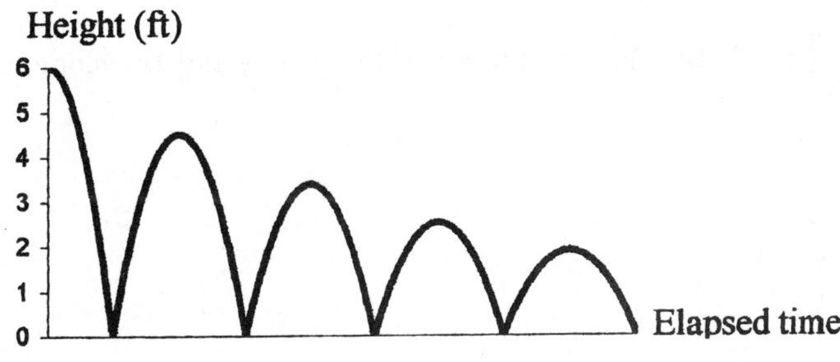

19. (a)

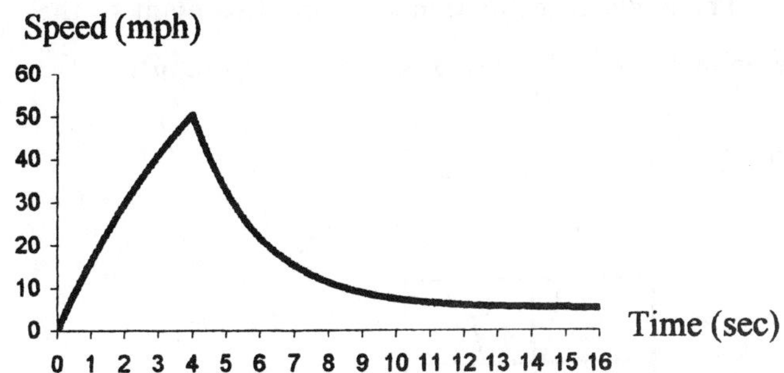

(b)

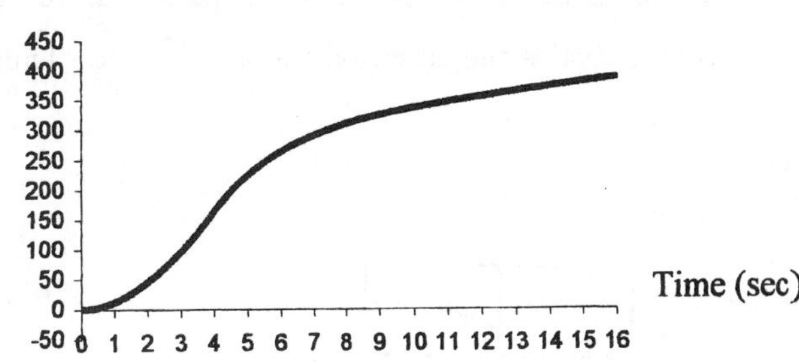

21. (a) The graph could represent the angle that a pendulum makes with the vertical as it swings back and forth. The variable y represents the angle and x represents time. Because of friction, the amplitude of the oscillation decays, as depicted by the graph. When y is positive, the pendulum is on one side of the vertical and when y is negative, the pendulum is on the other side.

(b) The graph could represent the angle the playground swing makes with the vertical as a child "pumps" on the swing to get it going. The variable y represents the angle and x represents time. Because the child puts mechanical energy into the system (swing + child), the amplitude of the oscillation grows with time, as depicted by the graph. When y is positive, the swing is on one side of the vertical and when y is negative, the swing is on the other side.

PRELIMINARY CHAPTER PRACTICE EXERCISES

1. $y = 3(x-1) + (-6)$
   $y = 3x - 9$

3. $x = 0$

5. $y = 2$

7. $y = -3x + 3$

9. Since $4x + 3y = 12$ is equivalent to $y = -\frac{4}{3}x + 4$, the slope of the given line (and hence the slope of the desired line) is $-\frac{4}{3}$.

$y = -\frac{4}{3}(x - 4) - 12$

$y = -\frac{4}{3}x - \frac{20}{3}$

11. Since $\frac{1}{2}x + \frac{1}{3}y = 1$ is equivalent to $y = -\frac{3}{2}x + 3$, the slope of the given line is $-\frac{3}{2}$ and the slope of the perpendicular line is $\frac{2}{3}$.

$y = \frac{2}{3}(x + 1) + 2$

$y = \frac{2}{3}x + \frac{8}{3}$

13. The area is $A = \pi r^2$ and the circumference is $C = 2\pi r$. Thus, $r = \frac{C}{2\pi} \Rightarrow A = \pi\left(\frac{C}{2\pi}\right)^2 = \frac{C^2}{4\pi}$.

15. The coordinates of a point on the parabola are (x, x^2). The angle of inclination θ joining this point to the origin satisfies the equation $\tan\theta = \frac{x^2}{x} = x$. Thus the point has coordinates $(x, x^2) = (\tan\theta, \tan^2\theta)$.

17.

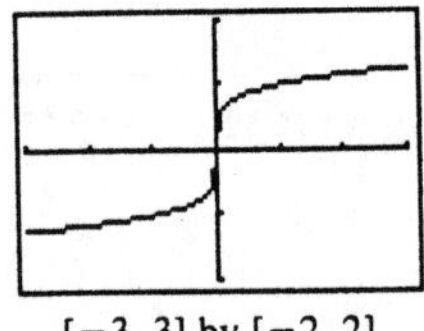

[−3, 3] by [−2, 2]

Symmetric about the origin.

19.

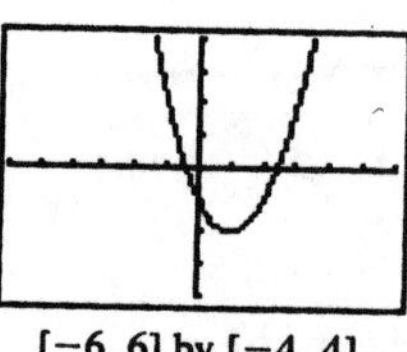

[−6, 6] by [−4, 4]

Neither

21. $y(-x) = (-x)^2 + 1 = x^2 + 1 = y(x)$
Even

23. $y(-x) = 1 - \cos(-x) = 1 - \cos x = y(x)$
Even

25. $y(-x) = \frac{(-x)^4 + 1}{(-x)^3 - 2(-x)} = \frac{x^4 + 1}{-x^3 + 2x} = -\frac{x^4 + 1}{x^3 - 2x} = -y(x)$

Odd

27. $y(-x) = -x + \cos(-x) = -x + \cos x$
Neither even nor odd

29. (a) The function is defined for all values of x, so the domain is $(-\infty, \infty)$.

(b) Since $|x|$ attains all nonnegative values, the range is $[-2, \infty)$.

31. (a) Since the square root requires $16 - x^2 \geq 0$, the domain is $]-4, 4]$.

(b) For values of x in the domain, $0 \leq 16 - x^2 \leq 16$, so $0 \leq \sqrt{16 - x^2} \leq 4$. The range is $[0, 4]$.

33. (a) The function is defined for all values of x, so the domain is $(-\infty, \infty)$.

(b) Since $2e^{-x}$ attains all positive values, the range is $(-3, \infty)$.

35. (a) The function is defined for all values of x, so the domain is $(-\infty, \infty)$.
(b) The sine function attains values from -1 to 1, so $-2 \le 2\sin(3x+\pi) \le 2$, and hence $-3 \le 2\sin(3x+\pi) - 1 \le 1$. The range is $[-3, 1]$.

37. (a) The logarithm requires $x - 3 > 0$, so the domain is $(3, \infty)$.
(b) The logarithm attains all real values, so the range is $(-\infty, \infty)$.

39. (a) The function is defined for $-4 \le x \le 4$, so the domain is $[-4, 4]$.
(b) The function is equivalent to $y = \sqrt{|x|}$, $-4 \le x \le 4$, which attains values from 0 to 2 for x in the domain. The range is $[0, 2]$.

41. First piece: Line through $(0, 1)$ and $(1, 0)$

$$m = \frac{0-1}{1-0} = \frac{-1}{1} = -1$$

$y = -x + 1$ or $1 - x$

Second piece: Line through $(1, 1)$ and $(2, 0)$

$$m = \frac{0-1}{2-1} = \frac{-1}{1} = -1$$

$y = -(x-1) + 1$

$y = -x + 2$ or $2 - x$

$$f(x) = \begin{cases} 1 - x, & 0 \le x < 1 \\ 2 - x, & 1 \le x \le 2 \end{cases}$$

43. (a) $(f \circ g)(-1) = f(g(-1)) = f\left(\frac{1}{\sqrt{-1+2}}\right) = f(1) = \frac{1}{1} = 1$

(b) $(g \circ f)(2) = g(f(2)) = g\left(\frac{1}{2}\right) = \frac{1}{\sqrt{1/2+2}} = \frac{1}{\sqrt{2.5}}$ or $\sqrt{\frac{2}{5}}$

(c) $(f \circ f)(x) = f(f(x)) = f\left(\frac{1}{x}\right) = \frac{1}{1/x} = x,\ x \ne 0$

(d) $(g \circ g)(x) = g(g(x)) = g\left(\frac{1}{\sqrt{x+2}}\right) = \frac{1}{\sqrt{1/\sqrt{x+2}+2}}$

$$= \frac{\sqrt[4]{x+2}}{\sqrt{1+2\sqrt{x+2}}}$$

45. (a) $(f \circ g)(x) = f(g(x))$

$= f\left(\sqrt{x+2}\right)$

$= 2 - \left(\sqrt{x+2}\right)^2$

$= -x,\ x \ge -2$

$(g \circ f)(x) = g(f(x))$

$= g\left(2 - x^2\right)$

$= \sqrt{(2-x^2)+2} = \sqrt{4-x^2}$

(b) Domain of $f \circ g$: $[-2, \infty)$

Domain of $g \circ f$: $[-2, 2]$

(c) Range of f ∘ g: $(-\infty, 2]$

Range of g ∘ f: $[0, 2]$

47.

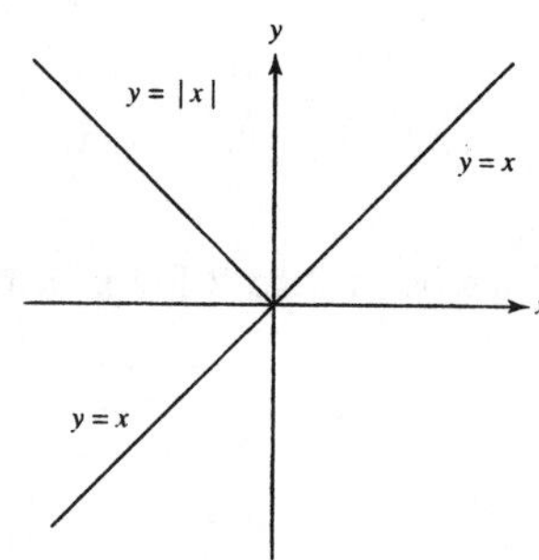

The graph of $f_2(x) = f_1(|x|)$ is the same as the graph of $f_1(x)$ to the right of the y-axis. The graph of $f_2(x)$ to the left of the y-axis is the reflection of $y = f_1(x)$, $x \geq 0$ across the y-axis.

49.

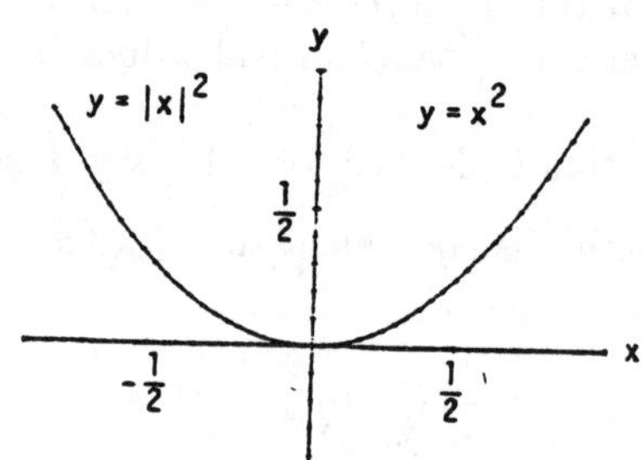

It does not change the graph.

51.

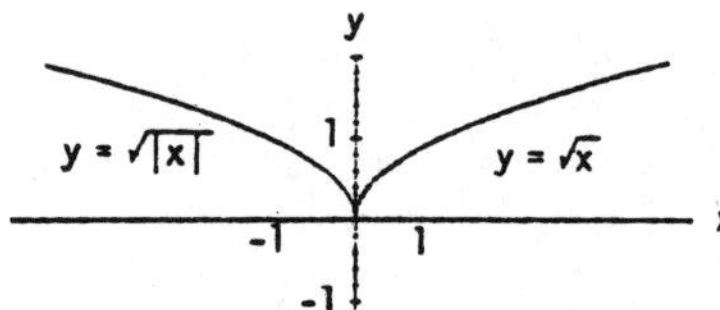

The graph of $f_2(x) = f_1(|x|)$ is the same as the graph of $f_1(x)$ to the right of the y-axis. The graph of $f_2(x)$ to the left of the y-axis is the reflection of $y = f_1(x)$, $x \geq 0$ across the y-axis.

53.

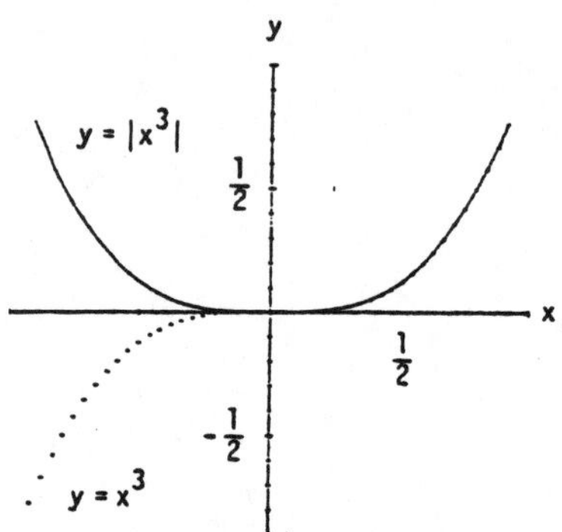

Whenever $g_1(x)$ is positive, the graph of $y = g_2(x) = |g_1(x)|$ is the same as the graph of $y = g_1(x)$. When $g_1(x)$ is negative, the graph of $y = g_2(x)$ is the reflection of the graph of $y = g_1(x)$ across the x-axis.

55.

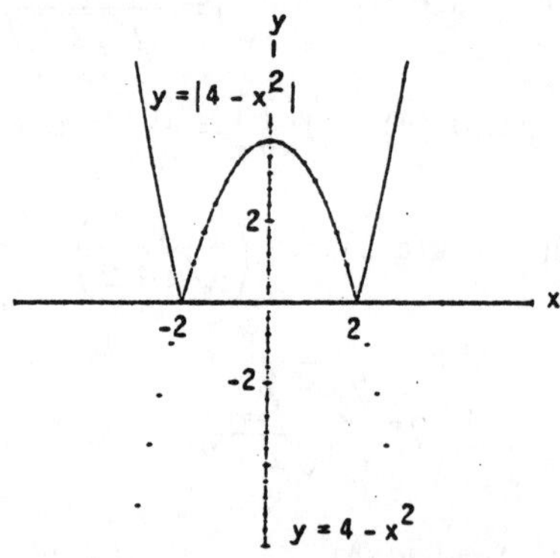

Whenever $g_1(x)$ is positive, the graph of $y = g_2(x) = |g_1(x)|$ is the same as the graph of $y = g_1(x)$. When $g_1(x)$ is negative, the graph of $y = g_2(x)$ is the reflection of the graph of $y = g_1(x)$ across the x-axis.

57. (a) The graph is symmetric about $y = x$.

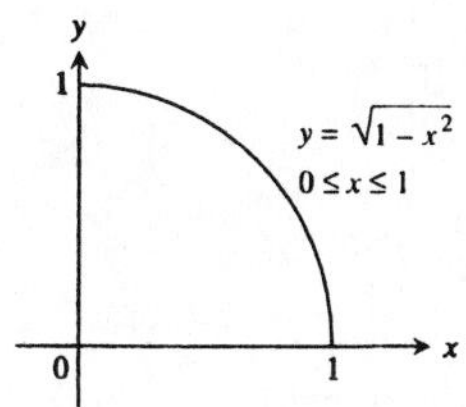

(b) $y = \sqrt{1-x^2} \Rightarrow y^2 = 1 - x^2 \Rightarrow x^2 = 1 - y^2 \Rightarrow x = \sqrt{1-y^2} \Rightarrow y = \sqrt{1-x^2} = f^{-1}(x)$

59. (a) $y = 2 - 3x \to 3x = 2 - y \to x = \frac{2-y}{3}$.

Interchange x and y: $y = \frac{2-x}{3} \to f^{-1}(x) = \frac{2-x}{3}$

Verify.

$(f \circ f^{-1})(x) = f(f^{-1}(x)) = f\left(\frac{2-x}{3}\right) = 2 - 3\left(\frac{2-x}{3}\right) = 2 - (2 - x) = x$

$(f^{-1} \circ f)(x) = f^{-1}(f(x)) = f^{-1}(2 - 3x) = \frac{2-(2-3x)}{3} = \frac{3x}{3} = x$

(b)

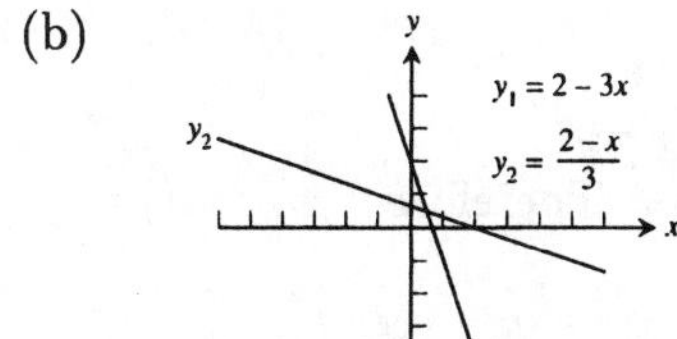

61. (a) $f(g(x)) = \left(\sqrt[3]{x}\right)^3 = x$, $g(f(x)) = \sqrt[3]{x^3} = x$

(b)

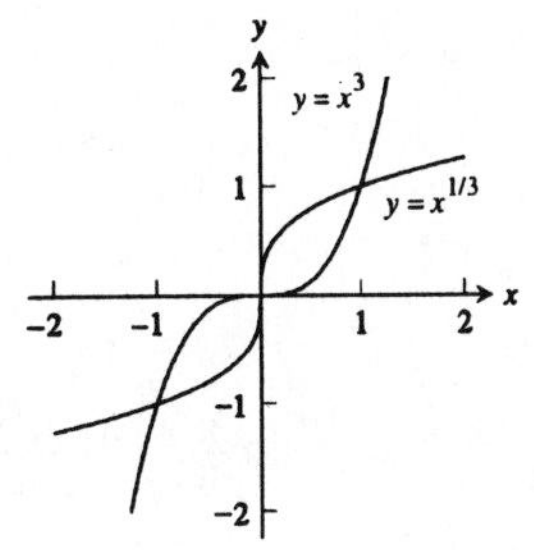

63. (a) $y = x + 1 \Rightarrow x = y - 1 \Rightarrow f^{-1}(x) = x - 1$

(b) $y = x + b \Rightarrow x = y - b \Rightarrow f^{-1}(x) = x - b$

(c) Their graphs will be parallel to one another and lie on opposite sides of the line $y = x$ equidistant from that line.

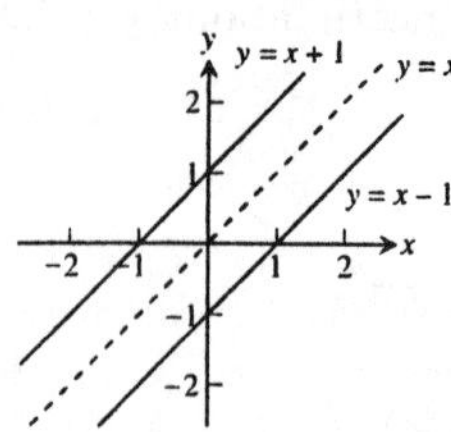

65. $x = 2.71828182846$ (using a TI-92 Plus calculator).

67. (a) $e^{\ln 7.2} = 7.2$ (b) $e^{-\ln x^2} = \frac{1}{e^{\ln x^2}} = \frac{1}{x^2}$ (c) $e^{\ln x - \ln y} = e^{\ln(x/y)} = \frac{x}{y}$

69. (a) $2 \ln \sqrt{e} = 2 \ln e^{1/2} = (2)\left(\frac{1}{2}\right) \ln e = 1$ (b) $\ln(\ln e^e) = \ln(e \ln e) = \ln e = 1$

(c) $\ln e^{\left(-x^2 - y^2\right)} = \left(-x^2 - y^2\right) \ln e = -x^2 - y^2$

71. Using a calculator, $\sin^{-1}(0.6) \approx 0.6435$ radians or $36.8699°$.

73. Since $\cos\theta = \frac{3}{7}$ and $0 \le \theta \le \pi$, $\sin\theta = \sqrt{1 - \cos^2\theta} = \sqrt{1 - \left(\frac{3}{7}\right)^2} = \sqrt{\frac{40}{49}} = \frac{\sqrt{40}}{7}$. Therefore,

$\sin\theta = \frac{\sqrt{40}}{7}$, $\cos\theta = \frac{3}{7}$, $\tan\theta = \frac{\sin\theta}{\cos\theta} = \frac{\sqrt{40}}{3}$, $\cot\theta = \frac{1}{\tan\theta}, = \frac{3}{\sqrt{40}}$, $\sec\theta = \frac{1}{\cos\theta} = \frac{7}{3}$, $\csc\theta = \frac{1}{\sin\theta} = \frac{7}{\sqrt{40}}$

75.

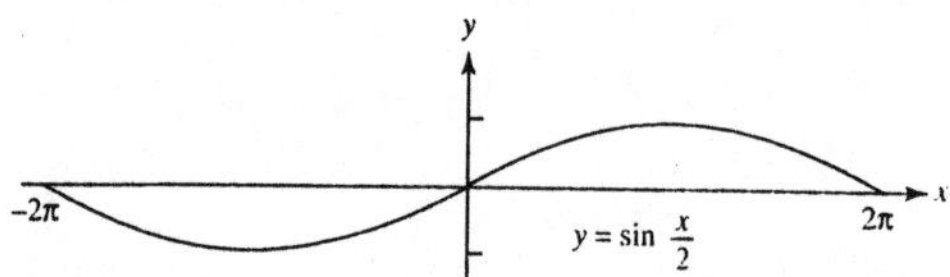

period $= 4\pi$

77.

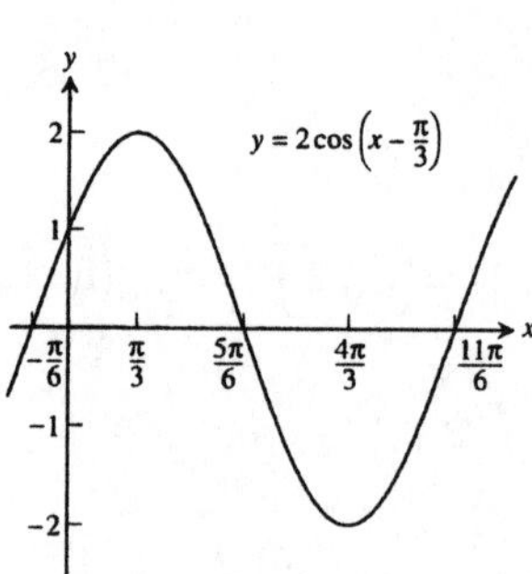

period $= 2\pi$

79. (a) $\sin B = \sin\frac{\pi}{3} = \frac{b}{c} = \frac{b}{2} \Rightarrow b = 2\sin\frac{\pi}{3} = 2\left(\frac{\sqrt{3}}{2}\right) = \sqrt{3}$. By the theorem of Pythagoras,

$a^2 + b^2 = c^2 \Rightarrow a = \sqrt{c^2 - b^2} = \sqrt{4 - 3} = 1$.

(b) $\sin B = \sin\frac{\pi}{3} = \frac{b}{c} = \frac{2}{c} \Rightarrow c = \frac{2}{\sin\frac{\pi}{3}} = \frac{2}{\left(\frac{\sqrt{3}}{2}\right)} = \frac{4}{\sqrt{3}}$. Thus, $a = \sqrt{c^2 - b^2} = \sqrt{\left(\frac{4}{\sqrt{3}}\right)^2 - (2)^2} = \sqrt{\frac{4}{3}} = \frac{2}{\sqrt{3}}$.

81. (a) $\tan B = \frac{b}{a} \Rightarrow a = \frac{b}{\tan B}$ (b) $\sin A = \frac{a}{c} \Rightarrow c = \frac{a}{\sin A}$

83. Since sin x has period 2π, $\sin^3(x + 2\pi) = \sin^3(x)$. This function has period 2π. A graph shows that no smaller number works for the period.

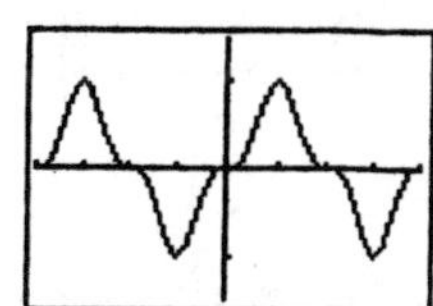

$[-2\pi, 2\pi]$ by $[-1.5, 1.5]$

85. $\cos\left(x + \frac{\pi}{2}\right) = \cos x \cos\left(\frac{\pi}{2}\right) - \sin x \sin\left(\frac{\pi}{2}\right) = (\cos x)(0) - (\sin x)(1) = -\sin x$

87. $\sin \frac{7\pi}{12} = \sin\left(\frac{\pi}{4} + \frac{\pi}{3}\right) = \sin \frac{\pi}{4} \cos \frac{\pi}{3} + \cos \frac{\pi}{4} \sin \frac{\pi}{3} = \left(\frac{\sqrt{2}}{2}\right)\left(\frac{1}{2}\right) + \left(\frac{\sqrt{2}}{2}\right)\left(\frac{\sqrt{3}}{2}\right) = \frac{\sqrt{6} + \sqrt{2}}{4}$

89. (a) $\frac{\pi}{6}$ (b) $-\frac{\pi}{4}$ (c) $\frac{\pi}{3}$

91. (a) $\frac{\pi}{4}$ (b) $\frac{5\pi}{6}$ (c) $\frac{\pi}{3}$

93. $\sec\left(\cos^{-1} \frac{1}{2}\right) = \sec\left(\frac{\pi}{3}\right) = 2$

95. $\tan\left(\sec^{-1} 1\right) + \sin\left(\csc^{-1}(-2)\right) = \tan\left(\cos^{-1} \frac{1}{1}\right) + \sin\left(\sin^{-1}\left(-\frac{1}{2}\right)\right) = \tan(0) + \sin\left(-\frac{\pi}{6}\right) = 0 + \left(-\frac{1}{2}\right) = -\frac{1}{2}$

97. $\alpha = \tan^{-1} 2x$ indicates the diagram

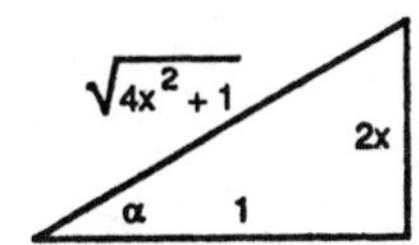

$\Rightarrow \sec\left(\tan^{-1} 2x\right) = \sec \alpha = \sqrt{4x^2 + 1}$

99. $\alpha = \cos^{-1} x$ indicates the diagram

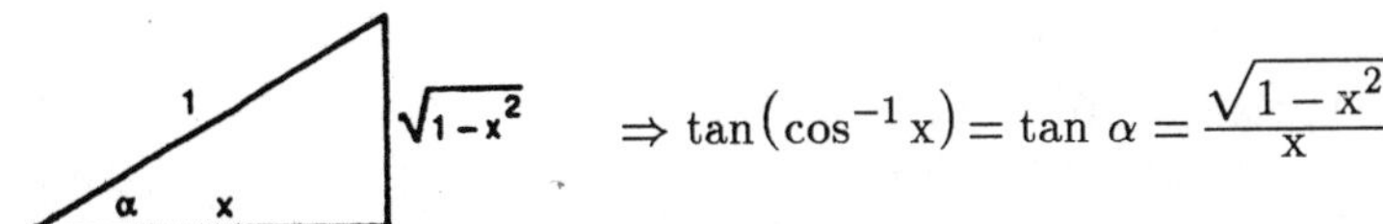

$\Rightarrow \tan\left(\cos^{-1} x\right) = \tan \alpha = \frac{\sqrt{1 - x^2}}{x}$

101. (a) Defined; there is an angle whose tangent is 2.
(b) Not defined; there is no angle whose cosine is 2.

103. (a) Not defined; there is no angle whose secant is 0.
(b) Not defined; there is no angle whose sine is $\sqrt{2}$.

105. Let h = height of vertical pole, and let b and c denote the distances of points B and C from the base of the pole, measured along the flat ground, respectively. Then, $\tan 50° = \frac{h}{c}$, $\tan 35° = \frac{h}{b}$, and $b - c = 10$. Thus, $h = c \tan 50°$ and $h = b \tan 35° = (c+10)\tan 35°$

$\Rightarrow c \tan 50° = (c+10)\tan 35° \Rightarrow c\,(\tan 50° - \tan 35°) = 10 \tan 35°$

$\Rightarrow c = \dfrac{10 \tan 35°}{\tan 50° - \tan 35°} \Rightarrow h = c \tan 50° = \dfrac{10 \tan 35° \tan 50°}{\tan 50° - \tan 35°}$

≈ 16.98 m.

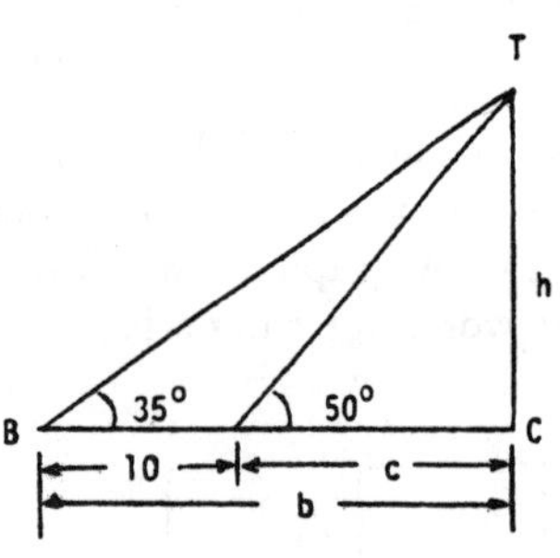

107. (a)

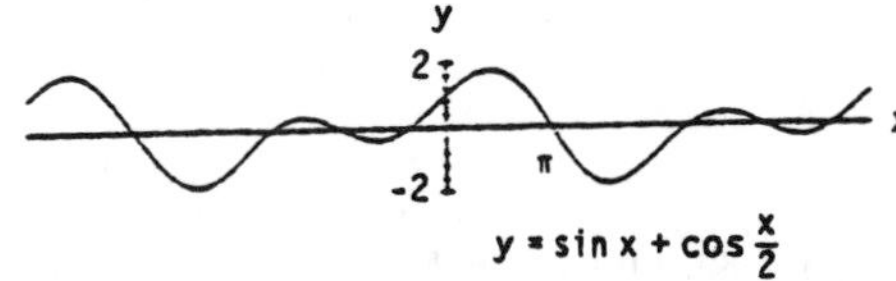

(b) The period appears to be 4π.

(c) $f(x+4\pi) = \sin(x+4\pi) + \cos\left(\frac{x+4\pi}{2}\right) = \sin(x+2\pi) + \cos\left(\frac{x}{2}+2\pi\right) = \sin x + \cos\frac{x}{2}$ since the period of sine and cosine is 2π. Thus, f(x) has period 4π.

109. (a) Substituting $\cos t = \frac{x}{5}$ and $\sin t = \frac{y}{2}$ in the identity $\cos^2 t + \sin^2 t = 1$ gives the Cartesian equation $\left(\frac{x}{5}\right)^2 + \left(\frac{y}{2}\right)^2 = 1$. The entire ellipse is traced by the curve.

(b)

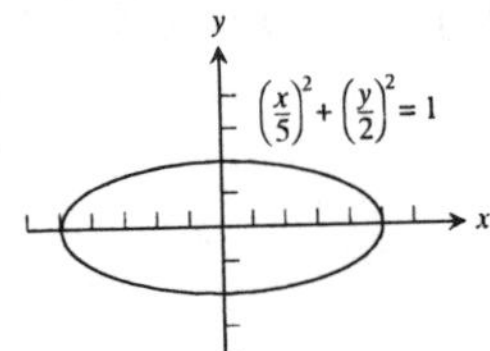

Initial point: (5, 0)
Terminal point: (5, 0)
The ellipse is traced exactly once in a counter-clockwise direction starting and ending at the point (5, 0).

111. (a) Substituting $t - x$ into $y = 11 - 2t$ gives the Cartesian equation $y = 11 - 2(2 - x)$, or $y = 2x + 7$. The part of the line from (4, 15) to (−2, 3) is traced by the parametrized curve. curve.

(b)

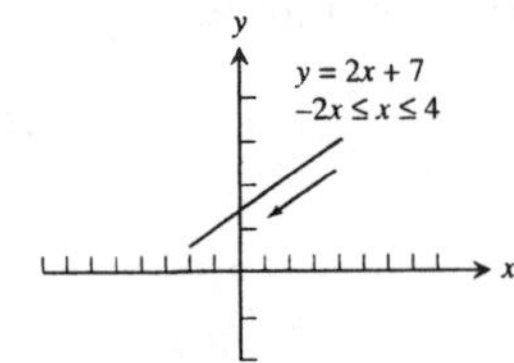

Initial point: (4, 15)
Terminal point: (−2, 3)
The line segment is traced from right to left starting at (4, 15) and ending at (−2, 3).

113. (a) For simplicity, we assume that x and y are linear functions of t, and that the point (x, y) starts at (−2, 5) for t = 0 and ends at (4, 3) for t = 1. Then x = f(t), where f(0) = −2 and f(1) = 4. Since slope $= \dfrac{\Delta x}{\Delta t} = \dfrac{4-(-2)}{1-0} = 6$, $x = f(t) = 6t - 2 = -2 + 6t$. Also, y = g(t), where g(0) = 5 and g(1) = 3.

Since slope $= \frac{\Delta y}{\Delta t} = \frac{3-5}{1-0} = -2$, $y = g(t) = -2t + 5 = 5 - 2t$. One possible parametrization is: $x = -2 + 6t$, $y = 5 - 2t$, $0 \le t \le 1$.

115. For simplicity, we assume that x and y are linear functions of t, and that the point (x, y) starts at (2, 5) for $t = 0$ and passes through $(-1, 0)$ for $t = 1$. Then $x = f(t)$, where $f(0) = 2$ and $f(1) = -1$. Since slope $= \frac{\Delta x}{\Delta t} = \frac{-1-2}{1-0} = -3$, $x = f(t) = -3t + 2 = 2 - 3t$. Also, $y = g(t)$, where $g(0) = 5$ and $g(1) = 0$. Since slope $= \frac{\Delta y}{\Delta t} = \frac{0-5}{1-0} = -5$, $y = g(t) = -5t + 5 = 5 - 5t$. One possible parametrization is: $x = 2 - 3t$, $y = 5 - 5t$, $t \ge 0$.

PRELIMINARY CHAPTER ADDITIONAL EXERCISES–THEORY, EXAMPLES, APPLICATIONS

1. (a) The given graph is reflected about the y-axis.

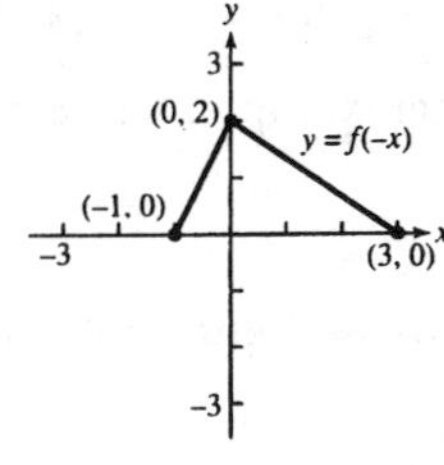

(b) The given graph is reflected about the x-axis.

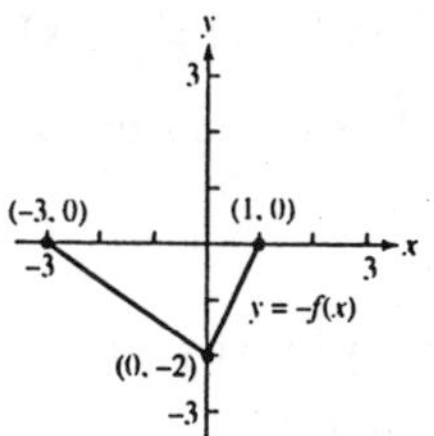

(c) The given graph is shifted left 1 unit, stretched vertically by a factor of 2, reflected about the x-axis, and then shifted upward 1 unit.

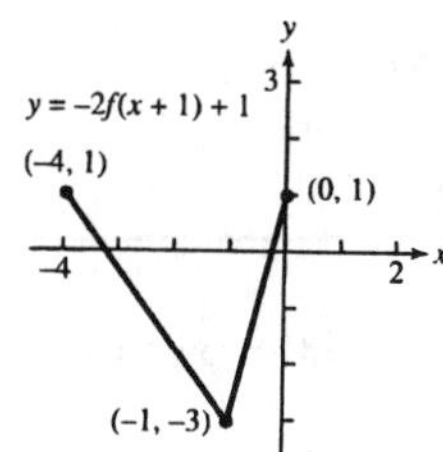

(d) The given graph is shifted right 2 units, stretched vertically by a factor of 3, and then shifted downward 2 units.

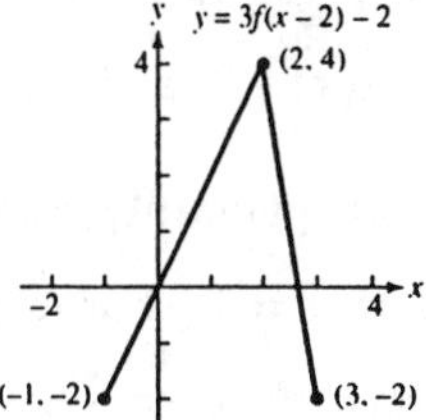

3. (a) $y = 100{,}000 - 10{,}000x$, $0 \le x \le 10$

(b)
$$\begin{aligned} y &= 55{,}000 \\ 100{,}000 - 10{,}000x &= 55{,}000 \\ -10{,}000x &= -45{,}000 \\ x &= 4.5 \end{aligned}$$
The value is \$55,000 after 4.5 years.

5. $1500(1.08)^t = 5000 \to 1.08^t = \frac{5000}{1500} = \frac{10}{3} \to \ln(1.08)^t = \ln\frac{10}{3} \to t \ln 1.08 = \ln\frac{10}{3} \to t = \frac{\ln(10/3)}{\ln 1.08} \approx 15.6439$
It will take about 15.6439 years. (If the bank only pays interest at the end of the year, it will take 16 years.)

7. As in the proof of the law of sines of Section P.5, Exercise 35, $ah = bc \sin A = ab \sin C = ac \sin B$ $\Rightarrow$ the area of ABC $= \frac{1}{2}(\text{base})(\text{height}) = \frac{1}{2}ah = \frac{1}{2}bc \sin A = \frac{1}{2}ab \sin C = \frac{1}{2}ac \sin B$.

9. Triangle ABD is an isosceles right triangle with its right angle at B and an angle of measure $\frac{\pi}{4}$ at A. We therefore have $\frac{\pi}{4} = \angle DAB = \angle DAE + \angle CAB = \tan^{-1}\frac{1}{3} + \tan^{-1}\frac{1}{2}$.

11. (a) If f is even, then $f(x) = f(-x)$ and $h(-x) = g(f(-x)) = g(f(x)) = h(x)$.
    If f is odd, then $f(-x) = -f(x)$ and $h(-x) = g(f(-x)) = g(-f(x)) = g(f(x)) = h(x)$ because g is even.
    If f is neither, then h may not be even. For example, if $f(x) = x^2 + x$ and $g(x) = x^2$, then $h(x) = x^4 + 2x^3 + x^2$ and $h(-x) = x^4 - 2x^3 + x^2 \neq h(x)$. Therefore, h need not be even.
    (b) No, h is not always odd. Let $g(t) = t$ and $f(x) = x^2$. Then, $h(x) = g(f(x)) = f(x) = x^2$ is even although g is odd.
    If f is odd, then $f(-x) = -f(x)$ and $h(-x) = g(f(-x)) = g(-f(x)) = -g(f(x)) = -h(x)$ because g is odd. In this case, h is odd. However, if f is even, as in the above counterexample, we see that h need not be odd.

13. There are (infinitely) many such function pairs. For example, $f(x) = 3x$ and $g(x) = 4x$ satisfy $f(g(x)) = f(4x) = 3(4x) = 12x = 4(3x) = g(3x) = g(f(x))$.

15. If f is odd and defined at x, then $f(-x) = -f(x)$. Thus $g(-x) = f(-x) - 2 = -f(x) - 2$ whereas $-g(x) = -(f(x) - 2) = -f(x) + 2$. Then g cannot be odd because $g(-x) = -g(x) \Rightarrow -f(x) - 2 = -f(x) + 2$ $\Rightarrow 4 = 0$, which is a contradiction. Also, g(x) is not even unless $f(x) = 0$ for all x. On the other hand, if f is even, then $g(x) = f(x) - 2$ is also even: $g(-x) = f(-x) - 2 = f(x) - 2 = g(x)$.

17. For (x, y) in the 1st quadrant, $|x| + |y| = 1 + x$ $\Leftrightarrow x + y = 1 + x \Leftrightarrow y = 1$. For (x, y) in the 2nd quadrant, $|x| + |y| = x + 1 \Leftrightarrow -x + y = x + 1$ $\Leftrightarrow y = 2x + 1$. In the 3rd quadrant, $|x| + |y| = x + 1$ $\Leftrightarrow -x - y = x + 1 \Leftrightarrow y = -2x - 1$. In the 4th quadrant, $|x| + |y| = x + 1 \Leftrightarrow x + (-y) = x + 1$ $\Leftrightarrow y = -1$. The graph is given at the right.

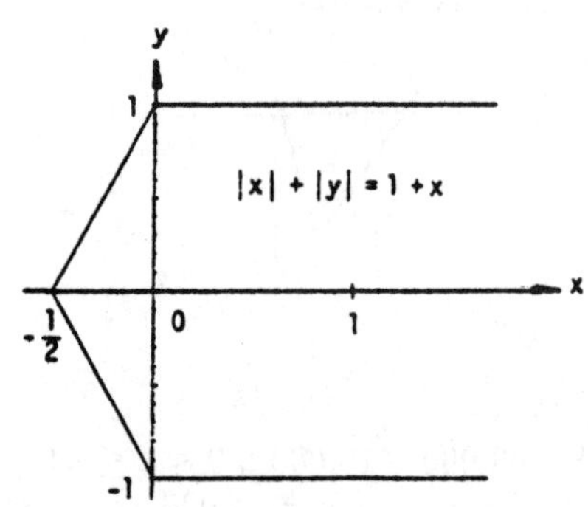

19. If f is even and odd, then $f(-x) = -f(x)$ and $f(-x) = f(x) \Rightarrow f(x) = -f(x)$ for all x in the domain of f. Thus $2f(x) = 0 \Rightarrow f(x) = 0$.

21. If the graph of f(x) passes the horizontal line test, so will the graph of $g(x) = -f(x)$ since it's the same graph reflected about the x-axis.
Alternate answer: If $g(x_1) = g(x_2)$ then $-f(x_1) = -f(x_2)$, $f(x_1) = f(x_2)$, and $x_1 = x_2$ since f is one-to-one.

23. (a) The expression $a(b^{c-x}) + d$ is defined for all values of x, so the domain is $(-\infty, \infty)$. Since b^{c-x} attains all positive values, the range is (d, ∞) if $a > 0$ and the range is $(-\infty, d)$ if $a < 0$.
    (b) The expression $a \log_b (x - c) + d$ is defined when $x - c > 0$, so the domain is (c, ∞).
    Since $a \log_b (x - c) + d$ attains every real value for some value of x, the range is $(-\infty, \infty)$.

25. (a)

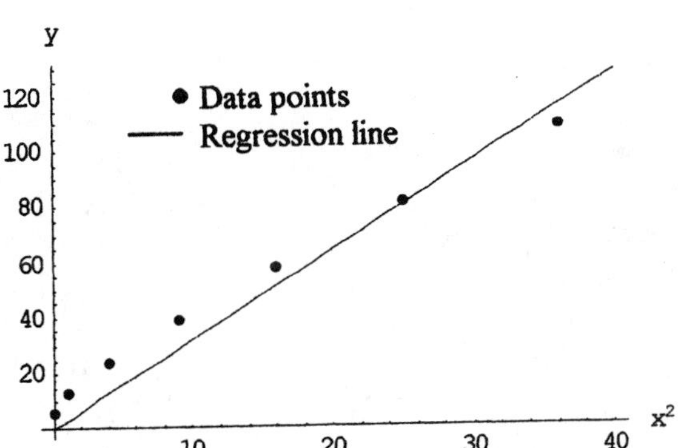

The graph does not support the assumption that $y \propto x^2$

(b)

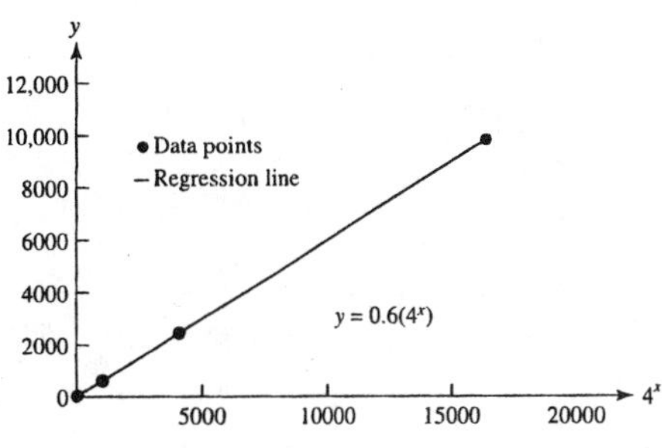

The graph supports the assumption that $y \propto 4^x$. The constant of proportionality is estimated from the slope of the regression line, which is 0.6, therefore, $y = 0.6(4^x)$.

27. (a) Since the elongation of the spring is zero when the stress is $5(10^{-3})(\text{lb/in.}^2)$, the data should be adjusted by subtracting this amount from each of the stress data values. This gives the following table, where $\bar{s} = s - 5(10^{-3})$.

$\bar{s} \times 10^{-3}$	0	5	15	25	35	45	55	65	75	85	95
$e \times 10^5$	0	19	57	94	134	173	216	256	297	343	390

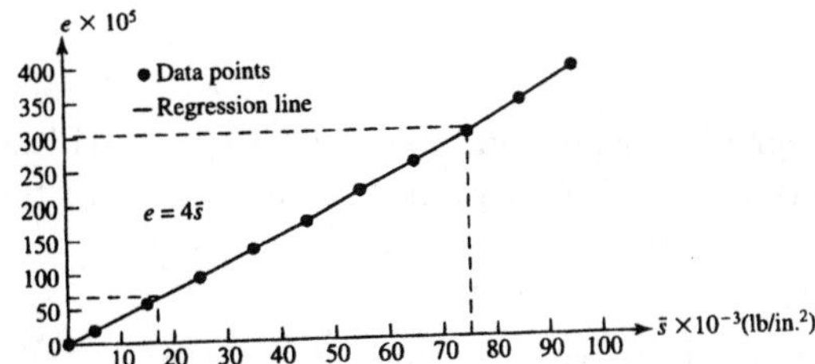

The slope of the graph is $\dfrac{(297 - 57)(10^5)}{(75 - 15)(10^{-3})} = 4.00(10^8)$ and the model is $e = 4(10^8)\bar{s}$ or $e = 4(10^8)(s - 5(10^{-3}))$.

(b) As show in the following graph, the largest relative error is about 6.4%

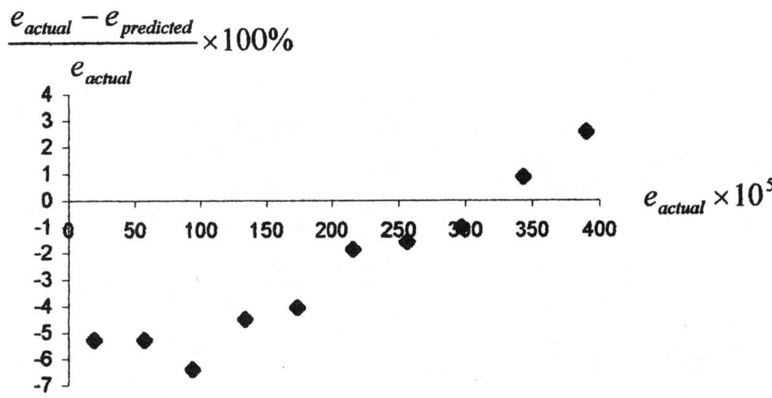

The model fits the data well. There does appear to be a pattern in the errors (i.e., they are not random) indicating that a refinement of the model is possible.

(c) $e = 4(10^8)(200 - 5)(10^{-3}) = 780(10^5)$(in./in.). Since $s = 200(10^{-3})$(lb/in.^{2}) is well outside the range of the data used for the model, one should not feel comfortable with this prediction without further testing of the spring.

29. (a) $y = 20.627x + 338.622$

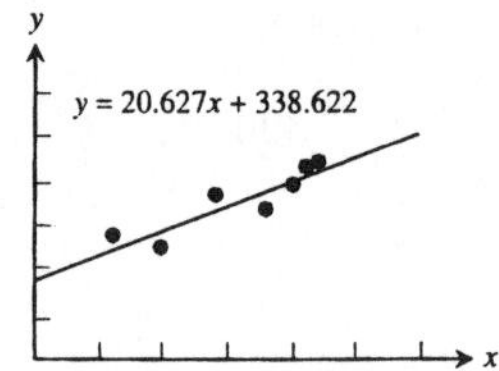

(b) When $x = 30$, $y = 957.445$. According to the regression equation, about 957 degrees will be earned.
(c) The slope is 20.627. It represents the approximate annual increase in the number of doctorates earned by Hispanic Americans per year.

31. (a) The TI-92 Plus calculator gives $f(x) = 2.000268 \sin(2.999187x - 1.000966) + 3.999881$.
(b) $f(x) = 2 \sin(3x - 1) + 4$

33. (a) The TI-93 Plus calculator gives $Q = 1.00(2.0138^x) = 1.00e^{0.7x}$

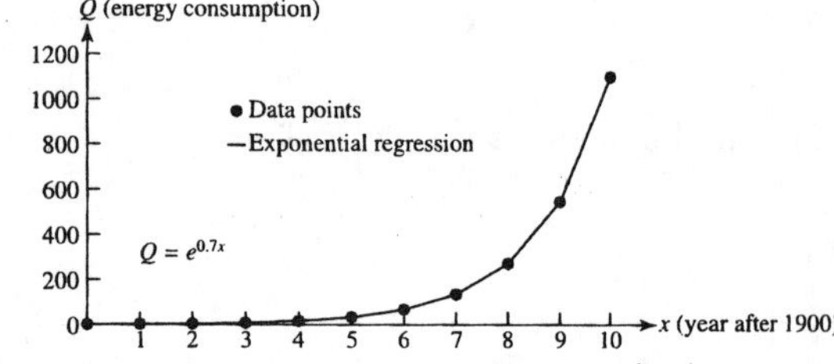

(b) For 1996, $x = 9.6 \Rightarrow Q(9.6) = e^{0.7(9.6)} = 828.82$ units of energy consumed that year as estimated by the exponential regression. The exponential regression shows that energy consumption has doubled (i.e., increased by 100%) each decade during the 20th century. The annual rate of increase during this time is $e^{0.7(0.1)} - e^{0.7(0)} = 0.0725 = 7.25\%$.

CHAPTER 1 LIMITS AND CONTINUITY

1.1 RATE OF CHANGE AND LIMITS

1. (a) $\frac{\Delta f}{\Delta x} = \frac{f(3) - f(2)}{3 - 2} = \frac{28 - 9}{1} = 19$ (b) $\frac{\Delta f}{\Delta x} = \frac{f(1) - f(-1)}{1 - (-1)} = \frac{2 - 0}{2} = 1$

3. (a) $\frac{\Delta h}{\Delta t} = \frac{h\left(\frac{3\pi}{4}\right) - h\left(\frac{\pi}{4}\right)}{\frac{3\pi}{4} - \frac{\pi}{4}} = \frac{-1 - 1}{\frac{\pi}{2}} = -\frac{4}{\pi}$ (b) $\frac{\Delta h}{\Delta t} = \frac{h\left(\frac{\pi}{2}\right) - h\left(\frac{\pi}{6}\right)}{\frac{\pi}{2} - \frac{\pi}{6}} = \frac{0 - \sqrt{3}}{\frac{\pi}{3}} = \frac{-3\sqrt{3}}{\pi}$

5. (a)

Q	Slope of PQ $= \frac{\Delta p}{\Delta t}$
$Q_1(10, 225)$	$\frac{650 - 225}{20 - 10} = 42.5$ m/sec
$Q_2(14, 375)$	$\frac{650 - 375}{20 - 14} = 45.83$ m/sec
$Q_3(16.5, 475)$	$\frac{650 - 475}{20 - 16.5} = 50.00$ m/sec
$Q_4(18, 550)$	$\frac{650 - 550}{20 - 18} = 50.00$ m/sec

(b) At $t = 20$, the Cobra was traveling approximately 50 m/sec or 180 km/h.

7. A plot of the data shows that the slope of the secant between $t = 0.8$ sec and $t = 1.0$ sec underestimates the instantaneous velocity (i.e., the slope of the tangent) at $t = 1.0$ sec, whereas the slope of the secant between $t = 1.0$ sec and $t = 1.2$ sec overestimates it.

Lower bound: $a = \frac{13.10 - 8.39}{1.0 - 0.8} = 23.55$ ft/sec

Upper bound: $b = \frac{18.87 - 13.10}{1.2 - 1.0} = 28.85$ ft/sec

$v(1) \approx \frac{a + b}{2} = \frac{23.55 + 28.85}{2} = 26.20$ ft/sec

9. (a) Does not exist. As x approaches 1 from the right, g(x) approaches 0. As x approaches 1 from the left, g(x) approaches 1. There is no single number L that all the values g(x) get arbitrarily close to as $x \to 1$.
(b) 1
(c) 0

11. (a) True (b) True (c) False
(d) False (e) False (f) True

13. $\lim_{x \to 0} \frac{x}{|x|}$ does not exist because $\frac{x}{|x|} = \frac{x}{x} = 1$ if $x > 0$ and $\frac{x}{|x|} = \frac{x}{-x} = -1$ if $x < 0$. As x approaches 0 from the left, $\frac{x}{|x|}$ approaches -1. As x approaches 0 from the right, $\frac{x}{|x|}$ approaches 1. There is no single number L that all the function values get arbitrarily close to as $x \to 0$.

15. Nothing can be said about $\lim_{x \to x_0} f(x)$ because the existence of a limit as $x \to x_0$ does not depend on how the function is defined at x_0. In order for a limit to exist, f(x) must be arbitrarily close to a single real number L when x is close enough to x_0. That is, the existence of a limit depends on the values of f(x) for x near x_0, not on the definition of f(x) at x_0 itself.

17. No, the definition does not require that f be defined at x = 1 in order for a limiting value to exist there. If f(1) is defined, it can be any real number, so we can conclude nothing about f(1) from $\lim_{x \to 1} f(x) = 5$.

19. (a) $f(x) = (x^2 - 9)/(x + 3)$

x	−3.1	−3.01	−3.001	−3.0001	−3.00001	−3.000001
f(x)	−6.1	−6.01	−6.001	−6.0001	−6.00001	−6.000001

x	−2.9	−2.99	−2.999	−2.9999	−2.99999	−2.999999
f(x)	−5.9	−5.99	−5.999	−5.9999	−5.99999	−5.999999

The estimate is $\lim_{x \to -3} f(x) = -6$.

(b)

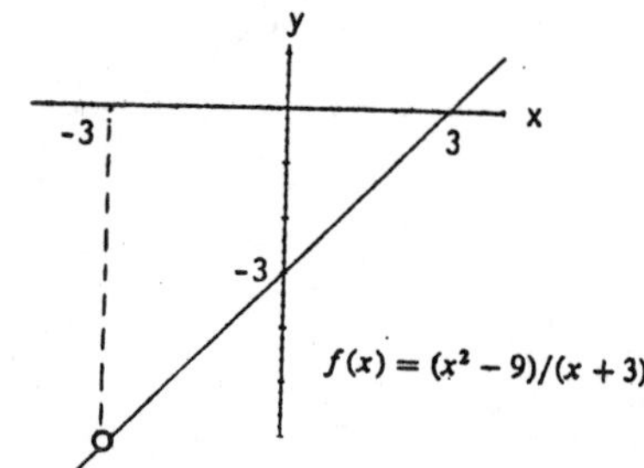

(c) $f(x) = \frac{x^2 - 9}{x + 3} = \frac{(x+3)(x-3)}{x+3} = x - 3$ if $x \neq -3$, and $\lim_{x \to -3} (x - 3) = -3 - 3 = -6$.

21. (a) $G(x) = (x + 6)/(x^2 + 4x - 12)$

x	−5.9	−5.99	−5.999	−5.9999	−5.99999	−5.999999
G(x)	−0.126582	−0.1251564	−0.1250156	−0.1250016	−0.12500016	−0.12500002

x	−6.1	−6.01	−6.001	−6.0001	−6.00001	−6.000001
G(x)	−0.123457	−0.1248439	−0.1249844	−0.1249984	−0.12499984	−0.12499998

The estimate is $\lim_{x \to -6} G(x) = -0.125$.

(b)

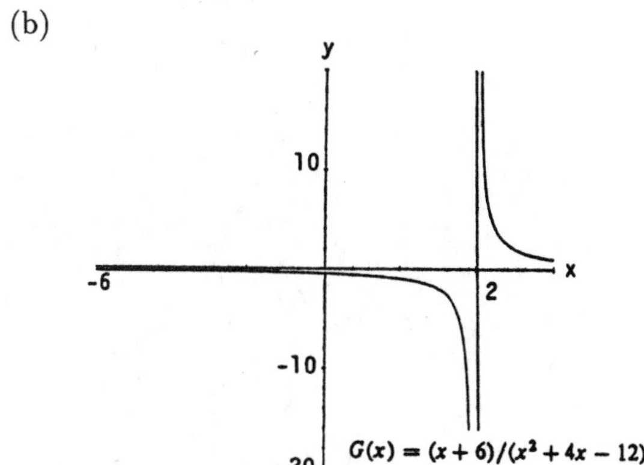

(c) $G(x) = \frac{x+6}{(x^2+4x-12)} = \frac{x+6}{(x+6)(x-2)} = \frac{1}{x-2}$ if $x \neq -6$, and $\lim_{x\to-6} \frac{1}{x-2} = \frac{1}{-6-2} = -\frac{1}{8} = -0.125$.

23. (a) $g(\theta) = (\sin\theta)/\theta$

θ	.1	.01	.001	.0001	.00001	.000001
$g(\theta)$	.998334	.999983	.999999	.999999	.999999	.999999

θ	−.1	−.01	−.001	−.0001	−.00001	−.000001
$g(\theta)$	.998334	.999983	.999999	.999999	.999999	.999999

$\lim_{\theta\to0} g(\theta) = 1$

(b)

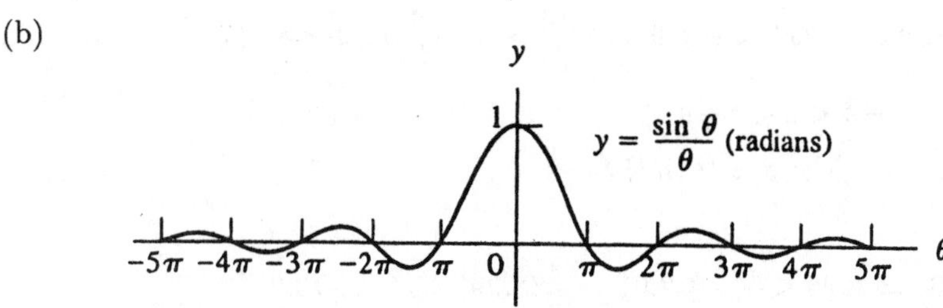

25. (a) $f(x) = x^{1/(1-x)}$

x	.9	.99	.999	.9999	.99999	.999999
f(x)	.348678	.366032	.367695	.367861	.367878	.367879

x	1.1	1.01	1.001	1.0001	1.00001	1.000001
f(x)	.385543	.369711	.368063	.367898	.367881	.367880

$\lim_{x\to1} f(x) \approx 0.36788$

(b)

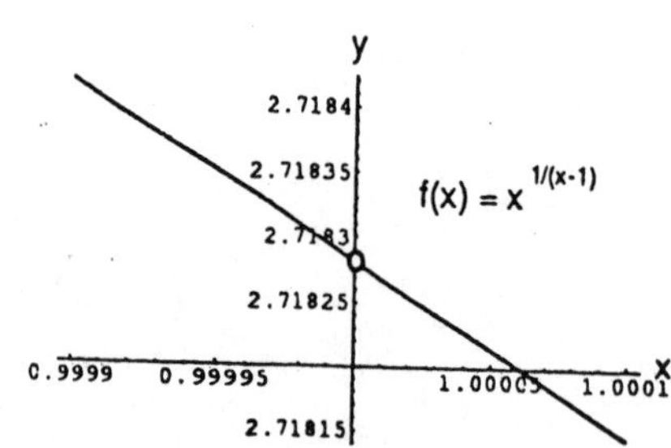

27. Step 1: $|x-5| < \delta \Rightarrow -\delta < x-5 < \delta \Rightarrow -\delta+5 < x < \delta+5$
    Step 2: From the graph, $-\delta+5 = 4.9 \Rightarrow \delta = 0.1$, or $\delta+5 = 5.1 \Rightarrow \delta = 0.1$; thus $\delta = 0.1$ in either case.

29. Step 1: $|x-1| < \delta \Rightarrow -\delta < x-1 < \delta \Rightarrow -\delta+1 < x < \delta+1$
    Step 2: From the graph, $-\delta+1 = \frac{9}{16} \Rightarrow \delta = \frac{7}{16}$, or $\delta+1 = \frac{25}{16} \Rightarrow \delta = \frac{9}{16}$; thus $\delta = \frac{7}{16}$.

31. Step 1: $\left|(x+1)-5\right| < 0.01 \Rightarrow |x-4| < 0.01 \Rightarrow -0.01 < x-4 < 0.01 \Rightarrow 3.99 < x < 4.01$
    Step 2: $|x-4| < \delta \Rightarrow -\delta < x-4 < \delta \Rightarrow -\delta+4 < x < \delta+4 \Rightarrow \delta = 0.01.$

33. Step 1: $\left|\sqrt{x+1}-1\right| < 0.1 \Rightarrow -0.1 < \sqrt{x+1}-1 < 0.1 \Rightarrow 0.9 < \sqrt{x+1} < 1.1 \Rightarrow 0.81 < x+1 < 1.21$
    $\Rightarrow -0.19 < x < 0.21$
    Step 2: $|x-0| < \delta \Rightarrow -\delta < x < \delta \Rightarrow \delta = 0.19.$

35. Step 1: $\left|\frac{1}{x}-\frac{1}{4}\right| < 0.05 \Rightarrow -0.05 < \frac{1}{x}-\frac{1}{4} < 0.05 \Rightarrow 0.2 < \frac{1}{x} < 0.3 \Rightarrow \frac{10}{2} > x > \frac{10}{3}$ or $\frac{10}{3} < x < 5.$
    Step 2: $|x-4| < \delta \Rightarrow -\delta < x-4 < \delta \Rightarrow -\delta+4 < x < \delta+4.$
    Then $-\delta+4 = \frac{10}{3}$ or $\delta = \frac{2}{3}$, or $\delta+4 = 5$ or $\delta = 1$; thus $\delta = \frac{2}{3}$.

37. $|A-9| \le 0.01 \Rightarrow -0.01 \le \pi\left(\frac{x}{2}\right)^2 - 9 \le 0.01 \Rightarrow 8.99 \le \frac{\pi x^2}{4} \le 9.01 \Rightarrow \frac{4}{\pi}(8.99) \le x^2 \le \frac{4}{\pi}(9.01)$
    $\Rightarrow 2\sqrt{\frac{8.99}{\pi}} \le x \le 2\sqrt{\frac{9.01}{\pi}}$ or $3.384 \le x \le 3.387$. To be safe, the left endpoint was rounded up and the right endpoint was rounded down.

39. (a) The limit can be found by substitution.
    $\lim_{x\to 2} f(x) = f(2) = \sqrt{3(2)-2} = \sqrt{4} = 2$
    (b) The graphs of $y_1 = f(x)$, $y_2 = 1.8$, and $y_3 = 2.2$ are shown.

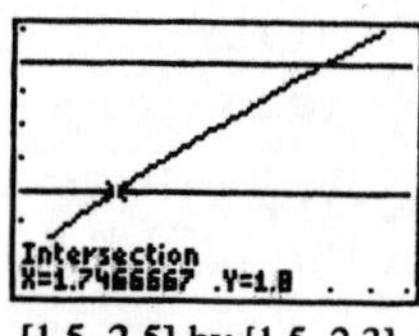

[1.5, 2.5] by [1.5, 2.3]

The intersections of y_1 with y_2 and y_3 are at $x \approx 1.7467$ and $x = 2.28$, respectively, so we may choose any value of a in $[1.7467, 2)$ (approximately) and any value of b in $[2, 2.28]$.
One possible answer: $a = 1.75$, $b = 2.28$.

(c) The graphs of $y_1 = f(x)$, $y_2 = 1.99$, and $y_3 = 2.01$ are shown.

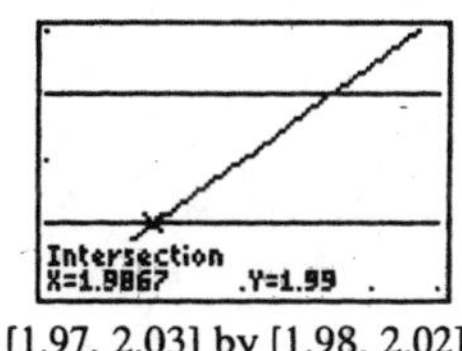

[1.97, 2.03] by [1.98, 2.02]

The intersections of y_1 with y_2 and y_3 are at $x = 1.9867$ and $x \approx 2.0134$, respectively, so we may choose any value of a in $[1.9867, 2)$ and any value of b in $[2, 2.0134]$ (approximately).
One possible answer: $a = 1.99$, $b = 2.01$.

41. (a) In three seconds, the ball falls $4.9(3)^2 = 44.1$ m, so its average speed is $\frac{44.1}{3} = 14.7$ m/sec.

(b) The average speed over the interval from time $t = 3$ to time $3 + h$ is

$$\frac{\Delta y}{\Delta t} = \frac{4.9(3+h)^2 - 4.9(3)^2}{(3+h) - 3} = \frac{4.9(6h + h^2)}{h} = 29.4 + 4.9h$$

Since $\lim_{h \to 0} (29.4 + 4.9h) = 29.4$, the instantaneous speed is 29.4 m/sec.

43. (a)

x	−0.1	−0.01	−0.001	−0.0001
f(x)	−0.054402	−0.005064	−0.000827	−0.000031

(b)

x	0.1	0.01	0.001	0.0001
f(x)	−0.054402	−0.005064	−0.000827	−0.000031

The limit appears to be 0.

45. (a)

x	−0.1	−0.01	−0.001	−0.0001
f(x)	2.0567	2.2763	2.2999	2.3023

(b)

x	0.1	0.01	0.001	0.0001
f(x)	2.5893	2.3293	2.3052	2.3029

The limit appears to be approximately 2.3.

1.2 RULES FOR FINDING LIMITS

1. (a) $\lim_{x \to 3^-} f(x) = 3$

3. (a) $\lim_{h \to 0^-} f(h) = -4$

(b) $\lim_{x\to 3^+} f(x) = -2$

(c) $\lim_{x\to 3} f(x)$ does not exist, because the left- and right-hand limits are not equal.

(d) $f(3) = 1$

(b) $\lim_{h\to 0^+} f(h) = -4$

(c) $\lim_{h\to 0} f(h) = -4$

(d) $f(0) = -4$

5. (a) $\lim_{x\to 0^-} F(x) = 4$

(b) $\lim_{x\to 0^+} F(x) = -3$

(c) $\lim_{x\to 0} F(x) =$ does not exist because the left- and right-hand limits are not equal.

(d) $F(0) = 4$

7. (a) quotient rule

(b) difference and power rules

(c) sum and constant multiple rules

9. (a) $\lim_{x\to c} f(x)\,g(x) = [\lim_{x\to c} f(x)][\lim_{x\to c} g(x)] = (5)(-2) = -10$

(b) $\lim_{x\to c} 2f(x)\,g(x) = 2[\lim_{x\to c} f(x)][\lim_{x\to c} g(x)] = 2(5)(-2) = -20$

(c) $\lim_{x\to c} [f(x) + 3g(x)] = \lim_{x\to c} f(x) + 3 \lim_{x\to c} g(x) = 5 + 3(-2) = -1$

(d) $\lim_{x\to c} \dfrac{f(x)}{f(x) - g(x)} = \dfrac{\lim_{x\to c} f(x)}{\lim_{x\to c} f(x) - \lim_{x\to c} g(x)} = \dfrac{5}{5-(-2)} = \dfrac{5}{7}$

11. (a) $\lim_{x\to -7} (2x+5) = 2(-7) + 5 = -14 + 5 = -9$

(b) $\lim_{t\to 6} 8(t-5)(t-7) = 8(6-5)(6-7) = -8$

(c) $\lim_{y\to 2} \dfrac{y+2}{y^2+5y+6} = \dfrac{2+2}{(2)^2+5(2)+6} = \dfrac{4}{4+10+6} = \dfrac{4}{20} = \dfrac{1}{5}$

(d) $\lim_{h\to 0} \dfrac{3}{\sqrt{3h+1}+1} = \dfrac{3}{\sqrt{3(0)+1}+1} = \dfrac{3}{\sqrt{1}+1} = \dfrac{3}{2}$

13. (a) $\lim_{t\to -5} \dfrac{t^2+3t-10}{t+5} = \lim_{t\to -5} \dfrac{(t+5)(t-2)}{t+5} = \lim_{t\to -5} (t-2) = -5-2 = -7$

(b) $\lim_{x\to -2} \dfrac{-2x-4}{x^3+2x^2} = \lim_{x\to -2} \dfrac{-2(x+2)}{x^2(x+2)} = \lim_{x\to -2} \dfrac{-2}{x^2} = \dfrac{-2}{4} = -\dfrac{1}{2}$

(c) $\lim_{y\to 1} \dfrac{y-1}{\sqrt{y+3}-2} = \lim_{y\to 1} \dfrac{(y-1)(\sqrt{y+3}+2)}{(\sqrt{y+3}-2)(\sqrt{y+3}+2)} = \lim_{y\to 1} \dfrac{(y-1)(\sqrt{y+3}+2)}{(y+3)-4} = \lim_{y\to 1} (\sqrt{y+3}+2)$

$= \sqrt{4}+2 = 4$

(d) $\lim_{x\to 3} \sin\left(\frac{1}{x} - \frac{1}{2}\right) = \sin\left(\frac{1}{3} - \frac{1}{2}\right) = -\sin\left(\frac{1}{6}\right) \approx -0.1659$

15. (a) $\lim_{x\to 0}\left(1-\frac{x^2}{6}\right)=1-\frac{0}{6}=1$ and $\lim_{x\to 0} 1=1$; by the sandwich theorem, $\lim_{x\to 0}\frac{x \sin x}{2-2\cos x}=1$

(b) For $x\neq 0$, $y=(x \sin x)/(2-2\cos x)$ lies between the other two graphs in the figure, and the graphs converge as $x\to 0$.

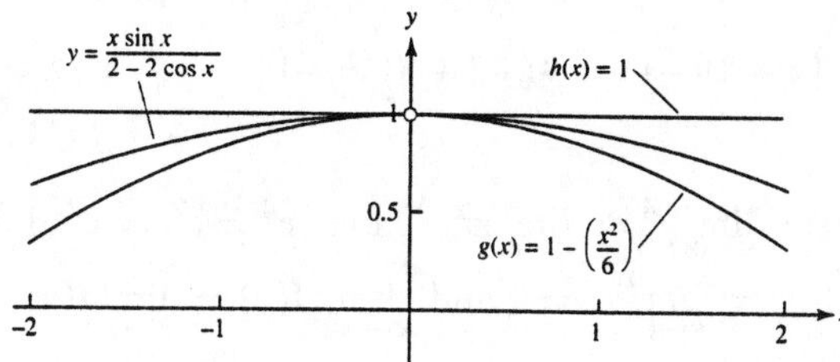

17. $\lim_{h\to 0}\frac{(1+h)^2-1^2}{h}=\lim_{h\to 0}\frac{1+2h+h^2-1}{h}=\lim_{h\to 0}\frac{h(2+h)}{h}=\lim_{h\to 0}(2+h)=2$

19. $\lim_{h\to 0}\frac{\left(\frac{1}{-2+h}\right)-\left(\frac{1}{-2}\right)}{h}=\lim_{h\to 0}\frac{\frac{-2}{-2+h}-1}{-2h}=\lim_{h\to 0}\frac{-2-(-2+h)}{-2h(-2+h)}=\lim_{h\to 0}\frac{-h}{h(4-2h)}=-\frac{1}{4}$

21. (a) False (b) True (c) False (d) True
(e) True (f) True (g) False (h) False
(i) False (j) False (k) True (l) False

23. (a) No, $\lim_{x\to 0^+} f(x)$ does not exist since $\sin\left(\frac{1}{x}\right)$ does not approach any single value as x approaches 0

(b) $\lim_{x\to 0^-} f(x)=\lim_{x\to 0^-} 0=0$

(c) $\lim_{x\to 0} f(x)$ does not exist because $\lim_{x\to 0^+} f(x)$ does not exist

25. (a) domain: $0\le x\le 2$
range: $0<y\le 1$ and $y=2$

(b) $\lim_{x\to c} f(x)$ exists for c belonging to

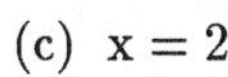

$(0,1)\cup(1,2)$

(c) $x=2$

(d) $x=0$

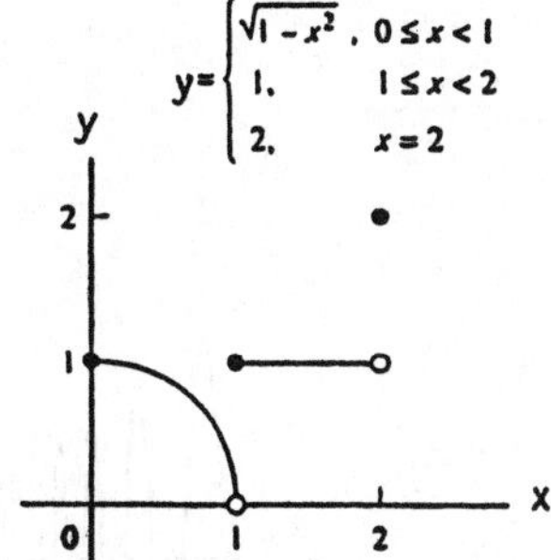

27. $\lim_{x\to -0.5^-}\sqrt{\frac{x+2}{x+1}}=\sqrt{\frac{-0.5+2}{-0.5+1}}=\sqrt{\frac{3/2}{1/2}}=\sqrt{3}$

29. $\lim_{h\to 0^+}\frac{\sqrt{h^2+4h+5}-\sqrt{5}}{h}=\lim_{h\to 0^+}\left(\frac{\sqrt{h^2+4h+5}-\sqrt{5}}{h}\right)\left(\frac{\sqrt{h^2+4h+5}+\sqrt{5}}{\sqrt{h^2+4h+5}+\sqrt{5}}\right)$

$=\lim_{h\to 0^+}\frac{(h^2+4h+5)-5}{h\left(\sqrt{h^2+4h+5}+\sqrt{5}\right)}=\lim_{h\to 0^+}\frac{h(h+4)}{h\left(\sqrt{h^2+4h+5}+\sqrt{5}\right)}=\frac{0+4}{\sqrt{5}+\sqrt{5}}=\frac{2}{\sqrt{5}}$

31. (a) $\lim_{x\to -2^+}(x+3)\frac{|x+2|}{x+2}=\lim_{x\to -2^+}(x+3)\frac{(x+2)}{(x+2)}$ $\quad(|x+2|=x+2 \text{ for } x>-2)$

$$= \lim_{x\to -2^+} (x+3) = (-2)+3 = 1$$

(b) $\lim\limits_{x\to -2^-} (x+3)\dfrac{|x+2|}{x+2} = \lim\limits_{x\to -2^-} (x+3)\left[\dfrac{-(x+2)}{(x+2)}\right]$ $\quad (|x+2| = -(x+2) \text{ for } x < -2)$

$$= \lim_{x\to -2^-} (x+3)(-1) = -(-2+3) = -1$$

33. $\lim\limits_{x\to c} f(x)$ exists at those points c where $\lim\limits_{x\to c} x^4 = \lim\limits_{x\to c} x^2$. Thus, $c^4 = c^2 \Rightarrow c^2(1-c^2) = 0$ $\Rightarrow c = 0,\ 1,$ or -1. Moreover, $\lim\limits_{x\to 0} f(x) = \lim\limits_{x\to 0} x^2 = 0$ and $\lim\limits_{x\to -1} f(x) = \lim\limits_{x\to 1} f(x) = 1$.

35. (a) $1 = \lim\limits_{x\to -2} \dfrac{f(x)}{x^2} = \dfrac{\lim\limits_{x\to -2} f(x)}{\lim\limits_{x\to -2} x^2} = \dfrac{\lim\limits_{x\to -2} f(x)}{4} \Rightarrow \lim\limits_{x\to -2} f(x) = 4.$

(b) $1 = \lim\limits_{x\to -2} \dfrac{f(x)}{x^2} = \left[\lim\limits_{x\to -2} \dfrac{f(x)}{x}\right]\left[\lim\limits_{x\to -2} \dfrac{1}{x}\right] = \left[\lim\limits_{x\to -2} \dfrac{f(x)}{x}\right]\left(\dfrac{1}{-2}\right) \Rightarrow \lim\limits_{x\to -2} \dfrac{f(x)}{x} = -2.$

37. Yes. If $\lim\limits_{x\to a^+} f(x) = L = \lim\limits_{x\to a^-} f(x)$, then $\lim\limits_{x\to a} f(x) = L$. If $\lim\limits_{x\to a^+} f(x) \neq \lim\limits_{x\to a^-} f(x)$, then $\lim\limits_{x\to a} f(x)$ does not exist.

39. $I = (5, 5+\delta) \Rightarrow 5 < x < 5+\delta$. Also, $\sqrt{x-5} < \epsilon \Rightarrow x-5 < \epsilon^2 \Rightarrow x < 5+\epsilon^2$. Choose $\delta = \epsilon^2$ $\Rightarrow \lim\limits_{x\to 5^+} \sqrt{x-5} = 0.$

41. If f is an odd function of x, then $f(-x) = -f(x)$. Given $\lim\limits_{x\to 0^+} f(x) = 3$, then $\lim\limits_{x\to 0^-} f(x) = -3$.

43. (a) $g(x) = x \sin\left(\frac{1}{x}\right)$

$-\frac{\pi}{20} \le x \le \frac{\pi}{20}$ $\qquad\qquad$ $-\frac{\pi}{80} \le x \le \frac{\pi}{180}$

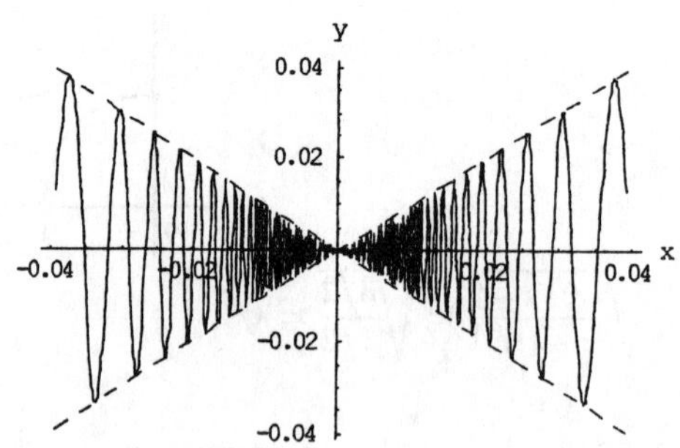

The graphs suggest that $\lim\limits_{x\to 0} g(x) = 0$.

(b) $k(x) = \sin\left(\frac{1}{x}\right)$

$-\frac{\pi}{20} \le x \le \frac{\pi}{20}$

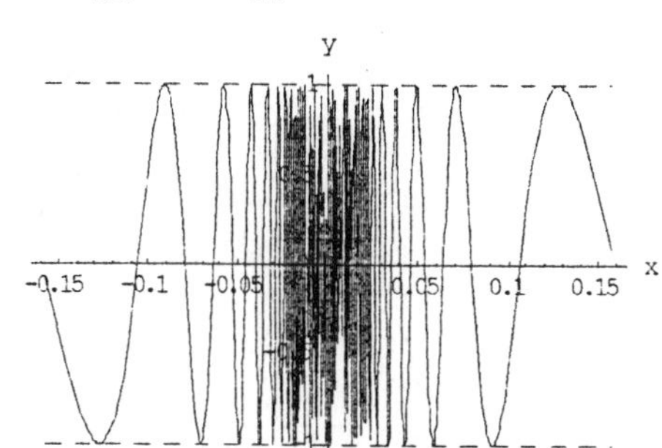

$-\frac{\pi}{80} \le x \le \frac{\pi}{80}$

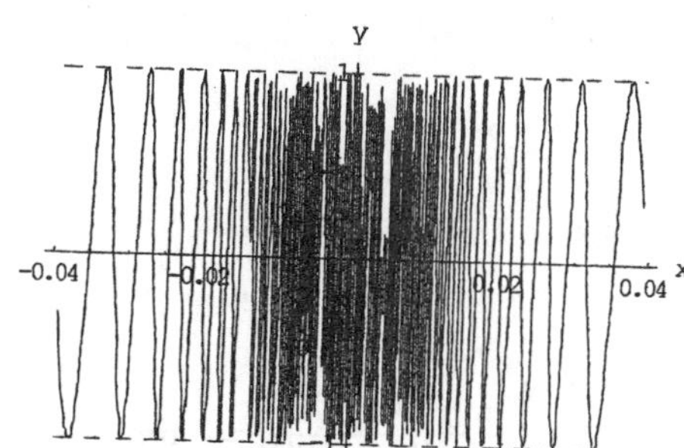

The graphs suggest that $\lim_{x\to 0} k(x)$ does not exist.

For both g(x) and k(x), the frequency of the oscillations increases without bound as $x \to 0$. For g(x), the sandwich theorem can be applied. If $x > 0$, $-x \le x \sin\left(\frac{1}{x}\right) \le x \Rightarrow \lim_{x\to 0^+} g(x) = 0$ and if $x < 0$, $x \le x \sin\left(\frac{1}{x}\right) \le -x \Rightarrow \lim_{x\to 0^-} g(x) = 0$. Therefore, $\lim_{x\to 0} g(x) = 0$ since the left- and right-hand limits are both 0. For k(x), the amplitude of the oscillations remains equal to one. Therefore, k(x) cannot be kept arbitrarily close to any number by keeping x sufficiently close to 0.

1.3 LIMITS INVOLVING INFINITY

Note: In these exercises we use the result $\lim_{x\to \pm\infty} \frac{1}{x^{m/n}} = 0$ whenever $\frac{m}{n} > 0$. This result follows immediately from Example 1 and the power rule in Theorem 7: $\lim_{x\to \pm\infty} \left(\frac{1}{x^{m/n}}\right) = \lim_{x\to \pm\infty} \left(\frac{1}{x}\right)^{m/n} = \left(\lim_{x\to \pm\infty} \frac{1}{x}\right)^{m/n} = 0^{m/n}$ $= 0$.

1. (a) π (b) π

3. (a) $-\frac{5}{3}$ (b) $-\frac{5}{3}$

5. $-\frac{1}{x} \le \frac{\sin 2x}{x} \le \frac{1}{x} \Rightarrow \lim_{x\to\infty} \frac{\sin 2x}{x} = 0$ by the Sandwich Theorem

7. (a) $\lim_{x\to\infty} \frac{2x+3}{5x+7} = \lim_{x\to\infty} \frac{2+\frac{3}{x}}{5+\frac{7}{x}} = \frac{2}{5}$ (b) $\frac{2}{5}$ (same process as part (a))

9. (a) $\lim_{x\to\infty} \frac{1-12x^3}{4x^2+12} = \lim_{x\to\infty} \frac{\frac{1}{x^2}-12x}{4+\frac{12}{x^2}} = -\infty$ (b) $\lim_{x\to-\infty} \frac{1-12x^3}{4x^2+12} = \lim_{x\to-\infty} \frac{\frac{1}{x^2}-12x}{4+\frac{12}{x^2}} = \infty$

11. (a) $\lim_{x\to\infty} \frac{3x^2-6x}{4x-8} = \lim_{x\to\infty} \frac{3x-6}{4-\frac{8}{x}} = \infty$ (b) $\lim_{x\to-\infty} \frac{3x^2-6x}{4x-8} = \lim_{x\to-\infty} \frac{3x-6}{4-\frac{8}{x}} = -\infty$

13. (a) $\lim\limits_{x\to\infty} \dfrac{-2x^3 - 2x + 3}{3x^3 + 3x^2 - 5x} = \lim\limits_{x\to\infty} \dfrac{-2 - \frac{2}{x^2} + \frac{3}{x^3}}{3 + \frac{3}{x} - \frac{5}{x^2}} = -\dfrac{2}{3}$

(b) $-\frac{2}{3}$ (same process as part (a))

15. $\lim\limits_{x\to\infty} \dfrac{2\sqrt{x} + x^{-1}}{3x - 7} = \lim\limits_{x\to\infty} \dfrac{\left(\frac{2}{x^{1/2}}\right) + \left(\frac{1}{x^2}\right)}{3 - \frac{7}{x}} = 0$

17. $\lim\limits_{x\to-\infty} \dfrac{\sqrt[3]{x} - \sqrt[5]{x}}{\sqrt[3]{x} + \sqrt[5]{x}} = \lim\limits_{x\to-\infty} \dfrac{1 - x^{(1/5)-(1/3)}}{1 + x^{(1/5)-(1/3)}} = \lim\limits_{x\to-\infty} \dfrac{1 - \left(\frac{1}{x^{2/15}}\right)}{1 + \left(\frac{1}{x^{2/15}}\right)} = 1$

19. $\lim\limits_{x\to\infty} \dfrac{2x^{5/3} - x^{1/3} + 7}{x^{8/5} + 3x + \sqrt{x}} = \lim\limits_{x\to\infty} \dfrac{2x^{1/15} - \frac{1}{x^{19/15}} + \frac{7}{x^{8/5}}}{1 + \frac{3}{x^{3/5}} + \frac{1}{x^{11/10}}} = \infty$

21. Here is one possibility.

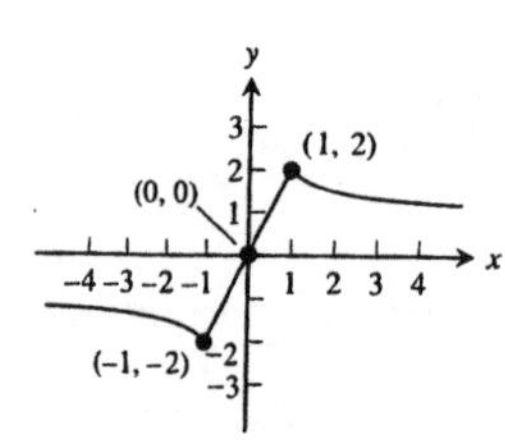

23. Here is one possibility.

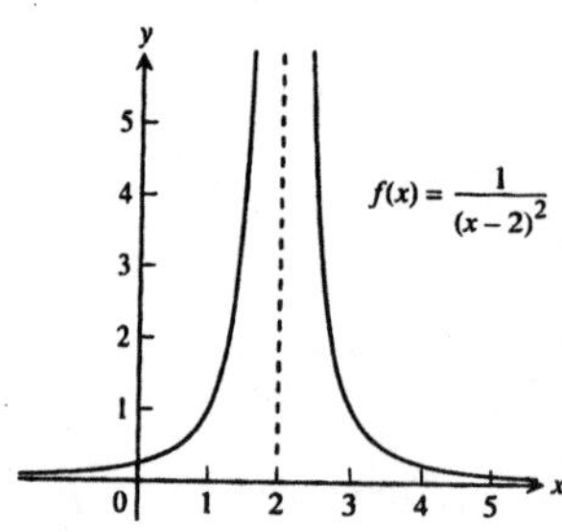

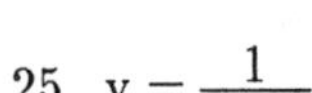

25. $y = \dfrac{1}{x-1}$

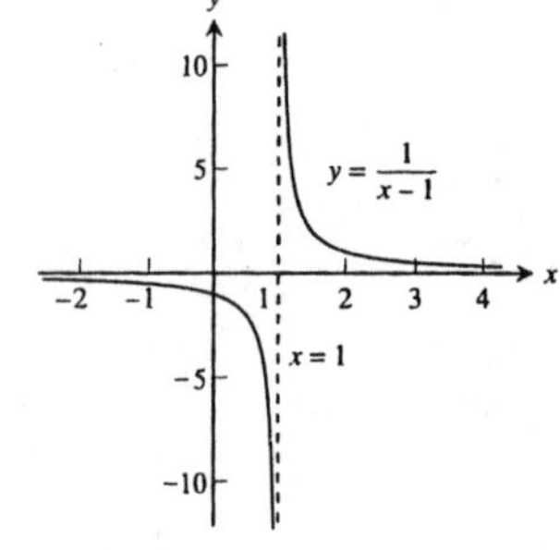

27. $y = \dfrac{2x^2 + x - 1}{x^2 - 1}$

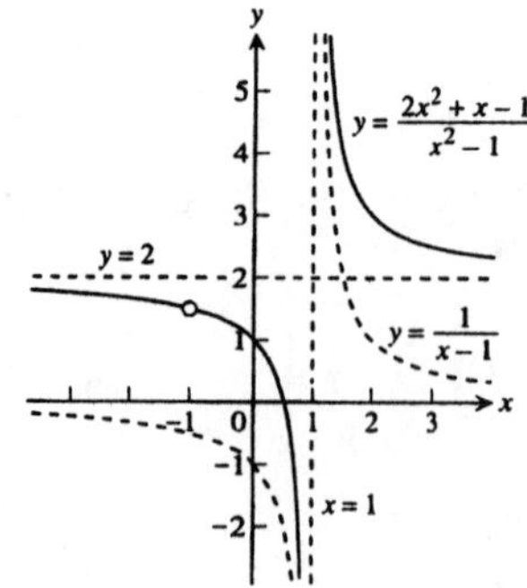

29. $y = \frac{x^4+1}{x^2} = x^2 + \frac{1}{x^2}$

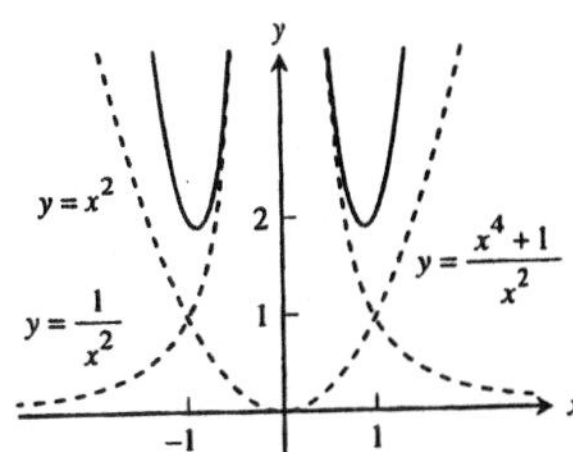

31. $y = \frac{x^2-x+1}{x-1} = x + \frac{1}{x-1}$

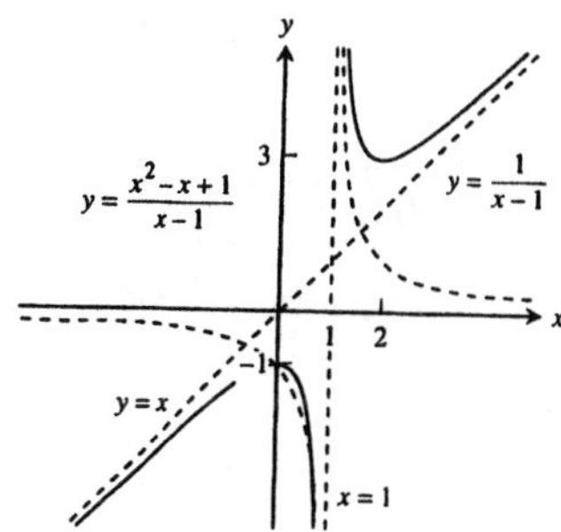

33. $y = \frac{8}{x^2+4}$

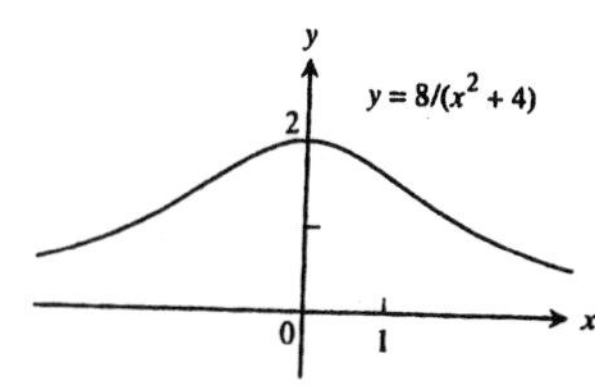

35. An end behavior model is $\frac{2x^3}{x} = 2x^2$. (a)

37. An end behavior model is $\frac{2x^4}{-x} = -2x^3$. (d)

39. (a) The function $y = e^x$ is a right end behavior model because $\lim_{x\to\infty} \frac{e^x - 2x}{e^x} = \lim_{x\to\infty} \left(1 - \frac{2x}{e^x}\right) = 1 - 0 = 1$.

(b) The function $y = -2x$ is a left end behavior model because $\lim_{x\to-\infty} \frac{e^x - 2x}{-2x} = \lim_{x\to-\infty} \left(-\frac{e^x}{2x} + 1\right) = 0 + 1 = 1$.

41. (a, b) The function $y = x$ is both a right end behavior model and a left end behavior model because

$$\lim_{x\to\pm\infty} \left(\frac{x + \ln|x|}{x}\right) = \lim_{x\to\pm\infty} \left(1 + \frac{\ln|x|}{x}\right) = 1 - 0 = 1.$$

43. $f(x) = \sqrt{x^2+x+1} - x$

(a)

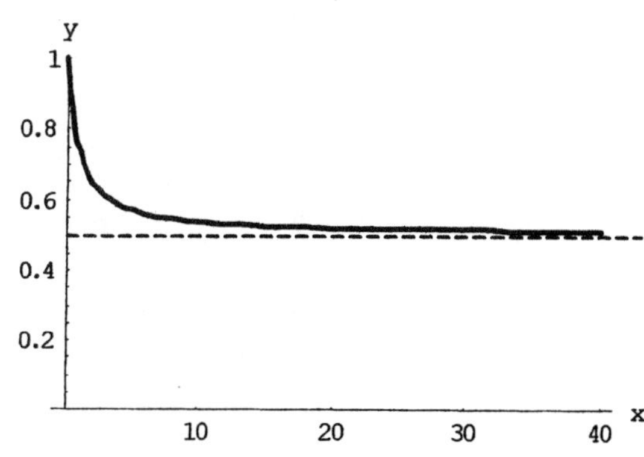

The graph suggests that $\lim_{x\to\infty} f(x) = \frac{1}{2}$.

(b)

x	f(x) to 6 decimal places
0	1.000000
10	0.535654
100	0.503731
1000	0.500375
10000	0.500037
100000	0.500004
1000000	0.500000

The table of values also suggest that $\lim_{x\to\infty} f(x) = \frac{1}{2}$

Proof: $\lim_{x\to\infty} \left(\sqrt{x^2+x+1}-x\right) = \lim_{x\to\infty} \left[\left(\sqrt{x^2+x+1}-x\right)\left(\frac{\sqrt{x^2+x+1}+x}{\sqrt{x^2+x+1}+x}\right)\right] = \lim_{x\to\infty} \left(\frac{1+x}{\sqrt{x^2+x+1}+x}\right)$

$$= \lim_{x\to\infty} \left(\frac{1+1/x}{\sqrt{1+1/x+1/x^2}+1}\right) = \frac{1}{2}$$

45. At most 2 horizontal asymptotes: one for $x \to \infty$ and possibly another for $x \to -\infty$.

47. $y = \frac{x}{\sqrt{4-x^2}}$

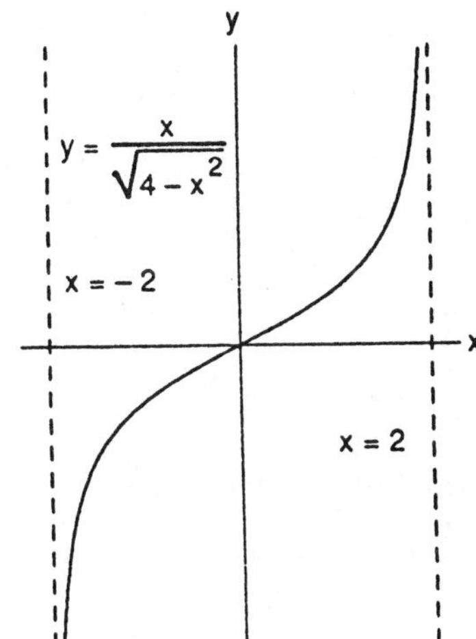

49. $y = x^{2/3} + \frac{1}{x^{1/3}}$

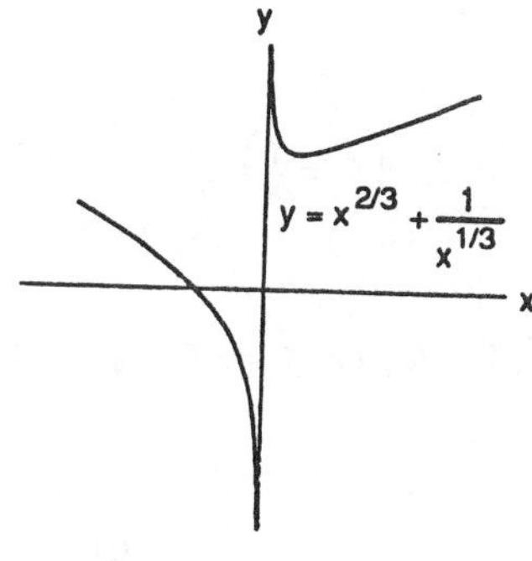

51.

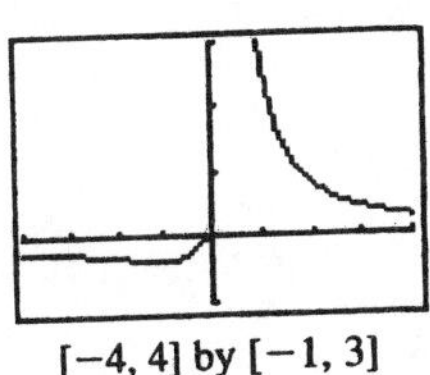

[−4, 4] by [−1, 3]

The graph of $y = f\left(\frac{1}{x}\right) = \frac{1}{x}e^{1/x}$ is shown.

$\lim_{x\to\infty} f(x) = \lim_{x\to 0^+} f\left(\frac{1}{x}\right) = \infty$

$\lim_{x\to -\infty} f(x) = \lim_{x\to 0^-} f\left(\frac{1}{x}\right) = 0$

53.

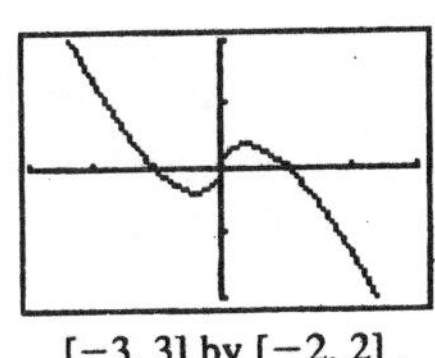

[−3, 3] by [−2, 2]

The graph of $y = f\left(\frac{1}{x}\right) = x \ln\left|\frac{1}{x}\right|$ is shown.

$$\lim_{x\to\infty} f(x) = \lim_{x\to 0^+} f\left(\frac{1}{x}\right) = 0$$

$$\lim_{x\to -\infty} f(x) = \lim_{x\to 0^-} f\left(\frac{1}{x}\right) = 0$$

55. $\lim_{x\to -\infty} \dfrac{\cos \frac{1}{x}}{1+\frac{1}{x}} = \lim_{\theta\to 0^-} \dfrac{\cos \theta}{1+\theta} = \frac{1}{1} = 1, \ \left(\theta = \frac{1}{x}\right)$

57. $\lim_{x\to \pm\infty} \left(3+\frac{2}{x}\right)\left(\cos \frac{1}{x}\right) = \lim_{\theta\to 0} (3+2\theta)(\cos \theta) = (3)(1) = 3, \ \left(\theta = \frac{1}{x}\right)$

59. $y = -\dfrac{x^2-4}{x+1} = 1 - x + \dfrac{3}{x+1}$

The graph of the function mimics each term as it becomes dominant.

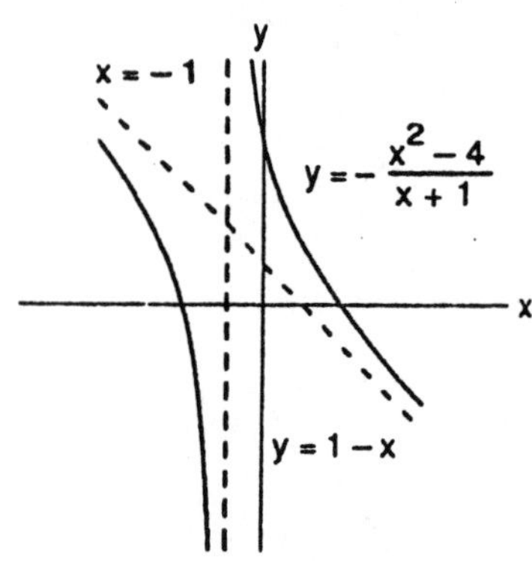

61. The graph of the function mimics each term as it becomes dominant.

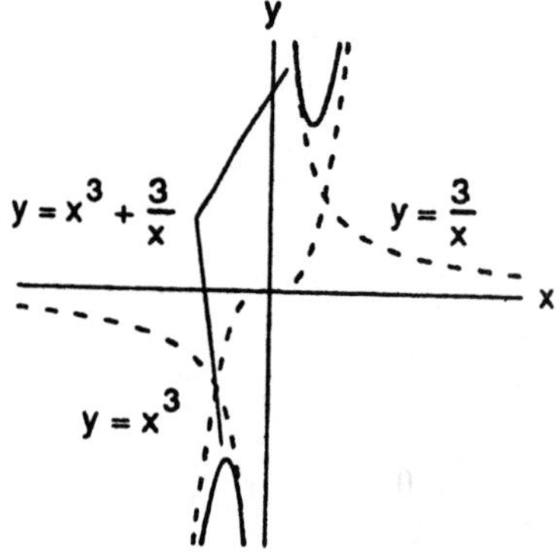

63. (a) $y \to \infty$ (see the accompanying graph)
(b) $y \to \infty$ (see the accompanying graph)
(c) cusps at $x = \pm 1$ (see the accompanying graph)

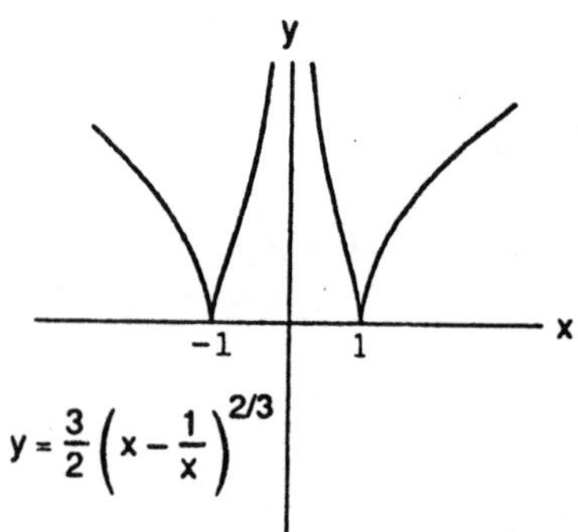

1.4 CONTINUITY

1. No, discontinuous at $x = 2$, not defined at $x = 2$

3. Continuous on $[-1, 3]$

5. (a) Yes (b) Yes, $\lim_{x\to -1^+} f(x) = 0$
   (c) Yes (d) Yes

7. (a) No (b) No

9. $f(2) = 0$, since $\lim_{x\to 2^-} f(x) = -2(2) + 4 = 0 = \lim_{x\to 2^+} f(x)$

11. The function f(x) is not continuous at $x = 0$ because $\lim_{x\to 0} f(x) = 0$, $f(0) = 1$ and, therefore, $\lim_{x\to 0} f(x) \neq f(0)$. The function f(x) is not continuous at $x = 1$ because $\lim_{x\to 1} f(x)$ does not exist since $\lim_{x\to 1^-} f(x) = -1$ and $\lim_{x\to 1^+} f(x) = 0$. The discontinuity at $x = 0$ is removable because the function would be continuous there if the value of f(0) were 0 instead of 1. The discontinuity at $x = 1$ is not removable because $\lim_{x\to 1} f(x)$ does not exist and the discontinuity cannot be removed by defining or redefining f(1).

13. Discontinuous only when $x - 2 = 0 \Rightarrow x = 2 \Rightarrow$ continuous on $(-\infty, 2) \cup (2, \infty)$

15. Discontinuous only when $t^2 - 4t + 3 = 0 \Rightarrow (t-3)(t-1) = 0 \Rightarrow t = 3$ or $t = 1 \Rightarrow$ continuous on $(-\infty, 1) \cup (1, 3) \cup (3, \infty)$

17. Discontinuous only at $\theta = 0 \Rightarrow$ continuous on $(-\infty, 0) \cup (0, \infty)$

19. Discontinuous when $2v + 3 < 0$ or $v < -\frac{3}{2} \Rightarrow$ continuous on the interval $\left[-\frac{3}{2}, \infty\right)$.

21. $\lim_{x\to\pi} \sin(x - \sin x) = \sin(\pi - \sin \pi) = \sin(\pi - 0) = \sin \pi = 0$; continuous at $x = \pi$

23. $\lim_{y\to 1} \sec\left(y \sec^2 y - \tan^2 y - 1\right) = \lim_{y\to 1} \sec\left(y \sec^2 y - \sec^2 y\right) = \lim_{y\to 1} \sec\left((y-1)\sec^2 y\right) = \sec\left((1-1)\sec^2 1\right)$ $= \sec 0 = 1$; continuous at $y = 1$

25. f(x) is continuous on $[0, 1]$ and $f(0) < 0$, $f(1) > 0$ $\Rightarrow$ by the Intermediate Value Theorem f(x) takes on every value between f(0) and f(1) $\Rightarrow$ the equation $f(x) = 0$ has at least one solution between $x = 0$ and $x = 1$.

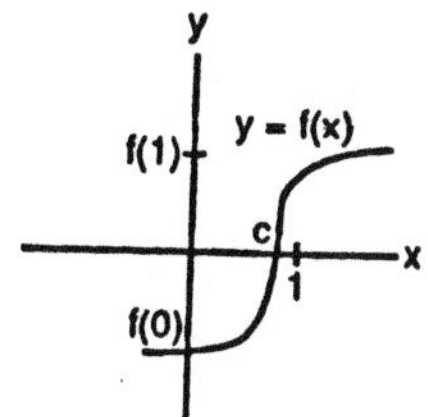

27. All five statements ask for the same information because of the intermediate value property of continuous functions.

    (a) A root of $f(x) = x^3 - 3x - 1$ is a point c where $f(c) = 0$. The roots are approximately $x_1 = -1.53$, $x_2 = -0.347$, $x_3 = 1.88$, the points where f(x) changes sign.

    (b) The points where $y = x^3$ crosses $y = 3x + 1$ have the same y-coordinate, or $y = x^3 = 3x + 1 \Rightarrow y = f(x)$ $= x^3 - 3x - 1 = 0$.

(c) $x^3 - 3x = 1 \Rightarrow x^3 - 3x - 1 = 0$. The solutions to the equation are the roots of $f(x) = x^3 - 3x - 1$.

(d) The points where $y = x^3 - 3x$ crosses $y = 1$ have common y-coordinates, or $y = x^3 - 3x = 1 \Rightarrow y = f(x) = x^3 - 3x - 1 = 0$.

(e) The solutions of $x^3 - 3x - 1 = 0$ are those points where $f(x) = x^3 - 3x - 1$ has value 0.

29. Answers may vary. For example, $f(x) = \frac{\sin(x-2)}{x-2}$ is discontinuous at $x = 2$ because it is not defined there. However, the discontinuity can be removed because f has a limit (namely 1) as $x \to 2$.

31. Noting that $r = 0$ is triple zero, the polynomial can be rewritten as $x^3(x^2 - x - 5)$. Therefore, the roots of the quintic polynomial are $r_1 = \frac{1 - \sqrt{21}}{2} \approx -1.791$, $r_2 = r_3 = r_4 = 0$, and $r_5 = \frac{1 + \sqrt{21}}{2} \approx 2.791$.

33. (a) Suppose x_0 is rational $\Rightarrow f(x_0) = 1$. Choose $\epsilon = \frac{1}{2}$. For any $\delta > 0$ there is an irrational number x (actually infinitely many) in the interval $(x_0 - \delta, x_0 + \delta) \Rightarrow f(x) = 0$. Then $0 < |x - x_0| < \delta$ but $|f(x) - f(x_0)| = 1 > \frac{1}{2} = \epsilon$, so $\lim_{x \to x_0} f(x)$ fails to exist $\Rightarrow$ f is discontinuous at x_0 rational. On the other hand, x_0 irrational $\Rightarrow f(x_0) = 0$ and there is a rational number x in $(x_0 - \delta, x_0 + \delta) \Rightarrow f(x) = 1$. Again $\lim_{x \to x_0} f(x)$ fails to exist $\Rightarrow$ f is discontinuous at x_0 irrational. That is, f is discontinuous at every point.

(b) f is neither right-continuous nor left-continuous at any point x_0 because in every interval $(x_0 - \delta, x_0)$ or $(x_0, x_0 + \delta)$ there exist both rational and irrational real numbers. Thus neither limits $\lim_{x \to x_0^-} f(x)$ and $\lim_{x \to x_0^+} f(x)$ exist by the same arguments used in part (a).

35. Yes, because of the Intermediate Value Theorem. If f(a) and f(b) did have different signs then f would have to equal zero at some point between a and b since f is continuous on [a, b].

37. If $f(0) = 0$ or $f(1) = 1$, we are done (i.e., $c = 0$ or $c = 1$ in those cases). Then let $f(0) = a > 0$ and $f(1) = b < 1$ because $0 \le f(x) \le 1$. Define $g(x) = f(x) - x \Rightarrow g$ is continuous on [0, 1]. Moreover, $g(0) = f(0) - 0 = a > 0$ and $g(1) = f(1) - 1 = b - 1 < 0 \Rightarrow$ by the Intermediate Value Theorem there is a number c in (0, 1) such that $g(c) = 0 \Rightarrow f(c) - c = 0$ or $f(c) = c$.

39. (a) Luisa's salary is $36,500 = $36,500(1.035)^0 for the first year ($0 \le t < 1$), $36,500(1.035) for the second year ($1 \le t < 2$), $36,500(1.035)^2 for the third year ($2 \le t < 3$), and so on. This corresponds to $y = 36{,}500(1.035)^{\text{int } t}$.

(b)

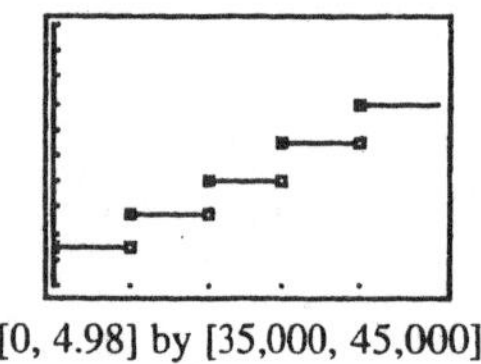

[0, 4.98] by [35,000, 45,000]

The function is continuous at all points in the domain [0, 5) except at t = 1, 2, 3, 4.

41. The function can be extended: $f(0) \approx 2.3$.

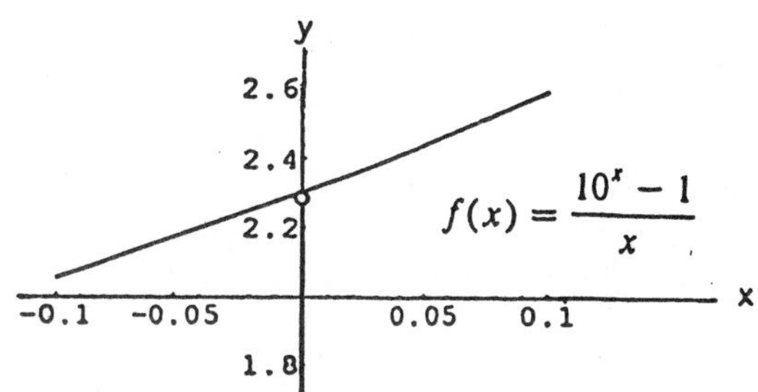

43. The function cannot be extended to be continuous at $x = 0$. If $f(0) = 1$, it will be continuous from the right. Or if $f(0) = -1$, it will be continuous from the left.

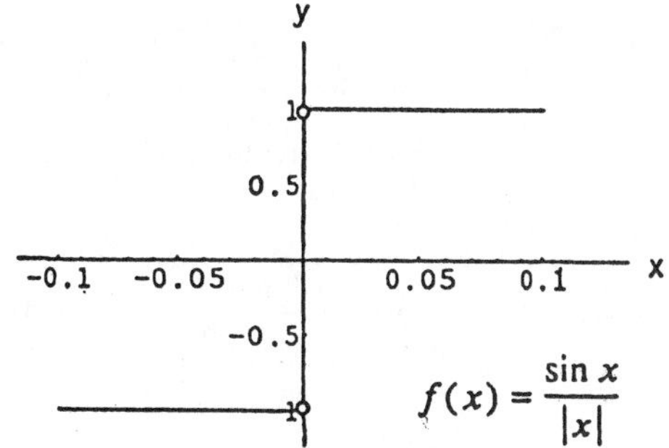

45. $x \approx 1.8794, -1.5321, -0.3473$

47. $x \approx 1.7549$

49. $x \approx 3.5156$

1.5 TANGENT LINES

1. P_1: $m_1 = 1$, P_2: $m_2 = 5$

3. P_1: $m_1 = \frac{5}{2}$, P_2: $m_2 = -\frac{1}{2}$

5. $$m = \lim_{h \to 0} \frac{[4-(-1+h)^2]-(4-(-1)^2)}{h}$$
$$= \lim_{h \to 0} \frac{-(1-2h+h^2)+1}{h} = \lim_{h \to 0} \frac{h(2-h)}{h} = 2;$$
at $(-1,3)$: $y = 3 + 2(x-(-1)) \Rightarrow y = 2x+5$, tangent line

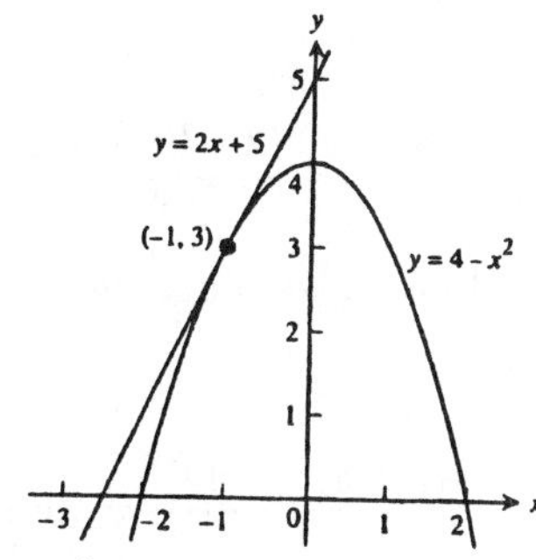

7. $$m = \lim_{h \to 0} \frac{(-2+h)^3-(-2)^3}{h} = \lim_{h \to 0} \frac{-8+12h-6h^2+h^3+8}{h}$$
$$= \lim_{h \to 0} \left(12 - 6h + h^2\right) = 12;$$
at $(-2,-8)$: $y = -8 + 12(x-(-2)) \Rightarrow y = 12x+16$, tangent line

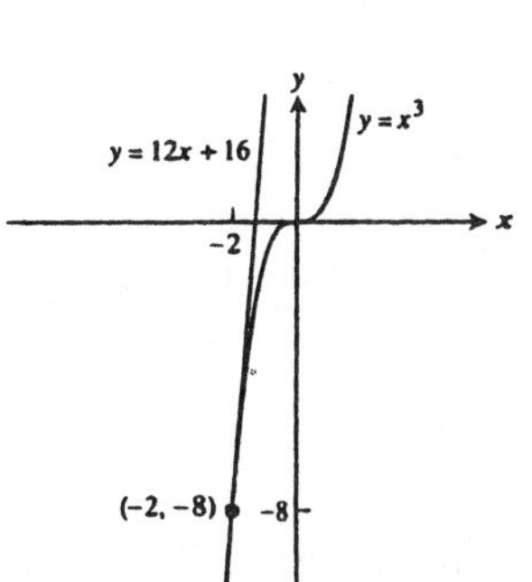

9. $m = \lim_{h\to 0} \frac{[(1+h)-2(1+h)^2]-(-1)}{h} = \lim_{h\to 0} \frac{(1+h-2-4h-2h^2)+1}{h} = \lim_{h\to 0} \frac{h(-3-2h)}{h} = -3;$
at $(1,-1)$: $y+1 = -3(x-1)$, tangent line

11. $m = \lim_{h\to 0} \frac{\frac{3+h}{(3+h)-2} - 3}{h} = \lim_{h\to 0} \frac{(3+h)-3(h+1)}{h(h+1)} = \lim_{h\to 0} \frac{-2h}{h(h+1)} = -2;$
at $(3,3)$: $y-3 = -2(u-3)$, tangent line

13. At $x = 3$, $y = \frac{1}{2} \Rightarrow m = \lim_{h\to 0} \frac{\frac{1}{(3+h)-1} - \frac{1}{2}}{h} = \lim_{h\to 0} \frac{2-(2+h)}{2h(2+h)} = \lim_{h\to 0} \frac{-h}{2h(2+h)} = -\frac{1}{4}$, slope

15. At a horizontal tangent the slope $m = 0 \Rightarrow 0 = m = \lim_{h\to 0} \frac{[(x+h)^2+4(x+h)-1]-(x^2+4x-1)}{h}$
$= \lim_{h\to 0} \frac{(x^2+2xh+h^2+4x+4h-1)-(x^2+4x-1)}{h} = \lim_{h\to 0} \frac{(2xh+h^2+4h)}{h} = \lim_{h\to 0} (2x+h+4) = 2x+4;$
$2x+4 = 0 \Rightarrow x = -2$. Then $f(-2) = 4-8-1 = -5 \Rightarrow (-2,-5)$ is the point on the graph where there is a horizontal tangent.

17. $-1 = m = \lim_{h\to 0} \frac{\frac{1}{(x+h)-1} - \frac{1}{x-1}}{h} = \lim_{h\to 0} \frac{(x-1)-(x+h-1)}{h(x-1)(x+h-1)} = \lim_{h\to 0} \frac{-h}{h(x-1)(x+h-1)} = -\frac{1}{(x-1)^2}$
$\Rightarrow (x-1)^2 = 1 \Rightarrow x^2-2x = 0 \Rightarrow x(x-2) = 0 \Rightarrow x = 0$ or $x = 2$. If $x = 0$, then $y = -1$ and $m = -1$
$\Rightarrow y = -1-(x-0) = -(x+1)$. If $x = 2$, then $y = 1$ and $m = -1 \Rightarrow y = 1-(x-2) = -(x-3)$.

19. $\lim_{h\to 0} \frac{f(2+h)-f(2)}{h} = \lim_{h\to 0} \frac{(100-4.9(2+h)^2)-(100-4.9(2)^2)}{h} = \lim_{h\to 0} \frac{-4.9(4+4h+h^2)+4.9(4)}{h}$
$= \lim_{h\to 0} (-19.6-4.9h) = -19.6$. The minus sign indicates the object is falling downward at a speed of 19.6 m/sec.

21. $\lim_{h\to 0} \frac{f(3+h)-f(3)}{h} = \lim_{h\to 0} \frac{\pi(3+h)^2-\pi(3)^2}{h} = \lim_{h\to 0} \frac{\pi[9+6h+h^2-9]}{h} = \lim_{h\to 0} \pi(6+h) = 6\pi$

23. $\lim_{h\to 0} \frac{s(1+h)-s(1)}{h} = \lim_{h\to 0} \frac{1.86(1+h)^2-1.86(1)^2}{h} = \lim_{h\to 0} \frac{1.86+3.72h+1.86h^2-1.86}{h} = \lim_{h\to 0} (3.72+1.86h)$
$= 3.72$

25. Slope at origin $= \lim_{h\to 0} \frac{f(0+h)-f(0)}{h} = \lim_{h\to 0} \frac{h^2 \sin\left(\frac{1}{h}\right)}{h} = \lim_{h\to 0} h \sin\left(\frac{1}{h}\right) = 0 \Rightarrow$ yes, f(x) does have a tangent at the origin with slope 0.

27. $\lim_{h\to 0^-} \frac{f(0+h)-f(0)}{h} = \lim_{h\to 0^-} \frac{-1-0}{h} = \infty$, and $\lim_{h\to 0^+} \frac{f(0+h)-f(0)}{h} = \lim_{h\to 0^+} \frac{1-0}{h} = \infty$. Therefore,

$$\lim_{h\to 0} \frac{f(0+h)-f(0)}{h} = \infty \Rightarrow \text{yes, the graph of f has a vertical tangent at the origin.}$$

29. (a) $\frac{\Delta f}{\Delta x} = \frac{f(0)-f(-2)}{0-(-2)} = \frac{1-e^{-2}}{2} \approx 0.432$ (b) $\frac{\Delta f}{\Delta x} = \frac{f(3)-f(1)}{3-1} = \frac{e^3-e}{2} \approx 8.684$

31. (a) $\frac{\Delta f}{\Delta t} = \frac{f(3\pi/4)-f(\pi/4)}{(3\pi/4)-(\pi/4)} = \frac{-1-1}{\pi/2} = -\frac{4}{\pi} \approx -1.273$

(b) $\frac{\Delta f}{\Delta t} = \frac{f(\pi/2)-f(\pi/6)}{(\pi/2)-(\pi/6)} = \frac{0-\sqrt{3}}{\pi/3} = -\frac{3\sqrt{3}}{\pi} \approx -1.654$

33. (a) $\frac{2.1-1.5}{1995-1993} = 0.3$

The rate of change was 0.3 billion dollars per year.

(b) $\frac{3.1-2.1}{1997-1995} = 0.5$

The rate of change was 0.5 billion dollars per year.

(c) $y = 0.0571x^2 - 0.1514x + 1.3943$

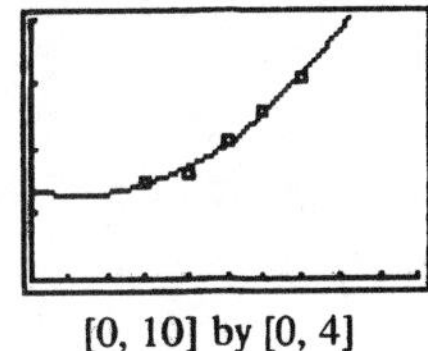

[0, 10] by [0, 4]

(d) $\frac{y(5)-y(3)}{5-3} \approx 0.31$

$\frac{y(7)-y(5)}{7-5} \approx 0.53$

According to the regression equation, the rates were 0.31 billion dollars per year and 0.53 billion dollars per year.

(e)
$$\begin{aligned}\lim_{h\to 0} \frac{y(7+h)-y(7)}{h} &= \lim_{h\to 0} \frac{[0.0571(7+h)^2 - 0.1514(7+h) + 1.3943] - [0.0571(7)^2 - 0.1514(7) + 1.3943]}{h} \\ &= \lim_{h\to 0} \frac{0.0571(14h+h^2) - 0.1514h}{h} \\ &= \lim_{h\to 0} [0.0571(14) - 0.1514 + 0.0571h] \\ &\approx 0.65\end{aligned}$$

The funding was growing at a rate of about 0.65 billion dollars per year.

35. (a) The graph appears to have a cusp at $x = 0$.

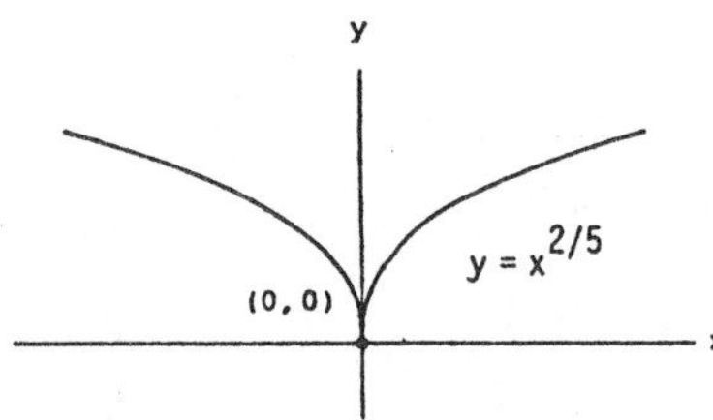

(b) $\lim\limits_{h\to 0^-} \frac{f(0+h)-f(0)}{h} = \lim\limits_{h\to 0^-} \frac{h^{2/5}-0}{h} = \lim\limits_{h\to 0^-} \frac{1}{h^{3/5}} = -\infty$ and $\lim\limits_{h\to 0^+} \frac{1}{h^{3/5}} = \infty \Rightarrow$ limit does not exist $\Rightarrow$ the graph of $y = x^{2/5}$ does not have a vertical tangent at $x = 0$.

37. (a) The graph appears to have a vertical tangent at $x = 0$.

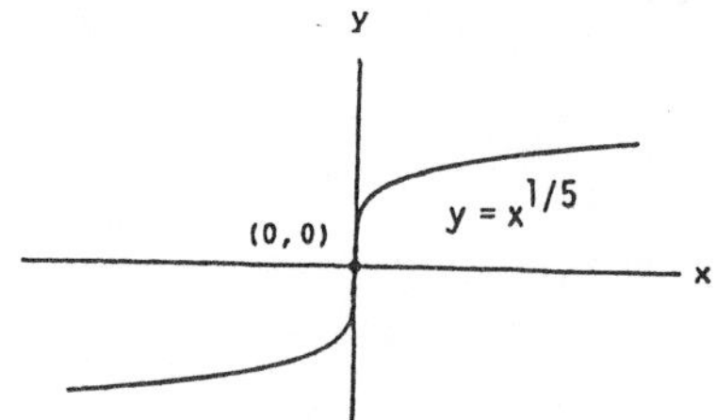

(b) $\lim\limits_{h\to 0} \frac{f(0+h)-f(0)}{h} = \lim\limits_{h\to 0} \frac{h^{1/5}-0}{h} = \lim\limits_{h\to 0} \frac{1}{h^{4/5}} = \infty \Rightarrow y = x^{1/5}$ has a vertical tangent at $x = 0$.

39. (a) The graph appears to have a cusp at $x = 0$.

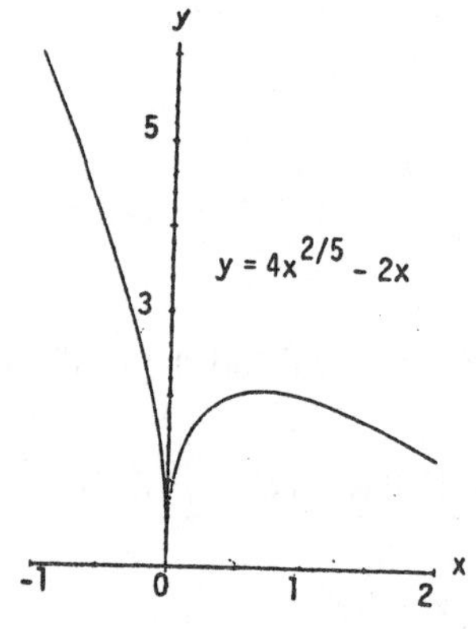

(b) $\lim\limits_{h\to 0^-} \frac{f(0+h)-f(0)}{h} = \lim\limits_{h\to 0^-} \frac{4h^{2/5}-2h}{h} = \lim\limits_{h\to 0^-} \frac{4h^{2/5}-2h}{h} = \lim\limits_{h\to 0^-} \frac{4}{h^{3/5}} - 2 = -\infty$ and $\lim\limits_{h\to 0^+} \frac{4}{h^{3/5}} - 2$ $= \infty \Rightarrow$ limit does not exist $\Rightarrow$ the graph of $y = 4x^{2/5} - 2x$ does not have a vertical tangent at $x = 0$.

41. (a) The graph appears to have a vertical tangent at $x = 1$ and a cusp at $x = 0$.

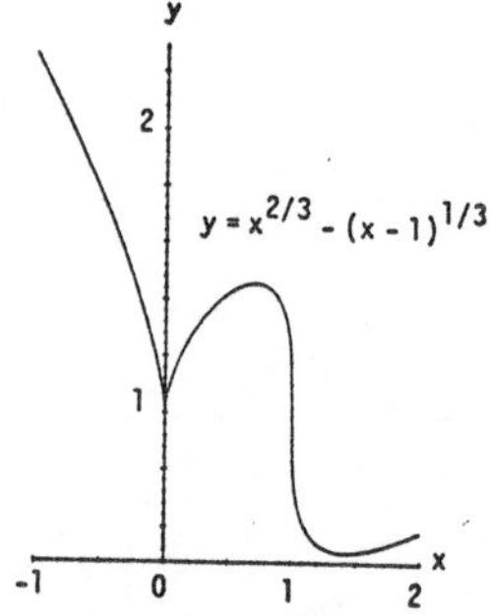

(b) x = 1: $\lim_{h\to 0} \frac{(1+h)^{2/3}-(1+h-1)^{1/3}-1}{h} = \lim_{h\to 0} \frac{(1+h)^{2/3}-h^{1/3}-1}{h} = -\infty$

$\Rightarrow y = x^{2/3}-(x-1)^{1/3}$ has a vertical tangent at x = 1;

x = 0: $\lim_{h\to 0} \frac{f(0+h)-f(0)}{h} = \lim_{h\to 0} \frac{h^{2/3}-(h-1)^{1/3}-(-1)^{1/3}}{h} = \lim_{h\to 0} \left[\frac{1}{h^{1/3}} - \frac{(h-1)^{1/3}}{h} + \frac{1}{h}\right]$

does not exist $\Rightarrow y = x^{2/3}-(x-1)^{1/3}$ does not have a vertical tangent at x = 0.

43. (a) The graph appears to have a vertical tangent at x = 0.

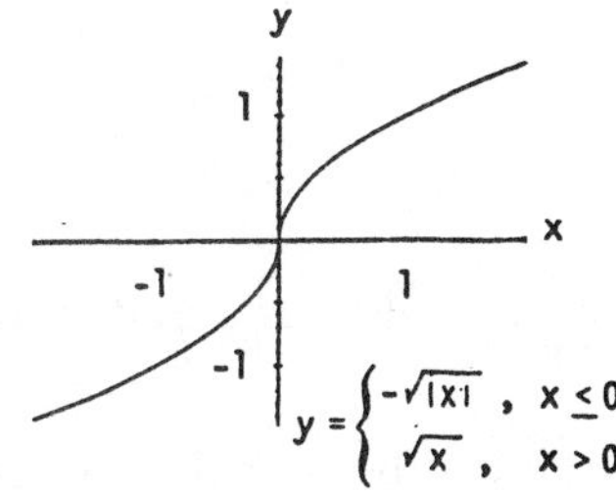

(b) $\lim_{h\to 0^+} \frac{f(0+h)-f(0)}{h} = \lim_{x\to 0^+} \frac{\sqrt{h}-0}{h} = \lim_{h\to 0^+} \frac{1}{\sqrt{h}} = \infty$;

$\lim_{h\to 0^-} \frac{f(0+h)-f(0)}{h} = \lim_{h\to 0^-} \frac{-\sqrt{|h|}-0}{h} = \lim_{h\to 0^-} \frac{-\sqrt{|h|}}{-|h|} = \lim_{h\to 0^-} \frac{1}{\sqrt{|h|}} = \infty$

$\Rightarrow$ y has a vertical tangent at x = 0.

CHAPTER 1 PRACTICE EXERCISES

1. At x = −1: $\lim_{x\to -1^-} f(x) = \lim_{x\to -1^+} f(x) = 1 \Rightarrow \lim_{x\to -1} f(x)$

$= 1 = f(-1) \Rightarrow$ f is continuous at x = −1.

At x = 0: $\lim_{x\to 0^-} f(x) = \lim_{x\to 0^+} f(x) = 0 \Rightarrow \lim_{x\to 0} f(x) = 0$.

But $f(0) = 1 \neq \lim_{x\to 0} f(x) \Rightarrow$ f is discontinuous at x = 0.

At x = 1: $\lim_{x\to 1^-} f(x) = -1$ and $\lim_{x\to 1^+} f(x) = 1 \Rightarrow \lim_{x\to 1} f(x)$

does not exist $\Rightarrow$ f is discontinuous at x = 1.

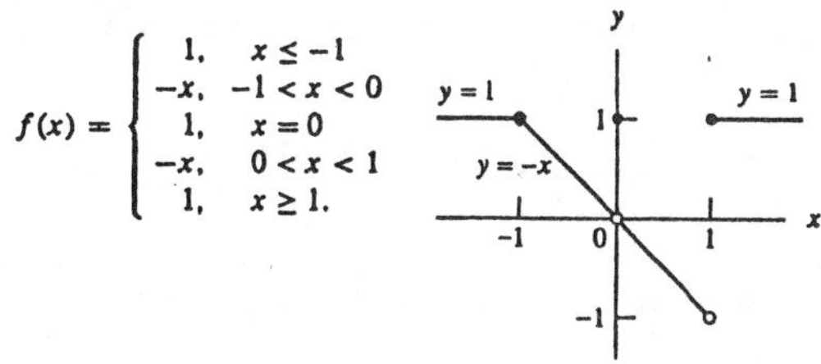

3. (a) $\lim_{t\to t_0} (3f(t)) = 3 \lim_{t\to t_0} f(t) = 3(-7) = -21$

(b) $\lim_{t\to t_0} (f(t))^2 = \left(\lim_{t\to t_0} f(t)\right)^2 = (-7)^2 = 49$

(c) $\lim_{t\to t_0} (f(t)\cdot g(t)) = \lim_{t\to t_0} f(t) \cdot \lim_{t\to t_0} g(t) = (-7)(0) = 0$

(d) $\lim\limits_{t\to t_0} \frac{f(t)}{g(t)-7} = \frac{\lim\limits_{t\to t_0} f(t)}{\lim\limits_{t\to t_0} (g(t)-7)} = \frac{-7}{\lim\limits_{t\to t_0} g(t) - \lim\limits_{t\to t_0} 7} = \frac{-7}{0-7} = 1$

(e) $\lim\limits_{t\to t_0} \cos(g(t)) = \cos\left(\lim\limits_{t\to t_0} g(t)\right) = \cos 0 = 1$

(f) $\lim\limits_{t\to t_0} |f(t)| = \left|\lim\limits_{t\to t_0} f(t)\right| = |-7| = 7$

(g) $\lim\limits_{t\to t_0} (f(t)+g(t)) = \lim\limits_{t\to t_0} f(t) + \lim\limits_{t\to t_0} g(t) = -7+0 = -7$

(h) $\lim\limits_{t\to t_0} (1/f(t)) = \frac{1}{\lim\limits_{t\to t_0} f(t)} = \frac{1}{-7} = -\frac{1}{7}$

5. Since $\lim\limits_{x\to 0} x = 0$ we must have that $\lim\limits_{x\to 0} (4-g(x)) = 0$. Otherwise, if $\lim\limits_{x\to 0} (4-g(x))$ is a finite positive number, we would have $\lim\limits_{x\to 0^-} \left[\frac{4-g(x)}{x}\right] = -\infty$ and $\lim\limits_{x\to 0^+} \left[\frac{4-g(x)}{x}\right] = \infty$ so the limit could not equal 1 as $x \to 0$. Similar reasoning holds if $\lim\limits_{x\to 0} (4-g(x))$ is a finite negative number. We conclude that $\lim\limits_{x\to 0} g(x) = 4$.

7. (a) $\lim\limits_{x\to c} f(x) = \lim\limits_{x\to c} x^{1/3} = c^{1/3} = f(c)$ for every real number $c \Rightarrow f$ is continuous on $(-\infty, \infty)$

(b) $\lim\limits_{x\to c} g(x) = \lim\limits_{x\to c} x^{3/4} = c^{3/4} = g(c)$ for every nonnegative real number $c \Rightarrow g$ is continuous on $[0, \infty)$

(c) $\lim\limits_{x\to c} h(x) = \lim\limits_{x\to c} x^{-2/3} = \frac{1}{c^{2/3}} = h(c)$ for every nonzero real number $c \Rightarrow h$ is continuous on $(-\infty, 0)$ and $(0, \infty)$

(d) $\lim\limits_{x\to c} k(x) = \lim\limits_{x\to c} x^{-1/6} = \frac{1}{c^{1/6}} = k(c)$ for every positive real number $c \Rightarrow h$ is continuous on $(0, \infty)$

9. (a) $\lim\limits_{x\to 0} \frac{x^2-4x+4}{x^3+5x^2-14x} = \lim\limits_{x\to 0} \frac{(x-2)(x-2)}{x(x+7)(x-2)} = \lim\limits_{x\to 0} \frac{x-2}{x(x+7)}$, $x \neq 2$; the limit does not exist because $\lim\limits_{x\to 0^-} \frac{x-2}{x(x+7)} = \infty$ and $\lim\limits_{x\to 0^+} \frac{x-2}{x(x+7)} = -\infty$

(b) $\lim\limits_{x\to 2} \frac{x^2-4x+4}{x^3+5x^2-14x} = \lim\limits_{x\to 2} \frac{(x-2)(x-2)}{x(x+7)(x-2)} = \lim\limits_{x\to 2} \frac{x-2}{x(x+7)}$, $x \neq 2 = \frac{0}{2(9)} = 0$

11. $\lim\limits_{x\to 1} \frac{1-\sqrt{x}}{1-x} = \lim\limits_{x\to 1} \frac{1-\sqrt{x}}{(1-\sqrt{x})(1+\sqrt{x})} = \lim\limits_{x\to 1} \frac{1}{1+\sqrt{x}} = \frac{1}{2}$

13. $\lim\limits_{h\to 0} \frac{(x+h)^2-x^2}{h} = \lim\limits_{h\to 0} \frac{(x^2+2hx+h^2)-x^2}{h} = \lim\limits_{h\to 0} (2x+h) = 2x$

15. $\lim\limits_{x\to 0} \frac{\frac{1}{2+x}-\frac{1}{2}}{x} = \lim\limits_{x\to 0} \frac{2-(2+x)}{2x(2+x)} = \lim\limits_{x\to 0} \frac{-1}{4+2x} = -\frac{1}{4}$

17. $\lim\limits_{x\to\infty} \frac{2x+3}{5x+7} = \lim\limits_{x\to\infty} \frac{2+\frac{3}{x}}{5+\frac{7}{x}} = \frac{2+0}{5+0} = \frac{2}{5}$

19. $\lim\limits_{x\to-\infty} \frac{x^2-4x+8}{3x^3} = \lim\limits_{x\to-\infty} \left(\frac{1}{3x} - \frac{4}{3x^2} + \frac{8}{3x^3}\right) = 0 - 0 + 0 = 0$

21. $\lim\limits_{x\to-\infty} \frac{x^2-7x}{x+1} = \lim\limits_{x\to-\infty} \left(\frac{x-7}{1+\frac{1}{x}}\right) = -\infty$

23. $\lim\limits_{x\to\infty} \frac{|\sin x|}{\text{int } x} \le \lim\limits_{x\to\infty} \frac{1}{\text{int } x} = 0$ since int $x \to \infty$ as $x \to \infty$

25. $\lim\limits_{x\to\infty} \frac{x + \sin x + 2\sqrt{x}}{x + \sin x} = \lim\limits_{x\to\infty} \left(\frac{1 + \frac{\sin x}{x} + \frac{2}{\sqrt{x}}}{1 + \frac{\sin x}{x}}\right) = \frac{1+0+0}{1+0} = 1$

27. $\lim\limits_{x\to\infty} e^{-x^2} = \lim\limits_{x\to\infty} \frac{1}{e^{x^2}} = 0$

29. (a) $f(-1) = -1$ and $f(2) = 5 \Rightarrow f$ has a root between -1 and 2 by the Intermediate Value Theorem.
(b), (c) root is 1.32471795724

CHAPTER 1 ADDITIONAL EXERCISES–THEORY, EXAMPLES, APPLICATIONS

1. (a)

x	0.1	0.01	0.001	0.0001	0.00001
x^x	0.7943	0.9550	0.9931	0.9991	0.9999

Apparently, $\lim\limits_{x\to0^+} x^x = 1$

(b)

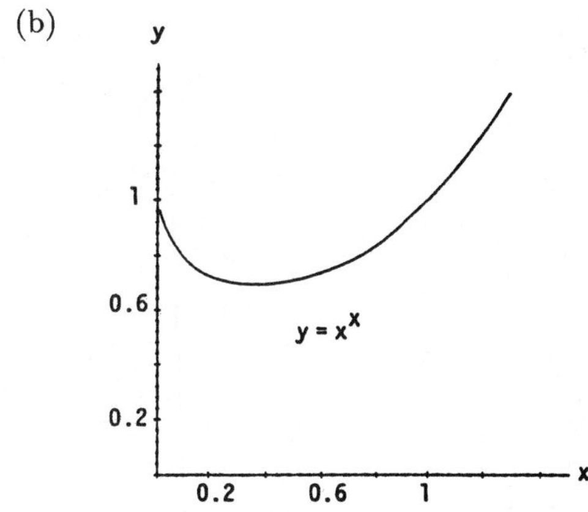

3. $\lim\limits_{v\to c^-} L = \lim\limits_{v\to c^-} L_0\sqrt{1-\frac{v^2}{c^2}} = L_0\sqrt{1-\frac{\lim\limits_{v\to c^-} v^2}{c^2}} = L_0\sqrt{1-\frac{c^2}{c^2}} = 0$

The left-hand limit was needed because the function L is undefined if $v > c$ (the rocket cannot move faster than the speed of light).

5. $\left|10+(t-70)\times 10^{-4}-10\right|<0.0005 \Rightarrow \left|(t-70)\times 10^{-4}\right|<0.0005 \Rightarrow -0.0005<(t-70)\times 10^{-4}<0.0005$
$\Rightarrow -5<t-70<5 \Rightarrow 65^\circ<t<75^\circ \Rightarrow$ Within 5° F.

7. (a) At $x=0$: $\lim\limits_{a\to 0} r_+(a) = \lim\limits_{a\to 0} \frac{-1+\sqrt{1+a}}{a} = \lim\limits_{a\to 0}\left(\frac{-1+\sqrt{1+a}}{a}\right)\left(\frac{-1-\sqrt{1+a}}{-1-\sqrt{1+a}}\right)$

$= \lim\limits_{a\to 0} \frac{1-(1+a)}{a\left(-1-\sqrt{1+a}\right)} = \frac{-1}{-1-\sqrt{1+0}} = \frac{1}{2}$

At $x=-1$: $\lim\limits_{a\to -1^+} r_+(a) = \lim\limits_{a\to -1^+} \frac{1-(1+a)}{a\left(-1-\sqrt{1+a}\right)} = \lim\limits_{a\to -1} \frac{-a}{a\left(-1-\sqrt{1+a}\right)} = \frac{-1}{-1-\sqrt{0}} = 1$

(b) At $x=0$: $\lim\limits_{a\to 0^-} r_-(a) = \lim\limits_{a\to 0^-} \frac{-1-\sqrt{1+a}}{a} = \lim\limits_{a\to 0^-}\left(\frac{-1-\sqrt{1+a}}{a}\right)\left(\frac{-1+\sqrt{1+a}}{-1+\sqrt{1+a}}\right)$

$= \lim\limits_{a\to 0^-} \frac{1-(1+a)}{a\left(-1+\sqrt{1+a}\right)} = \lim\limits_{a\to 0^-} \frac{a}{a\left(-1+\sqrt{1+a}\right)} = \lim\limits_{a\to 0^-} \frac{-1}{-1+\sqrt{1+a}} = \infty$ (because the

denominator is always negative); $\lim\limits_{a\to 0^+} r_-(a) = \lim\limits_{a\to 0^+} \frac{-1}{-1+\sqrt{1+a}} = -\infty$ (because the denominator

is always positive). Therefore, $\lim\limits_{a\to 0} r_-(a)$ does not exist.

At $x=-1$: $\lim\limits_{a\to -1^+} r_-(a) = \lim\limits_{a\to -1^+} \frac{-1-\sqrt{1+a}}{a} = \lim\limits_{a\to -1^+} \frac{-1}{-1+\sqrt{1+a}} = 1$

(c)

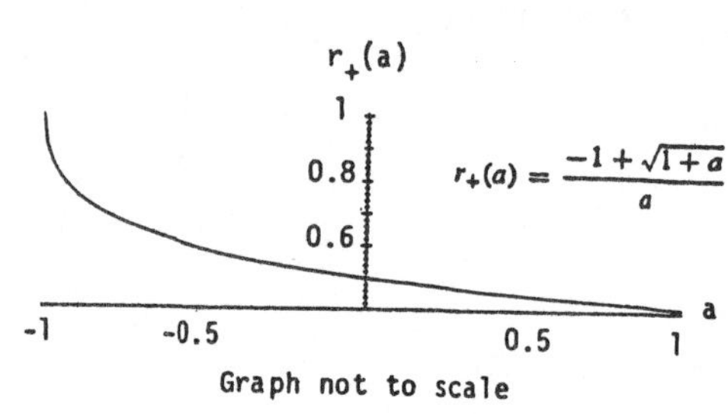

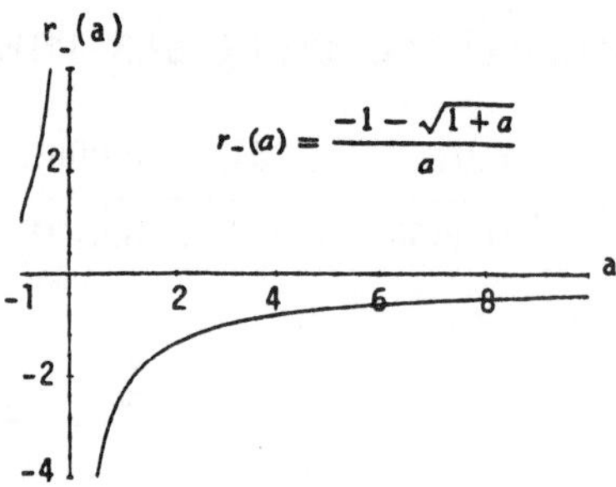

(d)

$f(x) = ax^2 + 2x - 1$

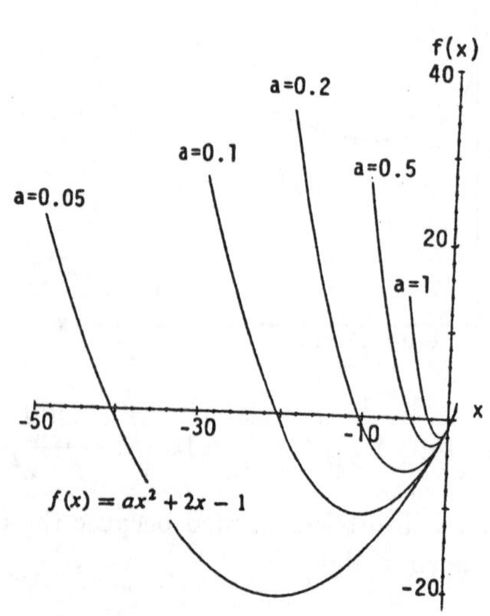

9. (a) True, because if $\lim_{x\to a} (f(x)+g(x))$ exists then $\lim_{x\to a} (f(x)+g(x)) - \lim_{x\to a} f(x) = \lim_{x\to a} [(f(x)+g(x)) - f(x)]$

$= \lim_{x\to a} g(x)$ exists, contrary to assumption.

(b) False; for example take $f(x) = \frac{1}{x}$ and $g(x) = -\frac{1}{x}$. Then neither $\lim_{x\to 0} f(x)$ nor $\lim_{x\to 0} g(x)$ exists, but

$\lim_{x\to 0} (f(x)+g(x)) = \lim_{x\to 0} \left(\frac{1}{x} - \frac{1}{x}\right) = \lim_{x\to 0} 0 = 0$ exists.

(c) True, because $g(x) = |x|$ is continuous $\Rightarrow g(f(x)) = |f(x)|$ is continuous (it is the composite of continuous functions).

(d) False; for example let $f(x) = \begin{cases} -1, & x \le 0 \\ 1, & x > 0 \end{cases} \Rightarrow f(x)$ is discontinuous at $x = 0$. However $|f(x)| = 1$ is continuous at $x = 0$.

11. Show $\lim_{x\to 1} f(x) = \lim_{x\to 1} (x^2 - 7) = -6 = f(1)$.

Step 1: $|(x^2-7)+6| < \epsilon \Rightarrow -\epsilon < x^2 - 1 < \epsilon \Rightarrow 1-\epsilon < x^2 < 1+\epsilon \Rightarrow \sqrt{1-\epsilon} < x < \sqrt{1+\epsilon}$.

Step 2: $|x-1| < \delta \Rightarrow -\delta < x - 1 < \delta \Rightarrow -\delta + 1 < x < \delta + 1$.

Then $-\delta + 1 = \sqrt{1-\epsilon}$ or $\delta + 1 = \sqrt{1+\epsilon}$. Choose $\delta = \min\left\{1 - \sqrt{1-\epsilon}, \sqrt{1+\epsilon} - 1\right\}$, then

$0 < |x-1| < \delta \Rightarrow |(x^2-7)-6| < \epsilon$ and $\lim_{x\to 1} f(x) = -6$. By the continuity test, f(x) is continuous at $x = 1$.

13. Show $\lim_{x\to 2} h(x) = \lim_{x\to 2} \sqrt{2x-3} = 1 = h(2)$.

Step 1: $\left|\sqrt{2x-3} - 1\right| < \epsilon \Rightarrow -\epsilon < \sqrt{2x-3} - 1 < \epsilon \Rightarrow 1-\epsilon < \sqrt{2x-3} < 1+\epsilon \Rightarrow \frac{(1-\epsilon)^2+3}{2} < x < \frac{(1+\epsilon)^2+3}{2}$.

Step 2: $|x-2| < \delta \Rightarrow -\delta < x - 2 < \delta$ or $-\delta + 2 < x < \delta + 2$.

Then $-\delta + 2 = \frac{(1-\epsilon)^2+3}{2} \Rightarrow \delta = 2 - \frac{(1-\epsilon)^2+3}{2} = \frac{1-(1-\epsilon)^2}{2} = \epsilon - \frac{\epsilon^2}{2}$, or $\delta + 2 = \frac{(1+\epsilon)^2+3}{2}$

$\Rightarrow \delta = \frac{(1+\epsilon)^2+3}{2} - 2 = \frac{(1+\epsilon)^2-1}{2} = \epsilon + \frac{\epsilon^2}{2}$. Choose $\delta = \epsilon - \frac{\epsilon^2}{2}$, the smaller of the two values. Then,

$0 < |x-2| < \delta \Rightarrow \left|\sqrt{2x-3} - 1\right| < \epsilon$, so $\lim_{x\to 2} \sqrt{2x-3} = 1$. By the continuity test, h(x) is continuous at $x = 2$.

15. (a) Let $\epsilon > 0$ be given. If x is rational, then $f(x) = x \Rightarrow |f(x) - 0| = |x - 0| < \epsilon \Leftrightarrow |x - 0| < \epsilon$; i.e., choose $\delta = \epsilon$. Then $|x - 0| < \delta \Rightarrow |f(x) - 0| < \epsilon$ for x rational. If x is irrational, then $f(x) = 0 \Rightarrow |f(x) - 0| < \epsilon$ $\Leftrightarrow 0 < \epsilon$ which is true no matter how close irrational x is to 0, so again we can choose $\delta = \epsilon$. In either case, given $\epsilon > 0$ there is a $\delta = \epsilon > 0$ such that $0 < |x - 0| < \delta \Rightarrow |f(x) - 0| < \epsilon$. Therefore, f is continuous at $x = 0$.

(b) Choose $x = c > 0$. Then within any interval $(c-\delta, c+\delta)$ there are both rational and irrational numbers. If c is rational, pick $\epsilon = \frac{c}{2}$. No matter how small we choose $\delta > 0$ there is an irrational number x in $(c-\delta, c+\delta) \Rightarrow |f(x) - f(c)| = |0 - c| = c > \frac{c}{2} = \epsilon$. That is, f is not continuous at any rational $c > 0$. On the other hand, suppose c is irrational $\Rightarrow f(c) = 0$. Again pick $\epsilon = \frac{c}{2}$. No matter how small we choose $\delta > 0$

there is a rational number x in $(c-\delta, c+\delta)$ with $|x-c| < \frac{c}{2} = \epsilon \Leftrightarrow \frac{c}{2} < x < \frac{3c}{2}$. Then $|f(x) - f(c)| = |x - 0|$ $= |x| > \frac{c}{2} = \epsilon \Rightarrow$ f is not continuous at any irrational $c > 0$.

If $x = c < 0$, repeat the argument picking $\epsilon = \frac{|c|}{2} = \frac{-c}{2}$. Therefore f fails to be continuous at any nonzero value $x = c$.

NOTES:

CHAPTER 2 DERIVATIVES

2.1 THE DERIVATIVE AS A FUNCTION

1. Step 1: $f(x) = 4 - x^2$ and $f(x+h) = 4 - (x+h)^2$

   Step 2: $\frac{f(x+h) - f(x)}{h} = \frac{[4-(x+h)^2]-(4-x^2)}{h} = \frac{(4-x^2-2xh-h^2)-4+x^2}{h} = \frac{-2xh-h^2}{h} = \frac{h(-2x-h)}{h}$
   $= -2x - h$ if $h \neq 0$

   Step 3: $f'(x) = \lim_{h\to 0} (-2x - h) = -2x$; $f'(-3) = 6$, $f'(0) = 0$

3. Step 1: $s(t) = t^3 - t^2$ and $s(t+h) = (t+h)^3 - (t+h)^3$

   Step 2: $\frac{s(t+h) - s(t)}{h} = \frac{[(t+h)^3-(t+h)^2]-(t^3-t^2)}{h}$

   $= \frac{(t^3+3t^2h+3th^2+h^3)+(t^2+2th+h^2)-(t^3-t^2)}{h}$

   $= \frac{h(3t^2+3th+h^2-2t-h)}{h} = 3t^2 - 2t + (3t-1)h + h^2$ if $h \neq 0$

   Step 3: $\frac{ds}{dt} = \lim_{h\to 0} (3t^2 - 2t + (3t-1)h + h^2) = 3t^2 - 2t$; $\left.\frac{ds}{dt}\right|_{t=-1} = 5$

5. Step 1: $p(\theta) = \sqrt{3\theta}$ and $p(\theta+h) = \sqrt{3(\theta+h)}$

   Step 2: $\frac{p(\theta+h) - p(\theta)}{h} = \frac{\sqrt{3(\theta+h)} - \sqrt{3\theta}}{h} = \frac{\left(\sqrt{3\theta+3h} - \sqrt{3\theta}\right)}{h} \cdot \frac{\left(\sqrt{3\theta+3h} + \sqrt{3\theta}\right)}{\left(\sqrt{3\theta+3h} + \sqrt{3\theta}\right)} = \frac{(3\theta+3h) - 3\theta}{h\left(\sqrt{3\theta+3h} + \sqrt{3\theta}\right)}$

   $= \frac{3h}{h\left(\sqrt{3\theta+3h} + \sqrt{3\theta}\right)} = \frac{3}{\sqrt{3\theta+3h} + \sqrt{3\theta}}$

   Step 3: $p'(\theta) = \lim_{h\to 0} \frac{3}{\sqrt{3\theta+3h} + \sqrt{3\theta}} = \frac{3}{\sqrt{3\theta} + \sqrt{3\theta}} = \frac{3}{2\sqrt{3\theta}}$; $p'(0.25) = \sqrt{3}$

7. $y = x^2 + x + 8 \Rightarrow \frac{dy}{dx} = 2x + 1 + 0 = 2x + 1 \Rightarrow \frac{d^2y}{dx^2} = 2$

9. $y = \frac{4x^3}{3} - 4 \Rightarrow \frac{dy}{dx} = \frac{d}{dx}\left(\frac{4}{3}x^3\right) - \frac{d}{dx}(4) = 4x^2 \Rightarrow \frac{d^2y}{dx^2} = \frac{d}{dx}(4x^2) = 8x$

11. $y = \frac{1}{2}x^4 - \frac{3}{2}x^2 - x \Rightarrow y' = 2x^3 - 3x - 1 \Rightarrow y'' = 6x^2 - 3 \Rightarrow y''' = 12x \Rightarrow y^{(4)} = 12 \Rightarrow y^{(n)} = 0$ for all $n \geq 5$

13. (a) $\frac{dy}{dx} = 3x^2 - 4 \Rightarrow m = \left.\frac{dy}{dx}\right|_{x=2} = 3(2)^2 - 4 = 8$

Therefore, the equation of the line tangent to the curve at the point $(2,1)$ is $y-1=8(x-2)$ or $y=8x-15$.

(b) Since $x^2 \geq 0$ for all real values of x, it follows that $3x^2 \geq 0$ and $3x^2-4 \geq -4$. In addition, $3x^2-4 \to +\infty$ as $x \to \pm\infty$. Therefore, the range of values of the curve's slope is $[-4,\infty)$. The graph of the derivative is a parabola that opens upward and its vertex is at the point $(0,-4)$.

(c) The equation of one such tangent line is found in part (a) when $x=2$. Also, $\frac{dy}{dx}=8 \Rightarrow 3x^2-4=8$ $\Rightarrow x^2=4 \Rightarrow x=2$ or $x=-2$. At $x=-2$, $y=(-2)^3-4(-2)+1=1$. Therefore, the equation of the line tangent to the curve at the point $(-2,1)$ is $y-1=8(x-(-2))$ or $y=8x+17$.

15. Note that as x increases, the slope of the tangent line to the curve is first negative, then zero (when $x=0$), then positive $\Rightarrow$ the slope is always increasing which matches (b).

17. $f_3(x)$ is an oscillating function like the cosine. Everywhere that the graph of f_3 has a horizontal tangent we expect f_3' to be zero, and (d) matches this condition.

19. (a) f' is not defined at $x=0, 1, 4$. At these points, the left-hand and right-hand derivatives do not agree. For example, $\lim_{x\to 0^-} \frac{f(x)-f(0)}{x-0}$ = slope of line joining $(-4,0)$ and $(0,2)=\frac{1}{2}$ but $\lim_{x\to 0^+} \frac{f(x)-f(0)}{x-0}$ = slope of line joining $(0,2)$ and $(1,-2)=-4$. Since these values are not equal, $f'(0)=\lim_{x\to 0} \frac{f(x)-f(0)}{x-0}$ does not exist.

(b)

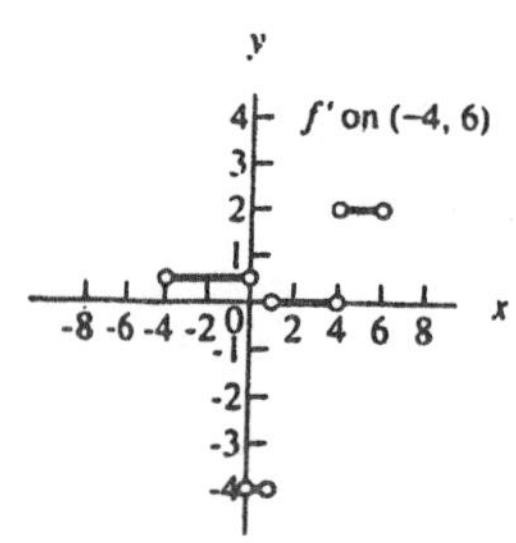

21. Left-hand derivative: For $h<0$, $f(0+h)=f(h)=h^2$ (using $y=x^2$ curve) $\Rightarrow \lim_{h\to 0^-} \frac{f(0+h)-f(0)}{h}$ $=\lim_{h\to 0^-} \frac{h^2-0}{h}=\lim_{h\to 0^-} h=0$;

Right-hand derivative: For $h>0$, $f(0+h)=f(h)=h$ (using $y=x$ curve) $\Rightarrow \lim_{h\to 0^+} \frac{f(0+h)-f(0)}{h}$ $=\lim_{h\to 0^+} \frac{h-0}{h}=\lim_{h\to 0^+} 1=1$;

Then $\lim_{h\to 0^-} \frac{f(0+h)-f(0)}{h} \neq \lim_{h\to 0^+} \frac{f(0+h)-f(0)}{h} \Rightarrow$ the derivative $f'(0)$ does not exist.

23. (a) The function is differentiable on its domain $-2 \leq x \leq 3$ (it is smooth)
(b) none
(c) none

25. (a) f is differentiable on $-1 \leq x < 0$ and $0 < x \leq 2$
(b) f is continuous but not differentiable at $x=0$: $\lim_{x\to 0} f(x)=0$ exists but there is a cusp at $x=0$, so $f'(0)=\lim_{h\to 0} \frac{f(0+h)-f(0)}{h}$ does not exist

(c) none

27. (a) $f'(x) = \lim_{h \to 0} \frac{f(x+h) - f(x)}{h} = \lim_{h \to 0} \frac{-(x+h)^2 - (-x^2)}{h} = \lim_{h \to 0} \frac{-x^2 - 2xh - h^2 + x^2}{h} = \lim_{h \to 0} (-2x - h) = -2x$

(b)

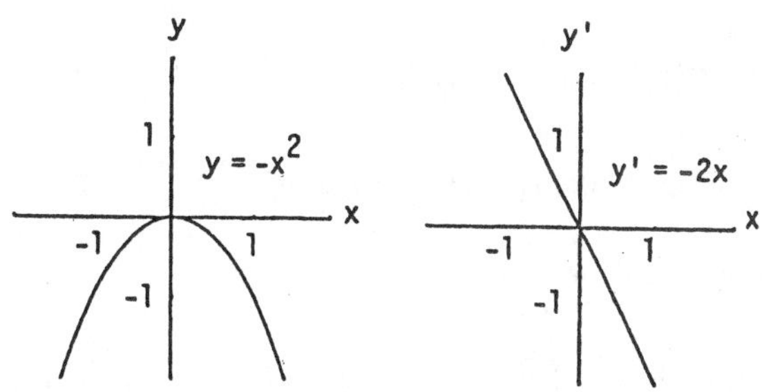

(c) $y' = -2x$ is positive for $x < 0$, y' is zero when $x = 0$, y' is negative when $x > 0$

(d) $y = -x^2$ is increasing for $-\infty < x < 0$ and decreasing for $0 < x < \infty$; the function is increasing on intervals where $y' > 0$ and decreasing on intervals where $y' < 0$

29. (a) Using the alternate formula for calculating derivatives: $f'(c) = \lim_{x \to c} \frac{f(x) - f(c)}{x - c} = \lim_{x \to c} \frac{\left(\frac{x^3}{3} - \frac{c^3}{3}\right)}{x - c}$

$= \lim_{x \to c} \frac{x^3 - c^3}{3(x - c)} = \lim_{x \to c} \frac{(x - c)(x^2 + xc + c^2)}{3(x - c)} = \lim_{x \to c} \frac{x^2 + xc + c^2}{3} = c^2 \Rightarrow f'(x) = x^2$

(b)

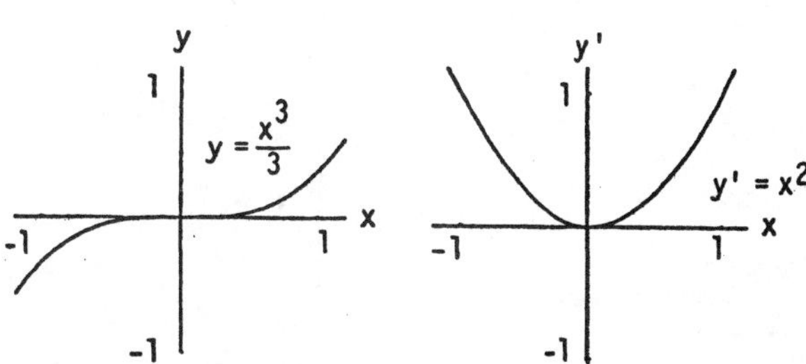

(c) y' is positive for all $x \neq 0$, and $y' = 0$ when $x = 0$; y' is never negative

(d) $y = \frac{x^3}{3}$ is increasing for all $x \neq 0$ (the graph is horizontal at $x = 0$) because y is increasing where $y' > 0$; y is never decreasing

31. $y' = \lim_{x \to c} \frac{f(x) - f(c)}{x - c} = \lim_{x \to c} \frac{x^3 - c^3}{x - c} = \lim_{x \to c} \frac{(x - c)(x^2 + xc + c^2)}{x - c} = \lim_{x \to c} (x^2 + xc + c^2) = 3c^2$.

The slope of the curve $y = x^3$ at $x = c$ is $y' = 3c^2$. Notice that $3c^2 \geq 0$ for all $c \Rightarrow y = x^3$ never has a negative slope.

33. $y' = \lim_{h \to 0} \frac{(2(x+h)^2 - 13(x+h) + 5) - (2x^2 - 13x + 5)}{h} = \lim_{h \to 0} \frac{2x^2 + 4xh + 2h^2 - 13x - 13h + 5 - 2x^2 + 13x - 5}{h}$

$= \lim_{h \to 0} \frac{4xh + 2h^2 - 13h}{h} = \lim_{h \to 0} (4x + 2h - 13) = 4x - 13$, slope at x. The slope is -1 when $4x - 13 = -1$

$\Rightarrow 4x = 12 \Rightarrow x = 3 \Rightarrow y = 2 \cdot 3^2 - 13 \cdot 3 + 5 = -16$. Thus the tangent line is $y + 16 = (-1)(x - 3)$ and the point of tangency is $(3, -16)$.

35. No. Derivatives of functions have the intermediate value property. The function $f(x) = \lfloor x \rfloor$ satisfies $f(0) = 0$ and $f(1) = 1$ but does not take on the value $\frac{1}{2}$ anywhere in $[0, 1] \Rightarrow f$ does not have the intermediate value property. Thus f cannot be the derivative of any function on $[0, 1] \Rightarrow f$ cannot be the derivative of any function on $(-\infty, \infty)$.

37. Yes; the derivative of $-f$ is $-f'$ so that $f'(x_0)$ exists $\Rightarrow -f'(x_0)$ exists as well.

39. Yes, $\lim\limits_{t \to 0} \frac{g(t)}{h(t)}$ can exist but it need not equal zero. For example, let $g(t) = mt$ and $h(t) = t$. Then $g(0) = h(0) = 0$, but $\lim\limits_{t \to 0} \frac{g(t)}{h(t)} = \lim\limits_{t \to 0} \frac{mt}{t} = \lim\limits_{t \to 0} m = m$, which need not be zero.

41. The graphs are shown below for $h = 1, 0.5, 0.1$. The function $y = \frac{1}{2\sqrt{x}}$ is the derivative of the function $y = \sqrt{x}$ so that $\frac{1}{2\sqrt{x}} = \lim\limits_{h \to 0} \frac{\sqrt{x+h} - \sqrt{x}}{h}$. The graphs reveal that $y = \frac{\sqrt{x+h} - \sqrt{x}}{h}$ gets closer to $y = \frac{1}{2\sqrt{x}}$ as h gets smaller and smaller.

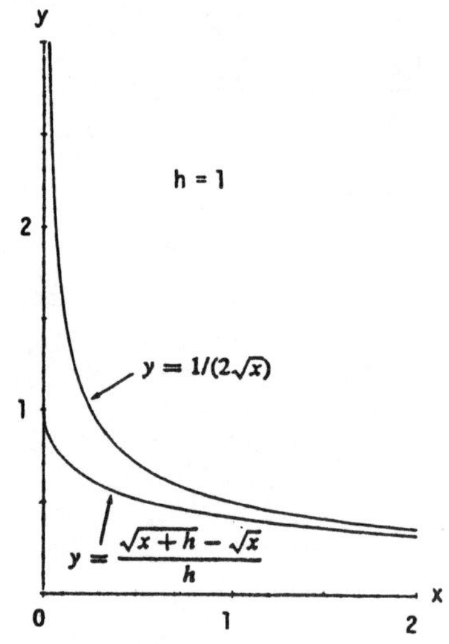

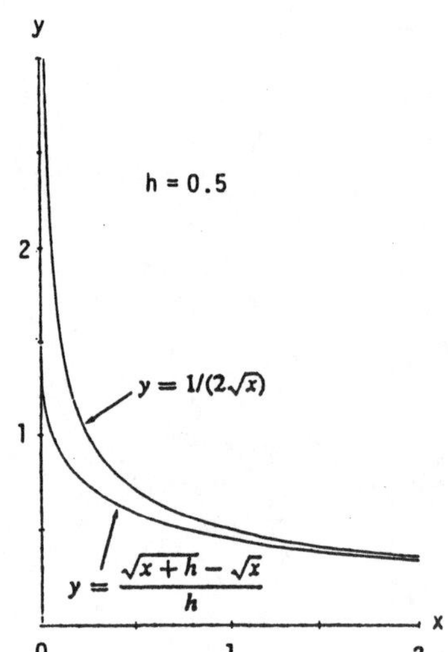

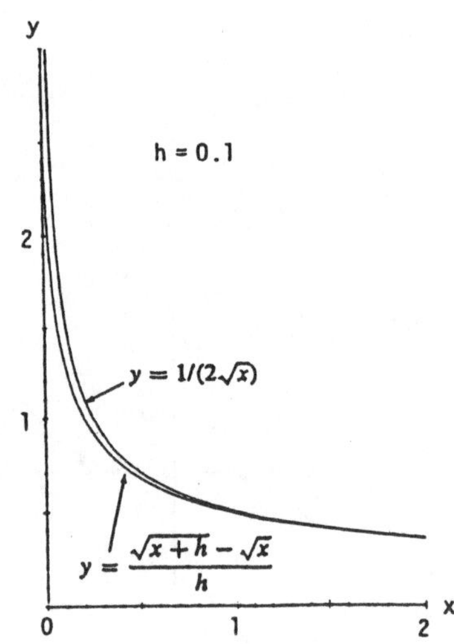

43. Weierstrass's nowhere differentiable continuous function.

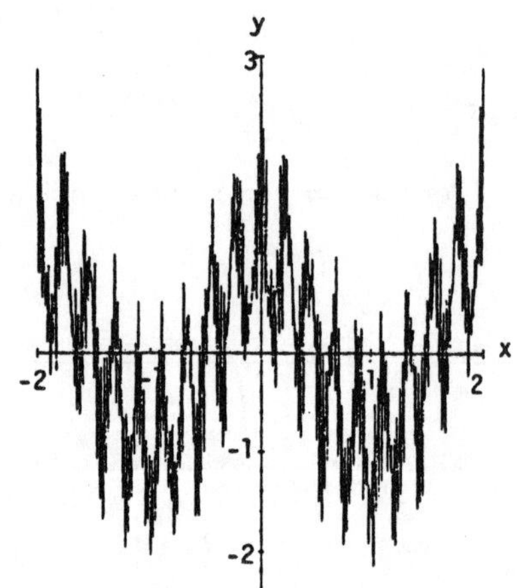

$$g(x) = \cos(\pi x) + \left(\frac{2}{3}\right)^1 \cos(9\pi x) + \left(\frac{2}{3}\right)^2 \cos(9^2\pi x) + \left(\frac{2}{3}\right)^3 \cos(9^3\pi x) + \cdots + \left(\frac{2}{3}\right)^7 \cos(9^7\pi x)$$

2.2 THE DERIVATIVE AS A RATE OF CHANGE

1. $s = t^2 - 3t + 2,\ 0 \le t \le 2$

(a) displacement $= \Delta s = s(2) - s(0) = -2\text{m},\ v_{av} = \frac{\Delta s}{\Delta t} = \frac{-2\text{ m}}{2\text{ sec}} = -1\text{ m/sec}$

(b) $v = \frac{ds}{dt} = 2t - 3,\ |v(0)| = |-3| = 3\text{ m/sec},\ |v(2)| = |1| = 1\text{ m/sec};\ a = \frac{dv}{dt} = \frac{d^2s}{dt^2} = 2,$

$a(0) = a(2) = 2\text{ m/sec}^2$

(c) $v = 0 \Rightarrow 2t - 3 = 0 \Rightarrow t = \frac{3}{2}$ sec. For $0 \le t < \frac{3}{2}$, v is negative and s is decreasing, whereas for $\frac{3}{2} < t \le 2$, v is positive and s is increasing. Therefore, the body changes direction at $t = \frac{3}{2}$.

3. $s = -t^3 + 3t^2 - 3t,\ 0 \le t \le 3$

(a) displacement $= \Delta s = s(3) - s(0) = -9\text{ m},\ v_{av} = \frac{\Delta s}{\Delta t} = \frac{-9}{3} = -3\text{ m/sec}$

(b) $v = \frac{ds}{dt} = -3t^2 + 6t - 3 \Rightarrow |v(0)| = |-3| = 3\text{ m/sec}$ and $|v(3)| = |-12| = 12\text{ m/sec};\ a = \frac{d^2s}{dt^2} = -6t + 6$

$\Rightarrow a(0) = 6\text{ m/sec}^2$ and $a(3) = -12\text{ m/sec}^2$

(c) $v = 0 \Rightarrow -3t^2 + 6t - 3 = 0 \Rightarrow t^2 - 2t + 1 = 0 \Rightarrow (t-1)^2 = 0 \Rightarrow t = 1$. For all other values of t in the interval the velocity v is negative (the graph of $v = -3t^2 + 6t - 3$ is a parabola with vertex at $t = 1$ which opens downward $\Rightarrow$ the body never changes direction.

5. $s = t^3 - 6t^2 + 9t$ and let the positive direction be to the right on the s-axis.

(a) $v = 3t^2 - 12t + 9$ so that $v = 0 \Rightarrow t^2 - 4t + 3 = (t-3)(t-1) = 0 \Rightarrow t = 1$ or 3; $a = 6t - 12 \Rightarrow a(1) = -6\text{ m/sec}^2$ and $a(3) = 6\text{ m/sec}^2$. Thus the body is motionless but being accelerated left when $t = 1$, and motionless but being accelerated right when $t = 3$.

(b) $a = 0 \Rightarrow 6t - 12 = 0 \Rightarrow t = 2$ with speed $|v(2)| = |12 - 24 + 9| = 3\text{ m/sec}$

(c) The body moves to the right or forward on $0 \le t < 1$, and to the left or backward on $1 < t < 2$. The positions are $s(0) = 0$, $s(1) = 4$ and $s(2) = 2 \Rightarrow$ total distance $= |s(1) - s(0)| + |s(2) - s(1)| = |4| + |-2| = 6$ m.

7. $s_m = 1.86t^2 \Rightarrow v_m = 3.72t$ and solving $3.72t = 27.8 \Rightarrow t \approx 7.5$ sec on Mars; $s_j = 11.44t^2 \Rightarrow v_j = 22.88t$ and solving $22.88t = 27.8 \Rightarrow t \approx 1.2$ sec on Jupiter.

9. $s = 15t - \frac{1}{2}g_s t^2 \Rightarrow v = 15 - g_s t$ so that $v = 0 \Rightarrow 15 - g_s t = 0 \Rightarrow t = \frac{15}{g_s}$. Therefore $\frac{15}{g_s} = 20 \Rightarrow g_s = \frac{3}{4} = 0.75\text{ m/sec}^2$

11. (a) $s = 179 - 16t^2 \Rightarrow v = -32t \Rightarrow$ speed $= |v| = 32t$ ft/sec and $a = -32\text{ ft/sec}^2$

(b) $s = 0 \Rightarrow 179 - 16t^2 = 0 \Rightarrow t = \sqrt{\frac{179}{16}} \approx 3.3$ sec

(c) When $t = \sqrt{\frac{179}{16}}$, $v = -32\sqrt{\frac{179}{16}} = -8\sqrt{179} \approx -107.0$ ft/sec

13. (a) at 2 and 7 seconds (b) between 3 and 6 seconds: $3 \le t \le 6$

(c)

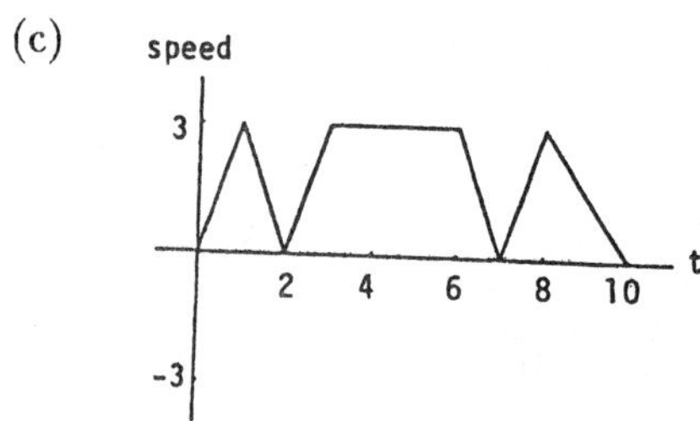

(d)

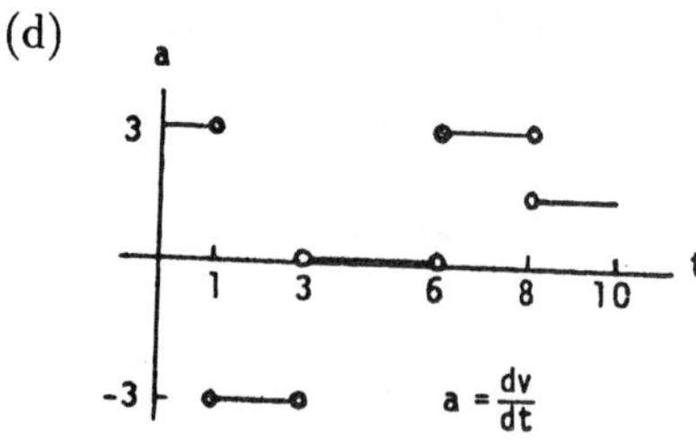

15. (a) 190 ft/sec (b) 2 sec
(c) at 8 sec, 0 ft/sec (d) 10.8 sec, 90 ft/sec
(e) From $t = 8$ until $t = 10.8$ sec, a total of 2.8 sec
(f) Greatest acceleration happens 2 sec after launch
(g) From $t = 2$ to $t = 10.8$ sec; during this period, $a = \dfrac{v(10.8) - v(2)}{10.8 - 2} \approx -32$ ft/sec^2

17. $s = 490t^2 \Rightarrow v = 980t \Rightarrow a = 980$
(a) Solving $160 = 490t^2 \Rightarrow t = \frac{4}{7}$ sec. The average velocity was $\dfrac{s(4/7) - s(0)}{4/7} = 280$ cm/sec.
(b) At the 160 cm mark the balls are falling at $v(4/7) = 560$ cm/sec. The acceleration at the 160 cm mark was 980 cm/sec^2.
(c) The light was flashing at a rate of $\dfrac{17}{4/7} = 29.75$ flashes per second.

19. C = position, B = velocity, and A = acceleration. Curve C cannot be the derivative of either A or B because C has only negative values while both A and B have some positive slopes. So, C represents position. Curve C has no positive slopes, so its derivative, the velocity, must be B. That leaves A for acceleration. Indeed, A is negative where B has negative slopes and positive where B has positive slopes.

21. (a) $r(x) = 2000\left(1 - \dfrac{1}{x+1}\right) \Rightarrow r'(x) = \dfrac{2000}{(x+1)^2}$, which is marginal revenue
(b) $r'(5) = \dfrac{2000}{36} \approx \$55.56$
(c) $\lim\limits_{x\to\infty} r'(x) = \lim\limits_{x\to\infty} \dfrac{2000}{(x+1)^2} = 0$. The increase in revenue as the number of items increases without bound will approach zero.

23. $Q(t) = 200(30 - t)^2 = 200(900 - 60t + t^2) \Rightarrow Q'(t) = 200(-60 + 2t) \Rightarrow Q'(10) = -8{,}000$ gallons/min is the rate the water is running at the end of 10 min. Then $\dfrac{Q(10) - Q(0)}{10} = -10{,}000$ gallons/min is the average rate the water flows during the first 10 min. The negative signs indicate water is <u>leaving</u> the tank.

25. (a) $V = \frac{4}{3}\pi r^3 \Rightarrow \dfrac{dV}{dr} = 4\pi r^2 \Rightarrow \left.\dfrac{dV}{dr}\right|_{r=2} = 4\pi(2)^2 = 16\pi$ ft^3/ft
(b) When $r = 2$, $\dfrac{dV}{dr} = 16\pi$ so that when r changes by 1 unit, we expect V to change by approximately 16π. Therefore when r changes by 0.2 units V changes by approximately $(16\pi)(0.2) = 3.2\pi \approx 10.05$ ft^3. Note that $V(2.2) - V(2) \approx 11.09$ ft^3.

27. $s = v_0t - 16t^2 \Rightarrow v = v_0 - 32t$; $v = 0 \Rightarrow t = \frac{v_0}{32}$; $1900 = v_0t - 16t^2$ so that $t = \frac{v_0}{32} \Rightarrow 1900 = \frac{v_0^2}{32} - \frac{v_0^2}{64}$

$\Rightarrow v_0 = \sqrt{(64)(1900)} = 80\sqrt{19}$ ft/sec and, finally, $\frac{80\sqrt{19}\text{ ft}}{\text{sec}} \cdot \frac{60\text{ sec}}{1\text{ min}} \cdot \frac{60\text{ min}}{1\text{ hr}} \cdot \frac{1\text{ mi}}{5280\text{ ft}} \approx 238$ mph.

29.

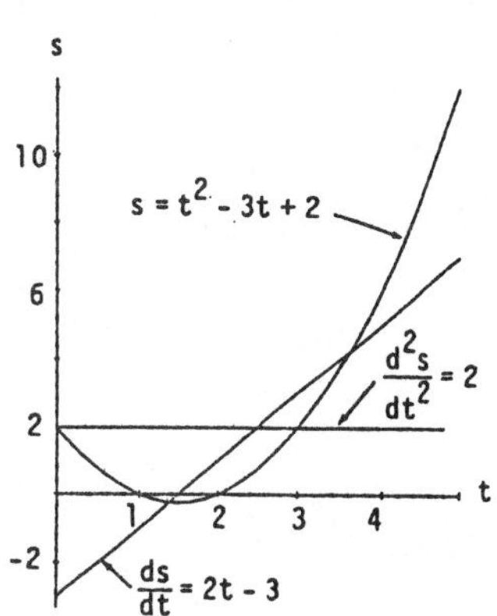

(a) $v = 0$ when $t = \frac{3}{2}$ sec

(b) $v < 0$ when $0 \le t < 1.5 \Rightarrow$ body moves down; $v > 0$ when $1.5 < t \le 5 \Rightarrow$ body moves up

(c) body changes direction at $t = \frac{3}{2}$ sec

(d) body speeds up on $\left(\frac{3}{2}, 5\right]$ and slows down on $\left[0, \frac{3}{2}\right)$

(e) body is moving fastest at $t = 5$ when the speed $= |v(5)| = 7$ units/sec; it is moving slowest at $t = \frac{3}{2}$ when the speed is 0

(f) When $t = 5$ the body is $s = 10$ units from the origin and farthest away.

31.

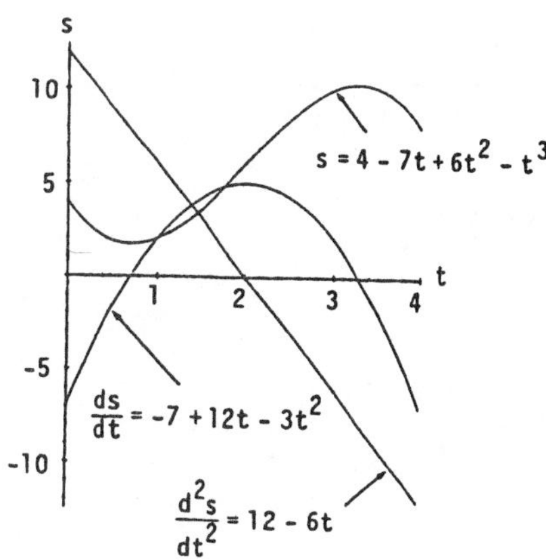

(a) $v = 0$ when $t = \frac{6 \pm \sqrt{15}}{3}$

(b) $v < 0$ when $0 \le t < \frac{6 - \sqrt{15}}{3}$ or $\frac{6 + \sqrt{15}}{3} < t \le 4 \Rightarrow$ body is moving left; $v > 0$ when $\frac{6 - \sqrt{15}}{3} < t < \frac{6 + \sqrt{15}}{3} \Rightarrow$ body is moving right

(c) body changes direction at $t = \frac{6 \pm \sqrt{15}}{3}$ sec

(d) body speeds up on $\left(\frac{6-\sqrt{15}}{3}, 2\right) \cup \left(\frac{6+\sqrt{15}}{3}, 4\right]$ and slows down on $\left[0, \frac{6-\sqrt{15}}{3}\right) \cup \left(2, \frac{6+\sqrt{15}}{3}\right)$

(e) The body is moving fastest at 7 units/sec when $t = 0$ and $t = 4$; it is moving slowest and stationary at $t = \frac{6 \pm \sqrt{15}}{3}$

(f) When $t = \frac{6+\sqrt{15}}{3}$ the position is $s \approx 10.303$ units and the body is farthest from the origin.

2.3 DERIVATIVES OF PRODUCTS, QUOTIENTS, AND NEGATIVE POWERS

1. $y = 6x^2 - 10x - 5x^{-2} \Rightarrow \frac{dy}{dx} = 12x - 10 + 10x^{-3} \Rightarrow \frac{d^2y}{dx^2} = 12 - 0 - 30x^{-4} = 12 - 30x^{-4}$

3. $r = \frac{1}{3}s^{-2} - \frac{5}{2}s^{-1} \Rightarrow \frac{dr}{ds} = -\frac{2}{3}s^{-3} + \frac{5}{2}s^{-2} = \frac{-2}{3s^3} + \frac{5}{2s^2} \Rightarrow \frac{d^2r}{ds^2} = 2s^{-4} - 5s^{-3} = \frac{2}{s^4} - \frac{5}{s^3}$

5. (a) $y = (3-x^2)(x^3-x+1) \Rightarrow y' = (3-x^2)\cdot\frac{d}{dx}(x^3-x+1) + (x^3-x+1)\cdot\frac{d}{dx}(3-x^2)$

$= (3-x^2)(3x^2-1) + (x^3-x+1)(-2x) = -5x^4 + 12x^2 - 2x - 3$

(b) $y = -x^5 + 4x^3 - x^2 - 3x + 3 \Rightarrow y' = -5x^4 + 12x^2 - 2x - 3$

7. $y = \frac{2x+5}{3x-2}$; use the quotient rule: $u = 2x+5$ and $v = 3x-2 \Rightarrow u' = 2$ and $v' = 3 \Rightarrow y' = \frac{vu' - uv'}{v^2}$

$= \frac{(3x-2)(2) - (2x+5)(3)}{(3x-2)^2} = \frac{6x - 4 - 6x - 15}{(3x-2)^2} = \frac{-19}{(3x-2)^2}$

9. $f(t) = \frac{t^2-1}{t^2+t-2} \Rightarrow f'(t) = \frac{(t^2+t-2)(2t) - (t^2-1)(2t+1)}{(t^2+t-2)^2} = \frac{(t-1)(t+2)(2t) - (t-1)(t+1)(2t+1)}{(t-1)^2(t+2)^2}$

$= \frac{(t+2)(2t) - (t+1)(2t+1)}{(t-1)(t+2)^2} = \frac{2t^2 + 4t - 2t^2 - 3t - 1}{(t-1)(t+2)^2} = \frac{t-1}{(t-1)(t+2)^2} = \frac{1}{(t+2)^2}$

11. $f(s) = \frac{\sqrt{s}-1}{\sqrt{s}+1} \Rightarrow f'(s) = \frac{(\sqrt{s}+1)\left(\frac{1}{2\sqrt{s}}\right) - (\sqrt{s}-1)\left(\frac{1}{2\sqrt{s}}\right)}{(\sqrt{s}+1)^2} = \frac{(\sqrt{s}+1) - (\sqrt{s}-1)}{2\sqrt{s}(\sqrt{s}+1)^2} = \frac{1}{\sqrt{s}(\sqrt{s}+1)^2}$

NOTE: $\frac{d}{ds}(\sqrt{s}) = \frac{1}{2\sqrt{s}}$ from Example 1 in Section 2.1

13. $y = \frac{1}{(x^2-1)(x^2+x+1)}$; use the quotient rule: $u = 1$ and $v = (x^2-1)(x^2+x+1) \Rightarrow u' = 0$ and

$v' = (x^2-1)(2x+1) + (x^2+x+1)(2x) = 2x^3 + x^2 - 2x - 1 + 2x^3 + 2x^2 + 2x = 4x^3 + 3x^2 - 1$

$\Rightarrow \frac{dy}{dx} = \frac{vu' - uv'}{v^2} = \frac{0 - 1(4x^3 + 3x^2 - 1)}{(x^2-1)^2(x^2+x+1)^2} = \frac{-4x^3 - 3x^2 + 1}{(x^2-1)^2(x^2+x+1)^2}$

15. $s = \frac{t^2+5t-1}{t^2} = 1 + \frac{5}{t} - \frac{1}{t^2} = 1 + 5t^{-1} - t^{-2} \Rightarrow \frac{ds}{dt} = 0 - 5t^{-2} + 2t^{-3} = -5t^{-2} + 2t^{-3} \Rightarrow \frac{d^2s}{dt^2} = 10t^{-3} - 6t^{-4}$

17. $w = \left(\frac{1+3z}{3z}\right)(3-z) = \left(\frac{1}{3}z^{-1} + 1\right)(3-z) = z^{-1} - \frac{1}{3} + 3 - z = z^{-1} + \frac{8}{3} - z \Rightarrow \frac{dw}{dz} = -z^{-2} + 0 - 1 = -z^{-2} - 1$

$\Rightarrow \frac{d^2w}{dz^2} = 2z^{-3} - 0 = 2z^{-3}$

19. $u(0) = 5,\ u'(0) = 3,\ v(0) = -1,\ v'(0) = 2$

(a) $\frac{d}{dx}(uv) = uv' + vu' \Rightarrow \frac{d}{dx}(uv)\Big|_{x=0} = u(0)v'(0) + v(0)u'(0) = 5 \cdot 2 + (-1)(3) = 7$

(b) $\frac{d}{dx}\left(\frac{u}{v}\right) = \frac{vu' - uv'}{v^2} \Rightarrow \frac{d}{dx}\left(\frac{u}{v}\right)\Big|_{x=0} = \frac{v(0)u'(0) - u(0)v'(0)}{(v(0))^2} = \frac{(-1)(3) - (5)(2)}{(-1)^2} = -13$

(c) $\frac{d}{dx}\left(\frac{v}{u}\right) = \frac{uv' - vu'}{u^2} \Rightarrow \frac{d}{dx}\left(\frac{v}{u}\right)\Big|_{x=0} = \frac{u(0)v'(0) - v(0)u'(0)}{(u(0))^2} = \frac{(5)(2) - (-1)(3)}{(5)^2} = \frac{13}{25}$

(d) $\frac{d}{dx}(7v - 2u) = 7v' - 2u' \Rightarrow \frac{d}{dx}(7v - 2u)\Big|_{x=0} = 7v'(0) - 2u'(0) = 7 \cdot 2 - 2(3) = 8$

21. $y = \frac{4x}{x^2+1} \Rightarrow \frac{dy}{dx} = \frac{(x^2+1)(4) - (4x)(2x)}{(x^2+1)^2} = \frac{4x^2 + 4 - 8x^2}{(x^2+1)^2} = \frac{4(-x^2+1)}{(x^2+1)^2}$. When $x = 0$, $y = 0$ and $y' = \frac{4(0+1)}{1}$ $= 4$, so the tangent to the curve at $(0,0)$ is the line $y = 4x$. When $x = 1$, $y = 2 \Rightarrow y' = 0$, so the tangent to the curve at $(1,2)$ is the line $y = 2$.

23. $y = ax^2 + bx + c$ passes through $(0,0) \Rightarrow 0 = a(0) + b(0) + c \Rightarrow c = 0$; $y = ax^2 + bx$ passes through $(1,2)$ $\Rightarrow 2 = a + b$; $y' = 2ax + b$ and since the curve is tangent to $y = x$ at the origin, its slope is 1 at $x = 0$ $\Rightarrow y' = 1$ when $x = 0 \Rightarrow 1 = 2a(0) + b \Rightarrow b = 1$. Then $a + b = 2 \Rightarrow a = 1$. In summary $a = b = 1$ and $c = 0$ so the curve is $y = x^2 + x$.

25. Let c be a constant $\Rightarrow \frac{dc}{dx} = 0 \Rightarrow \frac{d}{dx}(u \cdot c) = u \cdot \frac{dc}{dx} + c \cdot \frac{du}{dx} = u \cdot 0 + c\,\frac{du}{dx} = c\,\frac{du}{dx}$. Thus when one of the functions is a constant, the Product Rule is just the Constant Multiple Rule $\Rightarrow$ the Constant Multiple Rule is a special case of the Product Rule.

27. (a) $\frac{d}{dx}(uvw) = \frac{d}{dx}((uv) \cdot w) = (uv)\frac{dw}{dx} + w \cdot \frac{d}{dx}(uv) = uv\,\frac{dw}{dx} + w\left(u\,\frac{dv}{dx} + v\,\frac{du}{dx}\right) = uv\,\frac{dw}{dx} + wu\,\frac{dv}{dx} + wv\,\frac{du}{dx}$

$= uvw' + uv'w + u'vw$

(b) $\frac{d}{dx}(u_1u_2u_3u_4) = \frac{d}{dx}((u_1u_2u_3)u_4) = (u_1u_2u_3)\frac{du_4}{dx} + u_4\,\frac{d}{dx}(u_1u_2u_3) \Rightarrow \frac{d}{dx}(u_1u_2u_3u_4)$

$= u_1u_2u_3\,\frac{du_4}{dx} + u_4\left(u_1u_2\,\frac{du_3}{dx} + u_3u_1\,\frac{du_2}{dx} + u_3u_2\,\frac{du_1}{dx}\right)$ (using (a) above)

$\Rightarrow \frac{d}{dx}(u_1u_2u_3u_4) = u_1u_2u_3\,\frac{du_4}{dx} + u_1u_2u_4\,\frac{du_3}{dx} + u_1u_3u_4\,\frac{du_2}{dx} + u_2u_3u_4\,\frac{du_1}{dx}$

$= u_1u_2u_3u_4' + u_1u_2u_3'u_4 + u_1u_2'u_3u_4 + u_1'u_2u_3u_4$

(c) Generalizing (a) and (b) above, $\frac{d}{dx}(u_1 \cdots u_n) = u_1 u_2 \cdots u_{n-1} u_n' + u_1 u_2 \cdots u_{n-2} u_{n-1}' u_n + \ldots + u_1' u_2 \cdots u_n$

29. $P = \frac{nRT}{V - nb} - \frac{an^2}{V^2}$. We are holding T constant, and a, b, n, R are also constant so their derivatives are zero

$\Rightarrow \frac{dP}{dV} = \frac{(V - nb) \cdot 0 - (nRT)(1)}{(V - nb)^2} - \frac{V^2(0) - (an^2)(2V)}{(V^2)^2} = \frac{-nRT}{(V - nb)^2} + \frac{2an^2}{V^3}$

2.4 DERIVATIVES OF TRIGONOMETRIC FUNCTIONS

1. $y = -10x + 3 \cos x \Rightarrow \frac{dy}{dx} = -10 + 3 \frac{d}{dx}(\cos x) = -10 - 3 \sin x$

3. $y = \csc x - 4\sqrt{x} + 7 \Rightarrow \frac{dy}{dx} = -\csc x \cot x - \frac{4}{2\sqrt{x}} + 0 = -\csc x \cot x - \frac{2}{\sqrt{x}}$

5. $y = (\sec x + \tan x)(\sec x - \tan x) \Rightarrow \frac{dy}{dx} = (\sec x + \tan x) \frac{d}{dx}(\sec x - \tan x) + (\sec x - \tan x) \frac{d}{dx}(\sec x + \tan x)$

$= (\sec x + \tan x)(\sec x \tan x - \sec^2 x) + (\sec x - \tan x)(\sec x \tan x + \sec^2 x)$

$= (\sec^2 x \tan x + \sec x \tan^2 x - \sec^3 x - \sec^2 x \tan x) + (\sec^2 x \tan x - \sec x \tan^2 x + \sec^3 x - \tan x \sec^2 x) = 0.$

$\left(\text{Note also that } y = \sec^2 x - \tan^2 x = (\tan^2 x + 1) - \tan^2 x = 1 \Rightarrow \frac{dy}{dx} = 0.\right)$

7. $y = \frac{\cot x}{1 + \cot x} \Rightarrow \frac{dy}{dx} = \frac{(1 + \cot x) \frac{d}{dx}(\cot x) - (\cot x) \frac{d}{dx}(1 + \cot x)}{(1 + \cot x)^2} = \frac{(1 + \cot x)(-\csc^2 x) - (\cot x)(-\csc^2 x)}{(1 + \cot x)^2}$

$= \frac{-\csc^2 x - \csc^2 x \cot x + \csc^2 x \cot x}{(1 + \cot x)^2} = \frac{-\csc^2 x}{(1 + \cot x)^2}$

9. $y = \frac{4}{\cos x} + \frac{1}{\tan x} = 4 \sec x + \cot x \Rightarrow \frac{dy}{dx} = 4 \sec x \tan x - \csc^2 x$

11. $y = x^2 \sin x + 2x \cos x - 2 \sin x \Rightarrow \frac{dy}{dx} = (x^2 \cos x + (\sin x)(2x)) + ((2x)(-\sin x) + (\cos x)(2)) - 2 \cos x$

$= x^2 \cos x + 2x \sin x - 2x \sin x + 2 \cos x - 2 \cos x = x^2 \cos x$

13. $s = \tan t - t \Rightarrow \frac{ds}{dt} = \frac{d}{dt}(\tan t) - 1 = \sec^2 t - 1$

15. $s = \frac{1 + \csc t}{1 - \csc t} \Rightarrow \frac{ds}{dt} = \frac{(1 - \csc t)(-\csc t \cot t) - (1 + \csc t)(\csc t \cot t)}{(1 - \csc t)^2}$

$= \frac{-\csc t \cot t + \csc^2 t \cot t - \csc t \cot t - \csc^2 t \cot t}{(1 - \csc t)^2} = \frac{-2 \csc t \cot t}{(1 - \csc t)^2}$

17. $r = 4 - \theta^2 \sin \theta \Rightarrow \frac{dr}{d\theta} = -\left(\theta^2 \frac{d}{d\theta}(\sin \theta) + (\sin \theta)(2\theta)\right) = -(\theta^2 \cos \theta + 2\theta \sin \theta) = -\theta(\theta \cos \theta + 2 \sin \theta)$

19. $r = \sec\theta\csc\theta \Rightarrow \frac{dr}{d\theta} = (\sec\theta)(-\csc\theta\cot\theta) + (\csc\theta)(\sec\theta\tan\theta)$

$= \left(\frac{-1}{\cos\theta}\right)\left(\frac{1}{\sin\theta}\right)\left(\frac{\cos\theta}{\sin\theta}\right) + \left(\frac{1}{\sin\theta}\right)\left(\frac{1}{\cos\theta}\right)\left(\frac{\sin\theta}{\cos\theta}\right) = \frac{-1}{\sin^2\theta} + \frac{1}{\cos^2\theta} = \sec^2\theta - \csc^2\theta$

21. $p = 5 + \frac{1}{\cot q} = 5 + \tan q \Rightarrow \frac{dp}{dq} = \sec^2 q$

23. $p = \frac{\sin q + \cos q}{\cos q} \Rightarrow \frac{dp}{dq} = \frac{(\cos q)(\cos q - \sin q) - (\sin q + \cos q)(-\sin q)}{\cos^2 q}$

$= \frac{\cos^2 q - \cos q \sin q + \sin^2 q + \cos q \sin q}{\cos^2 q} = \frac{1}{\cos^2 q} = \sec^2 q$

25. (a) $y = \csc x \Rightarrow y' = -\csc x \cot x \Rightarrow y'' = -\Big((\csc x)\left(-\csc^2 x\right) + (\cot x)(-\csc x \cot x)\Big) = \csc^3 x + \csc x \cot^2 x$

$= (\csc x)\left(\csc^2 x + \cot^2 x\right) = (\csc x)\left(\csc^2 x + \csc^2 x - 1\right) = 2\csc^3 x - \csc x$

(b) $y = \sec x \Rightarrow y' = \sec x \tan x \Rightarrow y'' = (\sec x)\left(\sec^2 x\right) + (\tan x)(\sec x \tan x) = \sec^3 x + \sec x \tan^2 x$

$= (\sec x)\left(\sec^2 x + \tan^2 x\right) = (\sec x)\left(\sec^2 x + \sec^2 x - 1\right) = 2\sec^3 x - \sec x$

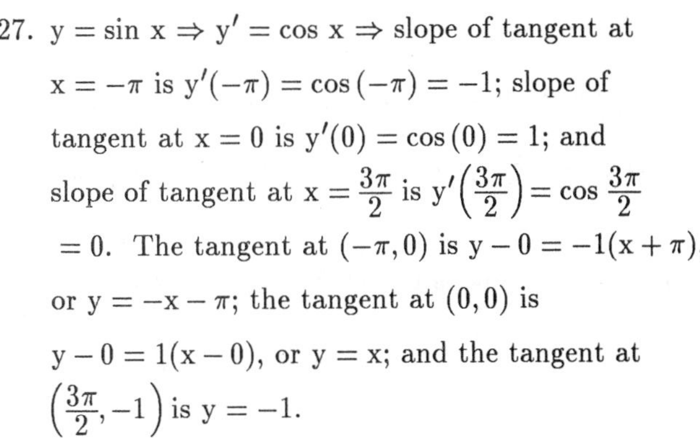

27. $y = \sin x \Rightarrow y' = \cos x \Rightarrow$ slope of tangent at $x = -\pi$ is $y'(-\pi) = \cos(-\pi) = -1$; slope of tangent at $x = 0$ is $y'(0) = \cos(0) = 1$; and slope of tangent at $x = \frac{3\pi}{2}$ is $y'\left(\frac{3\pi}{2}\right) = \cos\frac{3\pi}{2}$ $= 0$. The tangent at $(-\pi, 0)$ is $y - 0 = -1(x + \pi)$, or $y = -x - \pi$; the tangent at $(0,0)$ is $y - 0 = 1(x - 0)$, or $y = x$; and the tangent at $\left(\frac{3\pi}{2}, -1\right)$ is $y = -1$.

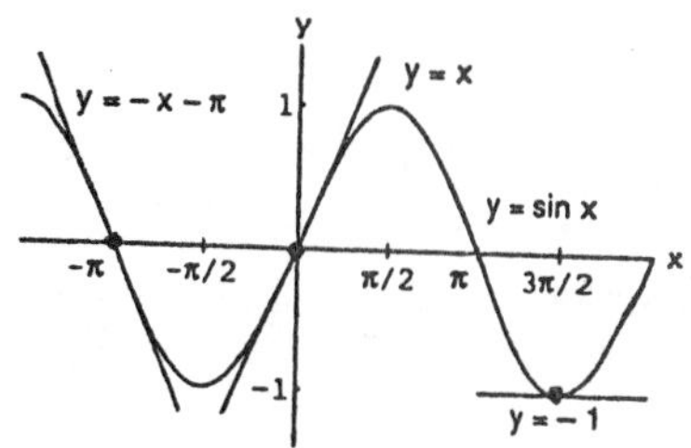

29. $y = \sec x \Rightarrow y' = \sec x \tan x \Rightarrow$ slope of tangent at $x = -\frac{\pi}{3}$ is $\sec\left(-\frac{\pi}{3}\right)\tan\left(-\frac{\pi}{3}\right) = -2\sqrt{3}$; slope of tangent at $x = \frac{\pi}{4}$ is $\sec\left(\frac{\pi}{4}\right)\tan\left(\frac{\pi}{4}\right) = \sqrt{2}$. The tangent at the point $\left(-\frac{\pi}{3}, \sec\left(-\frac{\pi}{3}\right)\right) = \left(-\frac{\pi}{3}, 2\right)$ is $y - 2 = -2\sqrt{3}\left(x + \frac{\pi}{3}\right)$; the tangent at the point $\left(\frac{\pi}{4}, \sec\left(\frac{\pi}{4}\right)\right) = \left(\frac{\pi}{4}, \sqrt{2}\right)$ is $y - \sqrt{2}$ $= \sqrt{2}\left(x - \frac{\pi}{4}\right)$.

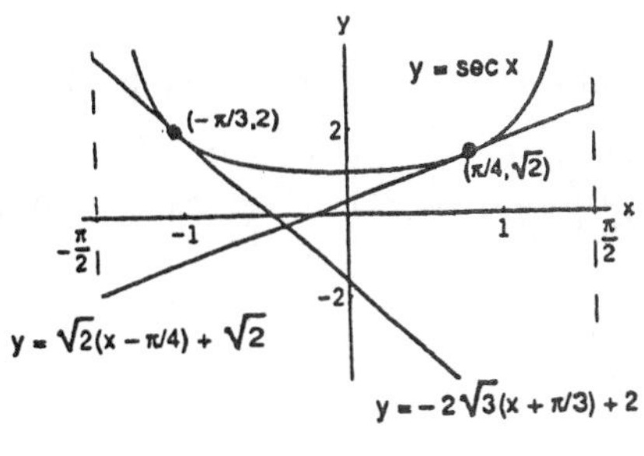

31. Yes, $y = x + \sin x \Rightarrow y' = 1 + \cos x$; horizontal tangent occurs where $1 + \cos x = 0 \Rightarrow \cos x = -1$ $\Rightarrow x = \pi$

33. No, $y = x - \cot x \Rightarrow y' = 1 + \csc^2 x$; horizontal tangent occurs where $1 + \csc^2 x = 0 \Rightarrow \csc^2 x = -1$. But there are no x-values for which $\csc^2 x = -1$.

35. We want all points on the curve where the tangent line has slope 2. Thus, $y = \tan x \Rightarrow y' = \sec^2 x$ so that $y' = 2 \Rightarrow \sec^2 x = 2 \Rightarrow \sec x = \pm\sqrt{2}$ $\Rightarrow x = \pm\frac{\pi}{4}$. Then the tangent line at $\left(\frac{\pi}{4}, 1\right)$ has equation $y - 1 = 2\left(x - \frac{\pi}{4}\right)$; the tangent line at $\left(-\frac{\pi}{4}, -1\right)$ has equation $y + 1 = 2\left(x + \frac{\pi}{4}\right)$.

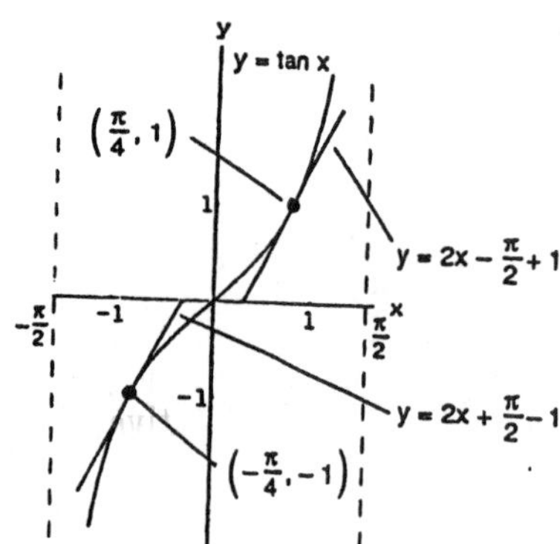

37. $y = 4 + \cot x - 2 \csc x \Rightarrow y' = -\csc^2 x + 2 \csc x \cot x = -\left(\frac{1}{\sin x}\right)\left(\frac{1 - 2 \cos x}{\sin x}\right)$

(a) When $x = \frac{\pi}{2}$, then $y' = -1$; the tangent line is $y = -x + \frac{\pi}{2} + 2$.

(b) To find the location of the horizontal tangent set $y' = 0 \Rightarrow 1 - 2 \cos x = 0 \Rightarrow x = \frac{\pi}{3}$ radians. When $x = \frac{\pi}{3}$, then $y = 4 - \sqrt{3}$ is the horizontal tangent.

39. $s = 2 - 2 \sin t \Rightarrow v = \frac{ds}{dt} = -2 \cos t \Rightarrow a = \frac{dv}{dt} = 2 \sin t \Rightarrow j = \frac{da}{dt} = 2 \cos t$. Therefore, velocity $= v\left(\frac{\pi}{4}\right)$ $= -\sqrt{2}$ m/sec; speed $= \left|v\left(\frac{\pi}{4}\right)\right| = \sqrt{2}$ m/sec; acceleration $= a\left(\frac{\pi}{4}\right) = \sqrt{2}$ m/sec^2; jerk $= j\left(\frac{\pi}{4}\right) = \sqrt{2}$ m/sec^3.

41. $\lim_{x \to 0} f(x) = \lim_{x \to 0} \frac{\sin^2 3x}{x^2} = \lim_{x \to 0} 9\left(\frac{\sin 3x}{3x}\right)\left(\frac{\sin 3x}{3x}\right) = 9$ so that f is continuous at $x = 0 \Rightarrow \lim_{x \to 0} f(x) = f(0)$ $\Rightarrow c = 9$.

43. $\frac{d^{999}}{dx^{999}}(\cos x) = \sin x$ because $\frac{d^4}{dx^4}(\cos x) = \cos x \Rightarrow$ the derivative of cos x any number of times that is a multiple of 4 is cos x. Thus, dividing 999 by 4 gives $999 = 249 \cdot 4 + 3 \Rightarrow \frac{d^{999}}{dx^{999}}(\cos x)$ $= \frac{d^3}{dx^3}\left[\frac{d^{249 \cdot 4}}{dx^{249 \cdot 4}}(\cos x)\right] = \frac{d^3}{dx^3}(\cos x) = \sin x$.

45.

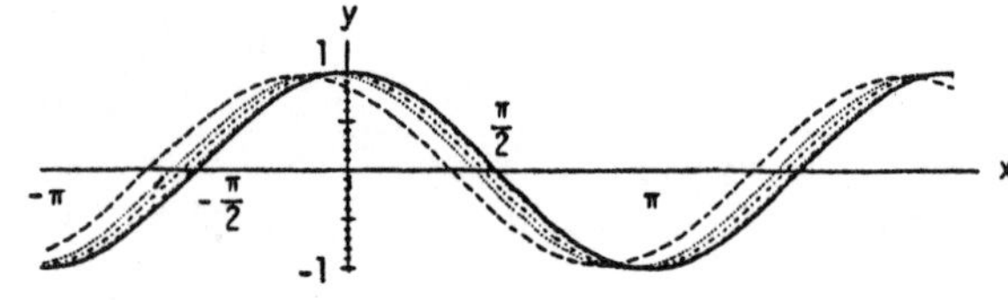

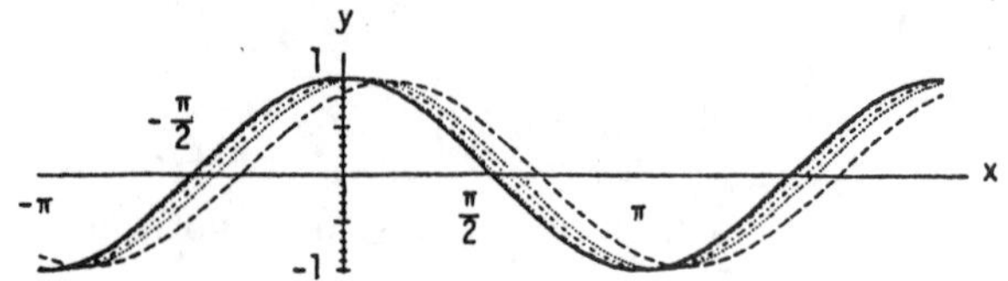

As h takes on the values of 1, 0.5, 0.3 and 0.1 the corresponding dashed curves of $y = \frac{\sin(x + h) - \sin x}{h}$ get closer and closer to the black curve $y = \cos x$ because $\frac{d}{dx}(\sin x) = \lim_{h \to 0} \frac{\sin(x + h) - \sin x}{h} = \cos x$. The same is true as h takes on the values of -1, -0.5, -0.3 and -0.1.

47. This is a grapher exercise. Compare your graphs with Exercise 45.

49. $y = \tan x \Rightarrow y' = \sec^2 x$, so the smallest value $y' = \sec^2 x$ takes on is $y' = 1$ when $x = 0$; y' has no maximum value since $\sec^2 x$ has no largest value on $\left(-\frac{\pi}{2}, \frac{\pi}{2}\right)$; y' is never negative since $\sec^2 x \geq 1$.

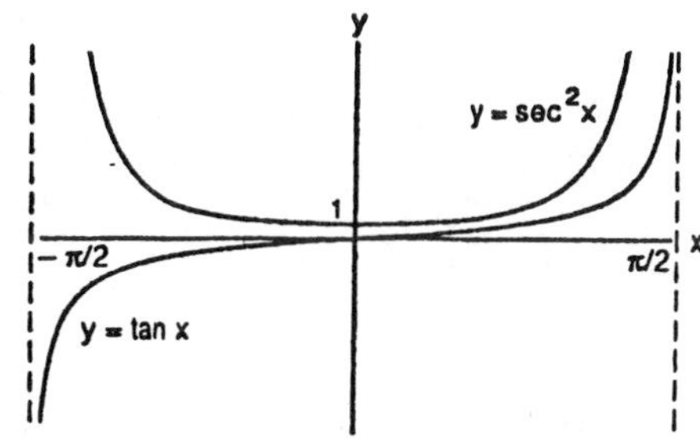

51. $y = \frac{\sin x}{x}$ appears to cross the y-axis at $y = 1$, since $\lim_{x \to 0} \frac{\sin x}{x} = 1$; $y = \frac{\sin 2x}{x}$ appears to cross the y-axis at $y = 2$, since $\lim_{x \to 0} \frac{\sin 2x}{x} = 2$; $y = \frac{\sin 4x}{x}$ appears to cross the y-axis at $y = 4$, since $\lim_{x \to 0} \frac{\sin 4x}{x} = 4$. However, none of these graphs actually cross the y-axis since $x = 0$ is not in the domain of the functions. Also, $\lim_{x \to 0} \frac{\sin 5x}{x} = 5$, $\lim_{x \to 0} \frac{\sin(-3x)}{x} = -3$, and $\lim_{x \to 0} \frac{\sin kx}{x} = k \Rightarrow$ the graphs of $y = \frac{\sin 5x}{x}$, $y = \frac{\sin(-3x)}{x}$, and $y = \frac{\sin kx}{x}$ approach 5, -3, and k, respectively, as $x \to 0$. However, the graphs do not actually cross the y-axis.

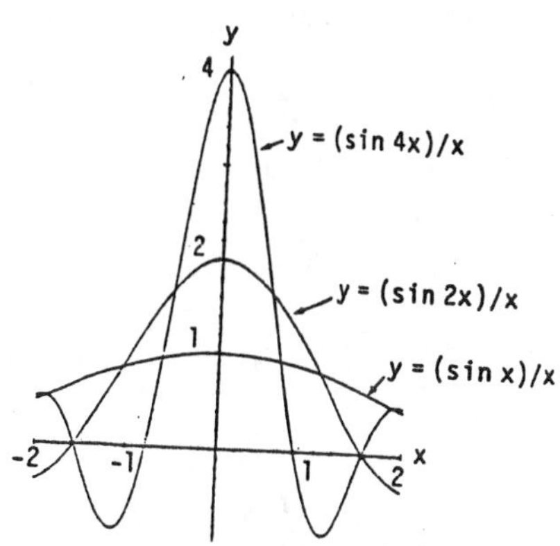

2.5 THE CHAIN RULE

1. $f(u) = 6u - 9 \Rightarrow f'(u) = 6 \Rightarrow f'(g(x)) = 6$; $g(x) = \frac{1}{2}x^4 \Rightarrow g'(x) = 2x^3$; therefore $\frac{dy}{dx} = f'(g(x))g'(x)$ $= 6 \cdot 2x^3 = 12x^3$

3. $f(u) = \sin u \Rightarrow f'(u) = \cos u \Rightarrow f'(g(x)) = \cos(3x+1)$; $g(x) = 3x + 1 \Rightarrow g'(x) = 3$; therefore $\frac{dy}{dx} = f'(g(x))g'(x)$ $= (\cos(3x+1))(3) = 3\cos(3x+1)$

5. $f(u) = \tan u \Rightarrow f'(u) = \sec^2 u \Rightarrow f'(g(x)) = \sec^2(10x - 5)$; $g(x) = 10x - 5 \Rightarrow g'(x) = 10$; therefore $\frac{dy}{dx} = f'(g(x))g'(x) = \left(\sec^2(10x-5)\right)(10) = 10\sec^2(10x-5)$

7. With $u = (4 - 3x)$, $y = u^9$: $\frac{dy}{dx} = \frac{dy}{du}\frac{du}{dx} = 9u^8 \cdot (-3) = -27(4-3x)^8$

9. With $u = \left(\frac{x^2}{8} + x - \frac{1}{x}\right)$, $y = u^4$: $\frac{dy}{dx} = \frac{dy}{du}\frac{du}{dx} = 4u^3 \cdot \left(\frac{x}{4} + 1 + \frac{1}{x^2}\right) = 4\left(\frac{x^2}{8} + x - \frac{1}{x}\right)^3\left(\frac{x}{4} + 1 + \frac{1}{x^2}\right)$

11. With $u = \pi - \frac{1}{x}$, $y = \cot u$: $\frac{dy}{dx} = \frac{dy}{du}\frac{du}{dx} = \left(-\csc^2 u\right)\left(\frac{1}{x^2}\right) = -\frac{1}{x^2}\csc^2\left(\pi - \frac{1}{x}\right)$

13. $q = \sqrt{2r - r^2} = \left(2r - r^2\right)^{1/2} \Rightarrow \frac{dq}{dr} = \frac{1}{2}\left(2r - r^2\right)^{-1/2} \cdot \frac{d}{dr}\left(2r - r^2\right) = \frac{1}{2}\left(2r - r^2\right)^{-1/2}(2 - 2r) = \frac{1 - r}{\sqrt{2r - r^2}}$

15. $r = (\csc\theta + \cot\theta)^{-1} \Rightarrow \frac{dr}{d\theta} = -(\csc\theta + \cot\theta)^{-2}\frac{d}{d\theta}(\csc\theta + \cot\theta) = \frac{\csc\theta\cot\theta + \csc^2\theta}{(\csc\theta + \cot\theta)^2} = \frac{\csc\theta(\cot\theta + \csc\theta)}{(\csc\theta + \cot\theta)^2}$
$= \frac{\csc\theta}{\csc\theta + \cot\theta}$

17. $y = x^2\sin^4 x + x\cos^{-2} x \Rightarrow \frac{dy}{dx} = x^2\frac{d}{dx}\left(\sin^4 x\right) + \sin^4 x \cdot \frac{d}{dx}\left(x^2\right) + x\frac{d}{dx}\left(\cos^{-2} x\right) + \cos^{-2} x \cdot \frac{d}{dx}(x)$
$= x^2\left(4\sin^3 x\frac{d}{dx}(\sin x)\right) + 2x\sin^4 x + x\left(-2\cos^{-3} x \cdot \frac{d}{dx}(\cos x)\right) + \cos^{-2} x$
$= x^2\left(4\sin^3 x\cos x\right) + 2x\sin^4 x + x\left(\left(-2\cos^{-3} x\right)(-\sin x)\right) + \cos^{-2} x$
$= 4x^2\sin^3 x\cos x + 2x\sin^4 x + 2x\sin x\cos^{-3} x + \cos^{-2} x$

19. $y = \frac{1}{21}(3x - 2)^7 + \left(4 - \frac{1}{2x^2}\right)^{-1} \Rightarrow \frac{dy}{dx} = \frac{7}{21}(3x - 2)^6 \cdot \frac{d}{dx}(3x - 2) + (-1)\left(4 - \frac{1}{2x^2}\right)^{-2} \cdot \frac{d}{dx}\left(4 - \frac{1}{2x^2}\right)$
$= \frac{7}{21}(3x - 2)^6 \cdot 3 + (-1)\left(4 - \frac{1}{2x^2}\right)^{-2}\left(\frac{1}{x^3}\right) = (3x - 2)^6 - \frac{1}{x^3\left(4 - \frac{1}{2x^2}\right)^2}$

21. $h(x) = x\tan\left(2\sqrt{x}\right) + 7 \Rightarrow h'(x) = x\frac{d}{dx}\left(\tan\left(2x^{1/2}\right)\right) + \tan\left(2x^{1/2}\right) \cdot \frac{d}{dx}(x) + 0$
$= x\sec^2\left(2x^{1/2}\right) \cdot \frac{d}{dx}\left(2x^{1/2}\right) + \tan\left(2x^{1/2}\right) = x\sec^2\left(2\sqrt{x}\right) \cdot \frac{1}{\sqrt{x}} + \tan\left(2\sqrt{x}\right) = \sqrt{x}\sec^2\left(2\sqrt{x}\right) + \tan\left(2\sqrt{x}\right)$

23. $f(\theta) = \left(\frac{\sin\theta}{1 + \cos\theta}\right)^2 \Rightarrow f'(\theta) = 2\left(\frac{\sin\theta}{1 + \cos\theta}\right) \cdot \frac{d}{d\theta}\left(\frac{\sin\theta}{1 + \cos\theta}\right) = \frac{2\sin\theta}{1 + \cos\theta} \cdot \frac{(1 + \cos\theta)(\cos\theta) - (\sin\theta)(-\sin\theta)}{(1 + \cos\theta)^2}$
$= \frac{(2\sin\theta)\left(\cos\theta + \cos^2\theta + \sin^2\theta\right)}{(1 + \cos\theta)^3} = \frac{(2\sin\theta)(\cos\theta + 1)}{(1 + \cos\theta)^3} = \frac{2\sin\theta}{(1 + \cos\theta)^2}$

25. $r = \left(\sec\sqrt{\theta}\right)\tan\left(\frac{1}{\theta}\right) \Rightarrow \frac{dr}{d\theta} = \left(\sec\sqrt{\theta}\right)\left(\sec^2\frac{1}{\theta}\right)\left(-\frac{1}{\theta^2}\right) + \tan\left(\frac{1}{\theta}\right)\left(\sec\sqrt{\theta}\tan\sqrt{\theta}\right)\left(\frac{1}{2\sqrt{\theta}}\right)$
$= -\frac{1}{\theta^2}\sec\sqrt{\theta}\sec^2\left(\frac{1}{\theta}\right) + \frac{1}{2\sqrt{\theta}}\tan\left(\frac{1}{\theta}\right)\sec\sqrt{\theta}\tan\sqrt{\theta} = \left(\sec\sqrt{\theta}\right)\left[\frac{\tan\sqrt{\theta}\tan\left(\frac{1}{\theta}\right)}{2\sqrt{\theta}} - \frac{\sec^2\left(\frac{1}{\theta}\right)}{\theta^2}\right]$

27. $y = \sin^2(\pi t - 2) \Rightarrow \frac{dy}{dt} = 2\sin(\pi t - 2) \cdot \frac{d}{dt}\sin(\pi t - 2) = 2\sin(\pi t - 2) \cdot \cos(\pi t - 2) \cdot \frac{d}{dt}(\pi t - 2)$
$= 2\pi\sin(\pi t - 2)\cos(\pi t - 2)$

29. $y = \left(1 + \cot\left(\frac{t}{2}\right)\right)^{-2} \Rightarrow \frac{dy}{dt} = -2\left(1 + \cot\left(\frac{t}{2}\right)\right)^{-3} \cdot \frac{d}{dt}\left(1 + \cot\left(\frac{t}{2}\right)\right) = -2\left(1 + \cot\left(\frac{t}{2}\right)\right)^{-3} \cdot \left(-\csc^2\left(\frac{t}{2}\right)\right) \cdot \frac{d}{dt}\left(\frac{t}{2}\right)$

$$= \frac{\csc^2\left(\frac{t}{2}\right)}{\left(1+\cot\left(\frac{t}{2}\right)\right)^3}$$

31. $y = \left[1+\tan^4\left(\frac{t}{12}\right)\right]^3 \Rightarrow \frac{dy}{dt} = 3\left[1+\tan^4\left(\frac{t}{12}\right)\right]^2 \cdot \frac{d}{dt}\left[1+\tan^4\left(\frac{t}{12}\right)\right] = 3\left[1+\tan^4\left(\frac{t}{12}\right)\right]^2\left[4\tan^3\left(\frac{t}{12}\right) \cdot \frac{d}{dt}\tan\left(\frac{t}{12}\right)\right]$

$= 12\left[1+\tan^4\left(\frac{t}{12}\right)\right]^2\left[\tan^3\left(\frac{t}{12}\right)\sec^2\left(\frac{t}{12}\right) \cdot \frac{1}{12}\right] = \left[1+\tan^4\left(\frac{t}{12}\right)\right]^2\left[\tan^3\left(\frac{t}{12}\right)\sec^2\left(\frac{t}{12}\right)\right]$

33. $t = \frac{\pi}{4} \Rightarrow x = 2\cos\frac{\pi}{4} = \sqrt{2}$, $y = 2\sin\frac{\pi}{4} = \sqrt{2}$; $\frac{dx}{dt} = -2\sin t$, $\frac{dy}{dt} = 2\cos t \Rightarrow \frac{dy}{dx} = \frac{dy/dt}{dx/dt} = \frac{2\cos t}{-2\sin t} = -\cot t$

$\Rightarrow \left.\frac{dy}{dx}\right|_{t=\frac{\pi}{4}} = -\cot\frac{\pi}{4} = -1$; tangent line is $y - \sqrt{2} = -1\left(x - \sqrt{2}\right)$ or $y = -x + 2\sqrt{2}$; $\frac{dy'}{dt} = \csc^2 t$

$\Rightarrow \frac{d^2y}{dx^2} = \frac{dy'/dt}{dx/dt} = \frac{\csc^2 t}{-2\sin t} = -\frac{1}{2\sin^3 t} \Rightarrow \left.\frac{d^2y}{dx^2}\right|_{t=\frac{\pi}{4}} = -\sqrt{2}$

35. $t = \frac{1}{4} \Rightarrow x = \frac{1}{4}$, $y = \frac{1}{2}$; $\frac{dx}{dt} = 1$, $\frac{dy}{dt} = \frac{1}{2\sqrt{t}} \Rightarrow \frac{dy}{dx} = \frac{dy/dt}{dx/dt} = \frac{1}{2\sqrt{t}} \Rightarrow \left.\frac{dy}{dx}\right|_{t=\frac{1}{4}} = \frac{1}{2\sqrt{\frac{1}{4}}} = 1$; tangent line is

$y - \frac{1}{2} = 1 \cdot \left(x - \frac{1}{4}\right)$ or $y = x + \frac{1}{4}$; $\frac{dy'}{dt} = -\frac{1}{4}t^{-3/2} \Rightarrow \frac{d^2y}{dx^2} = \frac{dy'/dt}{dx/dt} = -\frac{1}{4}t^{-3/2} \Rightarrow \left.\frac{d^2y}{dx^2}\right|_{t=\frac{1}{4}} = -2$

37. $t = -1 \Rightarrow x = 5$, $y = 1$; $\frac{dx}{dt} = 4t$, $\frac{dy}{dt} = 4t^3 \Rightarrow \frac{dy}{dx} = \frac{dy/dt}{dx/dt} = \frac{4t^3}{4t} = t^2 \Rightarrow \left.\frac{dy}{dx}\right|_{t=-1} = (-1)^2 = 1$; tangent line is

$y - 1 = 1 \cdot (x - 5)$ or $y = x - 4$; $\frac{dy'}{dt} = 2t \Rightarrow \frac{d^2y}{dx^2} = \frac{dy'/dt}{dx/dt} = \frac{2t}{4t} = \frac{1}{2} \Rightarrow \left.\frac{d^2y}{dx^2}\right|_{t=-1} = \frac{1}{2}$

39. $t = \frac{\pi}{2} \Rightarrow x = \cos\frac{\pi}{2} = 0$, $y = 1 + \sin\frac{\pi}{2} = 2$; $\frac{dx}{dt} = -\sin t$, $\frac{dy}{dt} = \cos t \Rightarrow \frac{dy}{dx} = \frac{\cos t}{-\sin t} = -\cot t$

$\Rightarrow \left.\frac{dy}{dx}\right|_{t=\frac{\pi}{2}} = -\cot\frac{\pi}{2} = 0$; tangent line is $y = 2$; $\frac{dy'}{dt} = \csc^2 t \Rightarrow \frac{d^2y}{dx^2} = \frac{\csc^2 t}{-\sin t} = -\csc^3 t \Rightarrow \left.\frac{d^2y}{dx^2}\right|_{t=\frac{\pi}{2}} = -1$

41. $y = \left(1+\frac{1}{x}\right)^3 \Rightarrow y' = 3\left(1+\frac{1}{x}\right)^2\left(-\frac{1}{x^2}\right) = -\frac{3}{x^2}\left(1+\frac{1}{x}\right)^2 \Rightarrow y'' = \left(-\frac{3}{x^2}\right) \cdot \frac{d}{dx}\left(1+\frac{1}{x}\right)^2 - \left(1+\frac{1}{x}\right)^2 \cdot \frac{d}{dx}\left(\frac{3}{x^2}\right)$

$= \left(-\frac{3}{x^2}\right)\left(2\left(1+\frac{1}{x}\right)\left(-\frac{1}{x^2}\right)\right) + \left(\frac{6}{x^3}\right)\left(1+\frac{1}{x}\right)^2 = \frac{6}{x^4}\left(1+\frac{1}{x}\right) + \frac{6}{x^3}\left(1+\frac{1}{x}\right)^2 = \frac{6}{x^3}\left(1+\frac{1}{x}\right)\left(\frac{1}{x}+1+\frac{1}{x}\right)$

$= \frac{6}{x^3}\left(1+\frac{1}{x}\right)\left(1+\frac{2}{x}\right)$

43. $y = \frac{1}{9}\cot(3x-1) \Rightarrow y' = -\frac{1}{9}\csc^2(3x-1)(3) = -\frac{1}{3}\csc^2(3x-1) \Rightarrow y'' = \left(-\frac{2}{3}\right)\left(\csc(3x-1) \cdot \frac{d}{dx}\csc(3x-1)\right)$

$= -\frac{2}{3}\csc(3x-1)\left(-\csc(3x-1)\cot(3x-1) \cdot \frac{d}{dx}(3x-1)\right) = 2\csc^2(3x-1)\cot(3x-1)$

45. $g(x) = \sqrt{x} \Rightarrow g'(x) = \frac{1}{2\sqrt{x}} \Rightarrow g(1) = 1$ and $g'(1) = \frac{1}{2}$; $f(u) = u^5 + 1 \Rightarrow f'(u) = 5u^4 \Rightarrow f'(g(1)) = f'(1) = 5$;

therefore, $(f\circ g)'(1) = f'(g(1))\cdot g'(1) = 5\cdot\frac{1}{2} = \frac{5}{2}$

47. $g(x) = 5\sqrt{x} \Rightarrow g'(x) = \frac{5}{2\sqrt{x}} \Rightarrow g(1) = 5$ and $g'(1) = \frac{5}{2}$; $f(u) = \cot\left(\frac{\pi u}{10}\right) \Rightarrow f'(u) = -\csc^2\left(\frac{\pi u}{10}\right)\left(\frac{\pi}{10}\right)$

$= \frac{-\pi}{10}\csc^2\left(\frac{\pi u}{10}\right) \Rightarrow f'(g(1)) = f'(5) = -\frac{\pi}{10}\csc^2\left(\frac{\pi}{2}\right) = -\frac{\pi}{10}$; therefore, $(f\circ g)'(1) = f'(g(1))g'(1) = -\frac{\pi}{10}\cdot\frac{5}{2}$

$= -\frac{\pi}{4}$

49. $g(x) = 10x^2 + x + 1 \Rightarrow g'(x) = 20x + 1 \Rightarrow g(0) = 1$ and $g'(0) = 1$; $f(u) = \frac{2u}{u^2+1} \Rightarrow f'(u) = \frac{(u^2+1)(2) - (2u)(2u)}{(u^2+1)^2}$

$= \frac{-2u^2+2}{(u^2+1)^2} \Rightarrow f'(g(0)) = f'(1) = 0$; therefore, $(f\circ g)'(0) = f'(g(0))g'(0) = 0\cdot 1 = 0$

51. (a) $y = 2f(x) \Rightarrow \frac{dy}{dx} = 2f'(x) \Rightarrow \left.\frac{dy}{dx}\right|_{x=2} = 2f'(2) = 2\left(\frac{1}{3}\right) = \frac{2}{3}$

(b) $y = f(x) + g(x) \Rightarrow \frac{dy}{dx} = f'(x) + g'(x) \Rightarrow \left.\frac{dy}{dx}\right|_{x=3} = f'(3) + g'(3) = 2\pi + 5$

(c) $y = f(x)\cdot g(x) \Rightarrow \frac{dy}{dx} = f(x)g'(x) + g(x)f'(x) \Rightarrow \left.\frac{dy}{dx}\right|_{x=3} = f(3)g'(3) + g(3)f'(3) = 3\cdot 5 + (-4)(2\pi) = 15 - 8\pi$

(d) $y = \frac{f(x)}{g(x)} \Rightarrow \frac{dy}{dx} = \frac{g(x)f'(x) - f(x)g'(x)}{[g(x)]^2} \Rightarrow \left.\frac{dy}{dx}\right|_{x=2} = \frac{g(2)f'(2) - f(2)g'(2)}{[g(2)]^2} = \frac{(2)\left(\frac{1}{3}\right) - (8)(-3)}{2^2} = \frac{37}{6}$

(e) $y = f(g(x)) \Rightarrow \frac{dy}{dx} = f'(g(x))g'(x) \Rightarrow \left.\frac{dy}{dx}\right|_{x=2} = f'(g(2))g'(2) = f'(2)(-3) = \frac{1}{3}(-3) = -1$

(f) $y = (f(x))^{1/2} \Rightarrow \frac{dy}{dx} = \frac{1}{2}(f(x))^{-1/2}\cdot f'(x) = \frac{f'(x)}{2\sqrt{f(x)}} \Rightarrow \left.\frac{dy}{dx}\right|_{x=2} = \frac{f'(2)}{2\sqrt{f(2)}} = \frac{\left(\frac{1}{3}\right)}{2\sqrt{8}} = \frac{1}{6\sqrt{8}} = \frac{1}{12\sqrt{2}} = \frac{\sqrt{2}}{24}$

(g) $y = (g(x))^{-2} \Rightarrow \frac{dy}{dx} = -2(g(x))^{-3}\cdot g'(x) \Rightarrow \left.\frac{dy}{dx}\right|_{x=3} = -2(g(3))^{-3}g'(3) = -2(-4)^{-3}\cdot 5 = \frac{5}{32}$

(h) $y = \left((f(x))^2 + (g(x))^2\right)^{1/2} \Rightarrow \frac{dy}{dx} = \frac{1}{2}\left((f(x))^2 + (g(x))^2\right)^{-1/2}(2f(x)\cdot f'(x) + 2g(x)\cdot g'(x))$

$\Rightarrow \left.\frac{dy}{dx}\right|_{x=2} = \frac{1}{2}\left((f(2))^2 + (g(2))^2\right)^{-1/2}(2f(2)f'(2) + 2g(2)g'(2)) = \frac{1}{2}\left(8^2 + 2^2\right)^{-1/2}\left(2\cdot 8\cdot\frac{1}{3} + 2\cdot 2\cdot(-3)\right)$

$= -\frac{5}{3\sqrt{17}}$

53. $\frac{ds}{dt} = \frac{ds}{d\theta}\cdot\frac{d\theta}{dt}$: $s = \cos\theta \Rightarrow \frac{ds}{d\theta} = -\sin\theta \Rightarrow \left.\frac{ds}{d\theta}\right|_{\theta=\frac{3\pi}{2}} = -\sin\left(\frac{3\pi}{2}\right) = 1$ so that $\frac{ds}{dt} = \frac{ds}{d\theta}\cdot\frac{d\theta}{dt} = 1\cdot 5 = 5$

55. With $y = x$, we should get $\frac{dy}{dx} = 1$ for both (a) and (b):

(a) $y = \frac{u}{5} + 7 \Rightarrow \frac{dy}{du} = \frac{1}{5}$; $u = 5x - 35 \Rightarrow \frac{du}{dx} = 5$; therefore, $\frac{dy}{dx} = \frac{dy}{du}\cdot\frac{du}{dx} = \frac{1}{5}\cdot 5 = 1$, as expected

(b) $y = 1 + \frac{1}{u} \Rightarrow \frac{dy}{du} = -\frac{1}{u^2}$; $u = (x-1)^{-1} \Rightarrow \frac{du}{dx} = -(x-1)^{-2}(1) = \frac{-1}{(x-1)^2}$; therefore $\frac{dy}{dx} = \frac{dy}{du} \cdot \frac{du}{dx}$

$= \frac{-1}{u^2} \cdot \frac{-1}{(x-1)^2} = \frac{-1}{\left((x-1)^{-1}\right)^2} \cdot \frac{-1}{(x-1)^2} = (x-1)^2 \cdot \frac{1}{(x-1)^2} = 1$, again as expected

57. $y = 2\tan\left(\frac{\pi x}{4}\right) \Rightarrow \frac{dy}{dx} = \left(2\sec^2\frac{\pi x}{4}\right)\left(\frac{\pi}{4}\right) = \frac{\pi}{2}\sec^2\frac{\pi x}{4}$

(a) $\left.\frac{dy}{dx}\right|_{x=1} = \frac{\pi}{2}\sec^2\left(\frac{\pi}{4}\right) = \pi \Rightarrow$ slope of tangent is 2; thus, $y(1) = 2\tan\left(\frac{\pi}{4}\right) = 2$ and $y'(1) = \pi \Rightarrow$ tangent line is given by $y - 2 = \pi(x-1) \Rightarrow y = \pi x + 2 - \pi$

(b) $y' = \frac{\pi}{2}\sec^2\left(\frac{\pi x}{4}\right)$ and the smallest value the secant function can have in $-2 < x < 2$ is $1 \Rightarrow$ the minimum value of y' is $\frac{\pi}{2}$ and that occurs when $\frac{\pi}{2} = \frac{\pi}{2}\sec^2\left(\frac{\pi x}{4}\right) \Rightarrow 1 = \sec^2\left(\frac{\pi x}{4}\right) \Rightarrow \pm 1 = \sec\left(\frac{\pi x}{4}\right) \Rightarrow x = 0$.

59. $s = A\cos(2\pi bt) \Rightarrow v = \frac{ds}{dt} = -A\sin(2\pi bt)(2\pi b) = -2\pi bA\sin(2\pi bt)$. If we replace b with 2b to double the frequency, the velocity formula gives $v = -4\pi bA\sin(4\pi bt) \Rightarrow$ doubling the frequency causes the velocity to double. Also $v = -2\pi bA\sin(2\pi bt) \Rightarrow a = \frac{dv}{dt} = -4\pi^2 b^2 A\cos(2\pi bt)$. If we replace b with 2b in the acceleration formula, we get $a = -16\pi^2 b^2 A\cos(4\pi bt) \Rightarrow$ doubling the frequency causes the acceleration to quadruple. Finally, $a = -4\pi^2 b^2 A\cos(2\pi bt) \Rightarrow j = \frac{da}{dt} = 8\pi^3 b^3 A\sin(2\pi bt)$. If we replace b with 2b in the jerk formula, we get $j = 64\pi^3 b^3 A\sin(2\pi bt) \Rightarrow$ doubling the frequency multiplies the jerk by a factor of 8.

61. $s = (1+4t)^{1/2} \Rightarrow v = \frac{ds}{dt} = \frac{1}{2}(1+4t)^{-1/2}(4) = 2(1+4t)^{-1/2} \Rightarrow v(6) = 2(1+4\cdot 6)^{-1/2} = \frac{2}{5}$ m/sec;

$v = 2(1+4t)^{-1/2} \Rightarrow a = \frac{dv}{dt} = -\frac{1}{2}\cdot 2(1+4t)^{-3/2}(4) = -4(1+4t)^{-3/2} \Rightarrow a(6) = -4(1+4\cdot 6)^{-3/2} = -\frac{4}{125}$ m/sec^2

63. v proportional to $\frac{1}{\sqrt{s}} \Rightarrow v = \frac{k}{\sqrt{s}}$ for some constant $k \Rightarrow \frac{dv}{ds} = -\frac{k}{2s^{3/2}}$. Thus, $a = \frac{dv}{dt} = \frac{dv}{ds}\cdot\frac{ds}{dt} = \frac{dv}{ds}\cdot v$

$= -\frac{k}{2s^{3/2}}\cdot\frac{k}{\sqrt{s}} = -\frac{k^2}{2}\left(\frac{1}{s^2}\right) \Rightarrow$ acceleration is a constant times $\frac{1}{s^2}$ so a is proportional to $\frac{1}{s^2}$.

65. $T = 2\pi\sqrt{\frac{L}{g}} \Rightarrow \frac{dT}{dL} = 2\pi\cdot\frac{1}{2\sqrt{\frac{L}{g}}}\cdot\frac{1}{g} = \frac{\pi}{g\sqrt{\frac{L}{g}}} = \frac{\pi}{\sqrt{gL}}$. Therefore, $\frac{dT}{du} = \frac{dT}{dL}\cdot\frac{dL}{du} = \frac{\pi}{\sqrt{gL}}\cdot kL = \frac{\pi k\sqrt{L}}{\sqrt{g}} = \frac{1}{2}\cdot 2\pi k\sqrt{\frac{L}{g}}$

$= \frac{kT}{2}$, as required.

67. The graph of $y = (f\circ g)(x)$ has a horizontal tangent at $x = 1$ provided that $(f\circ g)'(1) = 0 \Rightarrow f'(g(1))g'(1) = 0$ $\Rightarrow$ either $f'(g(1)) = 0$ or $g'(1) = 0$ (or both) $\Rightarrow$ either the graph of f has a horizontal tangent at $u = g(1)$, or the graph of g has a horizontal tangent at $x = 1$ (or both).

69. As $h \to 0$, the graph of $y = \frac{\sin 2(x+h) - \sin 2x}{h}$ approaches the graph of $y = 2 \cos 2x$ because $\lim\limits_{h \to 0} \frac{\sin 2(x+h) - \sin 2x}{h} = \frac{d}{dx}(\sin 2x) = 2 \cos 2x.$

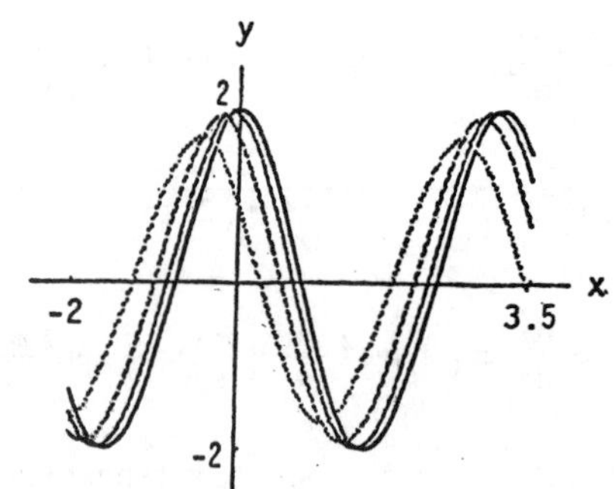

71. $\frac{dx}{dt} = \cos t$ and $\frac{dy}{dt} = 2 \cos 2t \Rightarrow \frac{dy}{dx} = \frac{dy/dt}{dx/dt} = \frac{2 \cos 2t}{\cos t} = \frac{2(2 \cos^2 t - 1)}{\cos t}$; then $\frac{dy}{dx} = 0 \Rightarrow \frac{2(2 \cos^2 t - 1)}{\cos t} = 0$ $\Rightarrow 2 \cos^2 t - 1 = 0 \Rightarrow \cos t = \pm\frac{1}{\sqrt{2}} \Rightarrow t = \frac{\pi}{4}, \frac{3\pi}{4}, \frac{5\pi}{4}, \frac{7\pi}{4}$. In the 1st quadrant: $t = \frac{\pi}{4} \Rightarrow x = \sin \frac{\pi}{4} = \frac{\sqrt{2}}{2}$ and $y = \sin 2\left(\frac{\pi}{4}\right) = 1 \Rightarrow \left(\frac{\sqrt{2}}{2}, 1\right)$ is the point where the tangent line is horizontal. At the origin: $x = 0$ and $y = 0$ $\Rightarrow \sin t = 0 \Rightarrow t = 0$ or $t = \pi$ and $\sin 2t = 0 \Rightarrow t = 0, \frac{\pi}{2}, \pi, \frac{3\pi}{2}$; thus $t = 0$ and $t = \pi$ give the tangent lines at the origin. Tangents at origin: $\left.\frac{dy}{dx}\right|_{t=0} = 2 \Rightarrow y = 2x$ and $\left.\frac{dy}{dx}\right|_{t=\pi} = -2 \Rightarrow y = -2x$

73. (a)

(b) $\frac{df}{dt} = 1.27324 \sin 2t + 0.42444 \sin 6t + 0.2546 \sin 10t + 0.18186 \sin 14t$

(c) The curve of $y = \frac{df}{dt}$ approximates $y = \frac{dg}{dt}$ the best when t is not $-\pi, -\frac{\pi}{2}, 0, \frac{\pi}{2}$, nor π.

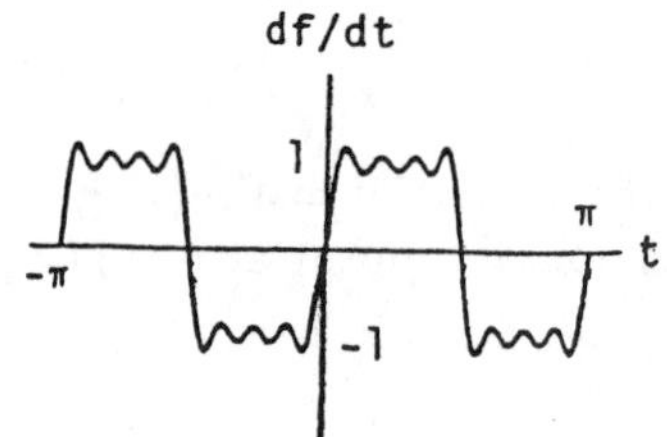

2.6 IMPLICIT DIFFERENTIATION

1. $y = x^{9/4} \Rightarrow \frac{dy}{dx} = \frac{9}{4}x^{5/4}$

3. $y = 7\sqrt{x+6} = 7(x+6)^{1/2} \Rightarrow \frac{dy}{dx} = \frac{7}{2}(x+6)^{-1/2} = \frac{7}{2\sqrt{x+6}}$

5. $y = x\left(x^2+1\right)^{1/2} \Rightarrow y' = (1)\left(x^2+1\right)^{1/2} + \left(\frac{x}{2}\right)\left(x^2+1\right)^{-1/2}(2x) = \frac{2x^2+1}{\sqrt{x^2+1}}$

7. $s = \sqrt[7]{t^2} = t^{2/7} \Rightarrow \frac{ds}{dt} = \frac{2}{7}t^{-5/7}$

9. $y = \sin\left((2t+5)^{-2/3}\right) \Rightarrow \frac{dy}{dt} = \cos\left((2t+5)^{-2/3}\right)\cdot\left(-\frac{2}{3}\right)(2t+5)^{-5/3}\cdot 2 = -\frac{4}{3}(2t+5)^{-5/3}\cos\left((2t+5)^{-2/3}\right)$

11. $g(x) = 2\left(2x^{-1/2}+1\right)^{-1/3} \Rightarrow g'(x) = -\frac{2}{3}\left(2x^{-1/2}+1\right)^{-4/3}\cdot(-1)x^{-3/2} = \frac{2}{3}\left(2x^{-1/2}+1\right)^{-4/3}x^{-3/2}$

13. $x^2y + xy^2 = 6$:

Step 1: $\left(x^2\frac{dy}{dx} + y\cdot 2x\right) + \left(x\cdot 2y\frac{dy}{dx} + y^2\cdot 1\right) = 0$

Step 2: $x^2\frac{dy}{dx} + 2xy\frac{dy}{dx} = -2xy - y^2$

Step 3: $\frac{dy}{dx}\left(x^2 + 2xy\right) = -2xy - y^2$

Step 4: $\frac{dy}{dx} = \frac{-2xy - y^2}{x^2 + 2xy}$

15. $x^3 - xy + y^3 = 1 \Rightarrow 3x^2 - y - x\frac{dy}{dx} + 3y^2\frac{dy}{dx} = 0 \Rightarrow \left(3y^2 - x\right)\frac{dy}{dx} = y - 3x^2 \Rightarrow \frac{dy}{dx} = \frac{y - 3x^2}{3y^2 - x}$

17. $y^2 = \frac{x-1}{x+1} \Rightarrow 2y\frac{dy}{dx} = \frac{(x+1)-(x-1)}{(x+1)^2} = \frac{2}{(x+1)^2} \Rightarrow \frac{dy}{dx} = \frac{1}{y(x+1)^2}$

19. $x = \tan y \Rightarrow 1 = \left(\sec^2 y\right)\frac{dy}{dx} \Rightarrow \frac{dy}{dx} = \frac{1}{\sec^2 y} = \cos^2 y$

21. $y\sin\left(\frac{1}{y}\right) = 1 - xy \Rightarrow y\left[\cos\left(\frac{1}{y}\right)\cdot(-1)\frac{1}{y^2}\cdot\frac{dy}{dx}\right] + \sin\left(\frac{1}{y}\right)\cdot\frac{dy}{dx} = -x\frac{dy}{dx} - y \Rightarrow \frac{dy}{dx}\left[-\frac{1}{y}\cos\left(\frac{1}{y}\right) + \sin\left(\frac{1}{y}\right) + x\right] = -y$

$\Rightarrow \frac{dy}{dx} = \frac{-y}{-\frac{1}{y}\cos\left(\frac{1}{y}\right) + \sin\left(\frac{1}{y}\right) + x} = \frac{-y^2}{y\sin\left(\frac{1}{y}\right) - \cos\left(\frac{1}{y}\right) + xy}$

23. $\theta^{1/2} + r^{1/2} = 1 \Rightarrow \frac{1}{2}\theta^{-1/2} + \frac{1}{2}r^{-1/2}\cdot\frac{dr}{d\theta} = 0 \Rightarrow \frac{dr}{d\theta}\left[\frac{1}{2\sqrt{r}}\right] = \frac{-1}{2\sqrt{\theta}} \Rightarrow \frac{dr}{d\theta} = -\frac{2\sqrt{r}}{2\sqrt{\theta}} = -\frac{\sqrt{r}}{\sqrt{\theta}}$

25. $\sin(r\theta) = \frac{1}{2} \Rightarrow [\cos(r\theta)]\left(r + \theta\frac{dr}{d\theta}\right) = 0 \Rightarrow \frac{dr}{d\theta}[\theta\cos(r\theta)] = -r\cos(r\theta) \Rightarrow \frac{dr}{d\theta} = \frac{-r\cos(r\theta)}{\theta\cos(r\theta)} = -\frac{r}{\theta}$, $\cos(r\theta) \neq 0$

27. $x^{2/3}+y^{2/3}=1 \Rightarrow \frac{2}{3}x^{-1/3}+\frac{2}{3}y^{-1/3}\frac{dy}{dx}=0 \Rightarrow \frac{dy}{dx}\left[\frac{2}{3}y^{-1/3}\right]=-\frac{2}{3}x^{-1/3} \Rightarrow y'=\frac{dy}{dx}=-\frac{x^{-1/3}}{y^{-1/3}}=-\left(\frac{y}{x}\right)^{1/3}$;

Differentiating again, $y''=\frac{x^{1/3}\cdot\left(-\frac{1}{3}y^{-2/3}\right)y'+y^{1/3}\left(\frac{1}{3}x^{-2/3}\right)}{x^{2/3}}=\frac{x^{1/3}\cdot\left(-\frac{1}{3}y^{-2/3}\right)\left(-\frac{y^{1/3}}{x^{1/3}}\right)+y^{1/3}\left(\frac{1}{3}x^{-2/3}\right)}{x^{2/3}}$

$\Rightarrow \frac{d^2y}{dx^2}=\frac{1}{3}x^{-2/3}y^{-1/3}+\frac{1}{3}y^{1/3}x^{-4/3}=\frac{y^{1/3}}{3x^{4/3}}+\frac{1}{3y^{1/3}x^{2/3}}$

29. $2\sqrt{y}=x-y \Rightarrow y^{-1/2}y'=1-y' \Rightarrow y'\left(y^{-1/2}+1\right)=1 \Rightarrow \frac{dy}{dx}=y'=\frac{1}{y^{-1/2}+1}=\frac{\sqrt{y}}{\sqrt{y}+1}$; we can differentiate the equation $y'\left(y^{-1/2}+1\right)=1$ again to find y'': $y'\left(-\frac{1}{2}y^{-3/2}y'\right)+\left(y^{-1/2}+1\right)y''=0$

$\Rightarrow \left(y^{-1/2}+1\right)y''=\frac{1}{2}[y']^2y^{-3/2} \Rightarrow \frac{d^2y}{dx^2}=y''=\frac{\frac{1}{2}\left(\frac{1}{y^{-1/2}+1}\right)^2y^{-3/2}}{\left(y^{-1/2}+1\right)}=\frac{1}{2y^{3/2}\left(y^{-1/2}+1\right)^3}=\frac{1}{2\left(1+\sqrt{y}\right)^3}$

31. $x^3+y^3=16 \Rightarrow 3x^2+3y^2y'=0 \Rightarrow 3y^2y'=-3x^2 \Rightarrow y'=-\frac{x^2}{y^2}$; we differentiate $y^2y'=-x^2$ to find y'':

$y^2y''+y'[2y\cdot y']=-2x \Rightarrow y^2y''=-2x-2y[y']^2 \Rightarrow y''=\frac{-2x-2y\left(-\frac{x^2}{y^2}\right)^2}{y^2}=\frac{-2x-\frac{2x^4}{y^3}}{y^2}$

$=\frac{-2xy^3-2x^4}{y^5} \Rightarrow \left.\frac{d^2y}{dx^2}\right|_{(2,2)}=\frac{-32-32}{32}=-2$

33. $x^2-2tx+2t^2=4 \Rightarrow 2x\frac{dx}{dt}-2x-2t\frac{dx}{dt}+4t=0 \Rightarrow (2x-2t)\frac{dx}{dt}=2x-4t \Rightarrow \frac{dx}{dt}=\frac{2x-4t}{2x-2t}=\frac{x-2t}{x-t}$;

$2y^3-3t^2=4 \Rightarrow 6y^2\frac{dy}{dt}-6t=0 \Rightarrow \frac{dy}{dt}=\frac{6t}{6y^2}=\frac{t}{y^2}$; thus $\frac{dy}{dx}=\frac{dy/dt}{dx/dt}=\frac{\left(\frac{t}{y^2}\right)}{\left(\frac{x-2t}{x-t}\right)}=\frac{t(x-t)}{y^2(x-2t)}$; $t=2$

$\Rightarrow x^2-2(2)x+2(2)^2=4 \Rightarrow x^2-4x+4=0 \Rightarrow (x-2)^2=0 \Rightarrow x=2$; $t=2 \Rightarrow 2y^3-3(2)^2=4$

$\Rightarrow 2y^3=16 \Rightarrow y^3=8 \Rightarrow y=2$; therefore $\left.\frac{dy}{dx}\right|_{t=2}=\frac{2(2-2)}{(2)^2(2-2(2))}=0$

35. $x+2x^{3/2}=t^2+t \Rightarrow \frac{dx}{dt}+3x^{1/2}\frac{dx}{dt}=2t+1 \Rightarrow \left(1+3x^{1/2}\right)\frac{dx}{dt}=2t+1 \Rightarrow \frac{dx}{dt}=\frac{2t+1}{1+3x^{1/2}}$; $y\sqrt{t+1}+2t\sqrt{y}=4$

$\Rightarrow \frac{dy}{dt}\sqrt{t+1}+y\left(\frac{1}{2}\right)(t+1)^{-1/2}+2\sqrt{y}+2t\left(\frac{1}{2}y^{-1/2}\right)\frac{dy}{dt}=0 \Rightarrow \frac{dy}{dt}\sqrt{t+1}+\frac{y}{2\sqrt{t+1}}+2\sqrt{y}+\left(\frac{t}{\sqrt{y}}\right)\frac{dy}{dt}=0$

$\Rightarrow \left(\sqrt{t+1}+\frac{t}{\sqrt{y}}\right)\frac{dy}{dt}=\frac{-y}{2\sqrt{t+1}}-2\sqrt{y} \Rightarrow \frac{dy}{dt}=\frac{\left(\frac{-y}{2\sqrt{t+1}}-2\sqrt{y}\right)}{\left(\sqrt{t+1}+\frac{t}{\sqrt{y}}\right)}=\frac{-y\sqrt{y}-4y\sqrt{t+1}}{2\sqrt{y}(t+1)+2t\sqrt{t+1}}$; thus

$$\frac{dy}{dx}=\frac{dy/dt}{dx/dt}=\frac{\left(\frac{-y\sqrt{y}-4y\sqrt{t+1}}{2\sqrt{y}\,(t+1)+2t\sqrt{t+1}}\right)}{\left(\frac{2t+1}{1+3x^{1/2}}\right)};\ t=0\Rightarrow x+2x^{3/2}=0\Rightarrow x\left(1+2x^{1/2}\right)=0\Rightarrow x=0;\ t=0$$

$$\Rightarrow y\sqrt{0+1}+2(0)\sqrt{y}=4\Rightarrow y=4;\text{ therefore }\left.\frac{dy}{dx}\right|_{t=0}=\frac{\left(\frac{-4\sqrt{4}-4(4)\sqrt{0+1}}{2\sqrt{4}(0+1)+2(0)\sqrt{0+1}}\right)}{\left(\frac{2(0)+1}{1+3(0)^{1/2}}\right)}=-6$$

37. $y^2+x^2=y^4-2x$ at $(-2,1)$ and $(-2,-1)\Rightarrow 2y\,\frac{dy}{dx}+2x=4y^3\,\frac{dy}{dx}-2\Rightarrow 2y\,\frac{dy}{dx}-4y^3\,\frac{dy}{dx}=-2-2x$

$\Rightarrow \frac{dy}{dx}\left(2y-4y^3\right)=-2-2x\Rightarrow \frac{dy}{dx}=\frac{x+1}{2y^3-y}\Rightarrow \left.\frac{dy}{dx}\right|_{(-2,1)}=-1$ and $\left.\frac{dy}{dx}\right|_{(-2,-1)}=1$

39. $x^2+xy-y^2=1\Rightarrow 2x+y+xy'-2yy'=0\Rightarrow (x-2y)y'=-2x-y\Rightarrow y'=\frac{2x+y}{2y-x}$;

(a) the slope of the tangent line $m=y'\big|_{(2,3)}=\frac{7}{4}\Rightarrow$ the tangent line is $y-3=\frac{7}{4}(x-2)\Rightarrow y=\frac{7}{4}x-\frac{1}{2}$

(b) the normal line is $y-3=-\frac{4}{7}(x-2)\Rightarrow y=-\frac{4}{7}x+\frac{29}{7}$

41. $y^2-2x-4y-1=0\Rightarrow 2yy'-2-4y'=0\Rightarrow 2(y-2)y'=2\Rightarrow y'=\frac{1}{y-2}$;

(a) the slope of the tangent line $m=y'\big|_{(-2,1)}=-1\Rightarrow$ the tangent line is $y-1=-1(x+2)\Rightarrow y=-x-1$

(b) the normal line is $y-1=1(x+2)\Rightarrow y=x+3$

43. $2xy+\pi\sin y=2\pi\Rightarrow 2xy'+2y+\pi(\cos y)y'=0\Rightarrow y'(2x+\pi\cos y)=-2y\Rightarrow y'=\frac{-2y}{2x+\pi\cos y}$;

(a) the slope of the tangent line $m=y'\big|_{\left(1,\frac{\pi}{2}\right)}=\left.\frac{-2y}{2x+\pi\cos y}\right|_{\left(1,\frac{\pi}{2}\right)}=-\frac{\pi}{2}\Rightarrow$ the tangent line is

$y-\frac{\pi}{2}=-\frac{\pi}{2}(x-1)\Rightarrow y=-\frac{\pi}{2}x+\pi$

(b) the normal line is $y-\frac{\pi}{2}=\frac{2}{\pi}(x-1)\Rightarrow y=\frac{2}{\pi}x-\frac{2}{\pi}+\frac{\pi}{2}$

45. $y=2\sin(\pi x-y)\Rightarrow y'=2\left[\cos(\pi x-y)\right]\cdot\left(\pi-y'\right)\Rightarrow y'[1+2\cos(\pi x-y)]=2\pi\cos(\pi x-y)$

$\Rightarrow y'=\frac{2\pi\cos(\pi x-y)}{1+2\cos(\pi x-y)}$;

(a) the slope of the tangent line $m=y'\big|_{(1,0)}=\left.\frac{2\pi\cos(\pi x-y)}{1+2\cos(\pi x-y)}\right|_{(1,0)}=2\pi\Rightarrow$ the tangent line is

$y-0=2\pi(x-1)\Rightarrow y=2\pi x-2\pi$

(b) the normal line is $y-0=-\frac{1}{2\pi}(x-1)\Rightarrow y=-\frac{x}{2\pi}+\frac{1}{2\pi}$

47. Solving $x^2+xy+y^2=7$ and $y=0 \Rightarrow x^2=7 \Rightarrow x=\pm\sqrt{7} \Rightarrow (-\sqrt{7},0)$ and $(\sqrt{7},0)$ are the points where the curve crosses the x-axis. Now $x^2+xy+y^2=7 \Rightarrow 2x+y+xy'+2yy'=0 \Rightarrow (x+2y)y'=-2x-y$ $\Rightarrow y'=-\frac{2x+y}{x+2y} \Rightarrow m=-\frac{2x+y}{x+2y} \Rightarrow$ the slope at $(-\sqrt{7},0)$ is $m=-\frac{-2\sqrt{7}}{-\sqrt{7}}=-2$ and the slope at $(\sqrt{7},0)$ is $m=-\frac{2\sqrt{7}}{\sqrt{7}}=-2$. Since the slope is -2 in each case, the corresponding tangents must be parallel.

49. $y^4=y^2-x^2 \Rightarrow 4y^3y'=2yy'-2x \Rightarrow 2(2y^3-y)y'=-2x \Rightarrow y'=\frac{x}{y-2y^3}$; the slope of the tangent line at $\left(\frac{\sqrt{3}}{4},\frac{\sqrt{3}}{2}\right)$ is $\left.\frac{x}{y-2y^3}\right|_{\left(\frac{\sqrt{3}}{4},\frac{\sqrt{3}}{2}\right)}=\frac{\frac{\sqrt{3}}{4}}{\frac{\sqrt{3}}{2}-\frac{6\sqrt{3}}{8}}=\frac{\frac{1}{4}}{\frac{1}{2}-\frac{3}{4}}=\frac{1}{2-3}=-1$; the slope of the tangent line at $\left(\frac{\sqrt{3}}{4},\frac{1}{2}\right)$ is $\left.\frac{x}{y-2y^3}\right|_{\left(\frac{\sqrt{3}}{4},\frac{1}{2}\right)}=\frac{\frac{\sqrt{3}}{4}}{\frac{1}{2}-\frac{2}{8}}=\frac{2\sqrt{3}}{4-2}=\sqrt{3}$

51. $y^4-4y^2=x^4-9x^2 \Rightarrow 4y^3y'-8yy'=4x^3-18x \Rightarrow y'(4y^3-8y)=4x^3-18x \Rightarrow y'=\frac{4x^3-18x}{4y^3-8y}=\frac{2x^3-9x}{2y^3-4y}$ $=\frac{x(2x^2-9)}{y(2y^2-4)}=m$; $(-3,2)$: $m=\frac{(-3)(18-9)}{2(8-4)}=-\frac{27}{8}$; $(-3,-2)$: $m=\frac{27}{8}$; $(3,2)$: $m=\frac{27}{8}$; $(3,-2)$: $m=-\frac{27}{8}$

53. (a) if $f(x)=\frac{3}{2}x^{2/3}-3$, then $f'(x)=x^{-1/3}$ and $f''(x)=-\frac{1}{3}x^{-4/3}$ so the claim $f''(x)=x^{-1/3}$ is false

(b) if $f(x)=\frac{9}{10}x^{5/3}-7$, then $f'(x)=\frac{3}{2}x^{2/3}$ and $f''(x)=x^{-1/3}$ is true

(c) $f''(x)=x^{-1/3} \Rightarrow f'''(x)=-\frac{1}{3}x^{-4/3}$ is true

(d) if $f'(x)=\frac{3}{2}x^{2/3}+6$, then $f''(x)=x^{-1/3}$ is true

55. $x^2+2xy-3y^2=0 \Rightarrow 2x+2xy'+2y-6yy'=0 \Rightarrow y'(2x-6y)=-2x-2y \Rightarrow y'=\frac{x+y}{3y-x} \Rightarrow$ the slope of the tangent line $m=y'|_{(1,1)}=\left.\frac{x+y}{3y-x}\right|_{(1,1)}=1 \Rightarrow$ the equation of the normal line at $(1,1)$ is $y-1=-1(x-1)$ $\Rightarrow y=-x+2$. To find where the normal line intersects the curve we substitute into its equation: $x^2+2x(2-x)-3(2-x)^2=0 \Rightarrow x^2+4x-2x^2-3(4-4x+x^2)=0 \Rightarrow -4x^2+16x-12=0 \Rightarrow x^2-4x+3=0$ $\Rightarrow (x-3)(x-1)=0 \Rightarrow x=3$ and $y=-x+2=-1$. Therefore, the normal to the curve at $(1,1)$ intersects the curve at the point $(3,-1)$. Note that it also intersects the curve at $(1,1)$.

57. $y^2=x \Rightarrow \frac{dy}{dx}=\frac{1}{2y}$. If a normal is drawn from $(a,0)$ to (x_1,y_1) on the curve its slope satisfies $\frac{y_1-0}{x_1-a}=-2y_1$ $\Rightarrow y_1=-2y_1(x_1-a)$ or $a=x_1+\frac{1}{2}$. Since $x_1\geq 0$ on the curve, we must have that $a\geq\frac{1}{2}$. By symmetry, the two points on the parabola are $(x_1,\sqrt{x_1})$ and $(x_1,-\sqrt{x_1})$. For the normal to be perpendicular,

$\left(\frac{\sqrt{x_1}}{x_1-a}\right)\left(\frac{\sqrt{x_1}}{a-x_1}\right)=-1 \Rightarrow \frac{x_1}{(a-x_1)^2}=1 \Rightarrow x_1=(a-x_1)^2 \Rightarrow x_1=\left(x_1+\frac{1}{2}-x_1\right)^2 \Rightarrow x_1=\frac{1}{4}$ and $y_1=\pm\frac{1}{2}$.

Therefore, $\left(\frac{1}{4}, \pm\frac{1}{2}\right)$ and $a=\frac{3}{4}$.

59. $xy^3+x^2y=6 \Rightarrow x\left(3y^2\frac{dy}{dx}\right)+y^3+x^2\frac{dy}{dx}+2xy=0 \Rightarrow \frac{dy}{dx}(3xy^2+x^2)=-y^3-2xy \Rightarrow \frac{dy}{dx}=\frac{-y^3-2xy}{3xy^2+x^2}$ $=-\frac{y^3+2xy}{3xy^2+x^2}$; also, $xy^3+x^2y=6 \Rightarrow x(3y^2)+y^3\frac{dx}{dy}+x^2+y\left(2x\frac{dx}{dy}\right)=0 \Rightarrow \frac{dx}{dy}(y^3+2xy)=-3xy^2-x^2$ $\Rightarrow \frac{dx}{dy}=-\frac{3xy^2+x^2}{y^3+2xy}$; thus $\frac{dx}{dy}$ appears to equal $\frac{1}{\frac{dy}{dx}}$. The two different treatments view the graphs as functions symmetric across the line $y=x$, so their slopes are reciprocals of one another at the corresponding points (a,b) and (b,a).

61. $x^4+4y^2=1$:

(a) $y^2=\frac{1-x^4}{4} \Rightarrow y=\pm\frac{1}{2}\sqrt{1-x^4} \Rightarrow \frac{dy}{dx}=\pm\frac{1}{4}(1-x^4)^{-1/2}(-4x^3)=\frac{\pm x^3}{(1-x^4)^{1/2}}$; differentiating implicitly, we find, $4x^3+8y\frac{dy}{dx}=0 \Rightarrow \frac{dy}{dx}=\frac{-4x^3}{8y}=\frac{-4x^3}{8\left(\pm\frac{1}{2}\sqrt{1-x^4}\right)}=\frac{\pm x^3}{(1-x^4)^{1/2}}$.

(b)

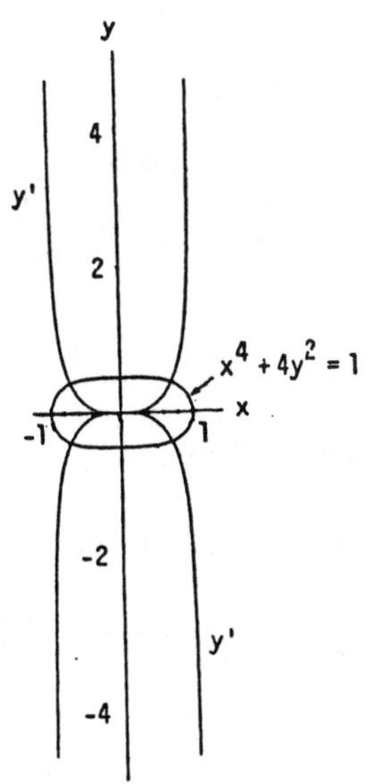

2.7 RELATED RATES

1. $A=\pi r^2 \Rightarrow \frac{dA}{dt}=2\pi r\frac{dr}{dt}$

3. (a) $V=\pi r^2h \Rightarrow \frac{dV}{dt}=\pi r^2\frac{dh}{dt}$ (b) $V=\pi r^2h \Rightarrow \frac{dV}{dt}=2\pi rh\frac{dr}{dt}$

(c) $V = \pi r^2 h \Rightarrow \frac{dV}{dt} = \pi r^2 \frac{dh}{dt} + 2\pi rh \frac{dr}{dt}$

5. (a) $\frac{dV}{dt} = 1$ volt/sec (b) $\frac{dI}{dt} = -\frac{1}{3}$ amp/sec

(c) $\frac{dV}{dt} = R\left(\frac{dI}{dt}\right) + I\left(\frac{dR}{dt}\right) \Rightarrow \frac{dR}{dt} = \frac{1}{I}\left(\frac{dV}{dt} - R\,\frac{dI}{dt}\right) \Rightarrow \frac{dR}{dt} = \frac{1}{I}\left(\frac{dV}{dt} - \frac{V}{I}\,\frac{dI}{dt}\right)$

(d) $\frac{dR}{dt} = \frac{1}{2}\left[1 - \frac{12}{2}\left(-\frac{1}{3}\right)\right] = \left(\frac{1}{2}\right)(3) = \frac{3}{2}$ ohms/sec, R is increasing

7. (a) $s = \sqrt{x^2+y^2} = \left(x^2+y^2\right)^{1/2} \Rightarrow \frac{ds}{dt} = \frac{x}{\sqrt{x^2+y^2}}\,\frac{dx}{dt}$

(b) $s = \sqrt{x^2+y^2} = \left(x^2+y^2\right)^{1/2} \Rightarrow \frac{ds}{dt} = \frac{x}{\sqrt{x^2+y^2}}\,\frac{dx}{dt} + \frac{y}{\sqrt{x^2+y^2}}\,\frac{dy}{dt}$

(c) $s = \sqrt{x^2+y^2} \Rightarrow s^2 = x^2 + y^2 \Rightarrow 2s\,\frac{ds}{dt} = 2x\,\frac{dx}{dt} + 2y\,\frac{dy}{dt} \Rightarrow 2s \cdot 0 = 2x\,\frac{dx}{dt} + 2y\,\frac{dy}{dt} \Rightarrow \frac{dx}{dt} = -\frac{y}{x}\,\frac{dy}{dt}$

9. (a) $A = \frac{1}{2}ab \sin\theta \Rightarrow \frac{dA}{dt} = \frac{1}{2}ab\cos\theta\,\frac{d\theta}{dt}$ (b) $A = \frac{1}{2}ab\sin\theta \Rightarrow \frac{dA}{dt} = \frac{1}{2}ab\cos\theta\,\frac{d\theta}{dt} + \frac{1}{2}b\sin\theta\,\frac{da}{dt}$

(c) $A = \frac{1}{2}ab\sin\theta \Rightarrow \frac{dA}{dt} = \frac{1}{2}ab\cos\theta\,\frac{d\theta}{dt} + \frac{1}{2}b\sin\theta\,\frac{da}{dt} + \frac{1}{2}a\sin\theta\,\frac{db}{dt}$

11. Given $\frac{d\ell}{dt} = -2$ cm/sec, $\frac{dw}{dt} = 2$ cm/sec, $\ell = 12$ cm and $w = 5$ cm.

(a) $A = \ell w \Rightarrow \frac{dA}{dt} = \ell\,\frac{dw}{dt} + w\,\frac{d\ell}{dt} \Rightarrow \frac{dA}{dt} = 12(2) + 5(-2) = 14\ \text{cm}^2/\text{sec}$, increasing

(b) $P = 2\ell + 2w \Rightarrow \frac{dP}{dt} = 2\,\frac{d\ell}{dt} + 2\,\frac{dw}{dt} = 2(-2) + 2(2) = 0$ cm/sec, constant

(c) $D = \sqrt{w^2+\ell^2} = \left(w^2+\ell^2\right)^{1/2} \Rightarrow \frac{dD}{dt} = \frac{1}{2}\left(w^2+\ell^2\right)^{-1/2}\left(2w\,\frac{dw}{dt} + 2\ell\,\frac{d\ell}{dt}\right) \Rightarrow \frac{dD}{dt} = \frac{w\,\frac{dw}{dt} + \ell\,\frac{d\ell}{dt}}{\sqrt{w^2+\ell^2}}$

$= \frac{(5)(2) + (12)(-2)}{\sqrt{25+144}} = -\frac{14}{13}$ cm/sec, decreasing

13. Given: $\frac{dx}{dt} = 5$ ft/sec, the ladder is 13 ft long, and $x = 12$, $y = 5$ at the instant of time

(a) Since $x^2 + y^2 = 169 \Rightarrow \frac{dy}{dt} = -\frac{x}{y}\,\frac{dx}{dt} = -\left(\frac{12}{5}\right)(5) = -12$ ft/sec, the ladder is sliding down the wall

(b) The area of the triangle formed by the ladder and walls is $A = \frac{1}{2}xy \Rightarrow \frac{dA}{dt} = \left(\frac{1}{2}\right)\left(x\,\frac{dy}{dt} + y\,\frac{dx}{dt}\right)$. The area is changing at $\frac{1}{2}[12(-12) + 5(5)] = -\frac{119}{2} = -59.5\ \text{ft}^2/\text{sec}$.

(c) $\cos\theta = \frac{x}{13} \Rightarrow -\sin\theta\,\frac{d\theta}{dt} = \frac{1}{13}\cdot\frac{dx}{dt} \Rightarrow \frac{d\theta}{dt} = -\frac{1}{13\sin\theta}\cdot\frac{dx}{dt} = -\left(\frac{1}{5}\right)(5) = -1$ rad/sec

15. Let s represent the distance between the girl and the kite and x represents the horizontal distance between the girl and kite $\Rightarrow s^2 = (300)^2 + x^2 \Rightarrow \frac{ds}{dt} = \frac{x}{s}\,\frac{dx}{dt} = \frac{400(25)}{500} = 20$ ft/sec.

17. $V = \frac{1}{3}\pi r^2 h$, $h = \frac{3}{8}(2r) = \frac{3r}{4} \Rightarrow r = \frac{4h}{3} \Rightarrow V = \frac{1}{3}\pi\left(\frac{4h}{3}\right)^2 h = \frac{16\pi h^3}{27} \Rightarrow \frac{dV}{dt} = \frac{16\pi h^2}{9}\,\frac{dh}{dt}$

(a) $\left.\frac{dh}{dt}\right|_{h=4} = \left(\frac{9}{16\pi 4^2}\right)(10) = \frac{90}{256\pi} \approx 0.1119$ m/sec $= 11.19$ cm/sec

(b) $r = \frac{4h}{3} \Rightarrow \frac{dr}{dt} = \frac{4}{3}\frac{dh}{dt} = \frac{4}{3}\left(\frac{90}{256\pi}\right) = \frac{15}{32\pi} \approx 0.1492$ m/sec $= 14.92$ cm/sec

19. (a) $V = \frac{\pi}{3}y^2(3R - y) \Rightarrow \frac{dV}{dt} = \frac{\pi}{3}[2y(3R - y) + y^2(-1)]\frac{dy}{dt} \Rightarrow \frac{dy}{dt} = \left[\frac{\pi}{3}(6Ry - 3y^2)\right]^{-1}\frac{dV}{dt} \Rightarrow$ at $R = 13$ and $y = 8$ we have $\frac{dy}{dt} = \frac{1}{144\pi}(-6) = \frac{-1}{24\pi}$ m/min

(b) The hemisphere is on the circle $r^2 + (13 - y)^2 = 169 \Rightarrow r = \sqrt{26y - y^2}$ m

(c) $r = (26y - y^2)^{1/2} \Rightarrow \frac{dr}{dt} = \frac{1}{2}(26y - y^2)^{-1/2}(26 - 2y)\frac{dy}{dt} \Rightarrow \frac{dr}{dt} = \frac{13 - y}{\sqrt{26y - y^2}}\frac{dy}{dt} \Rightarrow \left.\frac{dr}{dt}\right|_{y=8} = \frac{13 - 8}{\sqrt{26 \cdot 8 - 64}}\left(\frac{-1}{24\pi}\right)$ $= \frac{-5}{288\pi}$ m/min

21. If $V = \frac{4}{3}\pi r^3$, $r = 5$, and $\frac{dV}{dt} = 100\pi$ ft³/min, then $\frac{dV}{dt} = 4\pi r^2\frac{dr}{dt} \Rightarrow \frac{dr}{dt} = 1$ ft/min. Then $S = 4\pi r^2 \Rightarrow \frac{dS}{dt}$ $= 8\pi r\frac{dr}{dt} = 8\pi(5)(1) = 40\pi$ ft²/min, the rate at which the area is increasing.

23. Let s represent the distance between the bicycle and balloon, h the height of the balloon and x the horizontal distance between the balloon and the bicycle. The relationship between the variables is $s^2 = h^2 + x^2$ $\Rightarrow \frac{ds}{dt} = \frac{1}{s}\left(h\frac{dh}{dt} + x\frac{dx}{dt}\right) \Rightarrow \frac{ds}{dt} = \frac{1}{85}[68(1) + 51(17)] = 11$ ft/sec.

25. $y = QD^{-1} \Rightarrow \frac{dy}{dt} = D^{-1}\frac{dQ}{dt} - QD^{-2}\frac{dD}{dt} = \frac{1}{41}(0) - \frac{233}{(41)^2}(-2) = \frac{466}{1681}$ L/min $\Rightarrow$ increasing about 0.2772 L/min

27. Let $P(x, y)$ represent a point on the curve $y = x^2$ and θ the angle of inclination of a line containing P and the origin. Consequently, $\tan\theta = \frac{y}{x} \Rightarrow \tan\theta = \frac{x^2}{x} = x \Rightarrow \sec^2\theta\frac{d\theta}{dt} = \frac{dx}{dt} \Rightarrow \frac{d\theta}{dt} = \cos^2\theta\frac{dx}{dt}$. Since $\frac{dx}{dt} = 10$ m/sec and $\left.\cos^2\theta\right|_{x=3} = \frac{x^2}{y^2 + x^2} = \frac{3^2}{9^2 + 3^2} = \frac{1}{10}$, we have $\left.\frac{d\theta}{dt}\right|_{x=3} = 1$ rad/sec.

29. The distance from the origin is $s = \sqrt{x^2 + y^2}$ and we wish to find $\left.\frac{ds}{dt}\right|_{(5,12)}$

$$= \frac{1}{2}(x^2 + y^2)^{-1/2}\left.\left(2x\frac{dx}{dt} + 2y\frac{dy}{dt}\right)\right|_{(5,12)} = \frac{(5)(-1) + (12)(-5)}{\sqrt{25 + 144}} = -5 \text{ m/sec}$$

31. Let $s = 16t^2$ represent the distance the ball has fallen, h the distance between the ball and the ground, and I the distance between the shadow and the point directly beneath the ball. Accordingly, $s + h = 50$ and since the triangle LOQ and triangle PRQ are similar we have $I = \frac{30h}{50 - h} \Rightarrow h = 50 - 16t^2$ and $I = \frac{30(50 - 16t^2)}{50 - (50 - 16t^2)}$ $= \frac{1500}{16t^2} - 30 \Rightarrow \frac{dI}{dt} = -\frac{1500}{8t^3} \Rightarrow \left.\frac{dI}{dt}\right|_{t=\frac{1}{2}} = -1500$ ft/sec.

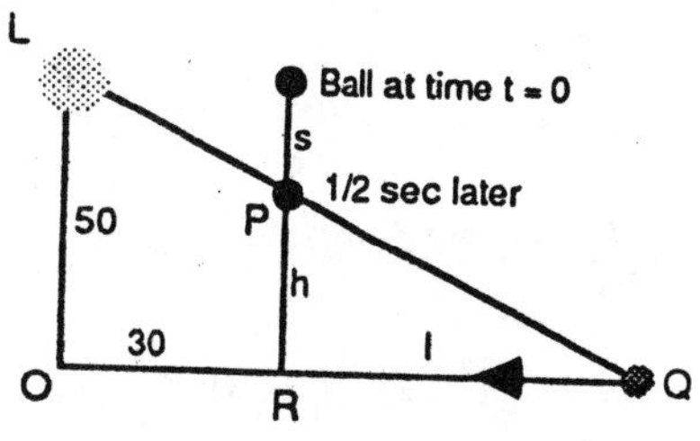

33. The volume of the ice is $V = \frac{4}{3}\pi r^3 - \frac{4}{3}\pi 4^3 \Rightarrow \frac{dV}{dt} = 4\pi r^2 \frac{dr}{dt} \Rightarrow \left.\frac{dr}{dt}\right|_{r=6} = \frac{5}{72\pi}$ in/min when $\frac{dV}{dt} = 10$ in^3/min.

The surface area is $S = 4\pi r^2 \Rightarrow \frac{dS}{dt} = 8\pi r \frac{dr}{dt} \Rightarrow \left.\frac{dS}{dt}\right|_{r=6} = 48\pi\left(\frac{5}{72\pi}\right) = \frac{10}{3}$ in^2/min.

35. When x represents the length of the shadow, then $\tan\theta = \frac{80}{x} \Rightarrow \sec^2\theta \frac{d\theta}{dt} = -\frac{80}{x^2}\frac{dx}{dt} \Rightarrow \frac{dx}{dt} = \frac{-x^2\sec^2\theta}{80}\frac{d\theta}{dt}$.

We are given that $\frac{d\theta}{dt} = 0.27° = \frac{3\pi}{2000}$ rad/min. At $x = 60$, $\cos\theta = \frac{3}{5} \Rightarrow$

$\left|\frac{dx}{dt}\right| = \left|\frac{-x^2\sec^2\theta}{80}\frac{d\theta}{dt}\right|\Bigg|_{\left(\frac{d\theta}{dt} = \frac{3\pi}{2000} \text{ and } \sec\theta = \frac{5}{3}\right)} = \frac{3\pi}{16}$ ft/min ≈ 0.589 ft/min ≈ 7.1 in/min.

37. Let x represent distance of the player from second base and s the distance to third base. Then $\frac{dx}{dt} = -16$ ft/sec

(a) $s^2 = x^2 + 8100 \Rightarrow 2s\frac{ds}{dt} = 2x\frac{dx}{dt} \Rightarrow \frac{ds}{dt} = \frac{x}{s}\frac{dx}{dt}$. When the player is 30 ft from first base, $x = 60$

$\Rightarrow s = 30\sqrt{13}$ and $\frac{ds}{dt} = \frac{60}{30\sqrt{13}}(-16) = \frac{-32}{\sqrt{13}} \approx -8.875$ ft/sec

(b) $\cos\theta_1 = \frac{90}{s} \Rightarrow -\sin\theta_1 \frac{d\theta_1}{dt} = -\frac{90}{s^2}\cdot\frac{ds}{dt} \Rightarrow \frac{d\theta_1}{dt} = \frac{90}{s^2\sin\theta_1}\cdot\frac{ds}{dt} = \frac{90}{sx}\cdot\frac{ds}{dt}$. Therefore, $x = 60$ and $s = 30\sqrt{13}$

$\Rightarrow \frac{d\theta_1}{dt} = \frac{90}{(30\sqrt{13})(60)}\cdot\left(\frac{-32}{\sqrt{13}}\right) = \frac{-8}{65}$ rad/sec; $\sin\theta_2 = \frac{90}{s} \Rightarrow \cos\theta_2\frac{d\theta_2}{dt} = -\frac{90}{s^2}\cdot\frac{ds}{dt} \Rightarrow \frac{d\theta_2}{dt} = \frac{-90}{s^2\cos\theta_2}\cdot\frac{ds}{dt}$

$= \frac{-90}{sx}\cdot\frac{ds}{dt}$. Therefore, $x = 60$ and $s = 30\sqrt{13} \Rightarrow \frac{d\theta_2}{dt} = \frac{8}{65}$ rad/sec.

(c) $\frac{d\theta_1}{dt} = \frac{90}{s^2\sin\theta_1}\cdot\frac{ds}{dt} = \frac{90}{\left(s^2\cdot\frac{x}{s}\right)}\cdot\left(\frac{x}{s}\right)\cdot\left(\frac{dx}{dt}\right) = \left(\frac{90}{s^2}\right)\left(\frac{dx}{dt}\right) = \left(\frac{90}{x^2+8100}\right)\frac{dx}{dt} \Rightarrow \lim_{x\to0}\frac{d\theta_1}{dt}$

$= \lim_{x\to0}\left(\frac{90}{x^2+8100}\right)(-15) = -\frac{1}{6}$ rad/sec; $\frac{d\theta_2}{dt} = \frac{-90}{s^2\cos\theta_2}\cdot\frac{ds}{dt} = \left(\frac{-90}{s^2\cdot\frac{x}{s}}\right)\left(\frac{x}{s}\right)\left(\frac{dx}{dt}\right) = \left(\frac{-90}{s^2}\right)\left(\frac{dx}{dt}\right)$

$= \left(\frac{-90}{x^2+8100}\right)\frac{dx}{dt} \Rightarrow \lim_{x\to0}\frac{d\theta_2}{dt} = \frac{1}{6}$ rad/sec

CHAPTER 2 PRACTICE EXERCISES

1. $y = x^5 - 0.125x^2 + 0.25x \Rightarrow \frac{dy}{dx} = 5x^4 - 0.25x + 0.25$

3. $y = x^3 - 3(x^2 + \pi^2) \Rightarrow \frac{dy}{dx} = 3x^2 - 3(2x + 0) = 3x^2 - 6x = 3x(x-2)$

5. $y = (x+1)^2(x^2+2x) \Rightarrow \frac{dy}{dx} = (x+1)^2(2x+2) + (x^2+2x)(2(x+1)) = 2(x+1)\left[(x+1)^2 + x(x+2)\right]$

$= 2(x+1)(2x^2+4x+1)$

7. $y = (\theta^2 + \sec\theta + 1)^3 \Rightarrow \frac{dy}{d\theta} = 3(\theta^2 + \sec\theta + 1)^2(2\theta + \sec\theta\tan\theta)$

9. $s = \frac{\sqrt{t}}{1+\sqrt{t}} \Rightarrow \frac{ds}{dt} = \frac{(1+\sqrt{t})\cdot\frac{1}{2\sqrt{t}} - \sqrt{t}\left(\frac{1}{2\sqrt{t}}\right)}{(1+\sqrt{t})^2} = \frac{(1+\sqrt{t})-\sqrt{t}}{2\sqrt{t}(1+\sqrt{t})^2} = \frac{1}{2\sqrt{t}(1+\sqrt{t})^2}$

11. $y = 2\tan^2 x - \sec^2 x \Rightarrow \frac{dy}{dx} = (4\tan x)(\sec^2 x) - (2\sec x)(\sec x \tan x) = 2\sec^2 x \tan x$

13. $s = \cos^4(1-2t) \Rightarrow \frac{ds}{dt} = 4\cos^3(1-2t)(-\sin(1-2t))(-2) = 8\cos^3(1-2t)\sin(1-2t)$

15. $s = (\sec t + \tan t)^5 \Rightarrow \frac{ds}{dt} = 5(\sec t + \tan t)^4(\sec t \tan t + \sec^2 t) = 5(\sec t)(\sec t + \tan t)^5$

17. $r = \sqrt{2\theta \sin\theta} = (2\theta\sin\theta)^{1/2} \Rightarrow \frac{dr}{d\theta} = \frac{1}{2}(2\theta\sin\theta)^{-1/2}(2\theta\cos\theta + 2\sin\theta) = \frac{\theta\cos\theta + \sin\theta}{\sqrt{2\theta\sin\theta}}$

19. $r = \sin\sqrt{2\theta} = \sin(2\theta)^{1/2} \Rightarrow \frac{dr}{d\theta} = \cos(2\theta)^{1/2}\left(\frac{1}{2}(2\theta)^{-1/2}(2)\right) = \frac{\cos\sqrt{2\theta}}{\sqrt{2\theta}}$

21. $y = \frac{1}{2}x^2\csc\frac{2}{x} \Rightarrow \frac{dy}{dx} = \frac{1}{2}x^2\left(-\csc\frac{2}{x}\cot\frac{2}{x}\right)\left(\frac{-2}{x^2}\right) + \left(\csc\frac{2}{x}\right)\left(\frac{1}{2}\cdot 2x\right) = \csc\frac{2}{x}\cot\frac{2}{x} + x\csc\frac{2}{x}$

23. $y = x^{-1/2}\sec(2x)^2 \Rightarrow \frac{dy}{dx} = x^{-1/2}\sec(2x)^2\tan(2x)^2(2(2x)\cdot 2) + \sec(2x)^2\left(-\frac{1}{2}x^{-3/2}\right)$

$= 8x^{1/2}\sec(2x)^2\tan(2x)^2 - \frac{1}{2}x^{-3/2}\sec(2x)^2 = \frac{1}{2}x^{1/2}\sec(2x)^2\left[16\tan(2x)^2 - x^{-2}\right]$

25. $y = 5\cot x^2 \Rightarrow \frac{dy}{dx} = 5(-\csc^2 x^2)(2x) = -10x\csc^2(x^2)$

27. $y = x^2\sin^2(2x^2) \Rightarrow \frac{dy}{dx} = x^2(2\sin(2x^2))(\cos(2x^2))(4x) + \sin^2(2x^2)(2x) = 8x^3\sin(2x^2)\cos(2x^2) + 2x\sin^2(2x^2)$

29. $s = \left(\frac{4t}{t+1}\right)^{-2} \Rightarrow \frac{ds}{dt} = -2\left(\frac{4t}{t+1}\right)^{-3}\left(\frac{(t+1)(4)-(4t)(1)}{(t+1)^2}\right) = -2\left(\frac{4t}{t+1}\right)^{-3}\frac{4}{(t+1)^2} = -\frac{(t+1)}{8t^3}$

31. $y = \left(\frac{\sqrt{x}}{x+1}\right)^2 \Rightarrow \frac{dy}{dx} = 2\left(\frac{\sqrt{x}}{x+1}\right)\cdot\frac{(x+1)\left(\frac{1}{2\sqrt{x}}\right) - (\sqrt{x})(1)}{(x+1)^2} = \frac{(x+1)-2x}{(x+1)^3} = \frac{1-x}{(x+1)^3}$

33. $y = \sqrt{\frac{x^2+x}{x^2}} = \left(1+\frac{1}{x}\right)^{1/2} \Rightarrow \frac{dy}{dx} = \frac{1}{2}\left(1+\frac{1}{x}\right)^{-1/2}\left(-\frac{1}{x^2}\right) = -\frac{1}{2x^2\sqrt{1+\frac{1}{x}}}$

35. $r = \left(\frac{\sin\theta}{\cos\theta - 1}\right)^2 \Rightarrow \frac{dr}{d\theta} = 2\left(\frac{\sin\theta}{\cos\theta-1}\right)\left[\frac{(\cos\theta-1)(\cos\theta)-(\sin\theta)(-\sin\theta)}{(\cos\theta-1)^2}\right]$

$= 2\left(\frac{\sin\theta}{\cos\theta-1}\right)\left(\frac{\cos^2\theta - \cos\theta + \sin^2\theta}{(\cos\theta-1)^2}\right) = \frac{(2\sin\theta)(1-\cos\theta)}{(\cos\theta-1)^3} = \frac{-2\sin\theta}{(\cos\theta-1)^2}$

37. $y = (2x+1)\sqrt{2x+1} = (2x+1)^{3/2} \Rightarrow \frac{dy}{dx} = \frac{3}{2}(2x+1)^{1/2}(2) = 3\sqrt{2x+1}$

39. $y = 3\left(5x^2 + \sin 2x\right)^{-3/2} \Rightarrow \frac{dy}{dx} = 3\left(-\frac{3}{2}\right)\left(5x^2 + \sin 2x\right)^{-5/2}[10x + (\cos 2x)(2)] = \frac{-9(5x + \cos 2x)}{\left(5x^2 + \sin 2x\right)^{5/2}}$

41. $xy + 2x + 3y = 1 \Rightarrow (xy' + y) + 2 + 3y' = 0 \Rightarrow xy' + 3y' = -2 - y \Rightarrow y'(x+3) = -2 - y \Rightarrow y' = -\frac{y+2}{x+3}$

43. $x^3 + 4xy - 3y^{4/3} = 2x \Rightarrow 3x^2 + \left(4x\frac{dy}{dx} + 4y\right) - 4y^{1/3}\frac{dy}{dx} = 2 \Rightarrow 4x\frac{dy}{dx} - 4y^{1/3}\frac{dy}{dx} = 2 - 3x^2 - 4y$

$\Rightarrow \frac{dy}{dx}\left(4x - 4y^{1/3}\right) = 2 - 3x^2 - 4y \Rightarrow \frac{dy}{dx} = \frac{2 - 3x^2 - 4y}{4x - 4y^{1/3}}$

45. $(xy)^{1/2} = 1 \Rightarrow \frac{1}{2}(xy)^{-1/2}\left(x\frac{dy}{dx} + y\right) = 0 \Rightarrow x^{1/2}y^{-1/2}\frac{dy}{dx} = -x^{-1/2}y^{1/2} \Rightarrow \frac{dy}{dx} = -x^{-1}y \Rightarrow \frac{dy}{dx} = -\frac{y}{x}$

47. $y^2 = \frac{x}{x+1} \Rightarrow 2y\frac{dy}{dx} = \frac{(x+1)(1) - (x)(1)}{(x+1)^2} \Rightarrow \frac{dy}{dx} = \frac{1}{2y(x+1)^2}$

49. $p^3 + 4pq - 3q^2 = 2 \Rightarrow 3p^2\frac{dp}{dq} + 4\left(p + q\frac{dp}{dq}\right) - 6q = 0 \Rightarrow 3p^2\frac{dp}{dq} + 4q\frac{dp}{dq} = 6q - 4p \Rightarrow \frac{dp}{dq}\left(3p^2 + 4q\right) = 6q - 4p$

$\Rightarrow \frac{dp}{dq} = \frac{6q - 4p}{3p^2 + 4q}$

51. $r\cos 2s + \sin^2 s = \pi \Rightarrow r(-\sin 2s)(2) + (\cos 2s)\left(\frac{dr}{ds}\right) + 2\sin s\cos s = 0 \Rightarrow \frac{dr}{ds}(\cos 2s) = 2r\sin 2s - 2\sin s\cos s$

$\Rightarrow \frac{dr}{ds} = \frac{2r\sin 2s - \sin 2s}{\cos 2s} = \frac{(2r-1)(\sin 2s)}{\cos 2s} = (2r-1)(\tan 2s)$

53. (a) $x^3 + y^3 = 1 \Rightarrow 3x^2 + 3y^2\frac{dy}{dx} = 0 \Rightarrow \frac{dy}{dx} = -\frac{x^2}{y^2} \Rightarrow \frac{d^2y}{dx^2} = \frac{y^2(-2x) - \left(-x^2\right)\left(2y\frac{dy}{dx}\right)}{y^4}$

$\Rightarrow \frac{d^2y}{dx^2} = \frac{-2xy^2 + \left(2yx^2\right)\left(-\frac{x^2}{y^2}\right)}{y^4} = \frac{-2xy^2 - \frac{2x^4}{y}}{y^4} = \frac{-2xy^3 - 2x^4}{y^5}$

(b) $y^2 = 1 - \frac{2}{x} \Rightarrow 2y\frac{dy}{dx} = \frac{2}{x^2} \Rightarrow \frac{dy}{dx} = \frac{1}{yx^2} \Rightarrow \frac{dy}{dx} = \left(yx^2\right)^{-1} \Rightarrow \frac{d^2y}{dx^2} = -\left(yx^2\right)^{-2}\left[y(2x) + x^2\frac{dy}{dx}\right]$

$\Rightarrow \frac{d^2y}{dx^2} = \frac{-2xy - x^2\left(\frac{1}{yx^2}\right)}{y^2x^4} = \frac{-2xy^2 - 1}{y^3x^4}$

55. (a) Let $h(x) = 6f(x) - g(x) \Rightarrow h'(x) = 6f'(x) - g'(x) \Rightarrow h'(1) = 6f'(1) - g'(1) = 6\left(\frac{1}{2}\right) - (-4) = 7$

(b) Let $h(x) = f(x)g^2(x) \Rightarrow h'(x) = f(x)\left(2g(x)\right)g'(x) + g^2(x)f'(x) \Rightarrow h'(0) = 2f(0)g(0)g'(0) + g^2(0)f'(0)$

$= 2(1)(1)\left(\frac{1}{2}\right) + (1)^2(-3) = -2$

(c) Let $h(x) = \frac{f(x)}{g(x)+1} \Rightarrow h'(x) = \frac{(g(x)+1)f'(x) - f(x)g'(x)}{(g(x)+1)^2} \Rightarrow h'(1) = \frac{(g(1)+1)f'(1) - f(1)g'(1)}{(g(1)+1)^2}$

$= \frac{(5+1)\left(\frac{1}{2}\right) - 3(-4)}{(5+1)^2} = \frac{5}{12}$

(d) Let $h(x) = f(g(x)) \Rightarrow h'(x) = f'(g(x))g'(x) \Rightarrow h'(0) = f'(g(0))g'(0) = f'(1)\left(\frac{1}{2}\right) = \left(\frac{1}{2}\right)\left(\frac{1}{2}\right) = \frac{1}{4}$

(e) Let $h(x) = g(f(x)) \Rightarrow h'(x) = g'(f(x))f'(x) \Rightarrow h'(0) = g'(f(0))f'(0) = h'(0) = g'(1)f'(0) = (-4)(-3) = 12$

(f) Let $h(x) = (x + f(x))^{3/2} \Rightarrow h'(x) = \frac{3}{2}(x + f(x))^{1/2}(1 + f'(x)) \Rightarrow h'(1) = \frac{3}{2}(1 + f(1))^{1/2}(1 + f'(1))$

$= \frac{3}{2}(1+3)^{1/2}\left(1 + \frac{1}{2}\right) = \frac{9}{2}$

(g) Let $h(x) = f(x + g(x)) \Rightarrow h'(x) = f'(x + g(x))(1 + g'(x)) \Rightarrow h'(0) = f'(g(0))(1 + g'(0))$

$= f'(1)\left(1 + \frac{1}{2}\right) = \left(\frac{1}{2}\right)\left(\frac{3}{2}\right) = \frac{3}{4}$

57. $x = t^2 + \pi \Rightarrow \frac{dx}{dt} = 2t$; $y = 3 \sin 2x \Rightarrow \frac{dy}{dx} = 3(\cos 2x)(2) = 6 \cos 2x = 6 \cos(2t^2 + 2\pi) = 6 \cos(2t^2)$; thus,

$\frac{dy}{dt} = \frac{dy}{dx} \cdot \frac{dx}{dt} = 6 \cos(2t^2) \cdot 2t \Rightarrow \left.\frac{dy}{dt}\right|_{t=0} = 6 \cos(0) \cdot 0 = 0$

59. $r = 8 \sin\left(s + \frac{\pi}{6}\right) \Rightarrow \frac{dr}{ds} = 8 \cos\left(s + \frac{\pi}{6}\right)$; $w = \sin(\sqrt{r} - 2) \Rightarrow \frac{dw}{dr} = \cos(\sqrt{r} - 2)\left(\frac{1}{2\sqrt{r}}\right)$

$= \frac{\cos \sqrt{8 \sin\left(s + \frac{\pi}{6}\right)} - 2}{2\sqrt{8 \sin\left(s + \frac{\pi}{6}\right)}}$; thus, $\frac{dw}{ds} = \frac{dw}{dr} \cdot \frac{dr}{ds} = \frac{\cos\left(\sqrt{8 \sin\left(s + \frac{\pi}{6}\right)} - 2\right)}{2\sqrt{8 \sin\left(s + \frac{\pi}{6}\right)}} \cdot \left[8 \cos\left(s + \frac{\pi}{6}\right)\right]$

$\Rightarrow \left.\frac{dw}{ds}\right|_{s=0} = \frac{\cos\left(\sqrt{8 \sin\left(\frac{\pi}{6}\right)} - 2\right) \cdot 8 \cos\left(\frac{\pi}{6}\right)}{2\sqrt{8 \sin\left(\frac{\pi}{6}\right)}} = \frac{(\cos 0)(8)\left(\frac{\sqrt{3}}{2}\right)}{2\sqrt{4}} = \sqrt{3}$

61. $y^3 + y = 2 \cos x \Rightarrow 3y^2 \frac{dy}{dx} + \frac{dy}{dx} = -2 \sin x \Rightarrow \frac{dy}{dx}(3y^2 + 1) = -2 \sin x \Rightarrow \frac{dy}{dx} = \frac{-2 \sin x}{3y^2 + 1} \Rightarrow \left.\frac{dy}{dx}\right|_{(0,1)}$

$= \frac{-2 \sin(0)}{3+1} = 0$; $\frac{d^2y}{dx^2} = \frac{(3y^2+1)(-2 \cos x) - (-2 \sin x)\left(6y \frac{dy}{dx}\right)}{(3y^2+1)^2}$

$\Rightarrow \left.\frac{d^2y}{dx^2}\right|_{(0,1)} = \frac{(3+1)(-2 \cos 0) - (-2 \sin 0)(6 \cdot 0)}{(3+1)^2} = -\frac{1}{2}$

63. $f(t) = \frac{1}{2t+1}$ and $f(t+h) = \frac{1}{2(t+h)+1} \Rightarrow \frac{f(t+h) - f(t)}{h} = \frac{\frac{1}{2(t+h)+1} - \frac{1}{2t+1}}{h} = \frac{2t+1-(2t+2h+1)}{(2t+2h+1)(2t+1)h}$

$= \frac{-2h}{(2t+2h+1)(2t+1)h} = \frac{-2}{(2t+2h+1)(2t+1)} \Rightarrow f'(t) = \lim_{h \to 0} \frac{f(t+h) - f(t)}{h} = \lim_{h \to 0} \frac{-2}{(2t+2h+1)(2t+1)}$

$= \frac{-2}{(2t+1)^2}$

65. (a)

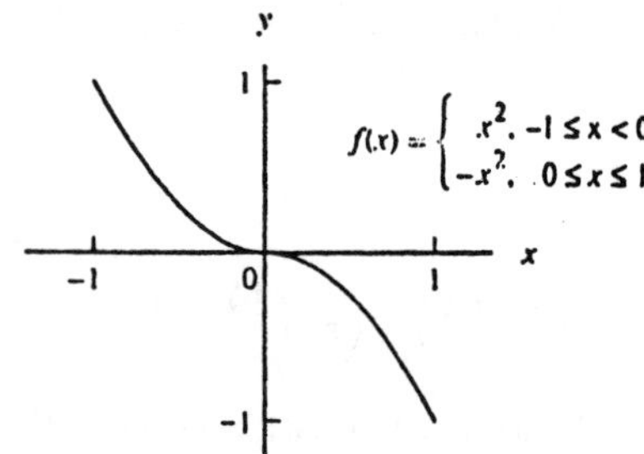

(b) $\lim_{x\to0^-} f(x) = \lim_{x\to0^-} x^2 = 0$ and $\lim_{x\to0^+} f(x) = \lim_{x\to0^+} -x^2 = 0 \Rightarrow \lim_{x\to0} f(x) = 0$. Since $\lim_{x\to0} f(x) = 0 = f(0)$ it follows that f is continuous at $x = 0$.

(c) $\lim_{x\to0^-} f'(x) = \lim_{x\to0^-} (2x) = 0$ and $\lim_{x\to0^+} f'(x) = \lim_{x\to0^+} (-2x) = 0 \Rightarrow \lim_{x\to0} f'(x) = 0$. Since this limit exists, it follows that f is differentiable at $x = 0$.

67. (a)

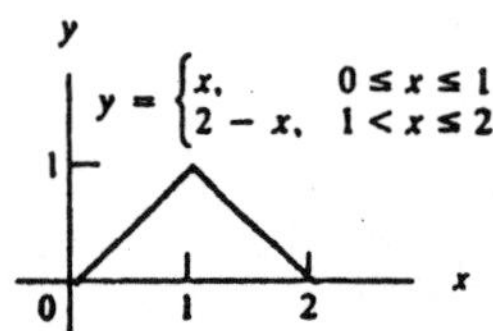

(b) $\lim_{x\to1^-} f(x) = \lim_{x\to1^-} x = 1$ and $\lim_{x\to1^+} f(x) = \lim_{x\to1^+} (2-x) = 1 \Rightarrow \lim_{x\to1} f(x) = 1$. Since $\lim_{x\to1} f(x) = 1 = f(1)$, it follows that f is continuous at $x = 1$.

(c) $\lim_{x\to1^-} f'(x) = \lim_{x\to1^-} 1 = 1$ and $\lim_{x\to1^+} f'(x) = \lim_{x\to1^+} -1 = -1 \Rightarrow \lim_{x\to1^-} f'(x) \neq \lim_{x\to1^+} f'(x)$, so $\lim_{x\to1} f'(x)$ does not exist $\Rightarrow$ f is not differentiable at $x = 1$.

69. $y = \frac{x}{2} + \frac{1}{2x-4} = \frac{1}{2}x + (2x-4)^{-1} \Rightarrow \frac{dy}{dx} = \frac{1}{2} - 2(2x-4)^{-2}$; the slope of the tangent is $-\frac{3}{2} \Rightarrow -\frac{3}{2}$ $= \frac{1}{2} - 2(2x-4)^{-2} \Rightarrow -2 = -2(2x-4)^{-2} \Rightarrow 1 = \frac{1}{(2x-4)^2} \Rightarrow (2x-4)^2 = 1 \Rightarrow 4x^2 - 16x + 16 = 1$ $\Rightarrow 4x^2 - 16x + 15 = 0 \Rightarrow (2x-5)(2x-3) = 0 \Rightarrow x = \frac{5}{2}$ or $x = \frac{3}{2} \Rightarrow \left(\frac{5}{2}, \frac{9}{4}\right)$ and $\left(\frac{3}{2}, -\frac{1}{4}\right)$ are points on the curve where the slope is $-\frac{3}{2}$.

71. $y = 2x^3 - 3x^2 - 12x + 20 \Rightarrow \frac{dy}{dx} = 6x^2 - 6x - 12$; the tangent is parallel to the x-axis when $\frac{dy}{dx} = 0$ $\Rightarrow 6x^2 - 6x - 12 = 0 \Rightarrow x^2 - x - 2 = 0 \Rightarrow (x-2)(x+1) = 0 \Rightarrow x = 2$ or $x = -1 \Rightarrow (2, 0)$ and $(-1, 27)$ are points on the curve where the tangent is parallel to the x-axis.

73. $y = 2x^3 - 3x^2 - 12x + 20 \Rightarrow \frac{dy}{dx} = 6x^2 - 6x - 12$

(a) The tangent is perpendicular to the line $y = 1 - \frac{x}{24}$ when $\frac{dy}{dx} = -\left(\frac{1}{-\left(\frac{1}{24}\right)}\right) = 24$; $6x^2 - 6x - 12 = 24$

$\Rightarrow x^2 - x - 2 = 4 \Rightarrow x^2 - x - 6 = 0 \Rightarrow (x-3)(x+2) = 0 \Rightarrow x = -2$ or $x = 3 \Rightarrow (-2, 16)$ and $(3, 11)$ are points where the tangent is perpendicular to $y = 1 - \frac{x}{24}$.

(b) The tangent is parallel to the line $y = \sqrt{2} - 12x$ when $\frac{dy}{dx} = -12 \Rightarrow 6x^2 - 6x - 12 = -12 \Rightarrow x^2 - x = 0$ $\Rightarrow x(x-1) = 0 \Rightarrow x = 0$ or $x = 1 \Rightarrow (0, 20)$ and $(1, 7)$ are points where the tangent is parallel to $y = \sqrt{2} - 12x$.

75. $y = \tan x,\ -\frac{\pi}{2} < x < \frac{\pi}{2} \Rightarrow \frac{dy}{dx} = \sec^2 x$; now the slope of $y = -\frac{x}{2}$ is $-\frac{1}{2} \Rightarrow$ the normal line is parallel to $y = -\frac{x}{2}$ when $\frac{dy}{dx} = 2$. Thus, $\sec^2 x = 2 \Rightarrow \frac{1}{\cos^2 x} = 2$ $\Rightarrow \cos^2 x = \frac{1}{2} \Rightarrow \cos x = \frac{\pm 1}{\sqrt{2}} \Rightarrow x = -\frac{\pi}{4}$ and $x = \frac{\pi}{4}$ for $-\frac{\pi}{2} < x < \frac{\pi}{2} \Rightarrow \left(-\frac{\pi}{4}, -1\right)$ and $\left(\frac{\pi}{4}, 1\right)$ are points where the normal is parallel to $y = -\frac{x}{2}$.

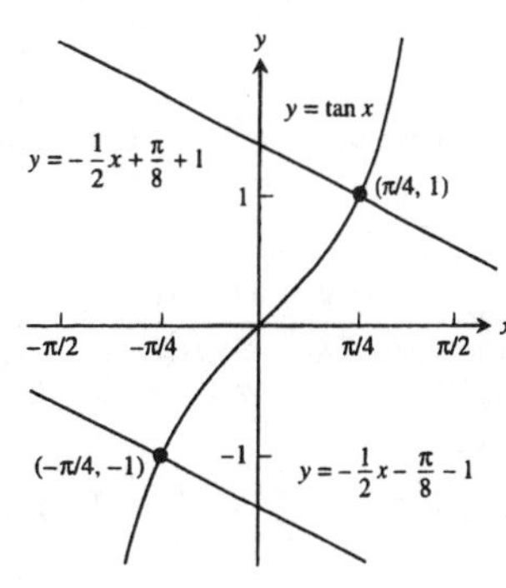

77. $y = x^2 + C \Rightarrow \frac{dy}{dx} = 2x$ and $y = x \Rightarrow \frac{dy}{dx} = 1$; the parabola is tangent to $y = x$ when $2x = 1 \Rightarrow x = \frac{1}{2} \Rightarrow y = \frac{1}{2}$; thus, $\frac{1}{2} = \left(\frac{1}{2}\right)^2 + C \Rightarrow C = \frac{1}{4}$

79. The line through $(0,3)$ and $(5,-2)$ has slope $m = \frac{3-(-2)}{0-5} = -1 \Rightarrow$ the line through $(0,3)$ and $(5,-2)$ is $y = -x + 3$; $y = \frac{c}{x+1} \Rightarrow \frac{dy}{dx} = \frac{-c}{(x+1)^2}$, so the curve is tangent to $y = -x + 3 \Rightarrow \frac{dy}{dx} = -1 = \frac{-c}{(x+1)^2}$ $\Rightarrow (x+1)^2 = c,\ x \neq -1$. Moreover, $y = \frac{c}{x+1}$ intersects $y = -x + 3 \Rightarrow \frac{c}{x+1} = -x + 3,\ x \neq -1$ $\Rightarrow c = (x+1)(-x+3),\ x \neq -1$. Thus $c = c \Rightarrow (x+1)^2 = (x+1)(-x+3) \Rightarrow (x+1)[x+1-(-x+3)]$ $= 0,\ x \neq -1 \Rightarrow (x+1)(2x-2) = 0 \Rightarrow x = 1$ (since $x \neq -1$) $\Rightarrow c = 4$.

81. $x^2 + 2y^2 = 9 \Rightarrow 2x + 4y\frac{dy}{dx} = 0 \Rightarrow \frac{dy}{dx} = -\frac{x}{2y} \Rightarrow \left.\frac{dy}{dx}\right|_{(1,2)} = -\frac{1}{4} \Rightarrow$ the tangent line is $y = 2 - \frac{1}{4}(x-1)$ $= -\frac{1}{4}x + \frac{9}{4}$ and the normal line is $y = 2 + 4(x-1) = 4x - 2$.

83. $xy + 2x - 5y = 2 \Rightarrow \left(x\frac{dy}{dx} + y\right) + 2 - 5\frac{dy}{dx} = 0 \Rightarrow \frac{dy}{dx}(x-5) = -y - 2 \Rightarrow \frac{dy}{dx} = \frac{-y-2}{x-5} \Rightarrow \left.\frac{dy}{dx}\right|_{(3,2)} = 2$ $\Rightarrow$ the tangent line is $y = 2 + 2(x-3) = 2x - 4$ and the normal line is $y = 2 + \frac{-1}{2}(x-3) = -\frac{1}{2}x + \frac{7}{2}$.

85. $x+\sqrt{xy}=6 \Rightarrow 1+\frac{1}{2\sqrt{xy}}\left(x\,\frac{dy}{dx}+y\right)=0 \Rightarrow x\,\frac{dy}{dx}+y=-2\sqrt{xy} \Rightarrow \frac{dy}{dx}=\frac{-2\sqrt{xy}-y}{x} \Rightarrow \left.\frac{dy}{dx}\right|_{(4,1)}=\frac{-5}{4}$
$\Rightarrow$ the tangent line is $y=1-\frac{5}{4}(x-4)=-\frac{5}{4}x+6$ and the normal line is $y=1+\frac{4}{5}(x-4)=\frac{4}{5}x-\frac{11}{5}$.

87. $x^3y^3+y^2=x+y \Rightarrow \left[x^3\left(3y^2\,\frac{dy}{dx}\right)+y^3(3x^2)\right]+2y\,\frac{dy}{dx}=1+\frac{dy}{dx} \Rightarrow 3x^3y^2\,\frac{dy}{dx}+2y\,\frac{dy}{dx}-\frac{dy}{dx}=1-3x^2y^3$
$\Rightarrow \frac{dy}{dx}(3x^3y^2+2y-1)=1-3x^2y^3 \Rightarrow \frac{dy}{dx}=\frac{1-3x^2y^3}{3x^3y^2+2y-1} \Rightarrow \left.\frac{dy}{dx}\right|_{(1,1)}=-\frac{2}{4}$, but $\left.\frac{dy}{dx}\right|_{(1,-1)}$ is undefined.
Therefore, the curve has slope $-\frac{1}{2}$ at $(1,1)$ but the slope is undefined at $(1,-1)$.

89. $x=\frac{1}{2}\tan t,\ y=\frac{1}{2}\sec t \Rightarrow \frac{dy}{dx}=\frac{dy/dt}{dx/dt}=\frac{\frac{1}{2}\sec t\tan t}{\frac{1}{2}\sec^2 t}=\frac{\tan t}{\sec t}=\sin t \Rightarrow \left.\frac{dy}{dx}\right|_{t=\pi/3}=\sin\frac{\pi}{3}=\frac{\sqrt{3}}{2};\ t=\frac{\pi}{3}$
$\Rightarrow x=\frac{1}{2}\tan\frac{\pi}{3}=\frac{\sqrt{3}}{2}$ and $y=\frac{1}{2}\sec\frac{\pi}{3}=1 \Rightarrow y=\frac{\sqrt{3}}{2}x+\frac{1}{4};\ \frac{d^2y}{dx^2}=\frac{dy'/dt}{dx/dt}=\frac{\cos t}{\frac{1}{2}\sec^2 t}=2\cos^3 t \Rightarrow \left.\frac{d^2y}{dx^2}\right|_{t=\pi/3}$
$=2\cos^3\left(\frac{\pi}{3}\right)=\frac{1}{4}$

91. B = graph of f, A = graph of f′. Curve B cannot be the derivative of A because A has only negative slopes while some of B's values are positive.

93.

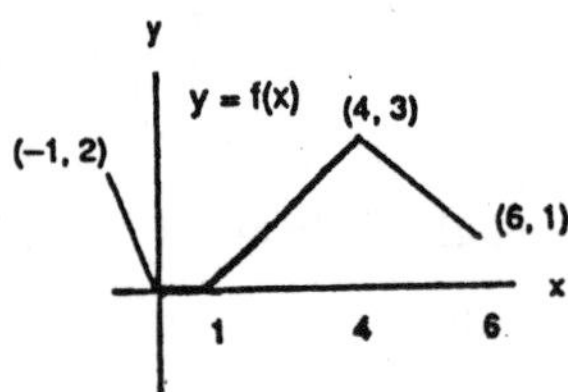

95. (a) 0, 0 (b) largest 1700, smallest about 1400

97. (a) $S=2\pi r^2+2\pi rh$ and h constant $\Rightarrow \frac{dS}{dt}=4\pi r\frac{dr}{dt}+2\pi h\,\frac{dr}{dt}=(4\pi r+2\pi h)\,\frac{dr}{dt}$
(b) $S=2\pi r^2+2\pi rh$ and r constant $\Rightarrow \frac{dS}{dt}=2\pi r\,\frac{dh}{dt}$
(c) $S=2\pi r^2+2\pi rh \Rightarrow \frac{dS}{dt}=4\pi r\,\frac{dr}{dt}+2\pi\left(r\,\frac{dh}{dt}+h\,\frac{dr}{dt}\right)=(4\pi r+2\pi h)\,\frac{dr}{dt}+2\pi r\,\frac{dh}{dt}$
(d S constant $\Rightarrow \frac{dS}{dt}=0 \Rightarrow 0=(4\pi r+2\pi h)\,\frac{dr}{dt}+2\pi r\,\frac{dh}{dt} \Rightarrow (2r+h)\,\frac{dr}{dt}=-r\,\frac{dh}{dt} \Rightarrow \frac{dr}{dt}=\frac{-r}{2r+h}\,\frac{dh}{dt}$

99. $A=\pi r^2 \Rightarrow \frac{dA}{dt}=2\pi r\,\frac{dr}{dt}$; so $r=10$ and $\frac{dr}{dt}=-\frac{2}{\pi}$ m/sec $\Rightarrow \frac{dA}{dt}=(2\pi)(10)\left(-\frac{2}{\pi}\right)=-40$ m^2/sec

101. $\frac{dR_1}{dt}=-1$ ohm/sec, $\frac{dR_2}{dt}=0.5$ ohm/sec; and $\frac{1}{R}=\frac{1}{R_1}+\frac{1}{R_2} \Rightarrow \frac{-1}{R^2}\,\frac{dR}{dt}=\frac{-1}{R_1^2}\,\frac{dR_1}{dt}-\frac{1}{R_2^2}\,\frac{dR_2}{dt}$. Also, $R_1=75$ ohms and $R_2=50$ ohms $\Rightarrow \frac{1}{R}=\frac{1}{75}+\frac{1}{50} \Rightarrow R=30$ ohms. Therefore, from the derivative equation,

$\frac{-1}{(30)^2}\frac{dR}{dt} = \frac{-1}{(75)^2}(-1) - \frac{1}{(50)^2}(0.5) = \left(\frac{1}{5625} - \frac{1}{5000}\right) \Rightarrow \frac{dR}{dt} = (-900)\left(\frac{5000-5625}{5625 \cdot 5000}\right) = \frac{9(625)}{50(5625)} = \frac{1}{50}$
$= 0.02$ ohm/sec.

103. Given $\frac{dx}{dt} = 10$ m/sec and $\frac{dy}{dt} = 5$ m/sec, let D be the distance from the origin $\Rightarrow D^2 = x^2 + y^2 \Rightarrow 2D\frac{dD}{dt}$ $= 2x\frac{dx}{dt} + 2y\frac{dy}{dt} \Rightarrow D\frac{dD}{dt} = x\frac{dx}{dt} + y\frac{dy}{dt}$. When $(x,y) = (3,-4)$, $D = \sqrt{3^2 + (-4)^2} = 5$ and $5\frac{dD}{dt} = (5)(10) + (12)(5) \Rightarrow \frac{dD}{dt} = \frac{110}{5} = 22$. Therefore, the particle is moving away from the origin at 22 m/sec (because the distance D is increasing).

105. (a) From the diagram we have $\frac{10}{h} = \frac{4}{r} \Rightarrow r = \frac{2}{5}h$.

(b) $V = \frac{1}{3}\pi r^2 h = \frac{1}{3}\pi\left(\frac{2}{5}h\right)^2 h = \frac{4\pi h^3}{75} \Rightarrow \frac{dV}{dt} = \frac{4\pi h^2}{25}\frac{dh}{dt}$, so $\frac{dV}{dt} = -5$ and $h = 6 \Rightarrow \frac{dh}{dt} = -\frac{125}{144\pi}$ ft/min.

107. (a) From the sketch in the text, $\frac{d\theta}{dt} = -0.6$ rad/sec and $x = \tan\theta$. Also $x = \tan\theta \Rightarrow \frac{dx}{dt} = \sec^2\theta\,\frac{d\theta}{dt}$; at point A, $x = 0 \Rightarrow \theta = 0 \Rightarrow \frac{dx}{dt} = (\sec^2 0)(-0.6) = -0.6$. Therefore the speed of the light is $0.6 = \frac{3}{5}$ km/sec when it reaches point A.

(b) $\frac{(3/5)\text{ rad}}{\text{sec}} \cdot \frac{1\text{ rev}}{2\pi\text{ rad}} \cdot \frac{60\text{ sec}}{\text{min}} = \frac{18}{\pi}$ revs/min

CHAPTER 2 ADDITIONAL EXERCISES–THEORY, EXAMPLES, APPLICATIONS

1. (a) $\sin 2\theta = 2\sin\theta\cos\theta \Rightarrow \frac{d}{d\theta}(\sin 2\theta) = \frac{d}{d\theta}(2\sin\theta\cos\theta) \Rightarrow 2\cos 2\theta = 2[(\sin\theta)(-\sin\theta) + (\cos\theta)(\cos\theta)]$ $\Rightarrow \cos 2\theta = \cos^2\theta - \sin^2\theta$

(b) $\cos 2\theta = \cos^2\theta - \sin^2\theta \Rightarrow \frac{d}{d\theta}(\cos 2\theta) = \frac{d}{d\theta}(\cos^2\theta - \sin^2\theta) \Rightarrow -2\sin 2\theta = (2\cos\theta)(-\sin\theta) - (2\sin\theta)(\cos\theta)$ $\Rightarrow \sin 2\theta = \cos\theta\sin\theta + \sin\theta\cos\theta \Rightarrow \sin 2\theta = 2\sin\theta\cos\theta$

3. (a) $f(x) = \cos x \Rightarrow f'(x) = -\sin x \Rightarrow f''(x) = -\cos x$, and $g(x) = a + bx + cx^2 \Rightarrow g'(x) = b + 2cx \Rightarrow g''(x) = 2c$; also, $f(0) = g(0) \Rightarrow \cos(0) = a \Rightarrow a = 1$; $f'(0) = g'(0) \Rightarrow -\sin(0) = b \Rightarrow b = 0$; $f''(0) = g''(0)$ $\Rightarrow -\cos(0) = 2c \Rightarrow c = -\frac{1}{2}$. Therefore, $g(x) = 1 - \frac{1}{2}x^2$.

(b) $f(x) = \sin(x + a) \Rightarrow f'(x) = \cos(x + a)$, and $g(x) = b\sin x + c\cos x \Rightarrow g'(x) = b\cos x - c\sin x$; also, $f(0) = g(0) \Rightarrow \sin(a) = b\sin(0) + c\cos(0) \Rightarrow c = \sin a$; $f'(0) = g'(0) \Rightarrow \cos(a) = b\cos(0) - c\sin(0)$ $\Rightarrow b = \cos a$. Therefore, $g(x) = \sin x\cos a + \cos x\sin a$.

(c) When $f(x) = \cos x$, $f'''(x) = \sin x$ and $f^{(4)}(x) = \cos x$; when $g(x) = 1 - \frac{1}{2}x^2$, $g'''(x) = 0$ and $g^{(4)}(x) = 0$. Thus $f'''(0) = 0 = g'''(0)$ so the third derivatives agree at $x = 0$. However, the fourth derivatives do not agree since $f^{(4)}(0) = 1$ but $g^{(4)}(0) = 0$. In case (b), when $f(x) = \sin(x + a)$ and $g(x)$ $= \sin x\cos a + \cos x\sin a$, notice that $f(x) = g(x)$ for all x, not just $x = 0$. Since this is an identity, we have $f^{(n)}(x) = g^{(n)}(x)$ for any x and any positive integer n.

5. If the circle $(x-h)^2+(y-k)^2=a^2$ and $y=x^2+1$ are tangent at $(1,2)$, then the slope of this tangent is $m=2x|_{(1,2)}=2$ and the tangent line is $y=2x$. The line containing (h,k) and $(1,2)$ is perpendicular to $y=2x \Rightarrow \frac{k-2}{h-1}=-\frac{1}{2} \Rightarrow h=5-2k \Rightarrow$ the location of the center is $(5-2k,k)$. Also, $(x-h)^2+(y-k)^2=a^2$ $\Rightarrow x-h+(y-k)y'=0 \Rightarrow 1+(y')^2+(y-k)y''=0 \Rightarrow y''=\frac{1+(y')^2}{k-y}$. At the point $(1,2)$ we know $y'=2$ from the tangent line and that $y''=2$ from the parabola. Since the second derivatives are equal at $(1,2)$ we obtain $2=\frac{1+(2)^2}{k-2} \Rightarrow k=\frac{9}{2}$. Then $h=5-2k=-4 \Rightarrow$ the circle is $(x+4)^2+\left(y-\frac{9}{2}\right)^2=a^2$. Since $(1,2)$ lies on the circle we have that $a=\frac{5\sqrt{5}}{2}$.

7. (a) $y=uv \Rightarrow \frac{dy}{dt}=\frac{du}{dt}v+u\frac{dv}{dt}=(0.04u)v+u(0.05v)=0.09uv=0.09y$

   (b) If $\frac{du}{dt}=-0.02u$ and $\frac{dv}{dt}=0.03v$, then $\frac{dy}{dt}=(-0.02u)v+(0.03v)u=0.01uv=0.01y$, increasing at 1% per year.

9. Answers will vary. Here is one possibility

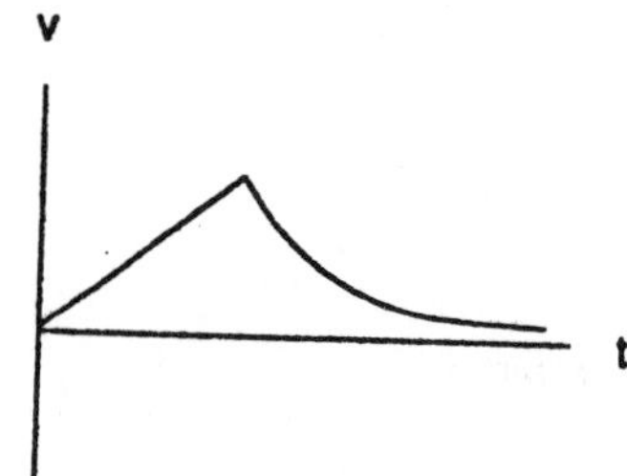

11. (a) $s(t)=64t-16t^2 \Rightarrow v(t)=\frac{ds}{dt}=64-32t=32(2-t)$. The maximum height is reached when $v(t)=0$ $\Rightarrow t=2$ sec. The velocity when it leaves the hand is $v(0)=64$ ft/sec.

    (b) $s(t)=64t-2.6t^2 \Rightarrow v(t)=\frac{ds}{dt}=64-5.2t$. The maximum height is reached when $v(t)=0 \Rightarrow t \approx 12.31$ sec. The maximum height is about $s(12.31)=393.85$ ft.

13. $m\left(v^2-v_0^2\right)=k\left(x_0^2-x^2\right) \Rightarrow m\left(2v\frac{dv}{dt}\right)=k\left(-2x\frac{dx}{dt}\right) \Rightarrow m\frac{dv}{dt}=k\left(-\frac{2x}{2v}\right)\frac{dx}{dt} \Rightarrow m\frac{dv}{dt}=-kx\left(\frac{1}{v}\right)\frac{dx}{dt}$. Then substituting $\frac{dx}{dt}=v \Rightarrow m\frac{dv}{dt}=-kx$, as claimed.

15. (a) To be continuous at $x=\pi$ requires that $\lim_{x\to\pi^-}\sin x=\lim_{x\to\pi^+}(mx+b) \Rightarrow 0=m\pi+b \Rightarrow m=-\frac{b}{\pi}$;

    (b) If $y'=\begin{cases}\cos x, & x<\pi\\ m, & x\ge\pi\end{cases}$ is differentiable at $x=\pi$, then $\lim_{x\to\pi^-}\cos x=m \Rightarrow m=-1$ and $b=\pi$.

17. (a) For all a, b and for all $x\neq 2$, f is differentiable at x. Next, f differentiable at $x=2 \Rightarrow$ f continuous at $x=2 \Rightarrow \lim_{x\to 2^-}f(x)=f(2) \Rightarrow 2a=4a-2b+3 \Rightarrow 2a-2b+3=0$. Also, f differentiable at $x\neq 2$

$\Rightarrow f'(x) = \begin{cases} a, & x < 2 \\ 2ax - b, & x > 2 \end{cases}$. In order that $f'(2)$ exist we must have $a = 2a(2) - b \Rightarrow a = 4a - b \Rightarrow 3a = b$.

Then $2a - 2b + 3 = 0$ and $3a = b \Rightarrow a = \frac{3}{4}$ and $b = \frac{9}{4}$.

(b) For $x < 2$, the graph of f is a straight line having a slope of $\frac{3}{4}$ and passing through the origin for $x \geq 2$, the graph of f is a parabola. At $x = 2$, the value of the y-coordinate on the parabola is $\frac{3}{2}$ which matches the y-coordinate of the point on the straight line at $x = 2$. In addition, the slope of the parabola at the match up point is $\frac{3}{4}$ which is equal to the slope of the straight line. Therefore, since the graph is differentiable at the match up point, the graph is smooth there.

19. f odd $\Rightarrow f(-x) = -f(x) \Rightarrow \frac{d}{dx}(f(-x)) = \frac{d}{dx}(-f(x)) \Rightarrow f'(-x)(-1) = -f'(x) \Rightarrow f'(-x) = f'(x) \Rightarrow f'$ is even.

21. Let $h(x) = (fg)(x) = f(x)\,g(x) \Rightarrow h'(x) = \lim\limits_{x \to x_0} \frac{h(x) - h(x_0)}{x - x_0} = \lim\limits_{x \to x_0} \frac{f(x)\,g(x) - f(x_0)\,g(x_0)}{x - x_0}$

$= \lim\limits_{x \to x_0} \frac{f(x)\,g(x) - f(x)\,g(x_0) + f(x)\,g(x_0) - f(x_0)\,g(x_0)}{x - x_0} = \lim\limits_{x \to x_0} \left[f(x)\left[\frac{g(x) - g(x_0)}{x - x_0}\right]\right] + \lim\limits_{x \to x_0} \left[g(x_0)\left[\frac{f(x) - f(x_0)}{x - x_0}\right]\right]$

$= f(x_0) \lim\limits_{x \to x_0} \left[\frac{g(x) - g(x_0)}{x - x_0}\right] + g(x_0)\,f'(x_0) = 0 \cdot \lim\limits_{x \to x_0} \left[\frac{g(x) - g(x_0)}{x - x_0}\right] + g(x_0)\,f'(x_0) = g(x_0)\,f'(x_0)$, if g is continuous at x_0. Therefore $(fg)(x)$ is differentiable at x_0 if $f(x_0) = 0$, and $(fg)'(x_0) = g(x_0)\,f'(x_0)$.

23. If $f(x) = x$ and $g(x) = x \sin\left(\frac{1}{x}\right)$, then $x^2 \sin\left(\frac{1}{x}\right)$ is differentiable at $x = 0$ because $f'(0) = 1$, $f(0) = 0$ and

$\lim\limits_{x \to 0} x \sin\left(\frac{1}{x}\right) = \lim\limits_{x \to 0} \frac{\sin\left(\frac{1}{x}\right)}{\frac{1}{x}} = \lim\limits_{t \to \infty} \frac{\sin t}{t} = 0$ (so g is continuous at $x = 0$). In fact, from Exercise 21,

$h'(0) = g(0)\,f'(0) = 0$. However, for $x \neq 0$, $h'(x) = \left[x^2 \cos\left(\frac{1}{x}\right)\right]\left(-\frac{1}{x^2}\right) + 2x \sin\left(\frac{1}{x}\right)$. But

$\lim\limits_{x \to 0} h'(x) = \lim\limits_{x \to 0} \left[-\cos\left(\frac{1}{x}\right) + 2x \sin\left(\frac{1}{x}\right)\right]$ does not exist because $\cos\left(\frac{1}{x}\right)$ has no limit as $x \to 0$. Therefore, the derivative is not continuous at $x = 0$ because it has no limit there.

25. Step 1: The formula holds for $n = 2$ (a single product) since $y = u_1 u_2 \Rightarrow \frac{dy}{dx} = \frac{du_1}{dx} u_2 + u_1 \frac{du_2}{dx}$.

Step 2: Assume the formula holds for $n = k$:

$$y = u_1 u_2 \cdots u_k \Rightarrow \frac{dy}{dx} = \frac{du_1}{dx} u_2 u_3 \cdots u_k + u_1 \frac{du_2}{dx} u_3 \cdots u_k + \ldots + u_1 u_2 \cdots u_{k-1} \frac{du_k}{dx}.$$

If $y = u_1 u_2 \cdots u_k u_{k+1} = (u_1 u_2 \cdots u_k) u_{k+1}$, then $\frac{dy}{dx} = \frac{d(u_1 u_2 \cdots u_k)}{dx} u_{k+1} + u_1 u_2 \cdots u_k \frac{du_{k+1}}{dx}$

$= \left(\frac{du_1}{dx} u_2 u_3 \cdots u_k + u_1 \frac{du_2}{dx} u_3 \cdots u_k + \cdots + u_1 u_2 \cdots u_{k-1} \frac{du_k}{dx}\right) u_{k+1} + u_1 u_2 \cdots u_k \frac{du_{k+1}}{dx}$

$$= \frac{du_1}{dx}\, u_2u_3\cdots u_{k+1} + u_1\,\frac{du_2}{dx}\, u_3\cdots u_{k+1} + \cdots + u_1u_2\cdots u_{k-1}\,\frac{du_k}{dx}\, u_{k+1} + u_1u_2\cdots u_k\,\frac{du_{k+1}}{dx}.$$

Thus the original formula holds for $n = (k+1)$ whenever it holds for $n = k$.

NOTES:

CHAPTER 3 APPLICATIONS OF DERIVATIVES

3.1 EXTREME VALUES OF FUNCTIONS

1. An absolute minimum at $x = c_2$, an absolute maximum at $x = b$. Theorem 1 guarantees the existence of such extreme values because h is continuous on $[a, b]$.

3. No absolute minimum. An absolute maximum at $x = c$. Since the function's domain is an open interval, the function does not satisfy the hypotheses of Theorem 1 and need not have absolute extreme values.

5. An absolute minimum at $x = a$ and an absolute maximum at $x = c$. Note that $y = g(x)$ is not continuous but still has extrema. When the hypothesis of Theorem 1 is satisfied then extrema are guaranteed, but when the hypothesis is not satisfied, absolute extrema may or may not occur.

7. Local minimum at $(-1, 0)$, local maximum at $(1, 0)$

9. Maximum at $(0, 5)$. Note that there is no minimum since the endpoint $(2, 0)$ is excluded from the graph.

11. Graph (c), since this is the only graph that has positive slope at c.

13. Graph (d), since this is the only graph representing a function that is differentiable at b but not at a.

15. $f(x) = \frac{2}{3}x - 5 \Rightarrow f'(x) = \frac{2}{3} \Rightarrow$ no critical points;

    $f(-2) = -\frac{19}{3}$, $f(3) = -3 \Rightarrow$ the absolute maximum is -3 at $x = 3$ and the absolute minimum is $-\frac{19}{3}$ at $x = -2$

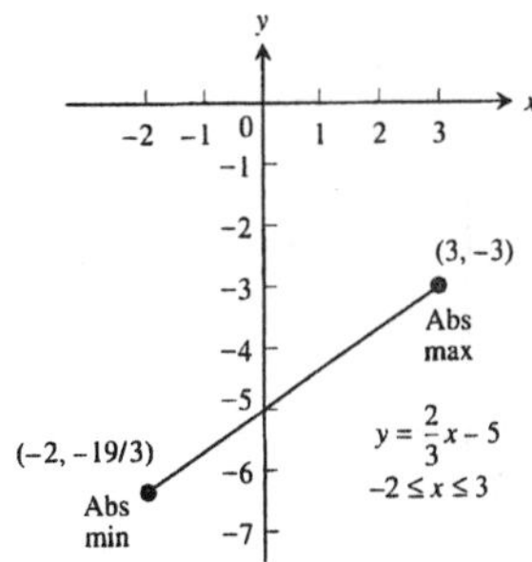

17. The first derivative of $f'(x) = \cos\left(x + \frac{\pi}{4}\right)$, has zeros at $x = \frac{\pi}{4}$, $x = \frac{5\pi}{4}$.

    Critical point values: $x = \frac{\pi}{4}$ $\quad f(x) = 1$

    $x = \frac{5\pi}{4}$ $\quad f(x) = -1$

    Endpoint values: $x = 0$ $\quad f(x) = \frac{1}{\sqrt{2}}$

    $x = \frac{7\pi}{4}$ $\quad f(x) = 0$

    Maximum value is 1 at $x = \frac{\pi}{4}$;

    minimum value is -1 at $x = \frac{5\pi}{4}$;

    local minimum at $\left(0, \frac{1}{\sqrt{2}}\right)$;

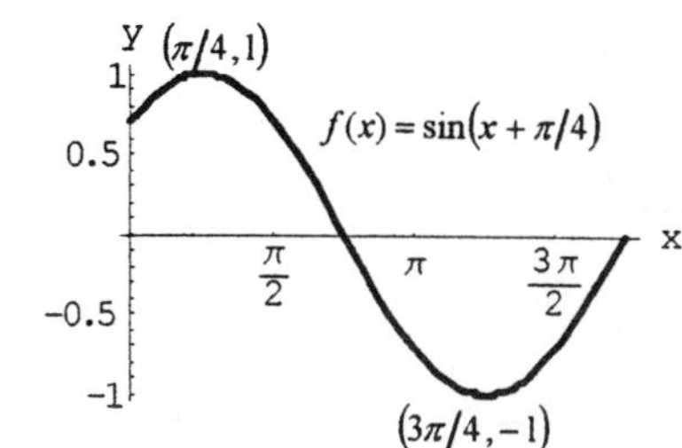

local maximum at $\left(\frac{7\pi}{4}, 0\right)$

19. $F(x) = -\frac{1}{x^2} = -x^{-2} \Rightarrow F'(x) = 2x^{-3} = \frac{2}{x^3}$, however $x = 0$ is not a critical point since 0 is not in the domain; $F(0.5) = -4$, $F(2) = -0.25 \Rightarrow$ the absolute maximum is -0.25 at $x = 2$ and the absolute minimum is -4 at $x = 0.5$

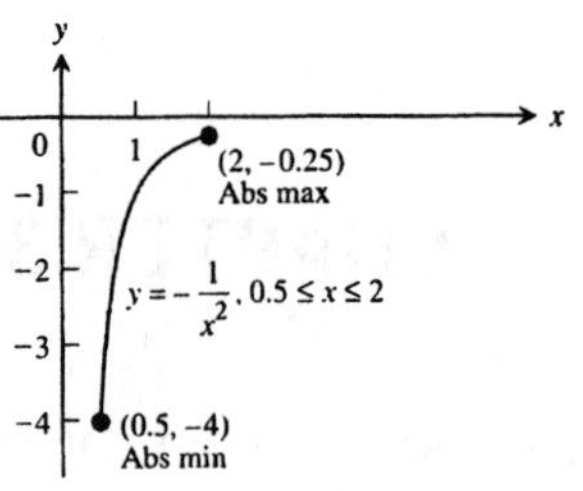

21.

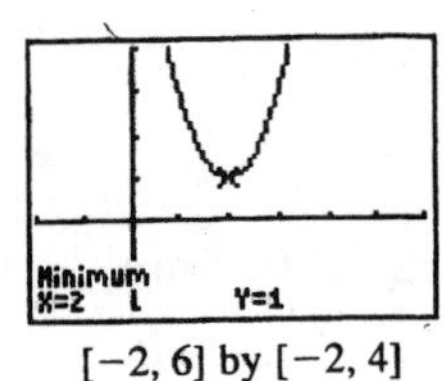

[−2, 6] by [−2, 4]

Minimum value is 1 at $x = 2$.

23.

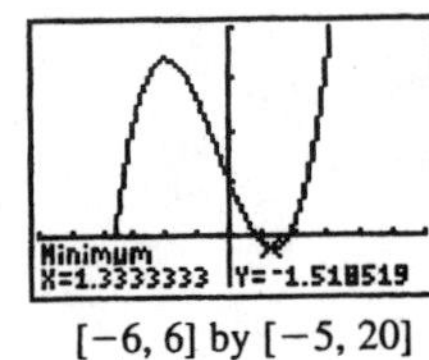

[−6, 6] by [−5, 20]

To find the exact values, note that $y' = 3x^2 + 2x - 8 = (3x - 4)(x + 2)$, which is zero when $x = -2$ or $x = \frac{4}{3}$. Local maximum at $(-2, 17)$; local minimum at $\left(\frac{4}{3}, -\frac{41}{27}\right)$

25.

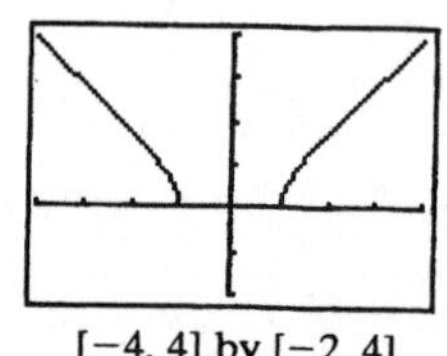
[−4, 4] by [−2, 4]

Minimum value is 0 at $x = -1$ and at $x = 1$.

27.

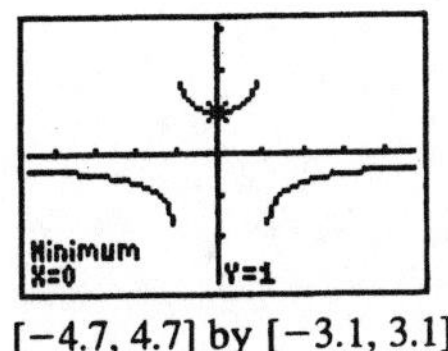

[−4.7, 4.7] by [−3.1, 3.1]

The actual graph of the function has asymptotes at $x = \pm 1$, so there are no extrema near these values. (This is an example of *grapher failure.*) There is a local minimum at $(0, 1)$.

29.

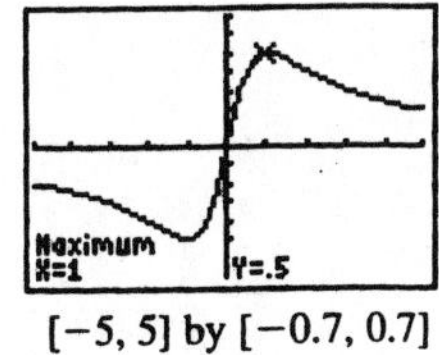

[−5, 5] by [−0.7, 0.7]

Maximum value is $\frac{1}{2}$ at $x = 1$;

minimum value is $-\frac{1}{2}$ at $x = -1$.

31.

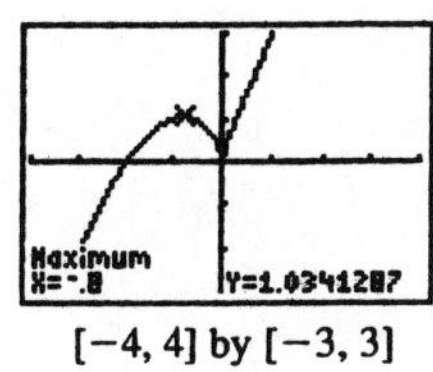

[−4, 4] by [−3, 3]

$$y' = x^{2/3}(1) + \frac{2}{3}x^{-1/3}(x+2) = \frac{5x+4}{3\sqrt[3]{x}}$$

crit. pt.	derivative	extremum	value
$x = -\frac{4}{5}$	0	local max	$\frac{12}{25}10^{1/3} = 1.034$
$x = 0$	undefined	local min	0

33.

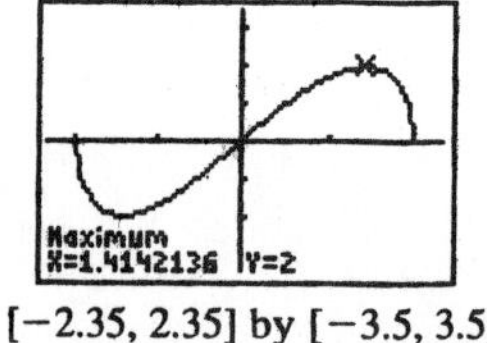

[−2.35, 2.35] by [−3.5, 3.5]

$$y' = x \cdot \frac{1}{2\sqrt{4-x^2}}(-2x) + (1)\sqrt{4-x^2} = \frac{-x^2 + (4-x^2)}{\sqrt{4-x^2}} = \frac{4-2x^2}{\sqrt{4-x^2}}$$

crit. pt.	derivative	extremum	value
$x = -2$	undefined	local max	0
$x = -\sqrt{2}$	0	minimum	−2
$x = \sqrt{2}$	0	maximum	2
$x = 2$	undefined	local min	0

35.

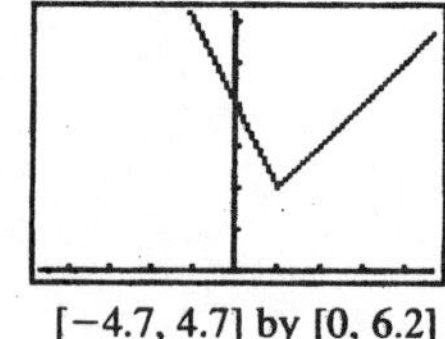
[−4.7, 4.7] by [0, 6.2]

$$y' = \begin{cases} -2, & x < 1 \\ 1, & x > 1 \end{cases}$$

crit. pt.	derivative	extremum	value
$x = 1$	undefined	minimum	2

37.

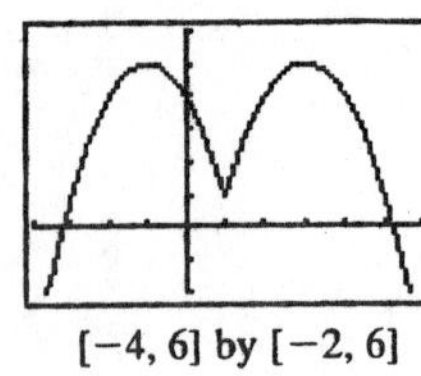
[−4, 6] by [−2, 6]

$$y' = \begin{cases} -2x - 2, & x < 1 \\ -2x + 6, & x > 1 \end{cases}$$

crit. pt.	derivative	extremum	value
$x=-1$	0	maximum	5
$x=1$	undefined	local min	1
$x=3$	0	maximum	5

39. (a) No, since $f'(x)=\frac{2}{3}(x-2)^{-1/3}$, which is undefined at $x=2$.

(b) The derivative is defined and nonzero for all $x \neq 2$. Also, $f(2)=0$ and $f(x)>0$ for all $x \neq 2$.

(c) No, f(x) need not have a global maximum because its domain is all real numbers. Any restriction of f to a closed interval of the form $[a,b]$ would have both a maximum value and a minimum value on the interval.

(d) The answers are the same as (a) and (b) with 2 replaced by a.

41.

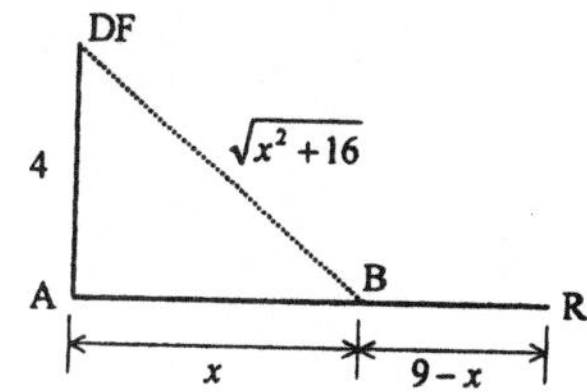

(a) The construction cost is $C(x)=0.3\sqrt{16+x^2}+0.2(9-x)$ million dollars, where $0 \le x \le 9$ miles. The following is a graph of C(x).

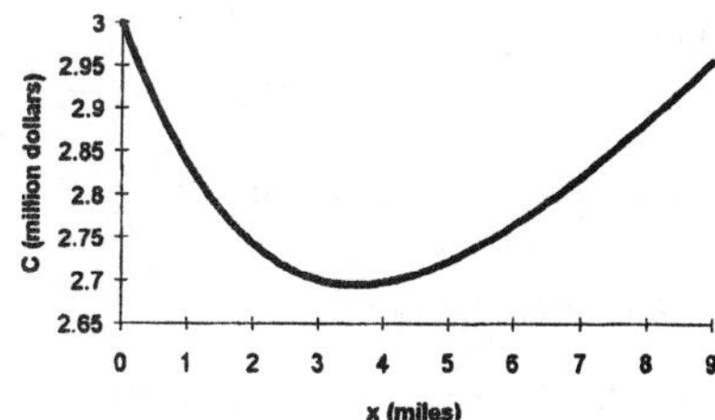

Solving $C'(x)=\dfrac{0.3x}{\sqrt{16+x^2}}-0.2=0$ gives $x=\pm\dfrac{8\sqrt{5}}{5}\approx \pm 3.58$ miles, but only $x=3.58$ miles is a critical point in the specified domain. Evaluating the costs at the critical and endpoints gives $C(0)=\$3$ million, $C(8\sqrt{5}/5)\approx \$2.694$ million, and $C(9)\approx \$2.955$ million. Therefore, to minimize the cost of construction, the pipeline should be placed from the docking facility to point B, 3.58 miles along the shore from point A, and then along the shore from B to the refinery.

(b) If the per mile cost of underwater construction is p, then $C(x)=p\sqrt{16+x^2}+0.2(9-x)$ and $C(x)=\dfrac{px}{\sqrt{16+x^2}}-0.2=0$ gives $x_c=\dfrac{0.8}{\sqrt{p^2-0.04}}$, which minimizes the construction cost provided $x_c \le 9$. The value of p that gives $x_c=9$ miles is 0.218864. Consequently, if the underwater construction costs \$218,864 per mile or less, then running the pipeline along a straight line directly from the docking facility to the refinery will minimize the cost of construction.

In theory, p would have to be infinite to justify running the pipe directly from the docking facility to point A (i.e., for x_c to be zero). For all values of $p > 0.218864$ there is always an $x_c \in (0,9)$ that will give a minimum value for C. This is proved by looking at $C''(x_c) = \frac{16p}{\left(16 + x_c^2\right)^{3/2}}$ which is always positive for $p > 0$.

43. 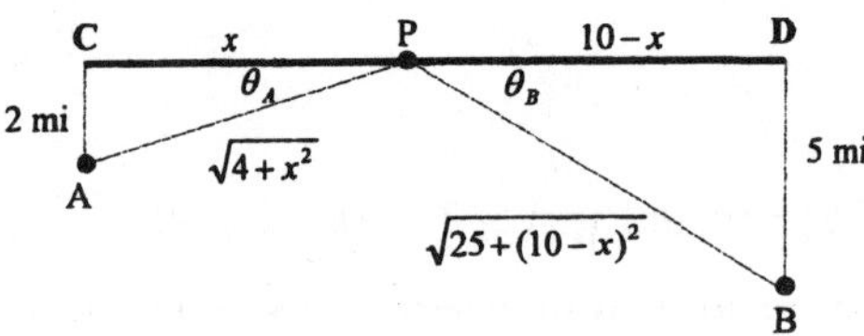

The length of pipeline is $L(x) = \sqrt{4+x^2} + \sqrt{25+(10-x)^2}$ for $0 \le x \le 10$. The following is a graph of L(x).

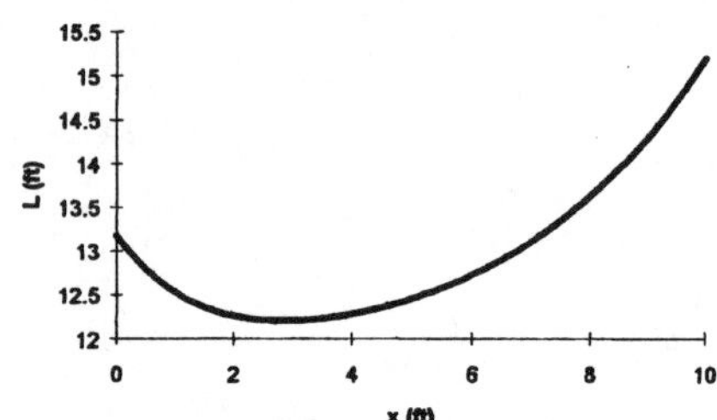

Setting the derivative of L(x) equal to zero gives $L'(x) = \frac{x}{\sqrt{4+x^2}} - \frac{(10-x)}{\sqrt{25+(10-x)^2}} = 0$. Note that $\frac{x}{\sqrt{4+x^2}} = \cos\theta_A$ and $\frac{10-x}{\sqrt{25-(10-x)^2}} = \cos\theta_B$, therefore, $L'(x) = 0$ when $\cos\theta_A = \cos\theta_B$, or $\theta_A = \theta_B$ and ΔACP is similar to ΔBDP. Use simple proportions to determine x as follows:

$$\frac{x}{2} = \frac{10-x}{5} \Rightarrow x = \frac{20}{7} \approx 2.857 \text{ miles along the coast from town A to town B.}$$

If the two towns were on opposite sides of the river, the obvious solution would be to place the pump station on a straight line (the shortest distance) between the two towns, again forcing $\theta_A = \theta_B$. The shortest length of pipe is the same regardless of whether the towns are on the same or opposite sides of the river.

45. (a) $V(x) = 160x - 52x^2 + 4x^3$
$V'(x) = 160 - 104x + 12x^2 = 4(x-2)(3x-20)$
The only critical point in the interval (0,5) is at $x = 2$. The maximum value of V(x) is 144 at $x = 2$.
(b) The largest possible volume of the box is 144 cubic units, and it occurs when $x = 2$.

47. Let x represent the length of the base and $\sqrt{25-x^2}$ the height of the triangle. The area of the triangle is represented by $A(x) = \frac{x}{2}\sqrt{25-x^2}$ where $0 \le x \le 5$. Consequently, solving $A'(x) = 0 \Rightarrow \frac{25-2x^2}{2\sqrt{25-x^2}} = 0$ $\Rightarrow x = \frac{5}{\sqrt{2}}$. Since $A(0) = A(5) = 0$, A(x) is maximized at $x = \frac{5}{\sqrt{2}}$. The largest possible area is $A\left(\frac{5}{\sqrt{2}}\right) = \frac{25}{4}$ cm².

49. $s = -\frac{1}{2}gt^2 + v_0t + s_0 \Rightarrow \frac{ds}{dt} = -gt + v_0 = 0 \Rightarrow t = \frac{v_0}{g}$. Then $s\left(\frac{v_0}{g}\right) = -\frac{1}{2}g\left(\frac{v_0}{g}\right)^2 + v_0\left(\frac{v_0}{g}\right) + s_0$ $= \frac{v_0^2}{2g} + s_0$ is the maximum height since $\frac{d^2s}{dt^2} = -g < 0$.

51. Yes, since $f(x) = |x| = \sqrt{x^2} = (x^2)^{1/2} \Rightarrow f'(x) = \frac{1}{2}(x^2)^{-1/2}(2x) = \frac{x}{(x^2)^{1/2}} = \frac{x}{|x|}$ is not defined at $x = 0$. Thus it is not required that f' be zero at a local extreme point since f' may be undefined there.

53. If g(c) is a local minimum value of g, then $g(x) \geq g(c)$ for all x in some open interval (a, b) containing c. Since g is odd, $g(-x) = -g(x) \leq -g(c) = g(-c)$ for all $-x$ in the open interval $(-b, -a)$ containing $-c$. That is, g assumes a local maximum at the point $-c$. This is also clear from the graph of g because the graph of an odd function is symmetric about the origin.

55. (a) $f'(x) = 3ax^2 + 2bx + c$ is a quadratic, so it can have 0, 1, or 2 zeros, which would be the critical points of f. Examples:

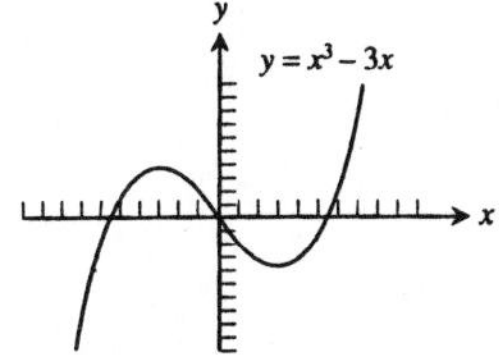

The function $f(x) = x^3 - 3x$ has two critical points at $x = -1$ and $x = 1$.

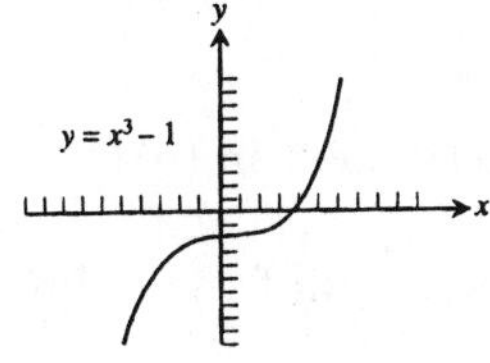

The function $f(x) = x^3 - 1$ has one critical point at $x = 0$.

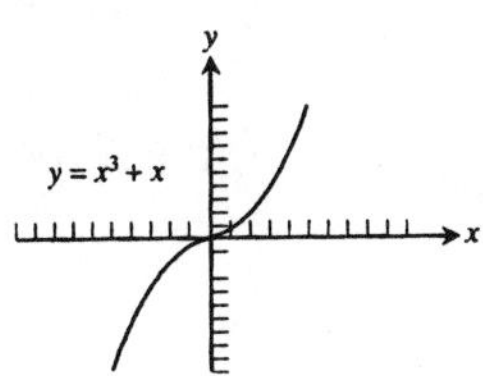

The function $f(x) = x^3 + x$ has no critical points.

(b) The function can have either two local extreme values or no extreme values. (If there is only one critical point, the cubic function has no extreme values.)

57.

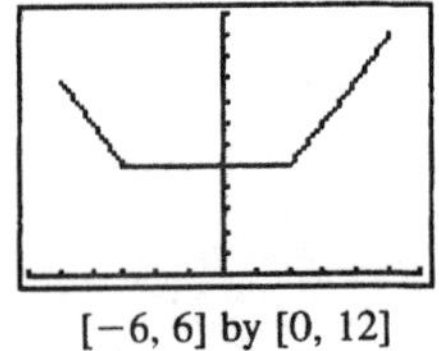

[−6, 6] by [0, 12]

Maximum value is 11 at $x = 5$;
minimum value is 5 on the interval $[-3, 2]$;
local maximum at $(-5, 9)$

59.

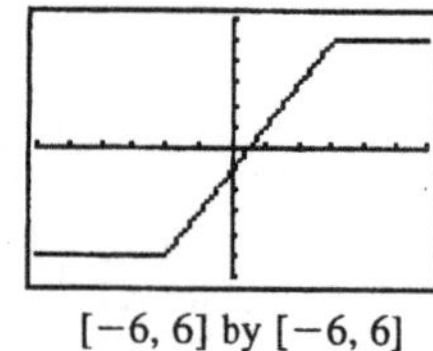

[−6, 6] by [−6, 6]

Maximum value is 5 on the interval $[3, \infty)$;
minimum value is -5 on the interval $(-\infty, -2]$.

3.2 THE MEAN VALUE THEOREM AND DIFFERENTIAL EQUATIONS

1. Does not; $f(x)$ is not differentiable at $x = 0$ in $(-1, 8)$.

3. Does; $s(t)$ is continuous for every point of $[0, 1]$ and differentiable for every point in $(0, 1)$.

5. Since $f(x)$ is not continuous on $0 \le x < 1$, Rolle's Theorem does not apply: $\lim\limits_{x \uparrow 1} f(x) = \lim\limits_{x \uparrow 1} x = 1$ $\neq 0 = f(1)$.

7. By Corollary 1, $f'(x) = 0$ for all $x \Rightarrow f(x) = C$, where C is a constant. Since $f(-1) = 3$ we have $C = 3$ $\Rightarrow f(x) = 3$ for all x.

9. (a) $y = \frac{x^2}{2} + C$ (b) $y = \frac{x^3}{3} + C$ (c) $y = \frac{x^4}{4} + C$

11. (a) $r' = -\theta^{-2} \Rightarrow r = \frac{1}{\theta} + C$ (b) $r = \theta + \frac{1}{\theta} + C$ (c) $r = 5\theta - \frac{1}{\theta} + C$

13. $f(x) = x^2 - x + C$; $0 = f(0) = 0^2 - 0 + C \Rightarrow C = 0 \Rightarrow f(x) = x^2 - x$

15. $r(\theta) = 8\theta + \cot\theta + C$; $0 = r\left(\frac{\pi}{4}\right) = 8\left(\frac{\pi}{4}\right) + \cot\left(\frac{\pi}{4}\right) + C \Rightarrow 0 = 2\pi + 1 + C \Rightarrow C = -2\pi - 1$ $\Rightarrow r(\theta) = 8\theta + \cot\theta - 2\pi - 1$

17. $v = \frac{ds}{dt} = 9.8t + 5 \Rightarrow s = 4.9t^2 + 5t + C$; at $s = 10$ and $t = 0$ we have $C = 10 \Rightarrow s = 4.9t^2 + 5t + 10$

19. $v = \frac{ds}{dt} = \sin(\pi t) \Rightarrow s = -\frac{1}{\pi}\cos(\pi t) + C$; at $s = 0$ and $t = 0$ we have $C = \frac{1}{\pi} \Rightarrow s = \frac{1 - \cos(\pi t)}{\pi}$

21. $a = 32 \Rightarrow v = 32t + C_1$; at $v = 20$ and $t = 0$ we have $C_1 = 20 \Rightarrow v = 32t + 20 \Rightarrow s = 16t^2 + 20t + C_2$; at $s = 5$ and $t = 0$ we have $C_2 = 5 \Rightarrow s = 16t^2 + 20t + 5$

23. $a = -4\sin(2t) \Rightarrow v = 2\cos(2t) + C_1$; at $v = 2$ and $t = 0$ we have $C_1 = 0 \Rightarrow v = 2\cos(2t)$ $\Rightarrow s = \sin(2t) + C_2$; at $s = -3$ and $t = 0$ we have $C_2 = -3 \Rightarrow s = \sin(2t) - 3$

25. $a(t) = v'(t) = 1.6 \Rightarrow v(t) = 1.6t + C$; at $(0,0)$ we have $C = 0 \Rightarrow v(t) = 1.6t$. When $t = 30$, then $v(30) = 48$ m/sec.

27. $a(t) = v'(t) = 9.8 \Rightarrow v(t) = 9.8t + C_1$; at $(0,0)$ we have $C_1 = 0 \Rightarrow s'(t) = v(t) = 9.8t \Rightarrow s(t) = 4.9t^2 + C_2$; at $(0,0)$ we have $C_2 = 0 \Rightarrow s(t) = 4.9t^2$. Then $s(t) = 10 \Rightarrow t^2 = \frac{10}{4.9} \Rightarrow t = \sqrt{\frac{10}{4.9}}$, and $v\left(\sqrt{\frac{10}{4.9}}\right) = 9.8\sqrt{\frac{10}{4.9}}$ $= \frac{2(4.9)\sqrt{10}}{\sqrt{4.9}} = (2)\sqrt{4.9}\sqrt{10} = 14$ m/sec.

29. (a) $v = \int a\,dt = \int \left(15t^{1/2} - 3t^{-1/2}\right) dt = 10t^{3/2} - 6t^{1/2} + C$; $\frac{ds}{dt}(1) = 4 \Rightarrow 4 = 10(1)^{3/2} - 6(1)^{1/2} + C \Rightarrow C = 0$ $\Rightarrow v = 10t^{3/2} - 6t^{1/2}$

(b) $s = \int v\,dt = \int \left(10t^{3/2} - 6t^{1/2}\right) dt = 4t^{5/2} - 4t^{3/2} + C$; $s(1) = 0 \Rightarrow 0 = 4(1)^{5/2} - 4(1)^{3/2} + C \Rightarrow C = 0$ $\Rightarrow s = 4t^{5/2} - 4t^{3/2}$

31. If $T(t)$ is the temperature of the thermometer at time t, then $T(0) = -19°$ C and $T(14) = 100°$ C. From the Mean Value Theorem there exists a $0 < t_0 < 14$ such that $\frac{T(14) - T(0)}{14 - 0} = 8.5°$ C/sec $= T'(t_0)$, the rate at which the temperature was changing at $t = t_0$ as measured by the rising mercury on the thermometer.

33. Because its average speed was approximately 7.667 knots, and by the Mean Value Theorem, it must have been going that speed at least once during the trip.

35. The conclusion of the Mean Value Theorem yields $\frac{\frac{1}{b} - \frac{1}{a}}{b - a} = -\frac{1}{c^2} \Rightarrow c^2\left(\frac{a - b}{ab}\right) = a - b \Rightarrow c = \sqrt{ab}$.

37. $f'(x) = [\cos x \sin(x + 2) + \sin x \cos(x + 2)] - 2\sin(x + 1)\cos(x + 1) = \sin(x + x + 2) - \sin 2(x + 1)$ $= \sin(2x + 2) - \sin(2x + 2) = 0$. Therefore, the function has the constant value $f(0) = -\sin^2 1 \approx -0.7081$ which explains why the graph is a horizontal line.

39. $f(x)$ must be zero at least once between a and b by the Intermediate Value Theorem. Now suppose that $f(x)$ is zero twice between a and b. Then by the Mean Value Theorem, $f'(x)$ would have to be zero at least once between the two zeros of $f(x)$, but this can't be true since we are given that $f'(x) \neq 0$ on this interval. Therefore, $f(x)$ is zero once and only once between a and b.

41. Yes. By Corollary 2 we have $f(x) = g(x) + c$ since $f'(x) = g'(x)$. If the graphs start at the same point $x = a$, then $f(a) = g(a) \Rightarrow c = 0 \Rightarrow f(x) = g(x)$.

43. Yes; f' is negative at some point c between a and b. By the Mean Value Theorem, $\frac{f(b) - f(a)}{b - a} = f'(c)$ for some point c between a and b. Since $b - a > 0$ and $f(b) < f(a)$, we have $f(b) - f(a) < 0 \Rightarrow f'(c) < 0$.

45. $f'(x) = \left(1 + x^4 \cos x\right)^{-1} \Rightarrow f''(x) = -\left(1 + x^4 \cos x\right)^{-2}\left(4x^3 \cos x - x^4 \sin x\right)$
$= -x^3\left(1 + x^4 \cos x\right)^{-2}(4 \cos x - x \sin x) < 0$ for $0 \le x \le 0.1 \Rightarrow f'(x)$ is decreasing when $0 \le x \le 0.1$
$\Rightarrow \min f' \approx 0.9999$ and $\max f' = 1$. Now we have $0.9999 \le \frac{f(0.1) - 1}{0.1} \le 1 \Rightarrow 0.09999 \le f(0.1) - 1 \le 0.1$
$\Rightarrow 1.09999 \le f(0.1) \le 1.1$.

3.3 THE SHAPE OF A GRAPH

1. The graph of $y = f''(x) \Rightarrow$ the graph of $y = f(x)$ is concave up on $(0, \infty)$, concave down on $(-\infty, 0) \Rightarrow$ a point of inflection at $x = 0$; the graph of $y = f'(x) \Rightarrow y' = +++\,|\,---\,|\,+++ \Rightarrow$ the graph $y = f(x)$ has both a local maximum and a local minimum

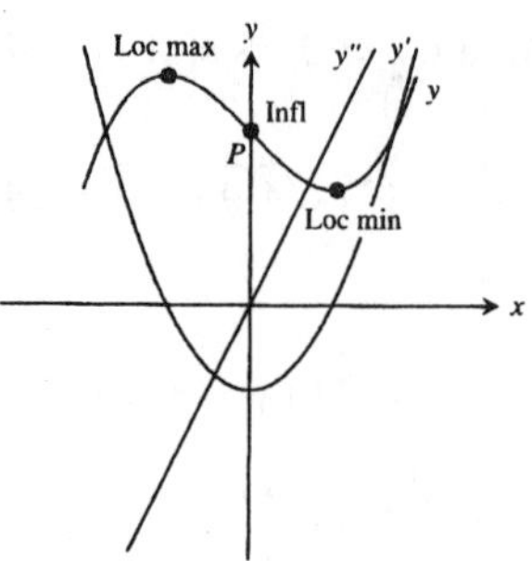

3. The graph of $y = f''(x) \Rightarrow y'' = ---\,|\,+++\,|\,--- \Rightarrow$ the graph of $y = f(x)$ has two points of inflection, the graph of $y = f'(x) \Rightarrow y' = ---\,|\,+++ \Rightarrow$ the graph of $y = f(x)$ has a local minimum

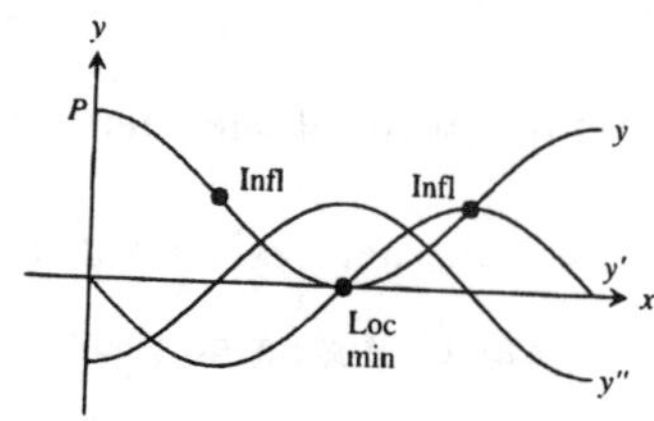

5.

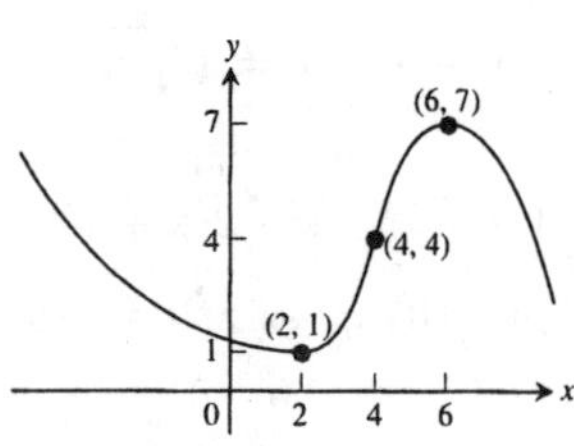

7. (a) Zero: $x = \pm 1$;
positive: $(-\infty, -1)$ and $(1, \infty)$;
negative: $(-1, 1)$
(b) Zero: $x = 0$;
positive: $(0, \infty)$;
negative: $(-\infty, 0)$

9. (a) $(-\infty, -2]$ and $[0, 2]$
(b) $[-2, 0]$ and $[2, \infty)$
(c) Local maxima: $x = -2$ and $x = 2$;
local minimum: $x = 0$

11. (a) $[0, 1]$, $[3, 4]$, and $[5.5, 6]$
(b) $[1, 3]$ and $[4, 5.5]$
(c) Local maxima: $x = 1$, $x = 4$ (if f is continuous at $x = 4$), and $x = 6$;
local minima: $x = 6$, $x = 3$, and $x = 5.5$

13. (a) $f'(x) = (x-1)(x+2) \Rightarrow$ critical points at -2 and 1
(b) $f' = +++\underset{-2}{|}---\underset{1}{|}+++ \Rightarrow$ increasing on $(-\infty, -2]$ and $[1, \infty)$, decreasing on $[-2, 1]$
(c) Local maximum at $x = -2$ and a local minimum at $x = 1$

15. (a) $f'(x) = (x-1)(x+2)(x-3) \Rightarrow$ critical points at -2, 1 and 3
(b) $f' = ---\underset{-2}{|}+++\underset{1}{|}---\underset{3}{|}+++ \Rightarrow$ increasing on $[-2, 1]$ and $[3, \infty)$, decreasing on $(-\infty, -2]$ and $[1, 3]$
(c) Local maximum at $x = 1$, local minima at $x = -2$ and $x = 3$

17. $y' = 2x - 1$

Intervals	$x < \frac{1}{2}$	$x > \frac{1}{2}$
Sign of y'	$-$	$+$
Behavior of y	Decreasing	Increasing

$y'' = 2$ (always positive: concave up)

Graphical support:

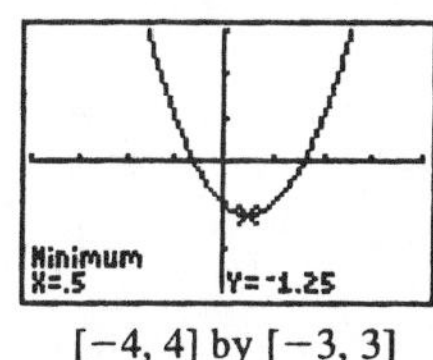

$[-4, 4]$ by $[-3, 3]$

(a) $\left[\frac{1}{2}, \infty\right)$

(b) $\left(-\infty, \frac{1}{2}\right]$

(c) $(-\infty, \infty)$

(d) Nowhere

(e) Local (and absolute) minimum at $\left(\frac{1}{2}, -\frac{5}{4}\right)$

(f) None

19. $y' = 8x^3 - 8x = 8x(x-1)(x+1)$

Intervals	$x < -1$	$-1 < x < 0$	$0 < x < 1$	$1 < x$
Sign of y'	$-$	$+$	$-$	$+$
Behavior of y	Decreasing	Increasing	Decreasing	Increasing

$y'' = 24x^2 - 8 = 8(\sqrt{3}x - 1)(\sqrt{3}x + 1)$

Intervals	$x < -\frac{1}{\sqrt{3}}$	$-\frac{1}{\sqrt{3}} < x < \frac{1}{\sqrt{3}}$	$\frac{1}{\sqrt{3}} < x$
Sign of y''	+	−	+
Behavior of y	Concave up	Concave down	Concave up

Graphical support:

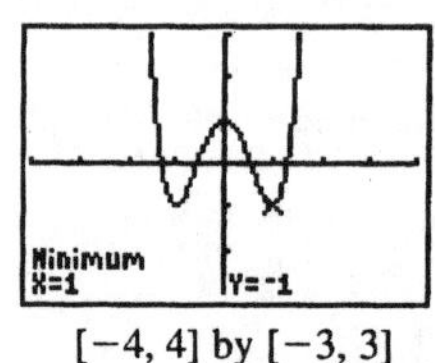

[−4, 4] by [−3, 3]

(a) $[-1, 0]$ and $[1, \infty)$

(b) $(-\infty, 1]$ and $[0, 1]$

(c) $\left(-\infty, -\frac{1}{\sqrt{3}}\right)$ and $\left(\frac{1}{\sqrt{3}}, \infty\right)$

(d) $\left(-\frac{1}{\sqrt{3}}, \frac{1}{\sqrt{3}}\right)$

(e) Local maximum: $(0, 1)$;
local (and absolute) minima: $(-1, -1)$ and $(1, -1)$

(f) $\left(\pm\frac{1}{\sqrt{3}}, -\frac{1}{9}\right)$

21. $y' = x\,\frac{1}{2\sqrt{8 - x^2}}(-2x) + (\sqrt{8 - x^2})(1) = \frac{8 - 2x^2}{\sqrt{8 - x^2}}$

Intervals	$-\sqrt{8} < x < -2$	$-2 < x < 2$	$2 < x < \sqrt{8}$
Sign of y'	−	−	−
Behavior of y	Decreasing	Decreasing	Decreasing

$$y'' = \frac{(\sqrt{8 - x^2})(-4x) - (8 - 2x^2)\frac{1}{2\sqrt{8 - x^2}}(-2x)}{(\sqrt{8 - x^2})^2} = \frac{2x^3 - 24x}{(8 - x^2)^{3/2}} = \frac{2x(x^2 - 12)}{(8 - x^2)^{3/2}}$$

Intervals	$-\sqrt{8} < x < 0$	$0 < x < \sqrt{8}$
Sign of y'	+	−
Behavior of y	Concave up	Concave down

Graphical support:

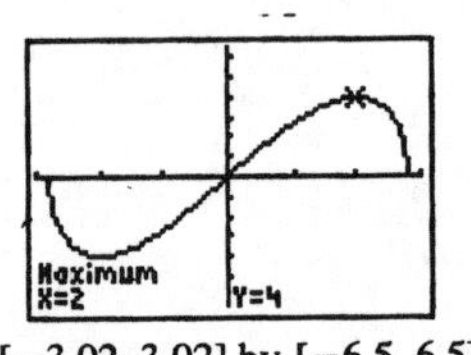

[−3.02, 3.02] by [−6.5, 6.5]

(a) $[-2,2]$ (b) $[-\sqrt{8},-2]$ and $[2,\sqrt{8}]$

(c) $(-\sqrt{8},0)$ (d) $(0,\sqrt{8})$

(e) Local maxima: $(-\sqrt{8},0)$ and $(2,4)$; local minima: $(-2,-4)$ and $(\sqrt{8},0)$
Note that the local extrema at $x = \pm 2$ are also absolute extrema

(f) $(0,0)$

23. $y' = 12x^2 + 42x + 36 = 6(x+2)(2x+3)$

Intervals	$x < -2$	$-2 < x < -\frac{3}{2}$	$-\frac{3}{2} < x$
Sign of y′	+	−	+
Behavior of y	Increasing	Decreasing	Increasing

$y'' = 24x + 42 = 6(4x+7)$

Intervals	$x < -\frac{7}{4}$	$-\frac{7}{4} < x$
Sign of y″	−	+
Behavior of y	Concave down	Concave up

Graphical support:

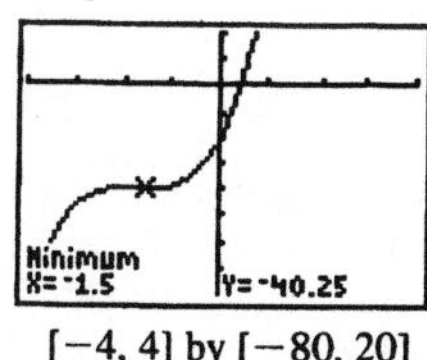

[−4, 4] by [−80, 20]

(a) $(-\infty,-2]$ and $\left[-\frac{3}{2},\infty\right)$ (b) $\left[-2,-\frac{3}{2}\right]$

(c) $\left(-\frac{7}{4},\infty\right)$ (d) $\left(-\infty,-\frac{7}{4}\right)$

(e) Local maximum: $(-2,-40)$; local minimum: $\left(-\frac{3}{2},-\frac{161}{4}\right)$

(f) $\left(-\frac{7}{4},-\frac{321}{8}\right)$

25. $y' = \frac{2}{5}x^{-4/5}$

Intervals	$x < 0$	$0 < x$
Sign of y′	+	+
Behavior of y	Increasing	Increasing

$y'' = -\frac{8}{25}x^{-9/5}$

Intervals	$x < 0$	$0 < x$
Sign of y''	$+$	$-$
Behavior of y	Concave up	Concave down

Graphical support:

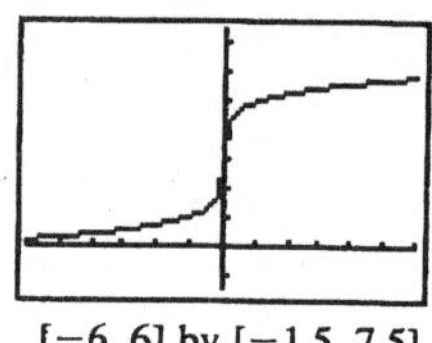

[−6, 6] by [−1.5, 7.5]

(a) $(-\infty, \infty)$

(b) None

(c) $(-\infty, 0)$

(d) $(0, \infty)$

(e) None

(f) $(0, 3)$

27. $y = x^{1/3}(x-4) = x^{4/3} - 4x^{1/3}$

$$y' = \frac{4}{3}x^{1/3} - \frac{4}{3}x^{-2/3} = \frac{4x-4}{3x^{2/3}}$$

Intervals	$x < 0$	$0 < x < 1$	$1 < x$
Sign of y'	$-$	$-$	$+$
Behavior of y	Decreasing	Decreasing	Increasing

$$y'' = \frac{4}{9}x^{-2/3} + \frac{8}{9}x^{-5/3} = \frac{4x+8}{9x^{5/3}}$$

Intervals	$x < -2$	$-2 < x < 0$	$0 < x$
Sign of y''	$+$	$-$	$+$
Behavior of y	Concave up	Concave down	Concave up

Graphical support:

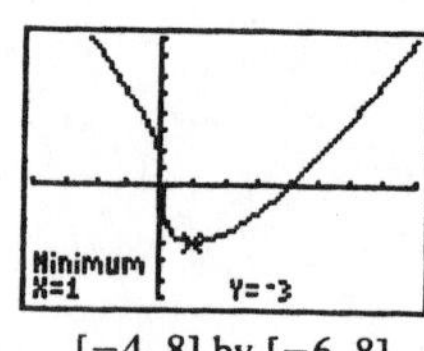

[−4, 8] by [−6, 8]

(a) $[1, \infty)$

(b) $(-\infty, 1]$

(c) $(-\infty, -2)$ and $(0, \infty)$

(d) $(-2, 0)$

(e) Local minimum: $(1, -3)$

(f) $(-2, 6\sqrt[3]{2}) \approx (-2, 7.56)$ and $(0, 0)$

29. We use a combination of analytic and grapher techniques to solve this problem. Depending on the viewing window chosen, graphs obtained using the nderiv function on a TI-92 calculator may exhibit strange behavior near $x = 2$ because, for example, $\text{nderiv}(y, x) \mid x = 2 \approx 1{,}000{,}000$ while y' is actually undefined at $x = 2$. The graph of $y = \dfrac{x^3 - 2x^2 + x - 1}{x - 2}$ is shown below:

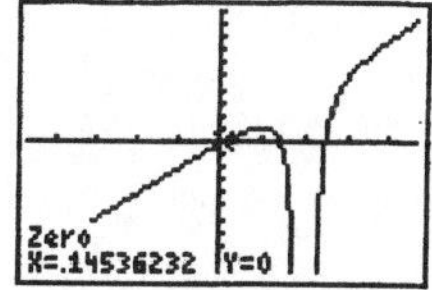

[−4,7, 4.7] by [−5, 15]

$$y' = \frac{(x-2)(3x^2 - 4x + 1) - (x^3 - 2x^2 + x - 1)(1)}{(x-2)^2} = \frac{2x^3 - 8x^2 + 8x - 1}{(x-2)^2}$$

The graph of y' is shown below:

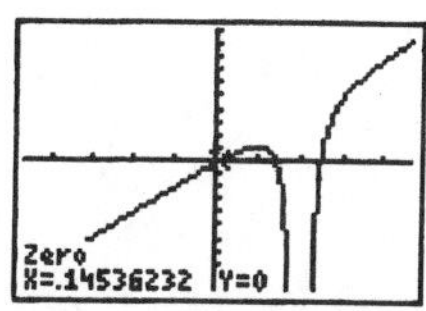

[−4,7, 4.7] by [−10, 10]

The zeros of y' are $x \approx 0.15$, $x \approx 1.40$, and $x \approx 2.45$.

Intervals	$x < 0.15$	$0.15 < x < 1.40$	$1.40 < x < 2$	$2 < x < 2.45$	$2.45 < x$
Sign of y'	−	+	−	−	+
Behavior of y	Decreasing	Increasing	Decreasing	Decreasing	Increasing

$$y'' = \frac{(x-2)^2(6x^2 - 16x + 8) - (2x^3 - 8x^2 + 8x - 1)(x-2)}{(x-2)^4}$$

$$= \frac{(x-2)(6x^2 - 16x + 8) - 2(2x^3 - 8x^2 - 1)}{(x-2)^3} = \frac{2x^3 - 12x^2 + 24x - 14}{(x-2)^3} = \frac{2(x-1)(x^2 - 5x + 7)}{(x-2)^3}$$

The graph of y'' is shown below.

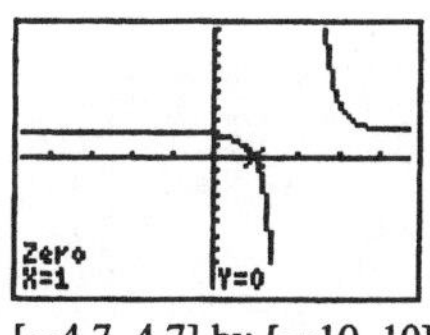

[−4,7, 4.7] by [−10, 10]

Note that the discriminant of $x^2 - 5x + 7$ is $(-5)^2 - 4(1)(7) = -3$, so the only solution of $y'' = 0$ is $x = 1$.

Intervals	$x < 1$	$1 < x < 2$	$2 < x$
Sign of y''	+	−	+
Behavior of y	Concave up	Concave down	Concave up

(a) Approximately $[0.15, 1.40]$ and $[2.45, \infty)$
(b) Approximately $(-\infty, 0.15]$, $[1.40, 2)$, and $(2, 2.45]$
(c) $(-\infty, 1)$ and $(2, \infty)$
(d) $(1, 2)$
(e) Local maximum: $\approx (1.40, 1.29)$; local minima: $\approx (0.15, 0.48)$ and $(2.45, 9.22)$
(f) $(1, 1)$

31. $y = x^{1/4}(x + 3) = x^{5/4} + 3x^{1/4}$

$$y' = \frac{5}{4}x^{1/4} + \frac{3}{4}x^{-3/4} = \frac{5x + 3}{4x^{3/4}}$$

Since $y' > 0$ for all $x > 0$, y is always increasing on its domain $x \geq 0$.

$$y'' = \frac{5}{16}x^{-3/4} - \frac{9}{16}x^{-7/4} = \frac{5x - 9}{16x^{7/4}}$$

Intervals	$0 < x < \frac{9}{5}$	$\frac{9}{5} < x$
Sign of y''	−	+
Behavior of y	Concave down	Concave up

Graphical support:

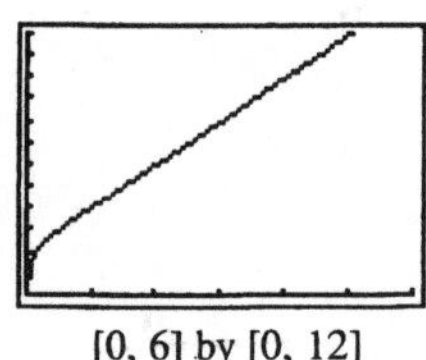

[0, 6] by [0, 12]

(a) $[0, \infty)$
(b) None
(c) $\left(\frac{9}{5}, \infty\right)$
(d) $\left(0, \frac{9}{5}\right)$
(e) Local (and absolute) minimum: $(0, 0)$
(f) $\left(\frac{9}{5}, \frac{24}{5}\cdot\sqrt[4]{\frac{9}{5}}\right) \approx (1.8, 5.56)$

33. $y' = (x - 1)^2(x - 2)$

Intervals	$x < 1$	$1 < x < 2$	$2 < x$
Sign of y'	−	−	+
Behavior of y	Decreasing	Decreasing	Increasing

$$y'' = (x - 1)^2(1) + (x - 2)(2)(x - 1) = (x - 1)[(x - 1) + 2(x - 2)] = (x - 1)(3x - 5)$$

Intervals	$x < 1$	$1 < x < \frac{5}{3}$	$\frac{5}{3} < x$
Sign of y''	$+$	$-$	$+$
Behavior of y	Concave up	Concave down	Concave up

(a) There are no local maxima.

(b) There is a local (and absolute) minimum at $x = 2$.

(c) There are points of inflection at $x = 1$ and at $x = \frac{5}{3}$.

35. (a) Absolute maximum at $(1, 2)$; absolute minimum at $(3, -2)$

(b) None

(c) One possible answer

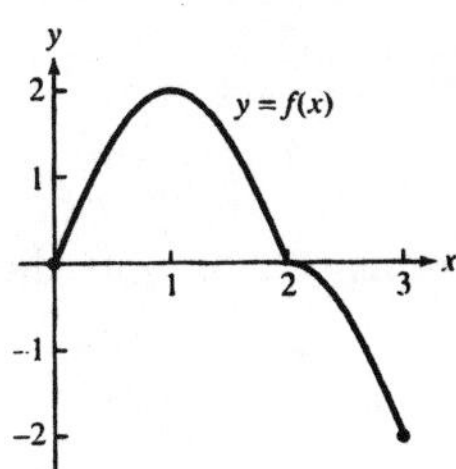

37. If $y = x^5 - 5x^4 - 240$, then $y' = 5x^3(x-4)$ and $y'' = 20x^2(x-3)$. The zeros of y' are extrema of y. The right-hand zero of y'' is a point of inflection for y. Inflection at $x = 3$, local maximum at $x = 0$, local minimum at $x = 4$.

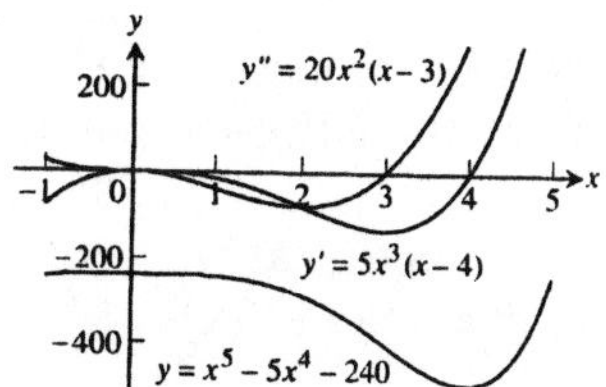

39. If $y = \frac{4}{5}x^5 + 16x^2 - 25$, then $y' = 4x\left(x^3 + 8\right)$ and $y'' = 16\left(x^3 + 2\right)$. The zeros of y' and y'' are extrema and points of inflection, respectively.

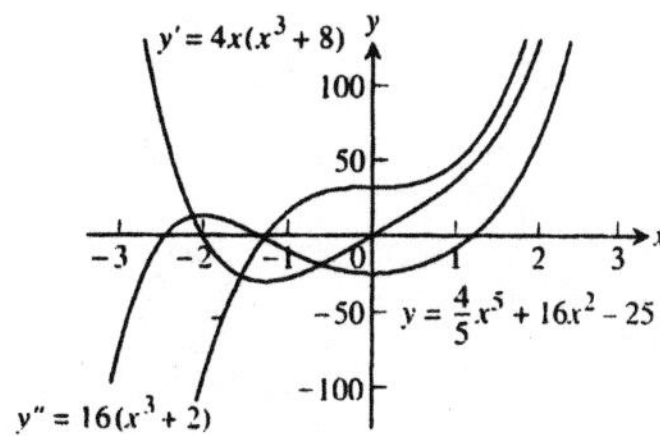

41. The graph of f falls where $f' < 0$, rises where $f' > 0$, and has horizontal tangents where $f' = 0$. It has local minima at points where f' changes from negative to positive and local maxima where f' changes from positive to negative. The graph of f is concave down where $f'' < 0$ and concave up where $f'' > 0$. It has points of inflection at values of x where f'' changes sign and a tangent line exists.

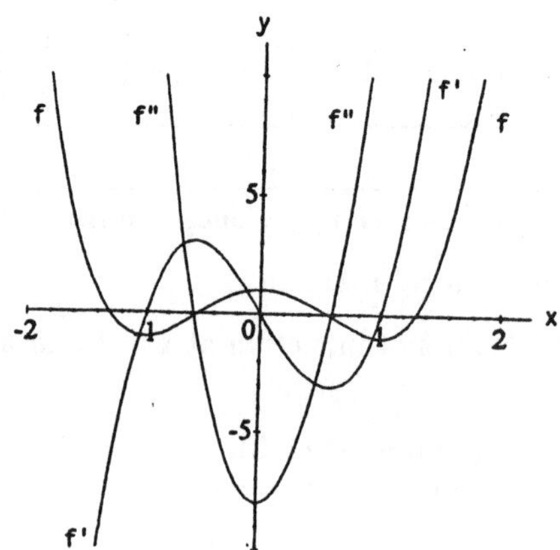

43. (a) $v(t) = s'(t) = 2t - 4$ (b) $a(t) = v'(t) = 2$
(c) It begins at position 3 moving in a negative direction. It moves to position -1 when $t = 2$, and then changes direction, moving in a positive direction thereafter.

45. (a) $v(t) = s'(t) = 3t^2 - 3$ (b) $a(t) = v'(t) = 6t$
(c) It begins at position 3 moving in a negative direction. It moves to position 1 when $t = 1$, and then changes direction, moving in a positive direction thereafter.

47. (a) The velocity is zero when the tangent line is horizontal, at approximately $t = 2.2$, $t = 6$, and $t = 9.8$.
(b) The acceleration is zero at the inflection points, approximately $t = 4$, $t = 8$, and $t = 12$.

49. No. f must have a horizontal tangent at that point, but f could be increasing (or decreasing), and there would be no local extremum. For example, if $f(x) = x^3$, $f'(0) = 0$ but there is no local extremum at $x = 0$.

51. One possible answer:

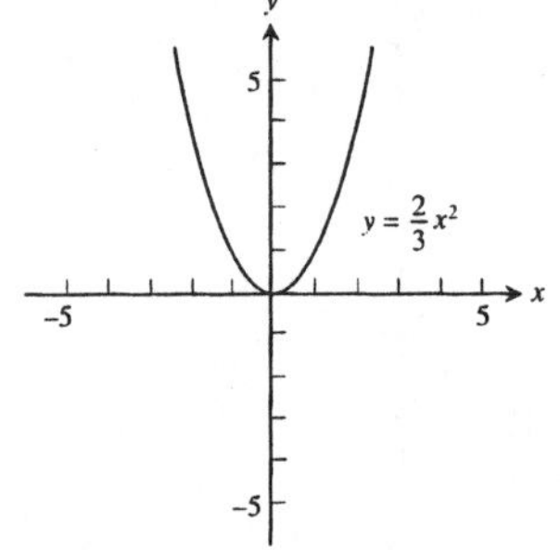

53. One possible answer:

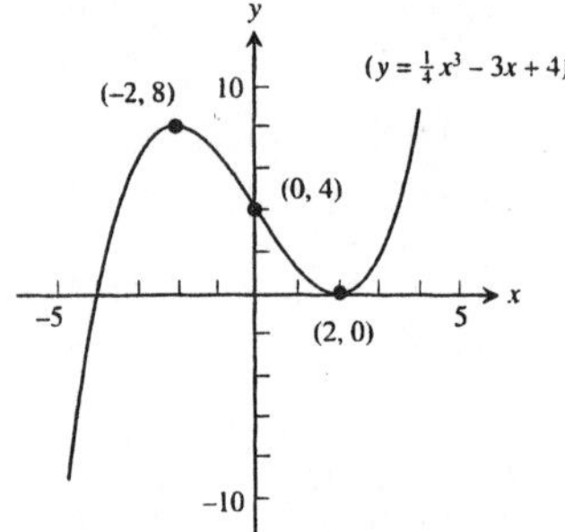

55. The graph must be concave down for $x > 0$ because

$f''(x) = -\frac{1}{x^2} < 0.$

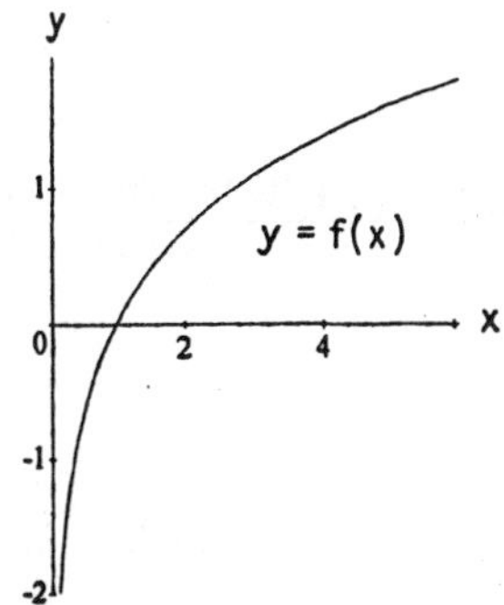

57. A quadratic curve never has an inflection point. If $y = ax^2 + bx + c$ where $a \neq 0$, then $y' = 2ax + b$ and $y'' = 2a$. Since 2a is a constant, it is not possible for y'' to change signs.

59. With $f(-2) = 11 > 0$ and $f(-1) = -1 < 0$ we conclude from the Intermediate Value Theorem that $f(x) = x^4 + 3x + 1$ has at least one zero between -2 and -1. Then $-2 < x < -1 \Rightarrow -8 < x^3 < -1$ $\Rightarrow -32 < 4x^3 < -4 \Rightarrow -29 < 4x^3 + 3 < -1 \Rightarrow f'(x) < 0$ for $-2 < x < -1 \Rightarrow f(x)$ is decreasing on $[-2, -1]$ $\Rightarrow f(x) = 0$ has exactly one solution in the interval $(-2, -1)$.

61. $r(\theta) = \theta + \sin^2\left(\frac{\theta}{3}\right) - 8 \Rightarrow r'(\theta) = 1 + \frac{2}{3}\sin\left(\frac{\theta}{3}\right)\cos\left(\frac{\theta}{3}\right) = 1 + \frac{1}{3}\sin\left(\frac{2\theta}{3}\right) > 0$ on $(-\infty, \infty) \Rightarrow r(\theta)$ is increasing on $(-\infty, \infty)$; $r(0) = -8$ and $r(8) = \sin^2\left(\frac{8}{3}\right) > 0 \Rightarrow r(\theta)$ has exactly one zero in $(-\infty, \infty)$.

63. (a) It appears to control the number and magnitude of the local extrema. If $k < 0$, there is a local maximum to the left of the origin and a local minimum to the right. The larger the magnitude of k ($k < 0$), the greater the magnitude of the extrema. If $k > 0$, the graph has only positive slopes and lies entirely in the first and third quadrants with no local extrema. The graph becomes increasingly steep and straight as $k \to \infty$.

(b) $f'(x) = 3x^2 + k \Rightarrow$ the discriminant $0^2 - 4(3)(k) = -12k$ is positive for $k < 0$, zero for $k = 0$, and negative for $k > 0$; f' has two zeros $x = \pm\sqrt{-\frac{k}{3}}$ when $k < 0$, one zero $x = 0$ when $k = 0$ and no real zeros when $k > 0$; the sign of k controls the number of local extrema.

(c) As $k \to \infty$, $f'(x) \to \infty$ and the graph becomes increasingly steep and straight. As $k \to -\infty$, the crest of the graph (local maximum) in the second quadrant becomes increasingly high and the trough (local minimum) in the fourth quadrant becomes increasingly deep.

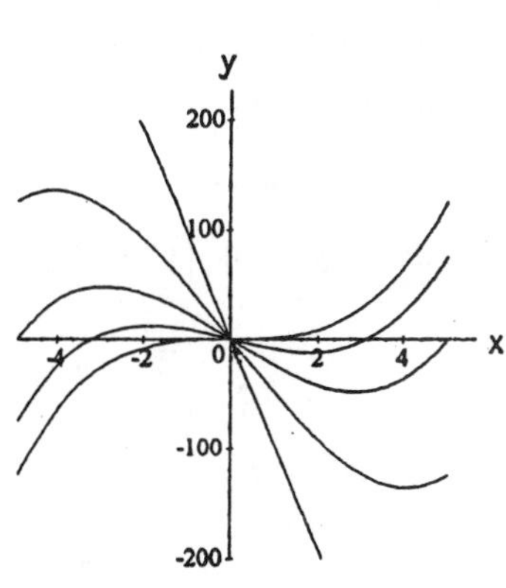

65. (a) $f'(x) = \frac{(1+ae^{-bx})(0)-(c)(-abe^{-bx})}{(1+ae^{-bx})^2} = \frac{abce^{-bx}}{(1+ae^{-bx})^2} = \frac{abce^{bx}}{(e^{bx}+a)^2}$

so the sign of $f'(x)$ is the same as the sign of abc.

(b) $f''(x) = \frac{(e^{bx}+a)^2(ab^2ce^{bx})-(abce^{bx})2(e^{bx}+a)(be^{bx})}{(e^{bx}+a)^4} = \frac{(e^{bx}+a)(ab^2ce^{bx})-(abce^{bx})(2be^{bx})}{(e^{bx}+a)^3 2}$

$= -\frac{ab^2ce^{bx}(e^{bx}-a)}{(e^{bx}+a)^3}$

Since $a > 0$, this changes sign when $x = \frac{\ln a}{b}$ due to the $e^{bx} - a$ factor in the numerator, and $f(x)$ has a point of inflection at that location.

3.4 GRAPHICAL SOLUTIONS TO DIFFERENTIAL EQUATIONS

1. $y' = (y+2)(y-3)$
   (a) $y = -2$ is a stable equilibrium value and $y = 3$ is an unstable equilibrium.
   (b) $y'' = (2y-1)y' = 2(y+2)(y-1/2)(y-3)$

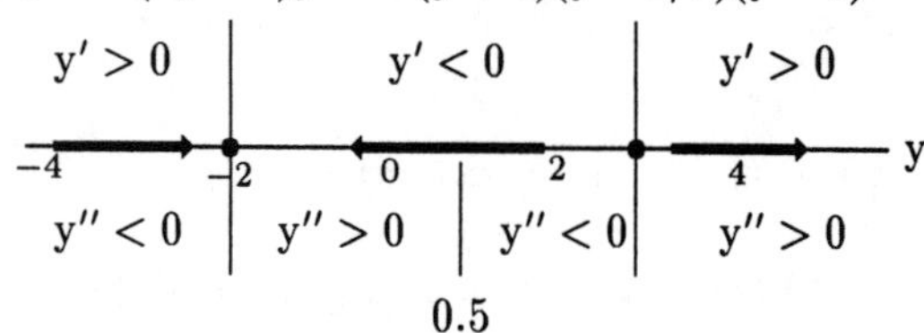

(c)

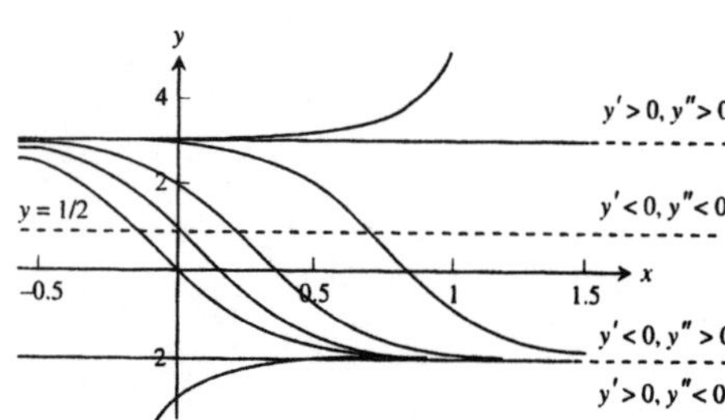

3. $y' = y^3 - y = (y+1)y(y-1)$
   (a) $y = -1$ and $y = 1$ are unstable equilibria and $y = 0$ is a stable equilibrium.
   (b) $y'' = (3y^2-1)y' = 3(y+1)(y+1/\sqrt{3})y(y-1/\sqrt{3})(y-1)$

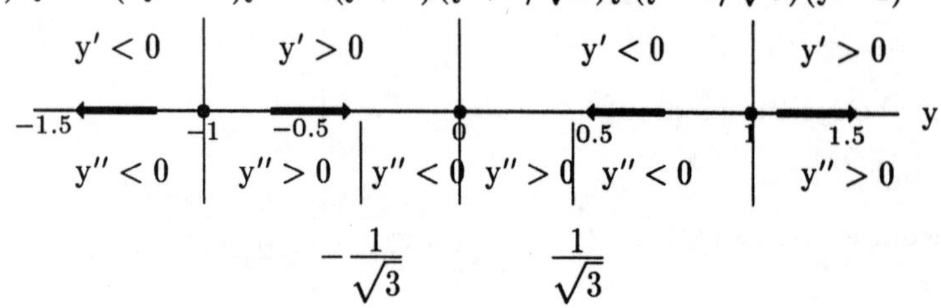

(c)

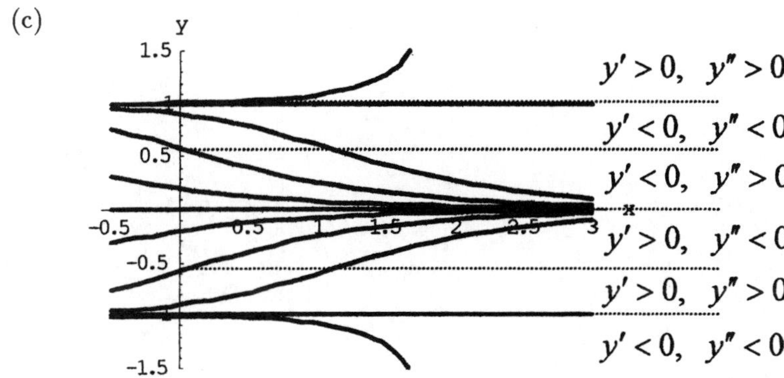

5. $y' = \sqrt{y}$, $y > 0$

(a) There are no equilibrium values.

(b) $y'' = \dfrac{1}{2\sqrt{y}} y' = \dfrac{1}{2\sqrt{y}} \cdot \sqrt{y} = \frac{1}{2}$

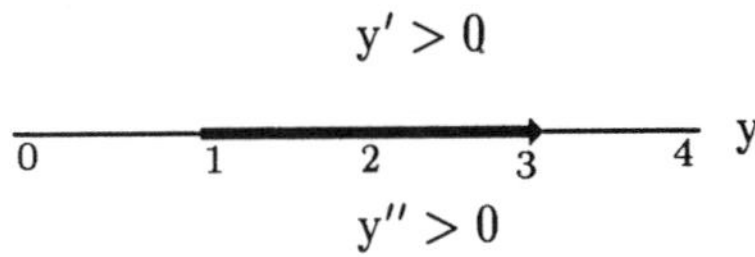

(c)

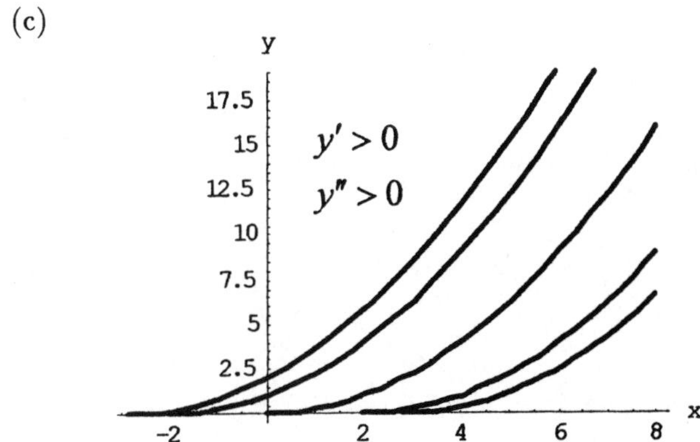

7. $y' = (y-1)(y-2)(y-3)$

(a) $y = 1$ and $y = 3$ are unstable equilibria and $y = 2$ is a stable equilibrium.

(b) $y'' = (3y^2 - 12y + 11)(y-1)(y-2)(y-3) = 3(y-1)\left(y - \dfrac{6-\sqrt{3}}{3}\right)(y-2)\left(y - \dfrac{6+\sqrt{3}}{3}\right)(y-3)$

$y' < 0$	$y' > 0$		$y' < 0$		$y' > 0$
0 — 1	1 —	— 2	2 —	— 3	3 — 4 y
$y'' < 0$	$y'' > 0$	$y'' < 0$	$y'' > 0$	$y'' < 0$	$y'' > 0$

$\dfrac{6-\sqrt{3}}{3} \approx 1.42 \qquad \dfrac{6+\sqrt{3}}{3} \approx 2.58$

(c)

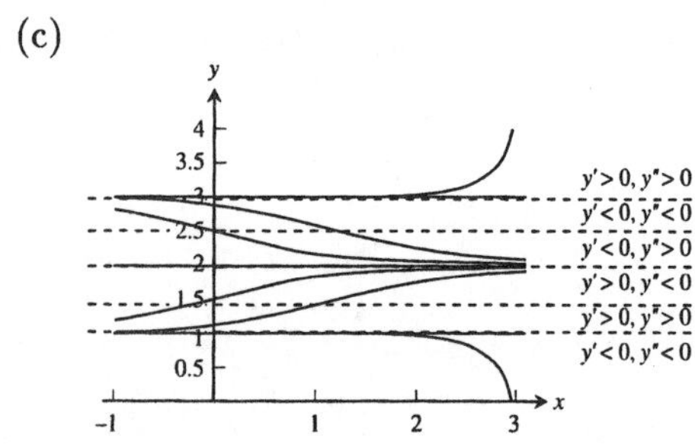

9. $\frac{dP}{dt} = 1 - 2P$ has a stable equilibrium at $P = \frac{1}{2}$. $\frac{d^2P}{dt^2} = -2\frac{dP}{dt} = -2(1-2P)$

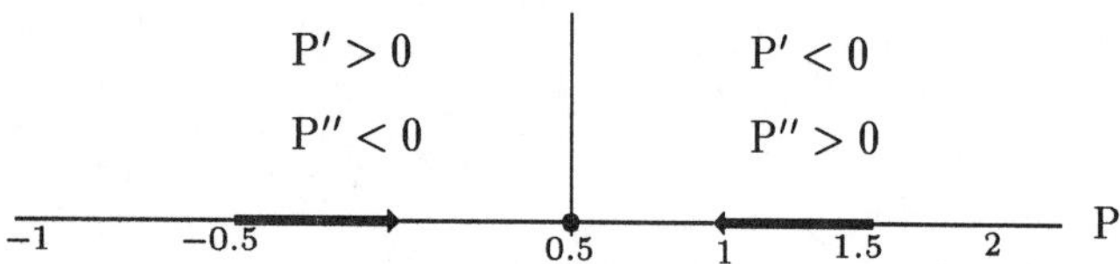

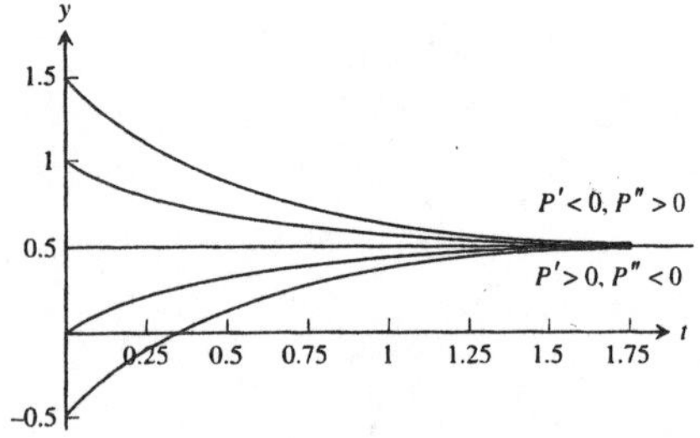

11. $\frac{dP}{dt} = 2P(P-3)$ has a stable equilibrium at $P = 0$ and an unstable equilibrium at $P = 3$.
$\frac{d^2P}{dt^2} = 2(2P-3)\frac{dP}{dt} = 4P(2P-3)(P-3)$

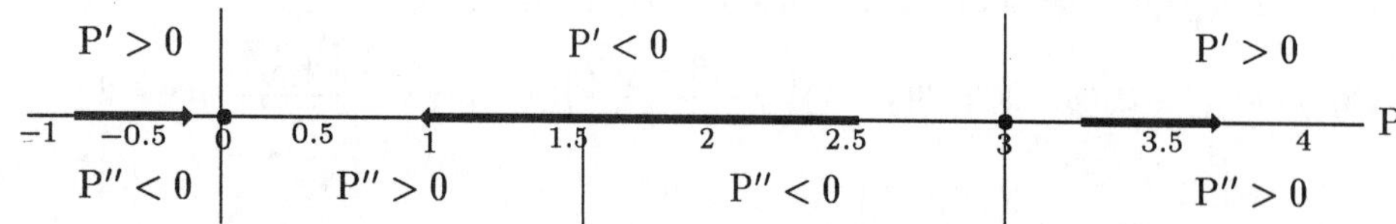

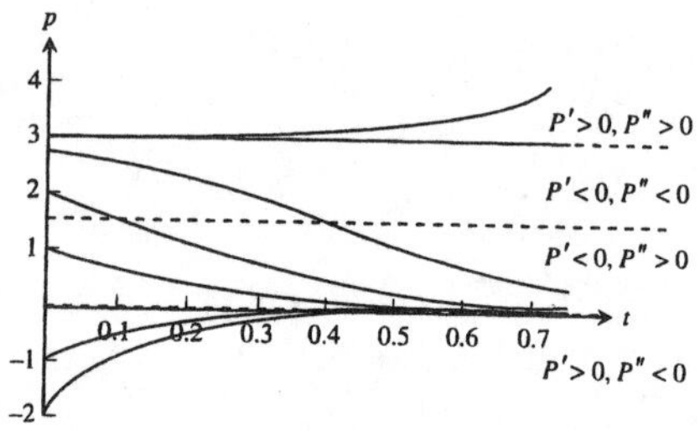

13.

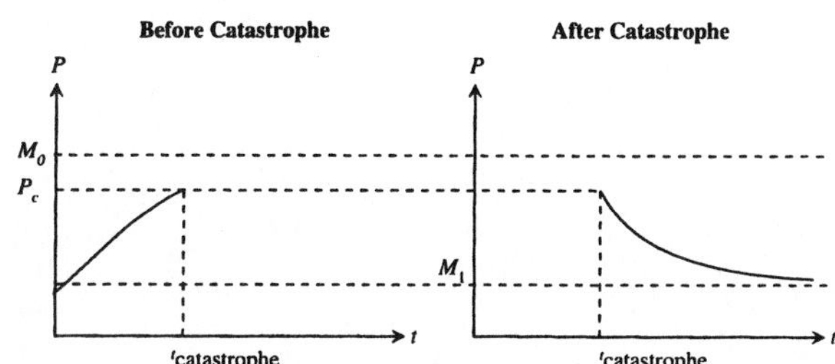

Before the catastrophe, the population exhibits logistic growth and $p(t) \to M_0$, the stable equilibrium.

After the catastrophe, the population declines logistically and $p(t) \to M_1$, the new stable equilibrium.

15. $\frac{dv}{dt} = g - \frac{k}{m}v^2$, g, k, m > 0 and $v(t) \geq 0$

Equilibrium: $\frac{dv}{dt} = g - \frac{k}{m}v^2 = 0 \Rightarrow v = \sqrt{\frac{mg}{k}}$

Concavity: $\frac{d^2v}{dt^2} = -2\left(\frac{k}{m}v\right)\frac{dv}{dt} = -2\left(\frac{k}{m}v\right)\left(g - \frac{k}{m}v^2\right)$

(a)

0 $\frac{dv}{dt} > 0$ $\frac{d^2v}{dt^2} < 0$ $\frac{dv}{dt} < 0$ $\frac{d^2v}{dt^2} > 0$ v

$v_{eq} = \sqrt{\frac{mg}{k}}$

(b)

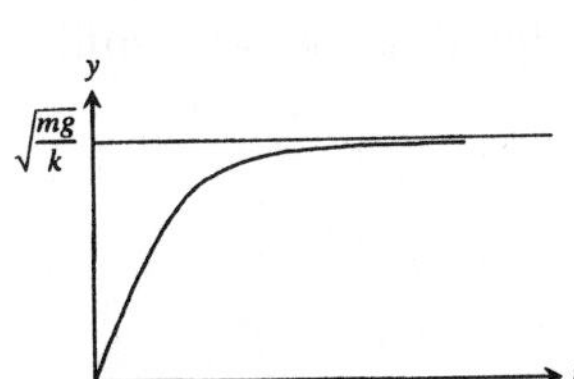

(c) $v_{terminal} = \sqrt{\frac{160}{0.005}} = 178.9\ \frac{ft}{s} = 122$ mph

17. $F = F_p - F_r$

$ma = 50 - 5|v|$

$\frac{dv}{dt} = \frac{1}{m}(50 - 5|v|)$.

The maximum velocity occurs when $\frac{dv}{dt} = 0$ or $v = 10$ ft/sec.

19. $L\frac{di}{dt} + Ri = V \Rightarrow \frac{di}{dt} = \frac{V}{L} - \frac{R}{L}i = \frac{R}{L}\left(\frac{V}{R} - i\right)$, V, L, R > 0

Equilibrium: $\frac{di}{dt} = \frac{R}{L}\left(\frac{V}{R} - i\right) = 0 \Rightarrow i = \frac{V}{R}$,

Concavity: $\frac{d^2i}{dt^2} = -\left(\frac{R}{L}\right)\frac{di}{dt} = -\left(\frac{R}{L}\right)^2\left(\frac{V}{R} - i\right)$

Phase Line:

Phase line:

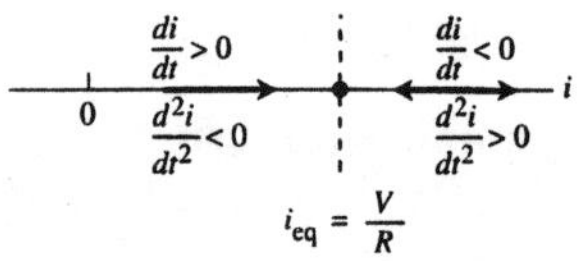

If the switch is closed at t = 0, then i(0) = 0, and the graph of the solution looks like this:

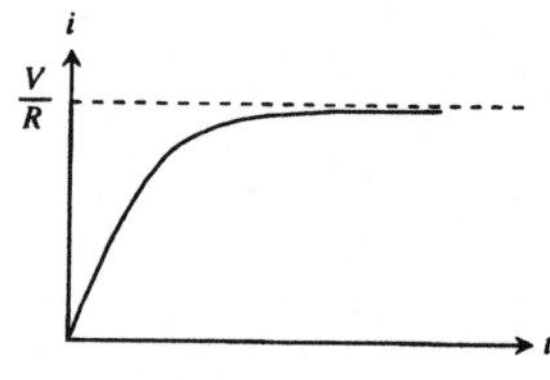

As $t \to \infty$, $i(t) \to i_{\text{steady state}} = \frac{V}{R}$.

As $t \to \infty$, $i(t) \to i_{\text{steady state}} = \frac{V}{R}$. (In the steady state condition, the self-inductance acts like a simple wire connector and, as a result, the current through the resistor can be calculated using the familiar version of Ohm's Law.)

3.5 MODELING AND OPTIMIZATION

1. Let ℓ and w represent the length and width of the rectangle, respectively. With an area of 16 in.^{2}, we have that $(\ell)(w) = 16 \Rightarrow w = 16\ell^{-1} \Rightarrow$ the perimeter is $P = 2\ell + 2w = 2\ell + 32\ell^{-1}$ and $P'(\ell) = 2 - \frac{32}{\ell^2} = \frac{2(\ell^2 - 16)}{\ell^2}$. Solving $P'(\ell) = 0 \Rightarrow \frac{2(\ell + 4)(\ell - 4)}{\ell^2} = 0 \Rightarrow \ell = -4, 4$. Since $\ell > 0$ for the length of a rectangle, ℓ must be 4 and $w = 4 \Rightarrow$ the perimeter is 16 in., a minimum since $P''(\ell) = \frac{16}{\ell^3} > 0$.

3. (a) The line containing point P also contains the points (0, 1) and (1, 0) $\Rightarrow$ the line containing P is $y = 1 - x$ $\Rightarrow$ a general point on that line is $(x, 1 - x)$.
   (b) The area $A(x) = 2x(1 - x)$, where $0 \le x \le 1$.

(c) When $A(x) = 2x - 2x^2$, then $A'(x) = 0 \Rightarrow 2 - 4x = 0 \Rightarrow x = \frac{1}{2}$. Since $A(0) = 0$ and $A(1) = 0$, we conclude that $A\left(\frac{1}{2}\right) = \frac{1}{2}$ sq units is the largest area.

5. The volume of the box is $V(x) = x(15 - 2x)(8 - 2x)$ $= 120x - 46x^2 + 4x^3$, where $0 \le x \le 4$. Solving $V'(x) = 0$ $\Rightarrow 120 - 92x + 12x^2 = 4(6 - x)(5 - 3x) = 0 \Rightarrow x = \frac{5}{3}$ or 6, but 6 is not in the domain. Since $V(0) = V(4) = 0$, $V\left(\frac{5}{3}\right)$ must be the maximum volume of the box with dimensions $\frac{14}{3} \times \frac{35}{3} \times \frac{5}{3}$ inches.

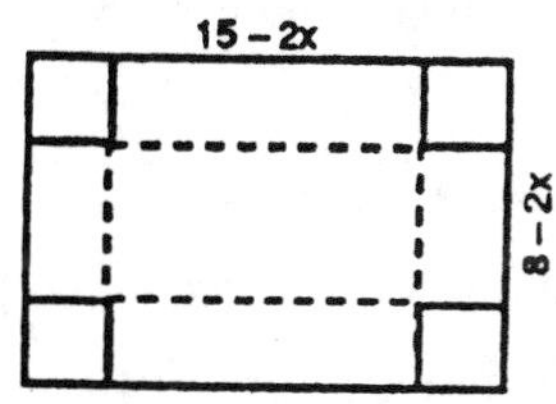

7. The area is $A(x) = x(800 - 2x)$, where $0 \le x \le 400$. Solving $A'(x) = 800 - 4x = 0 \Rightarrow x = 200$. With $A(0) = A(400) = 0$, the maximum area is $A(200) = 80{,}000\ m^2$.

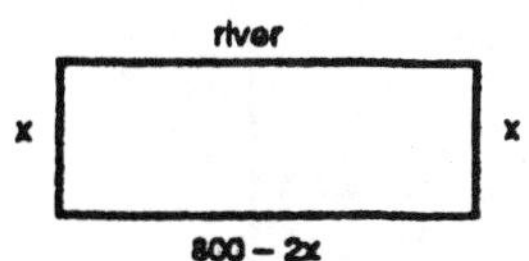

9. (a) We minimize the weight $= tS$ where S is the surface area, and t is the thickness of the steel walls of the tank. The surface area is $S = x^2 + 4xy$ where x is the length of a side of the square base of the tank, and y is its depth. The volume of the tank must be 500–ft^3 $\Rightarrow y = \frac{500}{x^2}$. Therefore, the weight of the tank is $w(x) = t\left(x^2 + \frac{2000}{x}\right)$. Treating the thickness as a constant gives $w'(x) = t\left(2x - \frac{2000}{x^2}\right)$ for $x > 0$. The critical value is at $x = 10$. Since $w''(10) = t\left(2 + \frac{4000}{10^3}\right) > 0$, there is a minimum at $x = 10$. Therefore, the optimum dimensions of the tank are 10–ft on the base edges and 5–ft deep.

(b) Minimizing the surface area of the tank minimizes its weight for a given wall thickness. The thickness of the steel walls would likely be determined by other considerations such as structural requirements.

11. The area of the printing is $(y - 4)(x - 8) = 50$. Consequently, $y = \left(\frac{50}{x - 8}\right) + 4$. The area of the paper is $A(x) = x\left(\frac{50}{x - 8} + 4\right)$, where $8 < x$. Then $A'(x) = \left(\frac{50}{x - 8} + 4\right) - x\left(\frac{50}{(x - 8)^2}\right)$ $= \frac{4(x - 8)^2 - 400}{(x - 8)^2} = 0 \Rightarrow$ the critical points are -2 and 18, but -2 is not in the domain. Thus $A''(18) > 0 \Rightarrow$ at $x = 18$ we have a minimum. Therefore the dimensions 18 by 9 inches minimize the amount of paper.

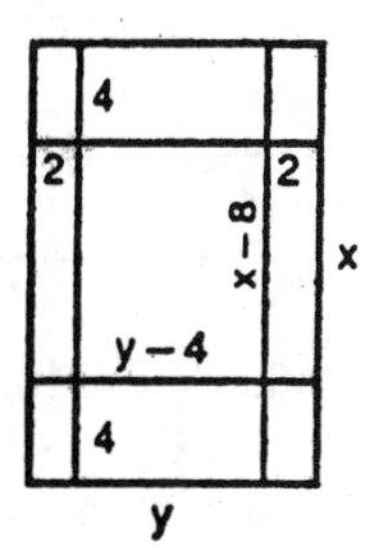

13. The area of the triangle is $A(\theta) = \frac{ab \sin \theta}{2}$, where $0 < \theta < \pi$. Solving $A'(\theta) = 0 \Rightarrow \frac{ab \cos \theta}{2} = 0 \Rightarrow \theta = \frac{\pi}{2}$. Since $A''(\theta)$ $= -\frac{ab \sin \theta}{2} \Rightarrow A''\left(\frac{\pi}{2}\right) < 0$, there is a maximum at $\theta = \frac{\pi}{2}$.

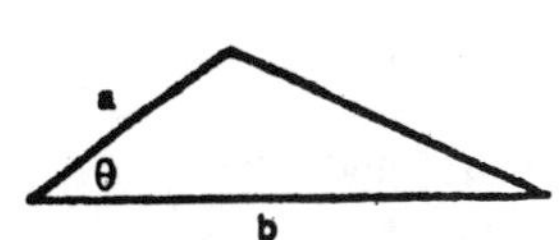

15. With a volume of 1000 cm and $V = \pi r^2 h$, then $h = \frac{1000}{\pi r^2}$. The amount of aluminum used per can is

$A = 8r^2 + 2\pi rh = 8r^2 + \frac{2000}{r}$. Then $A'(r) = 16r - \frac{2000}{r^2} = 0 \Rightarrow \frac{8r^3 - 1000}{r^2} = 0 \Rightarrow$ the critical points are 0 and 5,

but $r = 0$ results in no can. Since $A''(r) = 16 + \frac{1000}{r^3} > 0$ we have a minimum at $r = 5 \Rightarrow h = \frac{40}{\pi}$ and $h{:}r = 8{:}\pi$.

17. (a) The "sides" of the suitcase will measure $24 - 2x$ in. by $18 - 2x$ in. and will be $2x$ in. apart, so the volume formula is $V(x) = 2x(24 - 2x)(18 - 2x) = 8x^3 - 168x^2 + 864x$.

(b) We require $x > 0$, $2x < 18$, and $2x < 24$. Combining these requirements, the domain is the interval $(0, 9)$.

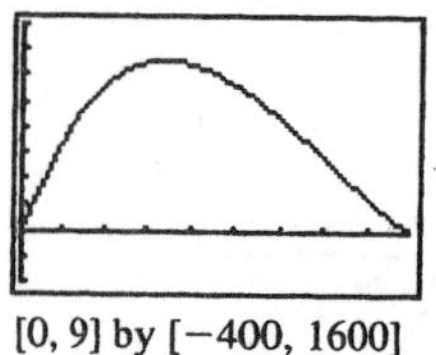

[0, 9] by [−400, 1600]

(c)

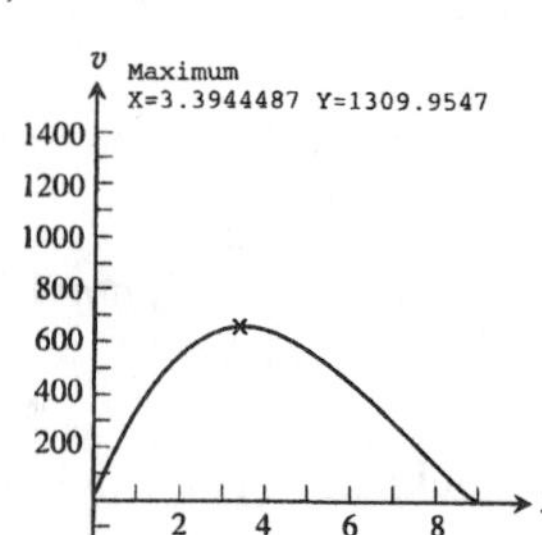

This maximum volume is approximately 1309.95 in^3 when $x \approx 3.39$ in.

(d) $V'(x) = 24x^2 - 336x + 864 = 24(x^2 - 14x + 36)$
The critical point is at

$$x = \frac{14 \pm \sqrt{(-14)^2 - 4(1)(36)}}{2(1)} = \frac{14 \pm \sqrt{52}}{2} = 7 \pm \sqrt{13},$$

that is, $x \approx 3.39$ or $x \approx 10.61$. We discard the larger value because it is not in the domain. Since $V''(x) = 24(2x - 14)$, which is negative when $x \approx 3.39$, the critical point corresponds to the maximum volume. The maximum value occurs at $x = 7 - \sqrt{13} \approx 3.39$, which confirms the results in (c).

(e) $8x^3 - 168x^2 + 864x = 1120$
$8(x^3 - 21x^2 + 108x - 140) = 0$
$8(x - 2)(x - 5)(x - 14) = 0$
Since 14 is not in the domain, the possible values of x are $x = 2$ in. or $x = 5$ in.

(f) The dimensions of the resulting box are $2x$ in., $(24 - 2x)$ in., and $(18 - 2x)$ in. Each of these measurements must be positive, so that gives the domain of $(0, 9)$.

19. Let the radius of the cylinder be r cm, $0 < r < 10$. Then the height is $2\sqrt{100 - r^2}$ and the volume is $V(r) = 2\pi r^2 \sqrt{100 - 4r^2}$ cm^3. Then

$$V'(r) = 2\pi r^2\left(\frac{1}{2\sqrt{100-r^2}}\right)(-2r) + \left(2\pi\sqrt{100-r^2}\right)(2r) = \frac{-2\pi r^3 + 4\pi r(100-r^2)}{\sqrt{100-r^2}} = \frac{2\pi r(200-3r^2)}{\sqrt{100-r^2}}$$

The critical point for $0 < r < 10$ occurs at $r = \sqrt{\frac{200}{3}} = 10\sqrt{\frac{2}{3}}$. Since $V'(r) > 0$ for $0 < r < 10\sqrt{\frac{2}{3}}$ and $V'(r) < 0$ for $10\sqrt{\frac{2}{3}} < r < 10$, the critical point corresponds to the maximum volume. The dimensions are $r = 10\sqrt{\frac{2}{3}} \approx 8.16$ cm and $h = \frac{20}{\sqrt{3}} \approx 11.55$ cm, and the volume is $\frac{4000\pi}{3\sqrt{3}} \approx 2418.40\ \text{cm}^3$.

21. (a) From the diagram we have $3h + 2w = 108$ and $V = h^2 w$ $\Rightarrow V(h) = h^2\left(54 - \frac{3}{2}h\right) = 54h^2 - \frac{3}{2}h^3$. Then $V'(h) = 108h - \frac{9}{2}h^2$ $= \frac{9}{2}h(24-h) = 0 \Rightarrow h = 0$ or $h = 24$, but $h = 0$ results in no box. Since $V''(h) = 108 - 9h < 0$ at $h = 24$, we have a maximum volume at $h = 24$ and $w = 54 - \frac{3}{2}h = 18$.

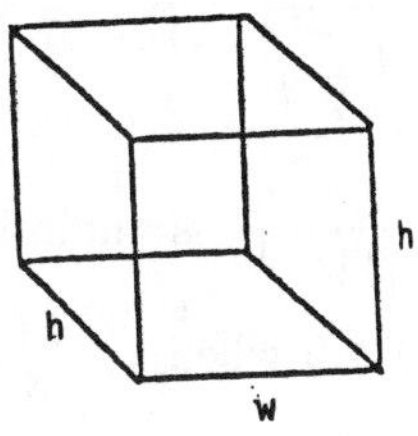

(b)

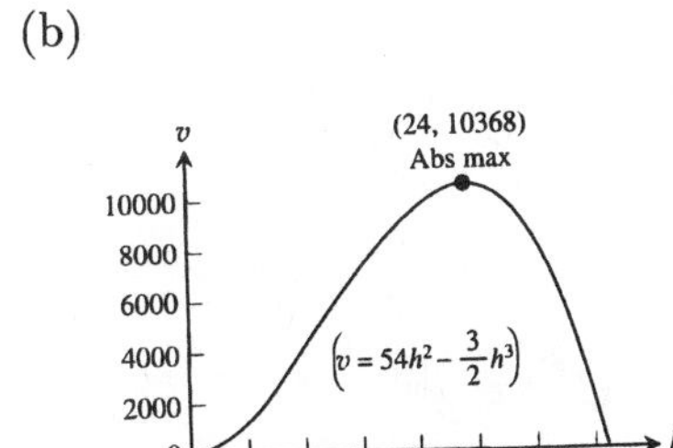

23. The fixed volume is $V = \pi r^2 h + \frac{2}{3}\pi r^3 \Rightarrow h = \frac{V}{\pi r^2} - \frac{2r}{3}$, where h is the height of the cylinder and r is the radius of the hemisphere. To minimize the cost we must minimize surface area of the cylinder added to twice the surface area of the hemisphere. Thus, we minimize $C = 2\pi rh + 4\pi r^2 = 2\pi r\left(\frac{V}{\pi r^2} - \frac{2r}{3}\right) + 4\pi r^2 = \frac{2V}{r} + \frac{8}{3}\pi r^2$. Then $\frac{dC}{dr} = -\frac{2V}{r^2} + \frac{16}{3}\pi r = 0 \Rightarrow V = \frac{8}{3}\pi r^3 \Rightarrow r = \left(\frac{3V}{8\pi}\right)^{1/3}$. From the volume equation, $h = \frac{V}{\pi r^2} - \frac{2r}{3}$ $= \frac{4V^{1/3}}{\pi^{1/3}\cdot 3^{2/3}} - \frac{2\cdot 3^{1/3}\cdot V^{1/3}}{3\cdot 2\cdot \pi^{1/3}} = \frac{3^{1/3}\cdot 2\cdot 4\cdot V^{1/3} - 2\cdot 3^{1/3}\cdot V^{1/3}}{3\cdot 2\cdot \pi^{1/3}} = \left(\frac{3V}{\pi}\right)^{1/3}$. Since $\frac{d^2C}{dr^2} = \frac{4V}{r^3} + \frac{16}{3}\pi > 0$, these dimensions do minimize the cost.

25. (a) From the diagram we have: $\overline{AP} = x$, $\overline{RA} = \sqrt{L - x^2}$, $\overline{PB} = 8.5 - x$, $\overline{CH} = \overline{DR} = 11 - \overline{RA} = 11 - \sqrt{L - x^2}$, $\overline{QB} = \sqrt{x^2 - (8.5-x)^2}$, $\overline{HQ} = 11 - \overline{CH} - \overline{QB}$ $= 11 - \left[11 - \sqrt{L - x^2} + \sqrt{x^2 - (8.5-x)^2}\right]$

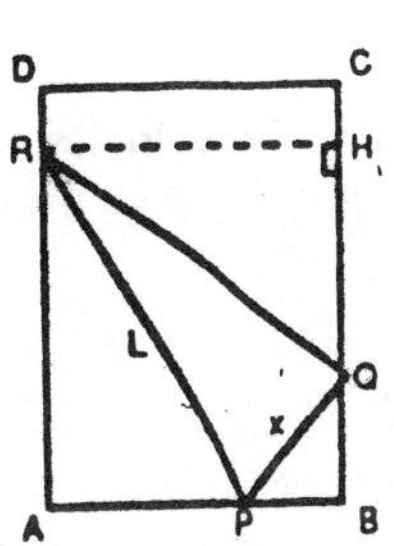

$= \sqrt{L - x^2} - \sqrt{x^2 - (8.5 - x)^2}$,

$\overline{RQ}^2 = \overline{RH}^2 + \overline{HQ}^2 = (8.5)^2 + \left(\sqrt{L - x^2} - \sqrt{x^2 - (8.5 - x)^2}\right)^2$. It follows that

$\overline{RP}^2 = \overline{PQ}^2 + \overline{RQ}^2 \Rightarrow L^2 = x^2 + \left(\sqrt{L^2 - x^2} - \sqrt{x^2 - (x - 8.5)^2}\right)^2 + (8.5)^2$

$\Rightarrow L^2 = x^2 + L^2 - x^2 - 2\sqrt{L^2 - x^2}\,\sqrt{17x - (8.5)^2} + 17x - (8.5)^2 + (8.5)^2$

$\Rightarrow 17^2x^2 = 4\left(L^2 - x^2\right)\left(17x - (8.5)^2\right) \Rightarrow L^2 = x^2 + \dfrac{17^2x^2}{4\left[17x - (8.5)^2\right]} = \dfrac{17x^3}{17x - (8.5)^2}$

$= \dfrac{17x^3}{17x - \left(\frac{17}{2}\right)^2} = \dfrac{4x^3}{4x - 17} = \dfrac{2x^3}{(2x - 8.5)}$.

(b) If $f(x) = \dfrac{4x^3}{4x - 17}$ is minimized, then L^2 is minimized. Now $f'(x) = \dfrac{4x^2(8x - 51)}{(4x - 17)^2} \Rightarrow f'(x) < 0$ when $x < \frac{51}{8}$ and $f'(x) > 0$ when $x > \frac{51}{8}$. Thus L^2 is minimized when $x = \frac{51}{8}$.

(c) When $x = \frac{51}{8}$, then $L \approx 11.0$ in.

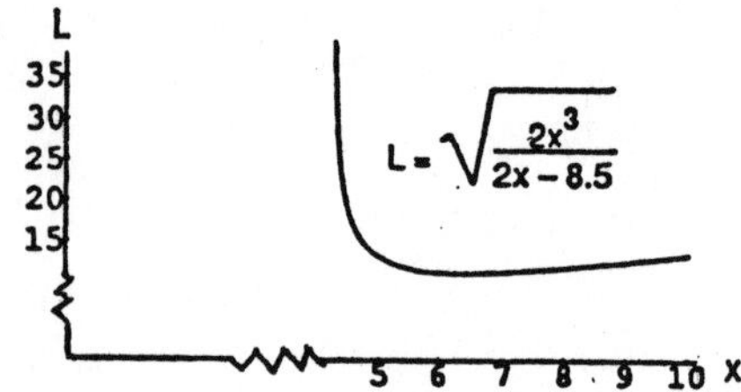

27. Note that $h^2 + r^2 = 3$ and so $r = \sqrt{3 - h^2}$. Then the volume is given by $V = \frac{\pi}{3}r^2h = \frac{\pi}{3}(3 - h^2)h = \pi h - \frac{\pi}{3}h^3$ for $0 < h < \sqrt{3}$, and so $\frac{dV}{dh} = \pi - \pi h^2 = \pi(1 - h^2)$. The critical point (for $h > 0$) occurs at $h = 1$. Since $\frac{dV}{dh} > 0$ for $0 < h < 1$ and $\frac{dV}{dh} < 0$ for $1 < h < \sqrt{3}$, the critical point corresponds to the maximum volume. The cone of greatest volume has radius $\sqrt{2}$ m, height 1 m, and volume $\frac{2\pi}{3}$ m^3.

29. If $f(x) = x^2 + \frac{a}{x}$, then $f'(x) = 2x - ax^{-2}$ and $f''(x) = 2 + 2ax^{-3}$. The critical points are 0 and $\sqrt[3]{\frac{a}{2}}$, but $x \neq 0$. Now $f''\left(\sqrt[3]{\frac{a}{2}}\right) = 6 > 0 \Rightarrow$ at $x = \sqrt[3]{\frac{a}{2}}$ there is a local minimum. However, no local maximum exists for any a.

31. (a) $s(t) = -16t^2 + 96t + 112$
$v(t) = s'(t) = -32t + 96$
At $t = 0$, the velocity is $v(0) = 96$ ft/sec.

(b) The maximum height occurs when $v(t) = 0$, when $t = 3$. The maximum height is $s(3) = 256$ ft and it occurs at $t = 3$ sec.

(c) Note that $s(t) = -16t^2 + 96t + 112 = -16(t + 1)(t - 7)$, so $s = 0$ at $t = -1$ or $t = 7$. Choosing the positive value of t, the velocity when $s = 0$ is $v(7) = -128$ ft/sec.

33. $\frac{8}{x} = \frac{h}{x+27} \Rightarrow h = 8 + \frac{216}{x}$ and $L(x) = \sqrt{h^2 + (x+27)^2}$

$= \sqrt{\left(8 + \frac{216}{x}\right)^2 + (x+27)^2}$ when $x \geq 0$. Note that L(x)

is minimized when $f(x) = \left(8 + \frac{216}{x}\right)^2 + (x+27)^2$ is minimized.

If $f'(x) = 0$, then $2\left(8 + \frac{216}{x}\right)\left(-\frac{216}{x^2}\right) + 2(x+27) = 0$

$\Rightarrow (x+27)\left(1 - \frac{1728}{x^3}\right) = 0 \Rightarrow x = -27$ (not acceptable since

distance is never negative) or $x = 12$. Then $L(12) = \sqrt{2197} \approx 46.87$ ft.

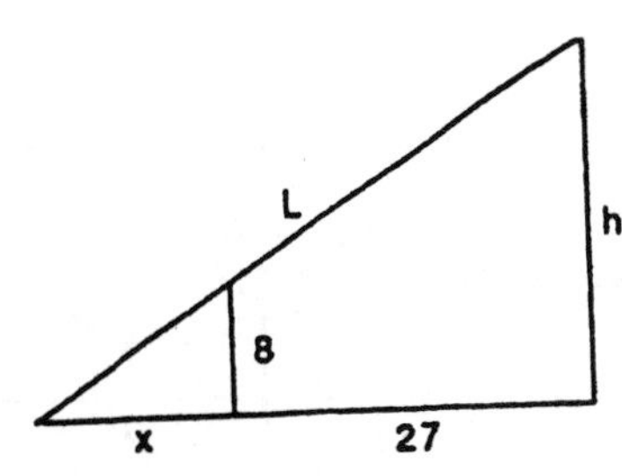

35. (a) From the situation we have $w^2 = 144 - d^2$. The stiffness of the beam is $S = kwd^3 = kd^3(144 - d^2)^{1/2}$, where $0 \leq d \leq 12$. Also, $S'(d) = \frac{4kd^2(108 - d^2)}{\sqrt{144 - d^2}} \Rightarrow$ critical points at 0, 12, and $6\sqrt{3}$. Both $d = 0$ and $d = 12$ cause $S = 0$. The maximum occurs at $d = 6\sqrt{3}$. The dimensions are 6 by $6\sqrt{3}$ inches.

(b)

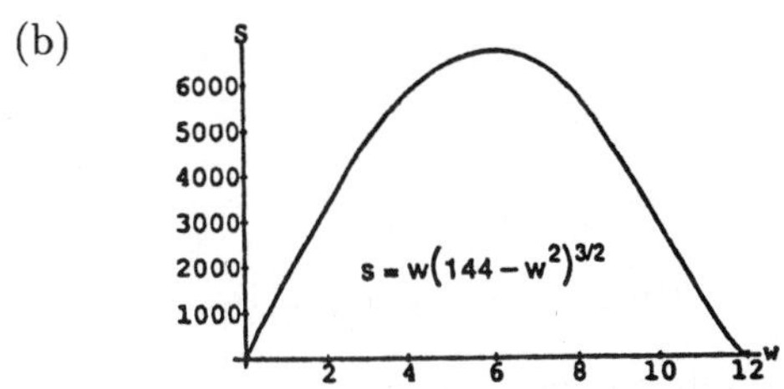

(c)

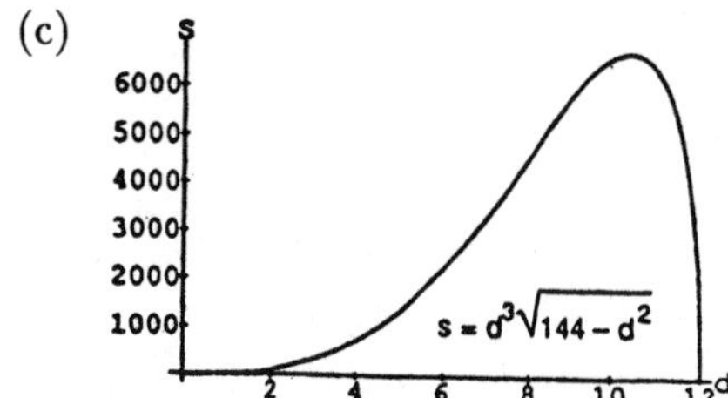

Both graphs indicate the same maximum value and are consistent with each other. The changing of k has no effect.

37. (a) $s = 10\cos(\pi t) \Rightarrow v = -10\pi\sin(\pi t) \Rightarrow$ speed $= |10\pi\sin(\pi t)| = 10\pi|\sin(\pi t)| \Rightarrow$ the maximum speed is $10\pi \approx 31.42$ cm/sec since the maximum value of $|\sin(\pi t)|$ is 1; the cart is moving the fastest at $t = 0.5$ sec, 1.5 sec, 2.5 sec and 3.5 sec when $|\sin(\pi t)|$ is 1. At these times the distance is $s = 10\cos\left(\frac{\pi}{2}\right) = 0$ cm and $a = -10\pi^2\cos(\pi t) \Rightarrow |a| = 10\pi^2|\cos(\pi t)| \Rightarrow |a| = 0$ cm/sec^2

(b) $|a| = 10\pi^2|\cos(\pi t)|$ is greatest at $t = 0.0$ sec, 1.0 sec, 2.0 sec, 3.0 sec and 4.0 sec, and at these times the magnitude of the cart's position is $|s| = 10$ cm from the rest position and the speed is 0 cm/sec.

39. (a) $s = \sqrt{(12 - 12t)^2 + (8t)^2} = ((12 - 12t)^2 + 64t^2)^{1/2}$

(b) $\frac{ds}{dt} = \frac{1}{2}((12 - 12t)^2 + 64t^2)^{-1/2}[2(12 - 12t)(-12) + 128t] = \frac{208t - 144}{\sqrt{(12 - 12t)^2 + 64t^2}}$

$\Rightarrow \left.\frac{ds}{dt}\right|_{t=0} = -12$ knots and $\left.\frac{ds}{dt}\right|_{t=1} = 8$ knots

(c) The graph indicates that the ships did not see each other because $s(t) > 5$ for all values of t.

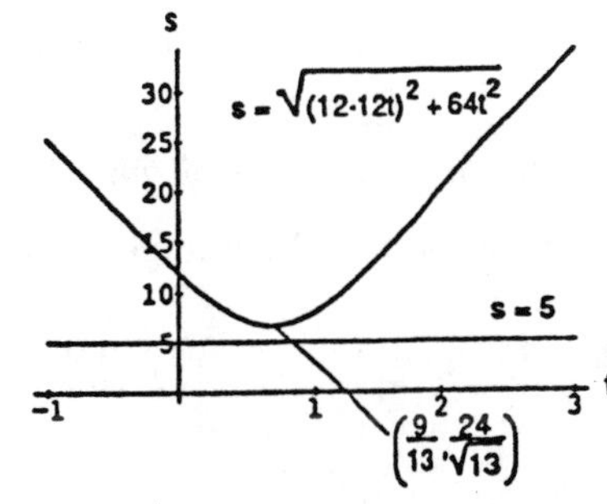

(d) The graph supports the conclusions in parts (b) and (c).

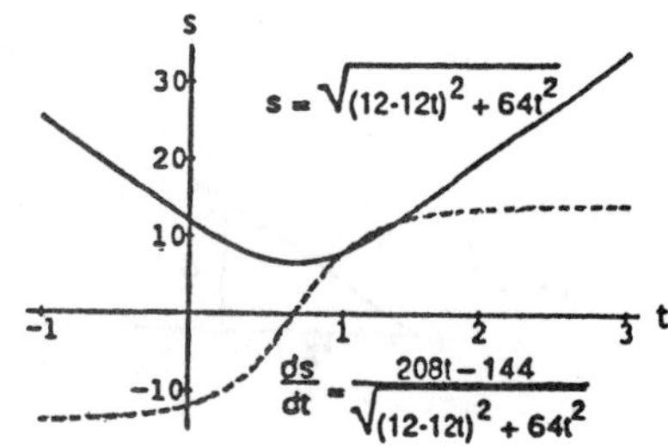

(e) $\lim\limits_{t\to\infty} \frac{ds}{dt} = \sqrt{\lim\limits_{t\to\infty} \frac{(208t-144)^2}{144(1-t)^2+64t^2}} = \sqrt{\lim\limits_{t\to\infty} \frac{\left(208-\frac{144}{t}\right)^2}{144\left(\frac{1}{t}-1\right)^2+64}} = \sqrt{\frac{208^2}{144+64}} = \sqrt{208} = 4\sqrt{13}$

which equals the square root of the sums of the squares of the individual speeds.

41. If $v = kax - kx^2$, then $v' = ka - 2kx$ and $v'' = -2k$, so $v' = 0 \Rightarrow x = \frac{a}{2}$. At $x = \frac{a}{2}$ there is a maximum since $v''\left(\frac{a}{2}\right) = -2k < 0$. The maximum value of v is $\frac{ka^2}{4}$.

43. The profit is $p = nx - nc = n(x-c) = \left[a(x-c)^{-1} + b(100-x)\right](x-c) = a + b(100-x)(x-c)$ $= a + (bc + 100b)x - 100bc - bx^2$. Then $p'(x) = bc + 100b - 2bx$ and $p''(x) = -2b$. Solving $p'(x) = 0 \Rightarrow$ $x = \frac{c}{2} + 50$. At $x = \frac{c}{2} + 50$ there is a maximum profit since $p''(x) = -2b < 0$ for all x.

45. (a) $A(q) = kmq^{-1} + cm + \frac{h}{2}q$, where $q > 0 \Rightarrow A'(q) = -kmq^{-2} + \frac{h}{2} = \frac{hq^2 - 2km}{2q^2}$ and $A''(q) = 2kmq^{-3}$. The critical points are $-\sqrt{\frac{2km}{h}}$, 0, and $\sqrt{\frac{2km}{h}}$, but only $\sqrt{\frac{2km}{h}}$ is in the domain. Then $A''\left(\sqrt{\frac{2km}{h}}\right) > 0 \Rightarrow$ at $q = \sqrt{\frac{2km}{h}}$ there is a minimum average weekly cost.

(b) $A(q) = \frac{(k+bq)m}{q} + cm + \frac{h}{2}q = kmq^{-1} + bm + cm + \frac{h}{2}q$, where $q > 0 \Rightarrow A'(q) = 0$ at $q = \sqrt{\frac{2km}{h}}$ as in (a). Also $A''(q) = 2kmq^{-3} > 0$ so the most economical quantity to order is still $q = \sqrt{\frac{2km}{h}}$ which minimizes the average weekly cost.

47. The profit $p(x) = r(x) - c(x) = 6x - \left(x^3 - 6x^2 + 15x\right) = -x^3 + 6x^2 - 9x$, where $x \ge 0$. Then $p'(x) = -3x^2 + 12x - 9 = -3(x-3)(x-1)$ and $p''(x) = -6x + 12$. The critical points are 1 and 3. Thus $p''(1) = 6 > 0 \Rightarrow$ at $x = 1$ there is a local minimum, and $p''(3) = -6 < 0 \Rightarrow$ at $x = 3$ there is a local maximum. But $p(3) = 0 \Rightarrow$ the best you can do is break even.

49. (a) The artisan should order px units of material in order to have enough until the next delivery.

(b) The average number of units in storage until the next delivery is $\frac{px}{2}$ and so the cost of storing them is $s\left(\frac{px}{2}\right)$ per day, and the total cost for x days is $\left(\frac{px}{2}\right)sx$. When added to the delivery cost, the total cost for delivery and storage for each cycle is: cost per cycle $= d + \frac{px}{2}sx$.

(c) The average cost per day for storage and delivery of materials is:

average cost per day $= \dfrac{\left(d + \frac{ps}{2}x^2\right)}{x} = \frac{d}{x} + \frac{ps}{2}x$. To minimize the average cost per day, set the derivative equal to zero. $\frac{d}{dx}\left(d(x)^{-1} + \frac{ps}{2}x\right) = -d(x)^{-2} + \frac{ps}{2} = 0 \Rightarrow x = \pm\sqrt{\frac{2d}{ps}}$. Only the positive root makes sense in this context so that $x^* = \sqrt{\frac{2d}{ps}}$. To verify that x^* gives a minimum, check the second derivative

$\left[\frac{d}{dx}\left(-dx^{-2} + \frac{ps}{2}\right)\right]\Bigg|_{\sqrt{\frac{2d}{ps}}} = \frac{2d}{x^3}\Bigg|_{\sqrt{\frac{2d}{ps}}} = \dfrac{2d}{\left(\sqrt{\frac{2d}{ps}}\right)^3} > 0 \Rightarrow$ a minimum. The amount to deliver is $px^* = \sqrt{\frac{2pd}{s}}$.

(d) The line and hyperbola intersect when $\frac{d}{x} = \frac{ps}{2}x$. Solving for x gives $x_{\text{intersection}} = \pm\sqrt{\frac{2d}{ps}}$. For $x > 0$, $x_{\text{intersection}} = \sqrt{\frac{2d}{ps}} = x^*$. From this result, the average cost per day is minimized when the average daily cost of delivery is equal to the average daily cost of storage.

51. From Section 2.2 we have $\frac{dR}{dM} = CM - M^2$. Solving $\frac{d^2R}{dM^2} = C - 2M = 0 \Rightarrow M = \frac{C}{2}$. Also, $\frac{d^3R}{dM^3} = -2 \Rightarrow$ at $M = \frac{C}{2}$ there is a maximum.

53. If $x > 0$, then $(x-1)^2 \geq 0 \Rightarrow x^2 + 1 \geq 2x \Rightarrow \frac{x^2+1}{x} \geq 2$. In particular if a, b, c and d are positive integers, then $\left(\frac{a^2+1}{a}\right)\left(\frac{b^2+1}{b}\right)\left(\frac{c^2+1}{c}\right)\left(\frac{d^2+1}{d}\right) \geq 16$.

55. At $x = c$, the tangents to the curves are parallel. Justification: The vertical distance between the curves is $D(x) = f(x) - g(x)$, so $D'(x) = f'(x) - g'(x)$. The maximum value of D will occur at a point c where $D' = 0$. At such a point, $f'(c) = g'(c) = 0$, or $f'(c) = g'(c)$.

57. (a) If $y = \cot x - \sqrt{2}\csc x$ where $0 < x < \pi$, then $y' = (\csc x)\left(\sqrt{2}\cot x - \csc x\right)$. Solving $y' = 0$ $\Rightarrow \cos x = \frac{1}{\sqrt{2}} \Rightarrow x = \frac{\pi}{4}$. For $0 < x < \frac{\pi}{4}$ we have $y' > 0$, and $y' < 0$ when $\frac{\pi}{4} < x < \pi$. Therefore, at $x = \frac{\pi}{4}$ there is a maximum value of $y = -1$.

(b)

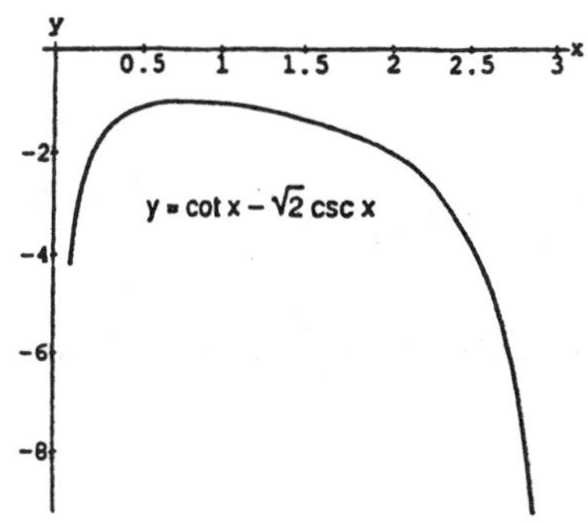

59. (a) The square of the distance is $D(x) = \left(x - \frac{3}{2}\right)^2 + (\sqrt{x} + 0)^2 = x^2 - 2x + \frac{9}{4}$, so $D'(x) = 2x - 2$ and the critical point occurs at $x = 1$. Since $D'(x) < 0$ for $x < 1$ and $D'(x) > 0$ for $x > 1$, the critical point corresponds to the minimum distance. The minimum distance is $\sqrt{D(1)} = \frac{\sqrt{5}}{2}$.

(b)

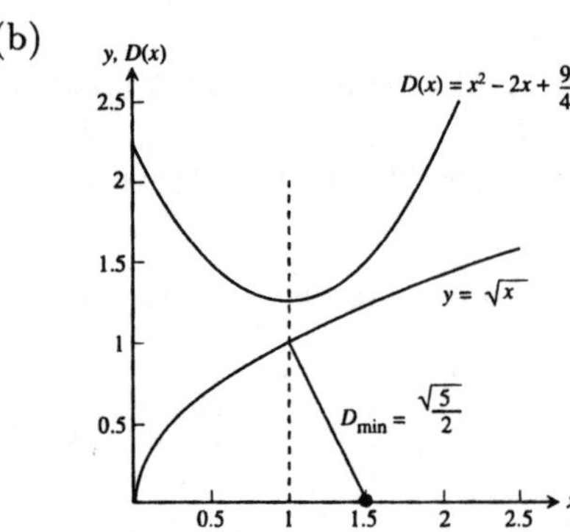

The minimum distance is from the point $(3/2, 0)$ to the point $(1, 1)$ on the graph of $y = \sqrt{x}$, and this occurs at the value $x = 1$ where $D(x)$, the distance squared, has its minimum value.

61. (a) The base radius of the cone is $r = \frac{2\pi a - x}{2\pi}$ and so the height is $h = \sqrt{a^2 - r^2} = \sqrt{a^2 - \left(\frac{2\pi a - x}{2\pi}\right)^2}$.

Therefore, $V(x) = \frac{\pi}{3} r^2 h = \frac{\pi}{3}\left(\frac{2\pi a - x}{2\pi}\right)^2 \sqrt{a^2 - \left(\frac{2\pi a - x}{2\pi}\right)^2}$.

(b) To simplify the calculations, we shall consider the volume as a function of r:

volume $= f(r) = \frac{\pi}{3} r^2 \sqrt{a^2 - r^2}$, where $0 < r < a$.

$$f'(r) = \frac{\pi}{3}\frac{d}{dr}\sqrt{r^2(a^2 - r^2)}$$

$$= \frac{\pi}{3}\left[r^2 \cdot \frac{1}{2\sqrt{a^2 - r^2}} \cdot (-2r) + \left(\sqrt{a^2 - r^2}\right)(2r)\right]$$

$$= \frac{\pi}{3}\left[\frac{-r^3 + 2r(a^2 - r^2)}{\sqrt{a^2 - r^2}}\right]$$

$$= \frac{\pi}{3}\left[\frac{2a^2 r - 3r^3}{\sqrt{a^2 - r^2}}\right]$$

$$= \frac{\pi r(2a^2 - 3r^2)}{3\sqrt{a^2 - r^2}}$$

The critical point occurs when $r^2 = \frac{2a^2}{3}$, which gives $r = a\sqrt{\frac{2}{3}} = \frac{a\sqrt{6}}{3}$. Then $h = \sqrt{a^2 - r^2} = \sqrt{a^2 - \frac{2a^2}{3}} = \sqrt{\frac{a^2}{3}} = \frac{a\sqrt{3}}{3}$. Using $r = \frac{a\sqrt{6}}{3}$ and $h = \frac{a\sqrt{3}}{3}$, we may now find the values of r and h for the given values of a.

When $a = 4$: $r = \frac{4\sqrt{6}}{3}$, $h = \frac{4\sqrt{3}}{3}$;

When $a = 5$: $r = \frac{5\sqrt{6}}{3}$, $h = \frac{5\sqrt{3}}{3}$;

When $a = 6$: $r = 2\sqrt{6}$, $h = 2\sqrt{3}$;

When $a = 8$: $r = \frac{8\sqrt{6}}{3}$, $h = \frac{8\sqrt{3}}{3}$

(c) Since $r = \frac{a\sqrt{6}}{3}$ and $h = \frac{a\sqrt{3}}{3}$, the relationship is $\frac{r}{h} = \sqrt{2}$.

3.6 LINEARIZATION AND DIFFERENTIALS

1. $f(x) = x^3 - 2x + 3 \Rightarrow f'(x) = 3x^2 - 2 \Rightarrow L(x) = f'(2)(x-2) + f(2) = 10(x-2) + 7 \Rightarrow L(x) = 10x - 13$ at $x = 2$

3. $f'(x) = 1 - x^{-2}$
We have $f(1) = 2$ and $f'(1) = 0$. $L(x) = f(1) + f'(1)(x-1) = 2 + 0(x-1) = 2$

5. $f'(x) = \sec^2 x$
We have $f(\pi) = 0$ and $f'(\pi) = 1$. $L(x) = f(\pi) + f'(\pi)(x-\pi) = 0 + 1(x-\pi) = x - \pi$

7. $f'(x) = k(1+x)^{k-1}$
We have $f(0) = 1$ and $f'(0) = k$. $L(x) = f(0) + f'(0)(x-0) = 1 + k(x-0) = 1 + kx$

9. Center $= -1$
$f'(x) = 4x + 4$
We have $f(-1) = -5$ and $f'(-1) = 0$
$L(x) = f(-1) + f'(-1)(x-(-1)) = -5 + 0(x+1) = -5$

11. Center $= 1$

$$f'(x) = \frac{(x+1)(1) - (x)(1)}{(x+1)^2} = \frac{1}{(x+1)^2}$$

We have $f(1) = \frac{1}{2}$ and $f'(1) = \frac{1}{4}$

$L(x) = f(1) + f'(1)(x-1) = \frac{1}{2} + \frac{1}{4}(x-1) = \frac{1}{4}x + \frac{1}{4}$

Alternate solution:

Using center $= \frac{3}{2}$, we have $f\left(\frac{3}{2}\right) = \frac{3}{5}$ and $f'\left(\frac{3}{2}\right) = \frac{4}{25}$.

$L(x) = f\left(\frac{3}{2}\right) + f'\left(\frac{3}{2}\right)\left(x - \frac{3}{2}\right) = \frac{3}{5} + \frac{4}{25}\left(x - \frac{3}{2}\right) = \frac{4}{25}x + \frac{9}{25}$

13. (a) $(1.0002)^{50} = (1 + 0.0002)^{50} \approx 1 + 50(0.0002) = 1 + .01 = 1.01$

(b) $\sqrt[3]{1.009} = (1 + 0.009)^{1/3} \approx 1 + \left(\frac{1}{3}\right)(0.009) = 1 + 0.003 = 1.003$

15. $y = x^3 - 3\sqrt{x} = x^3 - 3x^{1/2} \Rightarrow dy = \left(3x^2 - \frac{3}{2}x^{-1/2}\right)dx \Rightarrow dy = \left(3x^2 - \frac{3}{2\sqrt{x}}\right)dx$

17. $y = \frac{2x}{1+x^2} \Rightarrow dy = \left(\frac{(2)(1+x^2) - (2x)(2x)}{(1+x^2)^2} \right) dx = \frac{2-2x^2}{(1+x^2)^2} \, dx$

19. $2y^{3/2} + xy - x = 0 \Rightarrow 3y^{1/2}\,dy + y\,dx + x\,dy - dx = 0 \Rightarrow \left(3y^{1/2} + x\right) dy = (1-y)\,dx \Rightarrow dy = \frac{1-y}{3\sqrt{y}+x}\,dx$

21. $y = \sin(5\sqrt{x}) = \sin(5x^{1/2}) \Rightarrow dy = \left(\cos(5x^{1/2})\right)\left(\frac{5}{2}x^{-1/2}\right) dx \Rightarrow dy = \frac{5\cos(5\sqrt{x})}{2\sqrt{x}}\,dx$

23. $y = 4\tan\left(\frac{x^3}{3}\right) \Rightarrow dy = 4\left(\sec^2\left(\frac{x^3}{3}\right)\right)(x^2)\,dx \Rightarrow dy = 4x^2\sec^2\left(\frac{x^3}{3}\right) dx$

25. (a) $\Delta f = f(0.1) - f(0) = 0.21 - 0 = 0.21$
    (b) Since $f'(x) = 2x + 2$, $f'(0) = 2$.
    Therefore, $df = 2\,dx = 2(0.1) = 0.2$.
    (c) $|\Delta f - df| = |0.21 - 0.2| = 0.01$

27. (a) $\Delta f = f(0.55) - f(0.5) = \frac{20}{11} - 2 = -\frac{2}{11}$

    (b) Since $f'(x) = -x^{-2}$, $f'(0.5) = -4$.

    Therefore, $df = -4\,dx = -4(0.05) = -0.2 = -\frac{1}{5}$

    (c) $|\Delta f - df| = \left|-\frac{2}{11} + \frac{1}{5}\right| = \frac{1}{55}$

29. Note that $\frac{dV}{dr} = 4\pi r^2$, $dV = 4\pi r4^2\,dr$. When r changes from a to a + dr the change in volume is approximately $4\pi a^2\,dr$.

31. Note that $\frac{dV}{dx} = 3x^2$, so $dV = 3x^2\,dx$. When x changes from a to a + dx, the change in volume is approximately $3a^2\,dx$.

33. Given r = 2 m, dr = .02 m

    (a) $A = \pi r^2 \Rightarrow dA = 2\pi r\,dr = 2\pi(2)(.02) = .08\pi\ m^2$

    (b) $\left(\frac{.08\pi}{4\pi}\right)(100\%) = 2\%$

35. The volume of a cylinder is $V = \pi r^2 h$. When h is held fixed, we have $\frac{dV}{dr} = 2\pi rh$, and so $dV = 2\pi rh\,dr$. For h = 30 in., r = 6 in., and dr = 0.5 in., the thickness of the shell is approximately $dV = 2\pi rh\,dr = 2\pi(6)(30)(0.5)$ $= 180\pi \approx 565.5\ in^3$.

37. $V = \pi h^3 \Rightarrow dV = 3\pi h^2\,dh$; recall that $\Delta V \approx dV$. Then $|\Delta V| \le (1\%)(V) = \frac{(1)(\pi h^3)}{100} \Rightarrow |dV| \le \frac{(1)(\pi h^3)}{100}$ $\Rightarrow |3\pi h^2\,dh| \le \frac{(1)(\pi h^3)}{100} \Rightarrow |dh| \le \frac{1}{300}h = \left(\frac{1}{3}\%\right)h$. Therefore the greatest tolerated error in the measurement of h is $\frac{1}{3}\%$.

39. $V = \pi r^2 h$, h is constant $\Rightarrow dV = 2\pi rh\ dr$; recall that $\Delta V \approx dV$. We want $|\Delta V| \le \frac{1}{1000} V \Rightarrow |dV| \le \frac{\pi r^2 h}{1000}$

$\Rightarrow |2\pi rh\ dr| \le \frac{\pi r^2 h}{1000} \Rightarrow |dr| \le \frac{r}{2000} = (.05\%)r \Rightarrow$ a .05% variation in the radius can be tolerated.

41. $W = a + \frac{b}{g} = a + bg^{-1} \Rightarrow dW = -bg^{-2}\ dg = -\frac{b\ dg}{g^2} \Rightarrow \frac{dW_{moon}}{dW_{earth}} = \frac{\left(-\frac{b\ dg}{(5.2)^2}\right)}{\left(-\frac{b\ dg}{(32)^2}\right)} = \left(\frac{32}{5.2}\right)^2 = 37.87$, so a change of gravity on the moon has about 38 times the effect that a change of the same magnitude has on Earth.

43. (a) Window: $-0.00006 \le x \le 0.00006$, $0.9999 \le y \le 1.001$

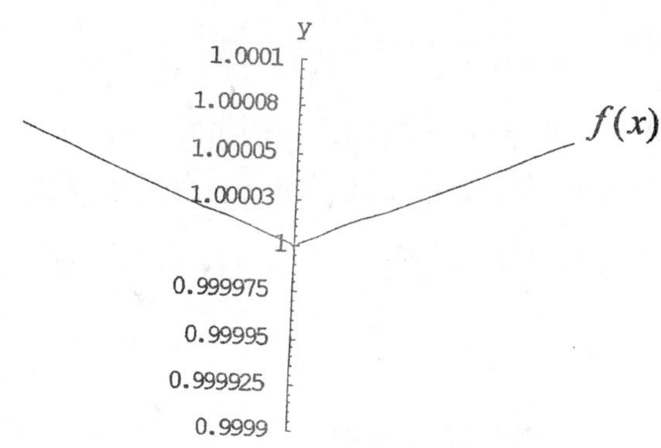

After zooming in seven times, starting with the window $-1 \le x \le 1$ and $0 \le y \le 2$ on a TI-02 Plus calculator, the graph of f(x) shows no signs of straightening out.

(b) Window: $-0.01 \le x \le 0.01$, $0.98 \le y \le 1.02$

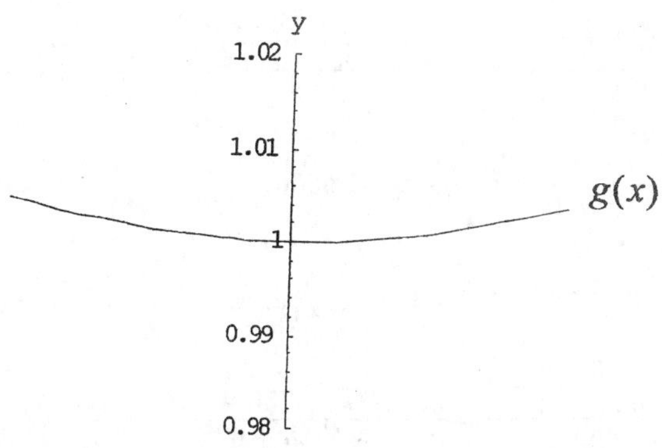

After zooming in only twice, starting with the window $-1 \le x \le 1$ and $0 \le y \le 2$ on a TI-02 Plus calculator, the graph of g(x) already appears to be smoothing toward a horizontal straight line.

(c) After seven zooms, starting with the window $-1 \le x \le 1$ and $0 \le y \le 2$ on a TI-02 Plus calculator, the graph of g(x) looks exactly like a horizontal straight line.

(d) Window: $-1 \le x \le 1$, $0 \le y \le 2$

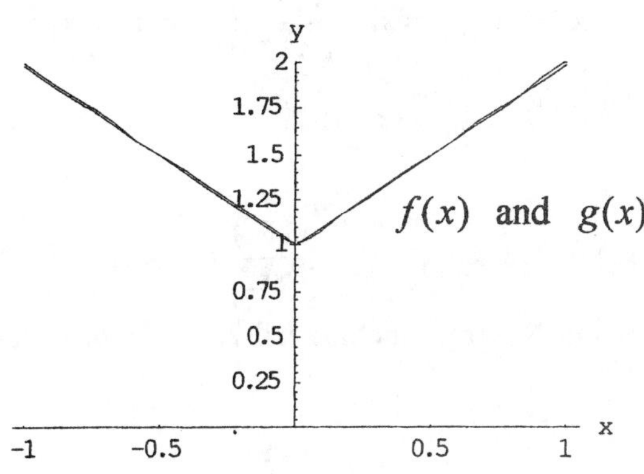

45. $E(x) = f(x) - g(x) \Rightarrow E(x) = f(x) - m(x-a) - c$. Then $E(a) = 0 \Rightarrow f(a) - m(a-a) - c = 0 \Rightarrow c = f(a)$. Next we calculate m: $\lim_{x\to a} \frac{E(x)}{x-a} = 0 \Rightarrow \lim_{x\to a} \frac{f(x) - m(x-a) - c}{x-a} = 0 \Rightarrow \lim_{x\to a} \left[\frac{f(x) - f(a)}{x-a} - m\right] = 0$ (since $c = f(a)$) $\Rightarrow f'(a) - m = 0 \Rightarrow m = f'(a)$. Therefore, $g(x) = m(x-a) + c = f'(a)(x-a) + f(a)$ is the linear approximation, as claimed.

47. $\lim_{x\to 0} \frac{\sqrt{1+x}}{1+\frac{x}{2}} = \frac{\sqrt{1+0}}{1+\frac{0}{2}} = 1$

49. $f(x) = \frac{4x}{x^2+1} \Rightarrow f'(x) = \frac{4(1-x^2)}{(x^2+1)^2}$;

At $x = 0$: $L(x) = f'(0)(x-0) + f(0) = 4x$;

At $x = \sqrt{3}$: $L(x) = f'(\sqrt{3})(x - \sqrt{3}) + f(\sqrt{3})$

$= \left(-\frac{1}{2}\right)(x - \sqrt{3}) + \sqrt{3} \Rightarrow L(x) = \frac{1}{2}(3\sqrt{3} - x)$

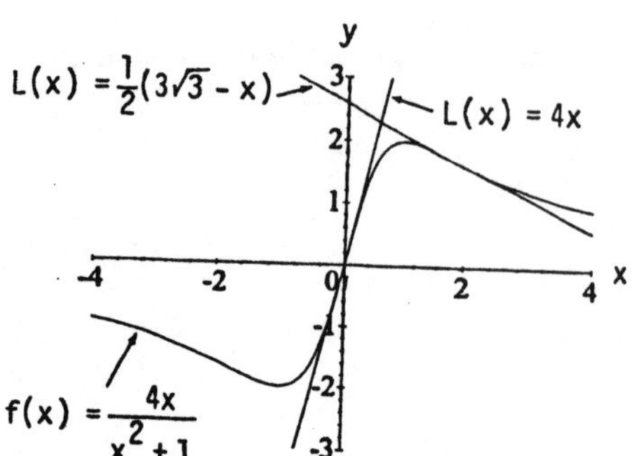

3.7 NEWTON'S METHOD

1. $y = x^2 + x - 1 \Rightarrow y' = 2x + 1 \Rightarrow x_{n+1} = x_n - \frac{x_n^2 + x_n - 1}{2x_n + 1}$; $x_0 = 1 \Rightarrow x_1 = 1 - \frac{1+1-1}{2+1} = \frac{2}{3}$ $\Rightarrow x_2 = \frac{2}{3} - \frac{\frac{4}{9} + \frac{2}{3} - 1}{\frac{4}{3} + 1} \Rightarrow x_2 = \frac{2}{3} - \frac{4+6-9}{12+9} = \frac{2}{3} - \frac{1}{21} = \frac{13}{21} \approx .61905$; $x_0 = -1 \Rightarrow x_1 = 1 - \frac{1-1-1}{-2+1} = -2$ $\Rightarrow x_2 = -2 - \frac{4-2-1}{-4+1} = -\frac{5}{3} \approx -1.66667$

3. $y = x^4 + x - 3 \Rightarrow y' = 4x^3 + 1 \Rightarrow x_{n+1} = x_n - \frac{x_n^4 + x_n - 3}{4x_n^3 + 1}$; $x_0 = 1 \Rightarrow x_1 = 1 - \frac{1+1-3}{4+1} = \frac{6}{5}$ $\Rightarrow x_2 = \frac{6}{5} - \frac{\frac{1296}{625} + \frac{6}{5} - 3}{\frac{864}{125} + 1} = \frac{6}{5} - \frac{1296 + 750 - 1875}{4320 + 625} = \frac{6}{5} - \frac{171}{4945} = \frac{5763}{4945} \approx 1.16542$; $x_0 = -1 \Rightarrow x_1 = -1 - \frac{1-1-3}{-4+1}$ $= -2 \Rightarrow x_2 = -2 - \frac{16-2-3}{-32+1} = -2 + \frac{11}{31} = -\frac{51}{31} \approx -1.64516$

5. $y = x^4 - 2 \Rightarrow y' = 4x^3 \Rightarrow x_{n+1} = x_n - \frac{x_n^4 - 2}{4x_n^3}$; $x_0 = 1 \Rightarrow x_1 = 1 - \frac{1-2}{4} = \frac{5}{4} \Rightarrow x_2 = \frac{5}{4} - \frac{\frac{625}{256} - 2}{\frac{125}{16}} = \frac{5}{4} - \frac{625-512}{2000}$ $= \frac{5}{4} - \frac{113}{2000} = \frac{2500 - 113}{2000} = \frac{2387}{2000} \approx 1.1935$

7. $f(x_0) = 0$ and $f'(x_0) \neq 0 \Rightarrow x_{n+1} = x_n - \frac{f(x_n)}{f'(x_n)}$ gives $x_1 = x_0 \Rightarrow x_2 = x_0 \Rightarrow x_n = x_0$ for all $n \geq 0$. That is, all of the approximations in Newton's method will be the root of $f(x) = 0$ as well as x_0.

9. If $x_0 = h > 0 \Rightarrow x_1 = x_0 - \frac{f(x_0)}{f'(x_0)} = h - \frac{f(h)}{f'(h)}$

$$= h - \frac{\sqrt{h}}{\left(\frac{1}{2\sqrt{h}}\right)} = h - (\sqrt{h})(2\sqrt{h}) = -h;$$

if $x_0 = -h < 0 \Rightarrow x_1 = x_0 - \frac{f(x_0)}{f'(x_0)} = -h - \frac{f(-h)}{f'(-h)}$

$$= -h - \frac{\sqrt{h}}{\left(\frac{-1}{2\sqrt{h}}\right)} = -h + (\sqrt{h})(2\sqrt{h}) = h.$$

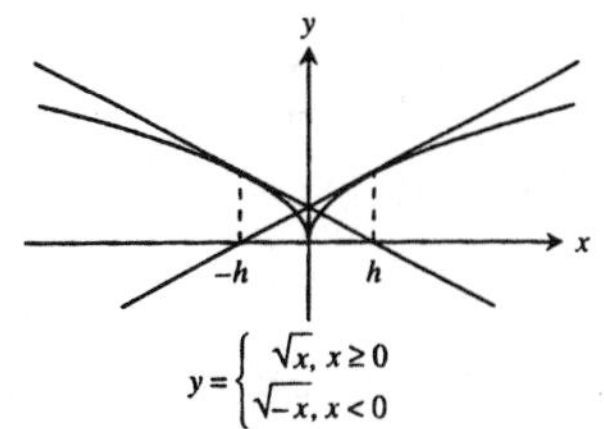

11. The points of intersection of $y = x^3$ and $y = 3x + 1$, or of $y = x^3 - 3x$ and $y = 1$, have the same x-values as the roots of $f(x) = x^3 - 3x - 1$ or the solutions of $g'(x) = 0$.

13. The following commands are for the TI-92 Plus calculator. (Be sure your calculator is in approximate mode.) Go to the home screen and type the following:
   (a) Define f(x) = x^3+3*x+1 (enter)
   f(x) STO> y0 (enter)
   nDeriv(f(x),x) STO> yp (enter)
   (b) −0.3 STO> x (enter)
   (c) x − y0 ÷ yp STO> x (enter)(enter)(enter)
   After executing the last command two times the value, x = −0.322185, does not change in the sixth decimal place thereafter.
   (d) Now try x0 = 0 by typing the following commands:
   0 STO> x (enter)
   x − y0 ÷ yp STO> x (enter)(enter)(enter)(enter)
   After executing the last command three times the value, x = −0.322185, does not change in the sixth decimal place thereafter.
   (e) Try f(x) = sin x to estimate the zero at x = π by typing the following:
   Define f(x) = sin(x) (enter)
   3 STO> x (enter)
   x − y0 ÷ yp STO> x (enter)(enter)(enter)

   After executing the last command two times the value, x = 3.14159, does not change in the fifth decimal place thereafter. The zeros (sin(x), x) command gives 3.14159 · @n1, which means any integer multiple of 3.14159. This matches the above result when @n1 = 1.

15. $f(x) = \tan x - 2x \Rightarrow f'(x) = \sec^2 x - 2 \Rightarrow x_{n+1} = x_n - \frac{\tan(x_n) - 2x_n}{\sec^2(x_n)}$; $x_0 = 1 \Rightarrow x_1 = 1.31047803$

$\Rightarrow x_2 = 1.223929097 \Rightarrow x_6 = x_7 = x_8 = 1.16556119$

17. (a) The graph of $f(x) = \sin 3x - 0.99 + x^2$ in the window $-2 \le x \le 2$, $-2 \le y \le 3$ suggests three roots. However, when you zoom in on the x-axis near x = 1.2, you can see that the graph lies above the axis there. There are only two roots, one near x = −1, the other near x = 0.4.

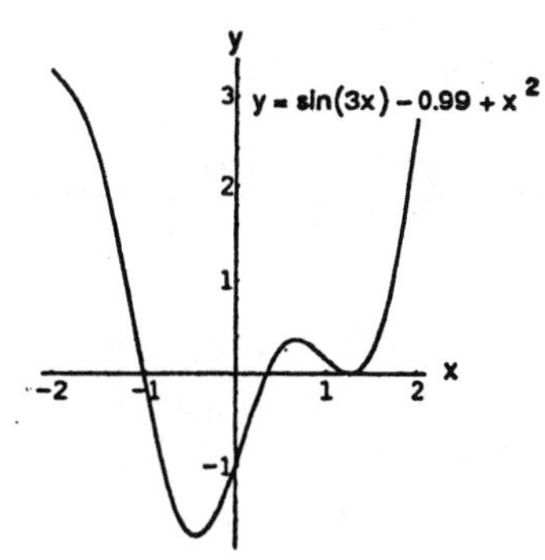

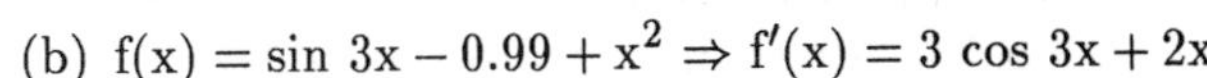
   (b) $f(x) = \sin 3x - 0.99 + x^2 \Rightarrow f'(x) = 3\cos 3x + 2x$

$\Rightarrow x_{n+1} = x_n - \frac{\sin(3x_n) - 0.99 + x_n^2}{3\cos(3x_n) + 2x_n}$ and the solutions

are approximately 0.35003501505249 and −1.0261731615301

19. $f(x) = 2x^4 - 4x^2 + 1 \Rightarrow f'(x) = 8x^3 - 8x \Rightarrow x_{n+1} = x_n - \frac{2x_n^4 - 4x_n^2 + 1}{8x_n^3 - 8x_n}$; if $x_0 = -2$, then $x_6 = -1.30656296$; if $x_0 = -0.5$, then $x_3 = -0.5411961$; the roots are approximately ± 0.5411961 and ± 1.30656296 because f(x) is an even function.

21. From the graph we let $x_0 = 0.5$ and $f(x) = \cos x - 2x$

$\Rightarrow x_{n+1} = x_n - \frac{\cos(x_n) - 2x_n}{-\sin(x_n) - 2} \Rightarrow x_1 = .45063$

$\Rightarrow x_2 = .45018 \Rightarrow$ at $x \approx 0.45$ we have $\cos x = 2x$.

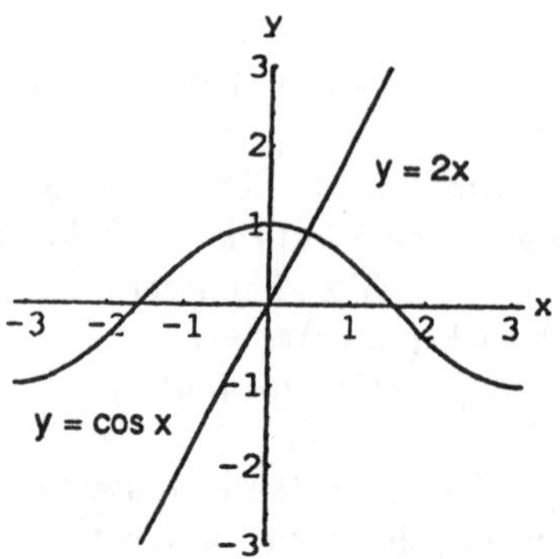

23. If $f(x) = x^3 + 2x - 4$, then $f(1) = -1 < 0$ and $f(2) = 8 > 0 \Rightarrow$ by the Intermediate Value Theorem the equation $x^3 + 2x - 4 = 0$ has a solution between 1 and 2. Consequently, $f'(x) = 3x^2 + 2$ and $x_{n+1} = x_n - \frac{x_n^3 + 2x_n - 4}{3x_n^2 + 2}$. Then $x_0 = 1 \Rightarrow x_1 = 1.2 \Rightarrow x_2 = 1.17975 \Rightarrow x_3 = 1.179509 \Rightarrow x_4 = 1.1795090 \Rightarrow$ the root is approximately 1.17951.

25. $f(x) = 4x^4 - 4x^2 \Rightarrow f'(x) = 16x^3 - 8x \Rightarrow x_{i+1} = x_i - \frac{f(x_i)}{f'(x_i)} = x_i - \frac{x_i^3 - x_i}{4x_i^2 - 2}$. Iterations are performed using the procedure in problem 13 in this section.

(a) For $x_0 = -2$ or $x_0 = -0.8$, $x_i \to -1$ as i gets large.

(b) For $x_0 = -0.5$ or $x_0 = 0.25$, $x_i \to 0$ as i gets large.

(c) For $x_0 = 0.8$ or $x_0 = 2$, $x_i \to 1$ as i gets large.

(d) (If your calculator has a CAS, put it in exact mode, otherwise approximate the radicals with a decimal value.) For $x_0 = -\frac{\sqrt{21}}{7}$ or $x_0 = \frac{\sqrt{21}}{7}$, Newton's method does not converge. The values of x_i alternate between $-\frac{\sqrt{21}}{7}$ and $\frac{\sqrt{21}}{7}$ as i increases.

27. $f(x) = (x-1)^{40} \Rightarrow f'(x) = 40(x-1)^{39} \Rightarrow x_{n+1} = x_n - \frac{(x_n - 1)^{40}}{40(x_n - 1)^{39}} = \frac{39x_n + 1}{40}$. With $x_0 = 2$, our computer gave $x_{87} = x_{88} = x_{89} = \cdots = x_{200} = 1.11051$, coming within 0.11051 of the root $x = 1$.

29. $f(x) = x^3 + 3.6x^2 - 36.4 \Rightarrow f'(x) = 3x^2 + 7.2x \Rightarrow x_{n+1} = x_n - \frac{x_n^3 + 3.6x_n^2 - 36.4}{3x_n^2 + 7.2x_n}$; $x_0 = 2 \Rightarrow x_1 = 2.53\overline{03}$ $\Rightarrow x_2 = 2.45418225 \Rightarrow x_3 = 2.45238021 \Rightarrow x_4 = 2.45237921$ which is 2.45 to two decimal places. Recall that $x = 10^4[H_3O^+] \Rightarrow [H_3O^+] = (x)(10^{-4}) = (2.45)(10^{-4}) = 0.000245$

CHAPTER 3 PRACTICE EXERCISES

1. The global minimum value of $\frac{1}{2}$ occurs at $x = 2$.

3. (a) The function is increasing on the intervals $[-3,-2]$ and $[1,2]$.
   (b) The function is decreasing on the intervals $[-2,0)$ and $(0,2]$.
   (c) Local maximum values occur only at $x = -2$ and at $x = 2$; local minimum values occur at $x = -3$ and at $x = 1$.

5. No, since $f(x) = x^3 + 2x + \tan x \Rightarrow f'(x) = 3x^2 + 2 + \sec^2 x > 0 \Rightarrow f(x)$ is always increasing on its domain.

7. No absolute minimum because $\lim_{x\to\infty} (7+x)(11-3x)^{1/3} = -\infty$. Next $f'(x) =$ $(11-3x)^{1/3} - (7+x)(11-3x)^{-2/3} = \frac{(11-3x)-(7+x)}{(11-3x)^{2/3}} = \frac{4(1-x)}{(11-3x)^{2/3}} \Rightarrow x = 1$ and $x = \frac{11}{3}$ are critical points.

   Since $f' > 0$ if $x < 1$ and $f' < 0$ if $x > 1$, $f(1) = 16$ is the absolute maximum.

9. Yes, because at each point of $[0,1]$ except $x = 0$, the function's value is a local minimum value as well as a local maximum value. At $x = 0$ the function's value, 0, is not a local minimum value because each open interval around $x = 0$ on the x-axis contains points to the left of 0 where f equals -1.

11. No, because the interval $0 < x < 1$ fails to be closed. The Extreme Value Theorem says that if the function is continuous throughout a finite closed interval $a \le x \le b$ then the existence of absolute extrema is guaranteed on that interval.

13. (a) $g(t) = \sin^2 t - 3t \Rightarrow g'(t) = 2 \sin t \cos t - 3 = \sin(2t) - 3 \Rightarrow g' < 0 \Rightarrow g(t)$ is always falling and hence must decrease on every interval in its domain.
    (b) One, since $\sin^2 t - 3t - 5 = 0$ and $\sin^2 t - 3t = 5$ have the same solutions: $f(t) = \sin^2 t - 3t - 5$ has the same derivative as $g(t)$ in part (a) and is always decreasing with $f(-3) > 0$ and $f(0) < 0$. The Intermediate Value Theorem guarantees the continuous function f has a root in $[-3,0]$.

15. (a) $f(x) = x^4 + 2x^2 - 2 \Rightarrow f'(x) = 4x^3 + 4x$. Since $f(0) = -2 < 0$, $f(1) = 1 > 0$ and $f'(x) \ge 0$ for $0 \le x \le 1$, we may conclude from the Intermediate Value Theorem that $f(x)$ has exactly one solution when $0 \le x \le 1$.
    (b) $x^2 = \frac{-2 \pm \sqrt{4+8}}{2} > 0 \Rightarrow x^2 = \sqrt{3} - 1$ and $x \ge 0 \Rightarrow x \approx \sqrt{.7320508076} \approx .8555996772$

17. Let $V(t)$ represent the volume of the water in the reservoir at time t, in minutes, let $V(0) = a_0$ be the initial amount and $V(1440) = a_0 + (1400)(43{,}560)(7.48)$ gallons be the amount of water contained in the reservoir after the rain, where 24 hr = 1440 min. Assume that $V(t)$ is continuous on $[0,1440]$ and differentiable on $(0,1440)$. The Mean Value Theorem says that for some t_0 in $(0,1440)$ we have $V'(t_0) = \frac{V(1440) - V(0)}{1440 - 0}$ $= \frac{a_0 + (1400)(43{,}560)(7.48) - a_0}{1440} = \frac{456{,}160{,}320 \text{ gal}}{1440 \text{ min}} = 316{,}778$ gal/min. Therefore at t_0 the reservoir's volume

was increasing at a rate in excess of 225,000 gal/min.

19. No, $\frac{x}{x+1} = 1 + \frac{-1}{x+1} \Rightarrow \frac{x}{x+1}$ differs from $\frac{-1}{x+1}$ by the constant 1. Both functions have the same derivative $\frac{d}{dx}\left(\frac{x}{x+1}\right) = \frac{(x+1) - x(1)}{(x+1)^2} = \frac{1}{(x+1)^2} = \frac{d}{dx}\left(\frac{-1}{x+1}\right)$.

21.

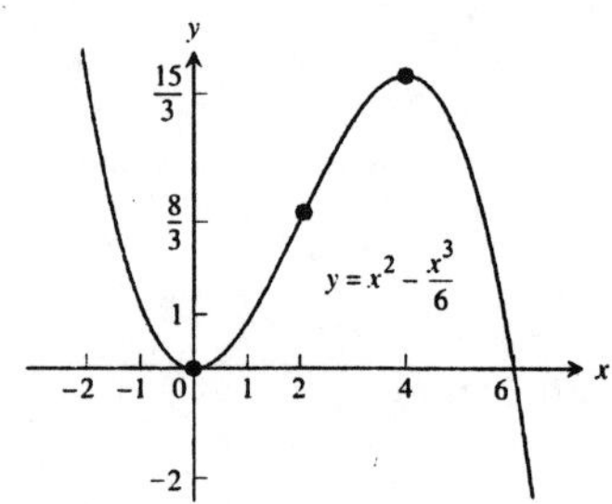

23.

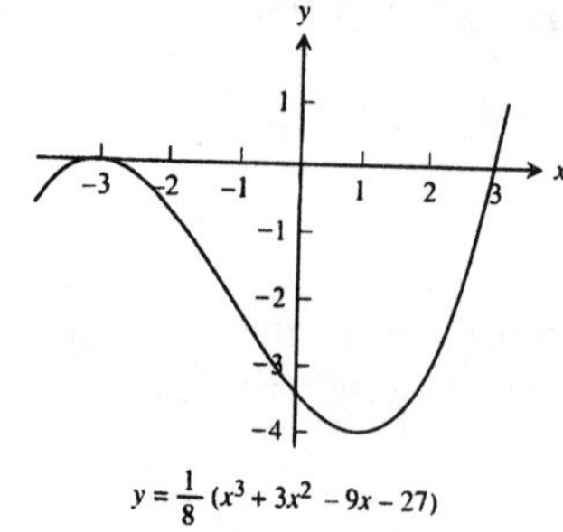

25.

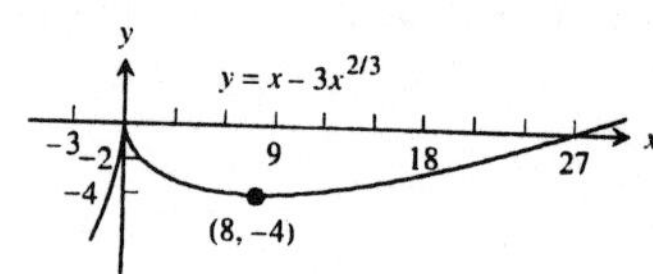

27. (a) $y' = 16 - x^2 \Rightarrow y' = \underset{-4}{---|}\underset{4}{+++|}---$ $\Rightarrow$ the curve is rising on $(-4,4)$, falling on $(-\infty,-4)$ and $(4,\infty)$ $\Rightarrow$ a local maximum at $x = 4$ and a local minimum at $x = -4$; $y'' = -2x \Rightarrow y'' = \underset{0}{+++|}---$ $\Rightarrow$ the curve is concave up on $(-\infty,0)$, concave down on $(0,\infty) \Rightarrow$ a point of inflection at $x = 0$

(b)

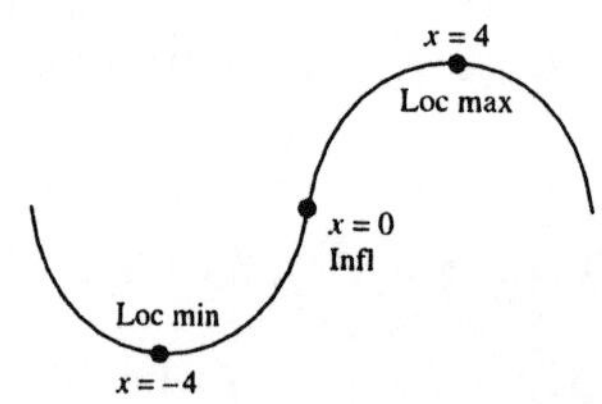

29. (a) $y' = 6x(x+1)(x-2) = 6x^3 - 6x^2 - 12x \Rightarrow y' = \underset{-1}{---|}\underset{0}{+++|}\underset{2}{---|}+++$ $\Rightarrow$ the graph is rising on $(-1,0)$ and $(2,\infty)$, falling on $(-\infty,-1)$ and $(0,2) \Rightarrow$ a local maximum at $x = 0$, local minima at $x = -1$ and $x = 2$; $y'' = 18x^2 - 12x - 12 = 6(3x^2 - 2x - 2) = 6\left(x - \frac{1-\sqrt{7}}{3}\right)\left(x - \frac{1+\sqrt{7}}{3}\right) \Rightarrow$

$y'' = +++ \;\big|\; --- \;\big|\; +++$ (sign changes at $\frac{1-\sqrt{7}}{3}$ and $\frac{1+\sqrt{7}}{3}$) $\Rightarrow$ the curve is concave up on $\left(-\infty, \frac{1-\sqrt{7}}{3}\right)$ and $\left(\frac{1+\sqrt{7}}{3}, \infty\right)$, concave down on $\left(\frac{1-\sqrt{7}}{3}, \frac{1+\sqrt{7}}{3}\right) \Rightarrow$ points of inflection at $x = \frac{1 \pm \sqrt{7}}{3}$

(b)

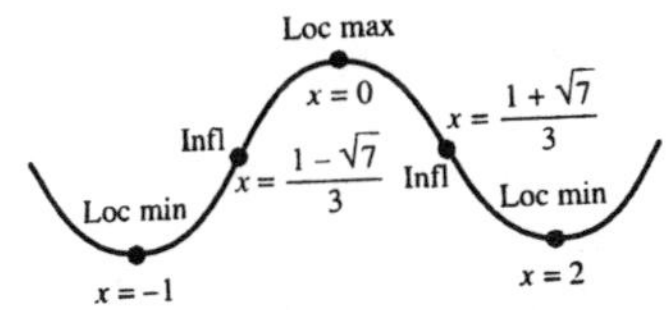

31. (a) $t = 0, 6, 12$ (b) $t = 3, 9$ (c) $6 < t < 12$ (d) $0 < t < 6,\ 12 < t < 14$

33. (a) $v(t) = s'(t) = 4 - 6t - 3t^2$
(b) $a(t) = v'(t) = -6 - 6t$
(c) The particle starts at position 3 moving in the positive direction, but decelerating. At approximately $t = 0.528$, it reaches a position 4.128 and changes direction, beginning to move in the negative direction. After that, it continues to accelerate while moving in the negative direction.

35. Since $\frac{d}{dx}\left(-\frac{1}{4}x^{-4} - \frac{1}{2}\cos 2x\right) = x^{-5} + \sin 2x$, $f(x) = -\frac{1}{4}x^{-4} - \frac{1}{2}\cos 2x + C$.

37. Since $\frac{d}{dx}\left(-\frac{2}{x} + \frac{1}{3}x^3 + x\right) = \frac{2}{x^2} + x^2 + 1$, $f(x) = -\frac{2}{x} + \frac{1}{3}x^3 + x + C$ for $x > 0$.

39. $v(t) = s'(t) = 9.8t + 5 \to s(t) = 4.9t^2 + 5t + C$; $s(0) = 10 \to C = 10 \to s(t) = 4.9t^2 + 5t + 10$

41. $\frac{dy}{dx} = y^2 - 1$

(a) $\frac{dy}{dx} = y^2 - 1 = 0 \Rightarrow y = \pm 1$; $y < -1 \Rightarrow \frac{dy}{dx} > 0$, $-1 < y < 1 \Rightarrow \frac{dy}{dx} < 0$, $y > 1 \Rightarrow \frac{dy}{dx} > 0$. Therefore, $y = -1$ is stable and $y = 1$ is unstable.

(b) $\frac{d^2y}{dx^2} = 2y\,\frac{dy}{dx} = 2y\left(y^2 - 1\right)$

y = −1 | y = 1
$\frac{dy}{dx} > 0$ | $\frac{dy}{dx} < 0$ | $\frac{dy}{dx} < 0$ | $\frac{dy}{dx} > 0$ | y
$\frac{d^2y}{dx^2} < 0$ | $\frac{d^2y}{dx^2} > 0$ | $\frac{d^2y}{dx^2} < 0$ | $\frac{d^2y}{dx^2} > 0$
y = 0

(c)

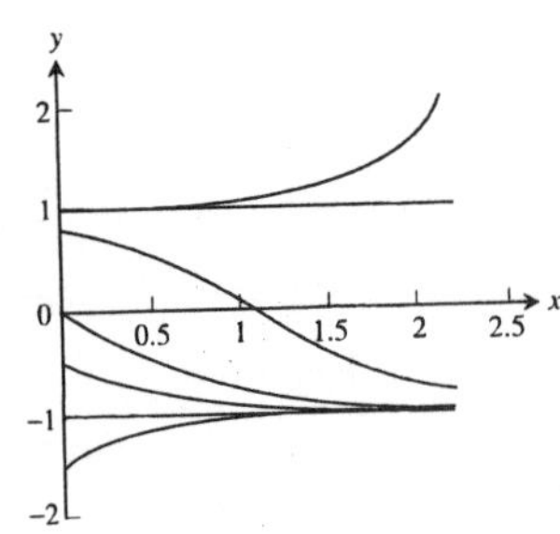

43. Note that $s = 100 - 2r$ and the sector area is given by $A = \pi r^2\left(\frac{s}{2\pi r}\right) = \frac{1}{2}rs = \frac{1}{2}r(100 - 2r) = 50r - r^2$. To find the domain of $A(r) = 50r - r^2$, note that $r > 0$ and $0 < s < 2\pi r$, which gives $12.1 \approx \frac{50}{\pi + 1} < r < 50$. Since $A'(r) = 50 - 2r$, the critical point occurs at $r = 25$. This value is in the domain and corresponds to the maximum area because $A''(r) = -2$, which is negative for all r. The greatest area is attained when $r = 25$ ft and $s = 50$ ft.

45. From the diagram we have $\left(\frac{h}{2}\right)^2 + r^2 = \left(\sqrt{3}\right)^2$ $\Rightarrow r^2 = \frac{12 - h^2}{4}$. The volume of the cylinder is $V = \pi r^2 h = \pi\left(\frac{12 - h^2}{4}\right)h = \frac{\pi}{4}\left(12h - h^3\right)$, where $0 \le h \le 2\sqrt{3}$. Then $V'(h) = \frac{3\pi}{4}(2 + h)(2 - h)$ $\Rightarrow$ the critical points are -2 and 2, but -2 is not in the domain. At $h = 2$ there is a maximum since $V''(2) = -3\pi < 0$. The dimensions of the largest cylinder are radius $= \sqrt{2}$ and height $= 2$.

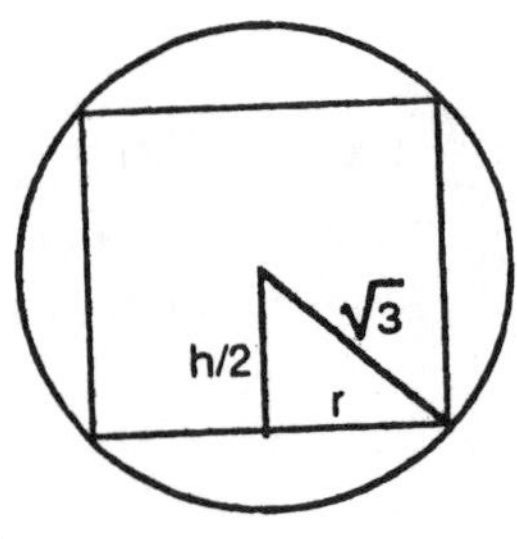

47. The profit $P = 2px + py = 2px + p\left(\frac{40 - 10x}{5 - x}\right)$, where p is the profit on grade B tires and $0 \le x \le 4$. Thus $P'(x) = \frac{2p}{(5 - x)^2}\left(x^2 - 10x + 20\right) \Rightarrow$ the critical points are $\left(5 - \sqrt{5}\right)$, 5, and $\left(5 + \sqrt{5}\right)$, but only $\left(5 - \sqrt{5}\right)$ is in the domain. Now $P'(x) > 0$ for $0 < x < \left(5 - \sqrt{5}\right)$ and $P'(x) < 0$ for $\left(5 - \sqrt{5}\right) < x < 4 \Rightarrow$ at $x = \left(5 - \sqrt{5}\right)$ there is a local maximum. Also $P(0) = 8p$, $P\left(5 - \sqrt{5}\right) = 4p\left(5 - \sqrt{5}\right) \approx 11p$, and $P(4) = 8p \Rightarrow$ at $x = \left(5 - \sqrt{5}\right)$ there is an absolute maximum. The maximum occurs when $x = \left(5 - \sqrt{5}\right)$ and $y = 2\left(5 - \sqrt{5}\right)$, the units are hundreds of tires, i.e., $x \approx 276$ tires and $y \approx 553$ tires.

49. The dimensions will be x in. by $10 - 2x$ in. by $16 - 2x$ in., so $V(x) = x(10 - 2x)(16 - 2x) = 4x^3 - 52x^2 + 160x$ for $0 < x < 5$. Then $V'(x) = 12x^2 - 104x + 160 = 4(x - 2)(3x - 20)$, so the critical point in the correct domain is $x = 2$. This critical point corresponds to the maximum possible volume because $V'(x) > 0$ for $0 < x < 2$ and $V'(x) < 0$ for $2 < x < 5$. The box of largest volume has a height of 2 in. and a base measuring 6 in. by 12 in.,

and its volume is 144 in^3.

Graphical support:

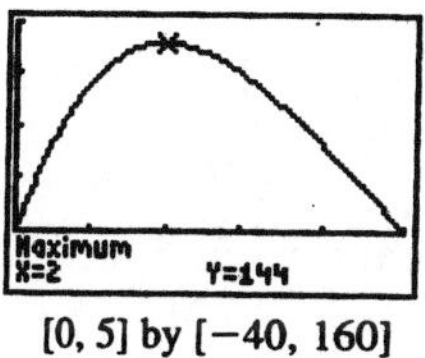

[0, 5] by [−40, 160]

51. (a) If $f(x) = \tan x$ and $x = -\frac{\pi}{4}$, then $f'(x) = \sec^2 x$,

$f\left(-\frac{\pi}{4}\right) = -1$ and $f'\left(-\frac{\pi}{4}\right) = 2$. The linearization of

$f(x)$ is $L(x) = 2\left(x + \frac{\pi}{4}\right) + (-1) = 2x + \frac{\pi - 2}{2}$.

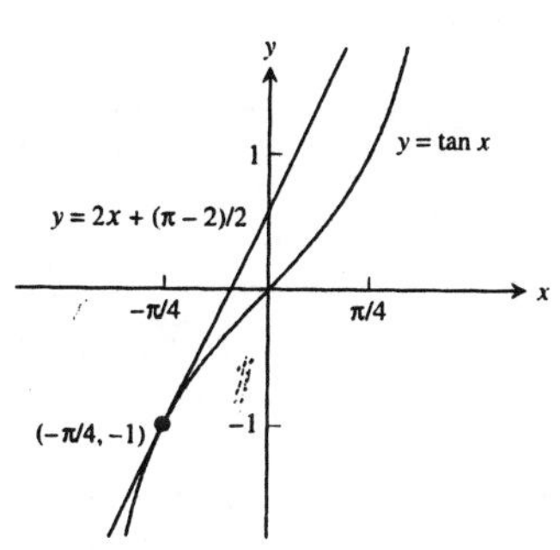

(b) If $f(x) = \sec x$ and $x = -\frac{\pi}{4}$, then $f'(x) = \sec x \tan x$,

$f\left(-\frac{\pi}{4}\right) = \sqrt{2}$ and $f'\left(-\frac{\pi}{4}\right) = -\sqrt{2}$. The linearization of

$f(x)$ is $L(x) = -\sqrt{2}\left(x + \frac{\pi}{4}\right) + \sqrt{2} = -\sqrt{2}x + \frac{\sqrt{2}(4-\pi)}{4}$.

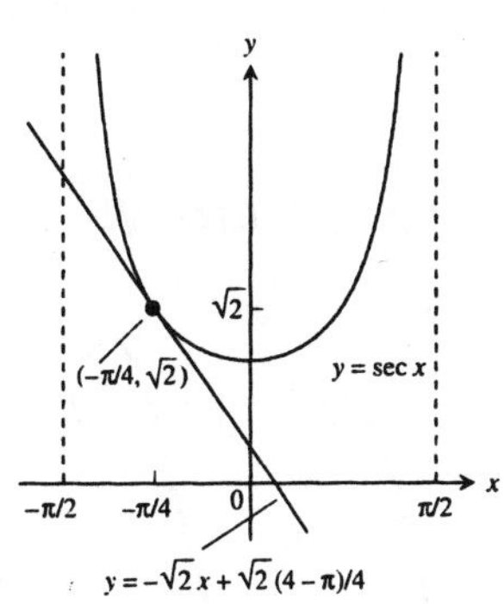

53. $f(x) = \sqrt{x+1} + \sin x - 0.5 = (x+1)^{1/2} + \sin x - 0.5 \Rightarrow f'(x) = \left(\frac{1}{2}\right)(x+1)^{-1/2} + \cos x$

$\Rightarrow L(x) = f'(0)(x-0) + f(0) = 1.5(x-0) + 0.5 \Rightarrow L(x) = 1.5x + 0.5$, the linearization of $f(x)$.

55. When the volume is $V = \frac{1}{3}\pi r^2 h$, then $dV = \frac{2}{3}\pi r_0 h\ dr$ estimates the change in the volume for fixed h.

57. $C = 2\pi r \Rightarrow r = \frac{C}{2\pi}$, $S = 4\pi r^2 = \frac{C^2}{\pi}$, and $V = \frac{4}{3}\pi r^3 = \frac{C^3}{6\pi^2}$. It also follows that $dr = \frac{1}{2\pi}\ dC$, $dS = \frac{2C}{\pi}\ dC$ and

$dV = \frac{C^2}{2\pi^2}\ dC$. Recall that $C = 10$ cm and $dC = 0.4$ cm.

(a) $dr = \frac{0.4}{2\pi} = \frac{0.2}{\pi}$ cm $\Rightarrow \left(\frac{dr}{r}\right)(100\%) = \left(\frac{0.2}{\pi}\right)\left(\frac{2\pi}{10}\right)(100\%) = (.04)(100\%) = 4\%$

(b) $dS = \frac{20}{\pi}(0.4) = \frac{8}{\pi}$ cm $\Rightarrow \left(\frac{dS}{S}\right)(100\%) = \left(\frac{8}{\pi}\right)\left(\frac{\pi}{100}\right)(100\%) = 8\%$

(c) $dV = \frac{10^2}{2\pi^2}(0.4) = \frac{20}{\pi^2}$ cm $\Rightarrow \left(\frac{dV}{V}\right)(100\%) = \left(\frac{20}{\pi^2}\right)\left(\frac{6\pi^2}{1000}\right)(100\%) = 12\%$

59. The graph of f(x) shows that for $1 \le x \le 2$, $f(x) = 0$ has one solution near $x = 1.7$. (Note: The exact solution is $x = \sqrt{3} \approx 1.732051$. Nonetheless, we use Newton's method to find an estimate for this solution.)

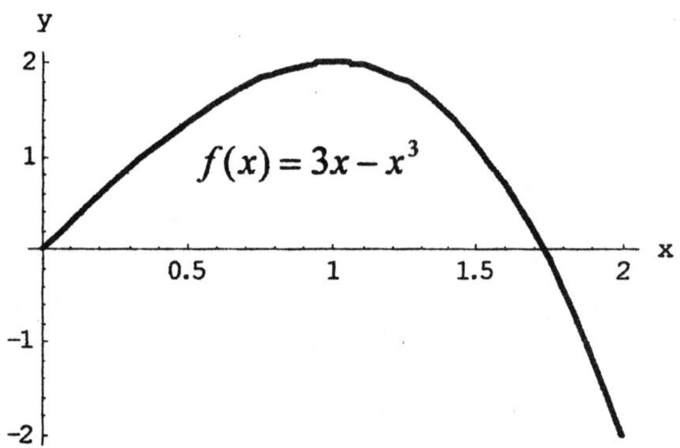

$f(x) = 3x - x^3 \Rightarrow f'(x) = 3 - 3x^2 \Rightarrow x_{n+1} = x_n - \frac{3x_n - x_n^3}{3 - 3x_n^2} \Rightarrow x_1 = 1.7$, $x_2 = 1.732981$, $x_3 = 1.732052$, $x_4 = 1.732051$, $x_5 = 1.732051$. Solution: $x \approx 1.732051$.

61. The domain of g(t) is $(-\infty, 1]$, and the graph of g(t) shows that $g(t) = 0$ has one solution near $t = -1$.

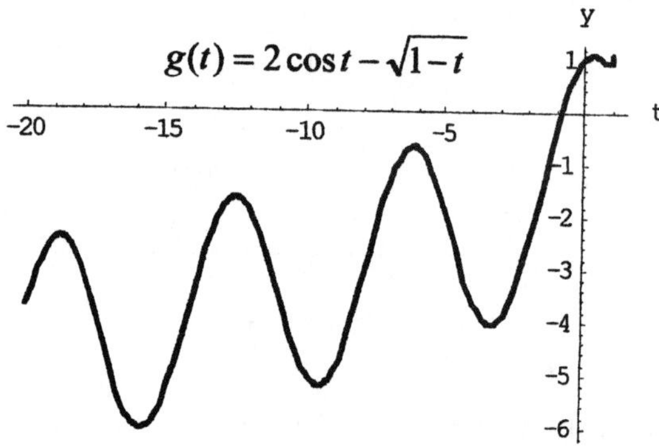

$g(t) = 2\cos t - \sqrt{1-t} \Rightarrow g'(t) = -2\sin t + \frac{1}{2\sqrt{1-t}} \Rightarrow t_{n+1} = t_n - \frac{2\cos t_n - \sqrt{1-t_n}}{-2\sin t_n + \frac{1}{\sqrt{1-t_n}}} \Rightarrow t_1 = -1$,

$t_2 = -0.836185$, $t_3 = -0.828381$, $t_4 = -0.828361$, $t_5 = -0.828361$. Solution: $t \approx -0.828361$.

CHAPTER 3 ADDITIONAL EXERCISES–THEORY, EXAMPLES, APPLICATIONS

1. If M and m are the maximum and minimum values, respectively, then $m \le f(x) \le M$ for all $x \in I$. If $m = M$ then f is constant on I.

3. On an open interval the extreme values of a continuous function (if any) must occur at an interior critical point. On a half-open interval the extreme values of a continuous function may be at a critical point or at the closed endpoint. Extreme values occur only where $f' = 0$, f' does not exist, or at the endpoints of the interval. Thus the extreme points will not be at the ends of an open interval.

5. (a) If $y' = 6(x+1)(x-2)^2$, then $y' < 0$ for $x < -1$ and $y' > 0$ for $x > -1$. The sign pattern is $f' = \underset{-1}{---|}\underset{2}{+++|}+++ \Rightarrow$ f has a local minimum at $x = -1$. Also $y'' = 6(x-2)^2 + 12(x+1)(x-2)$ $= 6(x-2)(3x) \Rightarrow y'' > 0$ for $x < 0$ or $x > 2$, while $y'' < 0$ for $0 < x < 2$. Therefore f has points of inflection at $x = 0$ and $x = 2$.

   (b) If $y' = 6x(x+1)(x-2)$, then $y' < 0$ for $x < -1$ and $0 < x < 2$; $y' > 0$ for $-1 < x < 0$ and $x > 2$. The sign sign pattern is $y' = \underset{-1}{---|}\underset{0}{+++|}\underset{2}{---|}+++$. Therefore f has a local maximum at $x = 0$ and local minima at $x = -1$ and $x = 2$. Also, $y'' = 6\left[x - \left(\frac{1-\sqrt{7}}{3}\right)\right]\left[x - \left(\frac{1+\sqrt{7}}{3}\right)\right]$, so $y'' < 0$ for $\frac{1-\sqrt{7}}{3} < x < \frac{1+\sqrt{7}}{3}$ and $y'' > 0$ for all other $x \Rightarrow$ f has points of inflection at $x = \frac{1 \pm \sqrt{7}}{3}$.

7. If f is continuous on $[a,c)$ and $f'(x) \le 0$ on $[a,c)$, then by the Mean Value Theorem for all $x \in [a,c)$ we have $\frac{f(c)-f(x)}{c-x} \le 0 \Rightarrow f(c) - f(x) \le 0 \Rightarrow f(x) \ge f(c)$. Also if f is continuous on $(c,b]$ and $f'(x) \ge 0$ on $(c,b]$, then for all $x \in (c,b]$ we have $\frac{f(x)-f(c)}{x-c} \ge 0 \Rightarrow f(x) - f(c) \ge 0 \Rightarrow f(x) \ge f(c)$. Therefore $f(x) \ge f(c)$ for all $x \in [a,b]$.

9. No. Corollary 1 requires that $f'(x) = 0$ for <u>all</u> x in some interval I, not $f'(x) = 0$ at a single point in I.

11. From (ii), $f(-1) = \frac{-1+a}{b-c+2} = 0 \Rightarrow a = 1$; from (iii), $1 = \lim_{x\to\infty} f(x) = \lim_{x\to\infty} \frac{x+1}{bx^2+cx+2} = \lim_{x\to\infty} \frac{1+\frac{1}{x}}{bx+c+\frac{2}{x}}$ $\Rightarrow b = 0$ (because $b = 1 \Rightarrow \lim_{x\to\infty} f(x) = 0$). Also, if $c = 0$ then $\lim_{x\to\infty} f(x) = \infty$ so we must have $c = 1$. In summary, $a = 1$, $b = 0$, and $c = 1$.

13. The area of the ΔABC is $A(x) = \frac{1}{2}(2)\sqrt{1-x^2} = (1-x^2)^{1/2}$, where $0 \le x \le 1$. Thus $A'(x) = \frac{-x}{\sqrt{1-x^2}} \Rightarrow 0$ and ± 1 are critical points. Also $A(\pm 1) = 0$ so $A(0) = 1$ is the maximum. When $x = 0$ the ΔABC is isosceles since $AC = BC = \sqrt{2}$.

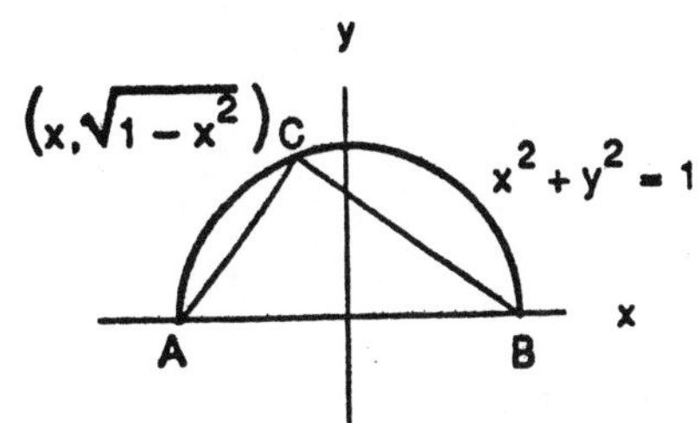

15. The time it would take the water to hit the ground from height y is $\sqrt{\frac{2y}{g}}$, where g is the acceleration of gravity. The product of time and exit velocity (rate) yields the distance the water travels:

$D(y) = \sqrt{\frac{2y}{g}}\sqrt{64(h-y)} = 8\sqrt{\frac{2}{g}}\left(hy - y^2\right)^{1/2}$, $0 \le y \le h \Rightarrow D'(y) = -4\sqrt{\frac{2}{g}}\left(hy - y^2\right)^{-1/2}(h - 2y) \Rightarrow 0, \frac{h}{2}$ and h are critical points. Now $D(0) = 0$, $D\left(\frac{h}{2}\right) = \frac{8h}{\sqrt{g}}$ and $D(h) = 0 \Rightarrow$ the best place to drill the hole is at $y = \frac{h}{2}$.

17. The surface area of the cylinder is $S = 2\pi r^2 + 2\pi rh$. From the diagram we have $\frac{r}{R} = \frac{H-h}{H} \Rightarrow h = \frac{RH - rH}{R}$ and

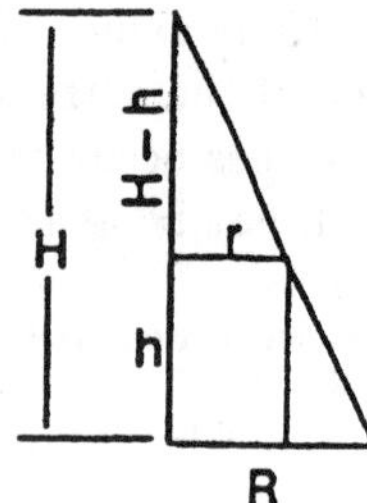

$S(r) = 2\pi r(r + h) = 2\pi r\left(r + H - r\frac{H}{R}\right) = 2\pi\left(1 - \frac{H}{R}\right)r^2 + 2\pi Hr$, where $0 \le r \le R$.

Case 1: $H < R \Rightarrow S(r)$ is a quadratic equation containing the origin and concave upward $\Rightarrow S(r)$ is maximum at $r = R$.

Case 2: $H = R \Rightarrow S(r)$ is a linear equation containing the origin with a positive slope $\Rightarrow S(r)$ is maximum at $r = R$.

Case 3: $H > R \Rightarrow S(r)$ is a quadratic equation containing the origin and concave downward. Then $\frac{dS}{dr} = 4\pi\left(1 - \frac{H}{R}\right)r + 2\pi H$ and $\frac{dS}{dr} = 0 \Rightarrow 4\pi\left(1 - \frac{H}{R}\right)r + 2\pi H = 0 \Rightarrow r = \frac{RH}{2(H-R)}$. For simplification we let $r^* = \frac{RH}{2(H-R)}$.

(a) If $R < H < 2R$, then $0 \ge H - 2R \Rightarrow H \ge 2(H - R) \Rightarrow \frac{RH}{2(H-R)} \ge R$ which is impossible.

(b) If $H = 2R$, then $r^* = \frac{2R^2}{2R} = R \Rightarrow S(r)$ is maximum at $r = R$.

(c) If $H > 2R$, then $2R + H \le 2H \Rightarrow H \le 2(H - R) \Rightarrow \frac{H}{2(H-R)} \le 1 \Rightarrow \frac{RH}{2(H-R)} \le R \Rightarrow r^* \le R$. Therefore, $S(r)$ is a maximum at $r = r^* = \frac{RH}{2(H-R)}$.

<u>Conclusion</u>: If $H \in (0, R]$ or $H = 2R$, then the maximum surface area is at $r = R$. If $H \in (R, 2R)$, then $r > R$ which is not possible. If $H \in (2R, \infty)$, then the maximum is at $r = r^* = \frac{RH}{2(H-R)}$.

19. $\lim\limits_{h\to 0} \frac{f'(c+h) - f'(c)}{h} = f''(c) \Leftrightarrow$ for $\epsilon = \frac{1}{2}|f''(c)| > 0$ there exists a $\delta > 0$ such that $0 < |h| < \delta$

$\Rightarrow \left|\frac{f'(c+h) - f'(c)}{h} - f''(c)\right| < \frac{1}{2}|f''(c)|$. Then $f'(c) = 0 \Rightarrow -\frac{1}{2}|f''(c)| < \frac{f'(c+h)}{h} - f''(c) < \frac{1}{2}|f''(c)|$

$\Rightarrow f''(c) - \frac{1}{2}|f''(c)| < \frac{f'(c+h)}{h} < f''(c) + \frac{1}{2}|f''(c)|$. If $f''(c) < 0$, then $|f''(c)| = -f''(c)$

$\Rightarrow \frac{3}{2}f''(c) < \frac{f'(c+h)}{h} < \frac{1}{2}f''(c) < 0$; likewise if $f''(c) > 0$, then $0 < \frac{1}{2}f''(c) < \frac{f'(c+h)}{h} < \frac{3}{2}f''(c)$.

(a) If $f''(c) < 0$, then $-\delta < h < 0 \Rightarrow f'(c+h) > 0$ and $0 < h < \delta \Rightarrow f'(c+h) < 0$. Therefore, f(c) is a local maximum.

(b) If $f''(c) > 0$, then $-\delta < h < 0 \Rightarrow f'(c+h) < 0$ and $0 < h < \delta \Rightarrow f'(c+h) > 0$. Therefore, f(c) is a local minimum.

21. (a) $(1)^2 = \frac{4\pi^2 L}{32.2} \Rightarrow L \approx 0.8156$ ft

(b) $2T\,dT = \frac{4\pi^2}{g}\,dL \Rightarrow dT = \frac{2\pi^2}{Tg}\,dL = \frac{2\pi^2}{\left(\frac{2\pi\sqrt{L}}{\sqrt{g}}\right)g}\,dL = \frac{\pi}{\sqrt{gL}}\,dL \approx \left(\frac{\pi}{\sqrt{32.2}\,\sqrt{0.8156}}\right)(0.01) = 0.00613$ sec.

(c) The original clock completes 1 swing every second or (24)(60)(60) = 86,400 swings per day. The new clock completes 1 swing every 1.00613 seconds. Therefore it takes (86,400)(1.00613) = 86,929.632 seconds for the new clock to complete the same number of swings. Thus the new clock loses $\frac{529.632}{60} \approx 8.83$ min/day.

NOTES.

CHAPTER 4 INTEGRATION

4.1 INDEFINITE INTEGRALS

1. (a) $3x^2$ (b) $\frac{x^8}{8}$ (c) $\frac{x^8}{8} - 3x^2 + 8x$

3. (a) $\frac{1}{x^2}$ (b) $\frac{-1}{4x^2}$ (c) $\frac{x^4}{4} + \frac{1}{2x^2}$

5. (a) $x^{2/3}$ (b) $x^{1/3}$ (c) $x^{-1/3}$

7. (a) $\tan x$ (b) $2 \tan\left(\frac{x}{3}\right)$ (c) $-\frac{2}{3}\tan\left(\frac{3x}{2}\right)$

9. $\int (x+1)\, dx = \frac{x^2}{2} + x + C$

11. $\int (2x^3 - 5x + 7)\, dx = \frac{1}{2}x^4 - \frac{5}{2}x^2 + 7x + C$

13. $\int x^{-1/3}\, dx = \frac{x^{2/3}}{\frac{2}{3}} + C = \frac{3}{2}x^{2/3} + C$

15. $\int \left(8y - \frac{2}{y^{1/4}}\right) dy = \int \left(8y - 2y^{-1/4}\right) dy = \frac{8y^2}{2} - 2\left(\frac{y^{3/4}}{\frac{3}{4}}\right) + C = 4y^2 - \frac{8}{3}y^{3/4} + C$

17. $\int \left(\frac{1}{7} - \frac{1}{y^{5/4}}\right) dy = \int \left(\frac{1}{7} - y^{-5/4}\right) dy = \frac{1}{7}y - \left(\frac{y^{-1/4}}{-\frac{1}{4}}\right) + C = \frac{y}{7} + \frac{4}{y^{1/4}} + C$

19. $\int \frac{t\sqrt{t} + \sqrt{t}}{t^2}\, dt = \int \left(\frac{t^{3/2}}{t^2} + \frac{t^{1/2}}{t^2}\right) dt = \int \left(t^{-1/2} + t^{-3/2}\right) dt = \frac{t^{1/2}}{\frac{1}{2}} + \left(\frac{t^{-1/2}}{-\frac{1}{2}}\right) + C = 2\sqrt{t} - \frac{2}{\sqrt{t}} + C$

21. $\int 7 \sin \frac{\theta}{3}\, d\theta = -21 \cos \frac{\theta}{3} + C$

23. $\int \left(1 + \tan^2 \theta\right) d\theta = \int \sec^2 \theta\, d\theta = \tan \theta + C$

25. $\int \cos \theta\, (\tan \theta + \sec \theta)\, d\theta = \int (\sin \theta + 1)\, d\theta = -\cos \theta + \theta + C$

27. $\frac{d}{dx}\left(\frac{(7x-2)^4}{28} + C\right) = \frac{4(7x-2)^3(7)}{28} = (7x-2)^3$

29. $\frac{d}{dx}\left(-3 \cot\left(\frac{x-1}{3}\right) + C\right) = -3\left(-\csc^2\left(\frac{x-1}{3}\right)\right)\left(\frac{1}{3}\right) = \csc^2\left(\frac{x-1}{3}\right)$

31. (a) Wrong: $\frac{d}{dx}\left(\frac{x^2}{2}\sin x + C\right) = \frac{2x}{2}\sin x + \frac{x^2}{2}\cos x = x\sin x + \frac{x^2}{2}\cos x$

(b) Wrong: $\frac{d}{dx}(-x\cos x + C) = -\cos x + x\sin x$

(c) Right: $\frac{d}{dx}(-x\cos x + \sin x + C) = -\cos x + x\sin x + \cos x = x\sin x$

33. Graph (b), because $\frac{dy}{dx} = 2x \Rightarrow y = x^2 + C$. Then $y(1) = 4 \Rightarrow C = 3$.

35. $\frac{dy}{dx} = 2x - 7 \Rightarrow y = x^2 - 7x + C$; at $x = 2$ and $y = 0$ we have $0 = 2^2 - 7(2) + C \Rightarrow C = 10 \Rightarrow y = x^2 - 7x + 10$

37. $\frac{dy}{dx} = 3x^{-2/3} \Rightarrow y = 9x^{1/3} + C$; at $x = -1$ and $y = -5$ we have $-5 = 9(-1) + C \Rightarrow C = 4 \Rightarrow y = 9x^{1/3} + 4$

39. $\frac{ds}{dt} = \cos t + \sin t \Rightarrow s = \sin t - \cos t + C$; at $t = \pi$ and $s = 1$ we have $1 = \sin\pi - \cos\pi + C \Rightarrow C = 0$
$\Rightarrow s = \sin t - \cos t$

41. $\frac{dv}{dt} = \frac{1}{2}\sec t\tan t \Rightarrow v = \frac{1}{2}\sec t + C$; at $v = 1$ and $t = 0$ we have $1 = \frac{1}{2}\sec(0) + C \Rightarrow C = \frac{1}{2} \Rightarrow v = \frac{1}{2}\sec t + \frac{1}{2}$

43. $\frac{d^2y}{dx^2} = 2 - 6x \Rightarrow \frac{dy}{dx} = 2x - 3x^2 + C_1$; at $\frac{dy}{dx} = 4$ and $x = 0$ we have $4 = 2(0) - 3(0)^2 + C_1 \Rightarrow C_1 = 4$
$\Rightarrow \frac{dy}{dx} = 2x - 3x^2 + 4 \Rightarrow y = x^2 - x^3 + 4x + C_2$; at $y = 1$ and $x = 0$ we have $1 = 0^2 - 0^3 + 4(0) + C_2 \Rightarrow C_2 = 1$
$\Rightarrow y = x^2 - x^3 + 4x + 1$

45. $\frac{d^3y}{dx^3} = 6 \Rightarrow \frac{d^2y}{dx^2} = 6x + C_1$; at $\frac{d^2y}{dx^2} = -8$ and $x = 0$ we have $-8 = 6(0) + C_1 \Rightarrow C_1 = -8 \Rightarrow \frac{d^2y}{dx^2} = 6x - 8$
$\Rightarrow \frac{dy}{dx} = 3x^2 - 8x + C_2$; at $\frac{dy}{dx} = 0$ and $x = 0$ we have $0 = 3(0)^2 - 8(0) + C_2 \Rightarrow C_2 = 0 \Rightarrow \frac{dy}{dx} = 3x^2 - 8x$
$\Rightarrow y = x^3 - 4x^2 + C_3$; at $y = 5$ and $x = 0$ we have $5 = 0^3 - 4(0)^2 + C_3 \Rightarrow C_3 = 5 \Rightarrow y = x^3 - 4x^2 + 5$

47. $v = \frac{ds}{dt} = 9.8t + 5 \Rightarrow s = 4.9t^2 + 5t + C$; at $s = 10$ and $t = 0$ we have $C = 10 \Rightarrow s = 4.9t^2 + 5t + 10$

49. $a = 32 \Rightarrow v = 32t + C_1$; at $v = 20$ and $t = 0$ we have $C_1 = 20 \Rightarrow v = 32t + 20 \Rightarrow s = 16t^2 + 20t + C_2$; at $s = 5$ and $t = 0$ we have $C_2 = 5 \Rightarrow s = 16t^2 + 20t + 5$

51. $m = y' = 3\sqrt{x} = 3x^{1/2} \Rightarrow y = 2x^{3/2} + C$; at $(9, 4)$ we have $4 = 2(9)^{3/2} + C \Rightarrow C = -50 \Rightarrow y = 2x^{3/2} - 50$

53. $\frac{dy}{dx} = 1 - \frac{4}{3}x^{1/3} \Rightarrow y = \int\left(1 - \frac{4}{3}x^{1/3}\right)dx = x - x^{4/3} + C$; at $(1, 0.5)$ on the curve we have $0.5 = 1 - 1^{4/3} + C$
$\Rightarrow C = 0.5 \Rightarrow y = x - x^{4/3} + \frac{1}{2}$

55. $\frac{dy}{dx} = \sin x - \cos x \Rightarrow y = \int(\sin x - \cos x)\,dx = -\cos x - \sin x + C$; at $(-\pi, -1)$ on the curve we have

$-1 = -\cos(-\pi) - \sin(-\pi) + C \Rightarrow C = -2 \Rightarrow y = -\cos x - \sin x - 2$

57. $a(t) = v'(t) = 1.6 \Rightarrow v(t) = 1.6t + C$; at $(0,0)$ we have $C = 0 \Rightarrow v(t) = 1.6t$. When $t = 30$, then $v(30) = 48$ m/sec.

59. Step 1: $\frac{d^2s}{dt^2} = -k \Rightarrow \frac{ds}{dt} = -kt + C_1$; at $\frac{ds}{dt} = 88$ and $t = 0$ we have $C_1 = 88 \Rightarrow \frac{ds}{dt} = -kt + 88 \Rightarrow$
$s = -k\left(\frac{t^2}{2}\right) + 88t + C_2$; at $s = 0$ and $t = 0$ we have $C_2 = 0 \Rightarrow s = -\frac{kt^2}{2} + 88t$

Step 2: $\frac{ds}{dt} = 0 \Rightarrow 0 = -kt + 88 \Rightarrow t = \frac{88}{k}$

Step 3: $242 = \frac{-k\left(\frac{88}{k}\right)^2}{2} + 88\left(\frac{88}{k}\right) \Rightarrow 242 = -\frac{(88)^2}{2k} + \frac{(88)^2}{k} \Rightarrow 242 = \frac{(88)^2}{2k} \Rightarrow k = 16$

61. (a) $v = \int a\,dt = \int \left(15t^{1/2} - 3t^{-1/2}\right) dt = 10t^{3/2} - 6t^{1/2} + C$; $\frac{ds}{dt}(1) = 4 \Rightarrow 4 = 10(1)^{3/2} - 6(1)^{1/2} + C \Rightarrow C = 0$
$\Rightarrow v = 10t^{3/2} - 6t^{1/2}$

(b) $s = \int v\,dt = \int \left(10t^{3/2} - 6t^{1/2}\right) dt = 4t^{5/2} - 4t^{3/2} + C$; $s(1) = 0 \Rightarrow 0 = 4(1)^{5/2} - 4(1)^{3/2} + C \Rightarrow C = 0$
$\Rightarrow s = 4t^{5/2} - 4t^{3/2}$

63. $\frac{d^2s}{dt^2} = a \Rightarrow \frac{ds}{dt} = \int a\,dt = at + C$; $\frac{ds}{dt} = v_0$ when $t = 0 \Rightarrow C = v_0 \Rightarrow \frac{ds}{dt} = at + v_0 \Rightarrow s = \frac{at^2}{2} + v_0 t + C_1$; $s = s_0$ when $t = 0 \Rightarrow s_0 = \frac{a(0)^2}{2} + v_0(0) + C_1 \Rightarrow C_1 = s_0 \Rightarrow s = \frac{at^2}{2} + v_0 t + s_0$

65. (a) $\frac{ds}{dt} = 9.8t - 3 \Rightarrow s = 4.9t^2 - 3t + C$; 1) at $s = 5$ and $t = 0$ we have $C = 5 \Rightarrow s = 4.9t^2 - 3t + 5$; displacement $= s(3) - s(1) = ((4.9)(9) - 9 + 5) - (4.9 - 3 + 5) = 33.2$ units; 2) at $s = -2$ and $t = 0$ we have $C = -2 \Rightarrow s = 4.9t^2 - 3t - 2$; displacement $= s(3) - s(1) = ((4.9)(9) - 9 - 2) - (4.9 - 3 - 2) = 33.2$ units; 3) at $s = s_0$ and $t = 0$ we have $C = s_0 \Rightarrow s = 4.9t^2 - 3t + s_0$; displacement $= s(3) - s(1)$ $= ((4.9)(9) - 9 + s_0) - (4.9 - 3 + s_0) = 33.2$ units

(b) True. Given an antiderivative $f(t)$ of the velocity function, we know that the body's position function is $s = f(t) + C$ for some constant C. Therefore, the displacement from $t = a$ to $t = b$ is $(f(b) + C) - (f(a) + C)$ $= f(b) - f(a)$. Thus we can find the displacement from any antiderivative f as the numerical difference $f(b) - f(a)$ without knowing the exact values of C and s.

4.2 INTEGRAL RULES; INTEGRATION BY SUBSTITUTION

1. Let $u = 2x^2 \Rightarrow du = 4x\,dx \Rightarrow \frac{1}{4}\,du = x\,dx$

$$\int x \sin\left(2x^2\right) dx = \int \frac{1}{4} \sin u\,du = -\frac{1}{4}\cos u + C = -\frac{1}{4}\cos 2x^2 + C$$

3. Let $u = 7x - 2 \Rightarrow du = 7\,dx \Rightarrow \frac{1}{7}\,du = dx$

$$\int 28(7x-2)^{-5}\,dx = \int \tfrac{1}{7}(28)u^{-5}\,du = \int 4u^{-5}\,du = -u^{-4} + C = -(7x-2)^{-4} + C$$

5. Let $u = 1 - r^3 \Rightarrow du = -3r^2\,dr \Rightarrow -3\,du = 9r^2\,dr$

$$\int \frac{9r^2\,dr}{\sqrt{1-r^3}} = \int -3u^{-1/2}\,du = -3(2)u^{1/2} + C = -6\left(1-r^3\right)^{1/2} + C$$

7. Let $u = x^{3/2} - 1 \Rightarrow du = \frac{3}{2}x^{1/2}\,dx \Rightarrow \frac{2}{3}\,du = \sqrt{x}\,dx$

$$\int \sqrt{x}\sin^2\left(x^{3/2}-1\right)dx = \int \tfrac{2}{3}\sin^2 u\,du = \tfrac{2}{3}\left(\tfrac{u}{2} - \tfrac{1}{4}\sin 2u\right) + C = \tfrac{1}{3}\left(x^{3/2}-1\right) - \tfrac{1}{6}\sin\left(2x^{3/2}-2\right) + C$$

9. (a) Let $u = \cot 2\theta \Rightarrow du = -2\csc^2 2\theta\,d\theta \Rightarrow -\frac{1}{2}\,du = \csc^2 2\theta\,d\theta$

$$\int \csc^2 2\theta \cot 2\theta\,d\theta = -\int \tfrac{1}{2}u\,du = -\tfrac{1}{2}\left(\tfrac{u^2}{2}\right) + C = -\tfrac{u^2}{4} + C = -\tfrac{1}{4}\cot^2 2\theta + C$$

(b) Let $u = \csc 2\theta \Rightarrow du = -2\csc 2\theta \cot 2\theta\,d\theta \Rightarrow -\frac{1}{2}\,du = \csc 2\theta \cot 2\theta\,d\theta$

$$\int \csc^2 2\theta \cot 2\theta\,d\theta = \int -\tfrac{1}{2}u\,du = -\tfrac{1}{2}\left(\tfrac{u^2}{2}\right) + C = -\tfrac{u^2}{4} + C = -\tfrac{1}{4}\csc^2 2\theta + C$$

11. Let $u = 3 - 2s \Rightarrow du = -2\,ds \Rightarrow -\frac{1}{2}\,du = ds$

$$\int \sqrt{3-2s}\,ds = \int \sqrt{u}\left(-\tfrac{1}{2}\,du\right) = -\tfrac{1}{2}\int u^{1/2}\,du = \left(-\tfrac{1}{2}\right)\left(\tfrac{2}{3}u^{3/2}\right) + C = -\tfrac{1}{3}(3-2s)^{3/2} + C$$

13. Let $u = 2 - x \Rightarrow du = -dx \Rightarrow -du = dx$

$$\int \frac{3}{(2-x)^2}\,dx = \int \frac{3(-du)}{u^2} = -3\int u^{-2}\,du = -3\left(\frac{u^{-1}}{-1}\right) + C = \frac{3}{2-x} + C$$

15. Let $u = 7 - 3y^2 \Rightarrow du = -6y\,dy \Rightarrow -\frac{1}{2}\,du = 3y\,dy$

$$\int 3y\sqrt{7-3y^2}\,dy = \int \sqrt{u}\left(-\tfrac{1}{2}\,du\right) = -\tfrac{1}{2}\int u^{1/2}\,du = \left(-\tfrac{1}{2}\right)\left(\tfrac{2}{3}u^{3/2}\right) + C = -\tfrac{1}{3}\left(7-3y^2\right)^{3/2} + C$$

17. Let $u = 1 + \sqrt{x} \Rightarrow du = \frac{1}{2\sqrt{x}}\,dx \Rightarrow 2\,du = \frac{1}{\sqrt{x}}\,dx$

$$\int \frac{\left(1+\sqrt{x}\right)^3}{\sqrt{x}}\,dx = \int u^3\,(2\,du) = 2\left(\tfrac{1}{4}u^4\right) + C = \tfrac{1}{2}\left(1+\sqrt{x}\right)^4 + C$$

19. Let $u = 3x + 2 \Rightarrow du = 3\,dx \Rightarrow \frac{1}{3}\,du = dx$

$$\int \sec^2(3x+2)\,dx = \int \left(\sec^2 u\right)\left(\tfrac{1}{3}\,du\right) = \tfrac{1}{3}\int \sec^2 u\,du = \tfrac{1}{3}\tan u + C = \tfrac{1}{3}\tan(3x+2) + C$$

21. Let $u = \tan\left(\frac{x}{2}\right) \Rightarrow du = \frac{1}{2}\sec^2\left(\frac{x}{2}\right) dx \Rightarrow 2\ du = \sec^2\left(\frac{x}{2}\right) dx$

$$\int \tan^7\left(\frac{x}{2}\right)\sec^2\left(\frac{x}{2}\right) dx = \int u^7\,(2\ du) = 2\left(\frac{1}{8}u^8\right) + C = \frac{1}{4}\tan^8\left(\frac{x}{2}\right) + C$$

23. Let $u = x^{3/2} + 1 \Rightarrow du = \frac{3}{2}x^{1/2}\,dx \Rightarrow \frac{2}{3}\,du = x^{1/2}\,dx$

$$\int x^{1/2}\sin\left(x^{3/2}+1\right) dx = \int (\sin u)\left(\frac{2}{3}\,du\right) = \frac{2}{3}\int \sin u\ du = \frac{2}{3}(-\cos u) + C = -\frac{2}{3}\cos\left(x^{3/2}+1\right) + C$$

25. Let $u = \cos(2t+1) \Rightarrow du = -2\sin(2t+1) \Rightarrow -\frac{1}{2}\,du = \sin(2t+1)$

$$\int \frac{\sin(2t+1)}{\cos^2(2t+1)}\,dt = \int -\frac{1}{2}\frac{du}{u^2} = \frac{1}{2u} + C = \frac{1}{2\cos(2t+1)} + C$$

27. Let $u = \cot y \Rightarrow du = -\csc^2 y\ dy \Rightarrow -du = \csc^2 y\ dy$

$$\int \sqrt{\cot y}\,\csc^2 y\ dy = \int \sqrt{u}\,(-du) = -\int u^{1/2}\,du = -\frac{2}{3}u^{3/2} + C = -\frac{2}{3}(\cot y)^{3/2} + C = -\frac{2}{3}\left(\cot^3 y\right)^{1/2} + C$$

29. Let $u = \sqrt{t} + 3 = t^{1/2} + 3 \Rightarrow du = \frac{1}{2}t^{-1/2}\,dt \Rightarrow 2\ du = \frac{1}{\sqrt{t}}\,dt$

$$\int \frac{1}{\sqrt{t}}\cos\left(\sqrt{t}+3\right) dt = \int (\cos u)(2\ du) = 2\int \cos u\ du = 2\sin u + C = 2\sin\left(\sqrt{t}+3\right) + C$$

31. Let $u = \csc\sqrt{\theta} \Rightarrow du = \left(-\csc\sqrt{\theta}\,\cot\sqrt{\theta}\right)\left(\frac{1}{2\sqrt{\theta}}\right) d\theta \Rightarrow -2\ du = \frac{1}{\sqrt{\theta}}\cot\sqrt{\theta}\,\csc\sqrt{\theta}\ d\theta$

$$\int \frac{\cos\sqrt{\theta}}{\sqrt{\theta}\,\sin^2\sqrt{\theta}}\,d\theta = \int \frac{1}{\sqrt{\theta}}\cot\sqrt{\theta}\,\csc\sqrt{\theta}\ d\theta = \int -2\ du = -2u + C = -2\csc\sqrt{\theta} + C = -\frac{2}{\sin\sqrt{\theta}} + C$$

33. (a) Let $u = \tan x \Rightarrow du = \sec^2 x\ dx$; $v = u^3 \Rightarrow dv = 3u^2\,du \Rightarrow 6\ dv = 18u^2\,du$; $w = 2 + v \Rightarrow dw = dv$

$$\int \frac{18\tan^2 x\sec^2 x}{\left(2+\tan^3 x\right)^2}\,dx = \int \frac{18u^2}{\left(2+u^3\right)^2}\,du = \int \frac{6\ dv}{(2+v)^2} = \int \frac{6\ dw}{w^2} = 6\int w^{-2}\,dw = -6w^{-1} + C = -\frac{6}{2+v} + C$$

$$= -\frac{6}{2+u^3} + C = -\frac{6}{2+\tan^3 x} + C$$

(b) Let $u = \tan^3 x \Rightarrow du = 3\tan^2 x\sec^2 x\ dx \Rightarrow 6\ du = 18\tan^2 x\sec^2 x\ dx$; $v = 2 + u \Rightarrow dv = du$

$$\int \frac{18\tan^2 x\sec^2 x}{\left(2+\tan^3 x\right)^2}\,dx = \int \frac{6\ du}{(2+u)^2} = \int \frac{6\ dv}{v^2} = -\frac{6}{v} + C = -\frac{6}{2+u} + C = -\frac{6}{2+\tan^3 x} + C$$

(c) Let $u = 2 + \tan^3 x \Rightarrow du = 3\tan^2 x\sec^2 x\ dx \Rightarrow 6\ du = 18\tan^2 x\sec^2 x\ dx$

$$\int \frac{18\tan^2 x\sec^2 x}{\left(2+\tan^3 x\right)^2}\,dx = \int \frac{6\ du}{u^2} = -\frac{6}{u} + C = -\frac{6}{2+\tan^3 x} + C$$

35. Let $u = 3(2r-1)^2 + 6 \Rightarrow du = 6(2r-1)(2)\,dr \Rightarrow \frac{1}{12}\,du = (2r-1)\,dr$; $v = \sqrt{u} \Rightarrow dv = \frac{1}{2\sqrt{u}}\,du \Rightarrow \frac{1}{6}\,dv$ $= \frac{1}{12\sqrt{u}}\,du$

$$\int \frac{(2r-1)\cos\sqrt{3(2r-1)^2+6}}{\sqrt{3(2r-1)^2+6}}\,dr = \int \left(\frac{\cos\sqrt{u}}{\sqrt{u}}\right)\left(\frac{1}{12}\,du\right) = \int (\cos v)\left(\frac{1}{6}\,dv\right) = \frac{1}{6}\sin v + C = \frac{1}{6}\sin\sqrt{u} + C$$

$$= \frac{1}{6}\sin\sqrt{3(2r-1)^2+6} + C$$

37. Let $u = 3t^2 - 1 \Rightarrow du = 6t\,dt \Rightarrow 2\,du = 12t\,dt$

$$s = \int 12t\left(3t^2-1\right)^3 dt = \int u^3\,(2\,du) = 2\left(\frac{1}{4}u^4\right) + C = \frac{1}{2}u^4 + C = \frac{1}{2}\left(3t^2-1\right)^4 + C;$$

$s = 3$ when $t = 1 \Rightarrow 3 = \frac{1}{2}(3-1)^4 + C \Rightarrow 3 = 8 + C \Rightarrow C = -5 \Rightarrow s = \frac{1}{2}\left(3t^2-1\right)^4 - 5$

39. Let $u = t + \frac{\pi}{12} \Rightarrow du = dt$

$$s = \int 8\sin^2\left(t+\frac{\pi}{12}\right) dt = \int 8\sin^2 u\,du = 8\left(\frac{u}{2} - \frac{1}{4}\sin 2u\right) + C = 4\left(t+\frac{\pi}{12}\right) - 2\sin\left(2t+\frac{\pi}{6}\right) + C;$$

$s = 8$ when $t = 0 \Rightarrow 8 = 4\left(\frac{\pi}{12}\right) - 2\sin\left(\frac{\pi}{6}\right) + C \Rightarrow C = 8 - \frac{\pi}{3} + 1 = 9 - \frac{\pi}{3} \Rightarrow s = 4t - 2\sin\left(2t+\frac{\pi}{6}\right) + 9$

41. Let $u = 2t - \frac{\pi}{2} \Rightarrow du = 2\,dt \Rightarrow -2\,du = -4\,dt$

$$\frac{ds}{dt} = \int -4\sin\left(2t-\frac{\pi}{2}\right) dt = \int (\sin u)(-2\,du) = 2\cos u + C_1 = 2\cos\left(2t-\frac{\pi}{2}\right) + C_1;$$

at $t = 0$ and $\frac{ds}{dt} = 100$ we have $100 = 2\cos\left(-\frac{\pi}{2}\right) + C_1 \Rightarrow C_1 = 100 \Rightarrow \frac{ds}{dt} = 2\cos\left(2t-\frac{\pi}{2}\right) + 100$

$$\Rightarrow s = \int \left(2\cos\left(2t-\frac{\pi}{2}\right) + 100\right) dt = \int (\cos u + 50)\,du = \sin u + 50u + C_2 = \sin\left(2t-\frac{\pi}{2}\right) + 50\left(2t-\frac{\pi}{2}\right) + C_2;$$

at $t = 0$ and $s = 0$ we have $0 = \sin\left(-\frac{\pi}{2}\right) + 50\left(-\frac{\pi}{2}\right) + C_2 \Rightarrow C_2 = 1 + 25\pi$

$\Rightarrow s = \sin\left(2t-\frac{\pi}{2}\right) + 100t - 25\pi + (1+25\pi) \Rightarrow s = \sin\left(2t-\frac{\pi}{2}\right) + 100t + 1$

43. Let $u = 2t \Rightarrow du = 2\,dt \Rightarrow 3\,du = 6\,dt$

$$s = \int 6\sin 2t\,dt = \int (\sin u)(3\,du) = -3\cos u + C = -3\cos 2t + C;$$

at $t = 0$ and $s = 0$ we have $0 = -3\cos 0 + C \Rightarrow C = 3 \Rightarrow s = 3 - 3\cos 2t \Rightarrow s\left(\frac{\pi}{2}\right) = 3 - 3\cos(\pi) = 6$ m

45. All three integrations are correct. In each case, the derivative of the function on the right is the integrand on the left, and each formula has an arbitrary constant for generating the remaining antiderivatives. Moreover, $\sin^2 x + C_1 = 1 - \cos^2 x + C_1 \Rightarrow C_2 = 1 + C_1$; also $-\cos^2 x + C_2 = -\frac{\cos 2x}{2} - \frac{1}{2} + C_2 \Rightarrow C_3 = C_2 - \frac{1}{2} = C_1 + \frac{1}{2}$.

4.3 ESTIMATING WITH FINITE SUMS

1. Using midpoints of the intervals, Area $\approx (0.25)(2)+(1.0)(2)+(2.0)(2)+(3.25)(2)+(4.0)(2)+(4.0)(2)$ $+(3.35)(2)+(2.25)(2)+(1.3)(2)+(0.75)(2)+(0.25)(2)=44.8$ mg·sec/L. Cardiac output $=\frac{\text{amount of dye}}{\text{area under curve}}\times 60 \approx \frac{5\text{ mg}}{44.5\text{ mg}\cdot\text{sec/L}}\times 60\ \frac{\text{sec}}{\text{min}} \approx 6.7$ L/min.

3. (a) $D \approx (0)(1)+(12)(1)+(22)(1)+(10)(1)+(5)(1)+(13)(1)+(11)(1)+(6)(1)+(2)(1)+(6)(1)=87$ inches

   (b) $D \approx (12)(1)+(22)(1)+(10)(1)+(5)(1)+(13)(1)+(11)(1)+(6)(1)+(2)(1)+(6)(1)+(0)(1)=87$ inches

5. (a) $D \approx (0)(10)+(44)(10)+(15)(10)+(35)(10)+(30)(10)+(44)(10)+(35)(10)+(15)(10)+(22)(10)$ $+(35)(10)+(44)(10)+(30)(10)=3490$ feet ≈ 0.73 miles

   (b) $D \approx (44)(10)+(15)(10)+(35)(10)+(30)(10)+(44)(10)+(35)(10)+(15)(10)+(22)(10)+(35)(10)$ $+(44)(10)+(30)(10)+(35)(10)=3840$ feet ≈ 0.66 miles

7. (a) $S_4=\pi\left[\sqrt{16-(-2)^2}\right]^2(2)+\pi\left[\sqrt{16-0^2}\right]^2(2)+\pi\left[\sqrt{16-(2)^2}\right]^2(2)=\pi[(16-4)+(16-0)+(16-4)](2)$ $=80\pi$

   (b) $\frac{|V-S_4|}{V}=\frac{\left|\left(\frac{256}{3}\right)\pi-80\pi\right|}{\left(\frac{256}{3}\right)\pi}=\frac{16}{256}\approx 6\%$

9. (a) $S_8=\pi\left[(16-0^2)+\left(16-\left(\frac{1}{2}\right)^2\right)+(16-(1)^2)+\left(16-\left(\frac{3}{2}\right)^2\right)+(16-(2)^2)+\left(16-\left(\frac{5}{2}\right)^2\right)\right.$ $\left.+(16-(3)^2)+\left(16-\left(\frac{7}{2}\right)\right)^2\right]\left(\frac{1}{2}\right)=\frac{\pi}{2}\left[128-\frac{1}{4}-1-\frac{9}{4}-4-\frac{25}{4}-9-\frac{49}{4}\right]=\frac{372\pi}{8}=\frac{93\pi}{2}$, overestimates

   (b) $V=\frac{2}{3}\pi r^3=\frac{128\pi}{3}\Rightarrow\frac{|V-S_8|}{V}=\frac{\left|\left(\frac{93}{2}\right)\pi-\left(\frac{128}{3}\right)\pi\right|}{\left(\frac{128}{3}\right)\pi}=\frac{23}{256}\approx 9\%$

11. (a) To have the same orientation as Exercise 11, tip the bowl sideways (assume the water is ice). The water covers the interval $[4,8]$. The function which will give us the values of the radii of the approximating cylinders is the equation of the upper semicircle formed by intersecting the hemisphere with the xy-plane, $f(x)=\sqrt{64-x^2}$. Using $\Delta x=\frac{1}{2}$ and left-endpoints for each interval $\Rightarrow S_8=\pi\left[(64-(4)^2)+\left(64-\left(\frac{9}{2}\right)^2\right)\right.$ $+(64-(5)^2)+\left(64-\left(\frac{11}{2}\right)^2\right)+(64-(6)^2)$ $\left.+\left(64-\left(\frac{13}{2}\right)^2\right)+(64-(7)^2)+\left(64-\left(\frac{15}{2}\right)^2\right)\right]\left(\frac{1}{2}\right)=\frac{\pi}{2}\left(512-16-\frac{81}{4}-25-\frac{121}{4}-36-\frac{169}{4}-49-\frac{225}{4}\right)$ $=\frac{\pi}{2}\left(386-\frac{596}{4}\right)=\frac{\pi}{8}(1544-596)=\frac{948}{8}\pi=118.5\pi;$

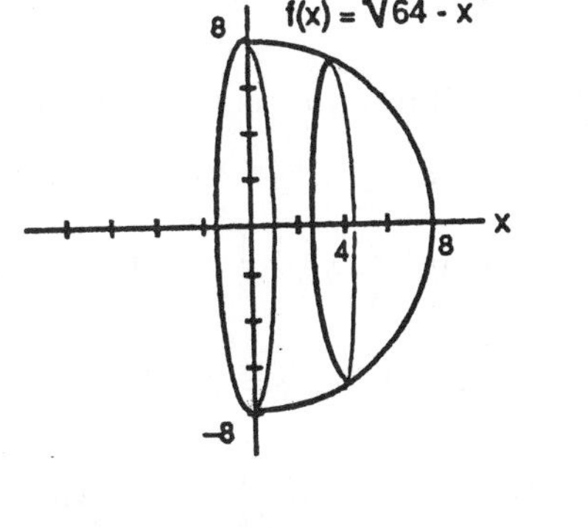

(b) $\frac{|V - S_8|}{V} = \frac{\left|\left(\frac{320}{3}\right)\pi - \left(\frac{948}{8}\right)\pi\right|}{\left(\frac{320}{3}\right)\pi} = \frac{2844 - 2560}{2560} \approx 11\%$

13. (a) $S_5 = \pi\left[\left(\sqrt{0}\right)^2 + \left(\sqrt{1}\right)^2 + \left(\sqrt{2}\right)^2 + \left(\sqrt{3}\right)^2 + \left(\sqrt{4}\right)^2\right](1) = 10\pi$, underestimates

(b) $\frac{|V - S_5|}{V} = \frac{\left(\frac{25}{2}\right)\pi - 10\pi}{\left(\frac{25}{2}\right)\pi} = \frac{5}{25} = 20\%$

15. (a) Because the acceleration is decreasing, an upper estimate is obtained using left end-points in summing acceleration $\cdot \Delta t$. Thus, $\Delta t = 1$ and speed $\approx [32.00 + 19.41 + 11.77 + 7.14 + 4.33](1) = 74.65$ ft/sec

(b) Using right end-points we obtain a lower estimate: speed $\approx [19.41 + 11.77 + 7.14 + 4.33 + 2.63](1)$ $= 45.28$ ft/sec

(c) Upper estimates for the speed at each second are:

t	0	1	2	3	4	5
v	0	32.00	51.41	63.18	70.32	74.65

Thus, the distance fallen when t = 3 seconds is s $\approx [32.00 + 51.41 + 63.18](1) = 146.59$ ft.

17. Partition [0, 2] into the four subintervals $\left[0, \frac{1}{2}\right]$, $\left[\frac{1}{2}, 1\right]$, $\left[1, \frac{3}{2}\right]$, and $\left[\frac{3}{2}, 2\right]$. The midpoints of these subintervals are $m_1 = \frac{1}{4}$, $m_2 = \frac{3}{4}$, $m_3 = \frac{5}{4}$, and $m_4 = \frac{7}{4}$. The heights of the four approximating rectangles are $f(m_1) = \left(\frac{1}{4}\right)^3 = \frac{1}{64}$, $f(m_2) = \left(\frac{3}{4}\right)^3 = \frac{27}{64}$, $f(m_3) = \left(\frac{5}{4}\right)^3 = \frac{125}{64}$, and $f(m_4) = \left(\frac{7}{4}\right)^3 = \frac{343}{64}$

$\Rightarrow$ Average value $\approx \frac{\frac{1}{64} + \frac{27}{64} + \frac{125}{64} + \frac{343}{64}}{4} = \frac{1 + 27 + 125 + 343}{4 \cdot 64} = \frac{496}{256} = \frac{31}{16}$. Notice that the average value is approximated by $\frac{\left(\frac{1}{4}\right)^3 + \left(\frac{3}{4}\right)^3 + \left(\frac{5}{4}\right)^3 + \left(\frac{7}{4}\right)^3}{4} = \frac{1}{2}\left[\left(\frac{1}{4}\right)^3\left(\frac{1}{2}\right) + \left(\frac{3}{4}\right)^3\left(\frac{1}{2}\right) + \left(\frac{5}{4}\right)^3\left(\frac{1}{2}\right) + \left(\frac{7}{4}\right)^3\left(\frac{1}{2}\right)\right]$

$= \frac{1}{\text{length of } [0,2]} \cdot \left[\begin{matrix}\text{approximate area under} \\ \text{curve } f(x) = x^3\end{matrix}\right]$. We use this observation in solving the next several exercises.

19. Partition [0, 2] into the four subintervals [0, 0.5], [0.5, 1], [1, 1.5], and [1.5, 2]. The midpoints of the subintervals are $m_1 = 0.25$, $m_2 = 0.75$, $m_3 = 1.25$, and $m_4 = 1.75$. The heights of the four approximating rectangles are $f(m_1) = \frac{1}{2} + \sin^2\frac{\pi}{4} = \frac{1}{2} + \frac{1}{2} = 1$, $f(m_2) = \frac{1}{2} + \sin^2\frac{3\pi}{4} = \frac{1}{2} + \frac{1}{2} = 1$, $f(m_3) = \frac{1}{2} + \sin^2\frac{5\pi}{4} = \frac{1}{2} + \left(-\frac{1}{\sqrt{2}}\right)^2$ $= \frac{1}{2} + \frac{1}{2} = 1$, and $f(m_4) = \frac{1}{2} + \sin^2\frac{7\pi}{4} = \frac{1}{2} + \left(-\frac{1}{\sqrt{2}}\right)^2 = 1$. The width of each rectangle is $\Delta x = \frac{1}{2}$. Thus, Area $\approx (1 + 1 + 1 + 1)\left(\frac{1}{2}\right) = 2 \Rightarrow$ average value $\approx \frac{\text{area}}{\text{length of } [0,2]} = \frac{2}{2} = 1$.

21. Since the leakage is increasing, an upper estimate uses right end-points and a lower estimate uses left end-points:

(a) upper estimate $= (70)(1) + (97)(1) + (136)(1) + (190)(1) + (265)(1) = 758$ gal,

lower estimate $= (50)(1) + (70)(1) + (97)(1) + (136)(1) + (190)(1) = 543$ gal.

(b) upper estimate $= (70 + 97 + 136 + 190 + 265 + 369 + 516 + 720) = 2363$ gal,

lower estimate $= (50 + 70 + 97 + 136 + 190 + 265 + 369 + 516) = 1693$ gal.

(c) worst case: $2363 + 720\text{t} = 25{,}000 \Rightarrow \text{t} \approx 31.4$ hrs;

best case: $1693 + 720\text{t} = 25{,}000 \Rightarrow \text{t} \approx 32.4$ hrs

23. (a) The diagonal of the square has length 2, so the side length is $\sqrt{2}$. Area $= \left(\sqrt{2}\right)^2 = 2$

(b) Think of the octagon as a collection of 16 right triangles with a hypotenuse of length 1 and an acute angle measuring $\frac{2\pi}{16} = \frac{\pi}{8}$.

Area $= 16\left(\frac{1}{2}\right)\left(\sin\frac{\pi}{8}\right)\left(\cos\frac{\pi}{8}\right) = 4\sin\frac{\pi}{4} = 2\sqrt{2} \approx 2.828$

(c) Think of the 16-gon as a collection of 32 right triangles with a hypotenuse of length 1 and an acute angle measuring $\frac{2\pi}{32} = \frac{\pi}{16}$.

Area $= 32\left(\frac{1}{2}\right)\left(\sin\frac{\pi}{16}\right)\left(\cos\frac{\pi}{16}\right) = 8\sin\frac{\pi}{8} \approx 3.061$

(d) Each area is less than the area of the circle, π. As n increases, the area approaches π.

4.4 RIEMANN SUMS AND DEFINITE INTEGRALS

1. $\sum_{k=1}^{2} \frac{6k}{k+1} = \frac{6(1)}{1+1} + \frac{6(2)}{2+1} = \frac{6}{2} + \frac{12}{3} = 7$

3. $\sum_{k=1}^{4} \cos k\pi = \cos(1\pi) + \cos(2\pi) + \cos(3\pi) + \cos(4\pi) = -1 + 1 - 1 + 1 = 0$

5. $\sum_{k=1}^{3} (-1)^{k+1} \sin\frac{\pi}{k} = (-1)^{1+1}\sin\frac{\pi}{1} + (-1)^{2+1}\sin\frac{\pi}{2} + (-1)^{3+1}\sin\frac{\pi}{3} = 0 - 1 + \frac{\sqrt{3}}{2} = \frac{\sqrt{3}-2}{2}$

7. (a) 

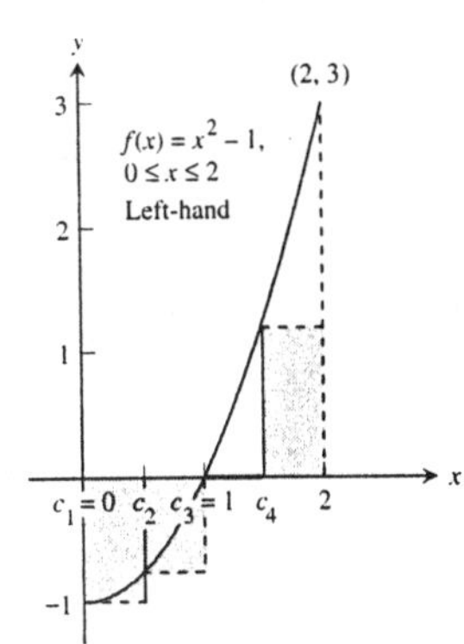

(b)

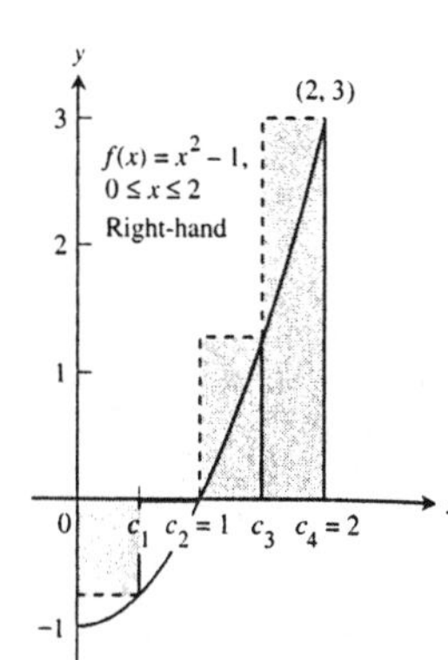

(c)

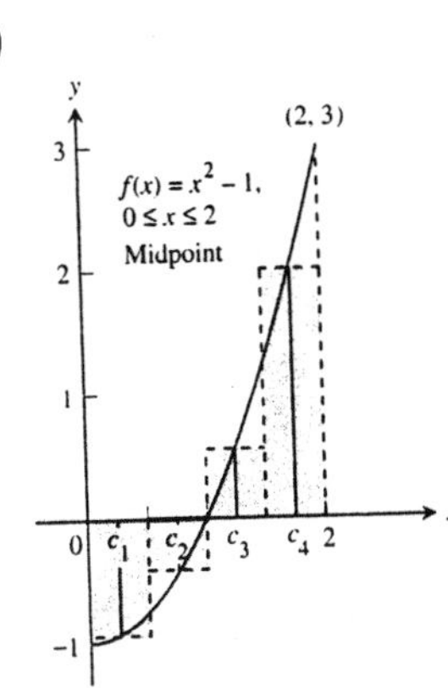

9. (a)

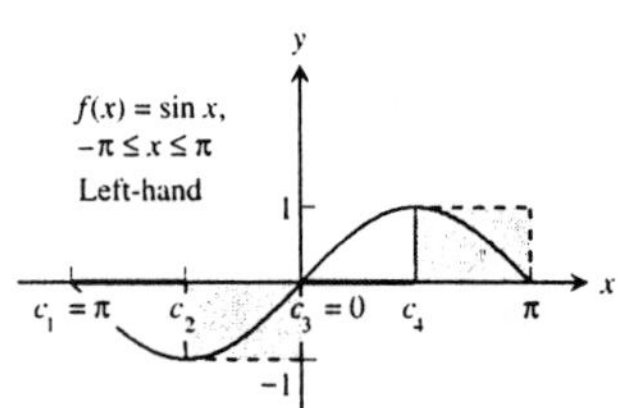

(b)

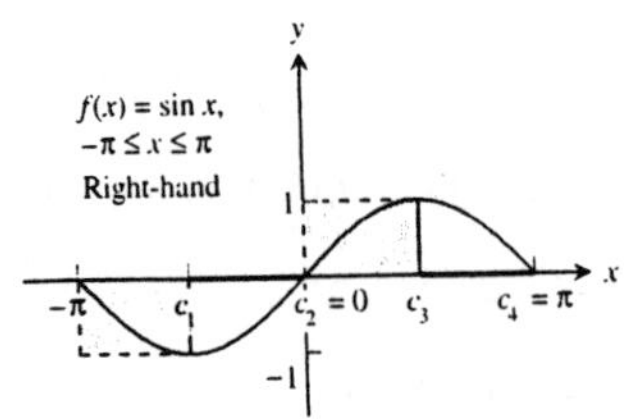

(c)

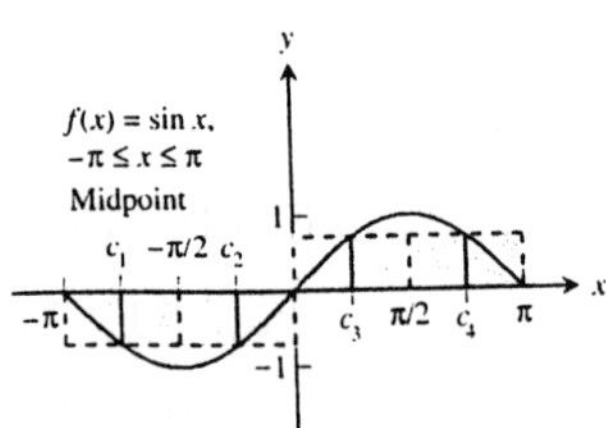

11. $\int_0^2 x^2\, dx$

13. $\int_{-7}^{5} (x^2 - 3x)\, dx$

15. $\int_0^1 \sqrt{4 - x^2}\, dx$

17. The area of the trapezoid is $A = \frac{1}{2}(B + b)h$

$= \frac{1}{2}(5 + 2)(6) = 21 \Rightarrow \int_{-2}^{4} \left(\frac{x}{2} + 3\right) dx$

$= 21$ square units

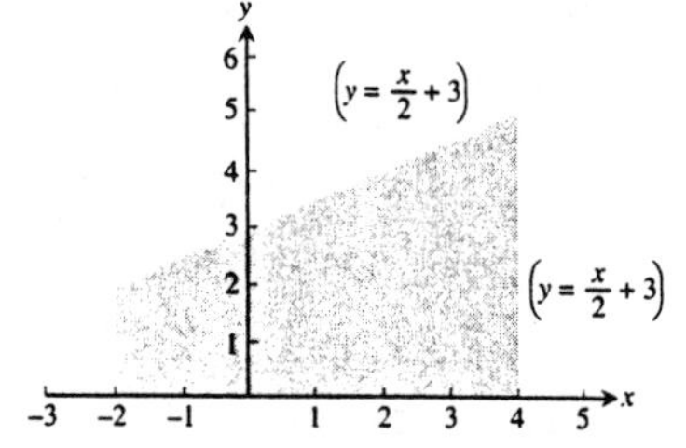

19. The area of the triangle on the left is $A = \frac{1}{2}bh = \frac{1}{2}(2)(2)$

$= 2$. The area of the triangle on the right is $A = \frac{1}{2}bh$

$= \frac{1}{2}(1)(1) = \frac{1}{2}$. Then, the total area is $2.5 \Rightarrow \int_{-2}^{1} |x|\, dx$

$= 2.5$ square units

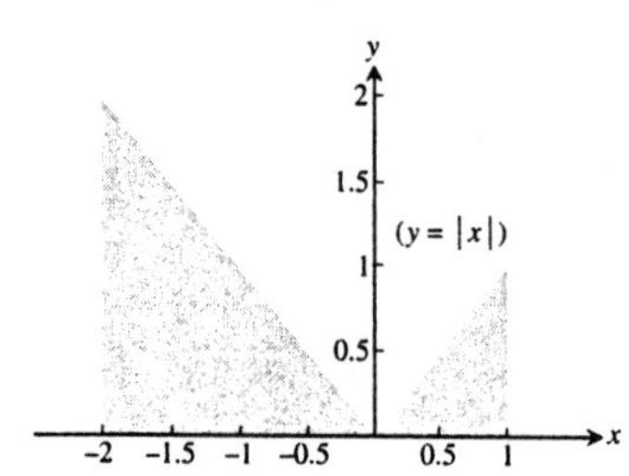

21. $\int_0^b x\, dx = \frac{1}{2}(b)(b) = \frac{b^2}{2}$

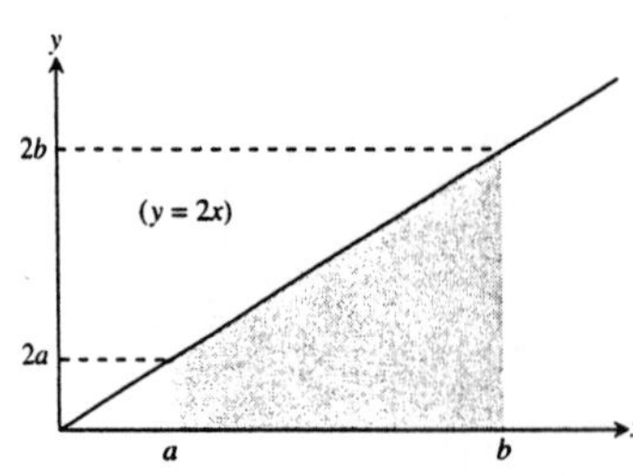

23. The graph of $f(x) = 1 - x$ on the interval $[0,1]$ forms a right isosceles triangle in the first quadrant with its two legs, each of length one, lying on the coordinate axes. The area of the triangle is $A = \frac{1}{2}bh = \frac{1}{2}(1)(1) = \frac{1}{2}$, which is also the value of the integral $\int_0^1 (1-x)\,dx = \frac{1}{2}$, therefore, $av(f) = \frac{1}{1-0}\int_0^1 (1-x)\,dx = (1)\left(\frac{1}{2}\right) = \frac{1}{2}$.

25. The function $f(x) = \sqrt{1-x^2}$ on the interval $[0,1]$ forms a quarter-circular area of radius 1 lying in the first quadrant with its center on the origin. The area of this quarter-circle is $A = \frac{\pi}{4}r^2 = \frac{\pi}{4}$, which is also the value of the integral $\int_0^1 \sqrt{1-x^2}\,dx$, therefore, $av(f) = \frac{1}{1-0}\int_0^1 \sqrt{1-x^2}\,dx = (1)\left(\frac{\pi}{4}\right) = \frac{\pi}{4}$.

27. (a) $\int_2^2 g(x)\,dx = 0$

(b) $\int_5^1 g(x)\,dx = -\int_1^5 g(x)\,dx = -8$

(c) $\int_1^2 3f(x)\,dx = 3\int_1^2 f(x)\,dx = 3(-4) = -12$

(d) $\int_2^5 f(x)\,dx = \int_1^5 f(x)\,dx - \int_1^2 f(x)\,dx = 6 - (-4) = 10$

(e) $\int_1^5 [f(x) - g(x)]\,dx = \int_1^5 f(x)\,dx - \int_1^5 g(x)\,dx = 6 - 8 = -2$

(f) $\int_1^5 [4f(x) - g(x)]\,dx = 4\int_1^5 f(x)\,dx - \int_1^5 g(x)\,dx = 4(6) - 8 = 16$

29. (a) $\int_1^2 f(u)\,du = \int_1^2 f(x)\,dx = 5$

(b) $\int_1^2 \sqrt{3}\,f(z)\,dz = \sqrt{3}\int_1^2 f(z)\,dz = 5\sqrt{3}$

(c) $\int_2^1 f(t)\,dt = -\int_1^2 f(t)\,dt = -5$

(d) $\int_1^2 [-f(x)]\,dx = -\int_1^2 f(x)\,dx = -5$

31. (a) $\int_3^4 f(z)\,dz = \int_0^4 f(z)\,dz - \int_0^3 f(z)\,dz = 7 - 3 = 4$

(b) $\int_4^3 f(t)\,dt = -\int_3^4 f(t)\,dt = -4$

33. To find where $x - x^2 \ge 0$, let $x - x^2 = 0 \Rightarrow x(1-x) = 0 \Rightarrow x = 0$ or $x = 1$. If $0 < x < 1$, then $x^2 < x$ $\Rightarrow 0 < x - x^2 \Rightarrow a = 0$ and $b = 1$ maximize the integral.

35. By the constant multiple rule, $\int_a^b k\,dx = k\int_a^b 1\,dx$. The Riemann sums definition of the definite integral gives

$\int_a^b 1\,dx = \lim_{\|P\|\to 0} \sum_{k=1}^{n} \Delta x_k$, and if $\Delta x_k = \frac{b-a}{n}$, then $\lim_{\|P\|\to 0} \sum_{k=1}^{n} \Delta x_k = \lim_{\|P\|\to 0} \sum_{k=1}^{n} \frac{b-a}{n}$

$= \lim_{\|P\|\to 0}\left(\frac{b-a}{n} \sum_{k=1}^{n} 1\right) = \lim_{\|P\|\to 0}\left(\frac{b-a}{n}\cdot n\right) = \lim_{\|P\|\to 0} (b-a) = b-a$. Therefore, $\int_a^b k\,dx = k(b-a)$,

for any k.

37. $f(x) = \frac{1}{1+x^2}$ is decreasing on $[0,1] \Rightarrow$ maximum value of f occurs at $0 \Rightarrow \max f = f(0) = 1$; minimum value of f occurs at $1 \Rightarrow \min f = f(1) = \frac{1}{1+1^2} = \frac{1}{2}$. Therefore, $(1-0)\min f \le \int_0^1 \frac{1}{1+x^2}\,dx \le (1-0)\max f$

$\Rightarrow \frac{1}{2} \le \int_0^1 \frac{1}{1+x^2}\,dx \le 1$. That is, an upper bound = 1 and a lower bound = $\frac{1}{2}$.

39. The car drove the first 150 miles in 5 hours and the second 150 miles in 3 hours, which means it drove 300 miles in 8 hours, for an average of $\frac{300}{8}$ mi/hr $= 37.5$ mi/hr. In terms of average values of functions, the function whose average value we seek is

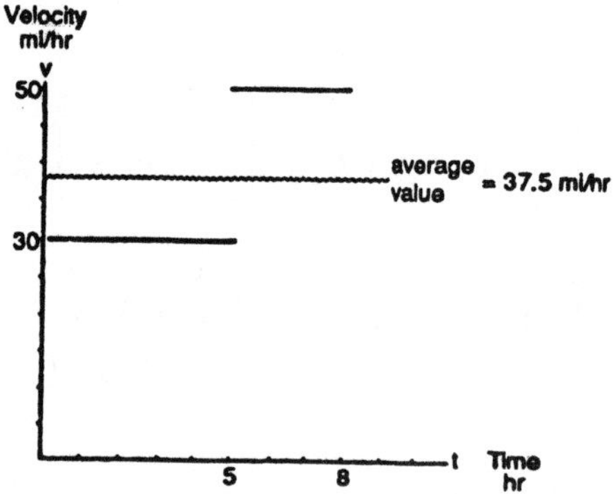

$v(t) = \begin{cases} 30, & 0 \le t \le 5 \\ 50, & 5 < t \le 8 \end{cases}$, and the average value is

$\frac{(30)(5)+(50)(3)}{8} = 37.5$ mph. It does not help to consider

$v(s) = \begin{cases} 30, & 0 \le s \le 150 \\ 50, & 150 < s \le 300 \end{cases}$ whose average value is $\frac{(30)(150)+(50)(150)}{300} = 40$ (mph)/mi because we want the average speed with respect to time, not distance.

4.5 THE MEAN VALUE AND FUNDAMENTAL THEOREMS

1. $\int_{-2}^{0} (2x+5)\,dx = \left[x^2+5x\right]_{-2}^{0} = \left(0^2+5(0)\right) - \left((-2)^2+5(-2)\right) = 6$

3. $\int_0^1 \left(x^2+\sqrt{x}\right) dx = \left[\frac{x^3}{3}+\frac{2}{3}x^{3/2}\right]_0^1 = \left(\frac{1}{3}+\frac{2}{3}\right) - 0 = 1$

5. $\int_0^{\pi} (1+\cos x)\,dx = [x+\sin x]_0^{\pi} = (\pi+\sin \pi) - (0+\sin 0) = \pi$

7. $\int_{\pi/4}^{3\pi/4} \csc\theta \cot\theta \, d\theta = [-\csc\theta]_{\pi/4}^{3\pi/4} = \left(-\csc\left(\frac{3\pi}{4}\right)\right) - \left(-\csc\left(\frac{\pi}{4}\right)\right) = -\sqrt{2} - (-\sqrt{2}) = 0$

9. $\int_{-\pi/2}^{\pi/2} (8y^2 + \sin y)\, dy = \left[\frac{8y^3}{3} - \cos y\right]_{-\pi/2}^{\pi/2} = \left(\frac{8\left(\frac{\pi}{2}\right)^3}{3} - \cos\frac{\pi}{2}\right) - \left(\frac{8\left(-\frac{\pi}{2}\right)^3}{3} - \cos\left(-\frac{\pi}{2}\right)\right) = \frac{2\pi^3}{3}$

11. $\int_1^{\sqrt{2}} \left(\frac{u^2}{2} - \frac{1}{u^5}\right) du = \int_1^{\sqrt{2}} \left(\frac{u^2}{2} - u^{-5}\right) du = \left(\frac{u^3}{6} + \frac{u^{-4}}{4}\right)\Big|_1^{\sqrt{2}} = \left[\left(\frac{(\sqrt{2})^3}{6} + \frac{1}{4(\sqrt{2})^4}\right) - \left(\frac{(1)^3}{6} + \frac{1}{4(1)^4}\right)\right]$

$= \frac{\sqrt{2}}{3} + \frac{1}{16} - \frac{1}{6} - \frac{1}{4} = \frac{16\sqrt{2} + 3 - 8 - 12}{48} = \frac{16\sqrt{2} - 17}{48}$

13. $\int_{-4}^{4} |x|\, dx = \int_{-4}^{0} |x|\, dx + \int_{0}^{4} |x|\, dx = -\int_{-4}^{0} x\, dx + \int_{0}^{4} x\, dx = \left[-\frac{x^2}{2}\right]_{-4}^{0} + \left[\frac{x^2}{2}\right]_0^4 = \left(-\frac{0^2}{2} + \frac{(-4)^2}{2}\right) + \left(\frac{4^2}{2} - \frac{0^2}{2}\right)$

$= 16$

15. (a) $\int_0^{\sqrt{x}} \cos t\, dt = [\sin t]_0^{\sqrt{x}} = \sin\sqrt{x} - \sin 0 = \sin\sqrt{x} \Rightarrow \frac{d}{dx}\left(\int_0^{\sqrt{x}} \cos t\, dt\right) = \frac{d}{dx}(\sin\sqrt{x}) = \cos\sqrt{x}\left(\frac{1}{2}x^{-1/2}\right)$

$= \frac{\cos\sqrt{x}}{2\sqrt{x}}$

(b) $\frac{d}{dx}\left(\int_0^{\sqrt{x}} \cos t\, dt\right) = (\cos\sqrt{x})\left(\frac{d}{dx}(\sqrt{x})\right) = (\cos\sqrt{x})\left(\frac{1}{2}x^{-1/2}\right) = \frac{\cos\sqrt{x}}{2\sqrt{x}}$

17. (a) $\int_0^{t^4} \sqrt{u}\, du = \int_0^{t^4} u^{1/2}\, du = \left[\frac{2}{3}u^{3/2}\right]_0^{t^4} = \frac{2}{3}(t^4)^{3/2} - 0 = \frac{2}{3}t^6 \Rightarrow \frac{d}{dt}\left(\int_0^{t^4} \sqrt{u}\, du\right) = \frac{d}{dt}\left(\frac{2}{3}t^6\right) = 4t^5$

(b) $\frac{d}{dt}\left(\int_0^{t^4} \sqrt{u}\, du\right) = \sqrt{t^4}\left(\frac{d}{dt}(t^4)\right) = t^2(4t^3) = 4t^5$

19. $y = \int_0^x \sqrt{1+t^2}\, dt \Rightarrow \frac{dy}{dx} = \sqrt{1+x^2}$

21. $y = \int_{\sqrt{x}}^0 \sin t^2\, dt \rightarrow y = -\int_0^{\sqrt{x}} \sin t^2\, dt \Rightarrow \frac{dy}{dx} = -\left(\sin(\sqrt{x})^2\right)\left(\frac{d}{dx}(\sqrt{x})\right) = -(\sin x)\left(\frac{1}{2}x^{-1/2}\right) = -\frac{\sin x}{2\sqrt{x}}$

23. $y = \int_0^{\sin x} \frac{dt}{\sqrt{1-t^2}}, |x| < \frac{\pi}{2} \Rightarrow \frac{dy}{dx} = \frac{1}{\sqrt{1-\sin^2 x}}\left(\frac{d}{dx}(\sin x)\right) = \frac{1}{\sqrt{\cos^2 x}}(\cos x) = \frac{\cos x}{|\cos x|} = \frac{\cos x}{\cos x} = 1$ since $|x| < \frac{\pi}{2}$

25. Let $u = 1 - 2x \Rightarrow du = -2\,dx$

$$\int (1-2x)^3\,dx = \int -\frac{1}{2}u^3\,du = -\frac{1}{8}u^4 + C \Rightarrow \int_0^1 (1-2x)^3\,dx = \left[-\frac{1}{8}(1-2x)^4\right]_0^1 = -\frac{1}{8}(-1)^4 - \left(-\frac{1}{8}\right)(1)^4 = 0$$

27. Let $u = 1 + \frac{\theta}{2} \Rightarrow du = \frac{1}{2}\,d\theta$

$$\int \sin^2\left(1+\frac{\theta}{2}\right)d\theta = \int 2\sin^2 u\,du = 2\left(\frac{u}{2} - \frac{1}{4}\sin 2u\right) + C \Rightarrow \int_0^{\pi} \sin^2\left(1+\frac{\theta}{2}\right)d\theta = \left[\left(1+\frac{\theta}{2}\right) - \frac{1}{2}\sin(2+\theta)\right]_0^{\pi}$$

$$= \left[\left(1+\frac{\pi}{2}\right) - \frac{1}{2}\sin(2+\pi)\right] - \left(1 - \frac{1}{2}\sin 2\right) = \frac{\pi}{2} + \sin 2$$

29. $y = \int_2^x \sec t\,dt + 3$

31. $y = \int_0^x \cos^2\theta \sin t\,dt - 1$

33. $-x^2 - 2x = 0 \Rightarrow -x(x+2) = 0 \Rightarrow x = 0$ or $x = -2$; Area

$$= -\int_{-3}^{-2} \left(-x^2 - 2x\right)dx + \int_{-2}^{0} \left(-x^2 - 2x\right)dx - \int_0^2 \left(-x^2 - 2x\right)dx$$

$$= -\left[-\frac{x^3}{3} - x^2\right]_{-3}^{-2} + \left[-\frac{x^3}{3} - x^2\right]_{-2}^{0} - \left[-\frac{x^3}{3} - x^2\right]_0^2$$

$$= -\left(\left(-\frac{(-2)^3}{3} - (-2)^2\right) - \left(-\frac{(-3)^3}{3} - (-3)^2\right)\right)$$

$$+\left(\left(-\frac{0^3}{3} - 0^2\right) - \left(-\frac{(-2)^3}{3} - (-2)^2\right)\right) - \left(\left(-\frac{2^3}{3} - 2^2\right) - \left(-\frac{0^3}{3} - 0^2\right)\right) = \frac{28}{3}$$

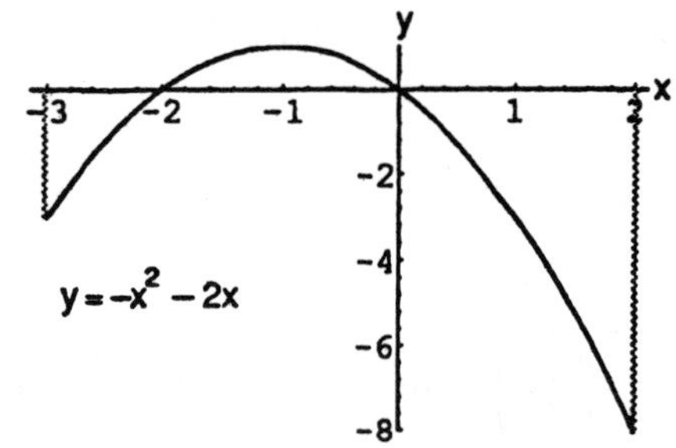

35. $x^3 - 4x = 0 \Rightarrow x\left(x^2 - 4\right) = 0 \Rightarrow x(x-2)(x+2) = 0$

$$\Rightarrow x = 0,\ 2,\ \text{or } -2;\ \text{Area} = \int_{-2}^{0} \left(x^3 - 4x\right)dx - \int_0^2 \left(x^3 - 4x\right)dx$$

$$= \left[\frac{x^4}{4} - 2x^2\right]_{-2}^{0} - \left[\frac{x^4}{4} - 2x^2\right]_0^2 = \left(\frac{0^4}{4} - 2(0)^2\right)$$

$$-\left(\frac{(-2)^4}{4} - 2(-2)^2\right) - \left[\left(\frac{2^4}{4} - 2(2)^2\right) - \left(\frac{0^4}{4} - 2(0)^2\right)\right] = 8$$

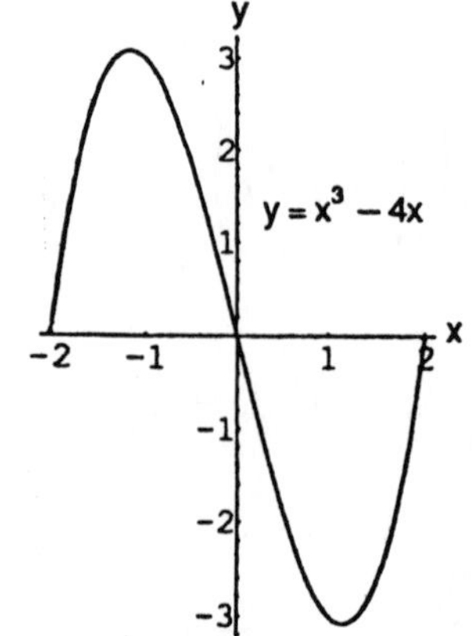

37. The area of the rectangle bounded by the lines $y = 2$, $y = 0$, $x = \pi$, and $x = 0$ is 2π. The area under the curve $y = 1 + \cos x$ on $[0, \pi]$ is $\int_0^{\pi} (1 + \cos x)\,dx = [x + \sin x]_0^{\pi} = (\pi + \sin \pi) - (0 + \sin 0) = \pi$. Therefore the area of

the shaded region is $2\pi - \pi = \pi$.

39. $\frac{dc}{dx} = \frac{1}{2\sqrt{x}} = \frac{1}{2}x^{-1/2} \Rightarrow c = \int_0^x \frac{1}{2}t^{-1/2}dt = \left[t^{1/2}\right]_0^x = \sqrt{x}$

(a) $c(100) - c(1) = \sqrt{100} - \sqrt{1} = \$9.00$

(b) $c(400) - c(100) = \sqrt{400} - \sqrt{100} = \$10.00$

41. (a) $v = \frac{ds}{dt} = \frac{d}{dt}\int_0^t f(x)\,dx = f(t) \Rightarrow v(5) = f(5) = 2$ m/sec

(b) $a = \frac{df}{dt}$ is negative since the slope of the tangent line at $t = 5$ is negative

(c) $s = \int_0^3 f(x)\,dx = \frac{1}{2}(3)(3) = \frac{9}{2}$ m since the integral is the area of the triangle formed by $y = f(x)$, the x-axis, and $x = 3$

(d) $t = 6$ since after $t = 6$ to $t = 9$, the region lies below the x-axis

(e) At $t = 4$ and $t = 7$, since there are horizontal tangents there

(f) Toward the origin between $t = 6$ and $t = 9$ since the velocity is negative on this interval. Away from the origin between $t = 0$ and $t = 6$ since the velocity is positive there.

(g) Right or positive side, because the integral of f from 0 to 9 is positive, there being more area above the x-axis than below it.

43. $\int_4^8 \pi\left(64 - x^2\right)dx = \pi\left[64x - \frac{x^3}{3}\right]_4^8 = \pi\left[\left(512 - \frac{512}{3}\right) - \left(256 - \frac{64}{3}\right)\right] = \pi\left(256 - \frac{448}{3}\right) = \frac{320\pi}{3}$

45. $\int_1^x f(t)\,dt = x^2 - 2x + 1 \Rightarrow f(x) = \frac{d}{dx}\int_1^x f(t)\,dt = \frac{d}{dx}\left(x^2 - 2x + 1\right) = 2x - 2$

47. $f(x) = 2 - \int_2^{x+1} \frac{9}{1+t}\,dt \Rightarrow f'(x) = -\frac{9}{1+(x+1)} = \frac{-9}{x+2} \Rightarrow f'(1) = -3;\ f(1) = 2 - \int_2^{1+1} \frac{9}{1+t}\,dt = 2 - 0 = 2;$

$L(x) = -3(x-1) + f(1) = -3(x-1) + 2 = -3x + 5$

49. (a) True: since f is continuous, g is differentiable by Part 1 of the Fundamental Theorem of Calculus.

(b) True: g is continuous because it is differentiable.

(c) True, since $g'(1) = f(1) = 0$.

(d) False, since $g''(1) = f'(1) > 0$.

(e) True, since $g'(1) = 0$ and $g''(1) = f'(1) > 0$.

(f) False: $g''(x) = f'(x) > 0$, so g'' never changes sign.

(g) True, since $g'(1) = f(1) = 0$ and $g'(x) = f(x)$ is an increasing function of x (because $f'(x) > 0$).

51. (a) $6 - x - x^2 = 0 \Rightarrow x^2 + x - 6 = 0$

$\Rightarrow (x+3)(x-2) = 0 \Rightarrow x = -3$ or $x = 2$;

$$\text{Area} = \int_{-3}^{2} (6 - x - x^2)\,dx = \left[6x - \frac{x^2}{2} - \frac{x^3}{3}\right]_{-3}^{2}$$

$$= \left(6(2) - \frac{2^2}{2} - \frac{2^3}{3}\right) - \left(6(-3) - \frac{(-3)^2}{2} - \frac{(-3)^3}{3}\right)$$

$$= \frac{125}{6}$$

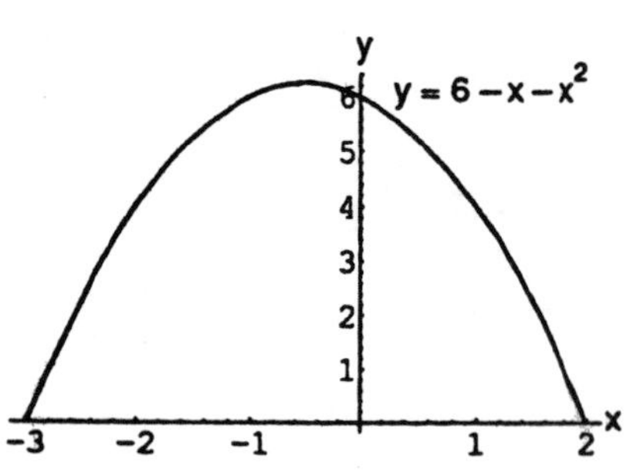

(b) $y' = -1 - 2x = 0 \Rightarrow x = -\frac{1}{2}$; $y' > 0$ for $x < -\frac{1}{2}$ and $y' < 0$ for $x > -\frac{1}{2} \Rightarrow x = -\frac{1}{2}$ yields a local maximum

$\Rightarrow$ height $= y\left(-\frac{1}{2}\right) = \frac{25}{4}$

(c) Base $= 2 - (-3) = 5$, height $= y\left(-\frac{1}{2}\right) = \frac{25}{4} \Rightarrow$ Area $= \frac{2}{3}$(Base)(Height) $= \frac{2}{3}(5)\left(\frac{25}{4}\right) = \frac{125}{6}$

(d) $$\text{Area} = \int_{-b/2}^{b/2} \left(h - \left(\frac{4h}{b^2}\right)x^2\right) dx = \left[hx - \frac{4hx^3}{3b^2}\right]_{-b/2}^{b/2}$$

$$= \left(h\left(\frac{b}{2}\right) - \frac{4h\left(\frac{b}{2}\right)^3}{3b^2}\right) - \left(h\left(-\frac{b}{2}\right) - \frac{4h\left(-\frac{b}{2}\right)^3}{3b^2}\right)$$

$$= \left(\frac{bh}{2} - \frac{bh}{6}\right) - \left(-\frac{bh}{2} + \frac{bh}{6}\right) = bh - \frac{bh}{3} = \frac{2}{3}bh$$

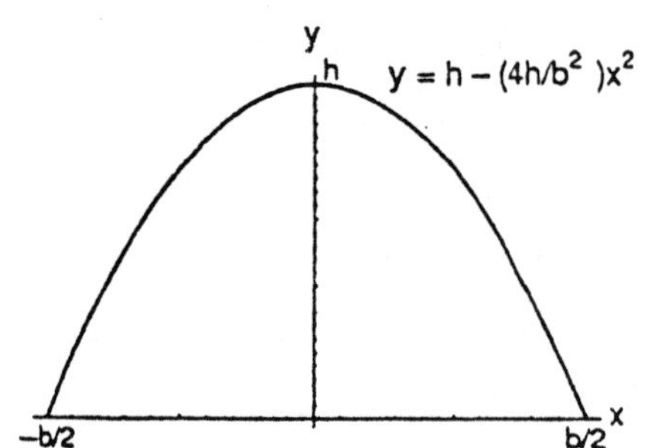

53.

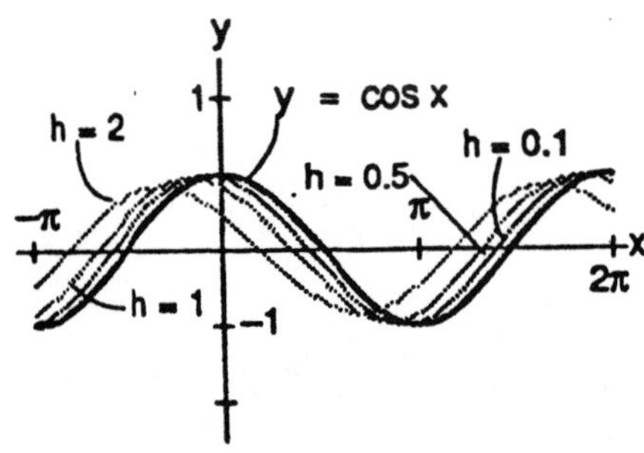

63. In Maple type diff(int(f(x),x=a..u(x)),x); or, in Mathematica type $\partial_x \int_a^{u[x]} f[t]\,dt$

4.6 SUBSTITUTION IN DEFINITE INTEGRALS

1. (a) Let $u = y + 1 \Rightarrow du = dy$; $y = 0 \Rightarrow u = 1$, $y = 3 \Rightarrow u = 4$

$$\int_0^3 \sqrt{y+1}\,dy = \int_1^4 u^{1/2}\,du = \left[\frac{2}{3}u^{3/2}\right]_1^4 = \left(\frac{2}{3}\right)(4)^{3/2} - \left(\frac{2}{3}\right)(1)^{3/2} = \left(\frac{2}{3}\right)(8) - \left(\frac{2}{3}\right)(1) = \frac{14}{3}$$

(b) Use the same substitution for u as in part (a); $y = -1 \Rightarrow u = 0$, $y = 0 \Rightarrow u = 1$

$$\int_{-1}^{0} \sqrt{y+1}\, dy = \int_{0}^{1} u^{1/2}\, du = \left[\frac{2}{3}u^{3/2}\right]_0^1 = \left(\frac{2}{3}\right)(1)^{3/2} - 0 = \frac{2}{3}$$

3. (a) Let $u = \cos x \Rightarrow du = -\sin x\, dx \Rightarrow -du = \sin x\, dx$; $x = 0 \Rightarrow u = 1$, $x = \pi \Rightarrow u = -1$

$$\int_{0}^{\pi} 3\cos^2 x \sin x\, dx = \int_{1}^{-1} -3u^2\, du = \left[-u^3\right]_1^{-1} = -(-1)^3 - \left(-(1)^3\right) = 2$$

(b) Use the same substitution as in part (a); $x = 2\pi \Rightarrow u = 1$, $x = 3\pi \Rightarrow u = -1$

$$\int_{2\pi}^{3\pi} 3\cos^2 x \sin x\, dx = \int_{1}^{-1} -3u^2\, du = 2$$

5. (a) Let $u = 4 + r^2 \Rightarrow du = 2r\, dr \Rightarrow \frac{1}{2}\, du = r\, dr$; $r = -1 \Rightarrow u = 5$, $r = 1 \Rightarrow u = 5$

$$\int_{-1}^{1} \frac{5r}{\left(4+r^2\right)^2}\, dr = 5\int_{5}^{5} \frac{1}{2}u^{-2}\, du = 0$$

(b) Use the same substitution as in part (a); $r = 0 \Rightarrow u = 4$, $r = 1 \Rightarrow u = 5$

$$\int_{0}^{1} \frac{5r}{\left(4+r^2\right)^2}\, dr = 5\int_{4}^{5} \frac{1}{2}u^{-2}\, du = 5\left[-\frac{1}{2}u^{-1}\right]_4^5 = 5\left(-\frac{1}{2}(5)^{-1}\right) - 5\left(-\frac{1}{2}(4)^{-1}\right) = \frac{1}{8}$$

7. (a) Let $u = 1 - \cos 3t \Rightarrow du = 3\sin 3t\, dt \Rightarrow \frac{1}{3}\, du = \sin 3t\, dt$; $t = 0 \Rightarrow u = 0$, $t = \frac{\pi}{6} \Rightarrow u = 1 - \cos\frac{\pi}{2} = 1$

$$\int_{0}^{\pi/6} (1 - \cos 3t)\sin 3t\, dt = \int_{0}^{1} \frac{1}{3}u\, du = \left[\frac{1}{3}\left(\frac{u^2}{2}\right)\right]_0^1 = \frac{1}{6}(1)^2 - \frac{1}{6}(0)^2 = \frac{1}{6}$$

(b) Use the same substitution as in part (a); $t = \frac{\pi}{6} \Rightarrow u = 1$, $t = \frac{\pi}{3} \Rightarrow u = 1 - \cos\pi = 2$

$$\int_{\pi/6}^{\pi/3} (1 - \cos 3t)\sin 3t\, dt = \int_{1}^{2} \frac{1}{3}u\, du = \left[\frac{1}{3}\left(\frac{u^2}{2}\right)\right]_1^2 = \frac{1}{6}(2)^2 - \frac{1}{6}(1)^2 = \frac{1}{2}$$

9. (a) Let $u = 4 + 3\sin z \Rightarrow du = 3\cos z\, dz \Rightarrow \frac{1}{3}\, du = \cos z\, dz$; $z = 0 \Rightarrow u = 4$, $z = 2\pi \Rightarrow u = 4$

$$\int_{0}^{2\pi} \frac{\cos z}{\sqrt{4 + 3\sin z}}\, dz = \int_{4}^{4} \frac{1}{\sqrt{u}}\left(\frac{1}{3}\, du\right) = 0$$

(b) Use the same substitution as in part (a); $z = -\pi \Rightarrow u = 4 + 3\sin(-\pi) = 4$, $z = \pi \Rightarrow u = 4$

$$\int_{-\pi}^{\pi} \frac{\cos z}{\sqrt{4 + 3\sin z}}\, dz = \int_{4}^{4} \frac{1}{\sqrt{u}}\left(\frac{1}{3}\, du\right) = 0$$

11. Let $u = t^5 + 2t \Rightarrow du = \left(5t^4 + 2\right) dt$; $t = 0 \Rightarrow u = 0$, $t = 1 \Rightarrow u = 3$

$$\int_0^1 \sqrt{t^5+2t}\,(5t^4+2)\,dt = \int_0^3 u^{1/2}\,du = \left[\tfrac{2}{3}u^{3/2}\right]_0^3 = \tfrac{2}{3}(3)^{3/2} - \tfrac{2}{3}(0)^{3/2} = 2\sqrt{3}$$

13. Let $u = \cos 2\theta \Rightarrow du = -2\sin 2\theta\,d\theta \Rightarrow -\frac{1}{2}\,du = \sin 2\theta\,d\theta$; $\theta = 0 \Rightarrow u = 1$, $\theta = \frac{\pi}{6} \Rightarrow u = \cos 2\left(\frac{\pi}{6}\right) = \frac{1}{2}$

$$\int_0^{\pi/6} \cos^{-3} 2\theta \sin 2\theta\,d\theta = \int_1^{1/2} u^{-3}\left(-\tfrac{1}{2}\,du\right) = -\tfrac{1}{2}\int_1^{1/2} u^{-3}\,du = \left[-\tfrac{1}{2}\left(\frac{u^{-2}}{-2}\right)\right]_1^{1/2} = \frac{1}{4\left(\frac{1}{2}\right)^2} - \frac{1}{4(1)^2} = \frac{3}{4}$$

15. Let $u = 1 - \sin 2t \Rightarrow du = -2\cos 2t\,dt \Rightarrow -\frac{1}{2}\,du = \cos 2t\,dt$; $t = 0 \Rightarrow u = 1$, $t = \frac{\pi}{4} \Rightarrow u = 0$

$$\int_0^{\pi/4} (1-\sin 2t)^{3/2} \cos 2t\,dt = \int_1^0 -\tfrac{1}{2}u^{3/2}\,du = \left[-\tfrac{1}{2}\left(\tfrac{2}{5}u^{5/2}\right)\right]_1^0 = \left(-\tfrac{1}{5}(0)^{5/2}\right) - \left(-\tfrac{1}{5}(1)^{5/2}\right) = \tfrac{1}{5}$$

17. $y(t) = \int \frac{1}{t^2}\sec^2 \frac{\pi}{t}\,dt$ with $y(4) = \frac{2}{\pi}$. Let $u = \frac{\pi}{t} \Rightarrow du = -\frac{\pi}{t^2}\,dt \Rightarrow \frac{1}{t^2}\,dt = -\frac{du}{\pi}$

$$\Rightarrow y(t) = \int \tfrac{1}{t^2}\sec^2 \tfrac{\pi}{t}\,dt = -\tfrac{1}{\pi}\int \sec^2 u\,du = -\tfrac{1}{\pi}\tan u + C = -\tfrac{1}{\pi}\tan\tfrac{\pi}{t} + C;$$

$$y(4) = \tfrac{2}{\pi} = -\tfrac{1}{\pi}\tan\tfrac{\pi}{4} + C \Rightarrow C = \tfrac{3}{\pi} \Rightarrow y(t) = \tfrac{1}{\pi}\left(3 - \tan\tfrac{\pi}{t}\right)$$

19. For the sketch given, $a = 0$, $b = \pi$; $f(x) - g(x) = 1 - \cos^2 x = \sin^2 x = \frac{1-\cos 2x}{2}$;

$$A = \int_0^{\pi} \frac{(1-\cos 2x)}{2}\,dx = \tfrac{1}{2}\int_0^{\pi} (1-\cos 2x)\,dx = \tfrac{1}{2}\left[x - \frac{\sin 2x}{2}\right]_0^{\pi} = \tfrac{1}{2}[(\pi - 0) - (0-0)] = \tfrac{\pi}{2}$$

21. For the sketch given, $a = -2$, $b = 2$; $f(x) - g(x) = 2x^2 - (x^4 - 2x^2) = 4x^2 - x^4$;

$$A = \int_{-2}^{2} (4x^2 - x^4)\,dx = \left[\frac{4x^3}{3} - \frac{x^5}{5}\right]_{-2}^{2} = \left(\tfrac{32}{3} - \tfrac{32}{5}\right) - \left[-\tfrac{32}{3} - \left(-\tfrac{32}{5}\right)\right] = \tfrac{64}{3} - \tfrac{64}{5} = \frac{320-192}{15} = \frac{128}{15}$$

23. AREA = A1 + A2

A1: For the sketch given, $a = -3$ and we find b by solving the equations $y = x^2 - 4$ and $y = -x^2 - 2x$ simultaneously for x: $x^2 - 4 = -x^2 - 2x \Rightarrow 2x^2 + 2x - 4 = 0 \Rightarrow 2(x+2)(x-1) \Rightarrow x = -2$ or $x = 1$ so $b = -2$: $f(x) - g(x) = (x^2 - 4) - (-x^2 - 2x) = 2x^2 + 2x - 4 \Rightarrow A1 = \int_{-3}^{-2} (2x^2 + 2x - 4)\,dx$

$$= \left[\frac{2x^3}{3} + \frac{2x^2}{2} - 4x\right]_{-3}^{-2} = \left(-\tfrac{16}{3} + 4 + 8\right) - (-18 + 9 + 12) = 9 - \tfrac{16}{3} = \tfrac{11}{3};$$

A2: For the sketch given, a = −2 and b = 1: $f(x) - g(x) = (-x^2 - 2x) - (x^2 - 4) = -2x^2 - 2x + 4$

$$\Rightarrow A2 = -\int_{-2}^{1} (2x^2 + 2x - 4)\,dx = -\left[\frac{2x^3}{3} + x^2 - 4x\right]_{-2}^{1} = -\left(\frac{2}{3} + 1 - 4\right) + \left(-\frac{16}{3} + 4 + 8\right)$$

$$= -\frac{2}{3} - 1 + 4 - \frac{16}{3} + 4 + 8 = 9;$$

Therefore, AREA = A1 + A2 = $\frac{11}{3} + 9 = \frac{38}{3}$

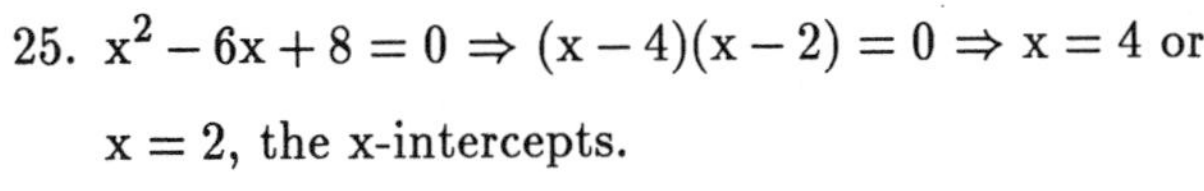

25. $x^2 - 6x + 8 = 0 \Rightarrow (x-4)(x-2) = 0 \Rightarrow x = 4$ or $x = 2$, the x-intercepts.

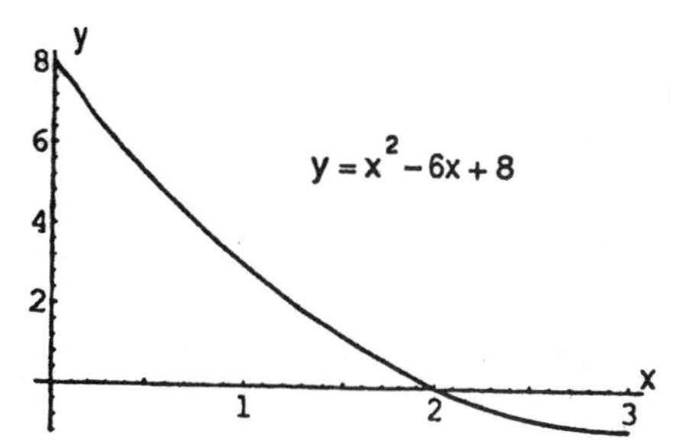

(a) $$\int_0^3 (x^2 - 6x + 8)\,dx = \int_0^3 x^2\,dx - 6\int_0^3 x\,dx + \int_0^3 8\,dx$$

$$= \frac{3^3}{3} - 6\left(\frac{3^2}{2} - \frac{0^2}{2}\right) + 8(3-0) = 6$$

(b) $$\text{Area} = \int_0^2 (x^2 - 6x + 8)\,dx + \left(-\int_2^3 (x^2 - 6x + 8)\,dx\right)$$

$$= \left(\int_0^2 x^2\,dx - 6\int_0^2 x\,dx + \int_0^2 8\,dx\right) - \left(\int_2^3 x^2\,dx - 6\int_2^3 x\,dx + \int_2^3 8\,dx\right)$$

$$= \left[\frac{2^3}{3} - 6\left(\frac{2^2}{2} - \frac{0^2}{2}\right) + 8(2-0)\right] - \left(\int_0^3 x^2\,dx - \int_0^2 x^2\,dx - 6\left(\frac{3^2}{2} - \frac{2^2}{2}\right) + 8(3-2)\right)$$

$$= \left(\frac{8}{3} - 12 + 16\right) - \left(\frac{3^3}{3} - \frac{2^3}{3} - 15 + 8\right) = \frac{22}{3} = 7\tfrac{1}{3}$$

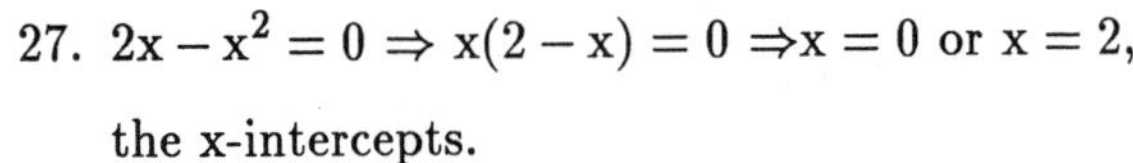

27. $2x - x^2 = 0 \Rightarrow x(2-x) = 0 \Rightarrow x = 0$ or $x = 2$, the x-intercepts.

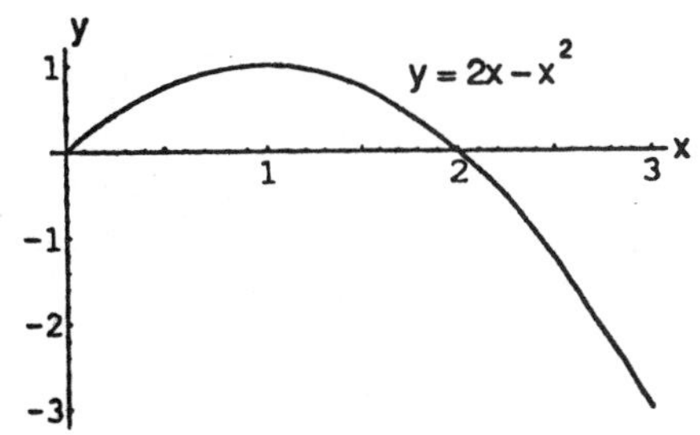

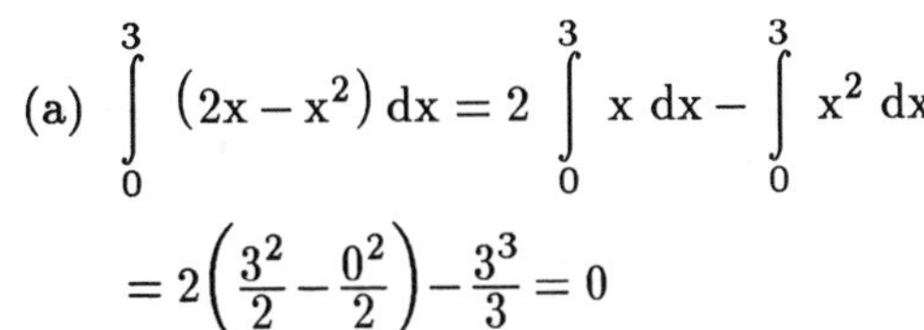

(a) $$\int_0^3 (2x - x^2)\,dx = 2\int_0^3 x\,dx - \int_0^3 x^2\,dx$$

$$= 2\left(\frac{3^2}{2} - \frac{0^2}{2}\right) - \frac{3^3}{3} = 0$$

(b) $$\text{Area} = \int_0^2 (2x - x^2)\,dx - \int_2^3 (2x - x^2)\,dx = 2\int_0^2 x\,dx - \int_0^2 x^2\,dx - \left(2\int_2^3 x\,dx - \int_2^3 x^2\,dx\right)$$

$$= 2\left(\frac{2^2}{2} - \frac{0^2}{2}\right) - \frac{2^3}{3} - 2\left(\frac{3^2}{2} - \frac{2^2}{2}\right) + \left(\int_0^3 x^2\,dx - \int_0^2 x^2\,dx\right) = 4 - \frac{8}{3} - 5 + \frac{3^3}{3} - \frac{2^3}{3} = \frac{8}{3}$$

29. $a = -2,\ b = 2;$

$f(x) - g(x) = 2 - (x^2 - 2) = 4 - x^2$

$$\Rightarrow A = \int_{-2}^{2} (4 - x^2)\,dx = \left[4x - \frac{x^3}{3}\right]_{-2}^{2} = \left(8 - \frac{8}{3}\right) - \left(-8 + \frac{8}{3}\right)$$

$$= 2 \cdot \left(\frac{24}{3} - \frac{8}{3}\right) = \frac{32}{3}$$

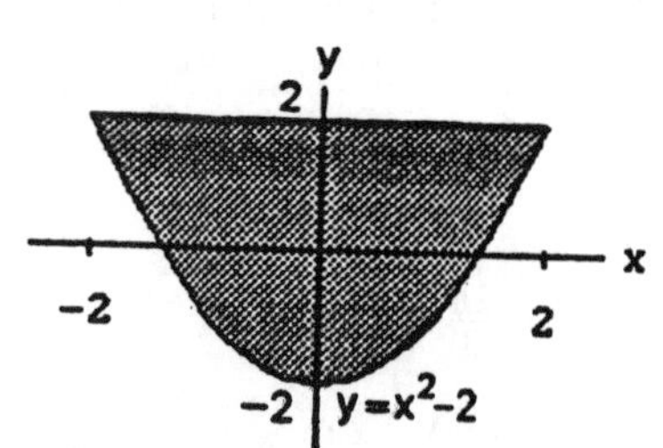

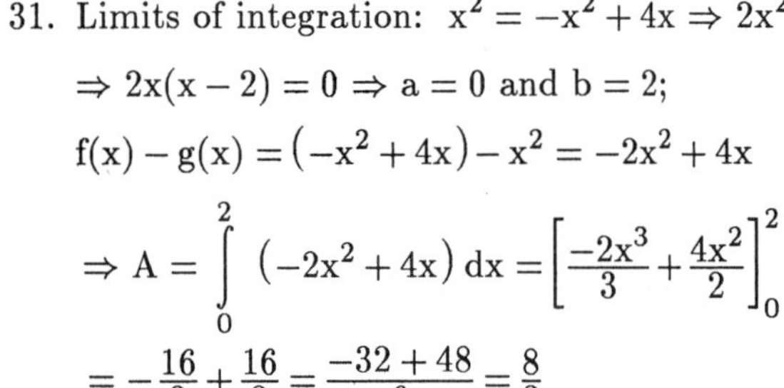

31. Limits of integration: $x^2 = -x^2 + 4x \Rightarrow 2x^2 - 4x = 0$

$\Rightarrow 2x(x - 2) = 0 \Rightarrow a = 0$ and $b = 2;$

$f(x) - g(x) = (-x^2 + 4x) - x^2 = -2x^2 + 4x$

$$\Rightarrow A = \int_{0}^{2} (-2x^2 + 4x)\,dx = \left[\frac{-2x^3}{3} + \frac{4x^2}{2}\right]_{0}^{2}$$

$$= -\frac{16}{3} + \frac{16}{2} = \frac{-32 + 48}{6} = \frac{8}{3}$$

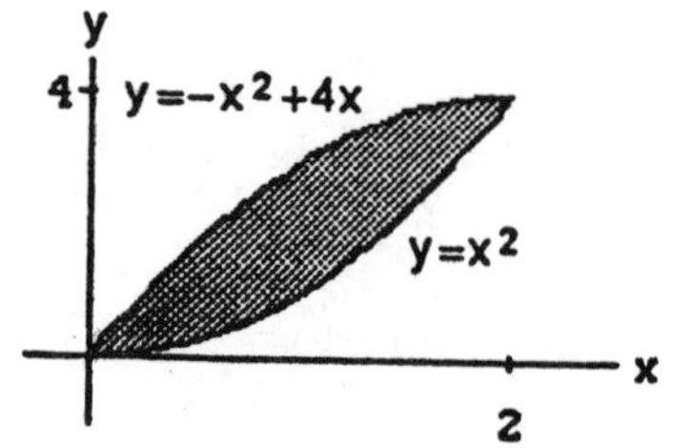

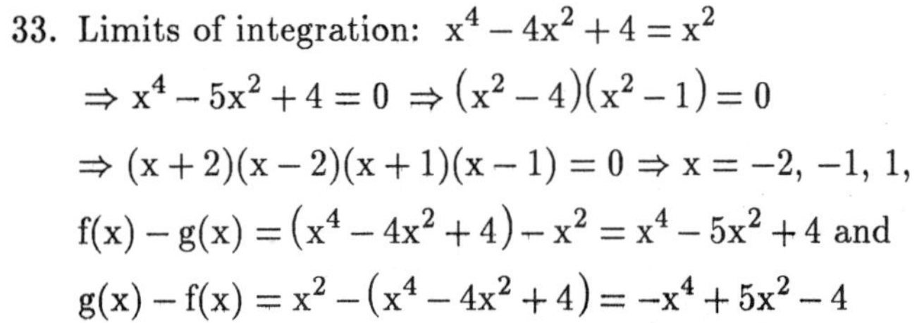

33. Limits of integration: $x^4 - 4x^2 + 4 = x^2$

$\Rightarrow x^4 - 5x^2 + 4 = 0 \Rightarrow (x^2 - 4)(x^2 - 1) = 0$

$\Rightarrow (x + 2)(x - 2)(x + 1)(x - 1) = 0 \Rightarrow x = -2, -1, 1, 2;$

$f(x) - g(x) = (x^4 - 4x^2 + 4) - x^2 = x^4 - 5x^2 + 4$ and

$g(x) - f(x) = x^2 - (x^4 - 4x^2 + 4) = -x^4 + 5x^2 - 4$

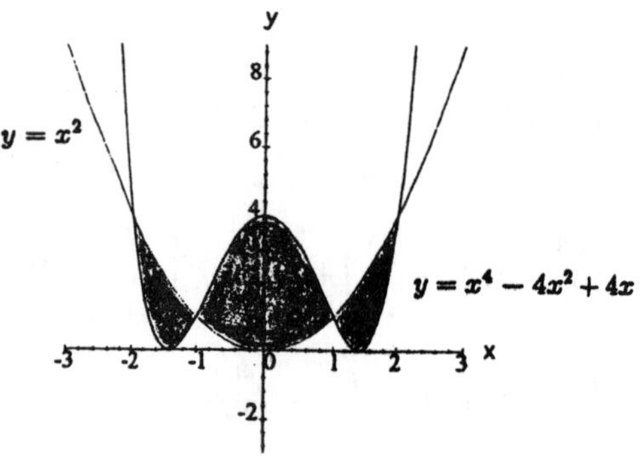

$$\Rightarrow A = \int_{-2}^{-1} (-x^4 + 5x^2 - 4)\,dx + \int_{-1}^{1} (x^4 - 5x^2 + 4)\,dx + \int_{1}^{2} (-x^4 + 5x^2 - 4)\,dx$$

$$= \left[-\frac{x^5}{5} + \frac{5x^3}{3} - 4x\right]_{-2}^{-1} + \left[\frac{x^5}{5} - \frac{5x^3}{3} + 4x\right]_{-1}^{1} + \left[\frac{-x^5}{5} + \frac{5x^3}{3} - 4x\right]_{1}^{2} = \left(\frac{1}{5} - \frac{5}{3} + 4\right) - \left(\frac{32}{5} - \frac{40}{3} + 8\right) + \left(\frac{1}{5} - \frac{5}{3} + 4\right)$$

$$- \left(-\frac{1}{5} + \frac{5}{3} - 4\right) + \left(-\frac{32}{5} + \frac{40}{3} - 8\right) - \left(-\frac{1}{5} + \frac{5}{3} - 4\right) = -\frac{60}{5} + \frac{60}{3} = \frac{300 - 180}{15} = 8$$

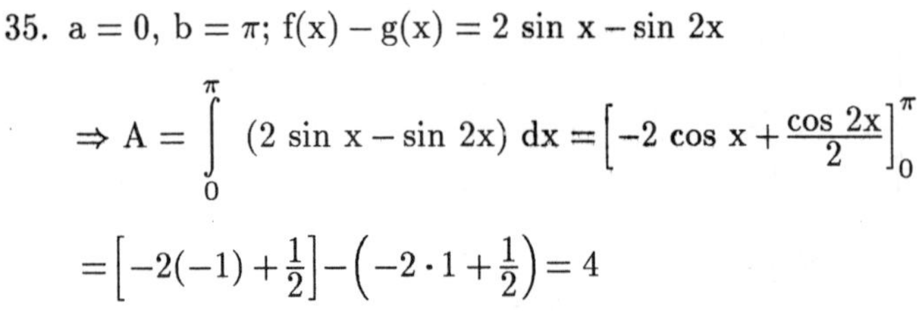

35. $a = 0,\ b = \pi;\ f(x) - g(x) = 2 \sin x - \sin 2x$

$$\Rightarrow A = \int_{0}^{\pi} (2 \sin x - \sin 2x)\,dx = \left[-2 \cos x + \frac{\cos 2x}{2}\right]_{0}^{\pi}$$

$$= \left[-2(-1) + \frac{1}{2}\right] - \left(-2 \cdot 1 + \frac{1}{2}\right) = 4$$

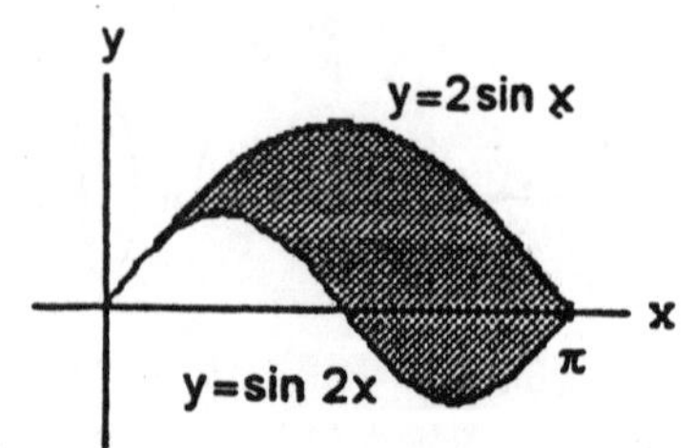

37. $A = A1 + A2$

$a_1 = -1,\ b_1 = 0$ and $a_2 = 0,\ b_2 = 1;$

$f_1(x) - g_1(x) = x - \sin\left(\frac{\pi x}{2}\right)$ and $f_2(x) - g_2(x) = \sin\left(\frac{\pi x}{2}\right) - x$

$\Rightarrow$ by symmetry about the origin,

$$A_1 + A_2 = 2A_1 \Rightarrow A = 2\int_0^1 \left[\sin\left(\frac{\pi x}{2}\right) - x\right] dx$$

$$= 2\left[-\frac{2}{\pi}\cos\left(\frac{\pi x}{2}\right) - \frac{x^2}{2}\right]_0^1 = 2\left[\left(-\frac{2}{\pi}\cdot 0 - \frac{1}{2}\right) - \left(-\frac{2}{\pi}\cdot 1 - 0\right)\right]$$

$$= 2\left(\frac{2}{\pi} - \frac{1}{2}\right) = 2\left(\frac{4-\pi}{2\pi}\right) = \frac{4-\pi}{\pi}$$

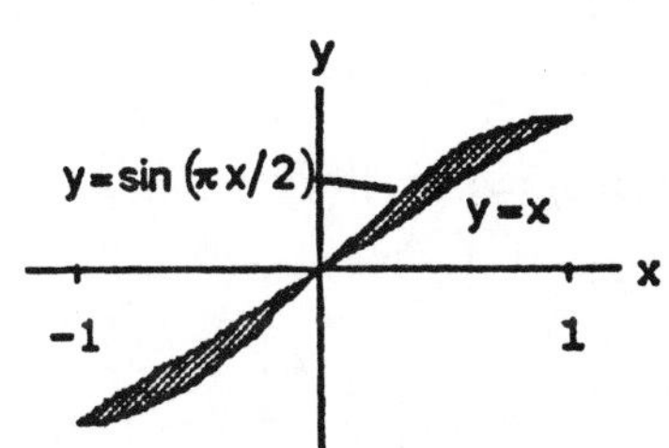

39. $A = A_1 + A_2$

Limits of integration: $y = x$ and $y = \frac{1}{x^2} \Rightarrow x = \frac{1}{x^2},\ x \neq 0$

$\Rightarrow x^3 = 1 \Rightarrow x = 1\ ,\ f_1(x) - g_1(x) = x - 0 = x$

$$\Rightarrow A_1 = \int_0^1 x\ dx = \left[\frac{x^2}{2}\right]_0^1 = \frac{1}{2};\ f_2(x) - g_2(x) = \frac{1}{x^2} - 0$$

$$= x^{-2} \Rightarrow A_2 = \int_1^2 x^{-2}\ dx = \left[\frac{-1}{x}\right]_1^2 = -\frac{1}{2} + 1 = \frac{1}{2};$$

$A = A_1 + A_2 = \frac{1}{2} + \frac{1}{2} = 1$

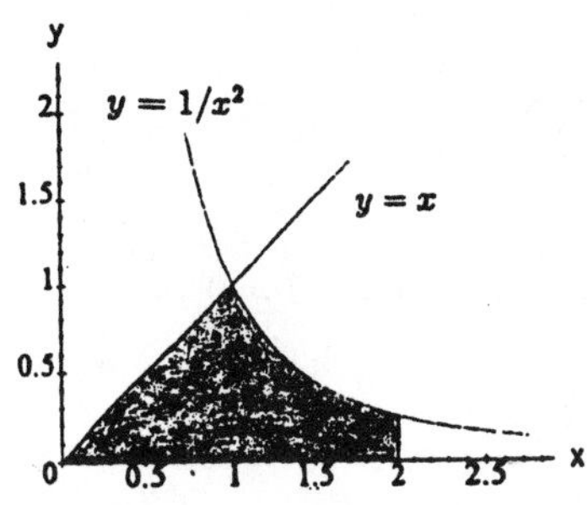

41. Limits of integration: $y = 3 - x^2$ and $y = -1$

$\Rightarrow 3 - x^2 = -1 \Rightarrow x^2 = 4 \Rightarrow a = -2$ and $b = 2;$

$f(x) - g(x) = (3 - x^2) - (-1) = 4 - x^2$

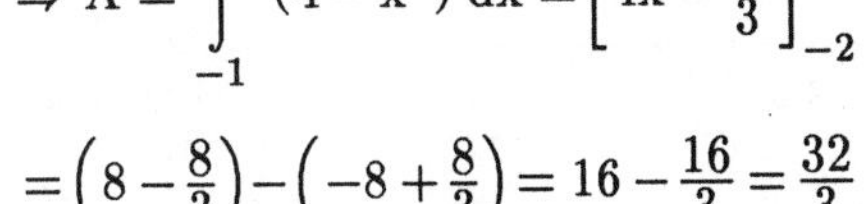

$$\Rightarrow A = \int_{-1}^{2} (4 - x^2)\ dx = \left[4x - \frac{x^3}{3}\right]_{-2}^{2}$$

$$= \left(8 - \frac{8}{3}\right) - \left(-8 + \frac{8}{3}\right) = 16 - \frac{16}{3} = \frac{32}{3}$$

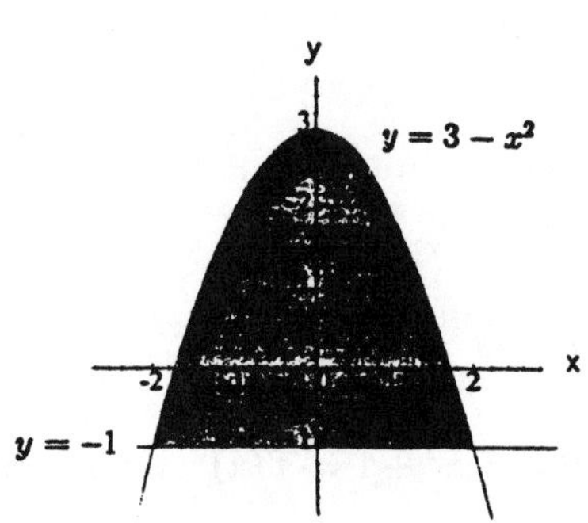

4.7 NUMERICAL INTEGRATION

1. $\int_1^2 x\ dx$

I. (a) For $n = 4$, $h = \frac{b-a}{n} = \frac{2-1}{4} = \frac{1}{4} \Rightarrow \frac{h}{2} = \frac{1}{8}$;

	x_i	$f(x_i)$	m	$mf(x_i)$
x_0	1	1	1	1
x_1	5/4	5/4	2	5/2
x_2	3/2	3/2	2	3
x_3	7/4	7/4	2	7/2
x_4	2	2	1	2

$\sum mf(x_i) = 12 \Rightarrow T = \frac{1}{8}(12) = \frac{3}{2}$;

$f(x) = x \Rightarrow f'(x) = 1 \Rightarrow f'' = 0 \Rightarrow M = 0$
$\Rightarrow |E_T| = 0$

(b) $\int_1^2 x\,dx = \left[\frac{x^2}{2}\right]_1^2 = 2 - \frac{1}{2} = \frac{3}{2} \Rightarrow |E_T| = \int_1^2 x\,dx - T = 0$

(c) $\frac{|E_T|}{\text{True Value}} \times 100 = 0\%$

II. (a) For $n = 4$, $h = \frac{b-a}{n} = \frac{2-1}{4} = \frac{1}{4} \Rightarrow \frac{h}{3} = \frac{1}{12}$;

	x_i	$f(x_i)$	m	$mf(x_i)$
x_0	1	1	1	1
x_1	5/4	5/4	4	5
x_2	3/2	3/2	2	3
x_3	7/4	7/4	4	7
x_4	2	2	1	2

$\sum mf(x_i) = 18 \Rightarrow S = \frac{1}{12}(18) = \frac{3}{2}$;

$f^{(4)}(x) = 0 \Rightarrow M = 0 \Rightarrow |E_S| = 0$

(b) $\int_1^2 x\,dx = \frac{3}{2} \Rightarrow |E_S| = \int_1^2 x\,dx - S = \frac{3}{2} - \frac{3}{2} = 0$

(c) $\frac{|E_S|}{\text{True Value}} \times 100 = 0\%$

3. $\int_{-1}^{1} (x^2 + 1)\,dx$

I. (a) For $n = 4$, $h = \frac{b-a}{n} = \frac{1-(-1)}{4} = \frac{2}{4} = \frac{1}{2} \Rightarrow \frac{h}{2} = \frac{1}{4}$;

	x_i	$f(x_i)$	m	$mf(x_i)$
x_0	−1	2	1	2
x_1	−1/2	5/4	2	5/2
x_2	0	1	2	2
x_3	1/2	5/4	2	5/2
x_4	1	2	1	2

$\sum mf(x_i) = 11 \Rightarrow T = \frac{1}{4}(11) = 2.75$;

$f(x) = x^2 + 1 \Rightarrow f'(x) = 2x \Rightarrow f''(x) = 2 \Rightarrow M = 2$

$\Rightarrow |E_T| \le \frac{1-(-1)}{12}\left(\frac{1}{2}\right)^2(2) = \frac{1}{12}$ or 0.08333

(b) $\int_{-1}^{1} (x^2 + 1)\,dx = \left[\frac{x^3}{3} + x\right]_{-1}^{1} = \left(\frac{1}{3} + 1\right) - \left(-\frac{1}{3} - 1\right) = \frac{8}{3} \Rightarrow E_T = \int_{-1}^{1} (x^2 + 1)\,dx - T = \frac{8}{3} - \frac{11}{4} = -\frac{1}{12}$

$\Rightarrow |E_T| = \left|-\frac{1}{12}\right| \approx 0.08333$

(c) $\frac{|E_T|}{\text{True Value}} \times 100 = \left(\frac{\frac{1}{12}}{\frac{8}{3}}\right) \times 100 \approx 3\%$

II. (a) For n = 4, $h = \frac{b-a}{n} = \frac{3-1}{4} = \frac{2}{4} = \frac{1}{2} \Rightarrow \frac{h}{3} = \frac{1}{6}$;

	x_i	$f(x_i)$	m	$mf(x_i)$
x_0	−1	2	1	2
x_1	−1/2	5/4	4	5
x_2	0	1	2	2
x_3	1/2	5/4	4	5
x_4	1	2	1	2

$\sum mf(x_i) = 16 \Rightarrow S = \frac{1}{6}(16) = \frac{8}{3} = 2.66667$;

$f^{(3)}(x) = 0 \Rightarrow f^{(4)}(x) = 0 \Rightarrow M = 0 \Rightarrow |E_S| = 0$

(b) $\int_{-1}^{1} (x^2+1)\,dx = \left[\frac{x^3}{3} + x\right]_{-1}^{1} = \frac{8}{3} \Rightarrow |E_S| = \int_{-1}^{1} (x^2+1)\,dx - S = \frac{8}{3} - \frac{8}{3} = 0$

(c) $\frac{|E_S|}{\text{True Value}} \times 100 = 0\%$

5. $\int_0^2 (t^3+t)\,dt$

I. (a) For n = 4, $h = \frac{b-a}{n} = \frac{2-0}{4} = \frac{2}{4} = \frac{1}{2} \Rightarrow \frac{h}{2} = \frac{1}{4}$;

	t_i	$f(t_i)$	m	$mf(t_i)$
t_0	0	0	1	0
t_1	1/2	5/8	2	5/4
t_2	1	2	2	4
t_3	3/2	39/8	2	39/4
t_4	2	10	1	10

$\sum mf(t_i) = 25 \Rightarrow T = \frac{1}{4}(25) = \frac{25}{4}$;

$f(t) = t^3 + t \Rightarrow f'(t) = 3t^2 + 1 \Rightarrow f''(t) = 6t \Rightarrow M = 12$

$= f''(2) \Rightarrow |E_T| \le \frac{2-0}{12}\left(\frac{1}{2}\right)^2 (12) = \frac{1}{2}$

(b) $\int_0^2 (t^3+t)\,dt = \left[\frac{t^4}{4} + \frac{t^2}{2}\right]_0^2 = \left(\frac{2^4}{4} + \frac{2^2}{2}\right) - 0 = 6 \Rightarrow |E_T| = \int_0^2 (t^3+t)\,dt - T = 6 - \frac{25}{4} = -\frac{1}{4} \Rightarrow |E_T| = \frac{1}{4}$

(c) $\frac{|E_T|}{\text{True Value}} \times 100 = \frac{\left|-\frac{1}{4}\right|}{6} \times 100 \approx 4\%$

II. (a) For n = 4, $h = \frac{b-a}{n} = \frac{2-0}{4} = \frac{2}{4} = \frac{1}{2} \Rightarrow \frac{h}{3} = \frac{1}{6}$;

	t_i	$f(t_i)$	m	$mf(t_i)$
t_0	0	0	1	0
t_1	1/2	5/8	4	5/2
t_2	1	2	2	4
t_3	3/2	39/8	4	39/2
t_4	2	10	1	10

$\sum mf(t_i) = 36 \Rightarrow S = \frac{1}{6}(36) = 6$;

$f^{(3)}(t) = 6 \Rightarrow f^{(4)}(t) = 0 \Rightarrow M = 0 \Rightarrow |E_S| = 0$

(b) $\int_0^2 (t^3+t)\,dt = 6 \Rightarrow |E_S| = \int_0^2 (t^3+t)\,dt - S = 6 - 6 = 0$

(c) $\frac{|E_S|}{\text{True Value}} \times 100 = 0\%$

7. $\int_1^2 \frac{1}{s^2}\,ds$

I. (a) For $n = 4$, $h = \frac{b-a}{n} = \frac{2-1}{4} = \frac{1}{4} \Rightarrow \frac{h}{2} = \frac{1}{8}$;

	s_i	$f(s_i)$	m	$mf(s_i)$
s_0	1	1	1	1
s_1	5/4	16/25	2	32/25
s_2	3/2	4/9	2	8/9
s_3	7/4	16/49	2	32/49
s_4	2	1/4	1	1/4

$\sum mf(s_i) = \frac{179{,}573}{44{,}100} \Rightarrow T = \frac{1}{8}\left(\frac{179{,}573}{44{,}100}\right) = \frac{179{,}573}{352{,}800}$

≈ 0.50899; $f(s) = \frac{1}{s^2} \Rightarrow f'(s) = -\frac{2}{s^3} \Rightarrow f''(s) = \frac{6}{s^4}$

$\Rightarrow M = 6 = f''(1) \Rightarrow |E_T| \le \frac{2-1}{12}\left(\frac{1}{4}\right)^2(6) = \frac{1}{32} = 0.03125$

(b) $\int_1^2 \frac{1}{s^2}\,ds = \int_1^2 s^{-2}\,ds = \left[-\frac{1}{s}\right]_1^2 = -\frac{1}{2} - \left(-\frac{1}{1}\right) = \frac{1}{2} \Rightarrow E_T = \int_1^2 \frac{1}{s^2}\,ds - T = \frac{1}{2} - 0.50899 = -0.00899$

$\Rightarrow |E_T| = 0.00899$

(c) $\frac{|E_T|}{\text{True Value}} \times 100 = \frac{0.00899}{0.5} \times 100 \approx 2\%$

II. (a) For $n = 4$, $h = \frac{b-a}{n} = \frac{2-1}{4} = \frac{1}{4} \Rightarrow \frac{h}{3} = \frac{1}{12}$;

	s_i	$f(s_i)$	m	$mf(s_i)$
s_0	1	1	1	1
s_1	5/4	16/25	4	64/25
s_2	3/2	4/9	2	8/9
s_3	7/4	16/49	4	64/49
s_4	2	1/4	1	1/4

$\sum mf(s_i) = \frac{264{,}821}{44{,}100} \Rightarrow S = \frac{1}{12}\left(\frac{264{,}821}{44{,}100}\right) = \frac{264{,}821}{529{,}200}$

≈ 0.50041; $f^{(3)}(s) = -\frac{24}{s^5} \Rightarrow f^{(4)}(s) = \frac{120}{s^6}$

$\Rightarrow M = 120 \Rightarrow |E_S| \le \left|\frac{2-1}{180}\right|\left(\frac{1}{4}\right)^4(120) = \frac{1}{384} \approx 0.002604$

(b) $\int_1^2 \frac{1}{s^2}\,ds = \frac{1}{2} \Rightarrow E_S = \int_1^2 \frac{1}{s^2}\,ds - S = \frac{1}{2} - 0.050041 = -0.00041 \Rightarrow |E_S| = 0.00041$

(c) $\frac{|E_S|}{\text{True Value}} \times 100 = \frac{0.0004}{0.5} \times 100 \approx 0.08\%$

9. $\int_0^\pi \sin t\,dt$

I. (a) For $n = 4$, $h = \frac{b-a}{n} = \frac{\pi - 0}{4} = \frac{\pi}{4} \Rightarrow \frac{h}{2} = \frac{\pi}{8}$;

	t_i	$f(t_i)$	m	$mf(t_i)$
t_0	0	0	1	0
t_1	$\pi/4$	$\sqrt{2}/2$	2	$\sqrt{2}$
t_2	$\pi/2$	1	2	2
t_3	$3\pi/4$	$\sqrt{2}/2$	2	$\sqrt{2}$
t_4	π	0	1	0

$\sum mf(t_i) = 2 + 2\sqrt{2} \approx 4.9294 \Rightarrow T = \frac{\pi}{8}(2 + 2\sqrt{2})$

≈ 1.89612; $f(t) = \sin t \Rightarrow f'(t) = \cos t \Rightarrow f''(t) = -\sin t$

$\Rightarrow M = 1 \Rightarrow |E_T| \le \frac{\pi - 0}{12}\left(\frac{\pi}{4}\right)^2(1) = \frac{\pi^3}{192} \approx 0.16149$

(b) $\int_0^\pi \sin t\, dt = [-\cos t]_0^\pi = (-\cos \pi) - (-\cos 0) = 2 \Rightarrow |E_T| = \int_0^\pi \sin t\, dt - T \approx 2 - 1.89612 = 0.10388$

(c) $\frac{|E_T|}{\text{True Value}} \times 100 = \frac{0.10388}{2} \times 100 \approx 5\%$

II. (a) For $n = 4$, $h = \frac{b-a}{n} = \frac{\pi - 0}{4} = \frac{\pi}{4} \Rightarrow \frac{h}{3} = \frac{\pi}{12}$;

	t_i	$f(t_i)$	m	$mf(t_i)$
t_0	0	0	1	0
t_1	$\pi/4$	$\sqrt{2}/2$	4	$2\sqrt{2}$
t_2	$\pi/2$	1	2	2
t_3	$3\pi/4$	$\sqrt{2}/2$	4	$2\sqrt{2}$
t_4	π	0	1	0

$\sum mf(t_i) = 2 + 4\sqrt{2} \approx 7.6569 \Rightarrow S = \frac{\pi}{12}(2 + 4\sqrt{2})$

≈ 2.00456; $f^{(3)}(t) = -\cos t \Rightarrow f^{(4)}(t) = \sin t$

$\Rightarrow M = 1 \Rightarrow |E_S| \le \frac{\pi - 0}{180}\left(\frac{\pi}{4}\right)^4(1) \approx 0.00664$

(b) $\int_0^\pi \sin t\, dt \Rightarrow E_S = \int_0^\pi \sin t\, dt - S \approx 2 - 2.00456 = -0.00456 \Rightarrow |E_S| \approx 0.00456$

(c) $\frac{|E_S|}{\text{True Value}} \times 100 = \frac{0.00456}{2} \times 100 \approx 0.23\%$

11. (a) $n = 8 \Rightarrow h = \frac{1}{8} \Rightarrow \frac{h}{2} = \frac{1}{16}$;

$\sum mf(x_i) = 1(0.0) + 2(0.12402) + 2(0.24206) + 2(0.34763) + 2(0.43301) + 2(0.48789) + 2(0.49608)$
$+ 2(0.42361) + 1(0) = 5.1086 \Rightarrow T = \frac{1}{16}(5.1086) = 0.31929$

(b) $n = 8 \Rightarrow h = \frac{1}{8} \Rightarrow \frac{h}{3} = \frac{1}{24}$;

$\sum mf(x_i) = 1(0.0) + 4(0.12402) + 2(0.24206) + 4(0.34763) + 2(0.43301) + 4(0.48789) + 2(0.49608)$
$+ 4(0.42361) + 1(0) = 7.8749 \Rightarrow S = \frac{1}{24}(7.8749) = 0.32812$

(c) Let $u = 1 - x^2 \Rightarrow du = -2x\, dx \Rightarrow -\frac{1}{2}\, du = x\, dx$; $x = 0 \Rightarrow u = 1$, $x = 1 \Rightarrow u = 0$

$$\int_0^1 x\sqrt{1-x^2}\, dx = \int_1^0 \sqrt{u}\left(-\frac{1}{2}\, du\right) = \frac{1}{2}\int_0^1 u^{1/2}\, du = \left[\frac{1}{2}\left(\frac{u^{3/2}}{\frac{3}{2}}\right)\right]_0^1 = \left[\frac{1}{3}u^{3/2}\right]_0^1 = \frac{1}{3}\left(\sqrt{1}\right)^3 - \frac{1}{3}\left(\sqrt{0}\right)^3 = \frac{1}{3};$$

$$E_T = \int_0^1 x\sqrt{1-x^2}\,dx - T \approx \tfrac{1}{3} - 0.31929 = 0.01404;\ E_S = \int_0^1 x\sqrt{1-x^2}\,dx - S \approx \tfrac{1}{3} - 0.32812 = 0.00521$$

13. (a) $n = 8 \Rightarrow h = \frac{\pi}{8} \Rightarrow \frac{h}{2} = \frac{\pi}{16}$;

$\sum mf(t_i) = 1(0.0) + 2(1.99138) + 2(1.26906) + 2(1.05961) + 2(0.75) + 2(0.48821) + 2(0.28946) + 2(0.13429)$
$+ 1(0) = 9.96402 \Rightarrow T = \frac{\pi}{16}(9.96402) \approx 1.95643$

(b) $n = 8 \Rightarrow h = \frac{\pi}{8} \Rightarrow \frac{h}{3} = \frac{\pi}{24}$;

$\sum mf(t_i) = 1(0.0) + 4(0.99138) + 2(1.26906) + 4(1.05961) + 2(0.75) + 4(0.48821) + 2(0.28946) + 4(0.13429)$
$+ 1(0) = 15.311 \Rightarrow S \approx \frac{\pi}{24}(15.311) \approx 2.00421$

(c) Let $u = 2 + \sin t \Rightarrow du = \cos t\,dt$; $t = -\frac{\pi}{2} \Rightarrow u = 2 + \sin\left(-\frac{\pi}{2}\right) = 1$, $t = \frac{\pi}{2} \Rightarrow u = 2 + \sin\frac{\pi}{2} = 3$

$$\int_{-\pi/2}^{\pi/2} \frac{3\cos t}{(2+\sin t)^2}\,dt = \int_1^3 \frac{3}{u^2}\,du = 3\int_1^3 u^{-2}\,du = \left[3\left(\frac{u^{-1}}{-1}\right)\right]_1^3 = 3\left(-\tfrac{1}{3}\right) - 3\left(-\tfrac{1}{1}\right) = 2;$$

$$E_T = \int_{-\pi/2}^{\pi/2} \frac{3\cos t}{(2+\sin t)^2}\,dt - T \approx 2 - 1.95643 = 0.04357;\ E_S = \int_{-\pi/2}^{\pi/2} \frac{3\cos t}{(2+\sin t)^2}\,dt - S$$
$$\approx 2 - 2.00421 = -0.00421$$

15. $\frac{5}{2}(6.0 + 2(8.2) + 2(9.1 + \ldots + 2(12.7) + 13.0)(30) = 15{,}990\ \text{ft}^3$

17. Use the conversion 30 mph = 44 fps (ft per sec) since time is measured in seconds. The distance traveled as the car accelerates from, say, 40 mph = 58.67 fps to 50 mph = 73.33 fps in (4.5 − 3.2) = 1.3 sec is the area of the trapezoid (see figure) associated with that time interval: $\frac{1}{2}(58.67 + 73.33)(1.3) = 85.8$ ft. The total distance traveled by the Ford Mustang Cobra is the sum of all these eleven trapezoids (using $\frac{\Delta t}{2}$ and the table below):

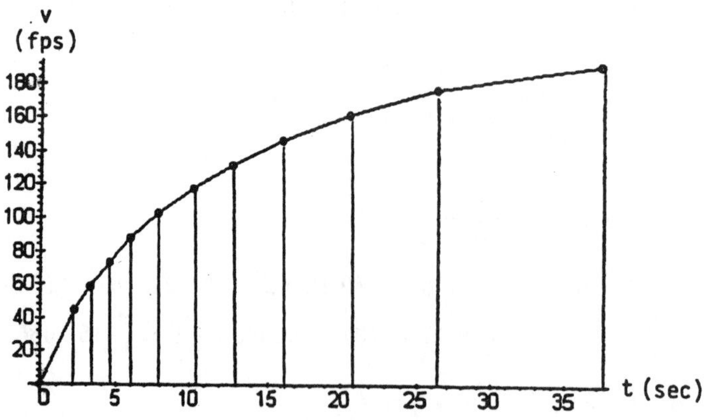

$s = (44)(1.1) + (102.67)(0.5) + (132)(0.65) + (161.33)(0.7) + (190.67)(0.95) + (220)(1.2) + (249.33)(1.25)$
$+ (278.67)(1.65) + (308)(2.3) + (337.33)(2.8) + (366.67)(5.45) = 5166.35\ \text{ft} \approx 0.978\ \text{mi}$

v (mph)	0	30	40	50	60	70	80	90	100	110	120	130
v (fps)	0	44	58.67	73.33	88	102.67	117.33	132	146.67	161.33	176	190.67
t (sec)	0	2.2	3.2	4.5	5.9	7.8	10.2	12.7	16	20.6	26.2	37.1
Δt/2	0	1.1	0.5	0.65	0.7	0.95	1.2	1.25	1.65	2.3	2.8	5.45

19. Using Simpson's Rule, $h = 1 \Rightarrow \frac{h}{3} = \frac{1}{3}$;

	x_i	y_i	m	my_i
x_0	0	1.5	1	1.5
x_1	1	1.6	4	6.4
x_2	2	1.8	2	3.6
x_3	3	1.9	4	7.6
x_4	4	2.0	2	4.0
x_5	5	2.1	4	8.4
x_6	6	2.1	1	2.1

$\sum my_i = 33.6 \Rightarrow$ Cross Section Area $\approx \frac{1}{3}(33.6) = 11.2\ \text{ft}^2$.
Let x be the length of the tank. Then the Volume V
= (Cross Sectional Area) x = 11.2x. Now 5000 lb of
gasoline at 42 lb/ft^3 $\Rightarrow V = \frac{5000}{42} = 119.05\ \text{ft}^3$
$\Rightarrow 119.05 = 11.2x \Rightarrow x \approx 10.63$ ft

21. $n = 2 \Rightarrow h = \frac{2-0}{2} = 1 \Rightarrow \frac{h}{3} = \frac{1}{3}$;

	x_i	$f(x_i)$	m	$mf(x_i)$
x_0	0	0	1	0
x_1	1	1	4	4
x_2	2	8	1	8

$\sum mf(x_i) = 12 \Rightarrow S = \frac{1}{3}(12) = 4$;

$$\int_0^2 x^3\,dx = \left[\frac{x^4}{4}\right]_0^2 = \frac{2^4}{4} - \frac{0^4}{4} = 4$$

23. (a) $h = \frac{b-a}{n} = \frac{1-0}{10} = 0.1 \Rightarrow \text{erf}(1) = \frac{2}{\sqrt{\pi}}\left(\frac{0.1}{3}\right)(y_0 + 4y_1 + 2y_2 + 4y_3 + \ldots + 4y_9 + y_{10})$

$$= \frac{2}{30\sqrt{\pi}}\left(e^0 + 4e^{-0.01} + 2e^{-0.04} + 4e^{-0.09} + \ldots + 4e^{-0.81} + e^{-1}\right) \approx 0.843$$

(b) $|E_s| \le \frac{1-0}{180}(0.1)^4(12) \approx 6.7 \times 10^{-6}$

25. $S_{50} \approx 3.13791$, $S_{100} \approx 3.14029$

27. 1.37076 (value obtained using a graphing calculator with n = 50)

29. (a) $T_{10} \approx 1.983523538$

$T_{100} \approx 1.999835504$

$T_{1000} \approx 1.999998355$

(b)

n	$\lvert E_T \rvert = 2 - T_n$
10	$0.016476462 = 1.6476462 \times 10^{-2}$
100	1.64496×10^{-4}
1000	1.645×10^{-6}

(c) $\left|E_{T_{10n}}\right| \approx 10^{-2}\left|E_{T_n}\right|$

(d) $b - a = \pi$, $h^2 = \frac{\pi^2}{n^2}$, $M = 1$

$$\left|E_{T_n}\right| \le \frac{\pi}{12}\left(\frac{\pi^2}{n^2}\right) = \frac{\pi^3}{12n^2}$$

$$\left|E_{T_{10n}}\right| \le \frac{\pi^3}{12(10n)^2} = 10^{-2}\left|E_{T_n}\right|$$

31. (a) $f'(x) = 2x\cos(x^2)$

$f''(x) = 2x \cdot -2x\sin(x^2) + 2\cos(x^2) = -4x^2\sin(x^2) + 2\cos(x^2)$

(b)

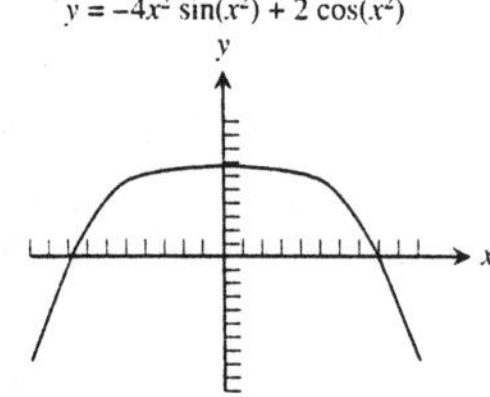

(c) The graph shows that $-3 \le f''(x) \le 2$ so $\left|f''(x)\right| \le 3$ for $-1 \le x \le 1$.

(d) $\left|E_T\right| \le \frac{1-(-1)}{12}(h^2)(3) = \frac{h^2}{2}$

(e) For $0 < h \le 0.1$, $\left|E_T\right| \le \frac{h^2}{2} \le \frac{0.1^2}{2} = 0.005 < 0.01$

(f) $n \ge \frac{1-(-1)}{h} \ge \frac{2}{0.1} = 20$

CHAPTER 4 PRACTICE EXERCISES

1. $\int (x^3 + 5x - 7)\,dx = \frac{x^4}{4} + \frac{5x^2}{2} - 7x + C$

3. $\int \left(3\sqrt{t} + \frac{4}{t^2}\right) dt = \int (3t^{1/2} + 4t^{-2})\,dt = \frac{3t^{3/2}}{\left(\frac{3}{2}\right)} + \frac{4t^{-1}}{-1} + C = 2t^{3/2} - \frac{4}{t} + C$

5. Let $u = r^2 + 5 \Rightarrow du = 2r\,dr \Rightarrow \frac{1}{2}\,du = r\,dr$

$$\int \frac{r\,dr}{\left(r^2+5\right)^2} = \int \frac{\left(\frac{1}{2}\right)du}{u^2} = \frac{1}{2}\int u^{-2}\,du = \frac{1}{2}\left(\frac{u^{-1}}{-1}\right) + C = -\frac{1}{2}u^{-1} + C = -\frac{1}{2\left(r^2+5\right)} + C$$

7. Let $u = 2 - \theta^2 \Rightarrow du = -2\theta\,d\theta \Rightarrow -\frac{1}{2}\,du = \theta\,d\theta$

$$\int 3\theta\sqrt{2-\theta^2}\,d\theta = \int \sqrt{u}\left(-\frac{3}{2}\,du\right) = -\frac{3}{2}\int u^{1/2}\,du = -\frac{3}{2}\left(\frac{u^{3/2}}{\frac{3}{2}}\right) + C = -u^{3/2} + C = -\left(2-\theta^2\right)^{3/2} + C$$

9. Let $u = 1 + x^4 \Rightarrow du = 4x^3\,dx \Rightarrow \frac{1}{4}\,du = x^3\,dx$

$$\int x^3(1+x^4)^{-1/4}\,dx = \int u^{-1/4}\left(\frac{1}{4}\,du\right) = \frac{1}{4}\int u^{-1/4}\,du = \frac{1}{4}\left(\frac{u^{3/4}}{\frac{3}{4}}\right) + C = \frac{1}{3}u^{3/4} + C = \frac{1}{3}(1+x^4)^{3/4} + C$$

11. Let $u = \frac{s}{10} \Rightarrow du = \frac{1}{10}\,ds \Rightarrow 10\,du = ds$

$$\int \sec^2 \frac{s}{10}\,ds = \int (\sec^2 u)(10\,du) = 10\int \sec^2 u\,du = 10\tan u + C = 10\tan\frac{s}{10} + C$$

13. Let $u = \sqrt{2}\,\theta \Rightarrow du = \sqrt{2}\,d\theta \Rightarrow \frac{1}{\sqrt{2}}\,du = d\theta$

$$\int \csc\sqrt{2}\theta\,\cot\sqrt{2}\theta\,d\theta = \int (\csc u\,\cot u)\left(\frac{1}{\sqrt{2}}\,du\right) = \frac{1}{\sqrt{2}}(-\csc u) + C = -\frac{1}{\sqrt{2}}\csc\sqrt{2}\theta + C$$

15. Let $u = \frac{x}{4} \Rightarrow du = \frac{1}{4}\,dx \Rightarrow 4\,du = dx$

$$\int \sin^2\frac{x}{4}\,dx = \int (\sin^2 u)(4\,du) = \int 4\left(\frac{1-\cos 2u}{2}\right)du = 2\int (1-\cos 2u)\,du = 2\left(u - \frac{\sin 2u}{2}\right) + C$$

$$= 2u - \sin 2u + C = 2\left(\frac{x}{4}\right) - \sin 2\left(\frac{x}{4}\right) + C = \frac{x}{2} - \sin\frac{x}{2} + C$$

17. Let $u = \cos x \Rightarrow du = -\sin x\,dx \Rightarrow -du = \sin x\,dx$

$$\int 2(\cos x)^{-1/2}\sin x\,dx = \int 2u^{-1/2}(-du) = -2\int u^{-1/2}\,du = -2\left(\frac{u^{1/2}}{\frac{1}{2}}\right) + C = -4u^{1/2} + C$$

$$= -4(\cos x)^{1/2} + C$$

19. $\int \left(t - \frac{2}{t}\right)\left(t + \frac{2}{t}\right)dt = \int \left(t^2 - \frac{4}{t^2}\right)dt = \int (t^2 - 4t^{-2})\,dt = \frac{t^3}{3} - 4\left(\frac{t^{-1}}{-1}\right) + C = \frac{t^3}{3} + \frac{4}{t} + C$

21. (a) Each time subinterval is of length $\Delta t = 0.4$ sec. The distance traveled over each subinterval, using the midpoint rule, is $\Delta h = \frac{1}{2}(v_i + v_{i+1})\Delta t$, where v_i is the velocity at the left, and v_{i+1} the velocity at the right, endpoint of the subinterval. We then add Δh to the height attained so far at the left endpoint v_i to arrive at the height associated with velocity v_{i+1} at the right endpoint. Using this methodology we build the following table based on the figure in the text:

t (sec)	0	0.4	0.8	1.2	1.6	2.0	2.4	2.8	3.2	3.6	4.0	4.4	4.8	5.2	5.6	6.0
v (fps)	0	10	25	55	100	190	180	170	155	140	130	120	105	90	80	65
h (ft)	0	2	9	25	56	114	188	258	323	382	436	486	531	570	604	633

t (sec)	6.4	6.8	7.2	7.6	8.0
v (fps)	52	40	30	15	0
h (ft)	656	674	688	697	700

NOTE: Your table values may vary slightly from ours depending on the v-values you read from the graph. Remember that some shifting of the graph occurs in the printing process.

The total height attained is about 700 ft.

(b) The graph is based on the table in part (a).

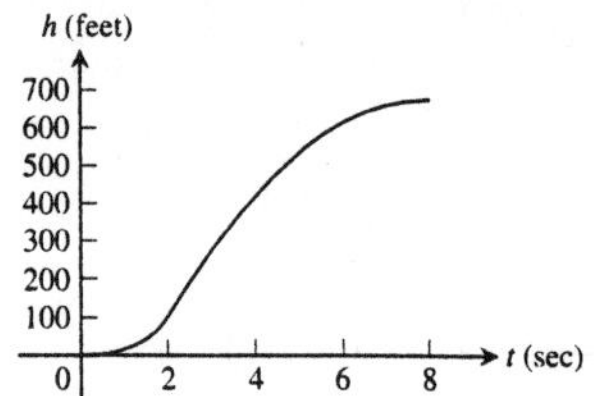

23. Let $u = 2x - 1 \Rightarrow du = 2\,dx \Rightarrow \frac{1}{2}\,du = dx$; $x = 1 \Rightarrow u = 1$, $x = 5 \Rightarrow u = 9$

$$\int_1^5 (2x-1)^{-1/2}\,dx = \int_1^9 u^{-1/2}\left(\frac{1}{2}\,du\right) = \left[u^{1/2}\right]_1^9 = 3 - 1 = 2$$

25. Let $u = \frac{x}{2} \Rightarrow 2\,du = dx$; $x = -\pi \Rightarrow u = -\frac{\pi}{2}$, $x = 0 \Rightarrow u = 0$

$$\int_{-\pi}^0 \cos\left(\frac{x}{2}\right)dx = \int_{-\pi/2}^0 (\cos u)(2\,du) = [2\sin u]_{-\pi/2}^0 = 2\sin 0 - 2\sin\left(-\frac{\pi}{2}\right) = 2(0-(-1)) = 2$$

27. (a) $\int_{-2}^2 f(x)\,dx = \frac{1}{3}\int_{-2}^2 3\,f(x)\,dx = \frac{1}{3}(12) = 4$

(b) $\int_2^5 f(x)\,dx = \int_{-2}^5 f(x)\,dx - \int_{-2}^2 f(x)\,dx = 6 - 4 = 2$

(c) $\int_5^{-2} g(x)\,dx = -\int_{-2}^5 g(x)\,dx = -2$

(d) $\int_{-2}^5 (-\pi\,g(x))\,dx = -\pi\int_{-2}^5 g(x)\,dx = -\pi(2) = -2\pi$

(e) $\int_{-2}^5 \left(\frac{f(x)+g(x)}{5}\right)dx = \frac{1}{5}\int_{-2}^5 f(x)\,dx + \frac{1}{5}\int_{-2}^5 g(x)\,dx = \frac{1}{5}(6) + \frac{1}{5}(2) = \frac{8}{5}$

29. $\int_{-1}^1 (3x^2 - 4x + 7)\,dx = \left[x^3 - 2x^2 + 7x\right]_{-1}^1 = \left[1^3 - 2(1)^2 + 7(1)\right] - \left[(-1)^3 - 2(-1)^2 + 7(-1)\right] = 6 - (-10) = 16$

31. $\int_1^2 \frac{4}{v^2}\,dv = \int_1^2 4v^{-2}\,dv = \left[-4v^{-1}\right]_1^2 = \left(\frac{-4}{2}\right) - \left(\frac{-4}{1}\right) = 2$

33. $\int_1^4 \frac{dt}{t\sqrt{t}} = \int_1^4 \frac{dt}{t^{3/2}} = \int_1^4 t^{-3/2}\,dt = \left[-2t^{-1/2}\right]_1^4 = \frac{-2}{\sqrt{4}} - \frac{(-2)}{\sqrt{1}} = 1$

35. Let $u = 2x + 1 \Rightarrow du = 2\,dx \Rightarrow 18\,du = 36\,dx$; $x = 0 \Rightarrow u = 1$, $x = 1 \Rightarrow u = 3$

$$\int_0^1 \frac{36\,dx}{(2x+1)^3} = \int_1^3 18u^{-3}\,du = \left[\frac{18u^{-2}}{-2}\right]_1^3 = \left[\frac{-9}{u^2}\right]_1^3 = \left(\frac{-9}{3^2}\right) - \left(\frac{-9}{1^2}\right) = 8$$

37. Let $u = 1 - x^{2/3} \Rightarrow du = -\frac{2}{3}x^{-1/3}\,dx \Rightarrow -\frac{3}{2}\,du = x^{-1/3}\,dx$; $x = \frac{1}{8} \Rightarrow u = 1 - \left(\frac{1}{8}\right)^{2/3} = \frac{3}{4}$, $x = 1 \Rightarrow u = 1 - 1^{2/3} = 0$

$$\int_{1/8}^1 x^{-1/3}\left(1 - x^{2/3}\right)^{3/2}\,dx = \int_{3/4}^0 u^{3/2}\left(-\frac{3}{2}\,du\right) = \left[\left(-\frac{3}{2}\right)\left(\frac{u^{5/2}}{\frac{5}{2}}\right)\right]_{3/4}^0 = \left[-\frac{3}{5}u^{5/2}\right]_{3/4}^0 = -\frac{3}{5}(0)^{5/2} - \left(-\frac{3}{5}\right)\left(\frac{3}{4}\right)^{5/2}$$

$$= \frac{27\sqrt{3}}{160}$$

39. Let $u = 5r \Rightarrow du = 5\,dr \Rightarrow \frac{1}{5}\,du = dr$; $r = 0 \Rightarrow u = 0$, $r = \pi \Rightarrow u = 5\pi$

$$\int_0^\pi \sin^2 5r\,dr = \int_0^{5\pi} \left(\sin^2 u\right)\left(\frac{1}{5}\,du\right) = \frac{1}{5}\left[\frac{u}{2} - \frac{\sin 2u}{4}\right]_0^{5\pi} = \left(\frac{\pi}{2} - \frac{\sin 10\pi}{20}\right) - \left(0 - \frac{\sin 0}{20}\right) = \frac{\pi}{2}$$

41. $$\int_0^{\pi/3} \sec^2\theta\,d\theta = [\tan\theta]_0^{\pi/3} = \tan\frac{\pi}{3} - \tan 0 = \sqrt{3}$$

43. Let $u = \frac{x}{6} \Rightarrow du = \frac{1}{6}\,dx \Rightarrow 6\,du = dx$; $x = \pi \Rightarrow u = \frac{\pi}{6}$, $x = 3\pi \Rightarrow u = \frac{\pi}{2}$

$$\int_\pi^{3\pi} \cot^2\frac{x}{6}\,dx = \int_{\pi/6}^{\pi/2} 6\cot^2 u\,du = 6\int_{\pi/6}^{\pi/2} \left(\csc^2 u - 1\right)du = [6(-\cot u - u)]_{\pi/6}^{\pi/2} = 6\left(-\cot\frac{\pi}{2} - \frac{\pi}{2}\right) - 6\left(-\cot\frac{\pi}{6} - \frac{\pi}{6}\right)$$

$$= 6\sqrt{3} - 2\pi$$

45. $$\int_{-\pi/3}^0 \sec x\tan x\,dx = [\sec x]_{-\pi/3}^0 = \sec 0 - \sec\left(-\frac{\pi}{3}\right) = 1 - 2 = -1$$

47. Let $u = \sin x \Rightarrow du = \cos x\,dx$; $x = 0 \Rightarrow u = 0$, $x = \frac{\pi}{2} \Rightarrow u = 1$

$$\int_0^{\pi/2} 5(\sin x)^{3/2}\cos x\,dx = \int_0^1 5u^{3/2}\,du = \left[5\left(\frac{2}{5}\right)u^{5/2}\right]_0^1 = \left[2u^{5/2}\right]_0^1 = 2(1)^{5/2} - 2(0)^{5/2} = 2$$

49. Let $u = 1 + 3\sin^2 x \Rightarrow du = 6\sin x\cos x\,dx \Rightarrow \frac{1}{2}\,du = 3\sin x\cos x\,dx$; $x = 0 \Rightarrow u = 1$, $x = \frac{\pi}{2}$ $\Rightarrow u = 1 + 3\sin^2\frac{\pi}{2} = 4$

$$\int_0^{\pi/2} \frac{3\sin x\cos x}{\sqrt{1 + 3\sin^2 x}}\,dx = \int_1^4 \frac{1}{\sqrt{u}}\left(\frac{1}{2}\,du\right) = \int_1^4 \frac{1}{2}u^{-1/2}\,du = \left[\frac{1}{2}\left(\frac{u^{1/2}}{\frac{1}{2}}\right)\right]_1^4 = \left[u^{1/2}\right]_1^4 = 4^{1/2} - 1^{1/2} = 1$$

51. Let $u = \sec\theta \Rightarrow du = \sec\theta\tan\theta\, d\theta$; $\theta = 0 \Rightarrow u = \sec 0 = 1$, $\theta = \frac{\pi}{3} \Rightarrow u = \sec\frac{\pi}{3} = 2$

$$\int_0^{\pi/3} \frac{\tan\theta}{\sqrt{2\sec\theta}}\, d\theta = \int_0^{\pi/3} \frac{\sec\theta\tan\theta}{\sec\theta\sqrt{2\sec\theta}}\, d\theta = \int_0^{\pi/3} \frac{\sec\theta\tan\theta}{\sqrt{2}(\sec\theta)^{3/2}}\, d\theta = \int_1^2 \frac{1}{\sqrt{2}\,u^{3/2}}\, du = \frac{1}{\sqrt{2}}\int_1^2 u^{-3/2}\, du$$

$$= \frac{1}{\sqrt{2}}\left[\frac{u^{-1/2}}{\left(-\frac{1}{2}\right)}\right]_1^2 = \left[-\frac{2}{\sqrt{2u}}\right]_1^2 = -\frac{2}{\sqrt{2(2)}} - \left(-\frac{2}{\sqrt{2(1)}}\right) = \sqrt{2} - 1$$

53. $x^2 - 4x + 3 = 0 \Rightarrow (x-3)(x-1) = 0 \Rightarrow x = 3$ or $x = 1$:

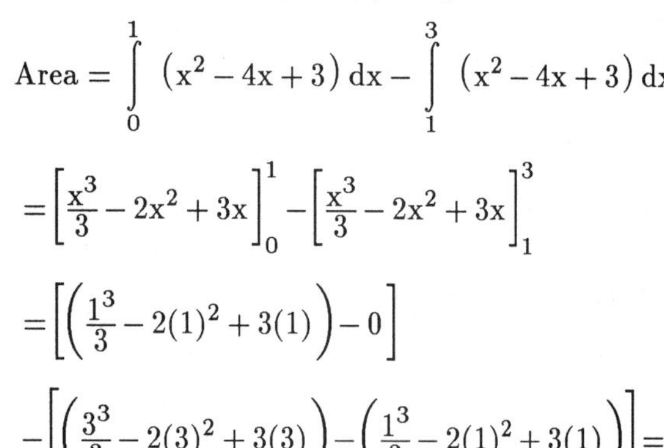

$$\text{Area} = \int_0^1 (x^2 - 4x + 3)\, dx - \int_1^3 (x^2 - 4x + 3)\, dx$$

$$= \left[\frac{x^3}{3} - 2x^2 + 3x\right]_0^1 - \left[\frac{x^3}{3} - 2x^2 + 3x\right]_1^3$$

$$= \left[\left(\frac{1^3}{3} - 2(1)^2 + 3(1)\right) - 0\right]$$

$$-\left[\left(\frac{3^3}{3} - 2(3)^2 + 3(3)\right) - \left(\frac{1^3}{3} - 2(1)^2 + 3(1)\right)\right] = \left(\frac{1}{3} + 1\right) - \left[0 - \left(\frac{1}{3} + 1\right)\right] = \frac{8}{3}$$

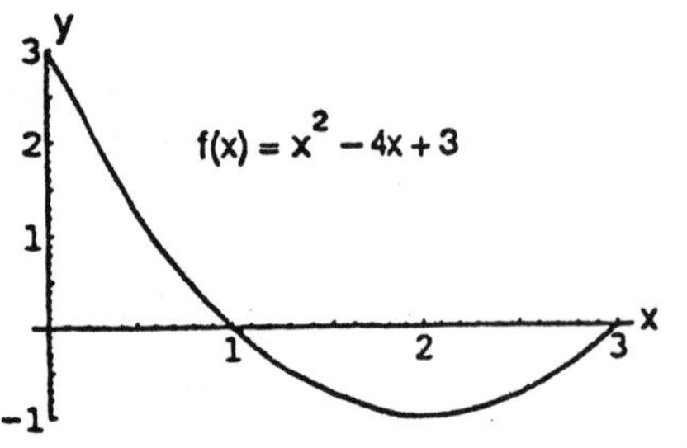

55. $5 - 5x^{2/3} = 0 \Rightarrow 1 - x^{2/3} = 0 \Rightarrow x = \pm 1$;

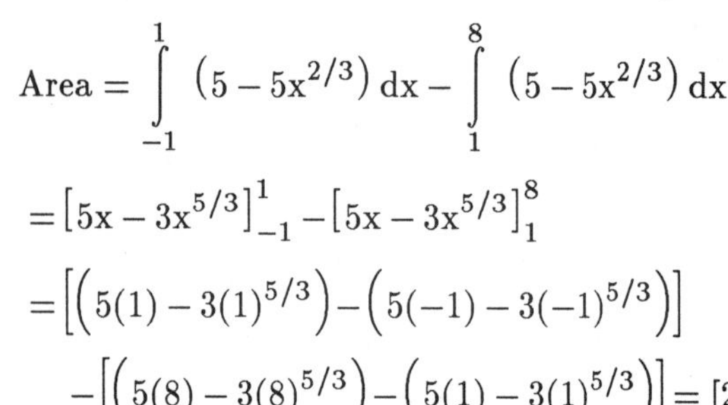

$$\text{Area} = \int_{-1}^1 (5 - 5x^{2/3})\, dx - \int_1^8 (5 - 5x^{2/3})\, dx$$

$$= \left[5x - 3x^{5/3}\right]_{-1}^1 - \left[5x - 3x^{5/3}\right]_1^8$$

$$= \left[\left(5(1) - 3(1)^{5/3}\right) - \left(5(-1) - 3(-1)^{5/3}\right)\right]$$

$$-\left[\left(5(8) - 3(8)^{5/3}\right) - \left(5(1) - 3(1)^{5/3}\right)\right] = [2 - (-2)] - [(40 - 96) - 2] = 62$$

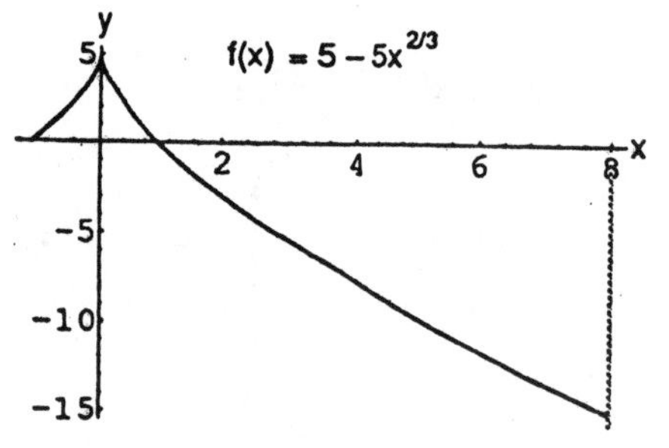

57. $f(x) = x$, $g(x) = \frac{1}{x^2}$, $a = 1$, $b = 2 \Rightarrow A = \int_a^b [f(x) - g(x)]\, dx$

$$= \int_1^2 \left(x - \frac{1}{x^2}\right) dx = \left[\frac{x^2}{2} + \frac{1}{x}\right]_1^2 = \left(\frac{4}{2} + \frac{1}{2}\right) - \left(\frac{1}{2} + 1\right) = 1$$

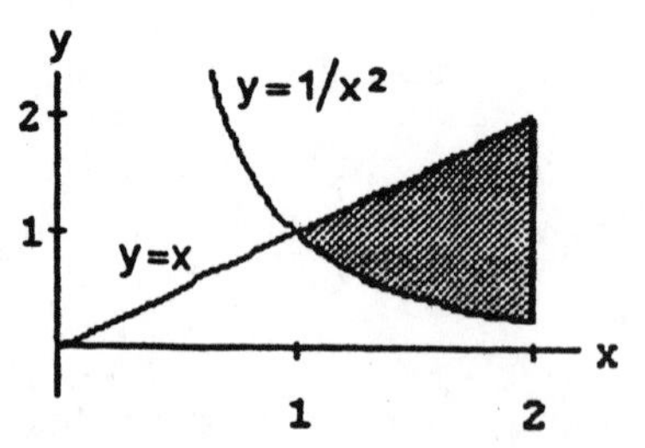

59. $f(x) = (1 - \sqrt{x})^2$, $g(x) = 0$, $a = 0$, $b = 1 \Rightarrow A = \int_a^b [f(x) - g(x)]\, dx = \int_0^1 (1 - \sqrt{x})^2\, dx = \int_0^1 (1 - 2\sqrt{x} + x)\, dx$

$$= \int_0^1 (1 - 2x^{1/2} + x)\, dx = \left[x - \frac{4}{3}x^{3/2} + \frac{x^2}{2}\right]_0^1 = 1 - \frac{4}{3} + \frac{1}{2} = \frac{1}{6}(6 - 8 + 3) = \frac{1}{6}$$

61. $f(x) = x$, $g(x) = \sin x$, $a = 0$, $b = \frac{\pi}{4}$

$$\Rightarrow A = \int_a^b [f(x) - g(x)]\,dx = \int_0^{\pi/4} (x - \sin x)\,dx$$

$$= \left[\frac{x^2}{2} + \cos x\right]_0^{\pi/4} = \left(\frac{\pi^2}{32} + \frac{\sqrt{2}}{2}\right) - 1$$

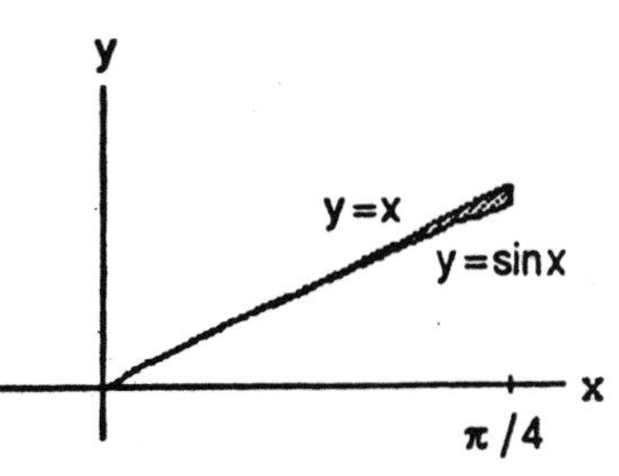

63. $a = 0$, $b = \pi$, $f(x) - g(x) = 2\sin x - \sin 2x$

$$\Rightarrow A = \int_0^{\pi} (2\sin x - \sin 2x)\,dx = \left[-2\cos x + \frac{\cos 2x}{2}\right]_0^{\pi}$$

$$= \left[-2\cdot(-1) + \frac{1}{2}\right] - \left(-2\cdot 1 + \frac{1}{2}\right) = 4$$

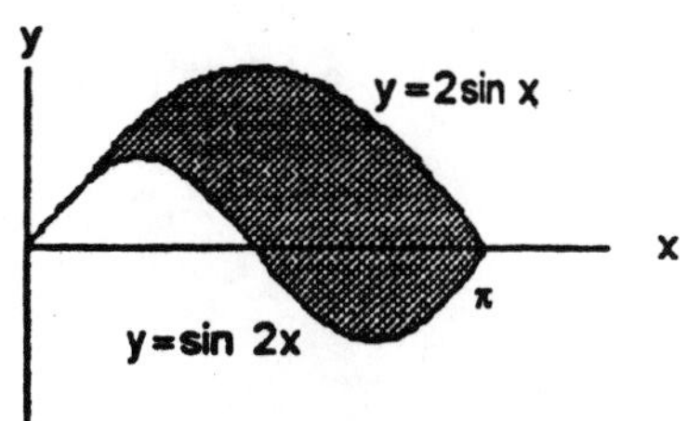

65. $f(x) = x^3 - 3x^2 = x^2(x-3) \Rightarrow f'(x) = 3x^2 - 6x = 3x(x-2) \Rightarrow f' = +++\underset{0}{|}----\underset{2}{|}+++$

$\Rightarrow f(0) = 0$ is a maximum and $f(2) = -4$ is a minimum. Then $A = -\int_0^3 (x^3 - 3x^2)\,dx = -\left[\frac{x^4}{4} - x^3\right]_0^3$

$= -\left(\frac{81}{4} - 27\right) = \frac{27}{4}$

67. $y = \int \frac{x^2+1}{x^2}\,dx = \int (1 + x^{-2})\,dx = x - x^{-1} + C = x - \frac{1}{x} + C$; $y = -1$ when $x = 1 \Rightarrow 1 - \frac{1}{1} + C = -1$

$\Rightarrow C = -1 \Rightarrow y = x - \frac{1}{x} - 1$

69. $\frac{dr}{dt} = \int \left(15\sqrt{t} + \frac{3}{\sqrt{t}}\right) dt = \int \left(15t^{1/2} + 3t^{-1/2}\right) dt = 10t^{3/2} + 6t^{1/2} + C$; $\frac{dr}{dt} = 8$ when $t = 1$

$\Rightarrow 10(1)^{3/2} + 6(1)^{1/2} + C = 8 \Rightarrow C = -8$. Thus $\frac{dr}{dt} = 10t^{3/2} + 6t^{1/2} - 8 \Rightarrow r = \int \left(10t^{3/2} + 6t^{1/2} - 8\right) dt$

$= 4t^{5/2} + 4t^{3/2} - 8t + C$; $r = 0$ when $t = 1 \Rightarrow 4(1)^{5/2} + 4(1)^{3/2} - 8(1) + C_1 = 0 \Rightarrow C_1 = 0$. Therefore,

$r = 4t^{5/2} + 4t^{3/2} - 8t$

71. $y = x^2 + \int_1^x \frac{1}{t}\,dt \Rightarrow \frac{dy}{dx} = 2x + \frac{1}{x} \Rightarrow \frac{d^2y}{dx^2} = 2 - \frac{1}{x^2}$; $y(1) = 1 + \int_1^1 \frac{1}{t}\,dt = 1$ and $y'(1) = 2 + 1 = 3$

73. $y = \int_5^x \frac{\sin t}{t}\,dt - 3 \Rightarrow \frac{dy}{dx} = \frac{\sin x}{x}$; $x = 5 \Rightarrow y = \int_5^5 \frac{\sin t}{t}\,dt - 3 = -3$

75. (a) $\text{av}(f) = \frac{1}{1-(-1)} \int_{-1}^{1} (mx + b)\,dx = \frac{1}{2}\left[\frac{mx^2}{2} + bx\right]_{-1}^{1} = \frac{1}{2}\left[\left(\frac{m(1)^2}{2} + b(1)\right) - \left(\frac{m(-1)^2}{2} + b(-1)\right)\right] = \frac{1}{2}(2b) = b$

(b) $\text{av}(f) = \frac{1}{k-(-k)} \int_{-k}^{k} (mx+b)\,dx = \frac{1}{2k}\left[\frac{mx^2}{2} + bx\right]_{-k}^{k} = \frac{1}{2k}\left[\left(\frac{m(k)^2}{2} + b(k)\right) - \left(\frac{m(-k)^2}{2} + b(-k)\right)\right]$

$= \frac{1}{2k}(2bk) = b$

77. $f'_{av} = \frac{1}{b-a}\int_a^b f'(x)\,dx = \frac{1}{b-a}[f(x)]_a^b = \frac{1}{b-a}[f(b) - f(a)] = \frac{f(b)-f(a)}{b-a}$ so the average value of f' over $[a,b]$ is the slope of the secant line joining the points $(a, f(a))$ and $(b, f(b))$.

79. $\frac{dy}{dx} = \sqrt{2 + \cos^3 x}$

81. $\frac{dy}{dx} = \frac{d}{dx}\left(-\int_1^x \frac{6}{3+t^4}\,dt\right) = -\frac{6}{3+x^4}$

83. $h = \frac{b-a}{n} = \frac{\pi - 0}{6} = \frac{\pi}{6} \Rightarrow \frac{h}{2} = \frac{\pi}{12}$;

	x_i	$f(x_i)$	m	$mf(x_i)$
x_0	0	0	1	0
x_1	$\pi/6$	1/2	2	1
x_2	$\pi/3$	3/2	2	3
x_3	$\pi/2$	2	2	4
x_4	$2\pi/3$	3/2	2	3
x_5	$5\pi/6$	1/2	2	1
x_6	π	0	1	0

$\sum_{i=0}^{6} mf(x_i) = 12 \Rightarrow T = \left(\frac{\pi}{12}\right)(12) = \pi$;

	x_i	$f(x_i)$	m	$mf(x_i)$
x_0	0	0	1	0
x_1	$\pi/6$	1/2	4	2
x_2	$\pi/3$	3/2	2	3
x_3	$\pi/2$	2	4	8
x_4	$2\pi/3$	3/2	2	3
x_5	$5\pi/6$	1/2	4	2
x_6	π	0	1	0

$\sum_{i=0}^{6} mf(x_i) = 18$ and $\frac{h}{3} = \frac{\pi}{18} \Rightarrow S = \left(\frac{\pi}{18}\right)(18) = \pi$.

85. $y_{av} = \frac{1}{365-0}\int_0^{365}\left[37\sin\left(\frac{2\pi}{365}(x-101)\right) + 25\right]dx = \frac{1}{365}\left[-37\left(\frac{365}{2\pi}\cos\left(\frac{2\pi}{365}(x-101)\right) + 25x\right)\right]_0^{365}$

$= \frac{1}{365}\left[\left(-37\left(\frac{365}{2\pi}\right)\cos\left[\frac{2\pi}{365}(365-101)\right] + 25(365)\right) - \left(-37\left(\frac{365}{2\pi}\right)\cos\left[\frac{2\pi}{365}(0-101)\right] + 25(0)\right)\right]$

$= -\frac{37}{2\pi}\cos\left(\frac{2\pi}{365}(264)\right) + 25 + \frac{37}{2\pi}\cos\left(\frac{2\pi}{365}(-101)\right) = -\frac{37}{2\pi}\left(\cos\left(\frac{2\pi}{365}(264)\right) - \cos\left(\frac{2\pi}{365}(-101)\right)\right) + 25$

$\approx -\frac{37}{2\pi}(0.16705 - 0.16705) + 25 = 25°\text{F}$

87. Using the trapezoidal rule, $h = 15 \Rightarrow \frac{h}{2} = 7.5$;

$\sum mf(x_i) = 794.8 \Rightarrow \text{Area} \approx (794.8)(7.5) = 5961 \text{ ft}^2$;

The cost is $\text{Area} \cdot (\$2.10/\text{ft}^2) \approx (5961 \text{ ft}^2)(\$2.10/\text{ft}^2)$
$= \$12{,}518.10 \Rightarrow$ the job cannot be done for \$11,000.

	x_i	$f(x_i)$	m	$mf(x_i)$
x_0	0	0	1	0
x_1	15	36	2	72
x_2	30	54	2	108
x_3	45	51	2	102
x_4	60	49.5	2	99
x_5	75	54	2	108
x_6	90	64.4	2	128.8
x_7	105	67.5	2	135
x_8	120	42	1	42

89. Yes. The function f, being differentiable on [a, b], is then continuous on [a, b]. The Fundamental Theorem of Calculus says that every continuous function on [a, b] is the derivative of a function on [a, b].

91. $y(x) = \int_5^x \frac{\sin t}{t}\, dt + 3$

93. (a) $g(1) = \int_1^1 f(t)\, dt = 0$

(b) $g(3) = \int_1^3 f(t)\, dt = -\frac{1}{2}(2)(1) = -1$

(c) $g(-1) = \int_1^{-1} f(t)\, dt = -\int_{-1}^{1} f(t)\, dt = -\frac{1}{4}\pi(2)^2 = -\pi$

(d) $g'(x) = f(x)$; Since $f(x) > 0$ for $-3 < x < 1$ and $f(x) < 0$ for $1 < x < 3$, g(x) has a relative maximum at $x = 1$.

(e) $g'(-1) = f(-1) = 2$
The equation of the tangent line is $y - (-\pi) = 2(x + 1)$ or $y = 2x + 2 - \pi$

(f) $g''(x) = f'(x)$, $f'(x) = 0$ at $x = -1$ and f' is not defined at $x = 2$. The inflection points are at $x = -1$ and $x = 2$. Note that $g''(x) = f'(x)$ is undefined at $x = 1$ as well, but since $g''(x) = f'(x)$ is negative on both sides of $x = 1$, $x = 1$ is not an inflection point.

(g) Note that the absolute maximum is $g(1) = 0$ and the absolute minimum is

$g(-3) = \int_1^{-3} f(t)\, dt = -\int_{-3}^{1} f(t)\, dt = -\frac{1}{2}\pi(2)^2 = -2\pi.$

The range of g is $[-2\pi, 0]$.

95. $av(I) = \frac{1}{30}\int_0^{30} (1200 - 40t)\, dt = \frac{1}{30}\left[1200t - 20t^2\right]_0^{30} = \frac{1}{30}[((1200(30) - 20(30)^2) - (1200(0) - 20(0)^2)]$

$= \frac{1}{30}(18{,}000) = 600$; Average Daily Holding Cost $= (600)(\$0.03) = \$18$

97. $av(I) = \frac{1}{30}\int_0^{30}\left(450 - \frac{t^2}{2}\right) dt = \frac{1}{30}\left[450t - \frac{t^3}{6}\right]_0^{30} = \frac{1}{30}\left[450(30) - \frac{30^3}{6} - 0\right] = 300$; Average Daily Holding Cost

$= (300)(\$0.02) = \$6$

CHAPTER 4 ADDITIONAL EXERCISES–THEORY, EXAMPLES, APPLICATIONS

1. (a) Yes, because $\int_0^1 f(x)\, dx = \frac{1}{7}\int_0^1 7f(x)\, dx = \frac{1}{7}(7) = 1$

   (b) No. For example, $\int_0^1 8x\, dx = \left[4x^2\right]_0^1 = 4$, but $\int_0^1 \sqrt{8x}\, dx = \left[2\sqrt{2}\left(\frac{x^{3/2}}{\frac{3}{2}}\right)\right]_0^1 = \frac{4\sqrt{2}}{3}\left(1^{3/2} - 0^{3/2}\right)$

   $= \frac{4\sqrt{2}}{3} \neq \sqrt{4}$

3. $y = \frac{1}{a}\int_0^x f(t)\sin a(x-t)\, dt = \frac{1}{a}\int_0^x f(t)\sin ax \cos at\, dt - \frac{1}{a}\int_0^x f(t)\cos ax \sin at\, dt$

   $= \frac{\sin ax}{a}\int_0^x f(t)\cos at\, dt - \frac{\cos ax}{a}\int_0^x f(t)\sin at\, dt \Rightarrow \frac{dy}{dx} = \cos ax \int_0^x f(t)\cos at\, dt$

   $+ \frac{\sin ax}{a}\left(\frac{d}{dx}\int_0^x f(t)\cos at\, dt\right) + \sin ax \int_0^x f(t)\sin at\, dt - \frac{\cos ax}{a}\left(\frac{d}{dx}\int_0^x f(t)\sin at\, dt\right)$

   $= \cos ax \int_0^x f(t)\cos at\, dt + \frac{\sin ax}{a}(f(x)\cos ax) + \sin ax \int_0^x f(t)\sin at\, dt - \frac{\cos ax}{a}(f(x)\sin ax)$

   $\Rightarrow \frac{dy}{dx} = \cos ax \int_0^x f(t)\cos at\, dt + \sin ax \int_0^x f(t)\sin at\, dt$. Next,

   $\frac{d^2y}{dx^2} = -a\sin ax \int_0^x f(t)\cos at\, dt + (\cos ax)\left(\frac{d}{dx}\int_0^x f(t)\cos at\, dt\right) + a\cos ax \int_0^x f(t)\sin at\, dt$

$$+(\sin ax)\left(\frac{d}{dx}\int_0^x f(t)\sin at\,dt\right) = -a\sin ax\int_0^x f(t)\cos at\,dt + (\cos ax)f(x)\cos ax$$

$$+a\cos ax\int_0^x f(t)\sin at\,dt + (\sin ax)f(x)\sin ax = -a\sin ax\int_0^x f(t)\cos at\,dt + a\cos ax\int_0^x f(t)\sin at\,dt + f(x).$$

Therefore, $y'' + a^2y = a\cos ax\int_0^x f(t)\sin at\,dt - a\sin ax\int_0^x f(t)\cos at\,dt + f(x)$

$$+a^2\left(\frac{\sin ax}{a}\int_0^x f(t)\cos at\,dt - \frac{\cos ax}{a}\int_0^x f(t)\sin at\,dt\right) = f(x).$$ Note also that $y'(0) = y(0) = 0$.

5. (a) $\int_0^{x^2} f(t)\,dt = x\cos \pi x \Rightarrow \frac{d}{dx}\int_0^{x^2} f(t)\,dt = \cos\pi x - \pi x\sin\pi x \Rightarrow f(x^2)(2x) = \cos\pi x - \pi x\sin\pi x$

$\Rightarrow f(x^2) = \frac{\cos\pi x - \pi x\sin\pi x}{2x}$. Thus, $x = 2 \Rightarrow f(4) = \frac{\cos 2\pi - 2\pi\sin 2\pi}{4} = \frac{1}{4}$

(b) $\int_0^{f(x)} t^2\,dt = \left[\frac{t^3}{3}\right]_0^{f(x)} = \frac{1}{3}(f(x))^3 \Rightarrow \frac{1}{3}(f(x))^3 = x\cos\pi x \Rightarrow (f(x))^3 = 3x\cos\pi x \Rightarrow f(x) = \sqrt[3]{3x\cos\pi x}$

$\Rightarrow f(4) = \sqrt[3]{3(4)\cos 4\pi} = \sqrt[3]{12}$

7. $\int_1^b f(x)\,dx = \sqrt{b^2+1} - \sqrt{2} \Rightarrow f(b) = \frac{d}{db}\int_1^b f(x)\,dx = \frac{1}{2}(b^2+1)^{-1/2}(2b) = \frac{b}{\sqrt{b^2+1}} \Rightarrow f(x) = \frac{x}{\sqrt{x^2+1}}$

9. $\frac{dy}{dx} = 3x^2 + 2 \Rightarrow y = \int (3x^2+2)\,dx = x^3 + 2x + C$. Then $(1,-1)$ on the curve $\Rightarrow 1^3 + 2(1) + C = -1 \Rightarrow C = -4$
$\Rightarrow y = x^3 + 2x - 4$

11. $\int_{-8}^{3} f(x)\,dx = \int_{-8}^{0} x^{2/3}\,dx + \int_0^3 -4\,dx$

$= \left[\frac{3}{5}x^{5/3}\right]_{-8}^{0} + [-4x]_0^3$

$= \left(0 - \frac{3}{5}(-8)^{5/3}\right) + (-4(3) - 0) = \frac{96}{5} - 12$

$= \frac{36}{5}$

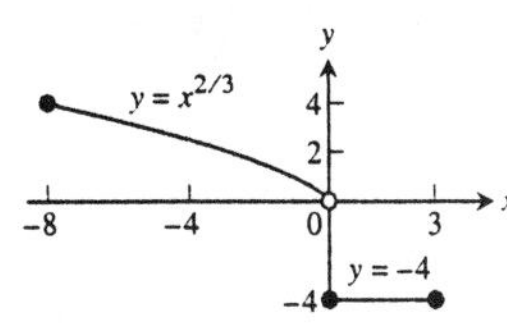

13. $\int_0^2 f(t)\,dt = \int_0^1 t\,dt + \int_1^2 \sin \pi t\,dt$

$= \left[\frac{t^2}{2}\right]_0^1 + \left[-\frac{1}{\pi}\cos \pi t\right]_1^2$

$= \left(\frac{1}{2} - 0\right) + \left[-\frac{1}{\pi}\cos 2\pi - \left(-\frac{1}{\pi}\cos \pi\right)\right]$

$= \frac{1}{2} - \frac{2}{\pi}$

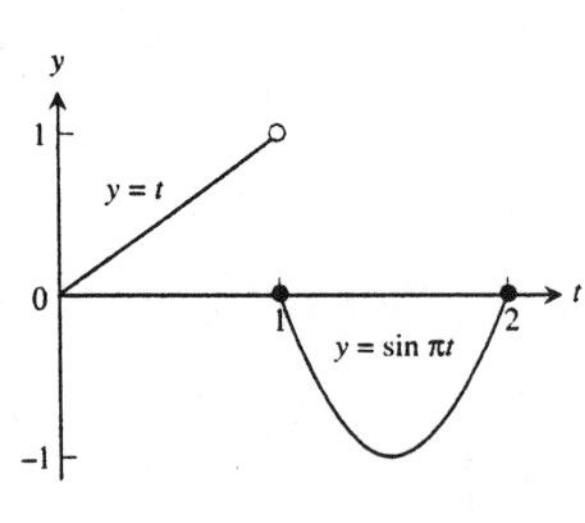

15. $\int_{-2}^2 f(x)\,dx = \int_{-2}^{-1} dx + \int_{-1}^1 (1-x^2)\,dx + \int_1^2 2\,dx$

$= [x]_{-2}^{-1} + \left[x - \frac{x^3}{3}\right]_{-1}^1 + [2x]_1^2$

$= -1 - (-2) + \left(1 - \frac{1^3}{3}\right) - \left(-1 - \frac{(-1)^3}{3}\right) + 2(2) - 2(1)$

$= 1 + \frac{2}{3} - \left(-\frac{2}{3}\right) + 4 - 2 = \frac{13}{3}$

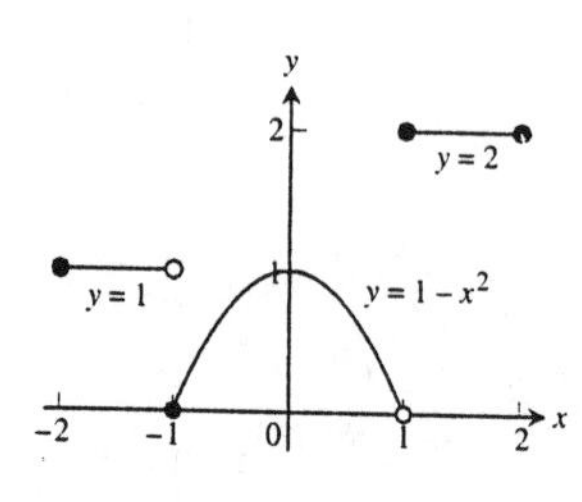

17. Ave. value $= \frac{1}{b-a}\int_a^b f(x)\,dx = \frac{1}{2-0}\int_0^2 f(x)\,dx = \frac{1}{2}\left[\int_0^1 x\,dx + \int_1^2 (x-1)\,dx\right] = \frac{1}{2}\left[\frac{x^2}{2}\right]_0^1 + \frac{1}{2}\left[\frac{x^2}{2} - x\right]_1^2$

$= \frac{1}{2}\left[\frac{1^2}{2} - 0 + \left(\frac{2^2}{2} - 2\right) - \left(\frac{1^2}{2} - 1\right)\right] = \frac{1}{2}$

19. $f(x) = \int_{1/x}^x \frac{1}{t}\,dt \Rightarrow f'(x) = \frac{1}{x}\left(\frac{dx}{dx}\right) - \left(\frac{1}{\frac{1}{x}}\right)\left(\frac{d}{dx}\left(\frac{1}{x}\right)\right) = \frac{1}{x} - x\left(-\frac{1}{x^2}\right) = \frac{1}{x} + \frac{1}{x} = \frac{2}{x}$

21. $g(y) = \int_{\sqrt{y}}^{2\sqrt{y}} \sin t^2\,dt \Rightarrow g'(y) = \left(\sin\left(2\sqrt{y}\right)^2\right)\left(\frac{d}{dy}\left(2\sqrt{y}\right)\right) - \left(\sin\left(\sqrt{y}\right)^2\right)\left(\frac{d}{dy}\left(\sqrt{y}\right)\right) = \frac{\sin 4y}{\sqrt{y}} - \frac{\sin y}{2\sqrt{y}}$

NOTES:

CHAPTER 5 APPLICATIONS OF INTEGRALS

5.1 VOLUMES BY SLICING AND ROTATION ABOUT AN AXIS

1. (a) $A = \pi(\text{radius})^2$ and radius $= \sqrt{1-x^2} \Rightarrow A(x) = \pi\left(1-x^2\right)$

   (b) $A = \text{width} \cdot \text{height}$, width $=$ height $= 2\sqrt{1-x^2} \Rightarrow A(x) = 4\left(1-x^2\right)$

   (c) $A = (\text{side})^2$ and diagonal $= \sqrt{2}(\text{side}) \Rightarrow A = \frac{(\text{diagonal})^2}{2}$; diagonal $= 2\sqrt{1-x^2} \Rightarrow A(x) = 2\left(1-x^2\right)$

   (d) $A = \frac{\sqrt{3}}{4}(\text{side})^2$ and side $= 2\sqrt{1-x^2} \Rightarrow A(x) = \sqrt{3}\left(1-x^2\right)$

3. $A(x) = \frac{(\text{diagonal})^2}{2} = \frac{\left(\sqrt{x}-\left(-\sqrt{x}\right)\right)^2}{2} = 2x$ (see Exercise 1c); $a = 0$, $b = 4$;

   $$V = \int_a^b A(x)\,dx = \int_0^4 2x\,dx = \left[x^2\right]_0^4 = 16$$

5. $A(x) = (\text{edge})^2 = \left[\sqrt{1-x^2} - \left(-\sqrt{1-x^2}\right)\right]^2 = \left(2\sqrt{1-x^2}\right)^2 = 4\left(1-x^2\right)$; $a = -1$, $b = 1$;

   $$V = \int_a^b A(x)\,dx = \int_{-1}^1 4\left(1-x^2\right)dx = 4\left[x - \frac{x^3}{3}\right]_{-1}^1 = 8\left(1-\tfrac{1}{3}\right) = \tfrac{16}{3}$$

7. (a) STEP 1) $A(x) = \frac{1}{2}(\text{side}) \cdot (\text{side}) \cdot \left(\sin\frac{\pi}{3}\right) = \frac{1}{2} \cdot \left(2\sqrt{\sin x}\right) \cdot \left(2\sqrt{\sin x}\right)\left(\sin\frac{\pi}{3}\right) = \sqrt{3}\sin x$

   STEP 2) $a = 0$, $b = \pi$

   STEP 3) $V = \int_a^b A(x)\,dx = \sqrt{3}\int_0^\pi \sin x\,dx = \left[-\sqrt{3}\cos x\right]_0^\pi = \sqrt{3}(1+1) = 2\sqrt{3}$

   (b) STEP 1) $A(x) = (\text{side})^2 = \left(2\sqrt{\sin x}\right)\left(2\sqrt{\sin x}\right) = 4\sin x$

   STEP 2) $a = 0$, $b = \pi$

   STEP 3) $V = \int_a^b A(x)\,dx = \int_0^\pi 4\sin x\,dx = \left[-4\cos x\right]_0^\pi = 8$

9. $A(y) = \frac{\pi}{4}(\text{diameter})^2 = \frac{\pi}{4}\left(\sqrt{5}y^2 - 0\right)^2 = \frac{5\pi}{4}y^4$;

$c = 0,\ d = 2;\ V = \int_c^d A(y)\,dy = \int_0^2 \frac{5\pi}{4}y^4\,dy$

$= \left[\left(\frac{5\pi}{4}\right)\left(\frac{y^5}{5}\right)\right]_0^2 = \frac{\pi}{4}\left(2^5 - 0\right) = 8\pi$

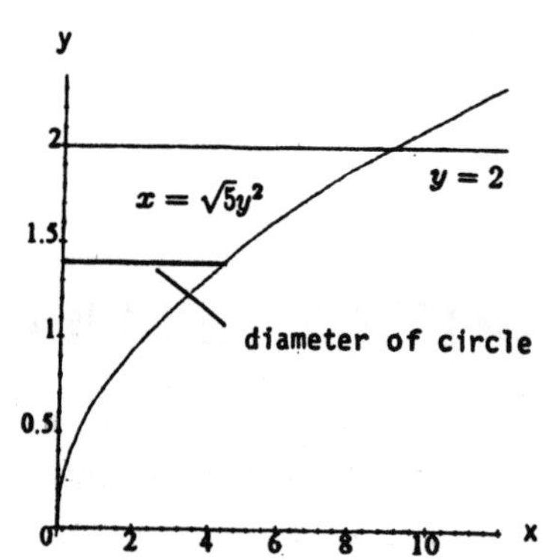

11. (a) It follows from Cavalieri's Theorem that the volume of a column is the same as the volume of a right prism with a square base of side length s and altitude h. Thus, STEP 1) $A(x) = (\text{side length})^2 = s^2$; STEP 2) $a = 0,\ b = h$; STEP 3) $V = \int_a^b A(x)\,dx = \int_0^h s^2\,dx = s^2h$

(b) From Cavalieri's Theorem we conclude that the volume of the column is the same as the volume of the prism described above, regardless of the number of turns $\Rightarrow V = s^2h$

13. $R(x) = y = 1 - \frac{x}{2} \Rightarrow V = \int_0^2 \pi[R(x)]^2\,dx = \pi \int_0^2 \left(1 - \frac{x}{2}\right)^2 dx = \pi \int_0^2 \left(1 - x + \frac{x^2}{4}\right) dx = \pi\left[x - \frac{x^2}{2} + \frac{x^3}{12}\right]_0^2$

$= \pi\left(2 - \frac{4}{2} + \frac{8}{12}\right) = \frac{2\pi}{3}$

15. $R(x) = \tan\left(\frac{\pi}{4}y\right)$; $u = \frac{\pi}{4}y \Rightarrow du = \frac{\pi}{4}\,dy \Rightarrow 4\,du = \pi\,dy$; $y = 0 \Rightarrow u = 0,\ y = 1 \Rightarrow u = \frac{\pi}{4}$;

$V = \int_0^1 \pi[R(y)]^2\,dy = \pi \int_0^1 \left[\tan\left(\frac{\pi}{4}y\right)\right]^2 dy = 4\int_0^{\pi/4} \tan^2 u\,du = 4\int_0^{\pi/4} \left(-1 + \sec^2 u\right) du = 4[-u + \tan u]_0^{\pi/4}$

$= 4\left(-\frac{\pi}{4} + 1 - 0\right) = 4 - \pi$

17. $R(x) = x^2 \Rightarrow V = \int_0^2 \pi[R(x)]^2\,dx = \pi \int_0^2 \left(x^2\right)^2 dx$

$= \pi \int_0^2 x^4\,dx = \pi\left[\frac{x^5}{5}\right]_0^2 = \frac{32\pi}{5}$

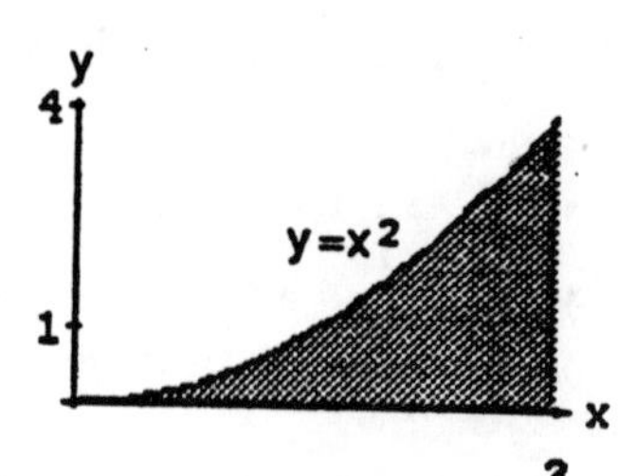

19. $R(x) = \sqrt{9 - x^2} \Rightarrow V = \int_{-3}^{3} \pi[R(x)]^2\, dx = \pi \int_{-3}^{3} (9 - x^2)\, dx$

$= \pi\left[9x - \frac{x^3}{3}\right]_{-3}^{3} = 2\pi\left[9(3) - \frac{27}{3}\right] = 2 \cdot \pi \cdot 18 = 36\pi$

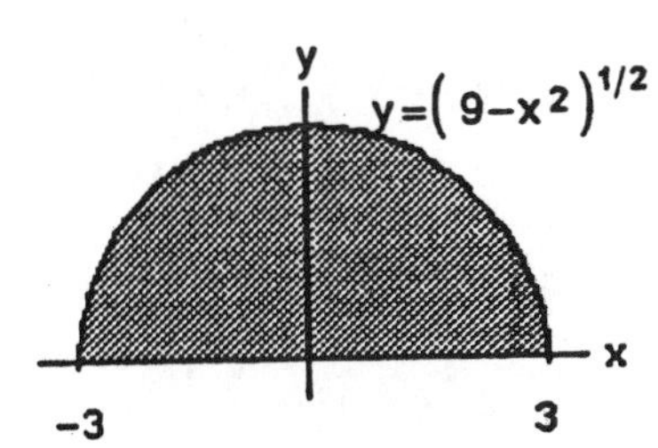

21. $R(x) = \sqrt{\cos x} \Rightarrow V = \int_{0}^{\pi/2} \pi[R(x)]^2\, dx = \pi \int_{0}^{\pi/2} \cos x\, dx$

$= \pi[\sin x]_0^{\pi/2} = \pi(1 - 0) = \pi$

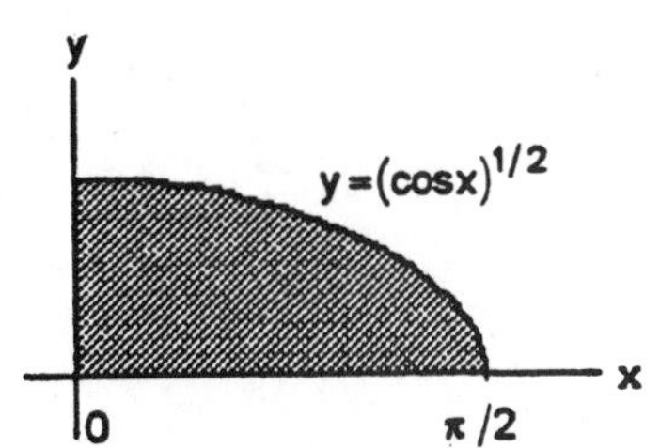

23. $R(x) = \sqrt{2} - \sec x \tan x \Rightarrow V = \int_{0}^{\pi/4} \pi[R(x)]^2\, dx$

$= \pi \int_{0}^{\pi/4} \left(\sqrt{2} - \sec x \tan x\right)^2 dx$

$= \pi \int_{0}^{\pi/4} \left(2 - 2\sqrt{2} \sec x \tan x + \sec^2 x \tan^2 x\right) dx$

$= \pi\left(\int_{0}^{\pi/4} 2\, dx - 2\sqrt{2} \int_{0}^{\pi/4} \sec x \tan x\, dx + \int_{0}^{\pi/4} (\tan x)^2 \sec^2 x\, dx\right)$

$= \pi\left([2x]_0^{\pi/4} - 2\sqrt{2}[\sec x]_0^{\pi/4} + \left[\frac{\tan^3 x}{3}\right]_0^{\pi/4}\right) = \pi\left[\left(\frac{\pi}{2} - 0\right) - 2\sqrt{2}\left(\sqrt{2} - 1\right) + \frac{1}{3}\left(1^3 - 0\right)\right] = \pi\left(\frac{\pi}{2} + 2\sqrt{2} - \frac{11}{3}\right)$

25. $R(y) = \sqrt{5} \cdot y^2 \Rightarrow V = \int_{-1}^{1} \pi[R(y)]^2\, dy = \pi \int_{-1}^{1} 5y^4\, dy$

$= \pi\left[y^5\right]_{-1}^{1} = \pi[1 - (-1)] = 2\pi$

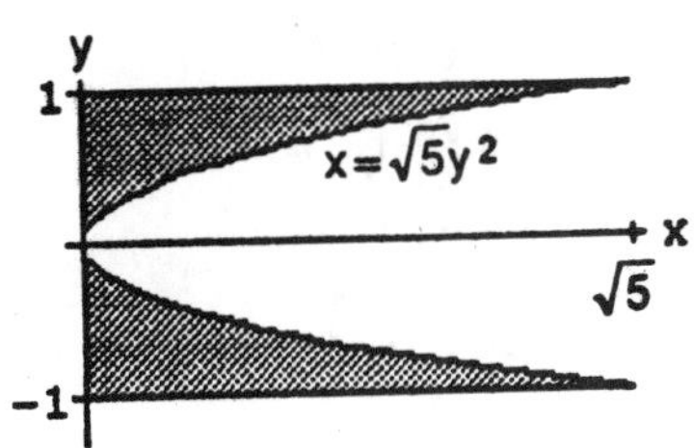

27. $R(y) = \sqrt{2 \sin 2y} \Rightarrow V = \int_0^{\pi/2} \pi[R(y)]^2\, dy = \pi \int_0^{\pi/2} 2 \sin 2y\, dy$

$= \pi[-\cos 2y]_0^{\pi/2} = \pi[1 - (-1)] = 2\pi$

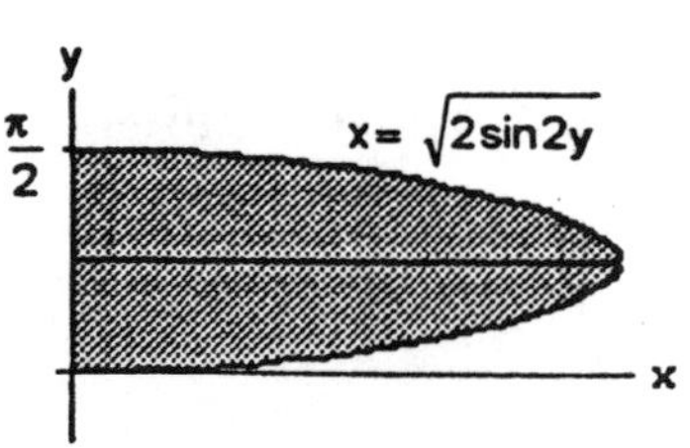

29. $R(y) = \frac{2}{y+1} \Rightarrow V = \int_0^3 \pi[R(y)]^2\, dy = 4\pi \int_0^3 \frac{1}{(y+1)^2}\, dy$

$= 4\pi\left[\frac{-1}{y+1}\right]_0^3 = 4\pi\left[-\frac{1}{4} - (-1)\right] = 3\pi$

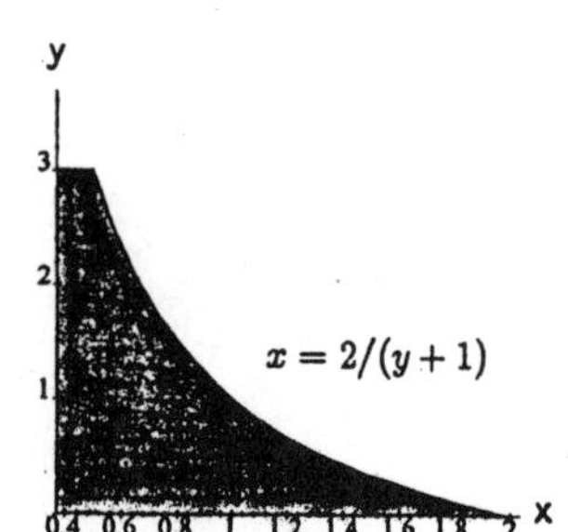

31. For the sketch given, $a = -\frac{\pi}{2}$, $b = \frac{\pi}{2}$; $R(x) = 1$, $r(x) = \sqrt{\cos x}$; $V = \int_a^b \pi\left([R(x)]^2 - [r(x)]^2\right) dx$

$= \int_{-\pi/2}^{\pi/2} \pi(1 - \cos x)\, dx = 2\pi \int_0^{\pi/2} (1 - \cos x)\, dx = 2\pi[x - \sin x]_0^{\pi/2} = 2\pi\left(\frac{\pi}{2} - 1\right) = \pi^2 - 2\pi$

33. $r(x) = x$ and $R(x) = 1 \Rightarrow V = \int_0^1 \pi\left([R(x)]^2 - [r(x)]^2\right) dx$

$= \int_0^1 \pi\left(1 - x^2\right) dx = \pi\left[x - \frac{x^3}{3}\right]_0^1 = \pi\left[\left(1 - \frac{1}{3}\right) - 0\right] = \frac{2\pi}{3}$

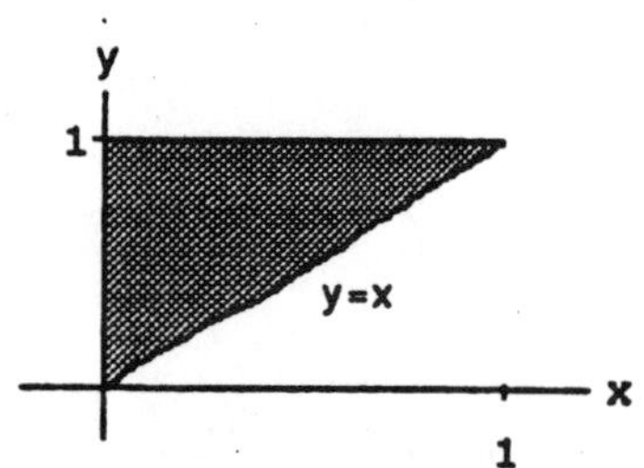

35. $r(x) = x^2 + 1$ and $R(x) = x + 3 \Rightarrow V = \int_{-1}^2 \pi\left([R(x)]^2 - [r(x)]^2\right) dx$

$= \pi \int_{-1}^2 \left[(x+3)^2 - \left(x^2+1\right)^2\right] dx$

$= \pi \int_{-1}^2 \left[\left(x^2 + 6x + 9\right) - \left(x^4 + 2x^2 + 1\right)\right] dx$

$= \pi \int_{-1}^2 \left(-x^4 - x^2 + 6x + 8\right) dx = \pi\left[-\frac{x^5}{5} - \frac{x^3}{3} + \frac{6x^2}{2} + 8x\right]_{-1}^2 = \pi\left[\left(-\frac{32}{5} - \frac{8}{3} + \frac{24}{2} + 16\right) - \left(\frac{1}{5} + \frac{1}{3} + \frac{6}{2} - 8\right)\right]$

$= \pi\left(-\frac{33}{5} - 3 + 28 - 3 + 8\right) = \pi\left(\frac{5 \cdot 30 - 33}{5}\right) = \frac{117\pi}{5}$

y
5
y=x+3
2
y=x^2+1
x
-1
2

37. $r(x) = \sec x$ and $R(x) = \sqrt{2} \Rightarrow V = \int_{-\pi/4}^{\pi/4} \pi\left([R(x)]^2 - [r(x)]^2\right) dx$

$$= \pi \int_{-\pi/4}^{\pi/4} \left(2 - \sec^2 x\right) dx = \pi[2x - \tan x]_{-\pi/4}^{\pi/4}$$

$$= \pi\left[\left(\frac{\pi}{2} - 1\right) - \left(-\frac{\pi}{2} + 1\right)\right] = \pi(\pi - 2)$$

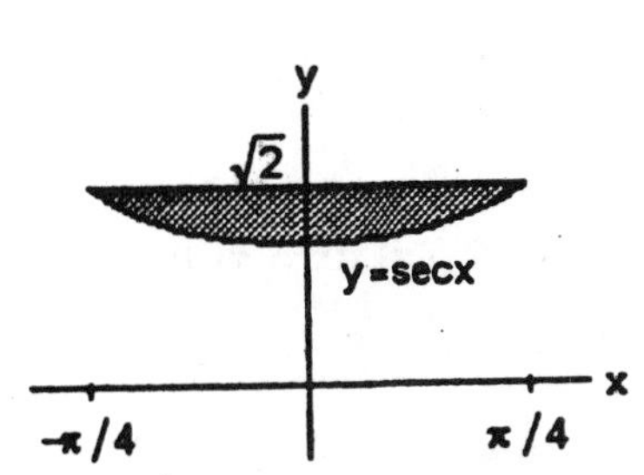

39. $r(y) = 1$ and $R(y) = 1 + y \Rightarrow V = \int_0^1 \pi\left([R(y)]^2 - [r(y)]^2\right) dy$

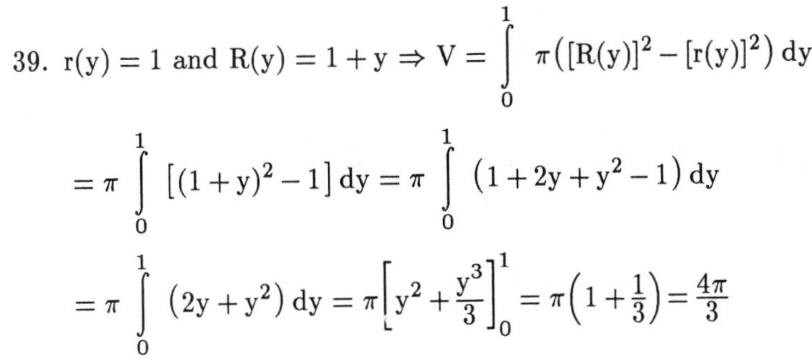

$$= \pi \int_0^1 \left[(1+y)^2 - 1\right] dy = \pi \int_0^1 \left(1 + 2y + y^2 - 1\right) dy$$

$$= \pi \int_0^1 \left(2y + y^2\right) dy = \pi\left[y^2 + \frac{y^3}{3}\right]_0^1 = \pi\left(1 + \frac{1}{3}\right) = \frac{4\pi}{3}$$

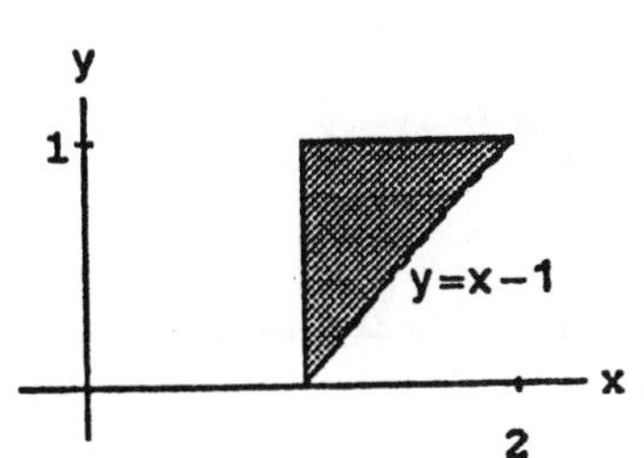

41. $R(y) = 2$ and $r(y) = \sqrt{y} \Rightarrow V = \int_0^4 \pi\left([R(y)]^2 - [r(y)]^2\right) dy$

$$= \pi \int_0^4 (4 - y)\, dy = \pi\left[4y - \frac{y^2}{2}\right]_0^4 = \pi(16 - 8) = 8\pi$$

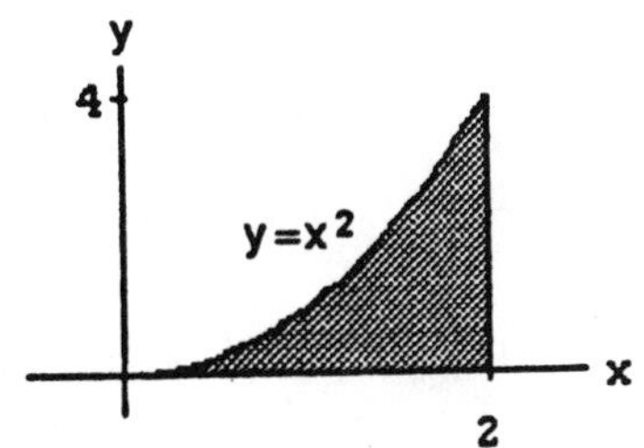

43. $R(y) = 2$ and $r(y) = 1 + \sqrt{y} \Rightarrow V = \int_0^1 \pi\left([R(y)]^2 - [r(y)]^2\right) dy$

$$= \pi \int_0^1 \left[4 - \left(1 + \sqrt{y}\right)^2\right] dy = \pi \int_0^1 \left(4 - 1 - 2\sqrt{y} - y\right) dy$$

$$= \pi \int_0^1 \left(3 - 2\sqrt{y} - y\right) dy = \pi\left[3y - \frac{4}{3}y^{3/2} - \frac{y^2}{2}\right]_0^1$$

$$= \pi\left(3 - \frac{4}{3} - \frac{1}{2}\right) = \pi\left(\frac{18 - 8 - 3}{6}\right) = \frac{7\pi}{6}$$

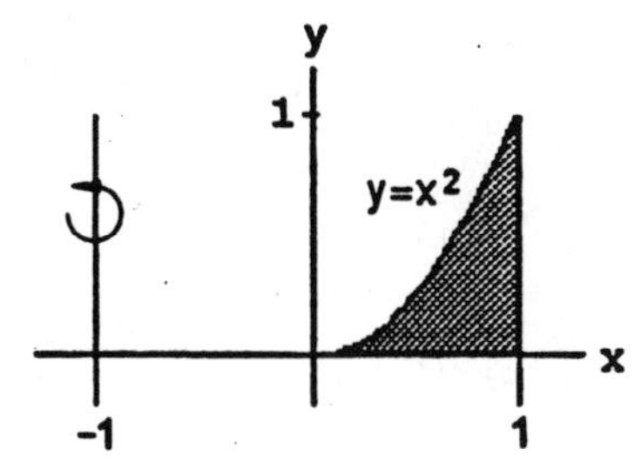

45. (a) $r(x) = \sqrt{x}$ and $R(x) = 2 \Rightarrow V = \int_0^4 \pi\left([R(x)]^2 - [r(x)]^2\right) dx$

$= \pi \int_0^4 (4 - x)\, dx = \pi\left[4x - \frac{x^2}{2}\right]_0^4 = \pi(16 - 8) = 8\pi$

(b) $r(y) = 0$ and $R(y) = y^2 \Rightarrow V = \int_0^2 \pi\left([R(y)]^2 - [r(y)]^2\right) dy$

$= \pi \int_0^2 y^4\, dy = \pi\left[\frac{y^5}{5}\right]_0^2 = \frac{32\pi}{5}$

(c) $r(x) = 0$ and $R(x) = 2 - \sqrt{x} \Rightarrow V = \int_0^4 \pi\left([R(x)]^2 - [r(x)]^2\right) dx = \pi \int_0^4 \left(2 - \sqrt{x}\right)^2 dx$

$= \pi \int_0^4 \left(4 - 4\sqrt{x} + x\right) dx = \pi\left[4x - \frac{8x^{3/2}}{3} + \frac{x^2}{2}\right]_0^4 = \pi\left(16 - \frac{64}{3} + \frac{16}{2}\right) = \frac{8\pi}{3}$

(d) $r(y) = 4 - y^2$ and $R(y) = 4 \Rightarrow V = \int_0^2 \pi\left([R(y)]^2 - [r(y)]^2\right) dy = \pi \int_0^2 \left[16 - \left(4 - y^2\right)^2\right] dy$

$= \pi \int_0^2 \left(16 - 16 + 8y^2 - y^4\right) dy = \pi \int_0^2 \left(8y^2 - y^4\right) dy = \pi\left[\frac{8}{3}y^3 - \frac{y^5}{5}\right]_0^2 = \pi\left(\frac{64}{3} - \frac{32}{5}\right) = \frac{224\pi}{15}$

47. (a) $r(x) = 0$ and $R(x) = 1 - x^2 \Rightarrow V = \int_{-1}^1 \pi\left([R(x)]^2 - [r(x)]^2\right) dx$

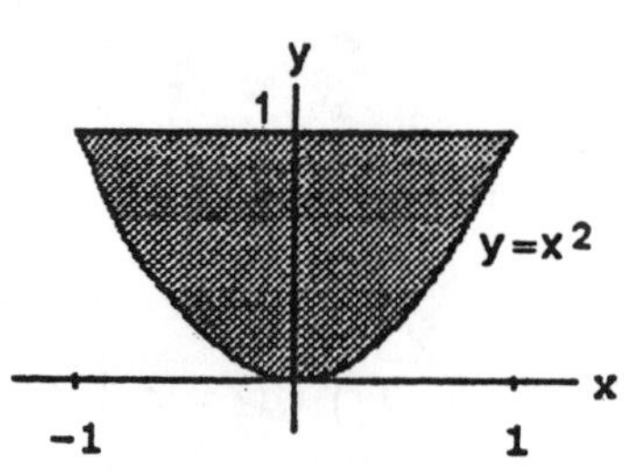

$= \pi \int_{-1}^1 \left(1 - x^2\right)^2 dx = \pi \int_{-1}^1 \left(1 - 2x^2 + x^4\right) dx$

$= \pi\left[x - \frac{2x^3}{3} + \frac{x^5}{5}\right]_{-1}^1 = 2\pi\left(1 - \frac{2}{3} + \frac{1}{5}\right) = 2\pi\left(\frac{15 - 10 + 3}{15}\right)$

$= \frac{16\pi}{15}$

(b) $r(x) = 1$ and $R(x) = 2 - x^2 \Rightarrow V = \int_{-1}^1 \pi\left([R(x)]^2 - [r(x)]^2\right) dx = \pi \int_{-1}^1 \left[\left(2 - x^2\right)^2 - 1\right] dx$

$= \pi \int_{-1}^1 \left(4 - 4x^2 + x^4 - 1\right) dx = \pi \int_{-1}^1 \left(3 - 4x^2 + x^4\right) dx = \pi\left[3x - \frac{4}{3}x^3 + \frac{x^5}{5}\right]_{-1}^1 = 2\pi\left(3 - \frac{4}{3} + \frac{1}{5}\right)$

$= \frac{2\pi}{15}(45 - 20 + 3) = \frac{56\pi}{15}$

(c) $r(x) = 1 + x^2$ and $R(x) = 2 \Rightarrow V = \int_{-1}^1 \pi\left([R(x)]^2 - [r(x)]^2\right) dx = \pi \int_{-1}^1 \left[4 - \left(1 + x^2\right)^2\right] dx$

$$= \pi \int_{-1}^{1} \left(4 - 1 - 2x^2 - x^4\right) dx = \pi \int_{-1}^{1} \left(3 - 2x^2 - x^4\right) dx = \pi\left[3x - \frac{2}{3}x^3 - \frac{x^5}{5}\right]_{-1}^{1} = 2\pi\left(3 - \frac{2}{3} - \frac{1}{5}\right)$$

$$= \frac{2\pi}{15}(45 - 10 - 3) = \frac{64\pi}{15}$$

49. $R(y) = b + \sqrt{a^2 - y^2}$ and $r(y) = b - \sqrt{a^2 - y^2}$

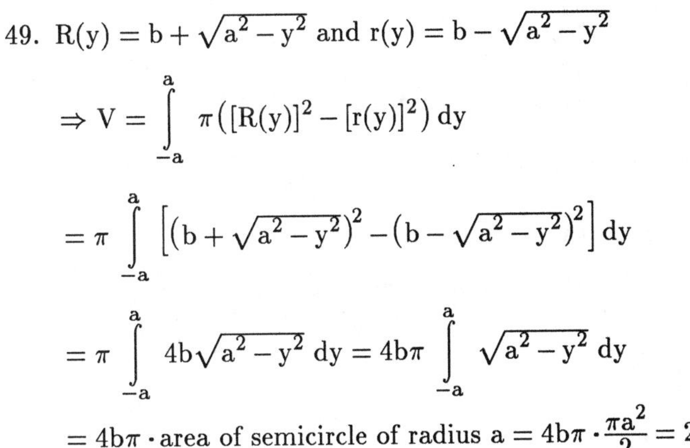

$$\Rightarrow V = \int_{-a}^{a} \pi\left([R(y)]^2 - [r(y)]^2\right) dy$$

$$= \pi \int_{-a}^{a} \left[\left(b + \sqrt{a^2 - y^2}\right)^2 - \left(b - \sqrt{a^2 - y^2}\right)^2\right] dy$$

$$= \pi \int_{-a}^{a} 4b\sqrt{a^2 - y^2}\, dy = 4b\pi \int_{-a}^{a} \sqrt{a^2 - y^2}\, dy$$

$$= 4b\pi \cdot \text{area of semicircle of radius } a = 4b\pi \cdot \frac{\pi a^2}{2} = 2a^2 b\pi^2$$

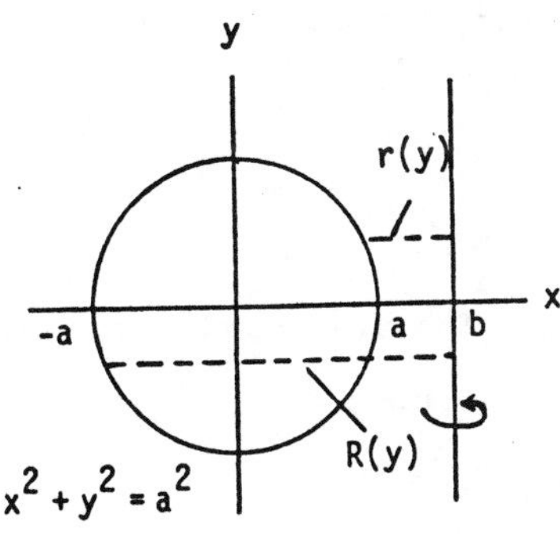

51. (a) $R(y) = \sqrt{a^2 - y^2} \Rightarrow V = \pi \int_{-a}^{h-a} \left(a^2 - y^2\right) dy = \pi\left[a^2 y - \frac{y^3}{3}\right]_{-a}^{h-a} = \pi\left[a^2 h - a^3 - \frac{(h-a)^3}{3} - \left(-a^3 + \frac{a^3}{3}\right)\right]$

$$= \pi\left[a^2 h - \frac{1}{3}\left(h^3 - 3h^2 a + 3ha^2 - a^3\right) - \frac{a^3}{3}\right] = \pi\left(a^2 h - \frac{h^3}{3} + h^2 a - ha^2\right) = \frac{\pi h^2(3a - h)}{3}$$

(b) Given $\frac{dV}{dt} = 0.2$ m^3/sec and $a = 5$ m, find $\left.\frac{dh}{dt}\right|_{h=4}$. From part (a), $V(h) = \frac{\pi h^2(15 - h)}{3} = 5\pi h^2 - \frac{\pi h^3}{3}$

$$\Rightarrow \frac{dV}{dh} = 10\pi h - \pi h^2 \Rightarrow \frac{dV}{dt} = \frac{dV}{dh} \cdot \frac{dh}{dt} = \pi h(10 - h)\frac{dh}{dt} \Rightarrow \left.\frac{dh}{dt}\right|_{h=4} = \frac{0.2}{4\pi(10 - 4)} = \frac{1}{(20\pi)(6)} = \frac{1}{120\pi} \text{ m/sec.}$$

53. The cross section of a solid right circular cylinder with a cone removed is a disk with radius R from which a disk of radius h has been removed (figure provided). Thus its area is $A_1 = \pi R^2 - \pi h^2 = \pi\left(R^2 - h^2\right)$. The cross section of the hemisphere is a disk of radius $\sqrt{R^2 - h^2}$ (figure provided). Therefore its area is $A_2 = \pi\left(\sqrt{R^2 - h^2}\right)^2 = \pi\left(R^2 - h^2\right)$. We can see that $A_1 = A_2$. The altitudes of both solids are R. Applying Cavalieri's Theorem we find Volume of Hemisphere = (Volume of Cylinder) − (Volume of Cone) $= \left(\pi R^2\right)R - \frac{1}{3}\pi\left(R^2\right)R = \frac{2}{3}\pi R^3$.

55. $R(y) = \sqrt{256 - y^2} \Rightarrow V = \int_{-16}^{-7} \pi[R(y)]^2\, dy = \pi \int_{-16}^{-7} \left(256 - y^2\right) dy = \pi\left[256y - \frac{y^3}{3}\right]_{-16}^{-7}$

$$= \pi\left[(256)(-7) + \frac{7^3}{3} - \left((256)(-16) + \frac{16^3}{3}\right)\right] = \pi\left(\frac{7^3}{3} + 256(16 - 7) - \frac{16^3}{3}\right) = 1053\pi \text{ cm}^3 \approx 3308 \text{ cm}^3$$

57. (a) $R(x) = |c - \sin x|$, so $V = \pi \int_0^{\pi} [R(x)]^2\, dx = \pi \int_0^{\pi} (c - \sin x)^2\, dx = \pi \int_0^{\pi} \left(c^2 - 2c \sin x + \sin^2 x\right) dx$

$$= \pi \int_0^{\pi} \left(c^2 - 2c \sin x + \frac{1 - \cos 2x}{2}\right) dx = \pi \int_0^{\pi} \left(c^2 + \frac{1}{2} - 2c \sin x - \frac{\cos 2x}{2}\right) dx$$

$$= \pi\left[\left(c^2 + \frac{1}{2}\right)x + 2c \cos x - \frac{\sin 2x}{4}\right]_0^{\pi} = \pi\left[\left(c^2\pi + \frac{\pi}{2} - 2c - 0\right) - (0 + 2c - 0)\right] = \pi\left(c^2\pi + \frac{\pi}{2} - 4c\right).$$ Let

$V(c) = \pi\left(c^2\pi + \frac{\pi}{2} - 4c\right)$. We find the extreme values of V(c): $\frac{dV}{dc} = \pi(2c\pi - 4) = 0 \Rightarrow c = \frac{2}{\pi}$ is a critical point, and $V\left(\frac{2}{\pi}\right) = \pi\left(\frac{4}{\pi} + \frac{\pi}{2} - \frac{8}{\pi}\right) = \pi\left(\frac{\pi}{2} - \frac{4}{\pi}\right) = \frac{\pi^2}{2} - 4$; Evaluate V at the endpoints: $V(0) = \frac{\pi^2}{2}$ and $V(1) = \pi\left(\frac{3}{2}\pi - 4\right) = \frac{\pi^2}{2} - (4 - \pi)\pi$. Now we see that the function's absolute minimum value is $\frac{\pi^2}{2} - 4$, taken on at the critical point $c = \frac{2}{\pi}$. (See also the accompanying graph.)

(b) From the discussion in part (a) we conclude that the function's absolute maximum value is $\frac{\pi^2}{2}$, taken on at the endpoint $c = 0$.

(c) The graph of the solid's volume as a function of c for $0 \le c \le 1$ is given at the right. As c moves away from $[0, 1]$ the volume of the solid increases without bound. If we approximate the solid as a set of solid disks, we can see that the radius of a typical disk increases without bounds as c moves away from $[0, 1]$.

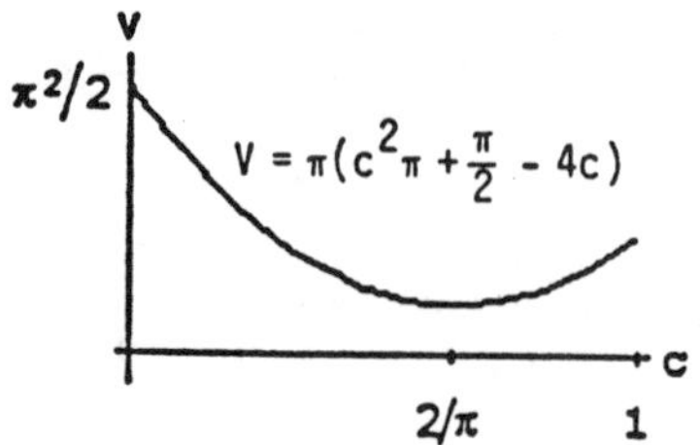

59. (a) Using $d = \frac{C}{\pi}$, and $A = \pi\left(\frac{d}{2}\right)^2 = \frac{C^2}{4\pi}$ yields the following areas (in square inches, rounded to the nearest tenth: 2.3, 1.6, 1.5, 2.1, 3.2, 4.8, 7.0, 9.3, 10.7, 10.7, 9.3, 6.4, 3.2.

(b) If C(y) is the circumference as a function of y, then the area of a cross section is

$A(y) = \pi\left(\frac{C(y)/\pi}{2}\right)^2 = \frac{C^2(y)}{4\pi}$, and the volume is $\frac{1}{4\pi}\int_0^6 C^2(y)\, dy$.

(c) $\int_0^6 A(y)\, dy = \frac{1}{4\pi}\int_0^6 C^2(y)\, dy \approx \frac{1}{4\pi}\left(\frac{6-0}{24}\right)[5.4^2 + 2(4.5^2 + 4.4^2 + 5.1^2 + 6.3^2 + 7.8^2 + 9.4^2 + 10.8^2 + 11.6^2 + 11.6^2 + 10.8^2 + 9.0^2) + 6.3^2] \approx 34.7 \text{ in.}^3$

(d) $V = \frac{1}{4\pi}\int_0^6 C^2(y)\, dy \approx \frac{1}{4\pi}\left(\frac{6-0}{36}\right)\Big[5.4^2 + 4(4.5^2) + 2(4.4^2) + 4(5.1^2) + 2(6.3^2) + 4(7.8^2) + 2(9.4^2) + 4(10.8^2) + 2(11.6^2) + 4(11.6^2) + 2(10.8^2) + 4(9.0^2) + 6.3^2\Big] = 34.792 \text{ in.}^3$

by Simpson's rule. The Simpson's rule estimate should be more accurate than the trapezoid estimate. The error in the Simpson's estimate is proportional to $h^4 = 0.0625$ whereas the error in the trapezoid estimate is proportional to $h^2 = 0.25$, a larger number when $h = 0.5$ in.

5.2 MODELING VOLUME USING CYLINDRICAL SHELLS

1. For the sketch given, a = 0, b = 2;

$$V = \int_a^b 2\pi\left(\begin{matrix}\text{shell}\\ \text{radius}\end{matrix}\right)\left(\begin{matrix}\text{shell}\\ \text{height}\end{matrix}\right) dx = \int_0^2 2\pi x\left(1+\frac{x^2}{4}\right) dx = 2\pi \int_0^2 \left(x+\frac{x^3}{4}\right) dx = 2\pi\left[\frac{x^2}{2}+\frac{x^4}{16}\right]_0^2 = 2\pi\left(\frac{4}{2}+\frac{16}{16}\right)$$

$$= 2\pi \cdot 3 = 6\pi$$

3. For the sketch given, c = 0, d = $\sqrt{2}$;

$$V = \int_c^d 2\pi\left(\begin{matrix}\text{shell}\\ \text{radius}\end{matrix}\right)\left(\begin{matrix}\text{shell}\\ \text{height}\end{matrix}\right) dy = \int_0^{\sqrt{2}} 2\pi y \cdot (y^2) dy = 2\pi \int_0^{\sqrt{2}} y^3\, dy = 2\pi\left[\frac{y^4}{4}\right]_0^{\sqrt{2}} = 2\pi$$

5. For the sketch given, a = 0, b = $\sqrt{3}$;

$$V = \int_a^b 2\pi\left(\begin{matrix}\text{shell}\\ \text{radius}\end{matrix}\right)\left(\begin{matrix}\text{shell}\\ \text{height}\end{matrix}\right) dx = \int_0^{\sqrt{3}} 2\pi x \cdot \left(\sqrt{x^2+1}\right) dx;$$

$\left[u = x^2+1 \Rightarrow du = 2x\,dx;\ x = 0 \Rightarrow u = 1,\ x = \sqrt{3} \Rightarrow u = 4\right]$

$$\to V = \pi \int_1^4 u^{1/2}\, du = \pi\left[\frac{2}{3}u^{3/2}\right]_1^4 = \frac{2\pi}{3}\left(4^{3/2}-1\right) = \left(\frac{2\pi}{3}\right)(8-1) = \frac{14\pi}{3}$$

7. a = 0, b = 2;

$$V = \int_a^b 2\pi\left(\begin{matrix}\text{shell}\\ \text{radius}\end{matrix}\right)\left(\begin{matrix}\text{shell}\\ \text{height}\end{matrix}\right) dx = \int_0^2 2\pi x\left[x-\left(-\frac{x}{2}\right)\right] dx$$

$$= \int_0^2 2\pi x^2 \cdot \frac{3}{2}\, dx = \pi \int_0^2 3x^2\, dx = \pi\left[x^3\right]_0^2 = 8\pi$$

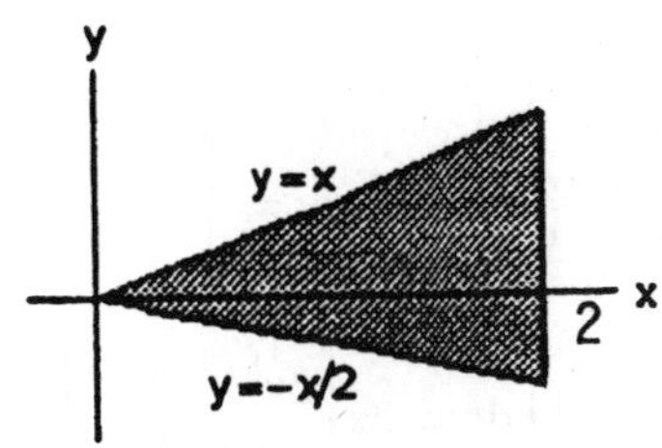

9. a = 0, b = 1;

$$V = \int_a^b 2\pi\left(\begin{matrix}\text{shell}\\ \text{radius}\end{matrix}\right)\left(\begin{matrix}\text{shell}\\ \text{height}\end{matrix}\right) dx = \int_0^1 2\pi x\left[(2-x)-x^2\right] dx$$

$$= 2\pi \int_0^1 \left(2x - x^2 - x^3\right) dx = 2\pi\left[x^2 - \frac{x^3}{3} - \frac{x^4}{4}\right]_0^1$$

$$= 2\pi\left(1-\frac{1}{3}-\frac{1}{4}\right) = 2\pi\left(\frac{12-4-3}{12}\right) = \frac{10\pi}{12} = \frac{5\pi}{6}$$

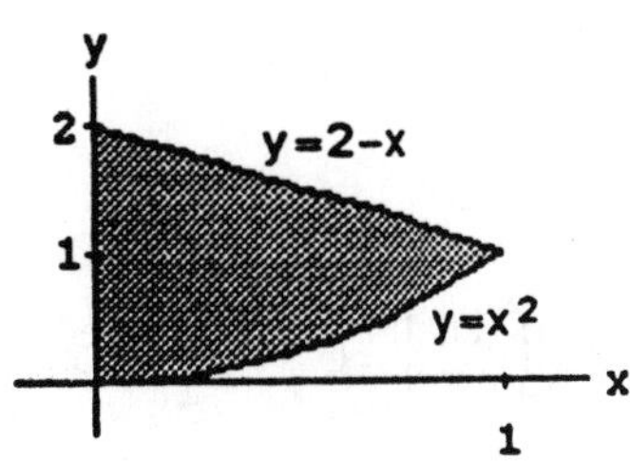

11. $a = 0,\ b = 1$;

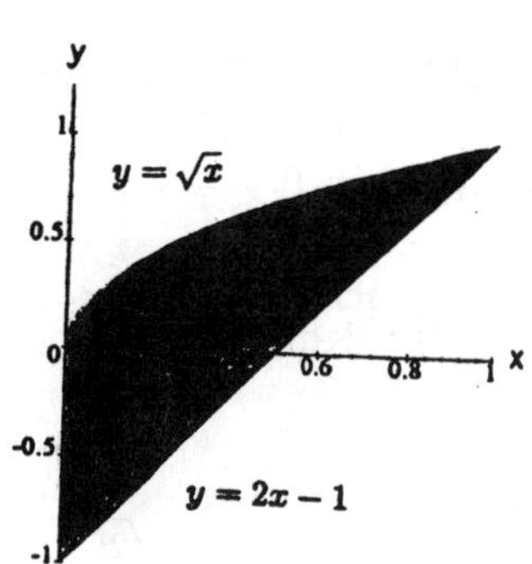

$$V = \int_a^b 2\pi\left(\begin{smallmatrix}\text{shell}\\ \text{radius}\end{smallmatrix}\right)\left(\begin{smallmatrix}\text{shell}\\ \text{height}\end{smallmatrix}\right) dx = \int_0^1 2\pi x\left[\sqrt{x} - (2x - 1)\right] dx$$

$$= 2\pi \int_0^1 \left(x^{3/2} - 2x^2 + x\right) dx = 2\pi\left[\frac{2}{5}x^{5/2} - \frac{2}{3}x^3 + \frac{1}{2}x^2\right]_0^1$$

$$= 2\pi\left(\frac{2}{5} - \frac{2}{3} + \frac{1}{2}\right) = 2\pi\left(\frac{12 - 20 + 15}{30}\right) = \frac{7\pi}{15}$$

13. (a) $xf(x) = \begin{cases} x \cdot \frac{\sin x}{x}, & 0 < x \le \pi \\ x, & x = 0 \end{cases} \Rightarrow xf(x) = \begin{cases} \sin x, & 0 < x \le \pi \\ 0, & x = 0 \end{cases}$; since $\sin 0 = 0$ we have

$$xf(x) = \begin{cases} \sin x, & 0 < x \le \pi \\ \sin x, & x = 0 \end{cases} \Rightarrow xf(x) = \sin x,\ 0 \le x \le \pi$$

(b) $V = \int_a^b 2\pi\left(\begin{smallmatrix}\text{shell}\\ \text{radius}\end{smallmatrix}\right)\left(\begin{smallmatrix}\text{shell}\\ \text{height}\end{smallmatrix}\right) dx = \int_0^\pi 2\pi x \cdot f(x)\, dx$ and $x \cdot f(x) = \sin x,\ 0 \le x \le \pi$ by part (a)

$$\Rightarrow V = 2\pi \int_0^\pi \sin x\, dx = 2\pi[-\cos x]_0^\pi = 2\pi(-\cos \pi + \cos 0) = 4\pi$$

15. $c = 0,\ d = 2$;

$$V = \int_c^d 2\pi\left(\begin{smallmatrix}\text{shell}\\ \text{radius}\end{smallmatrix}\right)\left(\begin{smallmatrix}\text{shell}\\ \text{height}\end{smallmatrix}\right) dy = \int_0^2 2\pi y\left[\sqrt{y} - (-y)\right] dy$$

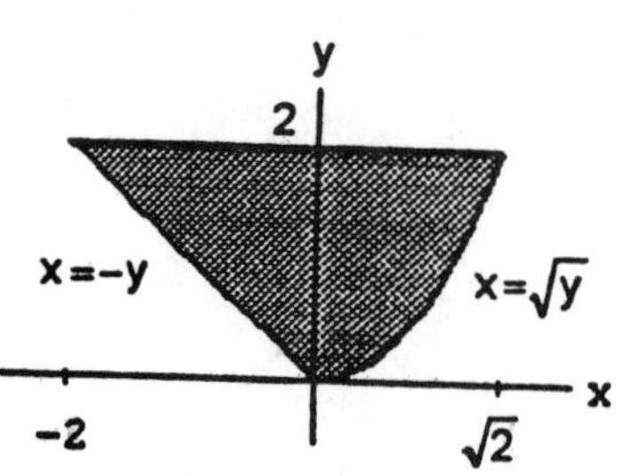

$$= 2\pi \int_0^2 \left(y^{3/2} + y^2\right) dy = 2\pi\left[\frac{2y^{5/2}}{5} + \frac{y^3}{3}\right]_0^2$$

$$= 2\pi\left[\frac{2}{5}\left(\sqrt{2}\right)^5 + \frac{2^3}{3}\right] = 2\pi\left(\frac{8\sqrt{2}}{5} + \frac{8}{3}\right) = 16\pi\left(\frac{\sqrt{2}}{5} + \frac{1}{3}\right)$$

$$= \frac{16\pi}{15}\left(3\sqrt{2} + 5\right)$$

17. $c = 0,\ d = 2$;

$$V = \int_c^d 2\pi\left(\begin{smallmatrix}\text{shell}\\ \text{radius}\end{smallmatrix}\right)\left(\begin{smallmatrix}\text{shell}\\ \text{height}\end{smallmatrix}\right) dy = \int_0^2 2\pi y\left(2y - y^2\right) dy$$

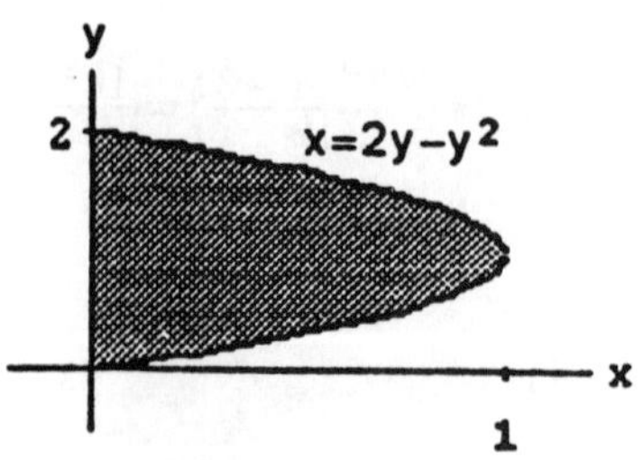

$$= 2\pi \int_0^2 \left(2y^2 - y^3\right) dy = 2\pi\left[\frac{2y^3}{3} - \frac{y^4}{4}\right]_0^2 = 2\pi\left(\frac{16}{3} - \frac{16}{4}\right)$$

$$= 32\pi\left(\frac{1}{3} - \frac{1}{4}\right) = \frac{32\pi}{12} = \frac{8\pi}{3}$$

19. $c = 0,\ d = 1;$

$$V = \int_c^d 2\pi\begin{pmatrix}\text{shell}\\ \text{radius}\end{pmatrix}\begin{pmatrix}\text{shell}\\ \text{height}\end{pmatrix} dy = 2\pi \int_0^1 y[y - (-y)]\, dy$$

$$= 2\pi \int_0^1 2y^2\, dy = \frac{4\pi}{3}\left[y^3\right]_0^1 = \frac{4\pi}{3}$$

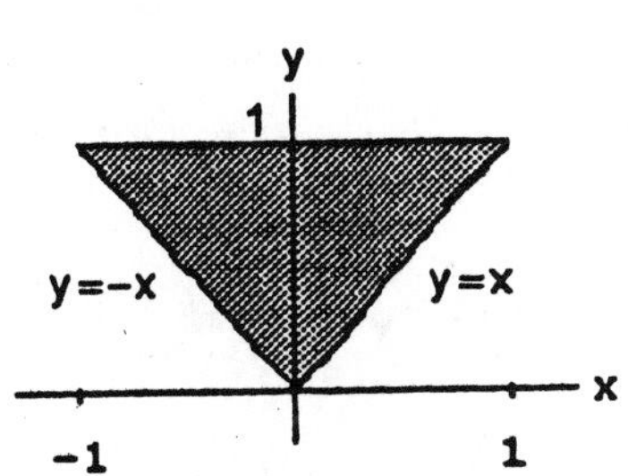

21. $c = 0,\ d = 2;$

$$V = \int_c^d 2\pi\begin{pmatrix}\text{shell}\\ \text{radius}\end{pmatrix}\begin{pmatrix}\text{shell}\\ \text{height}\end{pmatrix} dy = \int_0^2 2\pi y\left[(2 + y) - y^2\right] dy$$

$$= 2\pi \int_0^2 \left(2y + y^2 - y^3\right) dy = 2\pi\left[y^2 + \frac{y^3}{3} - \frac{y^4}{4}\right]_0^2$$

$$= 2\pi\left(4 + \frac{8}{3} - \frac{16}{4}\right) = \frac{\pi}{6}(48 + 32 - 48) = \frac{16\pi}{3}$$

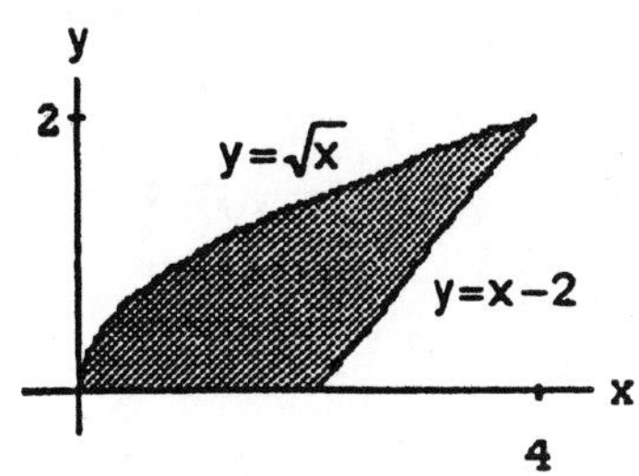

23. (a) $$V = \int_c^d 2\pi\begin{pmatrix}\text{shell}\\ \text{radius}\end{pmatrix}\begin{pmatrix}\text{shell}\\ \text{height}\end{pmatrix} dy = \int_0^1 2\pi y \cdot 12\left(y^2 - y^3\right) dy = 24\pi \int_0^1 \left(y^3 - y^4\right) dy = 24\pi\left[\frac{y^4}{4} - \frac{y^5}{5}\right]_0^1$$

$$= 24\pi\left(\frac{1}{4} - \frac{1}{5}\right) = \frac{24\pi}{20} = \frac{6\pi}{5}$$

(b) $$V = \int_c^d 2\pi\begin{pmatrix}\text{shell}\\ \text{radius}\end{pmatrix}\begin{pmatrix}\text{shell}\\ \text{height}\end{pmatrix} dy = \int_0^1 2\pi(1 - y)\left[12\left(y^2 - y^3\right)\right] dy = 24\pi \int_0^1 (1 - y)\left(y^2 - y^3\right) dy$$

$$= 24\pi \int_0^1 \left(y^2 - 2y^3 + y^4\right) dy = 24\pi\left[\frac{y^3}{3} - \frac{y^4}{2} + \frac{y^5}{5}\right]_0^1 = 24\pi\left(\frac{1}{3} - \frac{1}{2} + \frac{1}{5}\right) = 24\pi\left(\frac{1}{30}\right) = \frac{4\pi}{5}$$

(c) $$V = \int_c^d 2\pi\begin{pmatrix}\text{shell}\\ \text{radius}\end{pmatrix}\begin{pmatrix}\text{shell}\\ \text{height}\end{pmatrix} dy = \int_0^1 2\pi\left(\frac{8}{5} - y\right)\left[12\left(y^2 - y^3\right)\right] dy = 24\pi \int_0^1 \left(\frac{8}{5} - y\right)\left(y^2 - y^3\right) dy$$

$$= 24\pi \int_0^1 \left(\frac{8}{5}y^2 - \frac{13}{5}y^3 + y^4\right) dy = 24\pi\left[\frac{8}{15}y^3 - \frac{13}{20}y^4 + \frac{y^5}{5}\right]_0^1 = 24\pi\left(\frac{8}{15} - \frac{13}{20} + \frac{1}{5}\right) = \frac{24\pi}{60}(32 - 39 + 12)$$

$$= \frac{24\pi}{12} = 2\pi$$

(d) $$V = \int_c^d 2\pi\begin{pmatrix}\text{shell}\\ \text{radius}\end{pmatrix}\begin{pmatrix}\text{shell}\\ \text{height}\end{pmatrix} dy = \int_0^1 2\pi\left(y + \frac{2}{5}\right)\left[12\left(y^2 - y^3\right)\right] dy = 24\pi \int_0^1 \left(y + \frac{2}{5}\right)\left(y^2 - y^3\right) dy$$

$$= 24\pi \int_0^1 \left(y^3 - y^4 + \frac{2}{5}y^2 - \frac{2}{5}y^3\right) dy = 24\pi \int_0^1 \left(\frac{2}{5}y^2 + \frac{3}{5}y^3 - y^4\right) dy = 24\pi\left[\frac{2}{15}y^3 + \frac{3}{20}y^4 - \frac{y^5}{5}\right]_0^1$$

$$= 24\pi\left(\frac{2}{15} + \frac{3}{20} - \frac{1}{5}\right) = \frac{24\pi}{60}(8 + 9 - 12) = \frac{24\pi}{12} = 2\pi$$

25. (a) About the x-axis: $V = \int_a^b 2\pi\binom{\text{shell}}{\text{radius}}\binom{\text{shell}}{\text{height}}\,dy = \int_0^1 2\pi y\left(\sqrt{y} - y\right)dy = 2\pi \int_0^1 \left(y^{3/2} - y^2\right)dy$

$$= 2\pi\left(\frac{2}{5}y^{5/2} - \frac{1}{3}y^3\right)\Big|_{y=0}^{1} = 2\pi\left(\frac{2}{5} - \frac{1}{3}\right) = \frac{2\pi}{15}$$

About the y-axis: $V = \int_a^b 2\pi\binom{\text{shell}}{\text{radius}}\binom{\text{shell}}{\text{height}}\,dx = \int_0^1 2\pi x\left(x - x^2\right)dx = 2\pi \int_0^1 \left(x^2 - x^3\right)dx$

$$= 2\pi\left(\frac{x^3}{3} - \frac{x^4}{4}\right)\Big|_{x=0}^{1} = 2\pi\left(\frac{1}{3} - \frac{1}{4}\right) = \frac{\pi}{6}$$

(b) About the x-axis: $R(x) = x$ and $r(x) = x^2 \Rightarrow V = \int_a^b \pi\left[R(x)^2 - r(x)^2\right]dx = \int_0^1 \pi\left[x^2 - x^4\right]dx$

$$= \pi\left(\frac{x^3}{3} - \frac{x^5}{5}\right)\Big|_{r=0}^{1} = \pi\left(\frac{1}{3} - \frac{1}{5}\right) = \frac{2\pi}{15}$$

About the y-axis: $R(y) = \sqrt{y}$ and $r(y) = y \Rightarrow V = \int_a^b \pi\left[R(y)^2 - r(y)\right]dy = \int_0^1 \pi\left[y - y^2\right]dy$

$$= \pi\left(\frac{y^2}{2} - \frac{y^3}{3}\right)\Big|_{y=0}^{1} = \pi\left(\frac{1}{2} - \frac{1}{3}\right) = \frac{\pi}{6}$$

27. (a) $V = \int_c^d 2\pi\binom{\text{shell}}{\text{radius}}\binom{\text{shell}}{\text{height}}\,dy = \int_1^2 2\pi y(y-1)\,dy$

$$= 2\pi \int_1^2 \left(y^2 - y\right)dy = 2\pi\left[\frac{y^3}{3} - \frac{y^2}{2}\right]_1^2 = 2\pi\left[\left(\frac{8}{3} - \frac{4}{2}\right) - \left(\frac{1}{3} - \frac{1}{2}\right)\right]$$

$$= 2\pi\left(\frac{7}{3} - 2 + \frac{1}{2}\right) = \frac{\pi}{3}(14 - 12 + 3) = \frac{5\pi}{3}$$

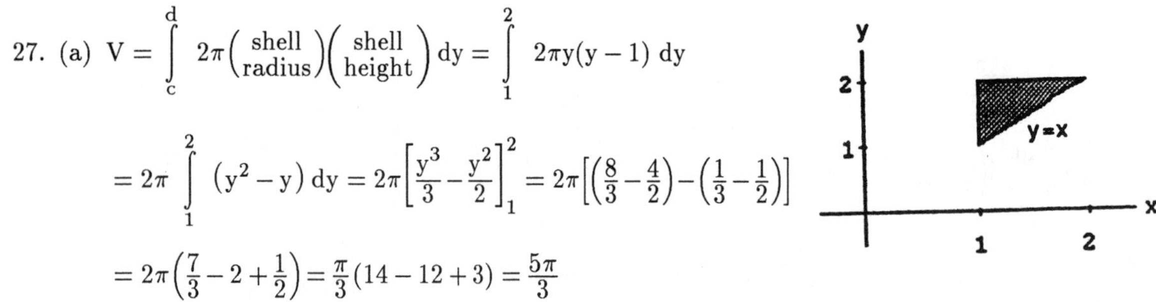

(b) $V = \int_a^b 2\pi\binom{\text{shell}}{\text{radius}}\binom{\text{shell}}{\text{height}}\,dx = \int_1^2 2\pi x(2-x)\,dx = 2\pi \int_1^2 \left(2x - x^2\right)dx = 2\pi\left[x^2 - \frac{x^3}{3}\right]_1^2$

$$= 2\pi\left[\left(4 - \frac{8}{3}\right) - \left(1 - \frac{1}{3}\right)\right] = 2\pi\left[\left(\frac{12-8}{3}\right) - \left(\frac{3-1}{3}\right)\right] = 2\pi\left(\frac{4}{3} - \frac{2}{3}\right) = \frac{4\pi}{3}$$

(c) $V = \int_a^b 2\pi\binom{\text{shell}}{\text{radius}}\binom{\text{shell}}{\text{height}}\,dx = \int_1^2 2\pi\left(\frac{10}{3} - x\right)(2-x)\,dx = 2\pi \int_1^2 \left(\frac{20}{3} - \frac{16}{3}x + x^2\right)dx$

$$= 2\pi\left[\frac{20}{3}x - \frac{8}{3}x^2 + \frac{1}{3}x^3\right]_1^2 = 2\pi\left[\left(\frac{40}{3} - \frac{32}{3} + \frac{8}{3}\right) - \left(\frac{20}{3} - \frac{8}{3} + \frac{1}{3}\right)\right] = 2\pi\left(\frac{3}{3}\right) = 2\pi$$

(d) $V = \int_c^d 2\pi\binom{\text{shell}}{\text{radius}}\binom{\text{shell}}{\text{height}}\,dy = \int_1^2 2\pi(y-1)(y-1)\,dy = 2\pi \int_1^2 (y-1)^2 = 2\pi\left[\frac{(y-1)^3}{3}\right]_1^2 = \frac{2\pi}{3}$

29. (a) $V = \int_c^d 2\pi\binom{\text{shell}}{\text{radius}}\binom{\text{shell}}{\text{height}}\,dy = \int_0^1 2\pi y(y - y^3)\,dy$

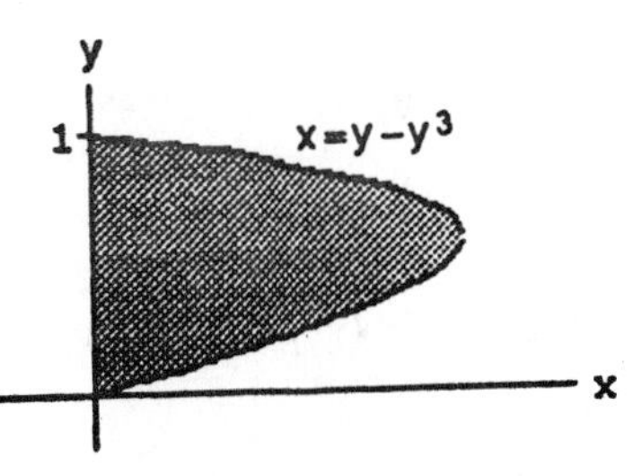

$= \int_0^1 2\pi(y^2 - y^4)\,dy = 2\pi\left[\frac{y^3}{3} - \frac{y^5}{5}\right]_0^1 = 2\pi\left(\frac{1}{3} - \frac{1}{5}\right)$

$= \frac{4\pi}{15}$

(b) $V = \int_c^d 2\pi\binom{\text{shell}}{\text{radius}}\binom{\text{shell}}{\text{height}}\,dy = \int_0^1 2\pi(1-y)(y - y^3)\,dy = 2\pi\int_0^1 (y - y^2 - y^3 + y^4)\,dy$

$= 2\pi\left[\frac{y^2}{2} - \frac{y^3}{3} - \frac{y^4}{4} + \frac{y^5}{5}\right]_0^1 = 2\pi\left(\frac{1}{2} - \frac{1}{3} - \frac{1}{4} + \frac{1}{5}\right) = \frac{2\pi}{60}(30 - 20 - 15 + 12) = \frac{7\pi}{30}$

31. (a) $V = \int_c^d 2\pi\binom{\text{shell}}{\text{radius}}\binom{\text{shell}}{\text{height}}\,dy = \int_0^2 2\pi y\left(\sqrt{8y} - y^2\right)dy$

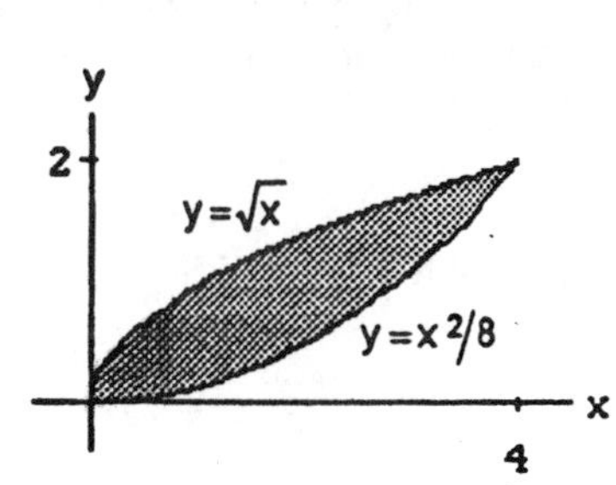

$= 2\pi\int_0^2 \left(2\sqrt{2}\,y^{3/2} - y^3\right)dy = 2\pi\left[\frac{4\sqrt{2}}{5}y^{5/2} - \frac{y^4}{4}\right]_0^2$

$= 2\pi\left(\frac{4\sqrt{2}\cdot\left(\sqrt{2}\right)^5}{5} - \frac{2^4}{4}\right) = 2\pi\left(\frac{4\cdot 2^3}{5} - \frac{4\cdot 4}{4}\right)$

$= 2\pi\cdot 4\left(\frac{8}{5} - 1\right) = \frac{8\pi}{5}(8 - 5) = \frac{24\pi}{5}$

(b) $V = \int_a^b 2\pi\binom{\text{shell}}{\text{radius}}\binom{\text{shell}}{\text{height}}\,dx = \int_0^4 2\pi x\left(\sqrt{x} - \frac{x^2}{8}\right)dx = 2\pi\int_0^4 \left(x^{3/2} - \frac{x^3}{8}\right)dx = 2\pi\left[\frac{2}{5}x^{5/2} - \frac{x^4}{32}\right]_0^4$

$= 2\pi\left(\frac{2\cdot 2^5}{5} - \frac{4^4}{32}\right) = 2\pi\left(\frac{2^6}{5} - \frac{2^8}{32}\right) = \frac{\pi\cdot 2^7}{160}(32 - 20) = \frac{\pi\cdot 2^9\cdot 3}{160} = \frac{\pi\cdot 2^4\cdot 3}{5} = \frac{48\pi}{5}$

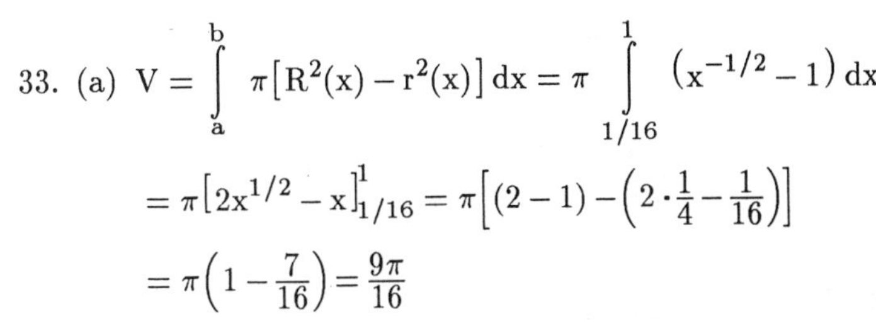

33. (a) $V = \int_a^b \pi\left[R^2(x) - r^2(x)\right]dx = \pi\int_{1/16}^1 \left(x^{-1/2} - 1\right)dx$

$= \pi\left[2x^{1/2} - x\right]_{1/16}^1 = \pi\left[(2-1) - \left(2\cdot\frac{1}{4} - \frac{1}{16}\right)\right]$

$= \pi\left(1 - \frac{7}{16}\right) = \frac{9\pi}{16}$

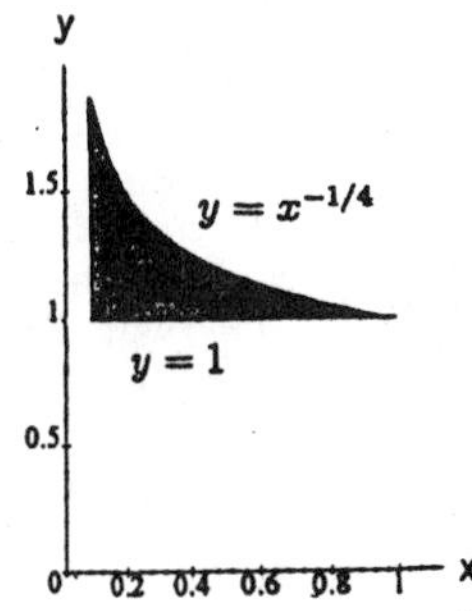

(b) $V = \int_a^b 2\pi\binom{\text{shell}}{\text{radius}}\binom{\text{shell}}{\text{height}}\,dy = \int_1^2 2\pi y\left(\frac{1}{y^4} - \frac{1}{16}\right)dy = 2\pi\int_1^2 \left(y^{-3} - \frac{y}{16}\right)dy = 2\pi\left[-\frac{1}{2}y^{-2} - \frac{y^2}{32}\right]_1^2$

$= 2\pi\left[\left(-\frac{1}{8} - \frac{1}{8}\right) - \left(-\frac{1}{2} - \frac{1}{32}\right)\right] = 2\pi\left(\frac{1}{4} + \frac{1}{32}\right) = \frac{2\pi}{32}(8 + 1) = \frac{9\pi}{16}$

35. (a) *Disc:* $V = V_1 - V_2$

$V_1 = \int_{a_1}^{b_1} \pi[R_1(x)]^2\,dx$ and $V_2 = \int_{a_2}^{b_2} \pi[R_2(x)]^2$ with $R_1(x) = \sqrt{\frac{x+2}{3}}$ and $R_2(x) = \sqrt{x}$,

$a_1 = -2$, $b_1 = 1$; $a_2 = 0$, $b_2 = 1 \Rightarrow$ two integrals are required

(b) *Washer*: $V = V_1 - V_2$

$$V_1 = \int_{a_1}^{b_1} \pi\left([R_1(x)]^2 - [r_1(x)]^2\right) dx \text{ with } R_1(x) = \sqrt{\frac{x+2}{3}} \text{ and } r_1(x) = 0;\ a_1 = -2 \text{ and } b_1 = 0;$$

$$V_2 = \int_{a_2}^{b_2} \pi\left([R_2(x)]^2 - [r_2(x)]^2\right) dx \text{ with } R_2(x) = \sqrt{\frac{x+2}{3}} \text{ and } r_2(x) = \sqrt{x};\ a_2 = 0 \text{ and } b_2 = 1$$

$\Rightarrow$ two integrals are required

(c) *Shell*: $V = \int_c^d 2\pi\binom{\text{shell}}{\text{radius}}\binom{\text{shell}}{\text{height}} dy = \int_c^d 2\pi y\binom{\text{shell}}{\text{height}} dy$ where shell height $= y^2 - (3y^2 - 2) = 2 - 2y^2$;

$c = 0$ and $d = 1$. Only *one* integral is required. It is, therefore preferable to use the *shell* method. However, whichever method you use, you will get $V = \pi$.

5.3 LENGTHS OF PLANE CURVES

1. $\frac{dy}{dx} = \frac{1}{3} \cdot \frac{3}{2}(x^2+2)^{1/2} \cdot 2x = \sqrt{(x^2+2)} \cdot x$

$$\Rightarrow L = \int_0^3 \sqrt{1 + (x^2+2)x^2}\, dx = \int_0^3 \sqrt{1 + 2x^2 + x^4}\, dx$$

$$= \int_0^3 \sqrt{(1+x^2)^2}\, dx = \int_0^3 (1+x^2)\, dx = \left[x + \frac{x^3}{3}\right]_0^3$$

$$= 3 + \frac{27}{3} = 12$$

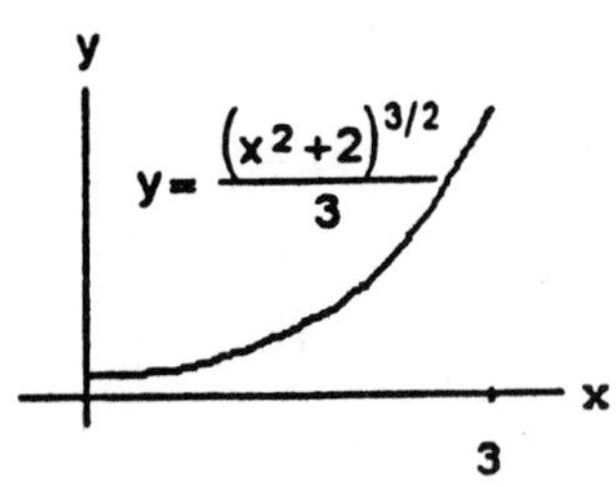

3. $\frac{dx}{dy} = y^2 - \frac{1}{4y^2} \Rightarrow \left(\frac{dx}{dy}\right)^2 = y^4 - \frac{1}{2} + \frac{1}{16y^4}$

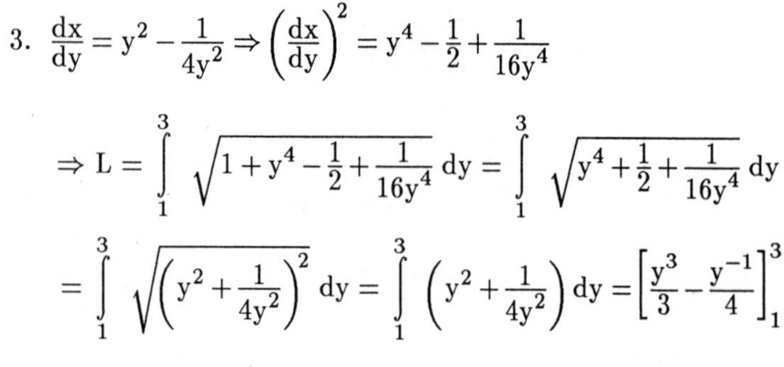

$$\Rightarrow L = \int_1^3 \sqrt{1 + y^4 - \frac{1}{2} + \frac{1}{16y^4}}\, dy = \int_1^3 \sqrt{y^4 + \frac{1}{2} + \frac{1}{16y^4}}\, dy$$

$$= \int_1^3 \sqrt{\left(y^2 + \frac{1}{4y^2}\right)^2}\, dy = \int_1^3 \left(y^2 + \frac{1}{4y^2}\right) dy = \left[\frac{y^3}{3} - \frac{y^{-1}}{4}\right]_1^3$$

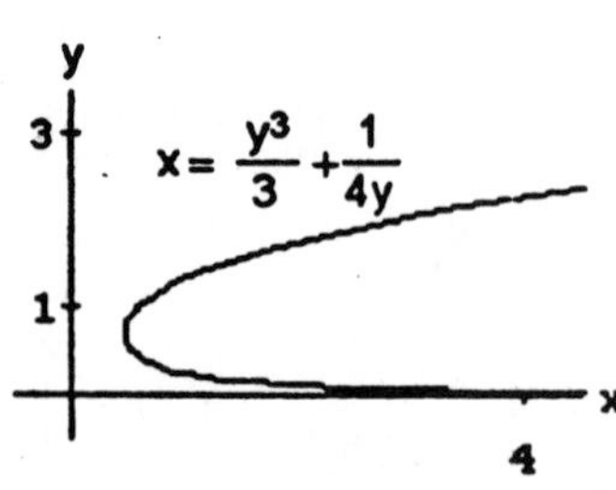

$$= \left(\frac{27}{3} - \frac{1}{12}\right) - \left(\frac{1}{3} - \frac{1}{4}\right) = 9 - \frac{1}{12} - \frac{1}{3} + \frac{1}{4} = 9 + \frac{(-1-4+3)}{12} = 9 + \frac{(-2)}{12} = \frac{53}{6}$$

5. $\frac{dx}{dy} = y^3 - \frac{1}{4y^3} \Rightarrow \left(\frac{dx}{dy}\right)^2 = y^6 - \frac{1}{2} + \frac{1}{16y^6}$

$$\Rightarrow L = \int_1^2 \sqrt{1 + y^6 - \frac{1}{2} + \frac{1}{16y^6}}\, dy = \int_1^2 \sqrt{y^6 + \frac{1}{2} + \frac{1}{16y^6}}\, dy$$

$$= \int_1^2 \sqrt{\left(y^3 + \frac{y^{-3}}{4}\right)^2}\, dy = \int_1^2 \left(y^3 + \frac{y^{-3}}{4}\right) dy$$

$$= \left[\frac{y^4}{4} - \frac{y^{-2}}{8}\right]_1^2 = \left(\frac{16}{4} - \frac{1}{(16)(2)}\right) - \left(\frac{1}{4} - \frac{1}{8}\right) = 4 - \frac{1}{32} - \frac{1}{4} + \frac{1}{8} = \frac{128 - 1 - 8 + 4}{32} = \frac{123}{32}$$

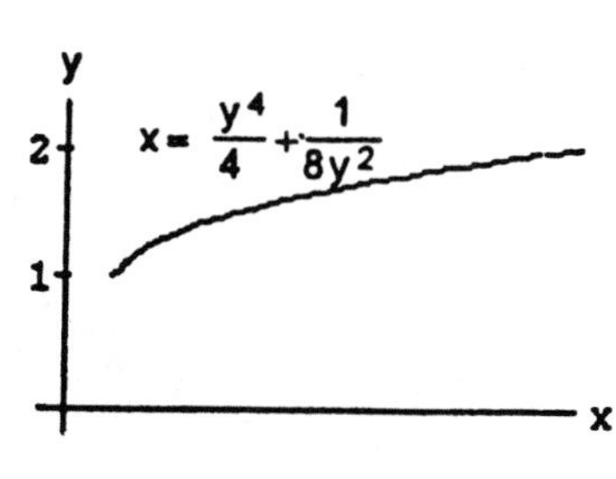

7. $\frac{dy}{dx} = x^{1/3} - \frac{1}{4}x^{-1/3} \Rightarrow \left(\frac{dy}{dx}\right)^2 = x^{2/3} - \frac{1}{2} + \frac{x^{-2/3}}{16}$

$$\Rightarrow L = \int_1^8 \sqrt{1 + x^{2/3} - \frac{1}{2} + \frac{x^{-2/3}}{16}}\, dx = \int_1^8 \sqrt{x^{2/3} + \frac{1}{2} + \frac{x^{-2/3}}{16}}\, dx$$

$$= \int_1^8 \sqrt{\left(x^{1/3} + \frac{1}{4}x^{-1/3}\right)^2}\, dx = \int_1^8 \left(x^{1/3} + \frac{1}{4}x^{-1/3}\right) dx$$

$$= \left[\frac{3}{4}x^{4/3} + \frac{3}{8}x^{2/3}\right]_1^8 = \frac{3}{8}\left[2x^{4/3} + x^{2/3}\right]_1^8 = \frac{3}{8}\left[\left(2 \cdot 2^4 + 2^2\right) - (2+1)\right] = \frac{3}{8}(32 + 4 - 3) = \frac{99}{8}$$

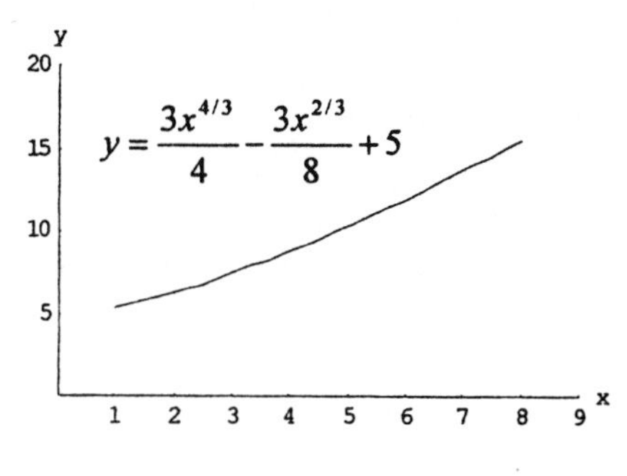

9. $\frac{dx}{dy} = \sqrt{\sec^4 y - 1} \Rightarrow \left(\frac{dx}{dy}\right)^2 = \sec^4 y - 1$

$$\Rightarrow L = \int_{-\pi/4}^{\pi/4} \sqrt{1 + \left(\sec^4 y - 1\right)}\, dy = \int_{-\pi/4}^{\pi/4} \sec^2 y\, dy$$

$$= \left[\tan y\right]_{-\pi/4}^{\pi/4} = 1 - (-1) = 2$$

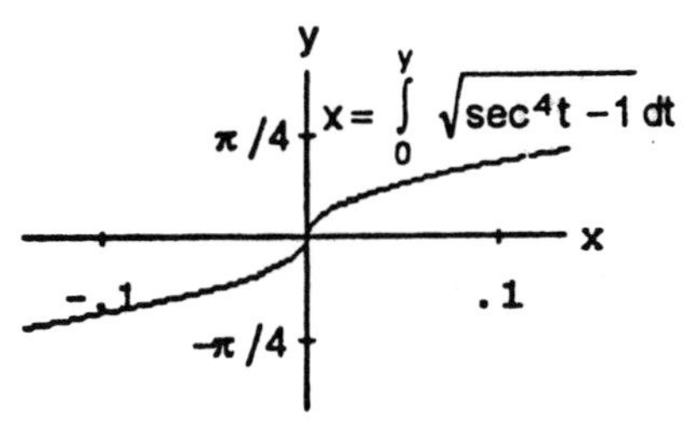

11. $\frac{dx}{dt} = -a \sin t$ and $\frac{dy}{dt} = a \cos t \Rightarrow \sqrt{\left(\frac{dx}{dt}\right)^2 + \left(\frac{dy}{dt}\right)^2} = \sqrt{(-a \sin t)^2 + (a \cos t)^2} = \sqrt{a^2\left(\sin^2 t + \cos^2 t\right)} = |a|$

$$\Rightarrow \text{Length} = \int_0^{2\pi} |a|\, dt = |a| \int_0^{2\pi} dt = 2\pi |a|.$$

13. $\frac{dx}{dt} = 3t^2$ and $\frac{dy}{dt} = 3t \Rightarrow \sqrt{\left(\frac{dx}{dt}\right)^2 + \left(\frac{dy}{dt}\right)^2} = \sqrt{\left(3t^2\right)^2 + (3t)^2} = \sqrt{9t^4 + 9t^2} = 3t\sqrt{t^2 + 1}$ (since $t \geq 0$ on $[0, \sqrt{3}]$)

$$\Rightarrow \text{Length} = \int_0^{\sqrt{3}} 3t\sqrt{t^2 + 1}\, dt;\ \left[u = t^2 + 1 \Rightarrow \frac{3}{2}\, du = 3t\, dt;\ t = 0 \Rightarrow u = 1,\ t = \sqrt{3} \Rightarrow u = 4\right]$$

$$\Rightarrow \int_1^4 \frac{3}{2}u^{1/2}\,du = \left[u^{3/2}\right]_1^4 = (8-1) = 7$$

15. $\frac{dx}{dt} = (2t+3)^{1/2}$ and $\frac{dy}{dt} = 1+t \Rightarrow \sqrt{\left(\frac{dx}{dt}\right)^2 + \left(\frac{dy}{dt}\right)^2} = \sqrt{(2t+3)+(1+t)^2} = \sqrt{t^2+4t+4} = |t+2| = t+2$ since $0 \le t \le 3 \Rightarrow \text{Length} = \int_0^3 (t+2)\,dt = \left[\frac{t^2}{2} + 2t\right]_0^3 = \frac{21}{2}$

17. $\sqrt{2}\,a = \int_0^a \sqrt{1+\left(\frac{dy}{dx}\right)^2}\,dx,\ a \ge 0 \Rightarrow \sqrt{2} = \sqrt{1+\left(\frac{dy}{dx}\right)^2} \Rightarrow \frac{dy}{dx} = \pm 1 \Rightarrow y = f(x) = \pm x + C$ where C is any real number.

19. (a) $\left(\frac{dy}{dx}\right)^2$ corresponds to $\frac{1}{4x}$ here, so take $\frac{dy}{dt}$ as $\frac{1}{2\sqrt{x}}$. Then $y = \sqrt{x} + C$, and since $(1,1)$ lies on the curve, $C = 0$. So $y = \sqrt{x}$.

(b) Only one. We know the derivative of the function and the value of the function at one value of x.

21. (a) $\frac{dy}{dx} = 2x \Rightarrow \left(\frac{dy}{dx}\right)^2 = 4x^2 \Rightarrow L = \int_{-1}^{2} \sqrt{1+\left(\frac{dy}{dx}\right)^2}\,dx$

$$= \int_{-1}^{2} \sqrt{1+4x^2}\,dx$$

(b)

(c) $L \approx 6.13$

23. (a) $\frac{dx}{dy} = \cos y \Rightarrow \left(\frac{dx}{dy}\right)^2 = \cos^2 y$

$$\Rightarrow L = \int_0^{\pi} \sqrt{1+\cos^2 y}\,dy$$

(c) $L \approx 3.82$

(b)

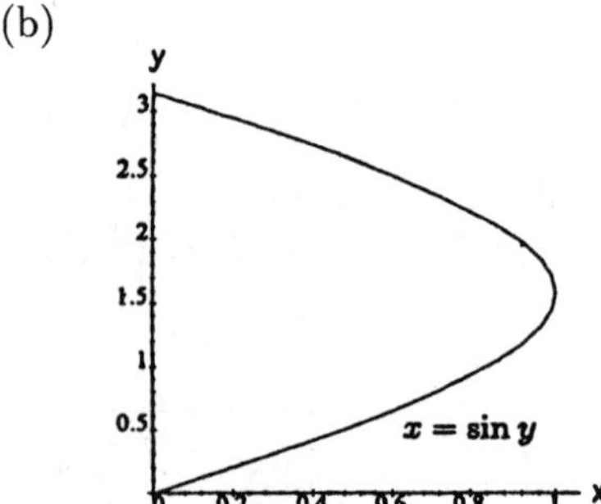

25. (a) $2y + 2 = 2\,\frac{dx}{dy} \Rightarrow \left(\frac{dx}{dy}\right)^2 = (y+1)^2$

$$\Rightarrow L = \int_{-1}^{3} \sqrt{1 + (y+1)^2}\,dy$$

(b)

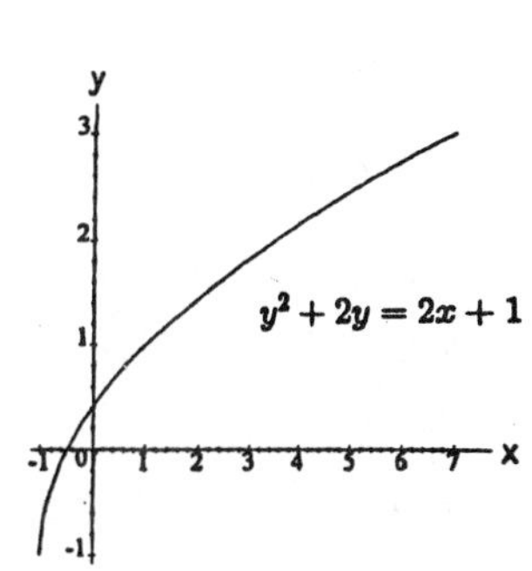

(c) $L \approx 9.29$

27. (a) $\frac{dy}{dx} = \tan x \Rightarrow \left(\frac{dy}{dx}\right)^2 = \tan^2 x$

$$\Rightarrow L = \int_0^{\pi/6} \sqrt{1 + \tan^2 x}\,dx = \int_0^{\pi/6} \sqrt{\frac{\sin^2 x + \cos^2 x}{\cos^2 x}}\,dx$$

$$= \int_0^{\pi/6} \frac{dx}{\cos x} = \int_0^{\pi/6} \sec x\,dx$$

(b)

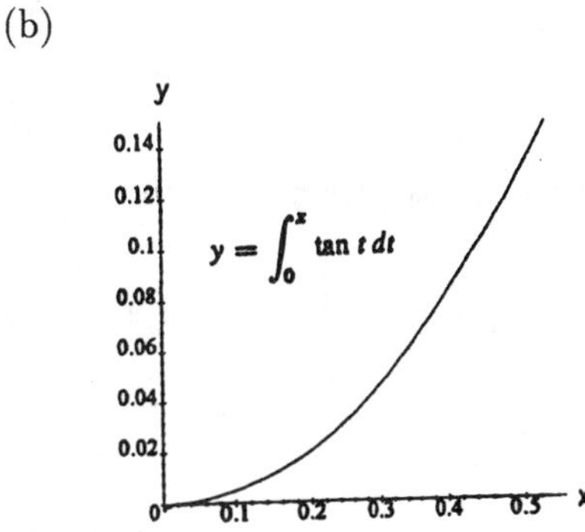

(c) $L \approx 0.55$

29. The length of the curve $y = \sin\left(\frac{3\pi}{20}x\right)$ from 0 to 20 is: $L = \int_0^{20} \sqrt{1 + \left(\frac{dy}{dx}\right)^2}\,dx$; $\frac{dy}{dx} = \frac{3\pi}{20}\cos\left(\frac{3\pi}{20}x\right) \Rightarrow \left(\frac{dy}{dx}\right)^2$ $= \frac{9\pi^2}{400}\cos^2\left(\frac{3\pi}{20}x\right) \Rightarrow L = \int_0^{20} \sqrt{1 + \frac{9\pi^2}{400}\cos^2\left(\frac{3\pi}{20}x\right)}\,dx$. Using numerical integration we find $L \approx 21.07$ in

5.4 SPRINGS, PUMPING AND LIFTING

1. The force required to lift the water is equal to the water's weight, which varies steadily from 40 lb to 0 lb over the 20-ft lift. When the bucket is x ft off the ground, the water weighs: $F(x) = 40\left(\frac{20-x}{20}\right) = 40\left(1 - \frac{x}{20}\right)$ $= 40 - 2x$ lb. The work done is: $W = \int_a^b F(x)\,dx = \int_0^{20} (40 - 2x)\,dx = \left[40x - x^2\right]_0^{20} = (40)(20) - 20^2$ $= 800 - 400 = 400$ ft · lb

3. The force required to haul up the rope is equal to the rope's weight, which varies steadily and is proportional to x, the length of the rope still hanging: $F(x) = 0.624x$. The work done is: $W = \int_0^{50} F(x)\,dx = \int_0^{50} 0.624x\,dx$ $= 0.624\left[\frac{x^2}{2}\right]_0^{50} = 780$ J

5. The force required to lift the cable is equal to the weight of the cable paid out: $F(x) = (4.5)(180 - x)$ where x is the position of the car off the first floor. The work done is: $W = \int_0^{180} F(x)\,dx = 4.5 \int_0^{180} (180 - x)\,dx$

$= 4.5\left[180x - \frac{x^2}{2}\right]_0^{180} = 4.5\left(180^2 - \frac{180^2}{2}\right) = \frac{4.5 \cdot 180^2}{2} = 72{,}900$ ft · lb

7. The force against the piston is $F = pA$. If $V = Ax$, where x is the height of the cylinder, then $dV = A\,dx$

$\Rightarrow \text{Work} = \int F\,dx = \int pA\,dx = \int_{(p_1,V_1)}^{(p_2,V_2)} p\,dV.$

9. The force required to stretch the spring from its natural length of 2 m to a length of 5 m is $F(x) = kx$. The work done by F is $W = \int_0^3 F(x)\,dx = k\int_0^3 x\,dx = \frac{k}{2}\left[x^2\right]_0^3 = \frac{9k}{2}$. This work is equal to 1800 J $\Rightarrow \frac{9}{2}k = 1800$

$\Rightarrow k = 400$ N/m

11. We find the force constant from Hooke's law: $F = kx$. A force of 2 N stretches the spring to 0.02 m $\Rightarrow 2 = k \cdot (0.02) \Rightarrow k = 100\,\frac{N}{m}$. The force of 4 N will stretch the rubber band y m, where $F = ky \Rightarrow y = \frac{F}{k}$

$\Rightarrow y = \frac{4N}{100\,\frac{N}{m}} \Rightarrow y = 0.04$ m $= 4$ cm. The work done to stretch the rubber band 0.04 m is $W = \int_0^{0.04} kx\,dx$

$= 100\int_0^{0.04} x\,dx = 100\left[\frac{x^2}{2}\right]_0^{0.04} = \frac{(100)(0.04)^2}{2} = 0.08$ J

13. (a) We find the spring's constant from Hooke's law: $F = kx \Rightarrow k = \frac{F}{x} = \frac{21{,}714}{8-5} = \frac{21{,}714}{3} \Rightarrow k = 7238\,\frac{\text{lb}}{\text{in}}$

(b) The work done to compress the assembly the first half inch is $W = \int_0^{0.5} kx\,dx = 7238\int_0^{0.5} x\,dx$

$= 7238\left[\frac{x^2}{2}\right]_0^{0.5} = (7238)\frac{(0.5)^2}{2} = \frac{(7238)(0.25)}{2} \approx 905$ in · lb. The work done to compress the assembly the second half inch is: $W = \int_{0.5}^{1.0} kx\,dx = 7238\int_{0.5}^{1.0} x\,dx = 7238\left[\frac{x^2}{2}\right]_{0.5}^{1.0} = \frac{7238}{2}\left[1 - (0.5)^2\right] = \frac{(7238)(0.75)}{2}$

≈ 2714 in · lb

15. We will use the coordinate system given.

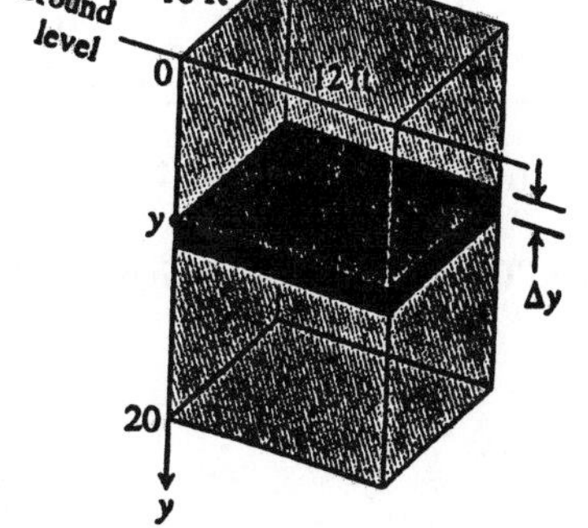

(a) The typical slab between the planes at y and $y + \Delta y$ has a volume of $\Delta V = (10)(12)\,\Delta y = 120\,\Delta y$ ft^3. The force F required to lift the slab is equal to its weight: $F = 62.4\,\Delta V = 62.4 \cdot 120\,\Delta y$ lb. The distance through which F must act is about y ft, so the work done lifting the slab is about $\Delta W = \text{force} \times \text{distance}$ $= 62.4 \cdot 120 \cdot y \cdot \Delta y$ ft·lb. The work it takes to lift all the water is approximately $W \approx \sum_0^{20} \Delta W$ $= \sum_0^{20} 62.4 \cdot 120y \cdot \Delta y$ ft·lb. This is a Riemann sum for the function $62.4 \cdot 120y$ over the interval $0 \le y \le 20$.

The work of pumping the tank empty is the limit of these sums: $W = \int_0^{20} 62.4 \cdot 120y\,dy$

$$= (62.4)(120)\left[\frac{y^2}{2}\right]_0^{20} = (62.4)(120)\left(\frac{400}{2}\right) = (62.4)(120)(200) = 1{,}497{,}600 \text{ ft}\cdot\text{lb}$$

(b) The time t it takes to empty the full tank with $\left(\frac{5}{11}\right)$-hp motor is $t = \dfrac{W}{250\,\frac{\text{ft}\cdot\text{lb}}{\text{sec}}} = \dfrac{1{,}497{,}600 \text{ ft}\cdot\text{lb}}{250\,\frac{\text{ft}\cdot\text{lb}}{\text{sec}}}$

$= 5990.4$ sec $= 1.664$ hr $\Rightarrow t \approx 1$ hr and 40 min

(c) Following all the steps of part (a), we find that the work it takes to lower the water level 10 ft is

$W = \int_0^{10} 62.4 \cdot 120y\,dy = (62.4)(120)\left[\frac{y^2}{2}\right]_0^{10} = (62.4)(120)\left(\frac{100}{2}\right) = 374{,}400$ ft·lb and the time is

$t = \dfrac{W}{250\,\frac{\text{ft}\cdot\text{lb}}{\text{sec}}} = 1497.6$ sec $= 0.416$ hr ≈ 25 min

(d) In a location where water weighs $62.26\,\frac{\text{lb}}{\text{ft}^3}$:

a) $W = (62.26)(24{,}000) = 1{,}494{,}240$ ft·lb.

b) $t = \dfrac{1{,}494{,}240}{250} = 5976.96$ sec ≈ 1.660 hr $\Rightarrow t \approx 1$ hr and 40 min

In a location where water weighs $62.59\,\frac{\text{lb}}{\text{ft}^3}$

a) $W = (62.59)(24{,}000) = 1{,}502{,}160$ ft·lb

b) $t = \dfrac{1{,}502{,}160}{250} = 6008.64$ sec ≈ 1.669 hr $\Rightarrow t \approx 1$ hr and 40.1 min

17. Using exactly the same procedure as done in Example 6 we change only the distance through which F must act: distance $\approx (10 - y)$ m. Then $\Delta W = 245{,}000\pi(10 - y)\,\Delta y$ J $\Rightarrow W \approx \sum_0^{10} \Delta W = \sum_0^{10} 245{,}000\pi(10 - y)\,\Delta y$

$$\Rightarrow W = \int_0^{10} 245{,}000\pi(10 - y)\,dy = 245{,}000\pi \int_0^{10} (10 - y)\,dy = 245{,}000\pi\left[10y - \frac{y^2}{2}\right]_0^{10} = 245{,}000\pi\left(100 - \frac{100}{2}\right)$$

$\approx (245{,}000\pi)(50) \approx 38{,}484{,}510$ J

19. The typical slab between the planes at y and and $y + \Delta y$ has a volume of $\Delta V = \pi(\text{radius})^2(\text{thickness})$ $= \pi\left(\frac{20}{2}\right)^2 \Delta y = \pi \cdot 100\ \Delta y\ \text{ft}^3$. The force F required to lift the slab is equal to its weight: $F = 51.2\,\Delta V = 51.2 \cdot 100\pi\,\Delta y$ lb $\Rightarrow F = 5120\pi\,\Delta y$ lb. The distance through which F must act is about $(30 - y)$ ft. The work it takes to lift all the kerosene is approximately $W \approx \sum_0^{30} \Delta W$ $= \sum_0^{30} 5120\pi(30 - y)\,\Delta y$ ft · lb which is a Riemann sum. The work to pump the tank dry is the limit of these sums: $W = \int_0^{30} 5120\pi(30 - y)\,dy = 5120\pi\left[30y - \frac{y^2}{2}\right]_0^{30} = 5120\pi\left(\frac{900}{2}\right) = (5120)(450\pi)$ $\approx 7{,}238{,}229.47$ ft · lb

21. (a) Follow all the steps of Example 7 but make the substitution of $64.5\ \frac{\text{lb}}{\text{ft}^3}$ for $57\ \frac{\text{lb}}{\text{ft}^3}$. Then,

$$W = \int_0^8 \frac{64.5\pi}{4}(10 - y)y^2\,dy = \frac{64.5\pi}{4}\left[\frac{10y^3}{3} - \frac{y^4}{4}\right]_0^8 = \frac{64.5\pi}{4}\left(\frac{10 \cdot 8^3}{3} - \frac{8^4}{4}\right) = \left(\frac{64.5\pi}{4}\right)(8^3)\left(\frac{10}{3} - 2\right)$$

$$= \frac{64.5\pi \cdot 8^3}{3} = 21.5\pi \cdot 8^3 \approx 34{,}582.65 \text{ ft} \cdot \text{lb}$$

(b) Exactly as done in Example 7 but change the distance through which F acts to distance $\approx (13 - y)$ ft.

$$\text{Then } W = \int_0^8 \frac{57\pi}{4}(13 - y)y^2\,dy = \frac{57\pi}{4}\left[\frac{13y^3}{3} - \frac{y^4}{4}\right]_0^8 = \frac{57\pi}{4}\left(\frac{13 \cdot 8^3}{3} - \frac{8^4}{4}\right) = \left(\frac{57\pi}{4}\right)(8^3)\left(\frac{13}{3} - 2\right) = \frac{57\pi \cdot 8^3 \cdot 7}{3 \cdot 4}$$

$$= (19\pi)(8^2)(7)(2) \approx 53.482.5 \text{ ft} \cdot \text{lb}$$

23. The typical slab between the planes at y and $y + \Delta y$ has a volume of about $\Delta V = \pi(\text{radius})^2(\text{thickness})$ $= \pi\left(\sqrt{25 - y^2}\right)^2 \Delta y\ \text{m}^3$. The force $F(y)$ required to lift this slab is equal to its weight: $F(y) = 9800 \cdot \Delta V$ $= 9800\pi\left(\sqrt{25 - y^2}\right)^2 \Delta y = 9800\pi(25 - y^2)\Delta y$ N. The distance through which $F(y)$ must act to lift the slab to the level of 4 m above the top of the reservoir is about $(4 - y)$ m, so the work done is approximately $\Delta W \approx 9800\pi(25 - y^2)(4 - y)\,\Delta y$ N · m. The work done lifting all the slabs from $y = -5$ m to $y = 0$ m is approximately $W \approx \sum_{-5}^{0} 9800\pi(25 - y^2)(4 - y)\,\Delta y$ N · m. Taking the limit of these Riemann sums, we get

$$W = \int_{-5}^{0} 9800\pi(25 - y^2)(4 - y)\,dy = 9800\pi \int_{-5}^{0} (100 - 25y - 4y^2 + y^3)\,dy = 9800\pi\left[100y - \frac{25}{2}y^2 - \frac{4}{3}y^3 + \frac{y^4}{4}\right]_{-5}^{0}$$

$$= -9800\pi\left(-500 - \frac{25 \cdot 25}{2} + \frac{4}{3} \cdot 125 + \frac{625}{4}\right) \approx 15{,}073{,}099.75 \text{ J}$$

25. $F = m\,\frac{dv}{dt} = mv\,\frac{dv}{dx}$ by the chain rule $\Rightarrow W = \int_{x_1}^{x_2} mv\,\frac{dv}{dx}\,dx = m\int_{x_1}^{x^2}\left(v\,\frac{dv}{dx}\right)dx = m\left[\frac{1}{2}v^2(x)\right]_{x_1}^{x_2}$ $= \frac{1}{2}m\left[v^2(x_2) - v^2(x_1)\right] = \frac{1}{2}mv_2^2 - \frac{1}{2}mv_1^2$, as claimed.

27. $90 \text{ mph} = \frac{90 \text{ mi}}{1 \text{ hr}} \cdot \frac{1 \text{ hr}}{60 \text{ min}} \cdot \frac{1 \text{ min}}{60 \text{ sec}} \cdot \frac{5280 \text{ ft}}{1 \text{ mi}} = 132 \text{ ft/sec}$; $m = \frac{0.3125 \text{ lb}}{32 \text{ ft/sec}^2} = \frac{0.3125}{32}$ slugs;

$$W = \left(\tfrac{1}{2}\right)\left(\frac{0.3125 \text{ lb}}{32 \text{ ft/sec}^2}\right)(132 \text{ ft/sec})^2 \approx 85.1 \text{ ft} \cdot \text{lb}$$

29. weight $= 2 \text{ oz} = \frac{1}{8} \text{ lb} \Rightarrow m = \frac{\frac{1}{8}}{32}$ slugs $= \frac{1}{256}$ slugs; $124 \text{ mph} = \frac{(124)(5280)}{(60)(60)} \approx 181.87$ ft/sec;

$$W = \left(\tfrac{1}{2}\right)\left(\tfrac{1}{256} \text{ slugs}\right)(181.87 \text{ ft/sec})^2 \approx 64.6 \text{ ft} \cdot \text{lb}$$

31. weight $= 6.5 \text{ oz} = \frac{6.5}{16} \text{ lb} \Rightarrow m = \frac{6.5}{(16)(32)}$ slugs; $W = \left(\tfrac{1}{2}\right)\left(\frac{6.5}{(16)(32)} \text{ slugs}\right)(132 \text{ ft/sec})^2 \approx 110.6 \text{ ft} \cdot \text{lb}$

33. (a) From the diagram,

$$r(y) = 60 - x = 60 - \sqrt{50^2 - (y - 325)^2}$$

for $325 \le y \le 375$ ft.

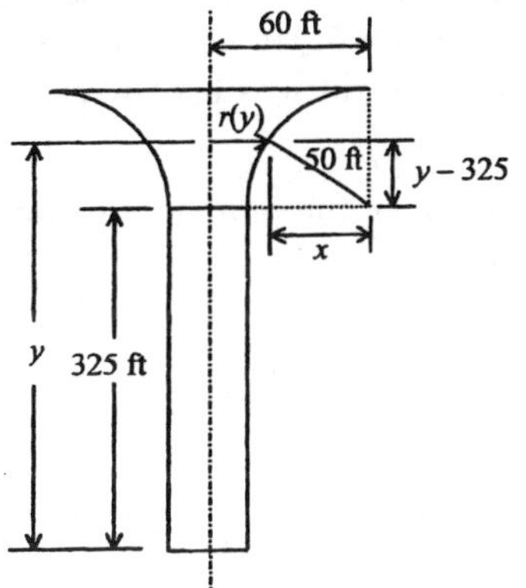

(b) The volume of a horizontal slice of the funnel is $\Delta V \approx \pi\,[r(y)]^2\,dy$

$$= \pi\left[60 - \sqrt{2500 - (y - 325)}\right]^2 \Delta y.$$

(c) The work required to lift the single slice of water is $\Delta W \approx 62.4 \Delta V(375 - y)$

$$= 62.4(375 - y)\pi\left[60 - \sqrt{2500 - (y - 325)}\right]^2 \Delta y.$$

The total work to pump out the funnel is

$$W = \int_{325}^{375} 62.4\pi(375 - y)\left[60 - \sqrt{2500 - (y - 325)}\right]^2 dy = 6.3358 \cdot 10^7 \text{ ft} \cdot \text{lb}.$$

35. We imagine the milkshake divided into thin slabs by planes perpendicular to the y-axis at the points of a partition of the interval $[0, 7]$. The typical slab between the planes at y and $y + \Delta y$ has a volume of about $\Delta V = \pi(\text{radius})^2(\text{thickness}) = \pi\left(\frac{y + 17.5}{14}\right)^2 \Delta y \text{ in}^3$. The force $F(y)$ required to lift this slab is equal to its weight: $F(y) = \frac{4}{9}\,\Delta V = \frac{4\pi}{9}\left(\frac{y + 17.5}{14}\right)^2 \Delta y$ oz. The distance through which $F(y)$ must act to lift this slab to the level of 1 inch above the top is about $(8 - y)$ in. The work done lifting the slab is about $\Delta W = \left(\frac{4\pi}{9}\right)\frac{(y + 17.5)^2}{14^2}(8 - y)\,\Delta y$ in·oz. The work done lifting all the slabs from $y = 0$ to $y = 7$ is approximately $W = \sum_0^7 \frac{4\pi}{9 \cdot 14^2}(y + 17.5)^2(8 - y)\,\Delta y$ in·oz which is a Riemann sum. The work is the limit of these sums as the norm of the partition goes to zero: $W = \int_0^7 \frac{4\pi}{9 \cdot 14^2}(y + 17.5)^2(8 - y)\,dy$

$$= \frac{4\pi}{9 \cdot 14^2}\int_0^7 \left(2450 - 26.25y - 27y^2 - y^3\right) dy = \frac{4\pi}{9 \cdot 14^2}\left[-\frac{y^4}{4} - 9y^3 - \frac{26.25}{2}\,y^2 + 2450y\right]_0^7$$

$$= \frac{4\pi}{9 \cdot 14^2}\left[-\frac{7^4}{4} - 9 \cdot 7^3 - \frac{26.25}{2} \cdot 7^2 + 2450 \cdot 7\right] \approx 91.32 \text{ in} \cdot \text{oz}$$

37. Work $= \int_{6,370,000}^{35,780,000} \frac{1000\,\text{MG}}{r^2}\,dr = 1000\,\text{MG} \int_{6,370,000}^{35,780,000} \frac{dr}{r^2} = 1000\,\text{MG}\left[-\frac{1}{r}\right]_{6,370,000}^{35,780,000}$

$$= (1000)\left(5.975 \cdot 10^{24}\right)\left(6.672 \cdot 10^{-11}\right)\left(\frac{1}{6{,}370{,}000} - \frac{1}{35{,}780{,}000}\right) \approx 5.144 \times 10^{10} \text{ J}$$

5.5 FLUID FORCES

1. To find the width of the plate at a typical depth y, we first find an equation for the line of the plate's right-hand edge: $y = x - 5$. If we let x denote the width of the right-hand half of the triangle at depth y, then $x = 5 + y$ and the total width is $L(y) = 2x = 2(5 + y)$. The depth of the strip is $(-y)$. The force exerted by the water against one side of the plate is therefore $F = \int_{-5}^{-2} w(-y) \cdot L(y)\,dy = \int_{-5}^{-2} 62.4 \cdot (-y) \cdot 2(5 + y)\,dy$

$$= 124.8 \int_{-5}^{-2} \left(-5y - y^2\right) dy = 124.8\left[-\frac{5}{2}y^2 - \frac{1}{3}y^3\right]_{-5}^{-2} = 124.8\left[\left(-\frac{5}{2} \cdot 4 + \frac{1}{3} \cdot 8\right) - \left(-\frac{5}{2} \cdot 25 + \frac{1}{3} \cdot 125\right)\right]$$

$$= (124.8)\left(\frac{105}{2} - \frac{117}{3}\right) = (124.8)\left(\frac{315 - 234}{6}\right) = 1684.8 \text{ lb}$$

3. Using the coordinate system of Exercise 2, we find the equation for the line of the plate's right-hand edge is $y = x - 3 \Rightarrow x = y + 3$. Thus the total width is $L(y) = 2x = 2(y + 3)$. The depth of the strip changes to $(4 - y)$

$$\Rightarrow F = \int_{-3}^{0} w(4 - y)L(y)\,dy = \int_{-3}^{0} 62.4 \cdot (4 - y) \cdot 2(y + 3)\,dy = 124.8 \int_{-3}^{0} \left(12 + y - y^2\right) dy$$

$$= 124.8\left[12y + \frac{y^2}{2} - \frac{y^3}{3}\right]_{-3}^{0} = (-124.8)\left(-36 + \frac{9}{2} + 9\right) = (-124.8)\left(-\frac{45}{2}\right) = 2808 \text{ lb}$$

5. Using the coordinate system of Exercise 2, we find the equation for the line of the plate's right-hand edge to be $y = 2x - 4 \Rightarrow x = \frac{y + 4}{2}$ and $L(y) = 2x = y + 4$. The depth of the strip is $(1 - y)$.

(a) $F = \int_{-4}^{0} w(1 - y)L(y)\,dy = \int_{-4}^{0} 62.4 \cdot (1 - y)(y + 4)\,dy = 62.4 \int_{-4}^{0} \left(4 - 3y - y^2\right) dy = 62.4\left[4y - \frac{3y^2}{2} - \frac{y^3}{3}\right]_{-4}^{0}$

$$= (-62.4)\left[(-4)(4) - \frac{(3)(16)}{2} + \frac{64}{3}\right] = (-62.4)\left(-16 - 24 + \frac{64}{3}\right) = \frac{(-62.4)(-120 + 64)}{3} = 1164.8 \text{ lb}$$

(b) $F = (-64.0)\left[(-4)(4) - \frac{(3)(16)}{2} + \frac{64}{3}\right] = \frac{(-64.0)(-120 + 64)}{3} \approx 1194.7 \text{ lb}$

7. Using the coordinate system given in the accompanying figure, we see that the total width is L(y) = 63 and the depth of the strip is (33.5 − y) ⇒ $F = \int_0^{33} w(33.5-y)L(y)\,dy$

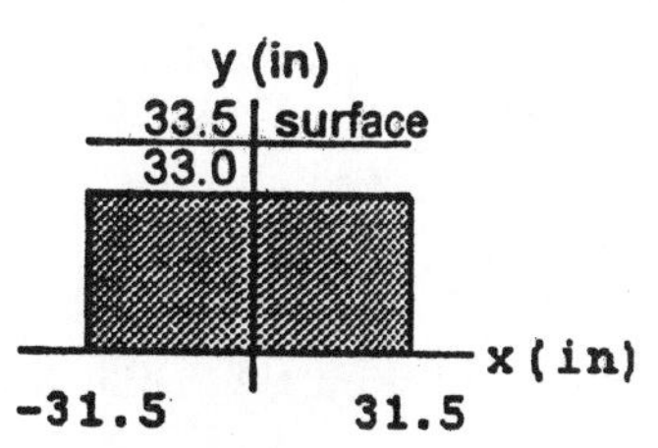

$= \int_0^{33} \frac{64}{12^3}\cdot(33.5-y)\cdot 63\,dy = \left(\frac{64}{12^3}\right)(63)\int_0^{33}(33.5-y)\,dy$

$= \left(\frac{64}{12^3}\right)(63)\left[33.5y - \frac{y^2}{2}\right]_0^{33} = \left(\frac{64\cdot 63}{12^3}\right)\left[(33.5)(33) - \frac{33^2}{2}\right] = \frac{(64)(63)(33)(67-33)}{(2)\left(12^3\right)} = 1309 \text{ lb}$

9. Using the coordinate system given in the accompanying figure, we see that the right-hand edge is $x = \sqrt{1-y^2}$ so the total width is $L(y) = 2x = 2\sqrt{1-y^2}$ and the depth of the strip is (−y). The force exerted by the water is therefore $F = \int_{-1}^{0} w\cdot(-y)\cdot 2\sqrt{1-y^2}\,dy$

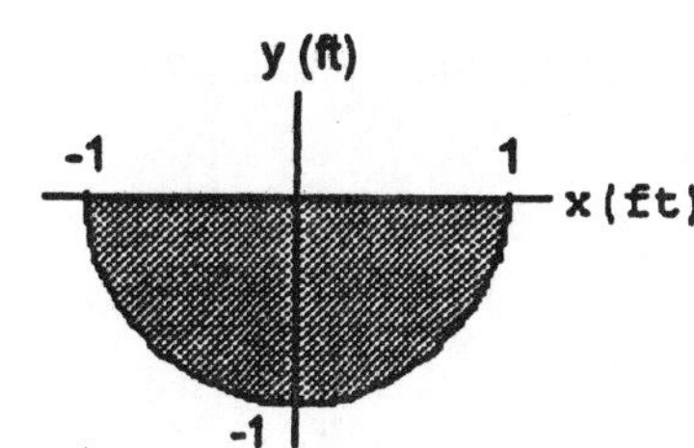

$= 62.4\int_{-1}^{0}\sqrt{1-y^2}\,d\left(1-y^2\right) = 62.4\left[\frac{2}{3}\left(1-y^2\right)^{3/2}\right]_{-1}^{0} = (62.4)\left(\frac{2}{3}\right)(1-0) = 41.6 \text{ lb}$

11. The coordinate system is given in the text. The right-hand edge is $x = \sqrt{y}$ and the total width is $L(y) = 2x = 2\sqrt{y}$.

(a) The depth of the strip is (2 − y) so the force exerted by the liquid on the gate is $F = \int_0^1 w(2-y)L(y)\,dy$

$= \int_0^1 50(2-y)\cdot 2\sqrt{y}\,dy = 100\int_0^1 (2-y)\sqrt{y}\,dy = 100\int_0^1\left(2y^{1/2} - y^{3/2}\right)dy = 100\left[\frac{4}{3}y^{3/2} - \frac{2}{5}y^{5/2}\right]_0^1$

$= 100\left(\frac{4}{3} - \frac{2}{5}\right) = \left(\frac{100}{15}\right)(20-6) = 93.33 \text{ lb}$

(b) Suppose that H is the maximum height to which the container can be filled without exceeding its design limitation. The depth of a typical strip is (H − y) and the force is $F = \int_0^1 w(H-y)L(y)\,dy = F_{max}$, where $F_{max} = 160$ lb. Therefore, $F_{max} = w\int_0^1 (H-y)\cdot 2\sqrt{y}\,dy = 100\int_0^1 (H-y)\sqrt{y}\,dy$

$= 100\int_0^1\left(Hy^{1/2} - y^{3/2}\right)dy = 100\left[\frac{2}{3}Hy^{3/2} - \frac{2}{5}y^{5/2}\right]_0^1 = 100\left(\frac{2H}{3} - \frac{2}{5}\right) = \left(\frac{100}{15}\right)(10H-6)$. When $F_{max} = 160$ lb we have $160 = \left(\frac{100}{15}\right)(10H-6) \Rightarrow 10H - 6 = 24 \Rightarrow H = 3$ ft

13. (a) The equation of the ellipse for the penstock gate is $\left(\frac{x}{8}\right)^2 + \left(\frac{y}{14}\right)^2 = 1$ or $49x^2 + 16y^2 = 3136 \Rightarrow x = \frac{\sqrt{3136 - 16y^2}}{7}$, where y is measured from the center of the ellipse.

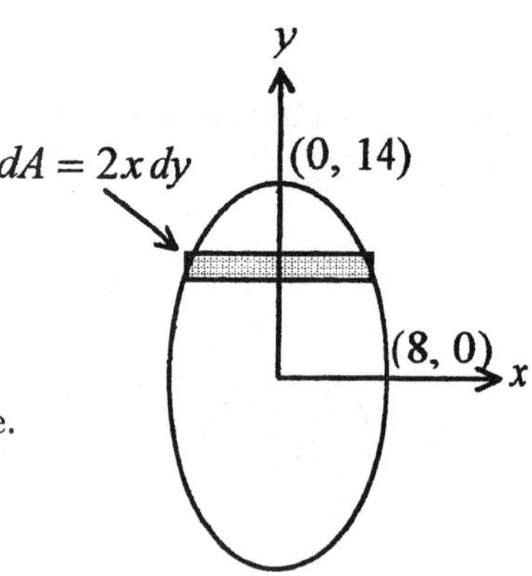

(b) $L(y) = 2x = \frac{2}{7}\sqrt{3136 - 16y^2}$

(c) $dF = 62.4[389 - (y + 115)](2x\ dy) = 124.8(274 - y)\ \frac{\sqrt{3136 - 16y^2}}{7}\ dy$. Therefore,

$$F = \int_{-14}^{14} 17.829(274 - y)\sqrt{3136 - 16y^2}\ dy = 6.0159 \cdot 10^6 \text{ lb} = 3008 \text{ tons.}$$

15. (a) The pressure at level y is $p(y) = w \cdot y \Rightarrow$ the average pressure is $\overline{p} = \frac{1}{b}\int_0^b p(y)\ dy = \frac{1}{b}\int_0^b w \cdot y\ dy = \frac{1}{b}w\left[\frac{y^2}{2}\right]_0^b$ $= \left(\frac{w}{b}\right)\left(\frac{b^2}{2}\right) = \frac{wb}{2}$. This is the pressure at level $\frac{b}{2}$, which is the pressure at the middle of the plate.

(b) The force exerted by the fluid is $F = \int_0^b w(\text{depth})(\text{length})\ dy = \int_0^b w \cdot y \cdot a\ dy$ $= (w \cdot a)\int_0^b y\ dy = (w \cdot a)\left[\frac{y^2}{2}\right]_0^b = w\left(\frac{ab^2}{2}\right) = \left(\frac{wb}{2}\right)(ab) = \overline{p} \cdot \text{Area}$, where $\overline{p}$ is the average value of the pressure (see part (a)).

17. (a) An equation of the right-hand edge is $y = \frac{3}{2}x \Rightarrow x = \frac{2}{3}y$ and $L(y) = 2x = \frac{4y}{3}$. The depth of the strip is $(3 - y) \Rightarrow F = \int_0^3 w(3 - y)L(y)\ dy = \int_0^3 (62.4)(3 - y)\left(\frac{4}{3}y\right) dy = (62.4) \cdot \left(\frac{4}{3}\right)\int_0^3 (3y - y^2)\ dy$

$$= (62.4)\left(\frac{4}{3}\right)\left[\frac{3}{2}y^2 - \frac{y^3}{3}\right]_0^3 = (62.4)\left(\frac{4}{3}\right)\left[\frac{27}{2} - \frac{27}{3}\right] = (62.4)\left(\frac{4}{3}\right)\left(\frac{27}{6}\right) = 374.4 \text{ lb}$$

(b) We want to find a new water level Y such that $F_Y = \frac{1}{2}(374.4) = 187.2$ lb. The new depth of the strip is $(Y - y)$, and Y is the new upper limit of integration. Thus, $F_Y = \int_0^Y w(Y - y)L(y)\ dy$

$$= 62.4\int_0^Y (Y-y)\left(\tfrac{4}{3}y\right)dy = (62.4)\left(\tfrac{4}{3}\right)\int_0^Y (Yy-y^2)\,dy = (62.4)\left(\tfrac{4}{3}\right)\left[Y\cdot\frac{y^2}{2}-\frac{y^3}{3}\right]_0^Y = (62.4)\left(\tfrac{4}{3}\right)\left(\frac{Y^3}{2}-\frac{Y^3}{3}\right)$$

$= (62.4)\left(\frac{2}{9}\right)Y^3$. Therefore $Y^3 = \frac{9F_Y}{2\cdot(62.4)} = \frac{(9)(187.2)}{124.8} \Rightarrow Y = \sqrt[3]{\frac{(9)(187.2)}{124.8}} = \sqrt[3]{13.5} \approx 2.3811$ ft. So,

$\Delta Y = 3 - Y \approx 3 - 2.3811 \approx 0.6189$ ft ≈ 7.5 in to the nearest half inch

(c) No, it does not matter how long the trough is. The fluid pressure and the resulting force depends only on depth of the water.

19. Use the same coordinate system as in Exercise 20 with $L(y) = 3.75$ and the depth of a typical strip being $(7.75 - y)$. Then $F = \int_0^{7.75} w(7.75-y)L(y)\,dy = \left(\frac{64.5}{12^3}\right)(3.75)\int_0^{7.75}(7.75-y)\,dy = \left(\frac{64.5}{12^3}\right)(3.75)\left[7.75y - \frac{y^2}{2}\right]_0^{7.75}$

$= \left(\frac{64.5}{12^3}\right)(3.75)\frac{(7.75)^2}{2} \approx 4.2$ lb

21. Suppose that h is the maximum height. Using the coordinate system given in the text, we find an equation for the line of the end plate's right-hand edge is $y = \frac{5}{2}x \Rightarrow x = \frac{2}{5}y$. The total width is $L(y) = 2x = \frac{4}{5}y$ and the depth of the typical horizontal strip at level y is $(h-y)$. Then the force is $F = \int_0^h w(h-y)L(y)\,dy = F_{max}$, where $F_{max} = 6667$ lb. Hence, $F_{max} = w\int_0^h (h-y)\cdot\frac{4}{5}y\,dy = (62.4)\left(\frac{4}{5}\right)\int_0^h (hy-y^2)\,dy$

$= (62.4)\left(\frac{4}{5}\right)\left[\frac{hy^2}{2}-\frac{y^3}{3}\right]_0^h = (62.4)\left(\frac{4}{5}\right)\left(\frac{h^3}{2}-\frac{h^3}{3}\right) = (62.4)\left(\frac{4}{5}\right)\left(\frac{1}{6}\right)h^3 = (10.4)\left(\frac{4}{5}\right)h^3 \Rightarrow h = \sqrt[3]{\left(\frac{5}{4}\right)\left(\frac{F_{max}}{10.4}\right)}$

$= \sqrt[3]{\left(\frac{5}{4}\right)\left(\frac{6667}{10.4}\right)} \approx 9.288$ ft. The volume of water which the tank can hold is $V = \frac{1}{2}(\text{Base})(\text{Height})\cdot 30$, where Height $= h$ and $\frac{1}{2}(\text{Base}) = \frac{2}{5}h \Rightarrow V = \left(\frac{2}{5}h^2\right)(30) = 12h^2 \approx 12(9.288)^2 \approx 1035$ ft^3.

5.6 MOMENTS AND CENTERS OF MASS

1. Because the children are balanced, the moment of the system about the origin must be equal to zero: $5\cdot 80 = x\cdot 100 \Rightarrow x = 4$ ft, the distance of the 100-lb child from the fulcrum.

3. The center of mass of each rod is in its center (see Example 1). The rod system is equivalent to two point masses located at the centers of the rods at coordinates $\left(\frac{L}{2}, 0\right)$ and $\left(0, \frac{L}{2}\right)$. Therefore $\overline{x} = \frac{m_y}{m}$

$= \frac{x_1m_1 + x_2m_2}{m_1+m_2} = \frac{\frac{L}{2}\cdot m + 0}{m+m} = \frac{L}{4}$ and $\overline{y} = \frac{m_x}{m} = \frac{y_1m_2 + y_2m_2}{m_1+m_2} = \frac{0 + \frac{L}{2}\cdot m}{m+m} = \frac{L}{4} \Rightarrow \left(\frac{L}{4}, \frac{L}{4}\right)$ is the center of mass location.

5. $M_0 = \int_0^2 x \cdot 4\,dx = \left[4\frac{x^2}{2}\right]_0^2 = 4 \cdot \frac{4}{2} = 8$; $M = \int_0^2 4\,dx = [4x]_0^2 = 4 \cdot 2 = 8 \Rightarrow \bar{x} = \frac{M_0}{M} = 1$

7. $M_0 = \int_0^3 x\left(1+\frac{x}{3}\right) dx = \int_0^3 \left(x+\frac{x^2}{3}\right) dx = \left[\frac{x^2}{2}+\frac{x^3}{9}\right]_0^3 = \left(\frac{9}{2}+\frac{27}{9}\right) = \frac{15}{2}$; $M = \int_0^3 \left(1+\frac{x}{3}\right) dx = \left[x+\frac{x^2}{6}\right]_0^3$
$= 3+\frac{9}{6} = \frac{9}{2} \Rightarrow \bar{x} = \frac{M_0}{M} = \frac{\left(\frac{15}{2}\right)}{\left(\frac{9}{2}\right)} = \frac{15}{9} = \frac{5}{3}$

9. $M_0 = \int_1^4 x\left(1+\frac{1}{\sqrt{x}}\right) dx = \int_1^4 \left(x+x^{1/2}\right) dx = \left[\frac{x^2}{2}+\frac{2x^{3/2}}{3}\right]_1^4 = \left(8+\frac{16}{3}\right)-\left(\frac{1}{2}+\frac{2}{3}\right) = \frac{15}{2}+\frac{14}{3} = \frac{45+28}{6} = \frac{73}{6}$;
$M = \int_1^4 \left(1+x^{-1/2}\right) dx = \left[x+2x^{1/2}\right]_1^4 = (4+4)-(1+2) = 5 \Rightarrow \bar{x} = \frac{M_0}{M} = \frac{\left(\frac{73}{6}\right)}{5} = \frac{73}{30}$

11. $M_0 = \int_0^1 x(2-x)\,dx + \int_1^2 x \cdot x\,dx = \int_0^1 \left(2x-x^2\right) dx + \int_1^2 x^2\,dx = \left[\frac{2x^2}{2}-\frac{x^3}{3}\right]_0^1 + \left[\frac{x^3}{3}\right]_1^2 = \left(1-\frac{1}{3}\right)+\left(\frac{8}{3}-\frac{1}{3}\right)$
$= \frac{9}{3} = 3$; $M = \int_0^1 (2-x)\,dx + \int_1^2 x\,dx = \left[2x-\frac{x^2}{2}\right]_0^1 + \left[\frac{x^2}{2}\right]_1^2 = \left(2-\frac{1}{2}\right)+\left(\frac{4}{2}-\frac{1}{2}\right) = 3 \Rightarrow \bar{x} = \frac{M_0}{M} = 1$

13. Since the plate is symmetric about the y-axis and its density is constant, the distribution of mass is symmetric about the y-axis and the center of mass lies on the y-axis. This means that $\bar{x} = 0$. It remains to find $\bar{y} = \frac{M_x}{M}$. We model the distribution of mass with *vertical* strips. The typical strip has center of mass: $(\tilde{x}, \tilde{y}) = \left(x, \frac{x^2+4}{2}\right)$, length: $4-x^2$, width: dx, area: $dA = \left(4-x^2\right) dx$, mass: $dm = \delta\,dA = \delta\left(4-x^2\right) dx$. The moment of the strip about the x-axis is $\tilde{y}\ dm = \left(\frac{x^2+4}{2}\right)\delta\left(4-x^2\right) dx = \frac{\delta}{2}\left(16-x^4\right) dx$. The moment of the plate about the x-axis is $M_x = \int \tilde{y}\,dm = \int_{-2}^2 \frac{\delta}{2}\left(16-x^4\right) dx = \frac{\delta}{2}\left[16x-\frac{x^5}{5}\right]_{-2}^2 = \frac{\delta}{2}\left[\left(16 \cdot 2-\frac{2^5}{5}\right)-\left(-16 \cdot 2+\frac{2^5}{5}\right)\right]$ 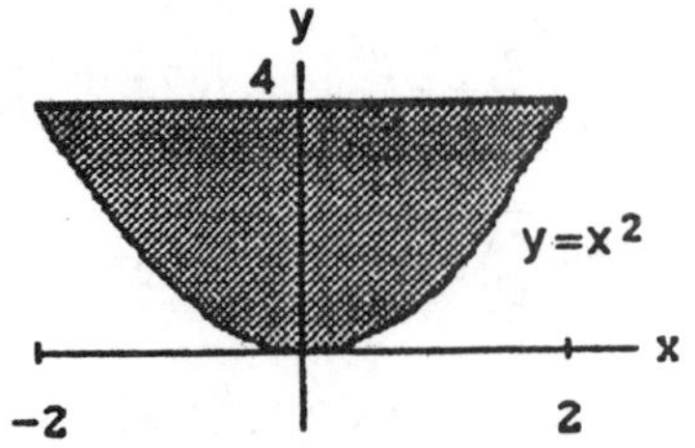
$= \frac{\delta \cdot 2}{2}\left(32-\frac{32}{5}\right) = \frac{128\delta}{5}$. The mass of the plate is $M = \int \delta\left(4-x^2\right) dx = \delta\left[4x-\frac{x^3}{3}\right]_{-2}^2 = 2\delta\left(8-\frac{8}{3}\right) = \frac{32\delta}{3}$.

Therefore $\bar{y} = \frac{M_x}{M} = \frac{\left(\frac{128\delta}{5}\right)}{\left(\frac{32\delta}{3}\right)} = \frac{12}{5}$. The plate's center of mass is the point $(\bar{x},\bar{y}) = \left(0, \frac{12}{5}\right)$.

15. Intersection points: $x - x^2 = -x \Rightarrow 2x - x^2 = 0$ $\Rightarrow x(2-x) = 0 \Rightarrow x = 0$ or $x = 2$. The typical *vertical* strip has center of mass: $(\tilde{x}, \tilde{y}) = \left(x, \frac{(x-x^2)+(-x)}{2}\right)$ $= \left(x, -\frac{x^2}{2}\right)$, length: $(x-x^2)-(-x) = 2x - x^2$, width: dx, area: $dA = (2x - x^2)\,dx$, mass: $dm = \delta\,dA = \delta(2x-x^2)\,dx$.

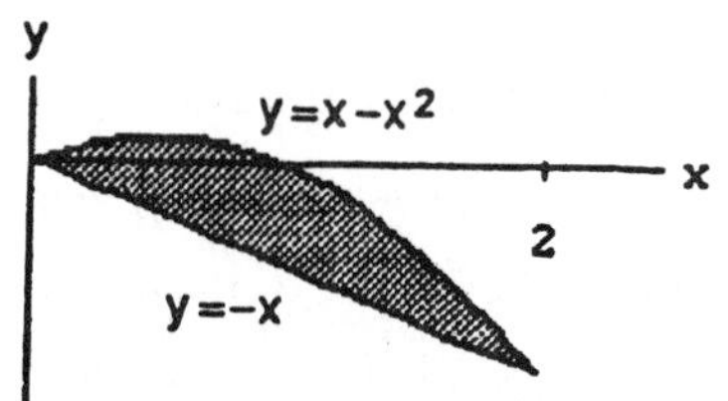

The moment of the strip about the x-axis is $\tilde{y}\,dm = \left(-\frac{x^2}{2}\right)\delta(2x-x^2)\,dx$; about the y-axis it is $\tilde{x}\,dm = x\cdot\delta(2x-x^2)\,dx$. Thus, $M_x = \int \tilde{y}\,dm = -\int_0^2 \left(\frac{\delta}{2}x^2\right)(2x-x^2)\,dx = -\frac{\delta}{2}\int_0^2 (2x^3 - x^4)\,dx$ $= -\frac{\delta}{2}\left[\frac{x^4}{2} - \frac{x^5}{5}\right]_0^2 = -\frac{\delta}{2}\left(2^3 - \frac{2^5}{5}\right) = -\frac{\delta}{2}\cdot 2^3\left(1 - \frac{4}{5}\right) = -\frac{4\delta}{5}$; $M_y = \int \tilde{x}\,dm = \int_0^2 x\cdot\delta(2x-x^2)\,dx$ $= \delta\int_0^2 (2x^2 - x^3) = \delta\left[\frac{2}{3}x^3 - \frac{x^4}{4}\right]_0^2 = \delta\left(2\cdot\frac{2^3}{3} - \frac{2^4}{4}\right) = \frac{\delta\cdot 2^4}{12} = \frac{4\delta}{3}$; $M = \int dm = \int_0^2 \delta(2x-x^2)\,dx$ $= \delta\int_0^2 (2x - x^2)\,dx = \delta\left[x^2 - \frac{x^3}{3}\right]_0^2 = \delta\left(4 - \frac{8}{3}\right) = \frac{4\delta}{3}$. Therefore, $\bar{x} = \frac{M_y}{M} = \left(\frac{4\delta}{3}\right)\left(\frac{3}{4\delta}\right) = 1$ and $\bar{y} = \frac{M_x}{M}$ $= \left(-\frac{4\delta}{5}\right)\left(\frac{3}{4\delta}\right) = -\frac{3}{5} \Rightarrow (\bar{x}, \bar{y}) = \left(1, -\frac{3}{5}\right)$ is the center of mass.

17. The typical *horizontal* strip has center of mass: $(\tilde{x}, \tilde{y}) = \left(\frac{y-y^3}{2}, y\right)$, length: $y - y^3$, width: dy, area: $dA = (y - y^3)\,dy$, mass: $dm = \delta\,dA = \delta(y-y^3)\,dy$. The moment of the strip about the y-axis is $\tilde{x}\,dm = \delta\left(\frac{y-y^3}{2}\right)(y-y^3)\,dy = \frac{\delta}{2}(y-y^3)^2\,dy$ $= \frac{\delta}{2}(y^2 - 2y^4 + y^6)\,dy$; the moment about the x-axis is $\tilde{y}\,dm = \delta y(y-y^3)\,dy = \delta(y^2 - y^4)\,dy$. Thus, $M_x = \int \tilde{y}\,dm = \delta\int_0^1 (y^2 - y^4)\,dy = \delta\left[\frac{y^3}{3} - \frac{y^5}{5}\right]_0^1 = \delta\left(\frac{1}{3} - \frac{1}{5}\right) = \frac{2\delta}{15}$; $M_y = \int \tilde{x}\,dm = \frac{\delta}{2}\int_0^1 (y^2 - 2y^4 + y^6)\,dy$ $= \frac{\delta}{2}\left[\frac{y^3}{3} - \frac{2y^5}{5} + \frac{y^7}{7}\right]_0^1 = \frac{\delta}{2}\left(\frac{1}{3} - \frac{2}{5} + \frac{1}{7}\right) = \frac{\delta}{2}\left(\frac{35 - 42 + 15}{3\cdot 5\cdot 7}\right) = \frac{4\delta}{105}$; $M = \int dm = \delta\int_0^1 (y - y^3)\,dy$ $= \delta\left[\frac{y^2}{2} - \frac{y^4}{4}\right]_0^1 = \delta\left(\frac{1}{2} - \frac{1}{4}\right) = \frac{\delta}{4}$. Therefore, $\bar{x} = \frac{M_y}{M} = \left(\frac{4\delta}{105}\right)\left(\frac{4}{\delta}\right) = \frac{16}{105}$ and $\bar{y} = \frac{M_x}{M} = \left(\frac{2\delta}{15}\right)\left(\frac{4}{\delta}\right) = \frac{8}{15}$ $\Rightarrow (\bar{x}, \bar{y}) = \left(\frac{16}{105}, \frac{8}{15}\right)$ is the center of mass.

19. Applying the symmetry argument analogous to the one used in Exercise 13, we find $\bar{x} = 0$. The typical *vertical* strip has center of mass: $(\tilde{x}, \tilde{y}) = \left(x, \frac{\cos x}{2}\right)$, length: $\cos x$, width: dx, area: $dA = \cos x\, dx$, mass: $dm = \delta\, dA = \delta \cos x\, dx$. The moment of the strip about the x-axis is $\tilde{y}\, dm = \delta \cdot \frac{\cos x}{2} \cdot \cos x\, dx$ $= \frac{\delta}{2} \cos^2 x\, dx = \frac{\delta}{2}\left(\frac{1 + \cos 2x}{2}\right) dx = \frac{\delta}{4}(1 + \cos 2x)\, dx$; thus,

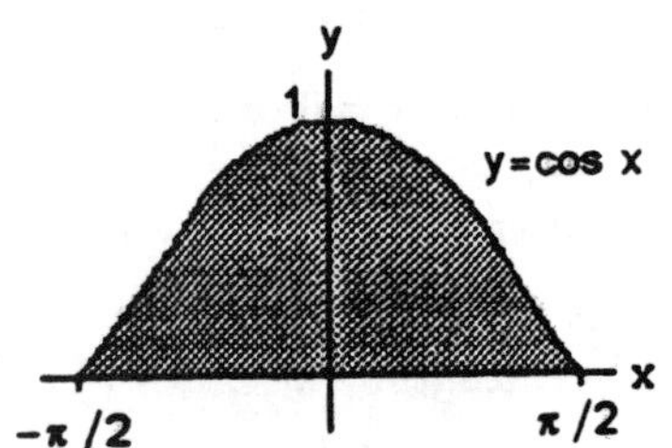

$$M_x = \int \tilde{y}\, dm = \int_{-\pi/2}^{\pi/2} \frac{\delta}{4}(1 + \cos 2x)\, dx = \frac{\delta}{4}\left[x + \frac{\sin 2x}{2}\right]_{-\pi/2}^{\pi/2} = \frac{\delta}{4}\left[\left(\frac{\pi}{2} + 0\right) - \left(-\frac{\pi}{2}\right)\right] = \frac{\delta\pi}{4};\ M = \int dm$$

$$= \delta \int_{-\pi/2}^{\pi/2} \cos x\, dx = \delta[\sin x]_{-\pi/2}^{\pi/2} = 2\delta. \text{ Therefore, } \bar{y} = \frac{M_x}{M} = \frac{\delta\pi}{4 \cdot 2\delta} = \frac{\pi}{8} \Rightarrow (\bar{x}, \bar{y}) = \left(0, \frac{\pi}{8}\right) \text{ is the center of mass.}$$

21. Since the plate is symmetric about the line $x = 1$ and its density is constant, the distribution of mass is symmetric about this line and the center of mass lies on it. This means that $\bar{x} = 1$. The typical *vertical* strip has center of mass: $(\tilde{x}, \tilde{y}) = \left(x, \frac{(2x - x^2) + (2x^2 - 4x)}{2}\right) = \left(x, \frac{x^2 - 2x}{2}\right)$, length: $(2x - x^2) - (2x^2 - 4x) = -3x^2 + 6x = 3(2x - x^2)$, width: dx, area: $dA = 3(2x - x^2)\, dx$, mass: $dm = \delta\, dA$ $= 3\delta(2x - x^2)\, dx$. The moment about the x-axis is

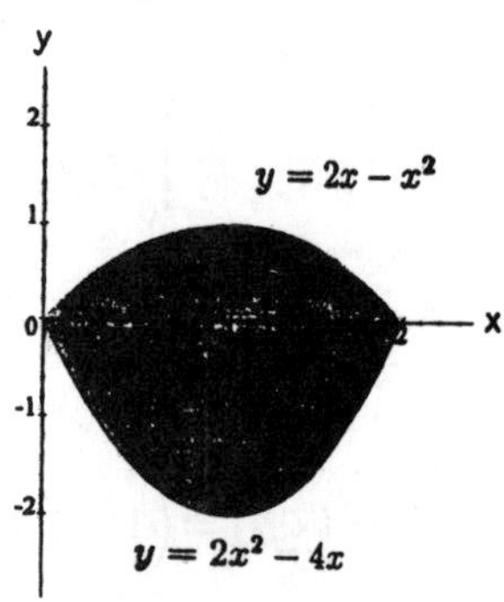

$\tilde{y}\, dm = \frac{3}{2}\delta(x^2 - 2x)(2x - x^2)\, dx = -\frac{3}{2}\delta(x^2 - 2x)^2\, dx = -\frac{3}{2}\delta(x^4 - 4x^3 + 4x^2)\, dx$. Thus, $M_x = \int \tilde{y}\, dm$

$$= -\int_0^2 \frac{3}{2}\delta(x^4 - 4x^3 + 4x^2)\, dx = -\frac{3}{2}\delta\left[\frac{x^5}{5} - x^4 + \frac{4}{3}x^3\right]_0^2 = -\frac{3}{2}\delta\left(\frac{2^5}{5} - 2^4 + \frac{4}{3} \cdot 2^3\right) = -\frac{3}{2}\delta \cdot 2^4\left(\frac{2}{5} - 1 + \frac{2}{3}\right)$$

$$= -\frac{3}{2}\delta \cdot 2^4\left(\frac{6 - 15 + 10}{15}\right) = -\frac{8\delta}{5};\ M = \int dm = \int_0^2 3\delta(2x - x^2)\, dx = 3\delta\left[x^2 - \frac{x^3}{3}\right]_0^2 = 3\delta\left(4 - \frac{8}{3}\right) = 4\delta.$$

Therefore, $\bar{y} = \frac{M_x}{M} = \left(-\frac{8\delta}{5}\right)\left(\frac{1}{4\delta}\right) = -\frac{2}{5} \Rightarrow (\bar{x}, \bar{y}) = \left(1, -\frac{2}{5}\right)$ is the center of mass.

23. Since the plate is symmetric about the line $x = y$ and its density is constant, the distribution of mass is symmetric about this line. This means that $\bar{x} = \bar{y}$. The typical *vertical* strip has

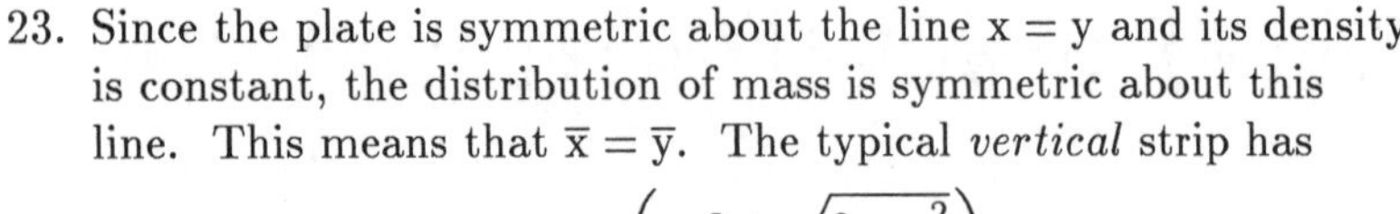

center of mass: $(\tilde{x}, \tilde{y}) = \left(x, \frac{3 + \sqrt{9 - x^2}}{2}\right)$,

length: $3 - \sqrt{9 - x^2}$, width: dx, area: $dA = (3 - \sqrt{9 - x^2})\, dx$, mass: $dm = \delta\, dA = \delta(3 - \sqrt{9 - x^2})\, dx$. The moment about the x-axis is $\tilde{y}\, dm = \delta \frac{(3 + \sqrt{9 - x^2})(3 - \sqrt{9 - x^2})}{2}\, dx$

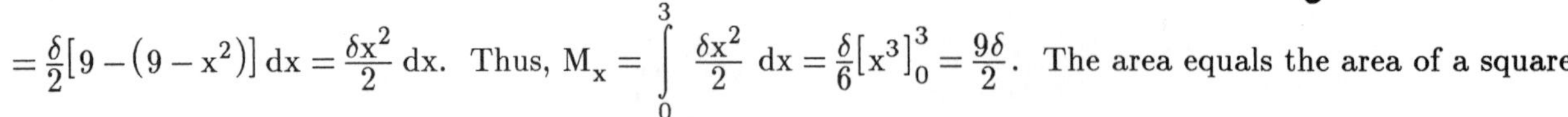

$= \frac{\delta}{2}[9 - (9 - x^2)]\, dx = \frac{\delta x^2}{2}\, dx$. Thus, $M_x = \int_0^3 \frac{\delta x^2}{2}\, dx = \frac{\delta}{6}[x^3]_0^3 = \frac{9\delta}{2}$. The area equals the area of a square with side length 3 minus one quarter the area of a disk with radius 3 $\Rightarrow A = 3^2 - \frac{\pi 9}{4} = \frac{9}{4}(4-\pi) \Rightarrow M = \delta A$ $= \frac{9\delta}{4}(4-\pi)$. Therefore, $\bar{y} = \frac{M_x}{M} = \left(\frac{9\delta}{2}\right)\left[\frac{4}{9\delta(4-\pi)}\right] = \frac{2}{4-\pi} \Rightarrow (\bar{x},\bar{y}) = \left(\frac{2}{4-\pi}, \frac{2}{4-\pi}\right)$ is the center of mass.

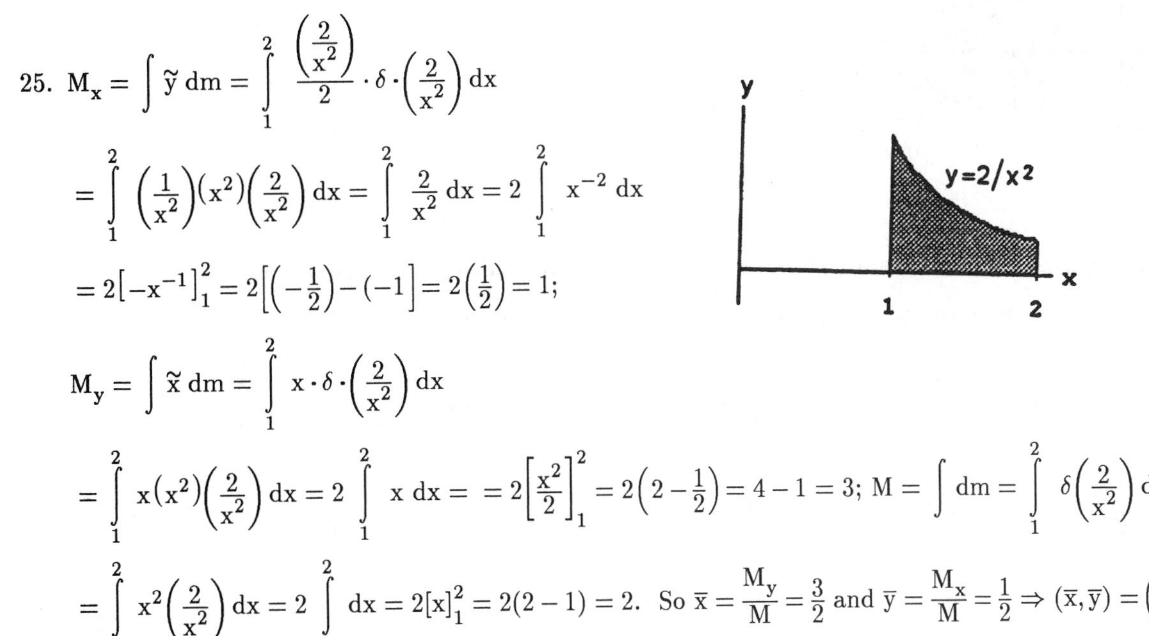

25. $M_x = \int \tilde{y}\,dm = \int_1^2 \frac{\left(\frac{2}{x^2}\right)}{2} \cdot \delta \cdot \left(\frac{2}{x^2}\right) dx$

$= \int_1^2 \left(\frac{1}{x^2}\right)(x^2)\left(\frac{2}{x^2}\right) dx = \int_1^2 \frac{2}{x^2}\,dx = 2\int_1^2 x^{-2}\,dx$

$= 2\left[-x^{-1}\right]_1^2 = 2\left[\left(-\frac{1}{2}\right) - (-1\right] = 2\left(\frac{1}{2}\right) = 1;$

$M_y = \int \tilde{x}\,dm = \int_1^2 x \cdot \delta \cdot \left(\frac{2}{x^2}\right) dx$

$= \int_1^2 x(x^2)\left(\frac{2}{x^2}\right) dx = 2\int_1^2 x\,dx = = 2\left[\frac{x^2}{2}\right]_1^2 = 2\left(2 - \frac{1}{2}\right) = 4 - 1 = 3;\ M = \int dm = \int_1^2 \delta\left(\frac{2}{x^2}\right) dx$

$= \int_1^2 x^2\left(\frac{2}{x^2}\right) dx = 2\int_1^2 dx = 2[x]_1^2 = 2(2-1) = 2.$ So $\bar{x} = \frac{M_y}{M} = \frac{3}{2}$ and $\bar{y} = \frac{M_x}{M} = \frac{1}{2} \Rightarrow (\bar{x},\bar{y}) = \left(\frac{3}{2},\frac{1}{2}\right)$ is the center of mass.

27. (a) We use the shell method:

$$V = \int_a^b 2\pi\left(\begin{smallmatrix}\text{shell}\\ \text{radius}\end{smallmatrix}\right)\left(\begin{smallmatrix}\text{shell}\\ \text{height}\end{smallmatrix}\right) dx = \int_1^4 2\pi x\left[\frac{4}{\sqrt{x}} - \left(-\frac{4}{\sqrt{x}}\right)\right] dx$$

$$= 16\pi \int_1^4 \frac{x}{\sqrt{x}}\,dx = 16\pi \int_1^4 x^{1/2}\,dx = 16\pi\left[\frac{2}{3}x^{3/2}\right]_1^4$$

$$= 16\pi\left(\frac{2}{3}\cdot 8 - \frac{2}{3}\right) = \frac{32\pi}{3}(8-1) = \frac{224\pi}{3}$$

(b) Since the plate is symmetric about the x-axis and its density $\delta(x) = \frac{1}{x}$ is a function of x alone, the distribution of its mass is symmetric about the x-axis. This means that $\bar{y} = 0$. We use the vertical strip approach to find $\bar{x}$: $M_y = \int \tilde{x}\,dm = \int_1^4 x \cdot \left[\frac{4}{\sqrt{x}} - \left(-\frac{4}{\sqrt{x}}\right)\right] \cdot \delta\,dx = \int_1^4 x \cdot \frac{8}{\sqrt{x}} \cdot \frac{1}{x}\,dx = 8\int_1^4 x^{-1/2}\,dx$

$= 8\left[2x^{1/2}\right]_1^4 = 8(2\cdot 2 - 2) = 16;\ M = \int dm = \int_1^4 \left[\frac{4}{\sqrt{x}} - \left(\frac{-4}{\sqrt{x}}\right)\right] \cdot \delta\,dx = 8\int_1^4 \left(\frac{1}{\sqrt{x}}\right)\left(\frac{1}{x}\right) dx = 8\int_1^4 x^{-3/2}\,dx$

$= 8\left[-2x^{-1/2}\right]_1^4 = 8[-1-(-2)] = 8.$ So $\bar{x} = \frac{M_y}{M} = \frac{16}{8} = 2 \Rightarrow (\bar{x},\bar{y}) = (2,0)$ is the center of mass.

(c)

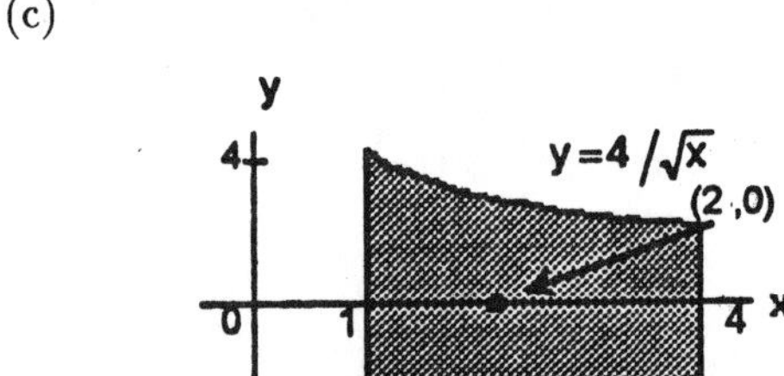

29. The mass of a horizontal strip is $dm = \delta\ dA = \delta L$, where L is the width of the triangle at a distance of y above its base on the x-axis as shown in the figure in the text. Also, by similar triangles we have $\frac{L}{b} = \frac{h-y}{h}$

$\Rightarrow L = \frac{b}{h}(h-y)$. Thus, $M_x = \int \tilde{y}\ dm = \int_0^h \delta y\left(\frac{b}{h}\right)(h-y)\ dy = \frac{\delta b}{h}\int_0^h (hy - y^2)\ dy = \frac{\delta b}{h}\left[\frac{hy^2}{2} - \frac{y^3}{3}\right]_0^h$

$= \frac{\delta b}{h}\left(\frac{h^3}{2} - \frac{h^3}{3}\right) = \delta bh^2\left(\frac{1}{2} - \frac{1}{3}\right) = \frac{\delta bh^2}{6}$; $M = \int dm = \int_0^h \delta\left(\frac{b}{h}\right)(h-y)\ dy = \frac{\delta b}{h}\int_0^h (h-y)\ dy = \frac{\delta b}{h}\left[hy - \frac{y^2}{2}\right]_0^h$

$= \frac{\delta b}{h}\left(h^2 - \frac{h^2}{2}\right) = \frac{\delta bh}{2}$. So $\bar{y} = \frac{M_x}{M} = \left(\frac{\delta bh^2}{6}\right)\left(\frac{2}{\delta bh}\right) = \frac{h}{3} \Rightarrow$ the center of mass lies above the base of the triangle one-third of the way toward the opposite vertex. Similarly the other two sides of the triangle can be placed on the x-axis and the same results will occur. Therefore the centroid does lie at the intersection of the medians, as claimed.

31. From the symmetry about the line $x = y$ it follows that $\bar{x} = \bar{y}$. It also follows that the line through the points $(0,0)$ and $\left(\frac{1}{2}, \frac{1}{2}\right)$ is a median $\Rightarrow \bar{y} = \bar{x} = \frac{2}{3}\cdot\left(\frac{1}{2} - 0\right) = \frac{1}{3}$ $\Rightarrow (\bar{x}, \bar{y}) = \left(\frac{1}{3}, \frac{1}{3}\right)$.

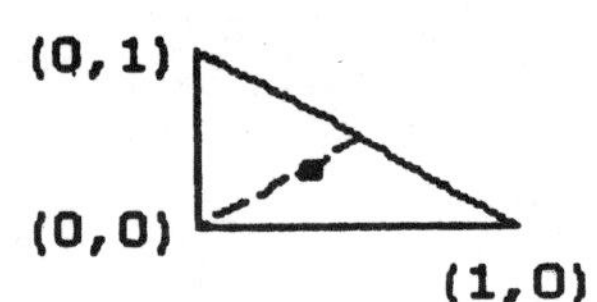

33. The point of intersection of the median from the vertex $(0,b)$ to the opposite side has coordinates $\left(0, \frac{a}{2}\right) \Rightarrow \bar{y} = (b-0)\cdot\frac{1}{3}$ $= \frac{b}{3}$ and $\bar{x} = \left(\frac{a}{2} - 0\right)\cdot\frac{2}{3} = \frac{a}{3} \Rightarrow (\bar{x}, \bar{y}) = \left(\frac{a}{3}, \frac{b}{3}\right)$.

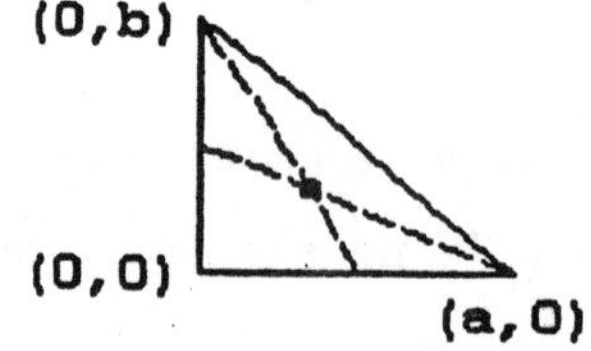

35. $y = x^{1/2} \Rightarrow dy = \frac{1}{2}x^{-1/2}\ dx \Rightarrow ds = \sqrt{(dx)^2 + (dy)^2} = \sqrt{1 + \frac{1}{4x}}\ dx$; $M_x = \delta\int_0^2 x\sqrt{1 + \frac{1}{4x}}\ dx$

$= \delta\int_0^2 x + \frac{1}{4}\ dx = \frac{2\delta}{3}\left[\left(x + \frac{1}{4}\right)^{3/2}\right]_0^2 = \frac{2\delta}{3}\left[\left(2 + \frac{1}{4}\right)^{3/2} - \left(\frac{1}{4}\right)^{3/2}\right] = \frac{2\delta}{3}\left[\left(\frac{9}{4}\right)^{3/2} - \left(\frac{1}{4}\right)^{3/2}\right] = \frac{2\delta}{3}\left(\frac{27}{8} - \frac{1}{8}\right) = \frac{13\delta}{6}$

37. From Example 6 we have $M_x = \int_0^{\pi} a(a \sin\theta)(k \sin\theta)\, d\theta = a^2k \int_0^{\pi} \sin^2\theta\, d\theta = \frac{a^2k}{2} \int_0^{\pi} (1 - \cos 2\theta)\, d\theta$

$= \frac{a^2k}{2}\left[\theta - \frac{\sin 2\theta}{2}\right]_0^{\pi} = \frac{a^2k\pi}{2}$; $M_y = \int_0^{\pi} a(a \cos\theta)(k \sin\theta)\, d\theta = a^2k \int_0^{\pi} \sin\theta \cos\theta\, d\theta = \frac{a^2k}{2}\left[\sin^2\theta\right]_0^{\pi} = 0$;

$M = \int_0^{\pi} ak \sin\theta\, d\theta = ak[-\cos\theta]_0^{\pi} = 2ak$. Therefore, $\bar{x} = \frac{M_y}{M} = 0$ and $\bar{y} = \frac{M_x}{M} = \left(\frac{a^2k\pi}{2}\right)\left(\frac{1}{2ak}\right) = \frac{a\pi}{4} \Rightarrow \left(0, \frac{a\pi}{4}\right)$

is the center of mass.

39. Consider the curve as an infinite number of line segments joined together. From the derivation of arc length we have that the length of a particular segment is $ds = \sqrt{(dx)^2 + (dy)^2}$. This implies that

$M_x = \int \delta y\, ds$, $M_y = \int \delta x\, ds$ and $M = \int \delta\, ds$. If δ is constant, then $\bar{x} = \frac{M_y}{M} = \frac{\int x\, ds}{\int ds} = \frac{\int x\, ds}{\text{length}}$ and

$\bar{y} = \frac{M_x}{M} = \frac{\int y\, ds}{\int ds} = \frac{\int y\, ds}{\text{length}}$.

41. A generalization of Example 6 yields $M_x = \int \tilde{y}\, dm = \int_{\pi/2 - \alpha}^{\pi/2 + \alpha} a^2 \sin\theta\, d\theta = a^2[-\cos\theta]_{\pi/2 - \alpha}^{\pi/2 + \alpha}$

$= a^2\left[-\cos\left(\frac{\pi}{2} + \alpha\right) + \cos\left(\frac{\pi}{2} - \alpha\right)\right] = a^2(\sin\alpha + \sin\alpha) = 2a^2 \sin\alpha$; $M = \int dm = \int_{\pi/2 - \alpha}^{\pi/2 + \alpha} a\, d\theta = a[\theta]_{\pi/2 - \alpha}^{\pi/2 + \alpha}$

$= a\left[\left(\frac{\pi}{2} + \alpha\right) - \left(\frac{\pi}{2} - \alpha\right)\right] = 2a\alpha$. Thus, $\bar{y} = \frac{M_x}{M} = \frac{2a^2 \sin\alpha}{2a\alpha} = \frac{a \sin\alpha}{\alpha}$. Now $s = a(2\alpha)$ and $a \sin\alpha = \frac{c}{2}$

$\Rightarrow c = 2a \sin\alpha$. Then $\bar{y} = \frac{a(2a \sin\alpha)}{2a\alpha} = \frac{ac}{s}$, as claimed.

CHAPTER 5 PRACTICE EXERCISES

1. $A(x) = \frac{\pi}{4}(\text{diameter})^2 = \frac{\pi}{4}\left(\sqrt{x} - x^2\right)^2 = \frac{\pi}{4}\left(x - 2\sqrt{x}\cdot x^2 + x^4\right)$;

$a = 0,\ b = 1 \Rightarrow V = \int_a^b A(x)\, dx = \frac{\pi}{4} \int_0^1 \left(x - 2x^{5/2} + x^4\right) dx$

$= \frac{\pi}{4}\left[\frac{x^2}{2} - \frac{4}{7}x^{7/2} + \frac{x^5}{5}\right]_0^1 = \frac{\pi}{4}\left(\frac{1}{2} - \frac{4}{7} + \frac{1}{5}\right) = \frac{\pi}{4 \cdot 70}(35 - 40 + 14)$

$= \frac{9\pi}{280}$

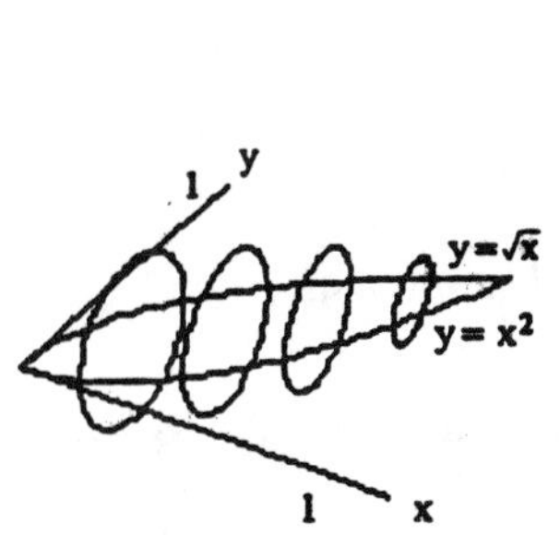

y
1
y = √x
y = x²
1
x

3. $A(x) = \frac{\pi}{4}(\text{diameter})^2 = \frac{\pi}{4}(2 \sin x - 2 \cos x)^2$

$= \frac{\pi}{4} \cdot 4(\sin^2 x - 2 \sin x \cos x + \cos^2 x) = \pi(1 - \sin 2x)$; $a = \frac{\pi}{4}$,

$b = \frac{5\pi}{4} \Rightarrow V = \int_a^b A(x)\, dx = \pi \int_{\pi/4}^{5\pi/4} (1 - \sin 2x)\, dx$

$= \pi\left[x + \frac{\cos 2x}{2}\right]_{\pi/4}^{5\pi/4} = \pi\left[\left(\frac{5\pi}{4} + \frac{\cos \frac{5\pi}{2}}{2}\right) - \left(\frac{\pi}{4} - \frac{\cos \frac{\pi}{2}}{2}\right)\right] = \pi^2$

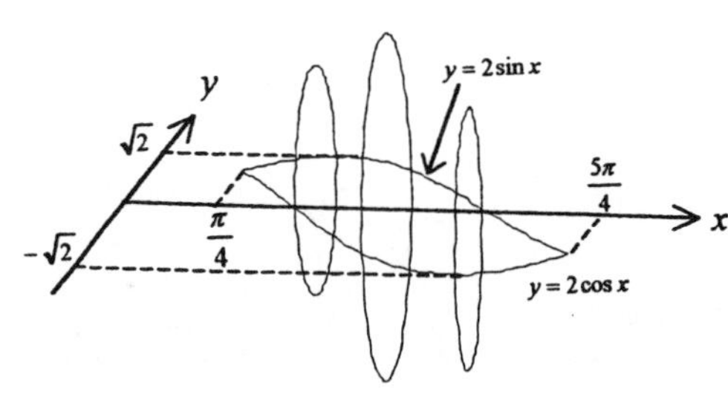

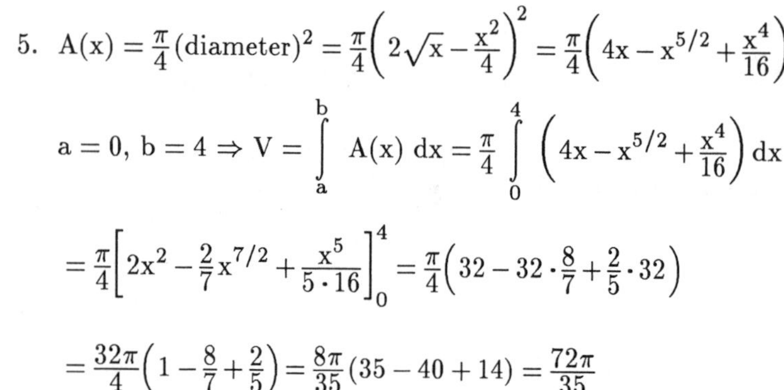

5. $A(x) = \frac{\pi}{4}(\text{diameter})^2 = \frac{\pi}{4}\left(2\sqrt{x} - \frac{x^2}{4}\right)^2 = \frac{\pi}{4}\left(4x - x^{5/2} + \frac{x^4}{16}\right)$;

$a = 0,\ b = 4 \Rightarrow V = \int_a^b A(x)\, dx = \frac{\pi}{4}\int_0^4 \left(4x - x^{5/2} + \frac{x^4}{16}\right) dx$

$= \frac{\pi}{4}\left[2x^2 - \frac{2}{7}x^{7/2} + \frac{x^5}{5 \cdot 16}\right]_0^4 = \frac{\pi}{4}\left(32 - 32 \cdot \frac{8}{7} + \frac{2}{5} \cdot 32\right)$

$= \frac{32\pi}{4}\left(1 - \frac{8}{7} + \frac{2}{5}\right) = \frac{8\pi}{35}(35 - 40 + 14) = \frac{72\pi}{35}$

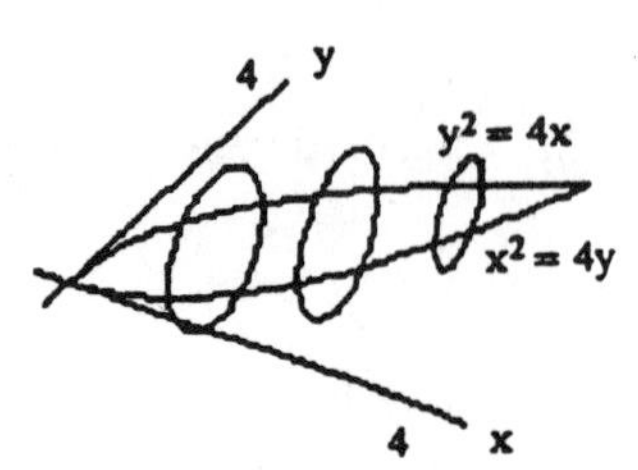

7. (a) *disk method*:

$V = \int_a^b \pi R^2(x)\, dx = \int_{-1}^{1} \pi\left(3x^4\right)^2 dx = \pi \int_{-1}^{1} 9x^8\, dx$

$= \pi\left[x^9\right]_{-1}^{1} = 2\pi$

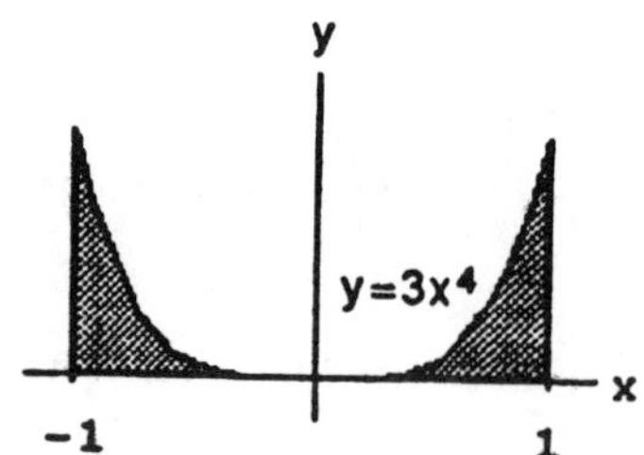

(b) *shell method*:

$V = \int_a^b 2\pi\left(\text{shell} \atop \text{radius}\right)\left(\text{shell} \atop \text{height}\right) dx = \int_0^1 2\pi x\left(3x^4\right) dx = 2\pi \cdot 3 \int_0^1 x^5\, dx = 2\pi \cdot 3\left[\frac{x^6}{6}\right]_0^1 = \pi$

Note: The lower limit of integration is 0 rather than −1.

(c) *shell method*:

$V = \int_a^b 2\pi\left(\text{shell} \atop \text{radius}\right)\left(\text{shell} \atop \text{height}\right) dx = 2\pi \int_{-1}^{1} (1 - x)\left(3x^4\right) dx = 2\pi\left[\frac{3x^5}{5} - \frac{x^6}{2}\right]_{-1}^{1} = 2\pi\left[\left(\frac{3}{5} - \frac{1}{2}\right) - \left(-\frac{3}{5} - \frac{1}{2}\right)\right] = \frac{12\pi}{5}$

(d) *washer method*:

$R(x) = 3,\ r(x) = 3 - 3x^4 = 3\left(1 - x^4\right) \Rightarrow V = \int_a^b \pi\left[R^2(x) - r^2(x)\right] dx = \int_{-1}^{1} \pi\left[9 - 9\left(1 - x^4\right)^2\right] dx$

$= 9\pi \int_{-1}^{1} \left[1 - \left(1 - 2x^4 + x^8\right)\right] dx = 9\pi \int_{-1}^{1} \left(2x^4 - x^8\right) dx = 9\pi\left[\frac{2x^5}{5} - \frac{x^9}{9}\right]_{-1}^{1} = 18\pi\left[\frac{2}{5} - \frac{1}{9}\right] = \frac{2\pi \cdot 13}{5} = \frac{26\pi}{5}$

9. (a) *disk method*:

$$V = \pi \int_1^5 \left(\sqrt{x-1}\right)^2 dx = \int_1^5 (x-1)\,dx = \pi\left[\frac{x^2}{2} - x\right]_1^5$$

$$= \pi\left[\left(\frac{25}{2} - 5\right) - \left(\frac{1}{2} - 1\right)\right] = \pi\left(\frac{24}{2} - 4\right) = 8\pi$$

(b) *washer method*:

$$R(y) = 5,\ r(y) = y^2 + 1 \Rightarrow V = \int_c^d \pi\left[R^2(y) - r^2(y)\right] dy = \pi \int_{-2}^{2} \left[25 - \left(y^2+1\right)^2\right] dy$$

$$= \pi \int_{-2}^{2} \left(25 - y^4 - 2y^2 - 1\right) dy = \pi \int_{-2}^{2} \left(24 - y^4 - 2y^2\right) dy = \pi\left[24y - \frac{y^5}{5} - \frac{2}{3}y^3\right]_{-2}^{2} = 2\pi\left(24 \cdot 2 - \frac{32}{5} - \frac{2}{3} \cdot 8\right)$$

$$= 32\pi\left(3 - \frac{2}{5} - \frac{1}{3}\right) = \frac{32\pi}{15}(45 - 6 - 5) = \frac{1088\pi}{15}$$

(c) *disk method*:

$$R(y) = 5 - \left(y^2+1\right) = 4 - y^2 \Rightarrow V = \int_c^d \pi R^2(y)\,dy = \int_{-2}^{2} \pi\left(4 - y^2\right)^2 dy = \pi \int_{-2}^{2} \left(16 - 8y^2 + y^4\right) dy$$

$$= \pi\left[16y - \frac{8y^3}{3} + \frac{y^5}{5}\right]_{-2}^{2} = 2\pi\left(32 - \frac{64}{3} + \frac{32}{5}\right) = 64\pi\left(1 - \frac{2}{3} + \frac{1}{5}\right) = \frac{64\pi}{15}(15 - 10 + 3) = \frac{512\pi}{15}$$

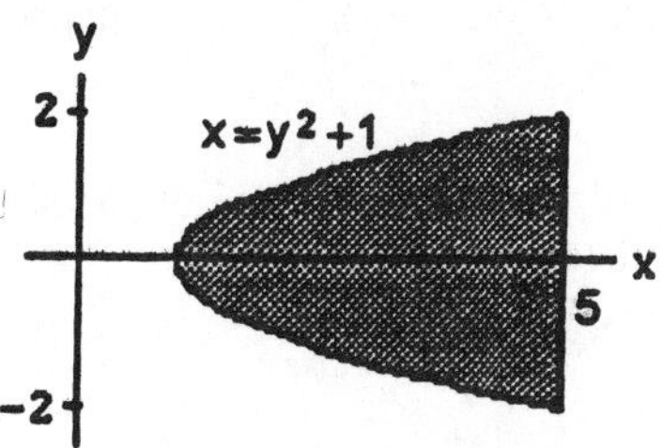

11. *disk method*:

$$R(x) = \tan x,\ a = 0,\ b = \frac{\pi}{3} \Rightarrow V = \pi \int_0^{\pi/3} \tan^2 x\,dx = \pi \int_0^{\pi/3} \left(\sec^2 x - 1\right) dx = \pi[\tan x - x]_0^{\pi/3} = \frac{\pi\left(3\sqrt{3} - \pi\right)}{3}$$

13. (a) *disk method*:

$$V = \pi \int_0^2 \left(x^2 - 2x\right)^2 dx = \pi \int_0^2 \left(x^4 - 4x^3 + 4x^2\right) dx = \pi\left[\frac{x^5}{5} - x^4 + \frac{4}{3}x^3\right]_0^2 = \pi\left(\frac{32}{5} - 16 + \frac{32}{3}\right)$$

$$= \frac{16\pi}{15}(6 - 15 + 10) = \frac{16\pi}{15}$$

(b) *disk method:*

$$V = 2\pi - \pi \int_0^2 \left[1+\left(x^2-2x\right)\right]^2 dx = 2\pi - \pi \int_0^2 \left[1+2\left(x^2-2x\right)+\left(x^2-2x\right)^2\right] dx$$

$$= 2\pi - \pi \int_0^2 \left(1+2x^2-4x+x^4-4x^3+4x^2\right) dx = 2\pi - \pi \int_0^2 \left(x^4-4x^3+6x^2-4x+1\right) dx$$

$$= 2\pi - \pi\left[\frac{x^5}{5} - x^4 + 2x^3 - 2x^2 + x\right]_0^2 = 2\pi - \pi\left(\frac{32}{5} - 16 + 16 - 8 + 2\right) = 2\pi - \frac{\pi}{5}(32-30) = 2\pi - \frac{2\pi}{5} = \frac{8\pi}{5}$$

(c) *shell method:*

$$V = \int_a^b 2\pi\begin{pmatrix}\text{shell}\\ \text{radius}\end{pmatrix}\begin{pmatrix}\text{shell}\\ \text{height}\end{pmatrix} dx = 2\pi \int_0^2 (2-x)\left[-\left(x^2-2x\right)\right] dx = 2\pi \int_0^2 (2-x)\left(2x-x^2\right) dx$$

$$= 2\pi \int_0^2 \left(4x-2x^2-2x^2+x^3\right) dx = 2\pi \int_0^2 \left(x^3-4x^2+4x\right) dx = 2\pi\left[\frac{x^4}{4} - \frac{4}{3}x^3 + 2x^2\right]_0^2 = 2\pi\left(4 - \frac{32}{3} + 8\right)$$

$$= \frac{2\pi}{3}(36-32) = \frac{8\pi}{3}$$

(d) *disk method:*

$$V = \pi \int_0^2 \left[2-\left(x^2-2x\right)\right]^2 dx - \pi \int_0^2 2^2\, dx = \pi \int_0^2 \left[4-4\left(x^2-2x\right)+\left(x^2-2x\right)^2\right] dx - 8\pi$$

$$= \pi \int_0^2 \left(4-4x^2+8x+x^4-4x^3+4x^2\right) dx - 8\pi = \pi \int_0^2 \left(x^4-4x^3+8x+4\right) dx - 8\pi$$

$$= \pi\left[\frac{x^5}{5} - x^4 + 4x^2 + 4x\right]_0^2 - 8\pi = \pi\left(\frac{32}{5} - 16 + 16 + 8\right) - 8\pi = \frac{\pi}{5}(32+40) - 8\pi = \frac{72\pi}{5} - \frac{40\pi}{5} = \frac{32\pi}{5}$$

15. The volume cut out is equivalent to the volume of the solid generated by revolving the region shown here about the x-axis. Using the *shell* method:

$$V = \int_c^d 2\pi\begin{pmatrix}\text{shell}\\ \text{radius}\end{pmatrix}\begin{pmatrix}\text{shell}\\ \text{height}\end{pmatrix} dy = \int_0^{\sqrt{3}} 2\pi y\left[\sqrt{4-y^2} - \left(-\sqrt{4-y^2}\right)\right] dy$$

$$= 2\pi \int_0^{\sqrt{3}} 2y\sqrt{4-y^2}\, dy = -2\pi \int_0^{\sqrt{3}} \sqrt{4-y^2}\, d\left(4-y^2\right)$$

$$= (-2\pi)\left(\frac{2}{3}\right)\left[\left(4-y^2\right)^{3/2}\right]_0^{\sqrt{3}} = -\frac{4\pi}{3}(1-8) = \frac{28\pi}{3}$$

17. $y = x^{1/2} - \frac{x^{3/2}}{3} \Rightarrow \frac{dy}{dx} = \frac{1}{2}x^{-1/2} - \frac{1}{2}x^{1/2} \Rightarrow \left(\frac{dy}{dx}\right)^2 = \frac{1}{4}\left(\frac{1}{x} - 2 + x\right) \Rightarrow L = \int_1^4 \sqrt{1+\frac{1}{4}\left(\frac{1}{x} - 2 + x\right)}\, dx$

$$\Rightarrow L = \int_1^4 \sqrt{\tfrac{1}{4}\left(\tfrac{1}{x} + 2 + x\right)}\, dx = \int_1^4 \sqrt{\tfrac{1}{4}\left(x^{-1/2} + x^{1/2}\right)^2}\, dx = \int_1^4 \tfrac{1}{2}\left(x^{-1/2} + x^{1/2}\right) dx = \tfrac{1}{2}\left[2x^{1/2} + \tfrac{2}{3}x^{3/2}\right]_1^4$$

$$= \tfrac{1}{2}\left[\left(4 + \tfrac{2}{3}\cdot 8\right) - \left(2 + \tfrac{2}{3}\right)\right] = \tfrac{1}{2}\left(2 + \tfrac{14}{3}\right) = \tfrac{10}{3}$$

19. $y = \frac{5}{12}x^{6/5} - \frac{5}{8}x^{4/5} \Rightarrow \frac{dy}{dx} = \frac{1}{2}x^{1/5} - \frac{1}{2}x^{-1/5} \Rightarrow \left(\frac{dy}{dx}\right)^2 = \frac{1}{4}\left(x^{2/5} - 2 + x^{-2/5}\right)$

$$\Rightarrow L = \int_1^{32} \sqrt{1 + \tfrac{1}{4}\left(x^{2/5} - 2 + x^{-2/5}\right)}\, dx \Rightarrow L = \int_1^{32} \sqrt{\tfrac{1}{4}\left(x^{2/5} + 2 + x^{-2/5}\right)}\, dx = \int_1^{32} \sqrt{\tfrac{1}{4}\left(x^{1/5} + x^{-1/5}\right)^2}\, dx$$

$$= \int_1^{32} \tfrac{1}{2}\left(x^{1/5} + x^{-1/5}\right) dx = \tfrac{1}{2}\left[\tfrac{5}{6}x^{6/5} + \tfrac{5}{4}x^{4/5}\right]_1^{32} = \tfrac{1}{2}\left[\left(\tfrac{5}{6}\cdot 2^6 + \tfrac{5}{4}\cdot 2^4\right) - \left(\tfrac{5}{6} + \tfrac{5}{4}\right)\right] = \tfrac{1}{2}\left(\tfrac{315}{6} + \tfrac{75}{4}\right)$$

$$= \tfrac{1}{48}(1260 + 450) = \tfrac{1710}{48} = \tfrac{285}{8}$$

21. $\frac{dx}{dt} = -5\sin t + 5\sin 5t$ and $\frac{dy}{dt} = 5\cos t - 5\cos 5t \Rightarrow \sqrt{\left(\frac{dx}{dt}\right)^2 + \left(\frac{dy}{dt}\right)^2}$

$$= \sqrt{(5\sin 5t - 5\sin t)^2 + (5\cos t - 5\cos 5t)^2}$$

$$= 5\sqrt{\sin^2 5t - 2\sin t\sin 5t + \sin^2 t + \cos^2 t - 2\cos t\cos 5t + \cos^2 5t} = 5\sqrt{2 - 2(\sin t\sin 5t + \cos t\cos 5t)}$$

$$= 5\sqrt{2(1 - \cos 4t)} = 5\sqrt{4\left(\tfrac{1}{2}\right)(1 - \cos 4t)} = 10\sqrt{\sin^2 2t} = 10|\sin 2t| = 10\sin 2t \text{ (since } 0 \le t \le \pi/2)$$

$$\Rightarrow \text{Length} = \int_0^{\pi/2} 10\sin 2t\, dt = (-5\cos 2t)\Big|_{t=0}^{\pi/2} = (-5)(-1) - (-5)(1) = 10$$

23. $\frac{dx}{d\theta} = -3\sin\theta$ and $\frac{dy}{d\theta} = 3\cos\theta \Rightarrow \sqrt{\left(\frac{dx}{d\theta}\right)^2 + \left(\frac{dy}{d\theta}\right)^2} = \sqrt{(-3\sin\theta)^2 + (3\cos\theta)^2} = \sqrt{3^2\left(\sin^2\theta + \cos^2\theta\right)} = 3$

$$\Rightarrow \text{Length} = \int_0^{3\pi/2} 3\, d\theta = 3\int_0^{3\pi/2} d\theta = 3\left(\tfrac{3\pi}{2} - 0\right) = \tfrac{9\pi}{2}$$

25. The equipment alone: the force required to lift the equipment is equal to its weight $\Rightarrow F_1(x) = 100$ N.

The work done is $W_1 = \int_a^b F_1(x)\, dx = \int_0^{40} 100\, dx = [100x]_0^{40} = 4000$ J; the rope alone: the force required to lift the rope is equal to the weight of the rope paid out at elevation $x \Rightarrow F_2(x) = 0.8(40 - x)$. The work done is $W_2 = \int_a^b F_2(x)\, dx = \int_0^{40} 0.8(40 - x)\, dx = 0.8\left[40x - \frac{x^2}{2}\right]_0^{40} = 0.8\left(40^2 - \frac{40^2}{2}\right) = \frac{(0.8)(1600)}{2} = 640$ J;

the total work is $W = W_1 + W_2 = 4000 + 640 = 4640$ J

27. Force constant: $F = kx \Rightarrow 20 = k \cdot 1 \Rightarrow k = 20$ lb/ft; the work to stretch the spring 1 ft is

$$W = \int_0^1 kx\,dx = k\int_0^1 x\,dx = \left[20\frac{x^2}{2}\right]_0^1 = 10 \text{ ft}\cdot\text{lb}$$; the work to stretch the spring an additional foot is

$$W = \int_1^2 kx\,dx = k\int_1^2 x\,dx = 20\left[\frac{x^2}{2}\right]_1^2 = 20\left(\frac{4}{2}-\frac{1}{2}\right) = 20\left(\frac{3}{2}\right) = 30 \text{ ft}\cdot\text{lb}$$

29. We imagine the water divided into thin slabs by planes perpendicular to the y-axis at the points of a partition of the interval $[0, 8]$. The typical slab between the planes at y and $y + \Delta y$ has a volume of about $\Delta V = \pi(\text{radius})^2(\text{thickness}) = \pi\left(\frac{5}{4}y\right)^2 = \frac{25\pi}{16}y^2\,\Delta y \text{ ft}^3$. The force $F(y)$ required to lift this slab is equal to its weight: $F(y) = 62.4\,\Delta V = \frac{(62.4)(25)}{16}\pi y^2\,\Delta y$ lb. The distance through which $F(y)$ must act to lift this slab to the level 6 ft above the top is about $(6 + 8 - y)$ ft, so the work done lifting the slab is about $\Delta W = \frac{(62.4)(25)}{16}\pi y^2(14 - y)\,\Delta y$ ft · lb. The work done lifting all the slabs from $y = 0$ to $y = 8$ to the level 6 ft above the top is approximately $W \approx \sum_0^8 \frac{(62.4)(25)}{16}\pi y^2\,(14 - y)\,\Delta y$ ft · lb so the work to pump the water is the limit of these Riemann sums as the norm of the partition goes to zero:

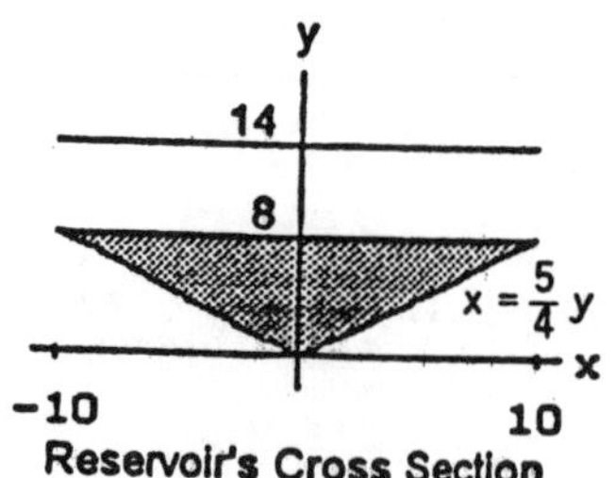

Reservoir's Cross Section

$$W = \int_0^8 \frac{(62.4)(25)}{(16)}\pi y^2(14 - y)\,dy = \frac{(62.4)(25)\pi}{16}\int_0^8 (14y^2 - y^3)\,dy = (62.4)\left(\frac{25\pi}{16}\right)\left[\frac{14}{3}y^3 - \frac{y^4}{4}\right]_0^8$$

$$= (62.4)\left(\frac{25\pi}{16}\right)\left(\frac{14}{3}\cdot 8^3 - \frac{8^4}{4}\right) \approx 418{,}208.81 \text{ ft}\cdot\text{lb}$$

31. The tank's cross section looks like the figure in Exercise 29 with right edge given by $x = \frac{5}{10}y = \frac{y}{2}$. A typical horizontal slab has volume $\Delta V = \pi(\text{radius})^2(\text{thickness}) = \pi\left(\frac{y}{2}\right)^2\Delta y = \frac{\pi}{4}y^2\,\Delta y$. The force required to lift this slab is its weight: $F(y) = 60 \cdot \frac{\pi}{4}y^2\,\Delta y$. The distance through which $F(y)$ must act is $(2 + 10 - y)$ ft, so the work to pump the liquid is $W = 60\int_0^{10} \pi(12 - y)\left(\frac{y^2}{4}\right)dy = 15\pi\left[\frac{12y^3}{3} - \frac{y^4}{4}\right]_0^{10} = 22{,}500\pi$ ft · lb; the time needed to empty the tank is $\frac{22{,}500 \text{ ft}\cdot\text{lb}}{275 \text{ ft}\cdot\text{lb/sec}} \approx 257$ sec

33. $$F = \int_a^b W\cdot\left(\begin{matrix}\text{strip}\\\text{depth}\end{matrix}\right)\cdot L(y)\,dy \Rightarrow F = 2\int_0^2 (62.4)(2 - y)(2y)\,dy = 249.6\int_0^2 (2y - y^2)\,dy = 249.6\left[y^2 - \frac{y^3}{3}\right]_0^2$$

$$= (249.6)\left(4 - \frac{8}{3}\right) = (249.6)\left(\frac{4}{3}\right) = 332.8 \text{ lb}$$

35. $F = \int_a^b W \cdot \begin{pmatrix}\text{strip}\\ \text{depth}\end{pmatrix} \cdot L(y)\,dy \Rightarrow F = 62.4 \int_0^4 (9-y)\left(2 \cdot \frac{\sqrt{y}}{2}\right) dy = 62.4 \int_0^4 \left(9y^{1/2} - y^{3/2}\right) dy$

$= 62.4\left[6y^{3/2} - \frac{2}{5}y^{5/2}\right]_0^4 = (62.4)\left(6 \cdot 8 - \frac{2}{5} \cdot 32\right) = \left(\frac{62.4}{5}\right)(48 \cdot 5 - 64) = \frac{(62.4)(176)}{5} = 2196.48 \text{ lb}$

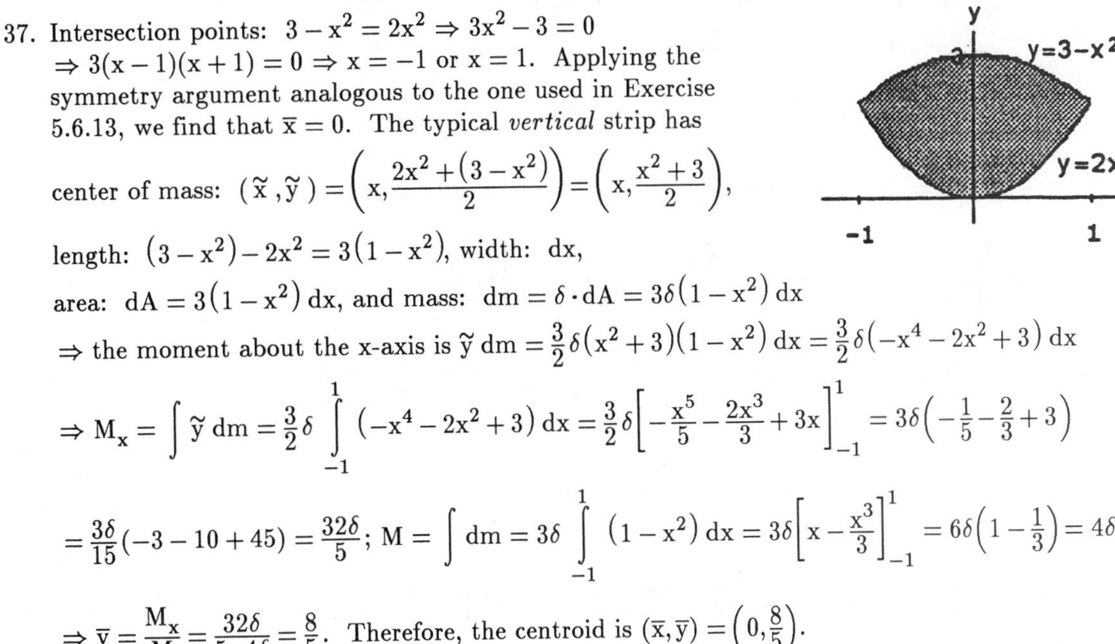

37. Intersection points: $3 - x^2 = 2x^2 \Rightarrow 3x^2 - 3 = 0$ $\Rightarrow 3(x-1)(x+1) = 0 \Rightarrow x = -1$ or $x = 1$. Applying the symmetry argument analogous to the one used in Exercise 5.6.13, we find that $\bar{x} = 0$. The typical *vertical* strip has center of mass: $(\tilde{x}, \tilde{y}) = \left(x, \frac{2x^2 + (3 - x^2)}{2}\right) = \left(x, \frac{x^2+3}{2}\right)$,

length: $(3 - x^2) - 2x^2 = 3(1 - x^2)$, width: dx,

area: $dA = 3(1 - x^2)\,dx$, and mass: $dm = \delta \cdot dA = 3\delta(1 - x^2)\,dx$

$\Rightarrow$ the moment about the x-axis is $\tilde{y}\,dm = \frac{3}{2}\delta(x^2+3)(1-x^2)\,dx = \frac{3}{2}\delta(-x^4 - 2x^2 + 3)\,dx$

$\Rightarrow M_x = \int \tilde{y}\,dm = \frac{3}{2}\delta \int_{-1}^{1} (-x^4 - 2x^2 + 3)\,dx = \frac{3}{2}\delta\left[-\frac{x^5}{5} - \frac{2x^3}{3} + 3x\right]_{-1}^{1} = 3\delta\left(-\frac{1}{5} - \frac{2}{3} + 3\right)$

$= \frac{3\delta}{15}(-3 - 10 + 45) = \frac{32\delta}{5}$; $M = \int dm = 3\delta \int_{-1}^{1} (1 - x^2)\,dx = 3\delta\left[x - \frac{x^3}{3}\right]_{-1}^{1} = 6\delta\left(1 - \frac{1}{3}\right) = 4\delta$

$\Rightarrow \bar{y} = \frac{M_x}{M} = \frac{32\delta}{5 \cdot 4\delta} = \frac{8}{5}$. Therefore, the centroid is $(\bar{x}, \bar{y}) = \left(0, \frac{8}{5}\right)$.

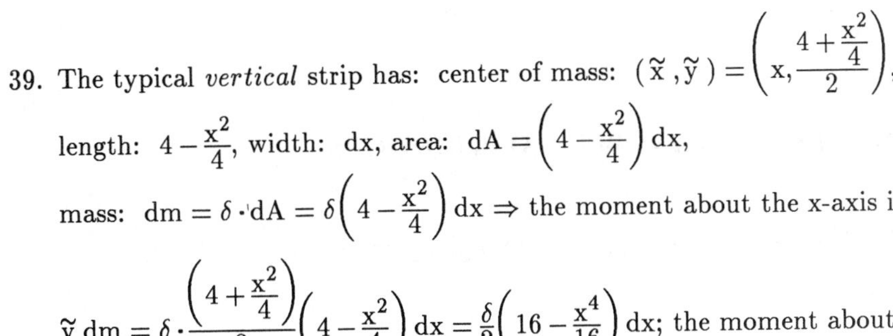

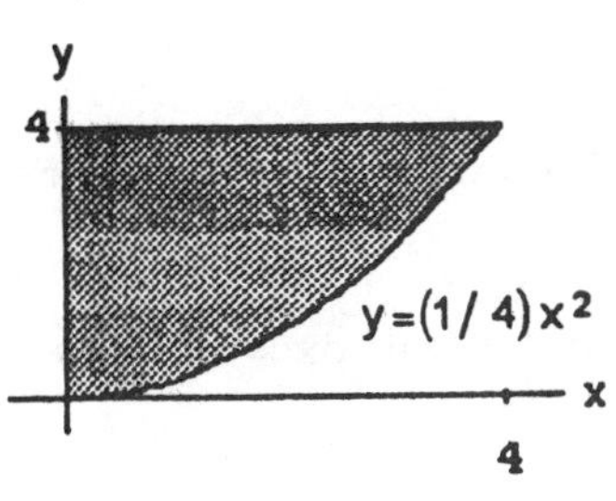

39. The typical *vertical* strip has: center of mass: $(\tilde{x}, \tilde{y}) = \left(x, \frac{4 + \frac{x^2}{4}}{2}\right)$,

length: $4 - \frac{x^2}{4}$, width: dx, area: $dA = \left(4 - \frac{x^2}{4}\right) dx$,

mass: $dm = \delta \cdot dA = \delta\left(4 - \frac{x^2}{4}\right) dx \Rightarrow$ the moment about the x-axis is

$\tilde{y}\,dm = \delta \cdot \frac{\left(4 + \frac{x^2}{4}\right)}{2}\left(4 - \frac{x^2}{4}\right) dx = \frac{\delta}{2}\left(16 - \frac{x^4}{16}\right) dx$; the moment about

the y-axis is $\tilde{x}\,dm = \delta\left(4 - \frac{x^2}{4}\right) \cdot x\,dx = \delta\left(4x - \frac{x^3}{4}\right) dx$. Thus, $M_x = \int \tilde{y}\,dm = \frac{\delta}{2} \int_0^4 \left(16 - \frac{x^4}{16}\right) dx$

$= \frac{\delta}{2}\left[16x - \frac{x^5}{5 \cdot 16}\right]_0^4 = \frac{\delta}{2}\left[64 - \frac{64}{5}\right] = \frac{128\delta}{5}$; $M_y = \int \tilde{x}\,dm = \delta \int_0^4 \left(4x - \frac{x^3}{4}\right) dx = \delta\left[2x^2 - \frac{x^4}{16}\right]_0^4$

$= \delta(32 - 16) = 16\delta$; $M = \int dm = \delta \int_{-0}^{4} \left(4 - \frac{x^2}{4}\right) dx = \delta\left[4x - \frac{x^3}{12}\right]_0^4 = \delta\left(16 - \frac{64}{12}\right) = \frac{32\delta}{3}$

$\Rightarrow \bar{x} = \frac{M_y}{M} = \frac{16 \cdot \delta \cdot 3}{32 \cdot \delta} = \frac{3}{2}$ and $\bar{y} = \frac{M_x}{M} = \frac{128 \cdot \delta \cdot 3}{5 \cdot 32 \cdot \delta} = \frac{12}{5}$. Therefore, the centroid is $(\bar{x}, \bar{y}) = \left(\frac{3}{2}, \frac{12}{5}\right)$.

41. A typical horizontal strip has: center of mass: $(\tilde{x}, \tilde{y}) = \left(\frac{y^2+2y}{2}, y\right)$, length: $2y - y^2$, width: dy, area: $dA = (2y - y^2)\,dy$, mass: $dm = \delta \cdot dA = (1+y)(2y-y^2)\,dy \Rightarrow$ the moment about the x-axis is $\tilde{y}\,dm = y(1+y)(2y-y^2)\,dy = (2y^2+2y^3-y^3-y^4)\,dy$ $= (2y^2+y^3-y^4)\,dy$; the moment about the y-axis is $\tilde{x}\,dm = \left(\frac{y^2+2y}{2}\right)(1+y)(2y-y^2)\,dy = \frac{1}{2}(4y^2-y^4)(1+y)\,dy$ $= \frac{1}{2}(4y^2+4y^3-y^4-y^5)\,dy \Rightarrow M_x = \int \tilde{y}\,dm = \int_0^2 (2y^2+y^3-y^4)\,dy = \left[\frac{2}{3}y^3+\frac{y^4}{4}-\frac{y^5}{5}\right]_0^2$ $= \left(\frac{16}{3}+\frac{16}{4}-\frac{32}{5}\right) = 16\left(\frac{1}{3}+\frac{1}{4}-\frac{2}{5}\right) = \frac{16}{60}(20+15-24) = \frac{4}{15}(11) = \frac{44}{15}$; $M_y = \int \tilde{x}\,dm$ $= \int_0^2 \frac{1}{2}(4y^2+4y^3-y^4-y^5)\,dy = \frac{1}{2}\left[\frac{4}{3}y^3+y^4-\frac{y^5}{5}-\frac{y^6}{6}\right]_0^2 = \frac{1}{2}\left(\frac{4\cdot 2^3}{3}+2^4-\frac{2^5}{5}-\frac{2^6}{6}\right)$ $= 4\left(\frac{4}{3}+2-\frac{4}{5}-\frac{8}{6}\right) = 4\left(2-\frac{4}{5}\right) = \frac{24}{5}$; $M = \int dm = \int_0^2 (1+y)(2y-y^2)\,dy = \int_0^2 (2y+y^2-y^3)\,dy$ $= \left[y^2+\frac{y^3}{3}-\frac{y^4}{4}\right]_0^2 = \left(4+\frac{8}{3}-\frac{16}{4}\right) = \frac{8}{3} \Rightarrow \bar{x} = \frac{M_y}{M} = \left(\frac{24}{5}\right)\left(\frac{3}{8}\right) = \frac{9}{5}$ and $\bar{y} = \frac{M_x}{M} = \left(\frac{44}{15}\right)\left(\frac{3}{8}\right) = \frac{44}{40} = \frac{11}{10}$. Therefore, the center of mass is $(\bar{x}, \bar{y}) = \left(\frac{9}{5}, \frac{11}{10}\right)$.

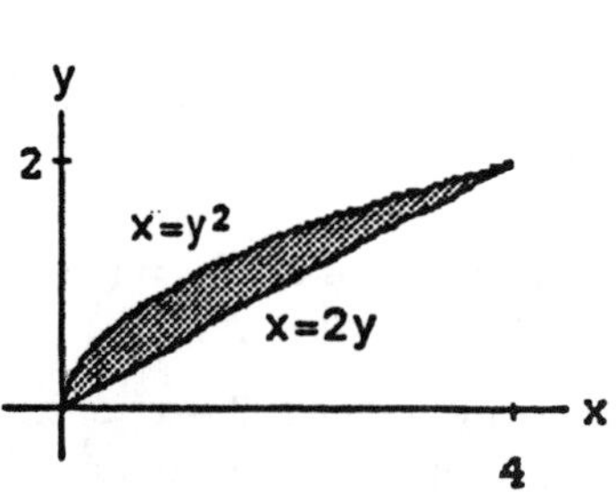

CHAPTER 5 ADDITIONAL EXERCISES–THEORY, EXAMPLES, APPLICATIONS

1. $V = \pi \int_0^a [f(x)]^2\,dx = a^2 + a \Rightarrow \pi \int_0^x [f(t)]^2\,dt = x^2 + x$ for all $x > a \Rightarrow \pi[f(x)]^2 = 2x + 1 \Rightarrow f(x) = \sqrt{\frac{2x+1}{\pi}}$

3. $s(x) = Cx \Rightarrow \int_0^x \sqrt{1+[f'(t)]^2}\,dt = Cx \Rightarrow \sqrt{1+[f'(x)]^2} = C \Rightarrow f'(x) = \sqrt{C^2-1}$ for $C \ge 1$

$\Rightarrow f(x) = \int_0^x \sqrt{C^2-1}\,dt + k$. Then $f(0) = a \Rightarrow a = 0 + k \Rightarrow f(x) = \int_0^x \sqrt{C^2-1}\,dt + a \Rightarrow f(x) = x\sqrt{C^2-1} + a$, where $C \ge 1$.

5. Converting to pounds and feet, 2 lb/in $= \frac{2\text{ lb}}{1\text{ in}} \cdot \frac{12\text{ in}}{1\text{ ft}} = 24$ lb/ft. Thus, $F = 24x \Rightarrow W = \int_0^{1/2} 24x\,dx$ $= \left[12x^2\right]_0^{1/2} = 3$ ft·lb. Since $W = \frac{1}{2}mv_0^2 - \frac{1}{2}mv_1^2$, where $W = 3$ ft·lb, $m = \left(\frac{1}{10}\text{ lb}\right)\left(\frac{1}{32\text{ ft/sec}^2}\right)$

$= \frac{1}{320}$ slugs, and $v_1 = 0$ ft/sec, we have $3 = \left(\frac{1}{2}\right)\left(\frac{1}{320}\,v_0^2\right) \Rightarrow v_0^2 = 3 \cdot 640$. For the projectile height, $s = -16t^2 + v_0 t$ (since $s = 0$ at $t = 0$) $\Rightarrow \frac{ds}{dt} = v = -32t + v_0$. At the top of the ball's path, $v = 0 \Rightarrow t = \frac{v_0}{32}$ and the height is $s = -16\left(\frac{v_0}{32}\right)^2 + v_0\left(\frac{v_0}{32}\right) = \frac{v_0^2}{64} = \frac{3 \cdot 640}{64} = 30$ ft.

7. Consider a rectangular plate of length ℓ and width w. The length is parallel with the surface of the fluid of weight density ω. The force on one side of the plate is $F = \omega \int_{-w}^{0} (-y)(\ell)\,dy = -\omega\ell\left[\frac{y^2}{2}\right]_{-w}^{0} = \frac{\omega\ell w^2}{2}$. The average force on one side of the plate is $F_{av} = \frac{\omega}{w}\int_{-w}^{0} (-y)\,dy = \frac{\omega}{w}\left[-\frac{y^2}{2}\right]_{-w}^{0} = \frac{\omega w}{2}$. Therefore the force $\frac{\omega\ell w^2}{2} = \left(\frac{\omega w}{2}\right)(\ell w)$ = (the average pressure up and down) $\cdot$ (the area of the plate).

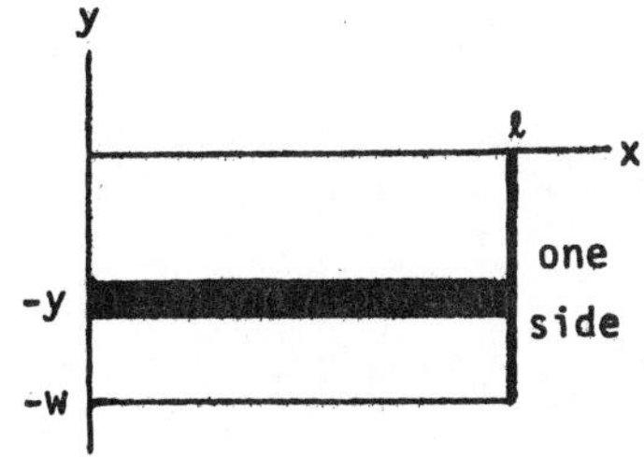

9. From the symmetry of $y = 1 - x^n$, n even, about the y-axis for $-1 \le x \le 1$, we have $\bar{x} = 0$. To find $\bar{y} = \frac{M_x}{M}$, we use the vertical strips technique. The typical strip has center of mass: $(\tilde{x}, \tilde{y}) = \left(x, \frac{1-x^n}{2}\right)$, length: $1 - x^n$, width: dx, area: $dA = (1 - x^n)\,dx$, mass: $dm = 1 \cdot dA = (1 - x^n)\,dx$. The moment of the strip about the x-axis is $\tilde{y}\,dm = \frac{(1-x^n)^2}{2}\,dx \Rightarrow M_x = \int_{-1}^{1} \frac{(1-x^n)^2}{2}\,dx = 2\int_0^1 \frac{1}{2}\left(1 - 2x^n + x^{2n}\right)dx = \left[x - \frac{2x^{n+1}}{n+1} + \frac{x^{2n+1}}{2n+1}\right]_0^1$

$$= 1 - \frac{2}{n+1} + \frac{1}{2n+1} = \frac{(n+1)(2n+1) - 2(2n+1) + (n+1)}{(n+1)(2n+1)} = \frac{2n^2 + 3n + 1 - 4n - 2 + n + 1}{(n+1)(2n+1)} = \frac{2n^2}{(n+1)(2n+1)}.$$

Also, $M = \int_{-1}^{1} dA = \int_{-1}^{1} (1 - x^n)\,dx = 2\int_0^1 (1 - x^n)\,dx = 2\left[x - \frac{x^{n+1}}{n+1}\right]_0^1 = 2\left(1 - \frac{1}{n+1}\right) = \frac{2n}{n+1}$. Therefore,

$\bar{y} = \frac{M_x}{M} = \frac{2n^2}{(n+1)(2n+1)} \cdot \frac{(n+1)}{2n} = \frac{n}{2n+1} \Rightarrow \left(0, \frac{n}{2n+1}\right)$ is the location of the centroid. As $n \to \infty$, $\bar{y} \to \frac{1}{2}$ so the limiting position of the centroid is $\left(0, \frac{1}{2}\right)$.

11. (a) Consider a single vertical strip with center of mass $(\tilde{x}, \tilde{y})$. If the plate lies to the right of the line, then the moment of this strip about the line $x = b$ is $(\tilde{x} - b)\,dm = (\tilde{x} - b)\,\delta\,dA \Rightarrow$ the plate's first moment about $x = b$ is the integral $\int (x - b)\delta\,dA = \int \delta x\,dA - \int \delta b\,dA = M_y - b\delta A$.

(b) If the plate lies to the left of the line, the moment of a vertical strip about the line $x = b$ is $(b - \tilde{x})\,dm = (b - \tilde{x})\,\delta\,dA \Rightarrow$ the plate's first moment about $x = b$ is $\int (b - x)\delta\,dA = \int b\delta\,dA - \int \delta x\,dA = b\delta A - M_y$.

13. (a) On $[0,a]$ a typical *vertical* strip has center of mass: $(\tilde{x},\tilde{y}) = \left(x, \frac{\sqrt{b^2-x^2}+\sqrt{a^2-x^2}}{2}\right)$, length: $\sqrt{b^2-x^2}-\sqrt{a^2-x^2}$, width: dx, area: $dA = \left(\sqrt{b^2-x^2}-\sqrt{a^2-x^2}\right)dx$, mass: $dm = \delta\, dA = \delta\left(\sqrt{b^2-x^2}-\sqrt{a^2-x^2}\right)dx$. On $[a,b]$ a typical *vertical* strip has center of mass: $(\tilde{x},\tilde{y}) = \left(x, \frac{\sqrt{b^2-x^2}}{2}\right)$, length: $\sqrt{b^2-x^2}$, width: dx, area: $dA = \sqrt{b^2-x^2}\,dx$, mass: $dm = \delta\, dA = \delta\sqrt{b^2-x^2}\,dx$. Thus, $M_x = \int \tilde{y}\, dm$

$$= \int_0^a \tfrac{1}{2}\left(\sqrt{b^2-x^2}+\sqrt{a^2-x^2}\right)\delta\left(\sqrt{b^2-x^2}-\sqrt{a^2-x^2}\right)dx + \int_a^b \tfrac{1}{2}\sqrt{b^2-x^2}\,\delta\sqrt{b^2-x^2}\,dx$$

$$= \tfrac{\delta}{2}\int_0^a \left[(b^2-x^2)-(a^2-x^2)\right]dx + \tfrac{\delta}{2}\int_a^b (b^2-x^2)\,dx = \tfrac{\delta}{2}\int_0^a (b^2-a^2)\,dx + \tfrac{\delta}{2}\int_a^b (b^2-x^2)\,dx$$

$$= \tfrac{\delta}{2}\left[(b^2-a^2)x\right]_0^a + \tfrac{\delta}{2}\left[b^2x-\tfrac{x^3}{3}\right]_a^b = \tfrac{\delta}{2}\left[(b^2-a^2)a\right] + \tfrac{\delta}{2}\left[\left(b^3-\tfrac{b^3}{3}\right)-\left(b^2a-\tfrac{a^3}{3}\right)\right]$$

$$= \tfrac{\delta}{2}\left(ab^2-a^3\right) + \tfrac{\delta}{2}\left(\tfrac{2}{3}b^3-ab^2+\tfrac{a^3}{3}\right) = \frac{\delta b^3}{3}-\frac{\delta a^3}{3} = \delta\left(\frac{b^3-a^3}{3}\right);\ M_y = \int \tilde{x}\, dm$$

$$= \int_0^a x\delta\left(\sqrt{b^2-x^2}-\sqrt{a^2-x^2}\right)dx + \int_a^b x\delta\sqrt{b^2-x^2}\,dx$$

$$= \delta\int_0^a x\left(b^2-x^2\right)^{1/2}dx - \delta\int_0^a x\left(a^2-x^2\right)^{1/2}dx + \delta\int_a^b x\left(b^2-x^2\right)^{1/2}dx$$

$$= \frac{-\delta}{2}\left[\frac{2\left(b^2-x^2\right)^{3/2}}{3}\right]_0^a + \frac{\delta}{2}\left[\frac{2\left(a^2-x^2\right)^{3/2}}{3}\right]_0^a - \frac{\delta}{2}\left[\frac{2\left(b^2-x^2\right)^{3/2}}{3}\right]_a^b$$

$$= -\tfrac{\delta}{3}\left[\left(b^2-a^2\right)^{3/2}-\left(b^2\right)^{3/2}\right] + \tfrac{\delta}{3}\left[0-\left(a^2\right)^{3/2}\right] - \tfrac{\delta}{3}\left[0-\left(b^2-a^2\right)^{3/2}\right] = \frac{\delta b^3}{3}-\frac{\delta a^3}{3} = \frac{\delta\left(b^3-a^3\right)}{3} = M_x;$$

We calculate the mass geometrically: $M = \delta A = \delta\left(\frac{\pi b^2}{4}\right)-\delta\left(\frac{\pi a^2}{4}\right) = \frac{\delta\pi}{4}\left(b^2-a^2\right)$. Thus, $\overline{x} = \frac{M_y}{M}$

$$= \frac{\delta\left(b^3-a^3\right)}{3}\cdot\frac{4}{\delta\pi\left(b^2-a^2\right)} = \frac{4}{3\pi}\left(\frac{b^3-a^3}{b^2-a^2}\right) = \frac{4}{3\pi}\,\frac{(b-a)\left(a^2+ab+b^2\right)}{(b-a)(b+a)} = \frac{4\left(a^2+ab+b^2\right)}{3\pi(a+b)};$$ likewise

$$\overline{y} = \frac{M_x}{M} = \frac{4\left(a^2+ab+b^2\right)}{3\pi(a+b)}.$$

(b) $\lim\limits_{b\to a} \frac{4}{3\pi}\left(\frac{a^2+ab+b^2}{a+b}\right) = \left(\frac{4}{3\pi}\right)\left(\frac{a^2+a^2+a^2}{a+a}\right) = \left(\frac{4}{3\pi}\right)\left(\frac{3a^2}{2a}\right) = \frac{2a}{\pi} \Rightarrow (\overline{x},\overline{y}) = \left(\frac{2a}{\pi},\frac{2a}{\pi}\right)$ is the limiting position of the centroid as $b \to a$. This is the centroid of a circle of radius a (and we note the two circles coincide when $b = a$).

CHAPTER 6 TRANSCENDENTAL FUNCTIONS AND DIFFERENTIAL EQUATIONS

6.1 LOGARITHMS

1. $y = \ln 3x \Rightarrow y' = \left(\frac{1}{3x}\right)(3) = \frac{1}{x}$

3. $y = \ln \frac{3}{x} = \ln 3x^{-1} \Rightarrow \frac{dy}{dx} = \left(\frac{1}{3x^{-1}}\right)\left(-3x^{-2}\right) = -\frac{1}{x}$

5. $y = \ln x^3 \Rightarrow \frac{dy}{dx} = \left(\frac{1}{x^3}\right)(3x^2) = \frac{3}{x}$

7. $y = t(\ln t)^2 \Rightarrow \frac{dy}{dt} = (\ln t)^2 + 2t(\ln t)\cdot\frac{d}{dt}(\ln t) = (\ln t)^2 + \frac{2t \ln t}{t} = (\ln t)^2 + 2 \ln t$

9. $y = \frac{x^4}{4}\ln x - \frac{x^4}{16} \Rightarrow \frac{dy}{dx} = x^3 \ln x + \frac{x^4}{4}\cdot\frac{1}{x} - \frac{4x^3}{16} = x^3 \ln x$

11. $y = \frac{1 + \ln t}{t} \Rightarrow \frac{dy}{dt} = \frac{t\left(\frac{1}{t}\right) - (1 + \ln t)(1)}{t^2} = \frac{1 - 1 - \ln t}{t^2} = -\frac{\ln t}{t^2}$

13. $y = \frac{x \ln x}{1 + \ln x} \Rightarrow y' = \frac{(1 + \ln x)\left(\ln x + x\cdot\frac{1}{x}\right) - (x \ln x)\left(\frac{1}{x}\right)}{(1 + \ln x)^2} = \frac{(1 + \ln x)^2 - \ln x}{(1 + \ln x)^2} = 1 - \frac{\ln x}{(1 + \ln x)^2}$

15. $y = \ln(\ln(\ln x)) \Rightarrow y' = \frac{1}{\ln(\ln x)}\cdot\frac{d}{dx}(\ln(\ln x)) = \frac{1}{\ln(\ln x)}\cdot\frac{1}{\ln x}\cdot\frac{d}{dx}(\ln x) = \frac{1}{x(\ln x)\ln(\ln x)}$

17. $y = \ln(\sec\theta + \tan\theta) \Rightarrow \frac{dy}{d\theta} = \frac{\sec\theta\tan\theta + \sec^2\theta}{\sec\theta + \tan\theta} = \frac{\sec\theta(\tan\theta + \sec\theta)}{\tan\theta + \sec\theta} = \sec\theta$

19. $y \equiv \frac{1 + \ln t}{1 - \ln t} \Rightarrow \frac{dy}{dt} = \frac{(1 - \ln t)\left(\frac{1}{t}\right) - (1 + \ln t)\left(\frac{-1}{t}\right)}{(1 - \ln t)^2} = \frac{\frac{1}{t} - \frac{\ln t}{t} + \frac{1}{t} + \frac{\ln t}{t}}{(1 - \ln t)^2} = \frac{2}{t(1 - \ln t)^2}$

21. $y = \ln(\sec(\ln\theta)) \Rightarrow \frac{dy}{d\theta} = \frac{1}{\sec(\ln\theta)}\cdot\frac{d}{d\theta}(\sec(\ln\theta)) = \frac{\sec(\ln\theta)\tan(\ln\theta)}{\sec(\ln\theta)}\cdot\frac{d}{d\theta}(\ln\theta) = \frac{\tan(\ln\theta)}{\theta}$

23. $y = \int_{x^2/2}^{x^2} \ln\sqrt{t}\,dt \Rightarrow \frac{dy}{dx} = \left(\ln\sqrt{x^2}\right)\cdot\frac{d}{dx}(x^2) - \left(\ln\sqrt{\frac{x^2}{2}}\right)\cdot\frac{d}{dx}\left(\frac{x^2}{2}\right) = 2x\ln|x| - x\ln\frac{|x|}{\sqrt{2}}$

25. $\int_{-3}^{-2} \frac{1}{x}\,dx = \left[\ln|x|\right]_{-3}^{-2} = \ln 2 - \ln 3 = \ln\frac{2}{3}$

27. $\int \frac{2y}{y^2 - 25}\,dy = \ln\left|y^2 - 25\right| + C$

29. $\int_0^{\pi} \frac{\sin t}{2-\cos t}\,dt = \left[\ln|2-\cos t|\right]_0^{\pi} = \ln 3 - \ln 1 = \ln 3$; or let $u = 2-\cos t \Rightarrow du = \sin t\,dt$ with $t = 0$

$\Rightarrow u = 1$ and $t = \pi \Rightarrow u = 3 \Rightarrow \int_0^{\pi} \frac{\sin t}{2-\cos t}\,dt = \int_1^3 \frac{1}{u}\,du = \left[\ln|u|\right]_1^3 = \ln 3 - \ln 1 = \ln 3$

31. Let $u = \ln x \Rightarrow du = \frac{1}{x}\,dx$; $x = 1 \Rightarrow u = 0$ and $x = 2 \Rightarrow u = \ln 2$;

$\int_1^2 \frac{2\ln x}{x}\,dx = \int_0^{\ln 2} 2u\,du = \left[u^2\right]_0^{\ln 2} = (\ln 2)^2$

33. Let $u = \ln x \Rightarrow du = \frac{1}{x}\,dx$; $x = 2 \Rightarrow u = \ln 2$ and $x = 4 \Rightarrow u = \ln 4$;

$\int_2^4 \frac{dx}{x(\ln x)^2} = \int_{\ln 2}^{\ln 4} u^{-2}\,du = \left[-\frac{1}{u}\right]_{\ln 2}^{\ln 4} = -\frac{1}{\ln 4} + \frac{1}{\ln 2} = -\frac{1}{\ln 2^2} + \frac{1}{\ln 2} = -\frac{1}{2\ln 2} + \frac{1}{\ln 2} = \frac{1}{2\ln 2} = \frac{1}{\ln 4}$

35. Let $u = 6 + 3\tan t \Rightarrow du = 3\sec^2 t\,dt$;

$\int \frac{3\sec^2 t}{6+3\tan t}\,dt = \int \frac{du}{u} = \ln|u| + C = \ln|6+3\tan t| + C$

37. Let $u = \cos\frac{x}{2} \Rightarrow du = -\frac{1}{2}\sin\frac{x}{2}\,dx \Rightarrow -2\,du = \sin\frac{x}{2}\,dx$; $x = 0 \Rightarrow u = 1$ and $x = \frac{\pi}{2} \Rightarrow u = \frac{1}{\sqrt{2}}$;

$\int_0^{\pi/2} \tan\frac{x}{2}\,dx = \int_0^{\pi/2} \frac{\sin\frac{x}{2}}{\cos\frac{x}{2}}\,dx = -2\int_1^{1/\sqrt{2}} \frac{du}{u} = \left[-2\ln|u|\right]_1^{1/\sqrt{2}} = -2\ln\frac{1}{\sqrt{2}} = 2\ln\sqrt{2} = \ln 2$

39. Let $u = \sin\frac{\theta}{3} \Rightarrow du = \frac{1}{3}\cos\frac{\theta}{3}\,d\theta \Rightarrow 6\,du = 2\cos\frac{\theta}{3}\,d\theta$; $\theta = \frac{\pi}{2} \Rightarrow u = \frac{1}{2}$ and $\theta = \pi \Rightarrow u = \frac{\sqrt{3}}{2}$;

$\int_{\pi/2}^{\pi} 2\cot\frac{\theta}{3}\,d\theta = \int_{\pi/2}^{\pi} \frac{2\cos\frac{\theta}{3}}{\sin\frac{\theta}{3}}\,d\theta = 6\int_{1/2}^{\sqrt{3}/2} \frac{du}{u} = 6\left[\ln|u|\right]_{1/2}^{\sqrt{3}/2} = 6\left(\ln\frac{\sqrt{3}}{2} - \ln\frac{1}{2}\right) = 6\ln\sqrt{3} = \ln 27$

41. $\int \frac{dx}{2\sqrt{x}+2x} = \int \frac{dx}{2\sqrt{x}(1+\sqrt{x})}$; let $u = 1+\sqrt{x} \Rightarrow du = \frac{1}{2\sqrt{x}}\,dx$; $\int \frac{dx}{2\sqrt{x}(1+\sqrt{x})} = \int \frac{du}{u} = \ln|u| + C$

$= \ln\left|1+\sqrt{x}\right| + C = \ln\left(1+\sqrt{x}\right) + C$

43. $y = \sqrt{x(x+1)} = (x(x+1))^{1/2} \Rightarrow \ln y = \frac{1}{2}\ln(x(x+1)) \Rightarrow 2\ln y = \ln(x) + \ln(x+1) \Rightarrow \frac{2y'}{y} = \frac{1}{x} + \frac{1}{x+1}$

$\Rightarrow y' = \left(\frac{1}{2}\right)\sqrt{x(x+1)}\left(\frac{1}{x} + \frac{1}{x+1}\right) = \frac{\sqrt{x(x+1)}\,(2x+1)}{2x(x+1)} = \frac{2x+1}{2\sqrt{x(x+1)}}$

45. $y = \sqrt{\theta+3}\,(\sin\theta) = (\theta+3)^{1/2}\sin\theta \Rightarrow \ln y = \frac{1}{2}\ln(\theta+3) + \ln(\sin\theta) \Rightarrow \frac{1}{y}\frac{dy}{d\theta} = \frac{1}{2(\theta+3)} + \frac{\cos\theta}{\sin\theta}$

$\Rightarrow \frac{dy}{d\theta} = \sqrt{\theta+3}\,(\sin\theta)\left[\frac{1}{2(\theta+3)} + \cot\theta\right]$

47. $y = t(t+1)(t+2) \Rightarrow \ln y = \ln t + \ln(t+1) + \ln(t+2) \Rightarrow \frac{1}{y}\frac{dy}{dt} = \frac{1}{t} + \frac{1}{t+1} + \frac{1}{t+2}$

$\Rightarrow \frac{dy}{dt} = t(t+1)(t+2)\left(\frac{1}{t} + \frac{1}{t+1} + \frac{1}{t+2}\right) = t(t+1)(t+2)\left[\frac{(t+1)(t+2) + t(t+2) + t(t+1)}{t(t+1)(t+2)}\right] = 3t^2 + 6t + 2$

49. $y = \frac{\theta \sin \theta}{\sqrt{\sec \theta}} \Rightarrow \ln y = \ln \theta + \ln(\sin \theta) - \frac{1}{2}\ln(\sec \theta) \Rightarrow \frac{1}{y}\frac{dy}{d\theta} = \left[\frac{1}{\theta} + \frac{\cos \theta}{\sin \theta} - \frac{(\sec \theta)(\tan \theta)}{2 \sec \theta}\right]$

$\Rightarrow \frac{dy}{d\theta} = \frac{\theta \sin \theta}{\sqrt{\sec \theta}}\left(\frac{1}{\theta} + \cot \theta - \frac{1}{2}\tan \theta\right)$

51. $y = \sqrt[3]{\frac{x(x-2)}{x^2+1}} \Rightarrow \ln y = \frac{1}{3}\left[\ln x + \ln(x-2) - \ln\left(x^2+1\right)\right] \Rightarrow \frac{y'}{y} = \frac{1}{3}\left(\frac{1}{x} + \frac{1}{x-2} - \frac{2x}{x^2+1}\right)$

$\Rightarrow y' = \frac{1}{3}\sqrt[3]{\frac{x(x-2)}{x^2+1}}\left(\frac{1}{x} + \frac{1}{x-2} - \frac{2x}{x^2+1}\right)$

53. $\frac{dy}{dx} = 1 + \frac{1}{x}$ at $(1,3) \Rightarrow y = x + \ln|x| + C$; $y = 3$ at $x = 1 \Rightarrow C = 2 \Rightarrow y = x + \ln|x| + 2$

55. $\int \frac{\log_{10} x}{x}\,dx = \int \left(\frac{\ln x}{\ln 10}\right)\left(\frac{1}{x}\right)dx$; $\left[u = \ln x \Rightarrow du = \frac{1}{x}\,dx\right]$

$\rightarrow \int \left(\frac{\ln x}{\ln 10}\right)\left(\frac{1}{x}\right)dx = \frac{1}{\ln 10}\int u\,du = \left(\frac{1}{\ln 10}\right)\left(\frac{1}{2}u^2\right) + C = \frac{(\ln x)^2}{2 \ln 10} + C$

57. $\int_0^2 \frac{\log_2(x+2)}{x+2}\,dx = \frac{1}{\ln 2}\int_0^2 [\ln(x+2)]\left(\frac{1}{x+2}\right)dx = \left(\frac{1}{\ln 2}\right)\left[\frac{(\ln(x+2))^2}{2}\right]_0^2 = \left(\frac{1}{\ln 2}\right)\left[\frac{(\ln 4)^2}{2} - \frac{(\ln 2)^2}{2}\right]$

$= \left(\frac{1}{\ln 2}\right)\left[\frac{4(\ln 2)^2}{2} - \frac{(\ln 2)^2}{2}\right] = \frac{3}{2}\ln 2$

59. $\int \frac{dx}{x \log_{10} x} = \int \left(\frac{\ln 10}{\ln x}\right)\left(\frac{1}{x}\right)dx = (\ln 10)\int \left(\frac{1}{\ln x}\right)\left(\frac{1}{x}\right)dx$; $\left[u = \ln x \Rightarrow du = \frac{1}{x}\,dx\right]$

$\rightarrow (\ln 10)\int \left(\frac{1}{\ln x}\right)\left(\frac{1}{x}\right)dx = (\ln 10)\int \frac{1}{u}\,du = (\ln 10)\ln|u| + C = (\ln 10)\ln|\ln x| + C$

61. $y = \log_2 5\theta = \frac{\ln 5\theta}{\ln 2} \Rightarrow \frac{dy}{d\theta} = \left(\frac{1}{\ln 2}\right)\left(\frac{1}{5\theta}\right)(5) = \frac{1}{\theta \ln 2}$

63. $y = \log_2 r \cdot \log_4 r = \left(\frac{\ln r}{\ln 2}\right)\left(\frac{\ln r}{\ln 4}\right) = \frac{\ln^2 r}{(\ln 2)(\ln 4)} \Rightarrow \frac{dy}{dr} = \left[\frac{1}{(\ln 2)(\ln 4)}\right](2 \ln r)\left(\frac{1}{r}\right) = \frac{2 \ln r}{r(\ln 2)(\ln 4)}$

65. $y = \theta \sin(\log_7 \theta) = \theta \sin\left(\frac{\ln \theta}{\ln 7}\right) \Rightarrow \frac{dy}{d\theta} = \sin\left(\frac{\ln \theta}{\ln 7}\right) + \theta\left[\cos\left(\frac{\ln \theta}{\ln 7}\right)\right]\left(\frac{1}{\theta \ln 7}\right) = \sin(\log_7 \theta) + \frac{1}{\ln 7}\cos(\log_7 \theta)$

67. (a) $f(x) = \ln(\cos x) \Rightarrow f'(x) = -\frac{\sin x}{\cos x} = -\tan x = 0 \Rightarrow x = 0$; $f'(x) > 0$ for $-\frac{\pi}{4} \le x < 0$ and $f'(x) < 0$ for $0 < x \le \frac{\pi}{3} \Rightarrow$ there is a relative maximum at $x = 0$ with $f(0) = \ln(\cos 0) = \ln 1 = 0$; $f\left(-\frac{\pi}{4}\right) = \ln\left(\cos\left(-\frac{\pi}{4}\right)\right) = \ln\left(\frac{1}{\sqrt{2}}\right) = -\frac{1}{2}\ln 2$ and $f\left(\frac{\pi}{3}\right) = \ln\left(\cos\left(\frac{\pi}{3}\right)\right) = \ln\frac{1}{2} = -\ln 2$. Therefore, the absolute minimum occurs at $x = \frac{\pi}{3}$ with $f\left(\frac{\pi}{3}\right) = -\ln 2$ and the absolute maximum occurs at $x = 0$ with $f(0) = 0$.

(b) $f(x) = \cos(\ln x) \Rightarrow f'(x) = \frac{-\sin(\ln x)}{x} = 0 \Rightarrow x = 1$; $f'(x) > 0$ for $\frac{1}{2} \le x < 1$ and $f'(x) < 0$ for $1 < x \le 2$ $\Rightarrow$ there is a relative maximum at $x = 1$ with $f(1) = \cos(\ln 1) = \cos 0 = 1$; $f\left(\frac{1}{2}\right) = \cos\left(\ln\left(\frac{1}{2}\right)\right)$ $= \cos(-\ln 2) = \cos(\ln 2)$ and $f(2) = \cos(\ln 2)$. Therefore, the absolute minimum occurs at $x = \frac{1}{2}$ and $x = 2$ with $f\left(\frac{1}{2}\right) = f(2) = \cos(\ln 2)$, and the absolute maximum occurs at $x = 1$ with $f(1) = 1$.

69. $\int_1^5 (\ln 2x - \ln x)\,dx = \int_1^5 (-\ln x + \ln 2 + \ln x)\,dx = (\ln 2)\int_1^5 dx = (\ln 2)(5-1) = \ln 2^4 = \ln 16$

71. $V = \pi \int_0^3 \left(\frac{2}{\sqrt{y+1}}\right)^2 dy = 4\pi \int_0^3 \frac{1}{y+1}\,dy = 4\pi\left[\ln|y+1|\right]_0^3 = 4\pi(\ln 4 - \ln 1) = 4\pi \ln 4$

73. $V = 2\pi \int_{1/2}^2 x\left(\frac{1}{x^2}\right) dx = 2\pi \int_{1/2}^2 \frac{1}{x}\,dx = 2\pi\left[\ln|x|\right]_{1/2}^2 = 2\pi\left(\ln 2 - \ln\frac{1}{2}\right) = 2\pi(2\ln 2) = \pi \ln 2^4 = \pi \ln 16$

75. (a) $L(x) = f(0) + f'(0)\cdot x$, and $f(x) = \ln(1+x) \Rightarrow f'(x)\big|_{x=0} = \frac{1}{1+x}\Big|_{x=0} = 1 \Rightarrow L(x) = \ln 1 + 1\cdot x \Rightarrow L(x) = x$

(b) On $[0, 0.1]$, $f(x)$ and $L(x)$are both increasing because $f'(x) = \frac{1}{1+x} > 0$ and $L'(x) = 1 > 0$ for $0 \le x \le 0.1$. In addition $0 \le x \le 0.1 \Rightarrow 1 \le 1+x \le 1.1 \Rightarrow \frac{1}{1.1} \le \frac{1}{x+1} \le 1 \Rightarrow L'(x) \ge f'(x) \Rightarrow E(x) = f(x) - L(x)$ is non-increasing on $[0, 0.1]$ because $E'(x) = f'(x) - L' \le 0$ on the interval. Therefore, the largest error is $|f(0.1) - L(0.1)| = |\ln(1.1) - 1.1| \approx 0.00469$.

(c) The approximation $y = x$ for $\ln(1+x)$ is best for smaller positive values of x on the interval $[0, 0.1]$ as seen on the graph. As x increases so does the magnitude of the error $|\ln(x) - x|$. From the graph, an upper bound for the magnitude of the error is $|\ln(1.1) - 0.1| \approx 0.00469$ which is consistent with the analytical result obtained in part (b).

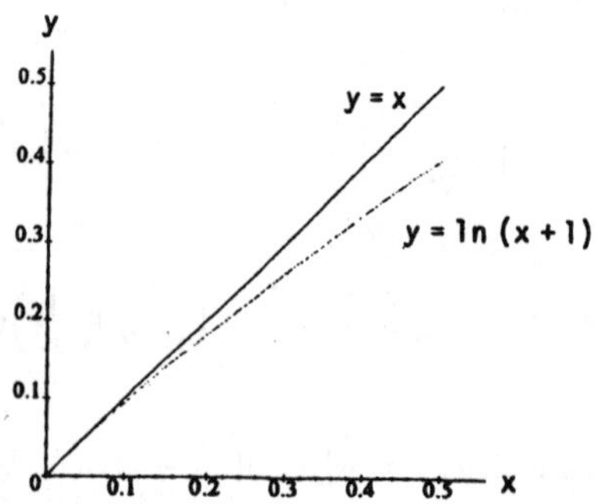

77. $\ln(1.2) = \ln(1+.2) \approx 0.2$, $\ln(.8) = \ln(1+(-0.2)) \approx -0.2$; with Simpson's rule for $n = 2$, $\ln(1.2) = \int_1^{1.2} \frac{1}{t}\,dt$

≈ 0.182323232 and $\ln(0.8) = \int_1^{0.8} \frac{1}{t}\,dt \approx -0.223148148$; alternatively, $\ln(1.2) = \ln(1+0.2) = \int_0^{0.2} \frac{1}{1+t}\,dt$

≈ 0.182323232 and $\ln(0.8) = \int_0^{-0.2} \frac{1}{1+t}\,dt \approx -0.223148148.$

79. (a) 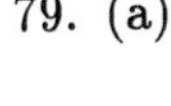

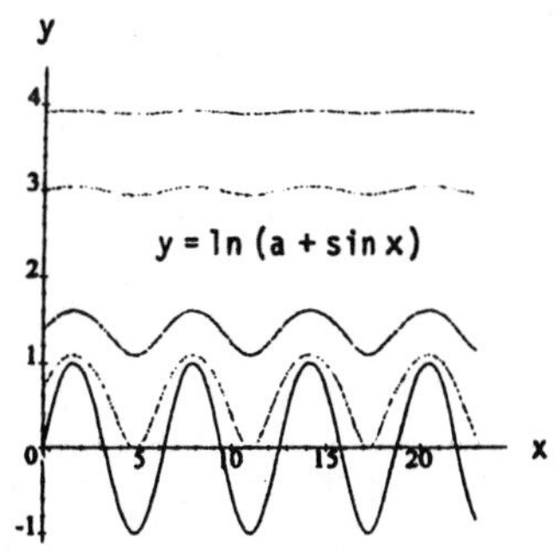

(b) As a increases, the value of $a + \sin x$ gets closer and closer to $a \pm 1$. Thus, $\ln(a + \sin x)$ looks more and more like the constant value $\ln a$ for larger and larger values of $a \Rightarrow$ the curves flatten as a increases.

6.2 EXPONENTIAL FUNCTIONS

1. $y = e^{-5x} \Rightarrow y' = e^{-5x}\frac{d}{dx}(-5x) \Rightarrow y' = -5e^{-5x}$

3. $y = e^{\left(4\sqrt{x}+x^2\right)} \Rightarrow y' = e^{\left(4\sqrt{x}+x^2\right)}\frac{d}{dx}\left(4\sqrt{x}+x^2\right) \Rightarrow y' = \left(\frac{2}{\sqrt{x}} + 2x\right)e^{\left(4\sqrt{x}+x^2\right)}$

5. $y = \left(x^2 - 2x + 2\right)e^x \Rightarrow y' = (2x-2)e^x + \left(x^2 - 2x + 2\right)e^x = x^2e^x$

7. $y = \ln\left(3\theta e^{-\theta}\right) = \ln 3 + \ln \theta + \ln e^{-\theta} = \ln 3 + \ln \theta - \theta \Rightarrow \frac{dy}{d\theta} = \frac{1}{\theta} - 1$

9. $y = \ln\left(2e^{-t}\sin t\right) = \ln 2 + \ln e^{-t} + \ln \sin t = \ln 2 - t + \ln \sin t \Rightarrow \frac{dy}{dt} = -1 + \left(\frac{1}{\sin t}\right)\frac{d}{dt}(\sin t) = -1 + \frac{\cos t}{\sin t}$

$= \frac{\cos t - \sin t}{\sin t}$

11. $y = \ln \frac{\sqrt{\theta}}{1+\sqrt{\theta}} = \ln \sqrt{\theta} - \ln(1+\sqrt{\theta}) \Rightarrow \frac{dy}{d\theta} = \left(\frac{1}{\sqrt{\theta}}\right)\frac{d}{d\theta}(\sqrt{\theta}) - \left(\frac{1}{1+\sqrt{\theta}}\right)\frac{d}{d\theta}(1+\sqrt{\theta})$

$= \left(\frac{1}{\sqrt{\theta}}\right)\left(\frac{1}{2\sqrt{\theta}}\right) - \left(\frac{1}{1+\sqrt{\theta}}\right)\left(\frac{1}{2\sqrt{\theta}}\right) = \frac{(1+\sqrt{\theta})-\sqrt{\theta}}{2\theta(1+\sqrt{\theta})} = \frac{1}{2\theta(1+\sqrt{\theta})} = \frac{1}{2\theta(1+\theta^{1/2})}$

13. $\int_0^{\ln x} \sin e^t \, dt \Rightarrow y' = (\sin e^{\ln x}) \cdot \frac{d}{dx}(\ln x) = \frac{\sin x}{x}$

15. $\ln y = e^y \sin x \Rightarrow \left(\frac{1}{y}\right)y' = (y'e^y)(\sin x) + e^y \cos x \Rightarrow y'\left(\frac{1}{y} - e^y \sin x\right) = e^y \cos x$

$\Rightarrow y'\left(\frac{1 - ye^y \sin x}{y}\right) = e^y \cos x \Rightarrow y' = \frac{ye^y \cos x}{1 - ye^y \sin x}$

17. $e^{2x} = \sin(x+3y) \Rightarrow 2e^{2x} = (1+3y')\cos(x+3y) \Rightarrow 1+3y' = \frac{2e^{2x}}{\cos(x+3y)} \Rightarrow 3y' = \frac{2e^{2x}}{\cos(x+3y)} - 1$

$\Rightarrow y' = \frac{2e^{2x} - \cos(x+3y)}{3\cos(x+3y)}$

19. $\int (e^{3x} + 5e^{-x}) \, dx = \frac{e^{3x}}{3} - 5e^{-x} + C$

21. $\int 8e^{(x+1)} \, dx = 8e^{(x+1)} + C$

23. Let $u = -r^{1/2} \Rightarrow du = -\frac{1}{2}r^{-1/2} \, dr \Rightarrow -2 \, du = r^{-1/2} \, dr$;

$\int \frac{e^{-\sqrt{r}}}{\sqrt{r}} \, dr = \int e^{-r^{1/2}} \cdot r^{-1/2} \, dr = -2 \int e^u \, du = -2e^{-r^{1/2}} + C = -2e^{-\sqrt{r}} + C$

25. Let $u = \frac{1}{x} \Rightarrow du = -\frac{1}{x^2} \, dx \Rightarrow -du = \frac{1}{x^2} \, dx$;

$\int \frac{e^{1/x}}{x^2} \, dx = \int -e^u \, du = -e^u + C = -e^{1/x} + C$

27. Let $u = \tan\theta \Rightarrow du = \sec^2\theta \, d\theta$; $\theta = 0 \Rightarrow u = 0$, $\theta = \frac{\pi}{4} \Rightarrow u = 1$;

$\int_0^{\pi/4} (1 + e^{\tan\theta}) \sec^2\theta \, d\theta = \int_0^{\pi/4} \sec^2\theta \, d\theta + \int_0^1 e^u \, du = [\tan\theta]_0^{\pi/4} + [e^u]_0^1 = \left[\tan\left(\frac{\pi}{4}\right) - \tan(0)\right] + (e^1 - e^0)$

$= (1-0) + (e-1) = e$

29. Let $u = \sec \pi t \Rightarrow du = \frac{1}{\pi} \sec \pi t \tan \pi t \, dt \Rightarrow \pi \, du = \sec \pi t \tan \pi t \, dt$;

$\int e^{\sec(\pi t)} \sec(\pi t) \tan(\pi t) \, dt = \frac{1}{\pi} \int e^u \, du = \frac{e^u}{\pi} + C = \frac{e^{\sec(\pi t)}}{\pi} + C$

31. Let $u = 1 + e^r \Rightarrow du = e^r \, dr$;

$\int \frac{e^r}{1+e^r} \, dr = \int \frac{1}{u} \, du = \ln|u| + C = \ln(1+e^r) + C$

33. $y = 2^x \Rightarrow y' = 2^x \ln 2$

35. $y = x^{\pi} \Rightarrow y' = \pi x^{(\pi - 1)}$

37. $y = 7^{\sec\theta} \ln 7 \Rightarrow \frac{dy}{d\theta} = \left(7^{\sec\theta} \ln 7\right)(\ln 7)(\sec\theta \tan\theta) = 7^{\sec\theta}(\ln 7)^2 (\sec\theta \tan\theta)$

39. $y = t^{1-e} \Rightarrow \frac{dy}{dt} = (1-e)\, t^{-e}$

41. $y = \log_3\left(\left(\frac{x+1}{x-1}\right)^{\ln 3}\right) = \frac{\ln\left(\frac{x+1}{x-1}\right)^{\ln 3}}{\ln 3} = \frac{(\ln 3)\ln\left(\frac{x+1}{x-1}\right)}{\ln 3} = \ln\left(\frac{x+1}{x-1}\right) = \ln(x+1) - \ln(x-1)$

$\Rightarrow \frac{dy}{dx} = \frac{1}{x+1} - \frac{1}{x-1} = \frac{-2}{(x+1)(x-1)}$

43. $y = \log_7\left(\frac{\sin\theta \cos\theta}{e^{\theta}2^{\theta}}\right) = \frac{\ln(\sin\theta) + \ln(\cos\theta) - \ln e^{\theta} - \ln 2^{\theta}}{\ln 7} = \frac{\ln(\sin\theta) + \ln(\cos\theta) - \theta - \theta\ln 2}{\ln 7}$

$\Rightarrow \frac{dy}{d\theta} = \frac{\cos\theta}{(\sin\theta)(\ln 7)} - \frac{\sin\theta}{(\cos\theta)(\ln 7)} - \frac{1}{\ln 7} - \frac{\ln 2}{\ln 7} = \left(\frac{1}{\ln 7}\right)(\cot\theta - \tan\theta - 1 - \ln 2)$

45. $y = (x+1)^x \Rightarrow \ln y = \ln(x+1)^x = x\ln(x+1) \Rightarrow \frac{y'}{y} = \ln(x+1) + x\cdot\frac{1}{(x+1)} \Rightarrow y' = (x+1)^x\left[\frac{x}{x+1} + \ln(x+1)\right]$

47. $y = (\sin x)^x \Rightarrow \ln y = \ln(\sin x)^x = x\ln(\sin x) \Rightarrow \frac{y'}{y} = \ln(\sin x) + x\left(\frac{\cos x}{\sin x}\right) \Rightarrow y' = (\sin x)^x[\ln(\sin x) + x\cot x]$

49. $y = x^{\ln x},\ x > 0 \Rightarrow \ln y = (\ln x)^2 \Rightarrow \frac{y'}{y} = 2(\ln x)\left(\frac{1}{x}\right) \Rightarrow y' = \left(x^{\ln x}\right)\left(\frac{\ln x^2}{x}\right)$

51. Let $u = x^2 \Rightarrow du = 2x\,dx \Rightarrow \frac{1}{2}du = x\,dx$; $x = 1 \Rightarrow u = 1$, $x = \sqrt{2} \Rightarrow u = 2$;

$$\int_1^{\sqrt{2}} x2^{\left(x^2\right)}\,dx = \int_1^2 \left(\frac{1}{2}\right)2^u\,du = \frac{1}{2}\left[\frac{2^u}{\ln 2}\right]_1^2 = \left(\frac{1}{2\ln 2}\right)\left(2^2 - 2^1\right) = \frac{1}{\ln 2}$$

53. Let $u = \ln x \Rightarrow du = \frac{1}{x}dx$; $x = 1 \Rightarrow u = 0$, $x = 2 \Rightarrow u = \ln 2$;

$$\int_1^2 \frac{2^{\ln x}}{x}\,dx = \int_0^{\ln 2} 2^u\,du = \left[\frac{2^u}{\ln 2}\right]_0^{\ln 2} = \left(\frac{1}{\ln 2}\right)\left(2^{\ln 2} - 2^0\right) = \frac{2^{\ln 2} - 1}{\ln 2}$$

55. $\int x^{\left(\sqrt{2}-1\right)}\,dx = \frac{x^{\sqrt{2}}}{\sqrt{2}} + C$

57. $\int_1^e x^{(\ln 2)-1}\,dx = \left[\frac{x^{\ln 2}}{\ln 2}\right]_1^e = \frac{e^{\ln 2} - 1^{\ln 2}}{\ln 2} = \frac{2-1}{\ln 2} = \frac{1}{\ln 2}$

59. $\frac{dy}{dt} = e^t\sin\left(e^t - 2\right) \Rightarrow y = \int e^t\sin\left(e^t - 2\right)dt$;

let $u = e^t - 2 \Rightarrow du = e^t\,dt \Rightarrow y = \int \sin u\,du = -\cos u + C = -\cos\left(e^t - 2\right) + C$; $y(\ln 2) = 0$

$\Rightarrow -\cos\left(e^{\ln 2} - 2\right) + C = 0 \Rightarrow -\cos(2-2) + C = 0 \Rightarrow C = \cos 0 = 1$; thus, $y = 1 - \cos\left(e^t - 2\right)$

61. $\frac{d^2y}{dx^2} = 2e^{-x} \Rightarrow \frac{dy}{dx} = -2e^{-x} + C$; $x = 0$ and $\frac{dy}{dx} = 0 \Rightarrow 0 = -2e^0 + C \Rightarrow C = 2$; thus $\frac{dy}{dx} = -2e^{-x} + 2$
$\Rightarrow y = 2e^{-x} + 2x + C_1$; $x = 0$ and $y = 1 \Rightarrow 1 = 2e^0 + C_1 \Rightarrow C_1 = -1 \Rightarrow y = 2e^{-x} + 2x - 1 = 2\left(e^{-x} + x\right) - 1$

63. $f(x) = e^x - 2x \Rightarrow f'(x) = e^x - 2$; $f'(x) = 0 \Rightarrow e^x = 2 \Rightarrow x = \ln 2$; $f(0) = 1$, the absolute maximum; $f(\ln 2) = 2 - 2\ln 2 \approx 0.613706$, the absolute minimum; $f(1) = e - 2 \approx 0.71828$, a relative or local maximum since $f''(x) = e^x$ is always positive

65. $f(x) = x^2 \ln \frac{1}{x} \Rightarrow f'(x) = 2x \ln \frac{1}{x} + x^2\left(\frac{1}{\frac{1}{x}}\right)\left(-x^{-2}\right) = 2x \ln \frac{1}{x} - x = -x(2\ln x + 1)$; $f'(x) = 0 \Rightarrow x = 0$ or $\ln x = -\frac{1}{2}$. Since $x = 0$ is not in the domain of f, $x = e^{-1/2} = \frac{1}{\sqrt{e}}$. Also, $f'(x) > 0$ for $0 < x < \frac{1}{\sqrt{e}}$ and $f'(x) < 0$ for $x > \frac{1}{\sqrt{e}}$. Therefore, $f\left(\frac{1}{\sqrt{e}}\right) = \frac{1}{e} \ln \sqrt{e} = \frac{1}{e} \ln e^{1/2} = \frac{1}{2e} \ln e = \frac{1}{2e}$ is the absolute maximum value of f assumed at $x = \frac{1}{\sqrt{e}}$.

67. Let $x = \frac{r}{k} \Rightarrow k = \frac{r}{x}$ and as $k \to \infty$, $x \to 0 \Rightarrow \lim\limits_{k\to\infty}\left(1 + \frac{r}{k}\right)^k = \lim\limits_{x\to 0}(1+x)^{r/x} = \lim\limits_{x\to 0}\left((1+x)^{1/x}\right)^r$
$= \left(\lim\limits_{x\to 0}(1+x)^{1/x}\right)^r$, since u^r is continuous. However, $\lim\limits_{x\to 0}(1+x)^{1/x} = e$ (by Theorem 2), therefore, $\lim\limits_{k\to\infty}\left(1 + \frac{r}{k}\right)^k = e^r$.

69. $y = mx + b \Rightarrow x = \frac{y}{m} - \frac{b}{m} \Rightarrow f^{-1}(x) = \frac{1}{m}x - \frac{b}{m}$; the graph of $f^{-1}(x)$ is a line with slope $\frac{1}{m}$ and y-intercept $-\frac{b}{m}$.

71. Note that $y = \ln x$ and $e^y = x$ are the same curve; $\int_1^a \ln x\, dx$ = area under the curve between 1 and a; $\int_0^{\ln a} e^y\, dy$ = area to the left of the curve. The sum of these areas is equal to the area of the rectangle
$\Rightarrow \int_1^a \ln x\, dx + \int_0^{\ln a} e^y\, dy = a \ln a$.

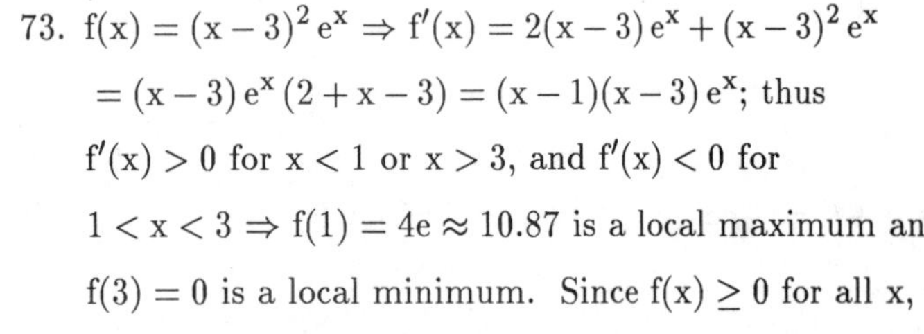

73. $f(x) = (x-3)^2 e^x \Rightarrow f'(x) = 2(x-3)e^x + (x-3)^2 e^x$
$= (x-3)e^x(2 + x - 3) = (x-1)(x-3)e^x$; thus
$f'(x) > 0$ for $x < 1$ or $x > 3$, and $f'(x) < 0$ for
$1 < x < 3 \Rightarrow f(1) = 4e \approx 10.87$ is a local maximum and
$f(3) = 0$ is a local minimum. Since $f(x) \geq 0$ for all x,
$f(3) = 0$ is also an absolute minimum.

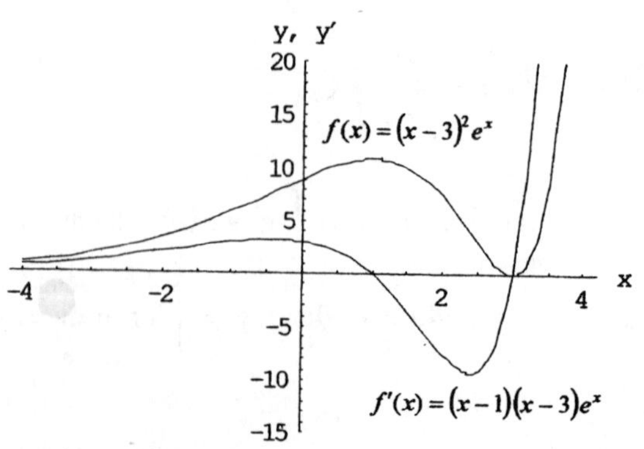

75. $e^{\ln x} = x$ for $x > 0$ and $\ln(e^x) = x$ for all x

77. (a) The point of tangency is $(p, \ln p)$ and $m_{tangent} = \frac{1}{p}$ since $\frac{dy}{dx} = \frac{1}{x}$. The tangent line passes through $(0,0)$ $\Rightarrow$ the equation of the tangent line is $y = \frac{1}{p}x$. The tangent line also passes through $(p, \ln p) \Rightarrow \ln p = \frac{1}{p}p = 1 \Rightarrow p = e$, and the tangent line equation is $y = \frac{1}{e}x$.

(b) $\frac{d^2y}{dx^2} = -\frac{1}{x^2} < 0$ for $x \neq 0 \Rightarrow y = \ln x$ is concave downward over its domain. Therefore, $y = \ln x$ lies below the graph of $y = \frac{1}{e}x$ for all $x > 0$, $x \neq e$ and $\ln x < \frac{x}{e}$ for $x > 0$, $x \neq e$.

(c) Multiplying by e, $e \ln x < x$ or $\ln x^e < x$.

(d) Exponentiating both sides of $\ln x^e < x$, we have $e^{\ln x^e} < e^x$, or $x^e < e^x$ for all positive $x \neq e$.

(e) Let $x = \pi$ to see that $\pi^e < e^\pi$. Therefore, e^π is bigger.

79. (a) $f(x) = 2^x \Rightarrow f'(x) = 2^x \ln 2$; $L(x) = (2^0 \ln 2)x + 2^0 = x \ln 2 + 1 \approx 0.69x + 1$

(b)

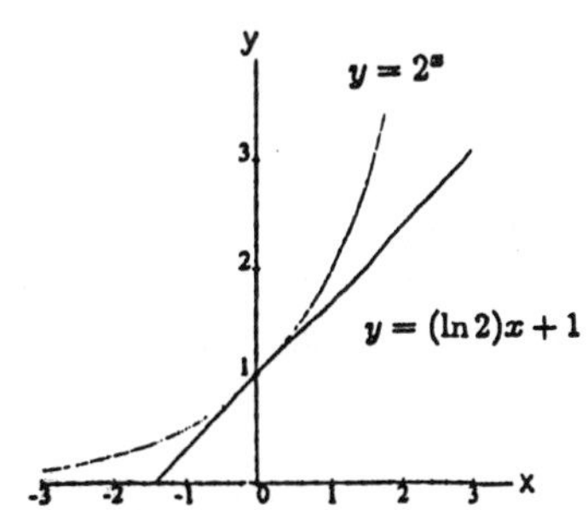

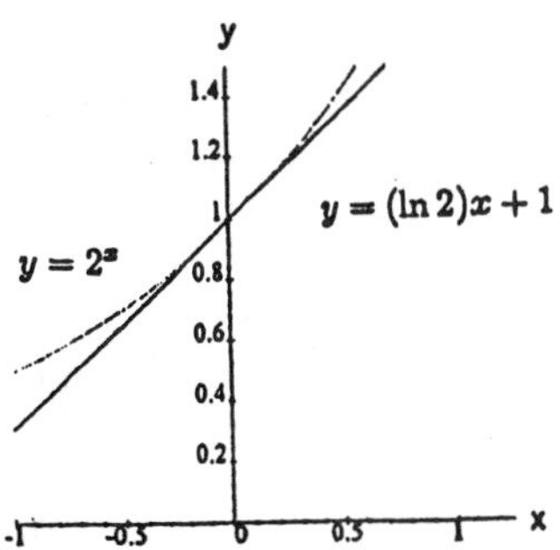

6.3 DERIVATIVES OF INVERSE TRIGONOMETRIC FUNCTIONS; INTEGRALS

1. $y = \cos^{-1}\left(\frac{1}{x}\right) = \sec^{-1} x \Rightarrow \frac{dy}{dx} = \frac{1}{|x|\sqrt{x^2 - 1}}$

3. $y = \sec^{-1}(2s + 1) \Rightarrow \frac{dy}{ds} = \frac{2}{|2s+1|\sqrt{(2s+1)^2 - 1}} = \frac{2}{|2s+1|\sqrt{4s^2 + 4s}} = \frac{1}{|2s+1|\sqrt{s^2 + s}}$

5. $y = \sec^{-1}\left(\frac{1}{t}\right) = \cos^{-1} t \Rightarrow \frac{dy}{dt} = \frac{-1}{\sqrt{1 - t^2}}$

7. $y = \ln(\tan^{-1} x) \Rightarrow \frac{dy}{dx} = \frac{\left(\frac{1}{1+x^2}\right)}{\tan^{-1} x} = \frac{1}{(\tan^{-1} x)(1 + x^2)}$

9. $y = \cos^{-1}(e^{-t}) \Rightarrow \frac{dy}{dt} = -\frac{-e^{-t}}{\sqrt{1 - (e^{-t})^2}} = \frac{e^{-t}}{\sqrt{1 - e^{-2t}}}$

11. $y = \tan^{-1}\sqrt{x^2-1} + \csc^{-1}x = \tan^{-1}\left(x^2-1\right)^{1/2} + \csc^{-1}x \Rightarrow \frac{dy}{dx} = \frac{\left(\frac{1}{2}\right)\left(x^2-1\right)^{-1/2}(2x)}{1+\left[\left(x^2-1\right)^{1/2}\right]^2} - \frac{1}{|x|\sqrt{x^2-1}}$

$= \frac{1}{x\sqrt{x^2-1}} - \frac{1}{|x|\sqrt{x^2-1}} = 0$, for $x > 1$

13. $y = x\sin^{-1}x + \sqrt{1-x^2} = x\sin^{-1}x + \left(1-x^2\right)^{1/2} \Rightarrow \frac{dy}{dx} = \sin^{-1}x + x\left(\frac{1}{\sqrt{1-x^2}}\right) + \left(\frac{1}{2}\right)\left(1-x^2\right)^{-1/2}(-2x)$

$= \sin^{-1}x + \frac{x}{\sqrt{1-x^2}} - \frac{x}{\sqrt{1-x^2}} = \sin^{-1}x$

15. $\int \frac{1}{\sqrt{1-4x^2}}\,dx = \frac{1}{2}\int \frac{2}{\sqrt{1-(2x)^2}}\,dx = \frac{1}{2}\int \frac{du}{\sqrt{1-u^2}}$, where $u = 2x$ and $du = 2\,dx$

$= \frac{1}{2}\sin^{-1}u + C = \frac{1}{2}\sin^{-1}(2x) + C$

17. $\int \frac{dx}{x\sqrt{25x^2-2}} = \int \frac{du}{u\sqrt{u^2-2}}$, where $u = 5x$ and $du = 5\,dx$

$= \frac{1}{\sqrt{2}}\sec^{-1}\left|\frac{u}{\sqrt{2}}\right| + C = \frac{1}{\sqrt{2}}\sec^{-1}\left|\frac{5x}{\sqrt{2}}\right| + C$

19. $\int_0^2 \frac{dt}{8+2t^2} = \frac{1}{\sqrt{2}}\int_0^{2\sqrt{2}} \frac{du}{8+u^2}$, where $u = \sqrt{2}t$ and $du = \sqrt{2}\,dt$; $t = 0 \Rightarrow u = 0$, $t = 2 \Rightarrow u = 2\sqrt{2}$

$= \left[\frac{1}{\sqrt{2}}\cdot\frac{1}{\sqrt{8}}\tan^{-1}\frac{u}{\sqrt{8}}\right]_0^{2\sqrt{2}} = \frac{1}{4}\left(\tan^{-1}\frac{2\sqrt{2}}{\sqrt{8}} - \tan^{-1}0\right) = \frac{1}{4}\left(\tan^{-1}1 - \tan^{-1}0\right) = \frac{1}{4}\left(\frac{\pi}{4} - 0\right) = \frac{\pi}{16}$

21. $\int \frac{3\,dr}{\sqrt{1-4(r-1)^2}} = \frac{3}{2}\int \frac{du}{\sqrt{1-u^2}}$, where $u = 2(r-1)$ and $du = 2\,dr$

$= \frac{3}{2}\sin^{-1}u + C = \frac{3}{2}\sin^{-1}[2(r-1)] + C$

23. $\int \frac{y\,dy}{\sqrt{1-y^4}} = \frac{1}{2}\int \frac{du}{\sqrt{1-u^2}}$, where $u = y^2$ and $du = 2y\,dy$

$= \frac{1}{2}\sin^{-1}u + C = \frac{1}{2}\sin^{-1}y^2 + C$

25. $\int_0^{\ln\sqrt{3}} \frac{e^x\,dx}{1+e^{2x}} = \int_1^{\sqrt{3}} \frac{du}{1+u^2}$, where $u = e^x$ and $du = e^x\,dx$; $x = 0 \Rightarrow u = 1$, $x = \ln\sqrt{3} \Rightarrow u = \sqrt{3}$

$= \left[\tan^{-1}u\right]_1^{\sqrt{3}} = \tan^{-1}\sqrt{3} - \tan^{-1}1 = \frac{\pi}{3} - \frac{\pi}{4} = \frac{\pi}{12}$

27. $\int \frac{dx}{\sqrt{-x^2+4x-3}} = \int \frac{dx}{\sqrt{1-(x^2-4x+4)}} = \int \frac{dx}{\sqrt{1-(x-2)^2}} = \sin^{-1}(x-2)+C$

29. $\int \frac{dy}{y^2-2y+5} = \int \frac{dy}{4+y^2-2y+1} = \int \frac{dy}{2^2+(y-1)^2} = \frac{1}{2}\tan^{-1}\left(\frac{y-1}{2}\right)+C$

31. $\int \frac{dx}{(x+1)\sqrt{x^2+2x}} = \int \frac{dx}{(x+1)\sqrt{x^2+2x+1-1}} = \int \frac{dx}{(x+1)\sqrt{(x+1)^2-1}}$

$= \int \frac{du}{u\sqrt{u^2-1}}$, where $u = x+1$ and $du = dx$

$= \sec^{-1}|u|+C = \sec^{-1}|x+1|+C$

33. $\int \frac{e^{\sin^{-1}x}}{\sqrt{1-x^2}}\,dx = \int e^u\,du$, where $u = \sin^{-1}x$ and $du = \frac{dx}{\sqrt{1-x^2}}$

$= e^u + C = e^{\sin^{-1}x} + C$

35. $\int \frac{1}{(\tan^{-1}y)(1+y^2)}\,dy = \int \frac{\left(\frac{1}{1+y^2}\right)}{\tan^{-1}y}\,dy = \int \frac{1}{u}\,du$, where $u = \tan^{-1}y$ and $du = \frac{dy}{1+y^2}$

$= \ln|u|+C = \ln\left|\tan^{-1}y\right|+C$

37. If $y = \ln x - \frac{1}{2}\ln(1+x^2) - \frac{\tan^{-1}x}{x} + C$, then $dy = \left[\frac{1}{x} - \frac{x}{1+x^2} - \frac{\left(\frac{x}{1+x^2}\right) - \tan^{-1}x}{x^2}\right]dx$

$= \left(\frac{1}{x} - \frac{x}{1+x^2} - \frac{1}{x(1+x^2)} + \frac{\tan^{-1}x}{x^2}\right)dx = \frac{x(1+x^2)-x^3-x+(\tan^{-1}x)(1+x^2)}{x^2(1+x^2)}\,dx = \frac{\tan^{-1}x}{x^2}\,dx,$

which verifies the formula

39. If $y = x(\sin^{-1}x)^2 - 2x + 2\sqrt{1-x^2}\,\sin^{-1}x + C$, then

$dy = \left[(\sin^{-1}x)^2 + \frac{2x(\sin^{-1}x)}{\sqrt{1-x^2}} - 2 + \frac{-2x}{\sqrt{1-x^2}}\sin^{-1}x + 2\sqrt{1-x^2}\left(\frac{1}{\sqrt{1-x^2}}\right)\right]dx = (\sin^{-1}x)^2\,dx$, which verifies

the formula

41. $\frac{dy}{dx} = \frac{1}{\sqrt{1-x^2}} \Rightarrow dy = \frac{dx}{\sqrt{1-x^2}} \Rightarrow y = \sin^{-1}x + C$; $x = 0$ and $y = 0 \Rightarrow 0 = \sin^{-1}0 + C \Rightarrow C = 0 \Rightarrow y = \sin^{-1}x$

43. $\frac{dy}{dx} = \frac{1}{x\sqrt{x^2-1}} \Rightarrow dy = \frac{dx}{x\sqrt{x^2-1}} \Rightarrow y = \sec^{-1}|x| + C$; $x = 2$ and $y = \pi \Rightarrow \pi = \sec^{-1}2 + C \Rightarrow C = \pi - \sec^{-1}2$

$= \pi - \frac{\pi}{3} = \frac{2\pi}{3} \Rightarrow y = \sec^{-1}(x) + \frac{2\pi}{3}$, $x > 1$

45. $\cos^{-1} u = \frac{\pi}{2} - \sin^{-1} u \Rightarrow \frac{d}{dx}\left(\cos^{-1} u\right) = \frac{d}{dx}\left(\frac{\pi}{2} - \sin^{-1} u\right) = 0 - \frac{\frac{du}{dx}}{\sqrt{1-u^2}}, |u| < 1$

47. $\csc^{-1} u = \frac{\pi}{2} - \sec^{-1} u \Rightarrow \frac{d}{dx}\left(\csc^{-1} u\right) = \frac{d}{dx}\left(\frac{\pi}{2} - \sec^{-1} u\right) = 0 - \frac{\frac{du}{dx}}{|u|\sqrt{u^2-1}} = -\frac{\frac{du}{dx}}{|u|\sqrt{u^2-1}}, |u| > 1$

49. $f(x) = \sin x \Rightarrow f'(x) = \cos x \Rightarrow \frac{df^{-1}}{dx} = \frac{1}{(\cos x)_{\sin^{-1} x}} \Rightarrow \frac{df^{-1}}{dx} = \frac{1}{\cos\left(\sin^{-1} x\right)} = \frac{1}{\sqrt{1-\sin^2\left(\sin^{-1} x\right)}} = \frac{1}{\sqrt{1-x^2}}$

51. $V = \pi \int_{-\sqrt{3}/3}^{\sqrt{3}} \left(\frac{1}{\sqrt{1+x^2}}\right)^2 dx = \pi \int_{-\sqrt{3}/3}^{\sqrt{3}} \frac{1}{1+x^2}\, dx = \pi\left[\tan^{-1} x\right]_{-\sqrt{3}/3}^{\sqrt{3}} = \pi\left[\tan^{-1}\sqrt{3} - \tan^{-1}\left(-\frac{\sqrt{3}}{3}\right)\right]$

$= \pi\left[\frac{\pi}{3} - \left(-\frac{\pi}{6}\right)\right] = \frac{\pi^2}{2}$

53. (a) $A(x) = \frac{\pi}{4}(\text{diameter})^2 = \frac{\pi}{4}\left[\frac{1}{\sqrt{1+x^2}} - \left(-\frac{1}{\sqrt{1+x^2}}\right)\right]^2 = \frac{\pi}{1+x^2} \Rightarrow V = \int_a^b A(x)\, dx = \int_{-1}^{1} \frac{\pi\, dx}{1+x^2}$

$= \pi\left[\tan^{-1} x\right]_{-1}^{1} = (\pi)(2)\left(\frac{\pi}{4}\right) = \frac{\pi^2}{2}$

(b) $A(x) = (\text{edge})^2 = \left[\frac{1}{\sqrt{1+x^2}} - \left(-\frac{1}{\sqrt{1+x^2}}\right)\right]^2 = \frac{\pi}{1+x^2} \Rightarrow V = \int_a^b A(x)\, dx = \int_{-1}^{1} \frac{4\, dx}{1+x^2}$

$= 4\left[\tan^{-1} x\right]_{-1}^{1} = 4\left[\tan^{-1}(1) - \tan^{-1}(-1)\right] = 4\left[\frac{\pi}{4} - \left(-\frac{\pi}{4}\right)\right] = 2\pi$

55. A calculator or computer numerical integrator yields $\sin^{-1} 0.6 \approx 0.643517104$.

57. The values of f increase over the interval $[-1, 1]$ because $f' > 0$, and the graph of f steepens as the values of f' increase towards the ends of the interval. The graph of f is concave down to the left of the origin where $f'' < 0$, and concave up to the right of the origin where $f'' > 0$. There is an inflection point at $x = 0$ where $f'' = 0$ and f' has a local minimum value.

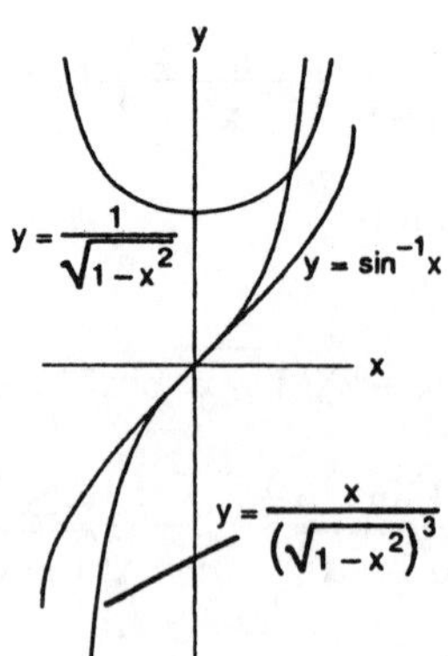

6.4 FIRST ORDER SEPARABLE DIFFERENTIAL EQUATIONS

1. (a) $y = e^{-x} \Rightarrow y' = -e^{-x} \Rightarrow 2y' + 3y = 2\left(-e^{-x}\right) + 3e^{-x} = e^{-x}$

(b) $y = e^{-x} + e^{-3x/2} \Rightarrow y' = -e^{-x} - \frac{3}{2}e^{-3x/2} \Rightarrow 2y' + 3y = 2\left(-e^{-x} - \frac{3}{2}e^{-3x/2}\right) + 3\left(e^{-x} + e^{-3x/2}\right) = e^{-x}$

(c) $y = e^{-x} + Ce^{-3x/2} \Rightarrow y' = -e^{-x} - \frac{3}{2}Ce^{-3x/2} \Rightarrow 2y' + 3y = 2\left(-e^{-x} - \frac{3}{2}Ce^{-3x/2}\right) + 3\left(e^{-x} + Ce^{-3x/2}\right) = e^{-x}$

3. $y = (x-2)\,e^{-x^2} \Rightarrow y' = e^{-x^2} + \left(-2xe^{-x^2}\right)(x-2) \Rightarrow y' = e^{-x^2} - 2xy;\ y(2) = (2-2)\,e^{-2^2} = 0$

5. $2\sqrt{xy}\,\frac{dy}{dx} = 1 \Rightarrow 2x^{1/2}y^{1/2}\,dy = dx \Rightarrow 2y^{1/2}\,dy = x^{-1/2}\,dx \Rightarrow \int 2y^{1/2}\,dy = \int x^{-1/2}\,dx \Rightarrow 2\left(\frac{2}{3}y^{3/2}\right)$
$= 2x^{1/2} + C_1 \Rightarrow \frac{2}{3}y^{3/2} - x^{1/2} = C$, where $C = \frac{1}{2}C_1$

7. $\frac{dy}{dx} = e^{x-y} \Rightarrow dy = e^x e^{-y}\,dx \Rightarrow e^y\,dy = e^x\,dx \Rightarrow \int e^y\,dy = \int e^x\,dx \Rightarrow e^y = e^x + C \Rightarrow e^y - e^x = C$

9. $\frac{dy}{dx} = \sqrt{y}\cos^2\sqrt{y} \Rightarrow dy = \left(\sqrt{y}\cos^2\sqrt{y}\right)dx \Rightarrow \frac{\sec^2\sqrt{y}}{\sqrt{y}}\,dy = dx \Rightarrow \int \frac{\sec^2\sqrt{y}}{\sqrt{y}}\,dy = \int dx$. In the integral on the left-hand side, substitute $u = \sqrt{y} \Rightarrow du = \frac{1}{2\sqrt{y}}\,dy \Rightarrow 2\,du = \frac{1}{\sqrt{y}}\,dy$, and we have

$2\int \sec^2 u\,du = \int dx \Rightarrow 2\tan u = x + C \Rightarrow -x + 2\tan\sqrt{y} = C$

11. $\sqrt{x}\,\frac{dy}{dx} = e^{y+\sqrt{x}} \Rightarrow \frac{dy}{dx} = \frac{e^y e^{\sqrt{x}}}{\sqrt{x}} \Rightarrow dy = \frac{e^y e^{\sqrt{x}}}{\sqrt{x}}\,dx \Rightarrow e^{-y}\,dy = \frac{e^{\sqrt{x}}}{\sqrt{x}}\,dx \Rightarrow \int e^{-y}\,dy = \int \frac{e^{\sqrt{x}}}{\sqrt{x}}\,dx.$

In the integral on the right-hand side, substitute $u = \sqrt{x} \Rightarrow du = \frac{1}{2\sqrt{x}}\,dx \Rightarrow 2\,du = \frac{1}{\sqrt{x}}\,dx.$ and we have

$\int e^{-y}\,dy = 2\int e^u\,du \Rightarrow -e^{-y} = 2e^u + C_1 \Rightarrow e^{-y} + 2e^{\sqrt{x}} = C$, where $C = -C_1$.

13. $\frac{dy}{dx} = 2x\sqrt{1-y^2} \Rightarrow dy = 2x\sqrt{1-y^2}\,dx \Rightarrow \frac{dy}{\sqrt{1-y^2}} = 2x\,dx \Rightarrow \int \frac{dy}{\sqrt{1-y^2}} = \int 2x\,dx \Rightarrow \sin^{-1} y = x^2 + C$

since $|y| < 1 \Rightarrow y = \sin(x^2 + C)$.

15. (a) $\frac{dp}{dh} = kp \Rightarrow p = p_0 e^{kh}$ where $p_0 = 1013$; $90 = 1013e^{20k} \Rightarrow k = \frac{\ln(90) - \ln(1013)}{20} \approx -0.121$

(b) $p = 1013e^{-6.05} \approx 2.389$ millibars

(c) $900 = 1013e^{(-0.121)h} \Rightarrow -0.121h = \ln\left(\frac{900}{1013}\right) \Rightarrow h = \frac{\ln(1013) - \ln(900)}{0.121} \approx 0.977$ km

17. $A = A_0 e^{kt} \Rightarrow 800 = 1000e^{10k} \Rightarrow k = \frac{\ln(0.8)}{10} \Rightarrow A = 1000e^{((\ln(0.8))/10)t}$, where A represents the amount of sugar that remains after time t. Thus after another 14 hrs, $A = 1000e^{((\ln(0.8)/10)14)} \approx 731.688$ kg

19. $V(t) = V_0 e^{-t/40} \Rightarrow 0.1V_0 = V_0 e^{-t/40}$ when the voltage is 10% of its original value $\Rightarrow t = -40\ln(0.1)$

≈ 92.1 sec

21. (a) $\frac{dQ}{dt} = -kQ + r$ where k is a positive constant and $Q = Q(t)$.

(b) $dQ = (-kQ + r)\ dt \Rightarrow dQ = -k\left(Q - \frac{r}{k}\right) dt \Rightarrow \frac{dQ}{\left(Q - \frac{r}{k}\right)} = -k\ dt \Rightarrow \int \frac{dQ}{\left(Q - \frac{r}{k}\right)} = -\int k\ dt$

$\Rightarrow \ln\left|Q - \frac{r}{k}\right| = -kt + C_1 \Rightarrow e^{\ln\left|Q - \frac{r}{k}\right|} = e^{-kt + C_1} \Rightarrow \left|Q - \frac{r}{k}\right| = e^{-kt}e^{C_1} \Rightarrow Q(t) = \frac{r}{k} \pm C_2 e^{-kt}$

$\Rightarrow Q(t) = \frac{r}{k} + Ce^{-kt}$ where $C_2 = e^{C_1}$ and $C = \pm C_2$. Apply the initial condition $Q(0) = Q_0 = \frac{r}{k} + Ce^0$

$\Rightarrow C = Q_0 - \frac{r}{k} \Rightarrow Q(t) = \frac{r}{k} + \left(Q_0 - \frac{r}{k}\right)e^{-kt}$

(c) $\lim\limits_{t\to\infty} Q(t) = \lim\limits_{t\to\infty} \left(\frac{r}{k} + \left(Q_0 - \frac{r}{k}\right)e^{-kt}\right) = \frac{r}{k} + \left(Q_0 - \frac{r}{k}\right)(0) = \frac{r}{k}$.

23. (a) $A_0 e^{(0.04)5} = A_0 e^{0.2}$

(b) $2A_0 = A_0 e^{(0.04)t} \Rightarrow \ln 2 = (0.04)t \Rightarrow t = \frac{\ln 2}{0.04} \approx 17.33$ years; $3A_0 = A_0 e^{(0.04)t} \Rightarrow \ln 3 = (0.04)t$

$\Rightarrow t = \frac{\ln 3}{0.04} \approx 27.47$ years

25. $y = y_0 e^{-0.18t}$ represents the decay equation; solving $(0.9)y_0 = y_0 e^{-0.18t} \Rightarrow t = \frac{\ln(0.9)}{-0.18} \approx 0.585$ days

27. $y = y_0 e^{-kt} = y_0 e^{-(k)(3/k)} = y_0 e^{-3} = \frac{y_0}{e^3} < \frac{y_0}{20} = (0.05)(y_0) \Rightarrow$ after three mean lifetimes less than 5% remains

29. $T - T_s = (T_0 - T_s)\, e^{-kt}$, $T_0 = 90°C$, $T_s = 20°C$, $T = 60°C \Rightarrow 60 - 20 = 70e^{-10k} \Rightarrow \frac{4}{7} = e^{-10k}$

$\Rightarrow k = \frac{\ln\left(\frac{7}{4}\right)}{10} \approx 0.05596$

(a) $35 - 20 = 70e^{-0.05596t} \Rightarrow t \approx 27.5$ min is the total time $\Rightarrow$ it will take $27.5 - 10 = 17.5$ to reach 35°C

(b) $T - T_s = (T_0 - T_s)\, e^{-kt}$, $T_0 = 90°C$, $T_s = -15°C \Rightarrow 35 + 15 = 105e^{-0.05596t} \Rightarrow t \approx 13.26$ min

31. $T - T_s = (T_0 - T_s)\, e^{-kt} \Rightarrow 39 - T_s = (46 - T_s)\, e^{-10k}$ and $33 - T_s = (46 - T_s)\, e^{-20k} \Rightarrow \frac{39 - T_s}{46 - T_s} = e^{-10k}$ and

$\frac{33 - T_s}{46 - T_s} = e^{-20k} = \left(e^{-10k}\right)^2 \Rightarrow \frac{33 - T_s}{46 - T_s} = \left(\frac{39 - T_s}{46 - T_s}\right)^2 \Rightarrow (33 - T_s)(46 - T_s) = (39 - T_s)^2 \Rightarrow 1518 - 79T_s + T_s^2$

$= 1521 - 78T_s + T_s^2 \Rightarrow -T_s = 3 \Rightarrow T_s = -3°C$

33. From Example 5, the half-life of carbon-14 is 5700 yr $\Rightarrow \frac{1}{2}c_0 = c_0 e^{-k(5700)} \Rightarrow k = \frac{\ln 2}{5700} \approx 0.0001216$

$\Rightarrow c = c_0 e^{-0.0001216t} \Rightarrow (0.445)c_0 = c_0 e^{-0.0001216t} \Rightarrow t = \frac{\ln(0.445)}{-0.0001216} \approx 6658$ years

35. From Exercise 33, $k \approx 0.0001216$ for carbon-14. Thus, $c = c_0 e^{-0.0001216t} \Rightarrow (0.995)c_0 = c_0 e^{-0.0001216t}$

$\Rightarrow t = \dfrac{\ln(0.995)}{-0.0001216} \approx 41$ years old

37. Note that the total mass is $66 + 7 = 73$ kg, therefore, $v = v_0 e^{-(k/m)t} \Rightarrow v = 9e^{-3.9t/73}$

(a) $s(t) = \int 9e^{-3.9t/73}\,dt = -\dfrac{2190}{13}e^{-3.9t/73} + C$

Since $s(0) = 0$ we have $C = \dfrac{2190}{13}$ and $\lim_{t\to\infty} s(t) = \lim_{t\to\infty} \dfrac{2190}{13}\left(1 - e^{-3.9t/3}\right) = \dfrac{2190}{13} \approx 168.5$

The cyclist will coast about 168.5 meters.

(b) $1 = 9e^{-3.9t/73} \Rightarrow \dfrac{3.9t}{73} = \ln 9 \Rightarrow t = \dfrac{73 \ln 9}{3.9} \approx 41.13$ sec

It will take about 41.13 seconds.

39. The total distance traveled $= \dfrac{v_0 m}{k} \Rightarrow \dfrac{(2.75)(39.92)}{k} = 4.91 \Rightarrow k = 22.36$. Therefore, the distance traveled is given by the function $s(t) = 4.91(1 - e^{-(22.36/39.92)t})$. The graphs shows s(t) and the data points.

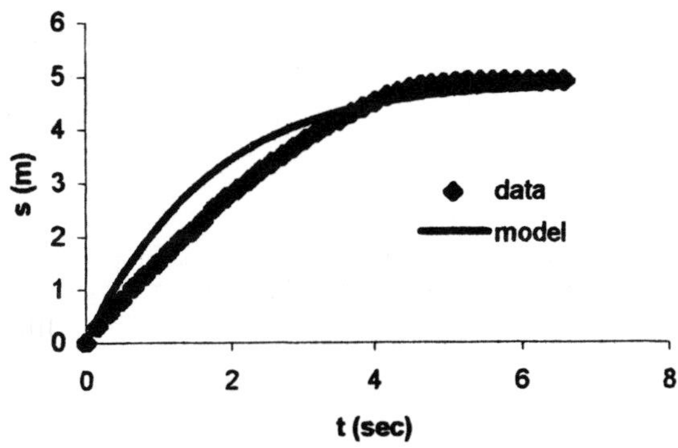

41.

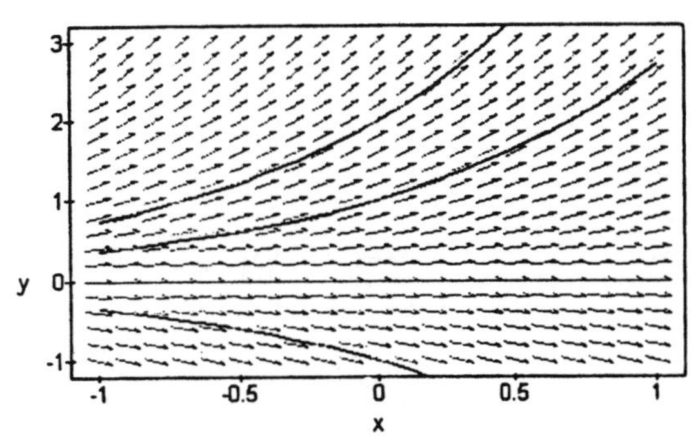

43.

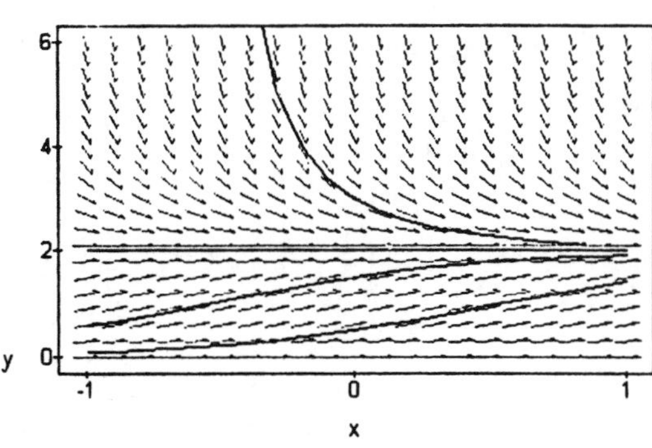

45.

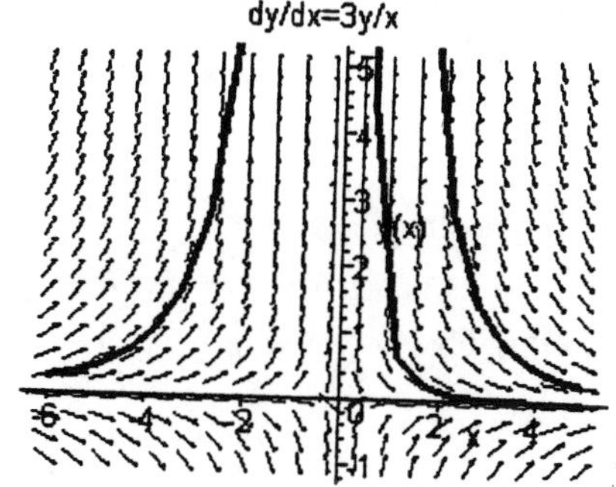

6.5 LINEAR FIRST ORDER DIFFERENTIAL EQUATIONS

1. $x\frac{dy}{dx} + y = e^x$

Step 1: $\frac{dy}{dx} + \left(\frac{1}{x}\right)y = \frac{e^x}{x},\ P(x) = \frac{1}{x},\ Q(x) = \frac{e^x}{x}$

Step 2: $\int P(x)\,dx = \int \frac{1}{x}\,dx = \ln|x| = \ln x,\ x > 0$

Step 3: $v(x) = e^{\int P(x)\,dx} = e^{\ln x} = x$

Step 4: $y = \frac{1}{v(x)}\int v(x)\,Q(x)\,dx = \frac{1}{x}\int x\left(\frac{e^x}{x}\right)dx = \frac{1}{x}(e^x + C) = \frac{e^x + C}{x}$

3. $xy' + 3y = \frac{\sin x}{x^2},\ x > 0$

Step 1: $\frac{dy}{dx} + \left(\frac{3}{x}\right)y = \frac{\sin x}{x^3},\ P(x) = \frac{3}{x},\ Q(x) = \frac{\sin x}{x^3}$

Step 2: $\int \frac{3}{x}\,dx = 3\ln|x| = \ln x^3,\ x > 0$

Step 3: $v(x) = e^{\ln x^3} = x^3$

Step 4: $y = \frac{1}{x^3}\int x^3\left(\frac{\sin x}{x^3}\right)dx = \frac{1}{x^3}\int \sin x\,dx = \frac{1}{x^3}(-\cos x + C) = \frac{C - \cos x}{x^3},\ x > 0$

5. $x\frac{dy}{dx} + 2y = 1 - \frac{1}{x},\ x > 0$

Step 1: $\frac{dy}{dx} + \left(\frac{2}{x}\right)y = \frac{1}{x} - \frac{1}{x^2},\ P(x) = \frac{2}{x},\ Q(x) = \frac{1}{x} - \frac{1}{x^2}$

Step 2: $\int \frac{2}{x}\,dx = 2\ln|x| = \ln x^2,\ x > 0$

Step 3: $v(x) = e^{\ln x^2} = x^2$

Step 4: $y = \frac{1}{x^2}\int x^2\left(\frac{1}{x} - \frac{1}{x^2}\right)dx = \frac{1}{x^2}\int (x-1)\,dx = \frac{1}{x^2}\left(\frac{x^2}{2} - x + C\right) = \frac{1}{2} - \frac{1}{x} + \frac{C}{x^2},\ x > 0$

7. $\frac{dy}{dx} - \frac{1}{2}y = \frac{1}{2}e^{x/2} \Rightarrow P(x) = -\frac{1}{2},\ Q(x) = \frac{1}{2}e^{x/2} \Rightarrow \int P(x)\,dx = -\frac{1}{2}x \Rightarrow v(x) = e^{-x/2}$

$\Rightarrow y = \frac{1}{e^{-x/2}}\int e^{-x/2}\left(\frac{1}{2}e^{x/2}\right)dx = e^{x/2}\int \frac{1}{2}\,dx = e^{x/2}\left(\frac{1}{2}x + C\right) = \frac{1}{2}xe^{x/2} + Ce^{x/2}$

9. $\frac{dy}{dx} - \left(\frac{1}{x}\right)y = 2\ln x \Rightarrow P(x) = -\frac{1}{x},\ Q(x) = 2\ln x \Rightarrow \int P(x)\,dx = -\int \frac{1}{x}\,dx = -\ln x,\ x > 0$

$\Rightarrow v(x) = e^{-\ln x} = \frac{1}{x} \Rightarrow y = x\int \left(\frac{1}{x}\right)(2\ln x)\,dx = x\left[(\ln x)^2 + C\right] = x(\ln x)^2 + Cx$

11. $\frac{ds}{dt}+\left(\frac{4}{t-1}\right)s=\frac{t+1}{(t-1)^3} \Rightarrow P(t)=\frac{4}{t-1},\ Q(t)=\frac{t+1}{(t-1)^3} \Rightarrow \int P(t)\,dt=\int \frac{4}{t-1}\,dt=4\ln|t-1|=\ln(t-1)^4$

$\Rightarrow v(t)=e^{\ln(t-1)^4}=(t-1)^4 \Rightarrow s=\frac{1}{(t-1)^4}\int (t-1)^4\left[\frac{t+1}{(t-1)^3}\right]dt=\frac{1}{(t-1)^4}\int \left(t^2-1\right)dt$

$=\frac{1}{(t-1)^4}\left(\frac{t^3}{3}-t+C\right)=\frac{t^3}{3(t-1)^4}-\frac{t}{(t-1)^4}+\frac{C}{(t-1)^4}$

13. $\frac{dr}{d\theta}+(\cot\theta)r=\sec\theta \Rightarrow P(\theta)=\cot\theta,\ Q(\theta)=\sec\theta \Rightarrow \int P(\theta)\,d\theta=\int \cot\theta\,d\theta=\ln|\sin\theta| \Rightarrow v(\theta)=e^{\ln|\sin\theta|}$

$=\sin\theta$ because $0<\theta<\frac{\pi}{2} \Rightarrow r=\frac{1}{\sin\theta}\int (\sin\theta)(\sec\theta)\,d\theta=\frac{1}{\sin\theta}\int \tan\theta\,d\theta=\frac{1}{\sin\theta}(\ln|\sec\theta|+C)$

$=(\csc\theta)(\ln|\sec\theta|+C)$

15. $\frac{dy}{dt}+2y=3 \Rightarrow P(t)=2,\ Q(t)=3 \Rightarrow \int P(t)\,dt=\int 2\,dt=2t \Rightarrow v(t)=e^{2t} \Rightarrow y=\frac{1}{e^{2t}}\int 3e^{2t}\,dt$

$=\frac{1}{e^{2t}}\left(\frac{3}{2}e^{2t}+C\right);\ y(0)=1 \Rightarrow \frac{3}{2}+C=1 \Rightarrow C=-\frac{1}{2} \Rightarrow y=\frac{3}{2}-\frac{1}{2}e^{-2t}$

17. $\frac{dy}{d\theta}+\left(\frac{1}{\theta}\right)y=\frac{\sin\theta}{\theta} \Rightarrow P(\theta)=\frac{1}{\theta},\ Q(\theta)=\frac{\sin\theta}{\theta} \Rightarrow \int P(\theta)\,d\theta=\ln|\theta| \Rightarrow v(\theta)=e^{\ln|\theta|}=|\theta|$

$\Rightarrow y=\frac{1}{|\theta|}\int |\theta|\left(\frac{\sin\theta}{\theta}\right)d\theta=\frac{1}{\theta}\int \theta\left(\frac{\sin\theta}{\theta}\right)d\theta$ for $\theta\neq 0 \Rightarrow y=\frac{1}{\theta}\int \sin\theta\,d\theta=\frac{1}{\theta}(-\cos\theta+C)$

$=-\frac{1}{\theta}\cos\theta+\frac{C}{\theta};\ y\left(\frac{\pi}{2}\right)=1 \Rightarrow C=\frac{\pi}{2} \Rightarrow y=-\frac{1}{\theta}\cos\theta+\frac{\pi}{2\theta}$

19. $(x+1)\frac{dy}{dx}-2\left(x^2+x\right)y=\frac{e^{x^2}}{x+1} \Rightarrow \frac{dy}{dx}-2\left[\frac{x(x+1)}{x+1}\right]y=\frac{e^{x^2}}{(x+1)^2} \Rightarrow \frac{dy}{dx}-2xy=\frac{e^{x^2}}{(x+1)^2} \Rightarrow P(x)=-2x,$

$Q(x)=\frac{e^{x^2}}{(x+1)^2} \Rightarrow \int P(x)\,dx=\int -2x\,dx=-x^2 \Rightarrow v(x)=e^{-x^2} \Rightarrow y=\frac{1}{e^{-x^2}}\int e^{-x^2}\left[\frac{e^{x^2}}{(x+1)^2}\right]dx$

$=e^{x^2}\int \frac{1}{(x+1)^2}\,dx=e^{x^2}\left[\frac{(x+1)^{-1}}{-1}+C\right]=-\frac{e^{x^2}}{x+1}+Ce^{x^2};\ y(0)=5 \Rightarrow -\frac{1}{0+1}+C=5 \Rightarrow -1+C=5$

$\Rightarrow C=6 \Rightarrow y=6e^{x^2}-\frac{e^{x^2}}{x+1}$

21. $\frac{dy}{dt}-ky=0 \Rightarrow P(t)=-k,\ Q(t)=0 \Rightarrow \int P(t)\,dt=\int -k\,dt=-kt \Rightarrow v(t)=e^{-kt}$

$\Rightarrow y=\frac{1}{e^{-kt}}\int \left(e^{-kt}\right)(0)\,dt=e^{kt}(0+C)=Ce^{kt};\ y(0)=y_0 \Rightarrow C=y_0 \Rightarrow y=y_0e^{kt}$

23. $x\int \frac{1}{x}\,dx=x(\ln|x|+C)=x\ln|x|+Cx \Rightarrow$ (b) is correct

25. Let y(t) = the amount of salt in the container and V(t) = the total volume of liquid in the tank at time t. Then, the departure rate is $\frac{y(t)}{V(t)}$ (the outflow rate).

(a) Rate entering $= \frac{2\text{ lb}}{\text{gal}} \cdot \frac{5\text{ gal}}{\text{min}} = 10$ lb/min

(b) Volume $= V(t) = 100\text{ gal} + (5t\text{ gal} - 4t\text{ gal}) = (100 + t)$ gal

(c) The volume at time t is $(100 + t)$ gal. The amount of salt in the tank at time t is y lbs. So the concentration at any time t is $\frac{y}{100+t}$ lbs/gal. Then, Rate leaving $= \frac{y}{100+t}$ (lbs/gal) $\cdot$ 4 (gal/min) $= \frac{4y}{100+t}$ lbs/min

(d) $\frac{dy}{dt} = 10 - \frac{4y}{100+t} \Rightarrow \frac{dy}{dt} + \left(\frac{4}{100+t}\right)y = 10 \Rightarrow P(t) = \frac{4}{100+t},\ Q(t) = 10 \Rightarrow \int P(t)\,dt = \int \frac{4}{100+t}\,dt$

$= 4\ln(100+t) \Rightarrow v(t) = e^{4\ln(100+t)} = (100+t)^4 \Rightarrow y = \frac{1}{(100+t)^4}\int (100+t)^4(10\,dt)$

$= \frac{10}{(100+t)^4}\left(\frac{(100+t)^5}{5} + C\right) = 2(100+t) + \frac{C}{(100+t)^4};\ y(0) = 50 \Rightarrow 2(100+0) + \frac{C}{(100+0)^4} = 50$

$\Rightarrow C = -(150)(100)^4 \Rightarrow y = 2(100+t) - \frac{(150)(100)^4}{(100+t)^4} \Rightarrow y = 2(100+t) - \frac{150}{\left(1+\frac{t}{100}\right)^4}$

(e) $y(25) = 2(100+25) - \frac{(150)(100)^4}{(100+25)^4} \approx 188.56$ lbs $\Rightarrow$ concentration $= \frac{y(25)}{\text{volume}} \approx \frac{188.6}{125} \approx 1.5$ lb/gal

27. Let y be the amount of fertilizer in the tank at time t. Then rate entering $= 1\,\frac{\text{lb}}{\text{gal}} \cdot 1\,\frac{\text{gal}}{\text{min}} = 1\,\frac{\text{lb}}{\text{min}}$ and the volume in the tank at time t is $V(t) = 100\text{ (gal)} + [1\text{ (gal/min)} - 3\text{ (gal/min)}]t\text{ min} = (100 - 2t)$ gal. Hence rate out $= \left(\frac{y}{100-2t}\right)3 = \frac{3y}{100-2t}$ lbs/min $\Rightarrow \frac{dy}{dt} = \left(1 - \frac{3y}{100-2t}\right)$ lbs/min $\Rightarrow \frac{dy}{dt} + \left(\frac{3}{100-2t}\right)y = 1$

$\Rightarrow P(t) = \frac{3}{100-2t},\ Q(t) = 1 \Rightarrow \int P(t)\,dt = \int \frac{3}{100-2t}\,dt = \frac{3\ln(100-2t)}{-2} \Rightarrow v(t) = e^{(-3\ln(100-2t))/2}$

$= (100-2t)^{-3/2} \Rightarrow y = \frac{1}{(100-2t)^{-3/2}}\int (100-2t)^{-3/2}\,dt = (100-2t)^{3/2}\left[\frac{-2(100-2t)^{-1/2}}{-2} + C\right]$

$= (100-2t) + C(100-2t)^{3/2};\ y(0) = 0 \Rightarrow [100 - 2(0)] + C[100 - 2(0)]^{3/2} \Rightarrow C(100)^{3/2} = -100$

$\Rightarrow C = -(100)^{-1/2} = -\frac{1}{10} \Rightarrow y = (100-2t) - \frac{(100-2t)^{3/2}}{10}$. Let $\frac{dy}{dt} = 0 \Rightarrow \frac{dy}{dt} = -2 - \frac{\left(\frac{3}{2}\right)(100-2t)^{1/2}(-2)}{10}$

$= -2 + \frac{3\sqrt{100-2t}}{10} = 0 \Rightarrow 20 = 3\sqrt{100-2t} \Rightarrow 400 = 9(100-2t) \Rightarrow 400 = 900 - 18t \Rightarrow -500 = -18t$

$\Rightarrow t \approx 27.8$ min, the time to reach the maximum. The maximum amount is then

$y(27.8) = [100 - 2(27.8)] - \frac{[100 - 2(27.8)]^{3/2}}{10} \approx 14.8$ lb

29. Steady State $= \frac{V}{R}$ and we want $i = \frac{1}{2}\left(\frac{V}{R}\right) \Rightarrow \frac{1}{2}\left(\frac{V}{R}\right) = \frac{V}{R}\left(1 - e^{-Rt/L}\right) \Rightarrow \frac{1}{2} = 1 - e^{-Rt/L} \Rightarrow -\frac{1}{2} = -e^{-Rt/L}$

$\Rightarrow \ln \frac{1}{2} = -\frac{Rt}{L} \Rightarrow -\frac{L}{R} \ln \frac{1}{2} = t \Rightarrow t = \frac{L}{R} \ln 2$ sec

31. (a) $t = \frac{3L}{R} \Rightarrow i = \frac{V}{R}\left(1 - e^{(-R/L)(3L/R)}\right) = \frac{V}{R}\left(1 - e^{-3}\right) \approx 0.9502 \frac{V}{R}$ amp, or about 95% of the steady state value

(b) $t = \frac{2L}{R} \Rightarrow i = \frac{V}{R}\left(1 - e^{(-R/L)(2L/R)}\right) = \frac{V}{R}\left(1 - e^{-2}\right) \approx 0.8647 \frac{V}{R}$ amp, or about 86% of the steady state value

6.6 EULER'S METHOD; POPULATION MODELS

1. $y_1 = y_0 + x_0(1 - y_0)\,dx = 0 + 1(1 - 0)(0.2) = 0.2,$

$y_2 = y_1 + x_1(1 - y_1)\,dx = 0.2 + 1.2(1 - 0.2)(0.2) = 0.392,$

$y_3 = y_2 + x_2(1 - y_2)\,dx = 0.392 + 1.4(1 - 0.392)(0.2) = 0.5622;$

$\frac{dy}{1-y} = x\,dx \Rightarrow -\ln|1-y| = \frac{x^2}{2} + C;\ x = 1,\ y = 0 \Rightarrow -\ln 1 = \frac{1}{2} + C \Rightarrow C = -\frac{1}{2} \Rightarrow \ln|1-y| = -\frac{x^2}{2} + \frac{1}{2}$

$\Rightarrow y = 1 - e^{\left(1-x^2\right)/2} \Rightarrow y(1.2) \approx 0.1975,\ y(1.4) \approx 0.3812,\ y(1.6) \approx 0.5416$

3. $y_1 = y_0 + (2x_0y_0 + 2y_0)\,dx = 3 + [2(0)(3) + 2(3)](0.2) = 4.2,$

$y_2 = y_1 + (2x_1y_1 + 2y_1)\,dx = 4.2 + [2(0.2)(4.2) + 2(4.2)](0.2) = 6.216,$

$y_3 = y_2 + (2x_2y_2 + 2y_2)\,dx = 6.216 + [2(0.4)(6.216) + 2(6.216)](0.2) = 9.6970;$

$\frac{dy}{dx} = 2y(x+1) \Rightarrow \frac{dy}{y} = 2(x+1)\,dx \Rightarrow \ln|y| = (x+1)^2 + C;\ x = 0,\ y = 3 \Rightarrow \ln 3 = 1 + C \Rightarrow C = \ln 3 - 1$

$\Rightarrow \ln y = (x+1)^2 + \ln 3 - 1 \Rightarrow y = e^{(x+1)^2 + \ln 3 - 1} = e^{\ln 3}e^{x^2+2x} = 3e^{x(x+2)} \Rightarrow y(0.2) \approx 4.6581,$

$y(0.4) \approx 7.8351,\ y(0.6) \approx 14.2765$

5. $y_1 = 1 + 1(0.2) = 1.2,$

$y_2 = 1.2 + (1.2)(0.2) = 1.44,$

$y_3 = 1.44 + (1.44)(0.2) = 1.728,$

$y_4 = 1.728 + (1.728)(0.2) = 2.0736,$

$y_5 = 2.0736 + (2.0736)(0.2) = 2.48832;$

$\frac{dy}{y} = dx \Rightarrow \ln y = x + C_1 \Rightarrow y = Ce^x;\ y(0) = 1 \Rightarrow 1 = Ce^0 \Rightarrow C = 1 \Rightarrow y = e^x \Rightarrow y(1) = e \approx 2.7183$

7. Let $z_n = y_{n-1} + 2y_{n-1}(x_{n-1} + 1)\,dx$ and $y_n = y_{n-1} + (y_{n-1}(x_{n-1} + 1) + z_n(x_n + 1))\,dx$ with $x_0 = 0$, $y_0 = 3$, and $dx = 0.2$. The exact solution is $y = 3e^{x(x+2)}$. Using a programmable calculator or a spreadsheet (I used a spreadsheet) gives the values in the following table.

x	z	y-approx	y-exact	Error
0	---	3	3	0
0.2	4.2	4.608	4.658122	0.050122
0.4	6.81984	7.623475	7.835089	0.211614
0.6	11.89262	13.56369	14.27646	0.712777

9. (a) $\frac{dP}{dt} = 0.0015P(150 - P) = \frac{0.225}{150}P(150 - P) = \frac{k}{M}P(M - P)$

Thus, $k = 0.225$ and $M = 150$, and $P = \frac{M}{1 + Ae^{-k/t}} = \frac{150}{1 + Ae^{-0.225t}}$

Initial condition: $P(0) = 6 \Rightarrow 6 = \frac{150}{1 + Ae^0} \Rightarrow 1 + A = 25 \Rightarrow A = 24$

Formula: $P = \frac{150}{1 + 24^{-0.225t}}$

(b) $100 = \frac{150}{1 + 24e^{-0.225t}} \Rightarrow 1 + 24e^{-0.2225t} = \frac{3}{2} \Rightarrow 24e^{-0.225t} = \frac{1}{2} \Rightarrow e^{-0.225t} = \frac{1}{48} \Rightarrow -0.225t = -\ln 48$

$\Rightarrow t = \frac{\ln 48}{0.225} \approx 17.21$ weeks

$125 = \frac{150}{1 + 24e^{-0.225t}} \Rightarrow 1 + 24e^{-0.2225t} = \frac{6}{5} \Rightarrow 24e^{-0.225t} = \frac{1}{5} \Rightarrow e^{-0.225t} = \frac{1}{120} \Rightarrow -0.225t = -\ln 120$

$\Rightarrow t = \frac{\ln 120}{0.225} \approx 21.28$

It will take about 17.21 weeks to reach 100 guppies, and about 21.28 weeks to reach 125 guppies.

11. (a) Using the general solution from Example 6, part (c),

$\frac{dy}{dt} = (0.08875 \times 10^{-7})(8 \times 10^7 - y)y \Rightarrow y(t) = \frac{M}{1 + Ae^{-rMt}} = \frac{8 \times 10^7}{1 + Ae^{-(0.08875)(8)t}} = \frac{8 \times 10^7}{1 + Ae^{-0.71t}}$

Apply the initial condition:

$y(0) = 1.6 \times 10^7 = \frac{8 \times 10^7}{1 + A} \Rightarrow A = \frac{8}{1.6} - 1 = 4 \Rightarrow y(1) = \frac{8 \times 10^7}{1 + 4e^{-0.71}} \approx 2.69671 \times 10^7$ kg.

(b) $y(t) = 4 \times 10^7 = \frac{8 \times 10^7}{1 + 4e^{-0.71t}} \Rightarrow 4e^{-0.71t} = 1 \Rightarrow t = -\frac{\ln(1/4)}{0.71} \approx 1.95253$ years.

13. (a) $\frac{dy}{dt} = 1 + y \Rightarrow dy = (1 + y)\,dt \Rightarrow \frac{dy}{1 + y} = dt \Rightarrow \ln|1 + y| = t + C_1 \Rightarrow e^{\ln|1+y|} = e^{t+C_1} \Rightarrow |1 + y| = e^t e^{C_1}$

$1 + y = \pm C_2 e^t \Rightarrow y = Ce^t - 1$, where $C_2 = e^{C_1}$ and $C = \pm C_2$. Apply the initial condition: $y(0) = 1$ $= Ce^0 - 1 \Rightarrow C = 2 \Rightarrow y = 2e^t - 1$.

(b) $\frac{dy}{dt} = 0.5(400 - y)y \Rightarrow dy = 0.5(400 - y)y\,dt \Rightarrow \frac{dy}{y(400 - y)} = 0.5\,dt$. Using the partial fraction decomposition in Example 6, part (c), we obtain $\frac{1}{400}\left(\frac{1}{y} + \frac{1}{400 - y}\right)dy = 0.5\,dt \Rightarrow \left(\frac{1}{y} + \frac{1}{400 - y}\right)dy$

$= 200\,dt \Rightarrow \int\left(\frac{1}{y} - \frac{1}{y - 400}\right)dy = \int 200\,dt \Rightarrow \ln|y| - \ln|y - 400| = 200t + C_1 \Rightarrow \ln\left|\frac{y}{y - 400}\right| = 200t + C_1$

$\Rightarrow e^{\ln\left|\frac{y}{y-400}\right|} = e^{200t + C_1} = e^{200t}e^{C_1} \Rightarrow \left|\frac{y}{y - 400}\right| = C_2 e^{200t}$ (where $C_2 = e^{C_1}$) $\Rightarrow \frac{y}{y - 400} = \pm C_2 e^{200t}$

$\Rightarrow \frac{y}{y - 400} = Ce^{200t}$ (where $C = \pm C_2$) $\Rightarrow y = Ce^{200t}y - 400Ce^{200t} \Rightarrow (1 - Ce^{200t})y = -400Ce^{200t}$

$\Rightarrow y = \frac{400Ce^{200t}}{Ce^{200t} - 1} \Rightarrow y = \frac{400}{1 - \frac{1}{C}e^{-200t}} = \frac{400}{1 + Ae^{-200t}}$, where $A = -\frac{1}{C}$. Apply the initial condition:

$y(0) = 2 = \frac{400}{1 + Ae^0} \Rightarrow A = 199 \Rightarrow y(t) = \frac{400}{1 + 199e^{-200t}}$.

15. (a) $\frac{dP}{dt} = kP^2 \Rightarrow \int P^{-2}\,dP = \int k\,dt \Rightarrow -P^{-1} = kt + C \Rightarrow P = \frac{-1}{kt + C}$

Initial condition: $P(0) = P_0 \Rightarrow P_0 = -\frac{1}{C} \Rightarrow C = -\frac{1}{P_0}$

Solution: $P = -\frac{1}{kt - (1/P_0)} = \frac{P_0}{1 - kP_0 t}$

(b) There is a vertical asymptote at $t = \frac{1}{kP_0}$

17. $\frac{dy}{dx} = 2xe^{x^2}$, $y(0) = 2 \Rightarrow y_{n+1} = y_n + 2x_n e^{x_n^2}\,dx = y_n + 2x_n e^{x_n^2}(0.1) = y_n + 0.2x_n e^{x_n^2}$

On a TI-92 Plus calculator home screen, type the following commands:

2 STO> y: 0 STO> x:y (enter)
y+0.2*x*e^(x^2) STO> y: x+0.1 STO>x: y (enter, 10 times)

The last value displayed gives $y_{Euler}(1) \approx 3.45835$

The exact solution: $dy = 2xe^{x^2}\,dx \Rightarrow y = e^{x^2} + C$; $y(0) = 2 = e^0 + C \Rightarrow C = 1 \Rightarrow y = 1 + e^{x^2}$

$\Rightarrow y_{exact}(1) = 1 + e \approx 3.71828$

19. $y_1 = -1 + \left[\frac{(-1)^2}{\sqrt{1}}\right](0.5) = -0.5,$

$y_2 = -0.5 + \left[\frac{(-0.5)^2}{\sqrt{1.5}}\right](0.5) = -0.39794,$

$y_3 = -0.39794 + \left[\frac{(-0.39794)^2}{\sqrt{2}}\right](0.5) = -0.34195,$

$y_4 = -0.34195 + \left[\frac{(-0.34195)^2}{\sqrt{2.5}}\right](0.5) = -0.30497,$

$y_5 = -0.27812$, $y_6 = -0.25745$, $y_7 = -0.24088$, $y_8 = -0.2272$;

$\frac{dy}{y^2} = \frac{dx}{\sqrt{x}} \Rightarrow -\frac{1}{y} = 2\sqrt{x} + C$; $y(1) = -1 \Rightarrow 1 = 2 + C \Rightarrow C = -1 \Rightarrow y = \frac{1}{1 - 2\sqrt{x}} \Rightarrow y(5) = \frac{1}{1 - 2\sqrt{5}} \approx -0.2880$

21. (a) $\frac{dy}{dx} = 2y^2(x - 1) \Rightarrow \frac{dy}{y^2} = 2(x - 1)\,dx \Rightarrow \int y^{-2}\,dy = \int (2x - 2)\,dx \Rightarrow -y^{-1} = x^2 - 2x + C$

Initial value: $y(2) = -\frac{1}{2} \Rightarrow 2 = 2^2 - 2(2) + C \Rightarrow C = 2$

Solution: $-y^{-1} = x^2 - 2x + 2$ or $y = -\frac{1}{x^2 - 2x + 2}$

$y(3) = -\frac{1}{3^2 - 2(3) + 2} = -\frac{1}{5} = -0.2$

(b) To find the approximation, set $y_1 = 2y^2(x - 1)$ and use EULERT with initial values $x = 2$ and $y = -\frac{1}{2}$ and step size 0.2 for 5 points. This gives $y(3) \approx -0.1851$; error ≈ 0.0149.

(c) Use step size 0.1 for 10 points. This gives $y(3) \approx -0.1929$; error ≈ 0.0071.

(d) Use step size 0.05 for 20 points. This gives $y(3) \approx -0.1965$; error ≈ 0.0035.

23. The exact solution is $y = \frac{1}{x^2 - 2x + 2}$, so $y(3) = -0.2$. To find the approximation, let

$z_n = y_{n-1} + 2y_{n-1}^2(x_n - 1)\,dx$ and $y_n = y_{n-1} + (y_{n-1}^2(x_{n-1} - 1) + z_n^2(x_n^2 - 1))\,dx$ with initial values $x_0 = 2$ and $y_0 = -\frac{1}{2}$. Use a spreadsheet, graphing calculator, or CAS as indicated in parts (a) through (d).

(a) Use $dx = 0.2$ with 5 steps to obtain $y(3) \approx -0.2024 \Rightarrow$ error ≈ 0.0024.

(b) Use $dx = 0.1$ with 10 steps to obtain $y(3) \approx -0.2005 \Rightarrow$ error ≈ 0.0005.

(c) Use $dx = 0.05$ with 20 steps to obtain $y(3) \approx -0.2001 \Rightarrow$ error ≈ 0.0001.

(d) Each time the step size is cut in half, the error is reduced to approximately one-fourth of what it was for the larger step size.

6.7 HYPERBOLIC FUNCTIONS

1. $\sinh x = -\frac{3}{4} \Rightarrow \cosh x = \sqrt{1 + \sinh^2 x} = \sqrt{1 + \left(-\frac{3}{4}\right)^2} = \sqrt{1 + \frac{9}{16}} = \sqrt{\frac{25}{16}} = \frac{5}{4}$, $\tanh x = \frac{\sinh x}{\cosh x} = \frac{\left(-\frac{3}{4}\right)}{\left(\frac{5}{4}\right)} = -\frac{3}{5}$,

$\coth x = \frac{1}{\tanh x} = -\frac{5}{3}$, $\operatorname{sech} x = \frac{1}{\cosh x} = \frac{4}{5}$, and $\operatorname{csch} x = \frac{1}{\sin x} = -\frac{4}{3}$

3. $\cosh x = \frac{17}{15}$, $x > 0 \Rightarrow \sinh x = \sqrt{\cosh^2 x - 1} = \sqrt{\left(\frac{17}{15}\right)^2 - 1} = \sqrt{\frac{289}{225} - 1} = \sqrt{\frac{64}{225}} = \frac{8}{15}$, $\tanh x = \frac{\sinh x}{\cosh x} = \frac{\left(\frac{8}{15}\right)}{\left(\frac{17}{15}\right)}$

$= \frac{8}{17}$, $\coth x = \frac{1}{\tanh x} = \frac{17}{8}$, $\operatorname{sech} x = \frac{1}{\cosh x} = \frac{15}{17}$, and $\operatorname{csch} x = \frac{1}{\sinh x} = \frac{15}{8}$

5. $2 \cosh(\ln x) = 2\left(\frac{e^{\ln x} + e^{-\ln x}}{2}\right) = e^{\ln x} + \frac{1}{e^{\ln x}} = x + \frac{1}{x}$

7. $\cosh 5x + \sinh 5x = \frac{e^{5x} + e^{-5x}}{2} + \frac{e^{5x} - e^{-5x}}{2} = e^{5x}$

9. $(\sinh x + \cosh x)^4 = \left(\frac{e^x - e^{-x}}{2} + \frac{e^x + e^{-x}}{2}\right)^4 = (e^x)^4 = e^{4x}$

11. (a) $\sinh 2x = \sinh(x + x) = \sinh x \cosh x + \cosh x \sinh x = 2 \sinh x \cosh x$

(b) $\cosh 2x = \cosh(x + x) = \cosh x \cosh x + \sinh x \sin x = \cosh^2 x + \sinh^2 x$

13. $y = 6 \sinh \frac{x}{3} \Rightarrow \frac{dy}{dx} = 6\left(\cosh \frac{x}{3}\right)\left(\frac{1}{3}\right) = 2 \cosh \frac{x}{3}$

15. $y = 2\sqrt{t} \tanh \sqrt{t} = 2t^{1/2} \tanh t^{1/2} \Rightarrow \frac{dy}{dt} = \left[\operatorname{sech}^2\left(t^{1/2}\right)\right]\left(\frac{1}{2}t^{-1/2}\right)\left(2t^{1/2}\right) + \left(\tanh t^{1/2}\right)\left(t^{-1/2}\right)$

$= \operatorname{sech}^2 \sqrt{t} + \frac{\tanh \sqrt{t}}{\sqrt{t}}$

17. $y = \ln(\sinh z) \Rightarrow \frac{dy}{dz} = \frac{\cosh z}{\sinh z} = \coth z$

19. $y = (\text{sech }\theta)(1 - \ln \text{sech }\theta) \Rightarrow \frac{dy}{d\theta} = \left(-\frac{-\text{sech }\theta \tanh \theta}{\text{sech }\theta}\right)(\text{sech }\theta) + (-\text{sech }\theta \tanh \theta)(1 - \ln \text{sech }\theta)$

$= \text{sech }\theta \tanh \theta - (\text{sech }\theta \tanh \theta)(1 - \ln \text{sech }\theta) = (\text{sech }\theta \tanh \theta)[1 - (1 - \ln \text{sech }\theta)]$

$= (\text{sech }\theta \tanh \theta)(\ln \text{sech }\theta)$

21. $y = \ln \cosh v - \frac{1}{2}\tanh^2 v \Rightarrow \frac{dy}{dv} = \frac{\sinh v}{\cosh v} - \left(\frac{1}{2}\right)(2 \tanh v)\left(\text{sech}^2 v\right) = \tanh v - (\tanh v)\left(\text{sech}^2 v\right)$

$= (\tanh v)\left(1 - \text{sech}^2 v\right) = (\tanh v)\left(\tanh^2 v\right) = \tanh^3 v$

23. $y = \left(x^2+1\right)\text{sech}(\ln x) = \left(x^2+1\right)\left(\frac{2}{e^{\ln x} + e^{-\ln x}}\right) = \left(x^2+1\right)\left(\frac{2}{x + x^{-1}}\right) = \left(x^2+1\right)\left(\frac{2x}{x^2+1}\right) = 2x \Rightarrow \frac{dy}{dx} = 2$

25. $y = \sinh^{-1}\sqrt{x} = \sinh^{-1}\left(x^{1/2}\right) \Rightarrow \frac{dy}{dx} = \frac{\left(\frac{1}{2}\right)x^{-1/2}}{\sqrt{1 + \left(x^{1/2}\right)^2}} = \frac{1}{2\sqrt{x}\sqrt{1+x}} = \frac{1}{2\sqrt{x(1+x)}}$

27. $y = (1-\theta)\tanh^{-1}\theta \Rightarrow \frac{dy}{d\theta} = (1-\theta)\left(\frac{1}{1-\theta^2}\right) + (-1)\tanh^{-1}\theta = \frac{1}{1+\theta} - \tanh^{-1}\theta$

29. $y = (1-t)\coth^{-1}\sqrt{t} = (1-t)\coth^{-1}\left(t^{1/2}\right) \Rightarrow \frac{dy}{dt} = (1-t)\left[\frac{\left(\frac{1}{2}\right)t^{-1/2}}{1 - \left(t^{1/2}\right)^2}\right] + (-1)\coth^{-1}\left(t^{1/2}\right) = \frac{1}{2\sqrt{t}} - \coth^{-1}\sqrt{t}$

31. $y = \cos^{-1}x - x\,\text{sech}^{-1}x \Rightarrow \frac{dy}{dx} = \frac{-1}{\sqrt{1-x^2}} - \left[x\left(\frac{-1}{x\sqrt{1-x^2}}\right) + (1)\,\text{sech}^{-1}x\right] = \frac{-1}{\sqrt{1-x^2}} + \frac{1}{\sqrt{1-x^2}} - \text{sech}^{-1}x$

$= -\text{sech}^{-1}x$

33. $y = \text{csch}^{-1}\left(\frac{1}{2}\right)^\theta \Rightarrow \frac{dy}{d\theta} = -\frac{\left[\ln\left(\frac{1}{2}\right)\right]\left(\frac{1}{2}\right)^\theta}{\left(\frac{1}{2}\right)^\theta\sqrt{1 + \left[\left(\frac{1}{2}\right)^\theta\right]^2}} = -\frac{\ln(1) - \ln(2)}{\sqrt{1 + \left(\frac{1}{2}\right)^{2\theta}}} = \frac{\ln 2}{\sqrt{1 + \left(\frac{1}{2}\right)^{2\theta}}}$

35. $y = \sinh^{-1}(\tan x) \Rightarrow \frac{dy}{dx} = \frac{\sec^2 x}{\sqrt{1 + (\tan x)^2}} = \frac{\sec^2 x}{\sqrt{\sec^2 x}} = \frac{\sec^2 x}{|\sec x|} = \frac{|\sec x||\sec x|}{|\sec x|} = |\sec x|$

37. (a) If $y = \tan^{-1}(\sinh x) + C$, then $\frac{dy}{dx} = \frac{\cosh x}{1 + \sinh^2 x} = \frac{\cosh x}{\cosh^2 x} = \text{sech } x$, which verifies the formula

(b) If $y = \sin^{-1}(\tanh x) + C$, then $\frac{dy}{dx} = \frac{\text{sech}^2 x}{\sqrt{1 - \tanh^2 x}} = \frac{\text{sech}^2 x}{\text{sech } x} = \text{sech } x$, which verifies the formula

39. If $y = \frac{x^2-1}{2}\coth^{-1}x + \frac{x}{2} + C$, then $\frac{dy}{dx} = x\coth^{-1}x + \left(\frac{x^2-1}{2}\right)\left(\frac{1}{1-x^2}\right) + \frac{1}{2} = x\coth^{-1}x$, which verifies the formula

41. $\int \sinh 2x\,dx = \frac{1}{2}\int \sinh u\,du$, where $u = 2x$ and $du = 2\,dx$

$= \frac{\cosh u}{2} + C = \frac{\cosh 2x}{2} + C$

43. $\int 6\cosh\left(\frac{x}{2}-\ln 3\right)dx = 12\int \cosh u\,du$, where $u=\frac{x}{2}-\ln 3$ and $du=\frac{1}{2}\,dx$

$= 12\sinh u + C = 12\sinh\left(\frac{x}{2}-\ln 3\right)+C$

45. $\int \tanh\frac{x}{7}\,dx = 7\int \frac{\sinh u}{\cosh u}\,du$, where $u=\frac{x}{7}$ and $du=\frac{1}{7}\,dx$

$= 7\ln|\cosh u| + C_1 = 7\ln\left|\cosh\frac{x}{7}\right| + C_1 = 7\ln\left|\frac{e^{x/7}+e^{-x/7}}{2}\right| + C_1 = 7\ln\left|e^{x/7}+e^{-x/7}\right| - 7\ln 2 + C_1$

$= 7\ln\left|e^{x/7}+e^{-x/7}\right| + C$

47. $\int \operatorname{sech}^2\left(x-\frac{1}{2}\right)dx = \int \sec^2 u\,du$, where $u=\left(x-\frac{1}{2}\right)$ and $du=dx$

$= \tanh u + C = \tan\left(x-\frac{1}{2}\right)+C$

49. $\int \frac{\operatorname{sech}\sqrt{t}\tanh\sqrt{t}}{\sqrt{t}}\,dt = 2\int \operatorname{sech} u\tanh u\,du$, where $u=\sqrt{t}=t^{1/2}$ and $du=\frac{dt}{2\sqrt{t}}$

$= 2(-\operatorname{sech} u)+C = -2\operatorname{sech}\sqrt{t}+C$

51. $\int_{\ln 2}^{\ln 4} \coth x\,dx = \int_{\ln 2}^{\ln 4} \frac{\cosh x}{\sinh x}\,dx = \int_{3/4}^{15/8} \frac{1}{u}\,du = [\ln|u|]_{3/4}^{15/8} = \ln\left|\frac{15}{8}\right| - \ln\left|\frac{3}{4}\right| = \ln\left|\frac{15}{8}\cdot\frac{4}{3}\right| = \ln\frac{5}{2}$,

where $u=\sinh x$, $du=\cosh x\,dx$, the lower limit is $\sinh(\ln 2) = \frac{e^{\ln 2}-e^{-\ln 2}}{2} = \frac{2-\left(\frac{1}{2}\right)}{2} = \frac{3}{4}$ and the upper limit is $\sinh(\ln 4) = \frac{e^{\ln 4}-e^{-\ln 4}}{2} = \frac{4-\left(\frac{1}{4}\right)}{2} = \frac{15}{8}$

53. $\int_{-\ln 4}^{-\ln 2} 2e^{\theta}\cosh\theta\,d\theta = \int_{-\ln 4}^{-\ln 2} 2e^{\theta}\left(\frac{e^{\theta}+e^{-\theta}}{2}\right)d\theta = \int_{-\ln 4}^{-\ln 2} \left(e^{2\theta}+1\right)d\theta = \left[\frac{e^{2\theta}}{2}+\theta\right]_{-\ln 4}^{-\ln 2}$

$= \left(\frac{e^{-2\ln 2}}{2}-\ln 2\right) - \left(\frac{e^{-2\ln 4}}{2}-\ln 4\right) = \left(\frac{1}{8}-\ln 2\right) - \left(\frac{1}{32}-\ln 4\right) = \frac{3}{32} - \ln 2 + 2\ln 2 = \frac{3}{32}+\ln 2$

55. $\int_{-\pi/4}^{\pi/4} \cosh(\tan\theta)\sec^2\theta\,d\theta = \int_{-1}^{1} \cosh u\,du = [\sinh u]_{-1}^{1} = \sinh(1) - \sinh(-1) = \left(\frac{e^1-e^{-1}}{2}\right) - \left(\frac{e^{-1}-e^1}{2}\right)$

$= \frac{e-e^{-1}-e^{-1}+e}{2} = e-e^{-1}$, where $u=\tan\theta$, $du=\sec^2\theta\,d\theta$, the lower limit is $\tan\left(-\frac{\pi}{4}\right)=-1$ and the upper limit is $\tan\left(\frac{\pi}{4}\right)=1$

57. $\int_{1}^{2} \frac{\cosh(\ln t)}{t}\,dt = \int_{0}^{\ln 2} \cosh u\,du = [\sinh u]_0^{\ln 2} = \sinh(\ln 2) - \sinh(0) = \frac{e^{\ln 2}-e^{-\ln 2}}{2} - 0 = \frac{2-\frac{1}{2}}{2} = \frac{3}{4}$, where $u=\ln t$, $du=\frac{1}{t}\,dt$, the lower limit is $\ln 1 = 0$ and the upper limit is $\ln 2$

59. $\int_{-\ln 2}^{0} \cosh^2\left(\frac{x}{2}\right) dx = \int_{-\ln 2}^{0} \frac{\cosh x + 1}{2}\, dx = \frac{1}{2}\int_{-\ln 2}^{0} (\cosh x + 1)\, dx = \frac{1}{2}[\sinh x + x]_{-\ln 2}^{0}$

$= \frac{1}{2}[(\sinh 0 + 0) - (\sinh(-\ln 2) - \ln 2)] = \frac{1}{2}\left[(0+0) - \left(\frac{e^{-\ln 2} - e^{\ln 2}}{2} - \ln 2\right)\right] = \frac{1}{2}\left[-\frac{\left(\frac{1}{2}\right) - 2}{2} + \ln 2\right]$

$= \frac{1}{2}\left(1 - \frac{1}{4} + \ln 2\right) = \frac{3}{8} + \frac{1}{2}\ln 2 = \frac{3}{8} + \ln\sqrt{2}$

61. $\sinh^{-1}\left(\frac{-5}{12}\right) = \ln\left(-\frac{5}{12} + \sqrt{\frac{25}{144} + 1}\right) = \ln\left(\frac{2}{3}\right)$

63. $\tanh^{-1}\left(-\frac{1}{2}\right) = \frac{1}{2}\ln\left(\frac{1 - (1/2)}{1 + (1/2)}\right) = -\frac{\ln 3}{2}$

65. $\operatorname{sech}^{-1}\left(\frac{3}{5}\right) = \ln\left(\frac{1 + \sqrt{1 - (9/25)}}{(3/5)}\right) = \ln 3$

67. (a) $\int_{0}^{2\sqrt{3}} \frac{dx}{\sqrt{4 + x^2}} = \left[\sinh^{-1}\frac{x}{2}\right]_0^{2\sqrt{3}} = \sinh^{-1}\sqrt{3} - \sinh 0 = \sinh^{-1}\sqrt{3}$

(b) $\sinh^{-1}\sqrt{3} = \ln\left(\sqrt{3} + \sqrt{3+1}\right) = \ln\left(\sqrt{3} + 2\right)$

69. (a) $\int_{5/4}^{2} \frac{1}{1 - x^2}\, dx = \left[\coth^{-1} x\right]_{5/4}^{2} = \coth^{-1} 2 - \coth^{-1}\frac{5}{4}$

(b) $\coth^{-1} 2 - \coth^{-1}\frac{5}{4} = \frac{1}{2}\left[\ln 3 - \ln\left(\frac{9/4}{1/4}\right)\right] = \frac{1}{2}\ln\frac{1}{3}$

71. (a) $\int_{1/5}^{3/13} \frac{dx}{x\sqrt{1 - 16x^2}} = \int_{4/5}^{12/13} \frac{du}{u\sqrt{a^2 - u^2}}$, where $u = 4x$, $du = 4\,dx$, $a = 1$

$= \left[-\operatorname{sech}^{-1} u\right]_{4/5}^{12/13} = -\operatorname{sech}^{-1}\frac{12}{13} + \operatorname{sech}^{-1}\frac{4}{5}$

(b) $-\operatorname{sech}^{-1}\frac{12}{13} + \operatorname{sech}^{-1}\frac{4}{5} = -\ln\left(\frac{1 + \sqrt{1 - (12/13)^2}}{(12/13)}\right) + \ln\left(\frac{1 + \sqrt{1 - (4/5)^2}}{(4/5)}\right)$

$= -\ln\left(\frac{13 + \sqrt{169 - 144}}{12}\right) + \ln\left(\frac{5 + \sqrt{25 - 16}}{4}\right) = \ln\left(\frac{5+3}{4}\right) - \ln\left(\frac{13+5}{12}\right) = \ln 2 - \ln\frac{3}{2}$

$= \ln\left(2 \cdot \frac{2}{3}\right) = \ln\frac{4}{3}$

73. (a) $\int_{0}^{\pi} \frac{\cos x}{\sqrt{1 + \sin^2 x}}\, dx = \int_{0}^{0} \frac{1}{\sqrt{1 + u^2}}\, du = \left[\sinh^{-1} u\right]_0^0 = \sinh^{-1} 0 - \sinh^{-1} 0 = 0$, where $u = \sin x$, $du = \cos x\, dx$

(b) $\sinh^{-1} 0 - \sinh^{-1} 0 = \ln\left(0 + \sqrt{0+1}\right) - \ln\left(0 + \sqrt{0+1}\right) = 0$

75. (a) Let $E(x) = \frac{f(x)+f(-x)}{2}$ and $O(x) = \frac{f(x)-f(-x)}{2}$. Then $E(x)+O(x) = \frac{f(x)+f(-x)}{2} + \frac{f(x)-f(-x)}{2}$ $= \frac{2f(x)}{2} = f(x)$. Also, $E(-x) = \frac{f(-x)+f(-(-x))}{2} = \frac{f(x)+f(-x)}{2} = E(x) \Rightarrow E(x)$ is even, and $O(-x) = \frac{f(-x)-f(-(-x))}{2} = -\frac{f(x)-f(-x)}{2} = -O(x) \Rightarrow O(x)$ is odd. Consequently, f(x) can be written as a sum of an even and an odd function.

(b) $f(x) = \frac{f(x)+f(-x)}{2}$ because $\frac{f(x)-f(-x)}{2} = 0$ and $f(x) = \frac{f(x)-f(-x)}{2}$ because $\frac{f(x)+f(-x)}{2} = 0$; thus $f(x) = \frac{2f(x)}{2} + 0$ and $f(x) = 0 + \frac{2f(x)}{2}$

77. (a) $m\frac{dv}{dt} = mg - kv^2 \Rightarrow \frac{m\frac{dv}{dt}}{mg-kv^2} = 1 \Rightarrow \frac{\frac{1}{g}\frac{dv}{dt}}{1-\frac{kv^2}{mg}} = 1 \Rightarrow \frac{\sqrt{\frac{k}{mg}}\,dv}{1-\left(\sqrt{v\left(\frac{k}{mg}\right)}\right)^2} = \sqrt{\frac{kg}{m}}\,dt \Rightarrow \int \frac{\sqrt{\frac{k}{mg}}\,dv}{1-\left(\sqrt{v\left(\frac{k}{mg}\right)}\right)^2}\,dv$ $= \int \sqrt{\frac{kg}{m}}\,dt \Rightarrow \tanh^{-1}\left(\sqrt{\frac{k}{mg}}\,v\right) = \sqrt{\frac{kg}{m}}\,t + C \Rightarrow v = \sqrt{\frac{mg}{k}}\tanh\left(\sqrt{\frac{kg}{m}}\,t + C\right)$; $v(0) = 0 \Rightarrow C = 0$ $\Rightarrow v = \sqrt{\frac{mg}{k}}\tanh\left(\sqrt{\frac{kg}{m}}\,t\right)$

(b) $\lim_{t\to\infty} v = \lim_{t\to\infty} \sqrt{\frac{mg}{k}}\tanh\left(\sqrt{\frac{kg}{m}}\,t\right) = \sqrt{\frac{mg}{k}}\lim_{t\to\infty}\tanh\left(\sqrt{\frac{kg}{m}}\,t\right) = \sqrt{\frac{mg}{k}}(1) = \sqrt{\frac{mg}{k}}$

(c) $\sqrt{\frac{160}{0.005}} = \sqrt{\frac{1{,}600{,}000}{5}} = \frac{400}{\sqrt{5}} = 80\sqrt{5} \approx 178.89$ ft/sec

79. $\frac{dy}{dx} = \frac{-1}{x\sqrt{1-x^2}} + \frac{x}{\sqrt{1-x^2}} \Rightarrow y = \int \frac{-1}{x\sqrt{1-x^2}}\,dx + \int \frac{x}{\sqrt{1-x^2}}\,dx \Rightarrow y = \operatorname{sech}^{-1}(x) - \sqrt{1-x^2} + C$; $x = 1$ and $y = 0 \Rightarrow C = 0 \Rightarrow y = \operatorname{sech}^{-1}(x) - \sqrt{1-x^2}$

81. $V = \pi\int_0^2 \left(\cosh^2 x - \sinh^2 x\right) dx = \pi\int_0^2 1\,dx = 2\pi$

83. $y = \frac{1}{2}\cosh 2x \Rightarrow y' = \sinh 2x \Rightarrow L = \int_0^{\ln\sqrt{5}} \sqrt{1+(\sinh 2x)^2}\,dx = \int_0^{\ln\sqrt{5}} \cosh 2x\,dx = \left[\frac{1}{2}\sinh 2x\right]_0^{\ln\sqrt{5}}$ $= \left[\frac{1}{2}\left(\frac{e^{2x}-e^{-2x}}{2}\right)\right]_0^{\ln\sqrt{5}} = \frac{1}{4}\left(5-\frac{1}{5}\right) = \frac{6}{5}$

85. (a) $y = \frac{H}{w}\cosh\left(\frac{w}{H}x\right) \Rightarrow \tan\phi = \frac{dy}{dx} = \left(\frac{H}{w}\right)\left[\frac{w}{H}\sinh\left(\frac{w}{H}x\right)\right] = \sinh\left(\frac{w}{H}x\right)$

(b) The tension at P is given by $T\cos\phi = H \Rightarrow T = H\sec\phi = H\sqrt{1+\tan^2\phi} = H\sqrt{1+\left(\sinh\frac{w}{H}x\right)^2}$ $= H\cosh\left(\frac{w}{H}x\right) = w\left(\frac{H}{w}\right)\cosh\left(\frac{w}{H}x\right) = wy$

87. (a) Since the cable is 32 ft long, s = 16 and x = 15. From Exercise 88, $x = \frac{1}{a}\sinh^{-1} as \Rightarrow 15a = \sinh^{-1} 16a$ $\Rightarrow \sinh 15a = 16a$.

(b) The intersection is near (0.042, 0.672).

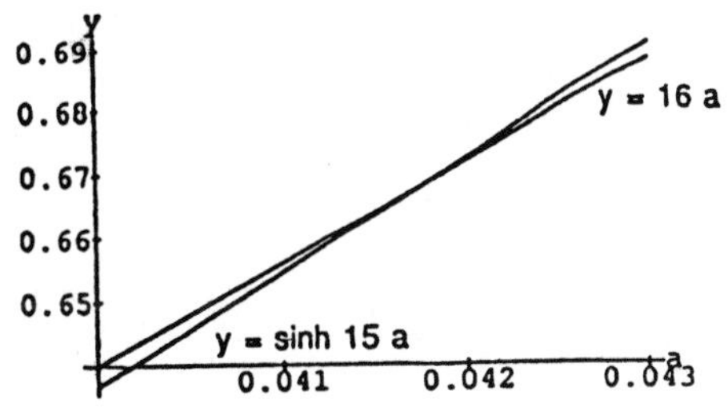

(c) Newton's method indicates that at $a \approx 0.0417525$ the curves $y = 16a$ and $y = \sinh 15a$ intersect.

(d) $T = wy \approx (2 \text{ lb})\left(\frac{1}{0.0417525}\right) \approx 47.90$ lb

(e) The sag is about 4.8 ft.

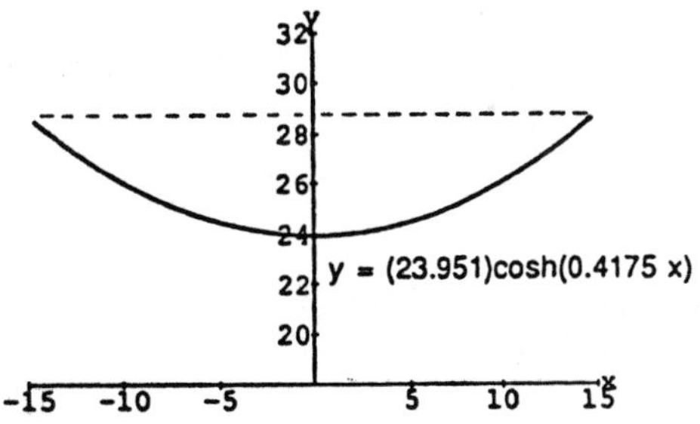

CHAPTER 6 PRACTICE EXERCISES

1. $y = 10e^{-x/5} \Rightarrow \frac{dy}{dx} = (10)\left(-\frac{1}{5}\right)e^{-x/5} = -2e^{-x/5}$

3. $y = \frac{1}{4}xe^{4x} - \frac{1}{16}e^{4x} \Rightarrow \frac{dy}{dx} = \frac{1}{4}\left[x\left(4e^{4x}\right) + e^{4x}(1)\right] - \frac{1}{16}\left(4e^{4}\right) = xe^{4x} + \frac{1}{4}e^{4x} - \frac{1}{4}e^{4x} = xe^{4x}$

5. $y = \ln\left(\sin^2\theta\right) \Rightarrow \frac{dy}{d\theta} = \frac{2(\sin\theta)(\cos\theta)}{\sin^2\theta} = \frac{2\cos\theta}{\sin\theta} = 2\cot\theta$

7. $y = \log_2\left(\frac{x^2}{2}\right) = \frac{\ln\left(\frac{x^2}{2}\right)}{\ln 2} \Rightarrow \frac{dy}{dx} = \frac{1}{\ln 2}\left(\frac{x}{\left(\frac{x^2}{2}\right)}\right) = \frac{2}{(\ln 2)x}$

9. $y = 8^{-t} \Rightarrow \frac{dy}{dt} = 8^{-t}(\ln 8)(-1) = -8^{-t}(\ln 8)$

11. $y = 5x^{3.6} \Rightarrow \frac{dy}{dx} = 5(3.6)x^{2.6} = 18x^{2.6}$

13. $y = (x+2)^{x+2} \Rightarrow \ln y = \ln (x+2)^{x+2} = (x+2) \ln (x+2) \Rightarrow \frac{y'}{y} = (x+2)\left(\frac{1}{x+2}\right) + (1) \ln (x+2)$
$\Rightarrow \frac{dy}{dx} = (x+2)^{x+2}[\ln (x+2) + 1]$

15. $y = \sin^{-1} \sqrt{1-u^2} = \sin^{-1}(1-u^2)^{1/2} \Rightarrow \frac{dy}{du} = \frac{\frac{1}{2}(1-u^2)^{-1/2}(-2u)}{\sqrt{1-\left[(1-u^2)^{1/2}\right]^2}} = \frac{-u}{\sqrt{1-u^2}\sqrt{1-(1-u^2)}} = \frac{-u}{|u|\sqrt{1-u^2}}$
$= \frac{-u}{u\sqrt{1-u^2}} = \frac{-1}{\sqrt{1-u^2}}$, $0 < u < 1$

17. $y = \ln(\cos^{-1} x) \Rightarrow y' = \frac{\left(\frac{-1}{\sqrt{1-x^2}}\right)}{\cos^{-1} x} = \frac{-1}{\sqrt{1-x^2} \cos^{-1} x}$

19. $y = t \tan^{-1} t - \left(\frac{1}{2}\right) \ln t \Rightarrow \frac{dy}{dt} = \tan^{-1} t + t\left(\frac{1}{1+t^2}\right) - \left(\frac{1}{2}\right)\left(\frac{1}{t}\right) = \tan^{-1} t + \frac{t}{1+t^2} - \frac{1}{2t}$

21. $y = z \sec^{-1} z - \sqrt{z^2-1} = z \sec^{-1} z - (z^2-1)^{1/2} \Rightarrow \frac{dy}{dz} = z\left(\frac{1}{|z|\sqrt{z^2-1}}\right) + (\sec^{-1} z)(1) - \frac{1}{2}(z^2-1)^{-1/2}(2z)$
$= \frac{z}{|z|\sqrt{z^2-1}} - \frac{z}{\sqrt{z^2-1}} + \sec^{-1} z = \frac{1-z}{\sqrt{z^2-1}} + \sec^{-1} z$, $z > 1$

23. $y = \csc^{-1} (\sec \theta) \Rightarrow \frac{dy}{d\theta} = \frac{-\sec \theta \tan \theta}{|\sec \theta|\sqrt{\sec^2 \theta - 1}} = -\frac{\tan \theta}{|\tan \theta|} = -1$, $0 < \theta < \frac{\pi}{2}$

25. $y = \frac{2(x^2+1)}{\sqrt{\cos 2x}} \Rightarrow \ln y = \ln\left(\frac{2(x^2+1)}{\sqrt{\cos 2x}}\right) = \ln (2) + \ln(x^2+1) - \frac{1}{2} \ln (\cos 2x) \Rightarrow \frac{y'}{y} = 0 + \frac{2x}{x^2+1} - \left(\frac{1}{2}\right)\frac{(-2 \sin 2x)}{\cos 2x}$
$\Rightarrow y' = \left(\frac{2x}{x^2+1} + \tan 2x\right)y = \frac{2(x^2+1)}{\sqrt{\cos 2x}}\left(\frac{2x}{x^2+1} + \tan 2x\right)$

27. $y = \left[\frac{(t+1)(t-1)}{(t-2)(t+3)}\right]^5 \Rightarrow \ln y = 5[\ln (t+1) + \ln (t-1) - \ln (t-2) - \ln (t+3)] \Rightarrow \left(\frac{1}{y}\right)\left(\frac{dy}{dt}\right)$
$= 5\left(\frac{1}{t+1} + \frac{1}{t-1} - \frac{1}{t-2} - \frac{1}{t+3}\right) \Rightarrow \frac{dy}{dt} = 5\left[\frac{(t+1)(t-1)}{(t-2)(t+3)}\right]^5\left(\frac{1}{t+1} + \frac{1}{t-1} - \frac{1}{t-2} - \frac{1}{t+3}\right)$

29. $y = (\sin \theta)^{\sqrt{\theta}} \Rightarrow \ln y = \sqrt{\theta} \ln (\sin \theta) \Rightarrow \left(\frac{1}{y}\right)\left(\frac{dy}{d\theta}\right) = \sqrt{\theta}\left(\frac{\cos \theta}{\sin \theta}\right) + \frac{1}{2}\theta^{-1/2} \ln (\sin \theta)$
$\Rightarrow \frac{dy}{d\theta} = (\sin \theta)^{\sqrt{\theta}}\left(\sqrt{\theta} \cot \theta + \frac{\ln (\sin \theta)}{2\sqrt{\theta}}\right) = \frac{1}{\sqrt{\theta}}(\sin \theta)^{\sqrt{\theta}}(\theta \cot \theta + \ln \sqrt{\sin \theta})$

31. $\int e^x \sin(e^x)\,dx = \int \sin u\,du$, where $u = e^x$ and $du = e^x\,dx$

$= -\cos u + C = -\cos(e^x) + C$

33. $\int e^x \sec^2(e^x - 7)\,dx = \int \sec^2 u\,du$, where $u = e^x - 7$ and $du = e^x\,dx$

$= \tan u + C = \tan(e^x - 7) + C$

35. $\int (\sec^2 x) e^{\tan x}\,dx = \int e^u\,du$, where $u = \tan x$ and $du = \sec^2 x\,dx$

$= e^u + C = e^{\tan x} + C$

37. $\int_{-1}^{1} \frac{1}{3x-4}\,dx = \frac{1}{3}\int_{-7}^{-1} \frac{1}{u}\,du$, where $u = 3x - 4$, $du = 3\,dx$; $x = -1 \Rightarrow u = -7$, $x = 1 \Rightarrow u = -1$

$= \frac{1}{3}\left[\ln|u|\right]_{-7}^{-1} = \frac{1}{3}\left[\ln|-1| - \ln|-7|\right] = \frac{1}{3}[0 - \ln 7] = -\frac{\ln 7}{3}$

39. $\int_0^{\pi} \tan\left(\frac{x}{3}\right)dx = \int_0^{\pi} \frac{\sin\left(\frac{x}{3}\right)}{\cos\left(\frac{x}{3}\right)}\,dx = -3\int_1^{1/2} \frac{1}{u}\,du$, where $u = \cos\left(\frac{x}{3}\right)$, $du = -\frac{1}{3}\sin\left(\frac{x}{3}\right)dx$; $x = 0 \Rightarrow u = 1$, $x = \pi$

$\Rightarrow u = \frac{1}{2}$

$= -3\left[\ln|u|\right]_1^{1/2} = -3\left[\ln\left|\frac{1}{2}\right| - \ln|1|\right] = -3\ln\frac{1}{2} = \ln 2^3 = \ln 8$

41. $\int_0^4 \frac{2t}{t^2 - 25}\,dt = \int_{-25}^{-9} \frac{1}{u}\,du$, where $u = t^2 - 25$, $du = 2t\,dt$; $t = 0 \Rightarrow u = -25$, $t = 4 \Rightarrow u = -9$

$= \left[\ln|u|\right]_{-25}^{-9} = \ln|-9| - \ln|-25| = \ln 9 - \ln 25 = \ln\frac{9}{25}$

43. $\int \frac{\tan(\ln v)}{v}\,dv = \int \tan u\,du = \int \frac{\sin u}{\cos u}\,du$, where $u = \ln v$ and $du = \frac{1}{v}\,dv$

$= -\ln|\cos u| + C = -\ln|\cos(\ln v)| + C$

45. $\int \frac{(\ln x)^{-3}}{x}\,dx = \int u^{-3}\,du$, where $u = \ln x$ and $du = \frac{1}{x}\,dx$

$= \frac{u^{-2}}{-2} + C = -\frac{1}{2}(\ln x)^{-2} + C$

47. $\int \frac{1}{r}\csc^2(1 + \ln r)\,dr = \int \csc^2 u\,du$, where $u = 1 + \ln r$ and $du = \frac{1}{r}\,dr$

$= -\cot u + C = -\cot(1 + \ln r) + C$

49. $\int x3^{x^2}\,dx = \frac{1}{2}\int 3^u\,du$, where $u = x^2$ and $du = 2x\,dx$

$= \frac{1}{2\ln 3}(3^u) + C = \frac{1}{2\ln 3}\left(3^{x^2}\right) + C$

51. $\int_1^7 \frac{3}{x}\,dx = 3\int_1^7 \frac{1}{x}\,dx = 3\left[\ln|x|\right]_1^7 = 3(\ln 7 - \ln 1) = 3\ln 7$

53. $\int_1^4 \left(\frac{x}{8} + \frac{1}{2x}\right)dx = \frac{1}{2}\int_1^4 \left(\frac{1}{4}x + \frac{1}{x}\right)dx = \frac{1}{2}\left[\frac{1}{8}x^2 + \ln|x|\right]_1^4 = \frac{1}{2}\left[\left(\frac{16}{8} + \ln 4\right) - \left(\frac{1}{8} + \ln 1\right)\right] = \frac{15}{16} + \frac{1}{2}\ln 4$

$= \frac{15}{16} + \ln\sqrt{4} = \frac{15}{16} + \ln 2$

55. $\int_{-2}^{-1} e^{-(x+1)}\,dx = -\int_1^0 e^u\,du$, where $u = -(x+1)$, $du = -dx$; $x = -2 \Rightarrow u = 1$, $x = -1 \Rightarrow u = 0$

$= -\left[e^u\right]_1^0 = -\left(e^0 - e^1\right) = e - 1$

57. $\int_0^{\ln 5} e^r\left(3e^r + 1\right)^{-3/2}\,dr = \frac{1}{3}\int_4^{16} u^{-3/2}\,du$, where $u = 3e^r + 1$, $du = 3e^r$; $r = 0 \Rightarrow u = 4$, $r = \ln 5 \Rightarrow u = 16$

$= -\frac{2}{3}\left[u^{-1/2}\right]_4^{16} = -\frac{2}{3}\left(16^{-1/2} - 4^{-1/2}\right) = \left(-\frac{2}{3}\right)\left(\frac{1}{4} - \frac{1}{2}\right) = \left(-\frac{2}{3}\right)\left(-\frac{1}{4}\right) = \frac{1}{6}$

59. $\int_1^e \frac{1}{x}(1 + 7\ln x)^{-1/3}\,dx = \frac{1}{7}\int_1^8 u^{-1/3}\,du$, where $u = 1 + 7\ln x$, $du = \frac{7}{x}\,dx$, $x = 1 \Rightarrow u = 1$, $x = e \Rightarrow u = 8$

$= \frac{3}{14}\left[u^{2/3}\right]_1^8 = \frac{3}{14}\left(8^{2/3} - 1^{2/3}\right) = \left(\frac{3}{14}\right)(4 - 1) = \frac{9}{14}$

61. $\int_1^3 \frac{[\ln(v+1)]^2}{v+1}\,dv = \int_1^3 [\ln(v+1)]^2\,\frac{1}{v+1}\,dv = \int_{\ln 2}^{\ln 4} u^2\,du$, where $u = \ln(v+1)$, $du = \frac{1}{v+1}dv$;
$v = 1 \Rightarrow u = \ln 2$, $v = 3 \Rightarrow u = \ln 4$;

$= \frac{1}{3}\left[u^3\right]_{\ln 2}^{\ln 4} = \frac{1}{3}\left[(\ln 4)^3 - (\ln 2)^3\right] = \frac{1}{3}\left[(2\ln 2)^3 - (\ln 2)^3\right] = \frac{(\ln 2)^3}{3}(8 - 1) = \frac{7}{3}(\ln 2)^3$

63. $\int_1^8 \frac{\log_4\theta}{\theta}\,d\theta = \frac{1}{\ln 4}\int_1^8 (\ln\theta)\left(\frac{1}{\theta}\right)d\theta = \frac{1}{\ln 4}\int_0^{\ln 8} u\,du$, where $u = \ln\theta$, $du = \frac{1}{\theta}\,d\theta$, $\theta = 1 \Rightarrow u = 0$, $\theta = 8 \Rightarrow u = \ln 8$

$= \frac{1}{2\ln 4}\left[u^2\right]_0^{\ln 8} = \frac{1}{\ln 16}\left[(\ln 8)^2 - 0^2\right] = \frac{(3\ln 2)^2}{4\ln 2} = \frac{9\ln 2}{4}$

65. $\int_{-3/4}^{3/4} \frac{6}{\sqrt{9-4x^2}}\,dx = 3\int_{-3/4}^{3/4} \frac{2}{\sqrt{3^2-(2x)^2}}\,dx = 3\int_{-3/2}^{3/2} \frac{1}{\sqrt{3^2-u^2}}\,du$, where $u = 2x$, $du = 2\,dx$; $x = -\frac{3}{4} \Rightarrow u = -\frac{3}{2}$, $x = \frac{3}{4} \Rightarrow u = \frac{3}{2}$

$= 3\left[\sin^{-1}\left(\frac{u}{3}\right)\right]_{-3/2}^{3/2} = 3\left[\sin^{-1}\left(\frac{1}{2}\right) - \sin^{-1}\left(-\frac{1}{2}\right)\right] = 3\left[\frac{\pi}{6} - \left(-\frac{\pi}{6}\right)\right] = 3\left(\frac{\pi}{3}\right) = \pi$

67. $\int_{-2}^{2} \frac{3}{4+3t^2}\,dt = \sqrt{3}\int_{-2}^{2} \frac{\sqrt{3}}{2^2+\left(\sqrt{3}t\right)^2}\,dt = \sqrt{3}\int_{-2\sqrt{3}}^{2\sqrt{3}} \frac{1}{2^2+u^2}\,du$, where $u = \sqrt{3}t$, $du = \sqrt{3}\,dt$; $t = -2 \Rightarrow u = -2\sqrt{3}$, $t = 2 \Rightarrow u = 2\sqrt{3}$

$= \sqrt{3}\left[\frac{1}{2}\tan^{-1}\left(\frac{u}{2}\right)\right]_{-2\sqrt{3}}^{2\sqrt{3}} = \frac{\sqrt{3}}{2}\left[\tan^{-1}\left(\sqrt{3}\right) - \tan^{-1}\left(-\sqrt{3}\right)\right] = \frac{\sqrt{3}}{2}\left[\frac{\pi}{3} - \left(-\frac{\pi}{3}\right)\right] = \frac{\pi}{\sqrt{3}}$

69. $\int \frac{1}{y\sqrt{4y^2-1}}\,dy = \int \frac{2}{(2y)\sqrt{(2y)^2-1}}\,dy = \int \frac{1}{u\sqrt{u^2-1}}\,du$, where $u = 2y$ and $du = 2\,dy$

$= \sec^{-1}|u| + C = \sec^{-1}|2y| + C$

71. $\int_{\sqrt{2}/3}^{2/3} \frac{1}{|y|\sqrt{9y^2-1}}\,dy = \int_{\sqrt{2}/3}^{2/3} \frac{3}{|3y|\sqrt{(3y)^2-1}}\,dy = \int_{\sqrt{2}}^{2} \frac{1}{|u|\sqrt{u^2-1}}\,du$, where $u = 3y$, $du = 3\,dy$; $y = \frac{\sqrt{2}}{3} \Rightarrow u = \sqrt{2}$, $y = \frac{2}{3} \Rightarrow u = 2$

$= \left[\sec^{-1}u\right]_{\sqrt{2}}^{2} = \left[\sec^{-1}2 - \sec^{-1}\sqrt{2}\right] = \frac{\pi}{3} - \frac{\pi}{4} = \frac{\pi}{12}$

73. $\int \frac{1}{\sqrt{-2x-x^2}}\,dx = \int \frac{1}{\sqrt{1-\left(x^2+2x+1\right)}}\,dx = \int \frac{1}{\sqrt{1-(x+1)^2}}\,dx = \int \frac{1}{\sqrt{1-u^2}}\,du$, where $u = x+1$ and $du = dx$

$= \sin^{-1}u + C = \sin^{-1}(x+1) + C$

75. $\int_{-2}^{-1} \frac{2}{v^2+4v+5}\,dv = 2\int_{-2}^{-1} \frac{1}{1+\left(v^2+4v+4\right)}\,dv = 2\int_{-2}^{-1} \frac{1}{1+(v+2)^2}\,dv = 2\int_{0}^{1} \frac{1}{1+u^2}\,du$,
where $u = v+2$, $du = dv$; $v = -2 \Rightarrow u = 0$, $v = -1 \Rightarrow u = 1$

$= 2\left[\tan^{-1}u\right]_0^1 = 2\left(\tan^{-1}1 - \tan^{-1}0\right) = 2\left(\frac{\pi}{4} - 0\right) = \frac{\pi}{2}$

77. $\int \frac{1}{(t+1)\sqrt{t^2+2t-8}}\,dt = \int \frac{1}{(t+1)\sqrt{\left(t^2+2t+1\right)-9}}\,dt = \int \frac{1}{(t+1)\sqrt{(t+1)^2-3^2}}\,dt = \int \frac{1}{u\sqrt{u^2-3^2}}\,du$
where $u = t+1$ and $du = dt$

$= \frac{1}{3}\sec^{-1}\left|\frac{u}{3}\right| + C = \frac{1}{3}\sec^{-1}\left|\frac{t+1}{3}\right| + C$

79. $\frac{df}{dx} = e^x + 1 \Rightarrow \left(\frac{df^{-1}}{dx}\right)_{x=f(\ln 2)} = \frac{1}{\left(\frac{df}{dx}\right)_{x=\ln 2}} \Rightarrow \left(\frac{df^{-1}}{dx}\right)_{x=f(\ln 2)} = \frac{1}{\left(e^x+1\right)_{x=\ln 2}} = \frac{1}{2+1} = \frac{1}{3}$

81. $y = x \ln 2x - x \Rightarrow y' = x\left(\frac{2}{2x}\right) + \ln(2x) - 1 = \ln 2x$; solving $y' = 0 \Rightarrow x = \frac{1}{2}$; $y' > 0$ for $x > \frac{1}{2}$ and $y' < 0$ for $x < \frac{1}{2} \Rightarrow$ relative minimum of $-\frac{1}{2}$ at $x = \frac{1}{2}$; $f\left(\frac{1}{2e}\right) = -\frac{1}{e}$ and $f\left(\frac{e}{2}\right) = 0 \Rightarrow$ absolute minimum is $-\frac{1}{2}$ at $x = \frac{1}{2}$ and the absolute maximum is 0 at $x = \frac{e}{2}$

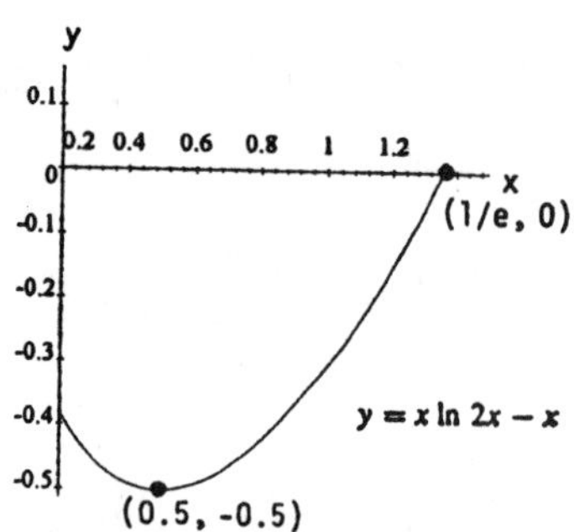

83. $A = \int_1^e \frac{2 \ln x}{x}\,dx = \int_0^1 2u\,du = \left[u^2\right]_0^1 = 1$, where $u = \ln x$ and $du = \frac{1}{x}\,dx$; $x = 1 \Rightarrow u = 0$, $x = e \Rightarrow u = 1$

85. $y = \ln x \Rightarrow \frac{dy}{dx} = \frac{1}{x}$; $\frac{dy}{dt} = \frac{dy}{dx}\frac{dx}{dt} \Rightarrow \frac{dy}{dt} = \left(\frac{1}{x}\right)\sqrt{x} = \frac{1}{\sqrt{x}} \Rightarrow \left.\frac{dy}{dt}\right|_{e^2} = \frac{1}{e}$ m/sec

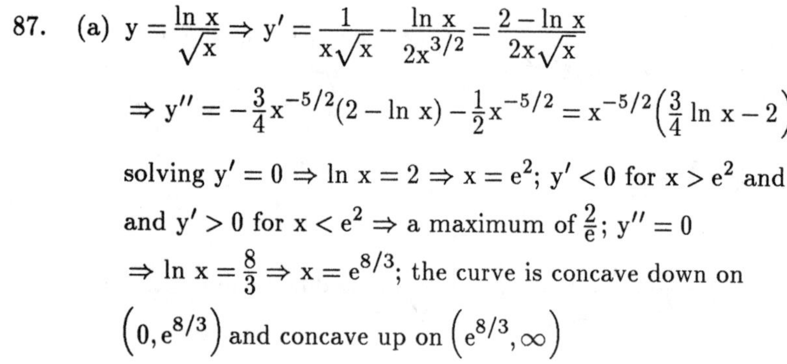

87. (a) $y = \frac{\ln x}{\sqrt{x}} \Rightarrow y' = \frac{1}{x\sqrt{x}} - \frac{\ln x}{2x^{3/2}} = \frac{2 - \ln x}{2x\sqrt{x}}$
$\Rightarrow y'' = -\frac{3}{4}x^{-5/2}(2 - \ln x) - \frac{1}{2}x^{-5/2} = x^{-5/2}\left(\frac{3}{4}\ln x - 2\right)$;
solving $y' = 0 \Rightarrow \ln x = 2 \Rightarrow x = e^2$; $y' < 0$ for $x > e^2$ and and $y' > 0$ for $x < e^2 \Rightarrow$ a maximum of $\frac{2}{e}$; $y'' = 0 \Rightarrow \ln x = \frac{8}{3} \Rightarrow x = e^{8/3}$; the curve is concave down on $\left(0, e^{8/3}\right)$ and concave up on $\left(e^{8/3}, \infty\right)$

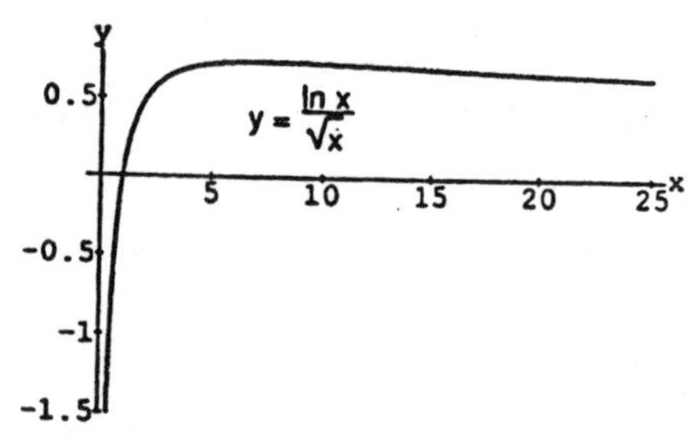

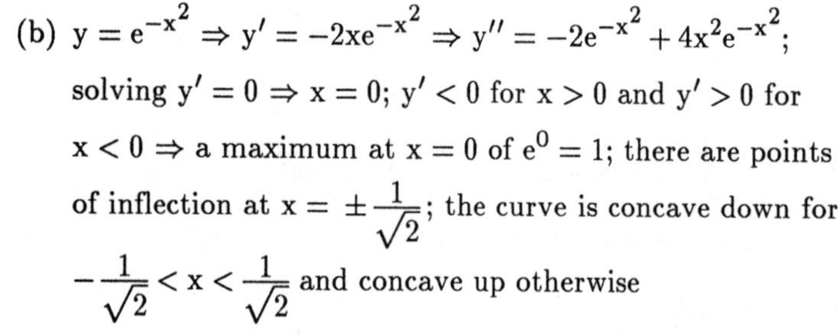

(b) $y = e^{-x^2} \Rightarrow y' = -2xe^{-x^2} \Rightarrow y'' = -2e^{-x^2} + 4x^2e^{-x^2}$;
solving $y' = 0 \Rightarrow x = 0$; $y' < 0$ for $x > 0$ and $y' > 0$ for $x < 0 \Rightarrow$ a maximum at $x = 0$ of $e^0 = 1$; there are points of inflection at $x = \pm\frac{1}{\sqrt{2}}$; the curve is concave down for $-\frac{1}{\sqrt{2}} < x < \frac{1}{\sqrt{2}}$ and concave up otherwise

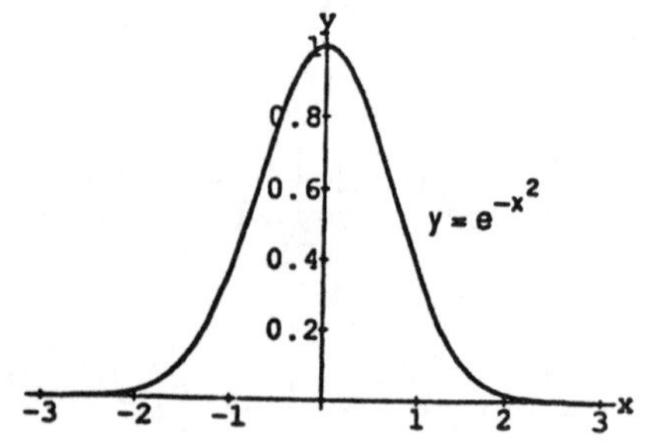

(c) $y = (1+x)e^{-x} \Rightarrow y' = e^{-x} - (1+x)e^{-x} = -xe^{-x}$ $\Rightarrow y'' = -e^{-x} + xe^{-x} = (x-1)e^{-x}$; solving $y' = 0$ $\Rightarrow -xe^{-x} = 0 \Rightarrow x = 0$; $y' < 0$ for $x > 0$ and $y' > 0$ for $x < 0 \Rightarrow$ a maximum at $x = 0$ of $(1+0)e^0 = 1$; there is a point of inflection at $x = 1$ and the curve is concave up for $x > 1$ and concave down for $x < 1$

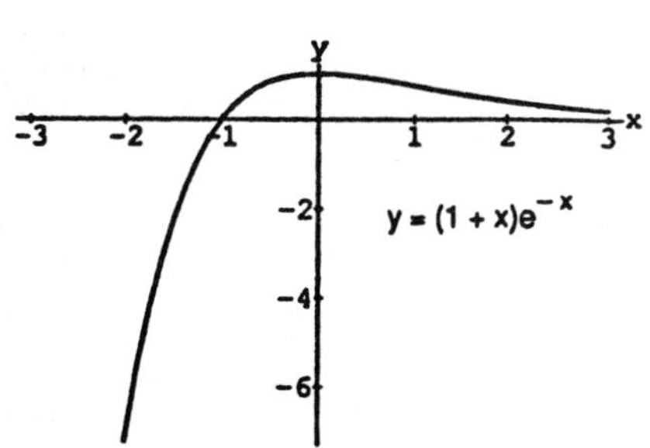

89. Since the half life is 5700 years and $A(t) = A_0e^{kt}$ we have $\frac{A_0}{2} = A_0e^{5700k} \Rightarrow \frac{1}{2} = e^{5700k} \Rightarrow \ln(0.5) = 5700k$ $\Rightarrow k = \frac{\ln(0.5)}{5700}$. With 10% of the original carbon-14 remaining we have $0.1A_0 = A_0e^{\frac{\ln(0.5)}{5700}t} \Rightarrow 0.1 = e^{\frac{\ln(0.5)}{5700}t}$ $\Rightarrow \ln(0.1) = \frac{\ln(0.5)}{5700}t \Rightarrow t = \frac{(5700)\ln(0.1)}{\ln(0.5)} \approx 18{,}935$ years (rounded to the nearest year).

91. $A = xy = xe^{-x^2} \Rightarrow \frac{dA}{dx} = e^{-x^2} + (x)(-2x)e^{-x^2} = e^{-x^2}(1-2x^2)$. Solving $\frac{dA}{dx} = 0 \Rightarrow 1 - 2x^2 = 0$ $\Rightarrow x = \frac{1}{\sqrt{2}}$; $\frac{dA}{dx} < 0$ for $x > \frac{1}{\sqrt{2}}$ and $\frac{dA}{dx} > 0$ for $0 < x < \frac{1}{\sqrt{2}} \Rightarrow$ absolute maximum of $\frac{1}{\sqrt{2}}e^{-1/2} = \frac{1}{\sqrt{2e}}$ at $x = \frac{1}{\sqrt{2}}$ units long by $y = e^{-1/2} = \frac{1}{\sqrt{e}}$ units high.

93. $K = \ln(5x) - \ln(3x) = \ln 5 + \ln x - \ln 3 - \ln x = \ln 5 - \ln 3 = \ln \frac{5}{3}$

95. $\theta = \pi - \cot^{-1}\left(\frac{x}{60}\right) - \cot^{-1}\left(\frac{5}{3} - \frac{x}{30}\right)$, $0 < x < 50 \Rightarrow \frac{d\theta}{dx} = \frac{\left(\frac{1}{60}\right)}{1+\left(\frac{x}{60}\right)^2} + \frac{\left(-\frac{1}{30}\right)}{1+\left(\frac{50-x}{30}\right)^2}$

$= 30\left[\frac{2}{60^2 + x^2} - \frac{1}{30^2 + (50-x)^2}\right]$; solving $\frac{d\theta}{dx} = 0 \Rightarrow x^2 - 200x + 3200 = 0 \Rightarrow x = 100 \pm 20\sqrt{17}$, but $100 + 20\sqrt{17}$ is not in the domain; $\frac{d\theta}{dx} > 0$ for $x < 20(5 - \sqrt{17})$ and $\frac{d\theta}{dx} < 0$ for $20(5-\sqrt{17}) < x < 50$ $\Rightarrow x = 20(5 - \sqrt{17}) \approx 17.54$ m maximizes θ

97. (a) Force = Mass times Acceleration (Newton's Second Law) or $F = ma$. Let $a = \frac{dv}{dt} = \frac{dv}{ds}\cdot\frac{ds}{dt} = v\frac{dv}{ds}$. Then $ma = -mgR^2s^{-2} \Rightarrow a = -gR^2s^{-2} \Rightarrow v\frac{dv}{ds} = -gR^2s^{-2} \Rightarrow v\,dv = -gR^2s^{-2}\,ds \Rightarrow \int v\,dv = \int -gR^2s^{-2}\,ds$ $\Rightarrow \frac{v^2}{2} = \frac{gR^2}{s} + C \Rightarrow v^2 = \frac{2gR^2}{s} + 2C_1 = \frac{2gR^2}{s} + C$. When $t = 0$, $v = v_0$ and $s = R \Rightarrow v_0^2 = \frac{2gR^2}{R} + C$ $\Rightarrow C = v_0^2 - 2gR \Rightarrow v^2 = \frac{2gR^2}{s} + v_0^2 - 2gR$

(b) If $v_0 = \sqrt{2gR}$, then $v^2 = \frac{2gR^2}{s} \Rightarrow v = \sqrt{\frac{2gR^2}{s}}$, since $v \ge 0$ if $v_0 \ge \sqrt{2gR}$. Then $\frac{ds}{dt} = \frac{\sqrt{2gR^2}}{\sqrt{s}}$ $\Rightarrow \sqrt{s}\,ds = \sqrt{2gR^2}\,dt \Rightarrow \int s^{1/2}\,ds = \int \sqrt{2gR^2}\,dt \Rightarrow \frac{2}{3}s^{3/2} = \left(\sqrt{2gR^2}\right)t + C_1 \Rightarrow s^{3/2} = \left(\frac{3}{2}\sqrt{2gR^2}\right)t + C$; $t = 0$ and $s = R \Rightarrow R^{3/2} = \left(\frac{3}{2}\sqrt{2gR^2}\right)(0) + C \Rightarrow C = R^{3/2} \Rightarrow s^{3/2} = \left(\frac{3}{2}\sqrt{2gR^2}\right)t + R^{3/2}$

$$= \left(\frac{3}{2}R\sqrt{2g}\right)t + R^{3/2} = R^{3/2}\left[\left(\frac{3}{2}R^{-1/2}\sqrt{2g}\right)t + 1\right] = R^{3/2}\left[\left(\frac{3\sqrt{2gR}}{2R}\right)t + 1\right]$$

$$= R^{3/2}\left[\left(\frac{3v_0}{2R}\right)t + 1\right] \Rightarrow s = R\left[1 + \left(\frac{3v_0}{2R}\right)t\right]^{2/3}$$

99. $\frac{dy}{dx} = e^{-x-y-2} \Rightarrow \frac{dy}{dx} = e^{-x-2} \cdot e^{-y} \Rightarrow e^y\,dy = e^{-x-2}\,dx = \int e^y\,dy = \int e^{-x-2}\,dx \Rightarrow e^y = -e^{-x-2} + C;\ x = 0$ and $y = -2 \Rightarrow e^{-2} = -e^{-2} + C \Rightarrow C = 2e^{-2} \Rightarrow e^y = -e^{-x-2} + 2e^{-2} \Rightarrow \ln(e^y) = \ln(-e^{-x-2} + 2e^{-2})$ $\Rightarrow y = \ln(-e^{-x-2} + 2e^{-2})$

101. $\frac{dy}{dx} + \left(\frac{2}{x+1}\right)y = \frac{x}{x+1} \Rightarrow P(x) = \frac{2}{x+1},\ Q(x) = \frac{x}{x+1} \Rightarrow \int P(x)\,dx = \int\left(\frac{2}{x+1}\right)dx = 2\ln|x+1| = \ln(x+1)^2$

$$\Rightarrow v(x) = e^{\ln(x+1)^2} = (x+1)^2 \Rightarrow y = \frac{1}{(x+1)^2}\int (x+1)^2\left(\frac{x}{x+1}\right)dx = \frac{1}{(x+1)^2}\int (x^2 + x)\,dx$$

$$= \left(\frac{1}{x+1}\right)^2\left(\frac{x^3}{3} + \frac{x^2}{2} + C\right);\ x = 0 \text{ and } y = 1 \Rightarrow 1 = 0 + 0 + C \Rightarrow C = 1 \Rightarrow y = \frac{1}{(x+1)^2}\cdot\left(\frac{x^3}{3} + \frac{x^2}{2} + 1\right)$$

103.

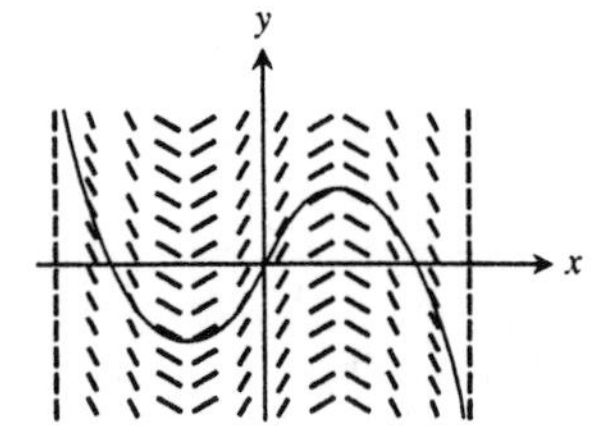

105. To find the approximate solution let $z_n = y_{n-1} + ((2 - y_{n-1})(2x_{n-1} + 3))(0.1)$ and $y_n = y_{n-1} + \left(\frac{(2 - y_{n-1})(2x_{n-1} + 3) + (2 - z_n)(2x_n + 3)}{2}\right)(0.1)$ with initial values $x_0 = -3$, $y_0 = 1$, and 20 steps. Use a spreadsheet, graphing calculator, or CAS to obtain the values in the following table.

x	y
−3	1
−2.9	0.6680
−2.8	0.2599
−2.7	−0.2294
−2.6	−0.8011
−2.5	−1.4509
−2.4	−2.1687
−2.3	−2.9374
−2.2	−3.7333
−2.1	−4.5268
−2.0	−5.2840

x	y
−1.9	−5.9686
−1.8	−6.5456
−1.7	−6.9831
−1.6	−7.2562
−1.5	−7.3488
−1.4	−7.2553
−1.3	−6.9813
−1.2	−6.5430
−1.1	−5.9655
−1.0	−5.2805

107. To estimate y(4), let $y_n = y_{n-1} + \left(\frac{x_{n-1}^2 - 2y_{n-1} + 1}{x_{n-1}}\right)(0.05)$ with initial values $x_0 = 1$, $y_0 = 1$, and 60 steps. Use a spreadsheet, programmable calculator, or CAS to obtain $y(4) \approx 4.4974$.

109. Let $z_n = y_{n-1} - \left(\frac{x_{n-1}^2 + y_{n-1}}{e^{y_{n-1}} + x_{n-1}}\right)(dx)$ and $y_n = y_{n-1} + \frac{1}{2}\left(\frac{x_{n-1}^2 + y_{n-1}}{e^{y_{n-1}} + x_{n-1}} + \frac{x_n^2 + z_n}{e^{z_n} + x_n}\right)(dx)$ with starting values $x_0 = 0$, $y_0 = 0$, and steps of 0.1 and -0.1. Use a spreadsheet, programmable calculator, or CAS to generate the following graphs.

(a)

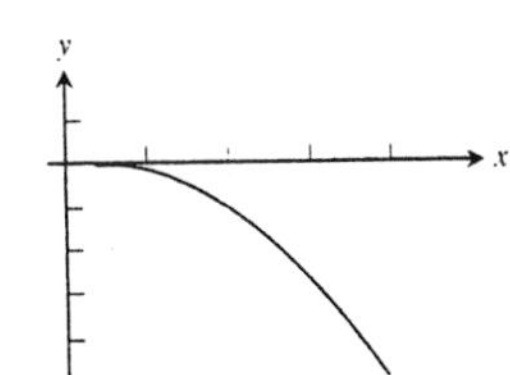

(b)

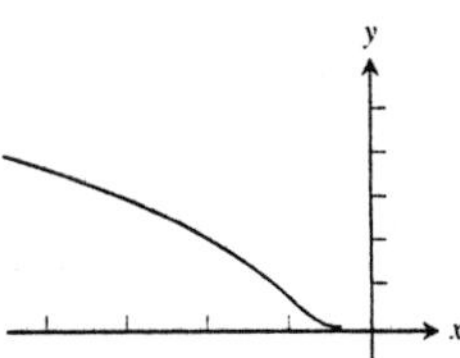

111.

x	1	1.2	1.4	1.6	1.8	2.0
y	−1	−0.8	−0.56	−0.28	0.04	0.4

$\frac{dy}{dx} = x \Rightarrow dy = x\,dx \Rightarrow y = \frac{x^2}{2} + C$; $x = 1$ and $y = -1$

$\Rightarrow -1 = \frac{1}{2} + C \Rightarrow C = -\frac{3}{2} \Rightarrow y\text{ (exact)} = \frac{x^2}{2} - \frac{3}{2}$

$\Rightarrow y(2) = \frac{2^2}{2} - \frac{3}{2} = \frac{1}{2}$ is the exact value

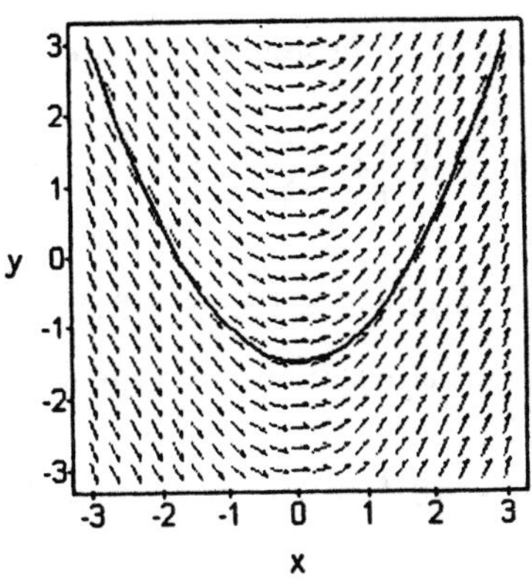

113.

x	1	1.2	1.4	1.6	1.8	2.0
y	−1	−1.2	−1.488	−1.9046	−2.5141	−3.4192

$\frac{dy}{dx} = xy \Rightarrow \frac{dy}{y} = x\,dx \Rightarrow \ln|y| = \frac{x^2}{2} + C \Rightarrow y = e^{\frac{x^2}{2} + C} = e^{x^2/2} \cdot e^C$

$= C_1 e^{x^2/2}$; $x = 1$ and $y = -1 \Rightarrow -1 = C_1 e^{1/2} \Rightarrow C_1 = -e^{-1/2}$

$\Rightarrow y\text{ (exact)} = -e^{-1/2} \cdot e^{x^2/2} = -e^{\left(x^2-1\right)/2} \Rightarrow y(2) = -e^{3/2}$

≈ -4.4817 is the exact value

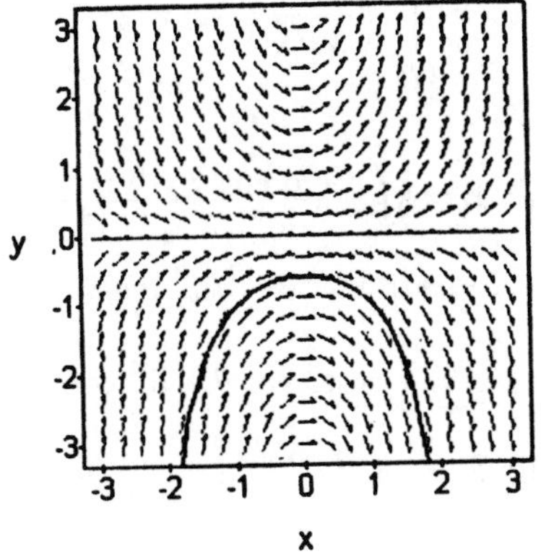

CHAPTER 6 ADDITIONAL EXERCISES–THEORY, EXAMPLES, APPLICATIONS

1. $A_1 = \int_1^e \frac{2\log_2 x}{x}\,dx = \frac{2}{\ln 2}\int_1^e \frac{\ln x}{x}\,dx = \left[\frac{(\ln x)^2}{\ln 2}\right]_1^e = \frac{1}{\ln 2}$; $A_2 = \int_1^e \frac{2\log_4 x}{4}\,dx = \frac{2}{\ln 4}\int_1^e \frac{\ln x}{x}\,dx$

$= \left[\frac{(\ln x)^2}{2\ln 2}\right]_1^e = \frac{1}{2\ln 2} \Rightarrow A_1 : A_2 = 2 : 1$

3. $f(x) = e^{g(x)} \Rightarrow f'(x) = e^{g(x)}\, g'(x)$, where $g'(x) = \frac{x}{1+x^4} \Rightarrow f'(2) = e^0\left(\frac{2}{1+16}\right) = \frac{2}{17}$

5. (a) The figure shows that $\frac{\ln e}{e} > \frac{\ln \pi}{\pi} \Rightarrow \pi \ln e > e \ln \pi \Rightarrow \ln e^\pi > \ln \pi^e \Rightarrow e^\pi > \pi^e$

(b) $y = \frac{\ln x}{x} \Rightarrow y' = \left(\frac{1}{x}\right)\left(\frac{1}{x}\right) - \frac{\ln x}{x^2} \Rightarrow \frac{1-\ln x}{x^2}$; solving $y' = 0 \Rightarrow \ln x = 1 \Rightarrow x = e$; $y' < 0$ for $x > e$ and $y' > 0$ for $0 < x < e \Rightarrow$ an absolute maximum occurs at $x = e$

7. (a) slope of $L_3 <$ slope of $L_2 <$ slope of $L_1 \Rightarrow \frac{1}{b} < \frac{\ln b - \ln a}{b-a} < \frac{1}{a}$

(b) area of small (shaded) rectangle < area under curve < area of large rectangle

$$\Rightarrow \frac{1}{b}(b-a) < \int_a^b \frac{1}{x}\,dx < \frac{1}{a}(b-a) \Rightarrow \frac{1}{b} < \frac{\ln b - \ln a}{b-a} < \frac{1}{a}$$

9. Use the Fundamental Theorem of Calculus.

$$y' = \frac{d}{dx}\left(\int_0^x \sin t^2\,dt\right) + \frac{d}{dx}(x^3 + x + 2) = (\sin x^2) + (3x^2 + 1)$$

$$y'' = \frac{d}{dx}(\sin x^2 + 3x^2 + 1) = (\cos x^2)(2x) + 6x = 2x\cos(x^2) + 6x$$

Thus, the differential equation is satisfied. Verify the initial conditions:

$$y'(0) = (\sin 0^2) + 3(0)^2 + 1 = 1 \text{ and } y(0) = \int_0^0 \sin(t^2)\,dt + 0^3 + 0 + 2 = 2$$

11. $V = \pi \int_{1/4}^{4} \left(\frac{1}{2\sqrt{x}}\right)^2 dx = \frac{\pi}{4}\int_{1/4}^{4} \frac{1}{x}\,dx = \frac{\pi}{4}\left[\ln|x|\right]_{1/4}^{4} = \frac{\pi}{4}\left(\ln 4 - \ln\frac{1}{4}\right) = \frac{\pi}{4}\ln 16 = \frac{\pi}{4}\ln(2^4) = \pi \ln 2$

13. (a) $L = k\left(\frac{a - b\cot\theta}{R^4} + \frac{b\csc\theta}{r^4}\right) \Rightarrow \frac{dL}{d\theta} = k\left(\frac{b\csc^2\theta}{R^4} - \frac{b\csc\theta\cot\theta}{r^4}\right)$; solving $\frac{dL}{d\theta} = 0$ $\Rightarrow r^4 b\csc^2\theta - bR^4\csc\theta\cot\theta = 0 \Rightarrow (b\csc\theta)(r^4\csc\theta - R^4\cot\theta) = 0$; but $b\csc\theta \neq 0$ since $\theta \neq \frac{\pi}{2} \Rightarrow r^4\csc\theta - R^4\cot\theta = 0 \Rightarrow \cos\theta = \frac{r^4}{R^4} \Rightarrow \theta = \cos^{-1}\left(\frac{r^4}{R^4}\right)$, the critical value of θ

(b) $\theta = \cos^{-1}\left(\frac{5}{6}\right)^4 \approx \cos^{-1}(0.48225) \approx 61°$

15. (a) $\frac{dy}{dt} = k\frac{A}{V}(c - y) \Rightarrow dy = -k\frac{A}{V}(y - c)\,dt \Rightarrow \frac{dy}{y-c} = -k\frac{A}{V}\,dt \Rightarrow \int \frac{dy}{y-c} = -\int k\frac{A}{V}\,dt \Rightarrow \ln|y - c|$ $= -k\frac{A}{V}t + C_1 \Rightarrow y - c = \pm e^{C_1}e^{-k\frac{A}{V}t}$. Apply the initial condition, $y(0) = y_0 \Rightarrow y_0 = c + C \Rightarrow C = y_0 - c$ $\Rightarrow y = c + (y_0 - c)e^{-k\frac{A}{V}t}$.

(b) Steady state solution: $y_\infty = \lim_{t\to\infty} y(t) = \lim_{t\to\infty}\left[c + (y_0 - c)e^{-k\frac{A}{V}t}\right] = c + (y_0 - c)(0) = c$

17. In the interval $\pi < x < 2\pi$ the function $\sin x < 0$ $\Rightarrow (\sin x)^{\sin x}$ is not defined for all values in that interval or its translation by 2π.

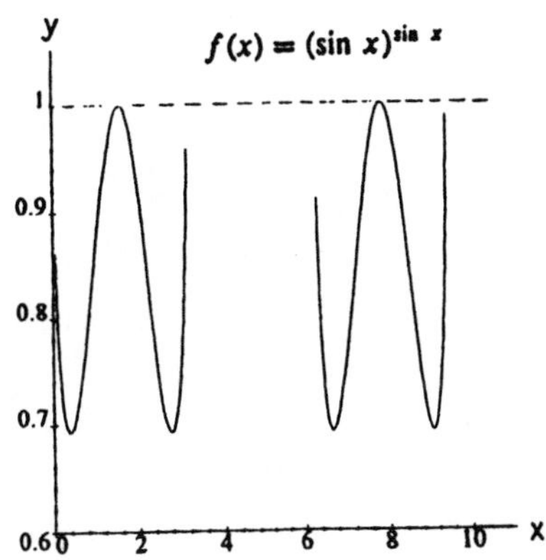

NOTES:

CHAPTER 7 TECHNIQUES OF INTEGRATION, L'HÔPITAL'S RULE, AND IMPROPER INTEGRALS

7.1 BASIC INTEGRATION FORMULAS

1. $\int \frac{16x\,dx}{\sqrt{8x^2+1}}; \begin{bmatrix} u = 8x^2+1 \\ du = 16x\,dx \end{bmatrix} \Rightarrow \int \frac{du}{\sqrt{u}} = 2\sqrt{u} + C = 2\sqrt{8x^2+1} + C$

3. $\int 3\sqrt{\sin v}\cos v\,dv; \begin{bmatrix} u = \sin v \\ du = \cos v\,dv \end{bmatrix} \Rightarrow \int 3\sqrt{u}\,du = 3 \cdot \frac{2}{3}u^{3/2} + C = 2(\sin v)^{3/2} + C$

5. $\int_0^1 \frac{16x\,dx}{8x^2+2}; \begin{bmatrix} u = 8x^2+2 \\ du = 16x\,dx \\ x = 0 \Rightarrow u = 2, \quad x = 1 \Rightarrow u = 10 \end{bmatrix} \Rightarrow \int_2^{10} \frac{du}{u} = [\ln|u|]_2^{10} = \ln 10 - \ln 2 = \ln 5$

7. $\int \frac{dx}{\sqrt{x}(\sqrt{x}+1)}; \begin{bmatrix} u = \sqrt{x} \\ du = \frac{1}{2\sqrt{x}}\,dx \\ 2\,du = \frac{dx}{\sqrt{x}} \end{bmatrix} \Rightarrow \int \frac{2\,du}{u} = 2\ln|u| + C = 2\ln(\sqrt{x}+1) + C$

9. $\int \cot(3-7x)\,dx; \begin{bmatrix} u = 3-7x \\ du = -7\,dx \end{bmatrix} \Rightarrow -\frac{1}{7}\int \cot u\,du = -\frac{1}{7}\ln|\sin u| + C = -\frac{1}{7}\ln|\sin(3-7x)| + C$

11. $\int e^\theta \csc(e^\theta+1)\,d\theta; \begin{bmatrix} u = e^\theta+1 \\ du = e^\theta\,d\theta \end{bmatrix} \Rightarrow \int \csc u\,du = -\ln|\csc u + \cot u| + C = -\ln\left|\csc(e^\theta+1) + \cot(e^\theta+1)\right| + C$

13. $\int \sec\frac{t}{3}\,dt; \begin{bmatrix} u = \frac{t}{3} \\ du = \frac{dt}{3} \end{bmatrix} \Rightarrow \int 3\sec u\,du = 3\ln|\sec u + \tan u| + C = 3\ln\left|\sec\frac{t}{3} + \tan\frac{t}{3}\right| + C$

15. $\int \csc(s-\pi)\,ds; \begin{bmatrix} u = s-\pi \\ du = ds \end{bmatrix} \Rightarrow \int \csc u\,du = -\ln|\csc u + \cot u| + C = -\ln|\csc(s-\pi) + \cot(s-\pi)| + C$

17. $\int_0^{\sqrt{\ln 2}} 2xe^{x^2}\,dx;\ \left[\begin{array}{c} u = x^2 \\ du = 2x\,dx \\ x = 0 \Rightarrow u = 0,\ x = \sqrt{\ln 2} \Rightarrow u = \ln 2 \end{array}\right] \Rightarrow \int_0^{\ln 2} e^u\,du = \left[e^u\right]_0^{\ln 2} = e^{\ln 2} - e^0 = 2 - 1 = 1$

19. $\int e^{\tan v}\sec^2 v\,dv;\ \left[\begin{array}{c} u = \tan v \\ du = \sec^2 v\,dv \end{array}\right] \Rightarrow \int e^u\,du = e^u + C = e^{\tan v} + C$

21. $\int 3^{x+1}\,dx;\ \left[\begin{array}{c} u = x+1 \\ du = dx \end{array}\right] \Rightarrow \int 3^u\,du = \left(\frac{1}{\ln 3}\right)3^u + C = \frac{3^{(x+1)}}{\ln 3} + C$

23. $\int \frac{2^{\sqrt{w}}\,dw}{2\sqrt{w}};\ \left[\begin{array}{c} u = \sqrt{w} \\ du = \frac{dw}{2\sqrt{w}} \end{array}\right] \Rightarrow \int 2^u\,du = \frac{2^u}{\ln 2} + C = \frac{2^{\sqrt{w}}}{\ln 2} + C$

25. $\int \frac{9\,du}{1+9u^2};\ \left[\begin{array}{c} x = 3u \\ dx = 3\,du \end{array}\right] \Rightarrow \int \frac{3\,dx}{1+x^2} = 3\tan^{-1}x + C = 3\tan^{-1}3u + C$

27. $\int_0^{1/6} \frac{dx}{\sqrt{1-9x^2}};\ \left[\begin{array}{c} u = 3x \\ du = 3\,dx \\ x = 0 \Rightarrow u = 0,\ x = \frac{1}{6} \Rightarrow u = \frac{1}{2} \end{array}\right] \Rightarrow \int_0^{1/2} \frac{1}{3}\frac{du}{\sqrt{1-u^2}} = \left[\frac{1}{3}\sin^{-1}u\right]_0^{1/2} = \frac{1}{3}\left(\frac{\pi}{6} - 0\right) = \frac{\pi}{18}$

29. $\int \frac{2s\,ds}{\sqrt{1-s^4}};\ \left[\begin{array}{c} u = s^2 \\ du = 2s\,ds \end{array}\right] \Rightarrow \int \frac{du}{\sqrt{1-u^2}} = \sin^{-1}u + C = \sin^{-1}s^2 + C$

31. $\int \frac{6\,dx}{x\sqrt{25x^2-1}} = \int \frac{6\,dx}{5x\sqrt{x^2-\frac{1}{25}}} = \frac{6}{5}\cdot 5\sec^{-1}|5x| + C = 6\sec^{-1}|5x| + C$

33. $\int \frac{dx}{e^x + e^{-x}} = \int \frac{e^x\,dx}{e^{2x}+1};\ \left[\begin{array}{c} u = e^x \\ du = e^x\,dx \end{array}\right] \Rightarrow \int \frac{du}{u^2+1} = \tan^{-1}u + C = \tan^{-1}e^x + C$

35. $\int_1^{e^{\pi/3}} \frac{dx}{x\cos(\ln x)};\ \left[\begin{array}{c} u = \ln x \\ du = \frac{dx}{x} \\ x = 1 \Rightarrow u = 0,\ x = e^{\pi/3} \Rightarrow u = \frac{\pi}{3} \end{array}\right] \Rightarrow \int_0^{\pi/3} \frac{du}{\cos u} = \int_0^{\pi/3} \sec u\,du = \left[\ln|\sec u + \tan u|\right]_0^{\pi/3}$

$= \ln\left|\sec\frac{\pi}{3} + \tan\frac{\pi}{3}\right| - \ln|\sec 0 + \tan 0| = \ln\left(2+\sqrt{3}\right) - \ln(1) = \ln\left(2+\sqrt{3}\right)$

37. $\int_1^2 \frac{8\,dx}{x^2-2x+2} = 8\int_1^2 \frac{dx}{1+(x-1)^2};\ \begin{bmatrix} u = x-1 \\ du = dx \\ x=1 \Rightarrow u=0,\ x=2 \Rightarrow u=1 \end{bmatrix} \Rightarrow 8\int_0^1 \frac{du}{1+u^2} = 8\left[\tan^{-1} u\right]_0^1$

$= 8\left(\tan^{-1} 1 - \tan^{-1} 0\right) = 8\left(\frac{\pi}{4} - 0\right) = 2\pi$

39. $\int \frac{dt}{\sqrt{-t^2+4t-3}} = \int \frac{dt}{\sqrt{1-(t-2)^2}};\ \begin{bmatrix} u = t-2 \\ du = dt \end{bmatrix} \Rightarrow \int \frac{du}{\sqrt{1-u^2}} = \sin^{-1} u + C = \sin^{-1}(t-2) + C$

41. $\int \frac{dx}{(x+1)\sqrt{x^2+2x}} = \int \frac{dx}{(x+1)\sqrt{(x+1)^2-1}};\ \begin{bmatrix} u = x+1 \\ du = dx \end{bmatrix} \Rightarrow \int \frac{du}{u\sqrt{u^2-1}} = \sec^{-1}|u| + C = \sec^{-1}|x+1| + C,$

$|u| = |x+1| > 1$

43. $\int (\sec x + \cot x)^2\,dx = \int \left(\sec^2 x + 2\sec x \cot x + \cot^2 x\right) dx = \int \sec^2 x\,dx + \int 2\csc x\,dx + \int \left(\csc^2 x - 1\right) dx$

$= \tan x - 2\ln|\csc x + \cot x| - \cot x - x + C$

45. $\int \csc x \sin 3x\,dx = \int (\csc x)(\sin 2x \cos x + \sin x \cos 2x)\,dx = \int (\csc x)\left(2\sin x \cos^2 x + \sin x \cos 2x\right) dx$

$= \int \left(2\cos^2 x + \cos 2x\right) dx = \int [(1+\cos 2x) + \cos 2x]\,dx = \int (1 + 2\cos 2x)\,dx = x + \sin 2x + C$

47. $\int \frac{x}{x+1}\,dx = \int \left(1 - \frac{1}{x+1}\right) dx = x - \ln|x+1| + C$

49. $\int_{\sqrt{2}}^3 \frac{2x^3}{x^2-1}\,dx = \int_{\sqrt{2}}^3 \left(2x + \frac{2x}{x^2-1}\right) dx = \left[x^2 + \ln|x^2-1|\right]_{\sqrt{2}}^3 = (9 + \ln 8) - (2 + \ln 1) = 7 + \ln 8$

51. $\int \frac{4t^3 - t^2 + 16t}{t^2+4}\,dt = \int \left[(4t-1) + \frac{4}{t^2+4}\right] dt = 2t^2 - t + 2\tan^{-1}\left(\frac{t}{2}\right) + C$

53. $\int \frac{1-x}{\sqrt{1-x^2}}\,dx = \int \frac{dx}{\sqrt{1-x^2}} - \int \frac{x\,dx}{\sqrt{1-x^2}} = \sin^{-1} x + \sqrt{1-x^2} + C$

55. $\int_0^{\pi/4} \frac{1+\sin x}{\cos^2 x}\,dx = \int_0^{\pi/4} \left(\sec^2 x + \sec x \tan x\right) dx = [\tan x + \sec x]_0^{\pi/4} = \left(1+\sqrt{2}\right) - (0+1) = \sqrt{2}$

57. $\int \frac{dx}{1+\sin x} = \int \frac{(1-\sin x)}{\left(1-\sin^2 x\right)}\,dx = \int \frac{(1-\sin x)}{\cos^2 x}\,dx = \int \left(\sec^2 x - \sec x \tan x\right) dx = \tan x - \sec x + C$

59. $\int \frac{1}{\sec\theta + \tan\theta}\, d\theta = \int \frac{\cos\theta}{1+\sin\theta}\, d\theta; \begin{bmatrix} u = 1+\sin\theta \\ du = \cos\theta\, d\theta \end{bmatrix} \Rightarrow \int \frac{du}{u} = \ln|u| + C = \ln|1+\sin\theta| + C$

61. $\int \frac{1}{1-\sec x}\, dx = \int \frac{\cos x}{\cos x - 1}\, dx = \int \left(1 + \frac{1}{\cos x - 1}\right) dx = \int \left(1 - \frac{1+\cos x}{\sin^2 x}\right) dx = \int \left(1 - \csc^2 x - \frac{\cos x}{\sin^2 x}\right) dx$

$= \int \left(1 - \csc^2 x - \csc x \cot x\right) dx = x + \cot x + \csc x + C$

63. $\int_0^{2\pi} \sqrt{\frac{1-\cos x}{2}}\, dx = \int_0^{2\pi} \left|\sin\frac{x}{2}\right| dx; \begin{bmatrix} \sin\frac{x}{2} \ge 0 \\ \text{for } 0 \le \frac{x}{2} \le \pi \end{bmatrix} \Rightarrow \int_0^{2\pi} \sin\left(\frac{x}{2}\right) dx = \left[-2\cos\frac{x}{2}\right]_0^{2\pi} = -2(\cos\pi - \cos 0)$

$= (-2)(-2) = 4$

65. $\int_{\pi/2}^{\pi} \sqrt{1+\cos 2t}\, dt = \int_{\pi/2}^{\pi} \sqrt{2}\,|\cos t|\, dt; \begin{bmatrix} \cos t \le 0 \\ \text{for } \frac{\pi}{2} \le t \le \pi \end{bmatrix} \Rightarrow \int_{\pi/2}^{\pi} -\sqrt{2}\cos t\, dt = \left[-\sqrt{2}\sin t\right]_{\pi/2}^{\pi}$

$= -\sqrt{2}\left(\sin\pi - \sin\frac{\pi}{2}\right) = \sqrt{2}$

67. $\int_{-\pi}^{0} \sqrt{1-\cos^2\theta}\, d\theta = \int_{-\pi}^{0} |\sin\theta|\, d\theta; \begin{bmatrix} \sin\theta \le 0 \\ \text{for } -\pi \le \theta \le 0 \end{bmatrix} \Rightarrow \int_{-\pi}^{0} -\sin\theta\, d\theta = [\cos\theta]_{-\pi}^{0} = \cos 0 - \cos(-\pi)$

$= 1 - (-1) = 2$

69. $\int_{-\pi/4}^{\pi/4} \sqrt{\tan^2 y + 1}\, dy = \int_{-\pi/4}^{\pi/4} |\sec y|\, dy; \begin{bmatrix} \sec y \ge 0 \\ \text{for } -\frac{\pi}{4} \le y \le \frac{\pi}{4} \end{bmatrix} \Rightarrow \int_{-\pi/4}^{\pi/4} \sec y\, dy = \left[\ln|\sec y + \tan y|\right]_{-\pi/4}^{\pi/4}$

$= \ln\left|\sqrt{2}+1\right| - \ln\left|\sqrt{2}-1\right|$

71. $\int_{\pi/4}^{3\pi/4} (\csc x - \cot x)^2\, dx = \int_{\pi/4}^{3\pi/4} \left(\csc^2 x - 2\csc x\cot x + \cot^2 x\right) dx = \int_{\pi/4}^{3\pi/4} \left(2\csc^2 x - 1 - 2\csc x\cot x\right) dx$

$= \left[-2\cot x - x + 2\csc x\right]_{\pi/4}^{3\pi/4} = \left(-2\cot\frac{3\pi}{4} - \frac{3\pi}{4} + 2\csc\frac{3\pi}{4}\right) - \left(-2\cot\frac{\pi}{4} - \frac{\pi}{4} + 2\csc\frac{\pi}{4}\right)$

$= \left[-2(-1) - \frac{3\pi}{4} + 2(\sqrt{2})\right] - \left[-2(1) - \frac{\pi}{4} + 2(\sqrt{2})\right] = 4 - \frac{\pi}{2}$

73. $\int \cos\theta\,\csc(\sin\theta)\, d\theta; \begin{bmatrix} u = \sin\theta \\ du = \cos\theta\, d\theta \end{bmatrix} \Rightarrow \int \csc u\, du = -\ln|\csc u + \cot u| + C$

$= -\ln\left|\csc(\sin\theta) + \cot(\sin\theta)\right| + C$

75. $\int (\csc x - \sec x)(\sin x + \cos x)\,dx = \int (1 + \cot x - \tan x - 1)\,dx = \int \cot x\,dx - \int \tan x\,dx$

$= \ln|\sin x| + \ln|\cos x| + C$

77. $\int \frac{6\,dy}{\sqrt{y}(1+y)};\ \left[\begin{array}{c} u = \sqrt{y} \\ du = \frac{1}{2\sqrt{y}}\,dy \end{array}\right] \Rightarrow \int \frac{12\,du}{1+u^2} = 12\tan^{-1} u + C = 12\tan^{-1}\sqrt{y} + C$

79. $\int \frac{7\,dx}{(x-1)\sqrt{x^2-2x-48}} = \int \frac{7\,dx}{(x-1)\sqrt{(x-1)^2-49}};\ \left[\begin{array}{c} u = x-1 \\ du = dx \end{array}\right] \Rightarrow \int \frac{7\,du}{u\sqrt{u^2-49}} = 7\cdot\frac{1}{7}\sec^{-1}\left|\frac{u}{7}\right| + C$

$= \sec^{-1}\left|\frac{x-1}{7}\right| + C$

81. $\int \sec^2 t \tan(\tan t)\,dt;\ \left[\begin{array}{c} u = \tan t \\ du = \sec^2 t\,dt \end{array}\right] \Rightarrow \int \tan u\,du = -\ln|\cos u| + C = \ln|\sec u| + C = \ln|\sec(\tan t)| + C$

83. (a) $\int \cos^3\theta\,d\theta = \int (\cos\theta)(1-\sin^2\theta)\,d\theta;\ \left[\begin{array}{c} u = \sin\theta \\ du = \cos\theta\,d\theta \end{array}\right] \Rightarrow \int (1-u^2)\,du = u - \frac{u^3}{3} + C = \sin\theta - \frac{1}{3}\sin^3\theta + C$

(b) $\int \cos^5\theta\,d\theta = \int (\cos\theta)(1-\sin^2\theta)^2\,d\theta = \int (1-u^2)^2\,du = \int (1-2u^2+u^4)\,du = u - \frac{2}{3}u^3 + \frac{u^5}{5} + C$

$= \sin\theta - \frac{2}{3}\sin^3\theta + \frac{1}{5}\sin^5\theta + C$

(c) $\int \cos^9\theta\,d\theta = \int (\cos^8\theta)(\cos\theta)\,d\theta = \int (1-\sin^2\theta)^4(\cos\theta)\,d\theta$

85. (a) $\int \tan^3\theta\,d\theta = \int (\sec^2\theta - 1)(\tan\theta)\,d\theta = \int \sec^2\theta\tan\theta\,d\theta - \int \tan\theta\,d\theta = \frac{1}{2}\tan^2\theta - \int \tan\theta\,d\theta$

$= \frac{1}{2}\tan^2\theta + \ln|\cos\theta| + C$

(b) $\int \tan^5\theta\,d\theta = \int (\sec^2\theta - 1)(\tan^3\theta)\,d\theta = \int \tan^3\theta\sec^2\theta\,d\theta - \int \tan^3\theta\,d\theta = \frac{1}{4}\tan^4\theta - \int \tan^3\theta\,d\theta$

(c) $\int \tan^7\theta\,d\theta = \int (\sec^2\theta - 1)(\tan^5\theta)\,d\theta = \int \tan^5\theta\sec^2\theta\,d\theta - \int \tan^5\theta\,d\theta = \frac{1}{6}\tan^6\theta - \int \tan^5\theta\,d\theta$

(d) $\int \tan^{2k+1}\theta\,d\theta = \int (\sec^2\theta - 1)(\tan^{2k-1}\theta)\,d\theta = \int \tan^{2k-1}\theta\sec^2\theta\,d\theta - \int \tan^{2k-1}\theta\,d\theta;$

$\left[\begin{array}{c} u = \tan\theta \\ du = \sec^2\theta\,d\theta \end{array}\right] \Rightarrow \int u^{2k-1}\,du - \int \tan^{2k-1}\theta\,d\theta = \frac{1}{2k}u^{2k} - \int \tan^{2k-1}\theta\,d\theta = \frac{1}{2k}\tan^{2k}\theta - \int \tan^{2k-1}\theta\,d\theta$

87. $A = \int_{-\pi/4}^{\pi/4} (2\cos x - \sec x)\,dx = [2\sin x - \ln|\sec x + \tan x|]_{-\pi/4}^{\pi/4}$

$= [\sqrt{2} - \ln(\sqrt{2}+1)] - [-\sqrt{2} - \ln(\sqrt{2}-1)]$

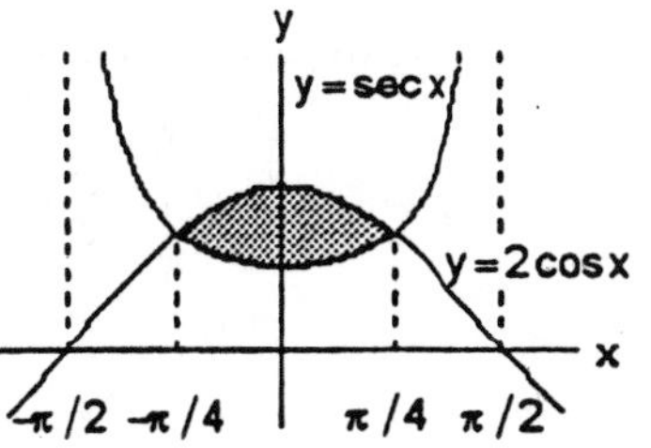

$$= 2\sqrt{2} - \ln\left(\frac{\sqrt{2}+1}{\sqrt{2}-1}\right) = 2\sqrt{2} - \ln\left(\frac{(\sqrt{2}+1)^2}{2-1}\right)$$
$$= 2\sqrt{2} - \ln(3 + 2\sqrt{2})$$

89. $V = \int_{-\pi/4}^{\pi/4} \pi(2\cos x)^2\,dx - \int_{-\pi/4}^{\pi/4} \pi\sec^2 x\,dx = 4\pi \int_{-\pi/4}^{\pi/4} \cos^2 x\,dx - \pi \int_{-\pi/4}^{\pi/4} \sec^2 x\,dx$

$$= 2\pi \int_{-\pi/4}^{\pi/4} (1 + \cos 2x)\,dx - \pi[\tan x]_{-\pi/4}^{\pi/4} = 2\pi\left[x + \tfrac{1}{2}\sin 2x\right]_{-\pi/4}^{\pi/4} - \pi[1 - (-1)]$$
$$= 2\pi\left[\left(\tfrac{\pi}{4} + \tfrac{1}{2}\right) - \left(-\tfrac{\pi}{4} - \tfrac{1}{2}\right)\right] - 2\pi = 2\pi\left(\tfrac{\pi}{2} + 1\right) - 2\pi = \pi^2$$

91. $y = \ln(\cos x) \Rightarrow \frac{dy}{dx} = -\frac{\sin x}{\cos x} \Rightarrow \left(\frac{dy}{dx}\right)^2 = \tan^2 x = \sec^2 x - 1;\ L = \int_a^b \sqrt{1 + \left(\frac{dy}{dx}\right)^2}\,dx$

$$= \int_0^{\pi/3} \sqrt{1 + (\sec^2 x - 1)}\,dx = \int_0^{\pi/3} \sec x\,dx = [\ln|\sec x + \tan x|]_0^{\pi/3} = \ln|2 + \sqrt{3}| - \ln|1 + 0| = \ln(2 + \sqrt{3})$$

93. $\int \csc x\,dx = \int (\csc x)(1)\,dx = \int (\csc x)\left(\frac{\csc x + \cot x}{\csc x + \cot x}\right)dx = \int \frac{\csc^2 x + \csc x\cot x}{\csc x + \cot x}\,dx;$

$$\left[\begin{array}{c} u = \csc x + \cot x \\ du = (-\csc x\cot x - \csc^2 x)\,dx \end{array}\right] \Rightarrow \int \frac{-du}{u} = -\ln|u| + C = -\ln|\csc x + \cot x| + C$$

7.2 INTEGRATION BY PARTS

1. $u = x,\ du = dx;\ dv = \sin\frac{x}{2}\,dx,\ v = -2\cos\frac{x}{2};$

$$\int x\sin\tfrac{x}{2}\,dx = -2x\cos\tfrac{x}{2} - \int\left(-2\cos\tfrac{x}{2}\right)dx = -2x\cos\left(\tfrac{x}{2}\right) + 4\sin\left(\tfrac{x}{2}\right) + C$$

3.
$$\begin{array}{lcl} & & \cos t \\ t^2 & \xrightarrow{(+)} & \sin t \\ 2t & \xrightarrow{(-)} & -\cos t \\ 2 & \xrightarrow{(+)} & -\sin t \\ 0 & & \end{array}$$

$$\int t^2\cos t\,dt = t^2\sin t + 2t\cos t - 2\sin t + C$$

5. $u = \ln x,\ du = \frac{dx}{x};\ dv = x\,dx,\ v = \frac{x^2}{2};$

$$\int_1^2 x\ln x\,dx = \left[\frac{x^2}{2}\ln x\right]_1^2 - \int_1^2 \frac{x^2}{2}\frac{dx}{x} = 2\ln 2 - \left[\frac{x^2}{4}\right]_1^2 = 2\ln 2 - \frac{3}{4} = \ln 4 - \frac{3}{4}$$

7. $u = \tan^{-1} y$, $du = \frac{dy}{1+y^2}$; $dv = dy$, $v = y$;

$$\int \tan^{-1} y\ dy = y \tan^{-1} y - \int \frac{y\ dy}{(1+y^2)} = y \tan^{-1} y - \frac{1}{2}\ln(1+y^2) + C = y \tan^{-1} y - \ln \sqrt{1+y^2} + C$$

9. $u = x$, $du = dx$; $dv = \sec^2 x\ dx$, $v = \tan x$;

$$\int x \sec^2 x\ dx = x \tan x - \int \tan x\ dx = x \tan x + \ln|\cos x| + C$$

11.

$$\begin{array}{lcl} & & e^x \\ x^3 & \xrightarrow{(+)} & e^x \\ 3x^2 & \xrightarrow{(-)} & e^x \\ 6x & \xrightarrow{(+)} & e^x \\ 6 & \xrightarrow{(-)} & e^x \\ 0 & & \end{array}$$

$$\int x^3 e^x\ dx = x^3 e^x - 3x^2 e^x + 6xe^x - 6e^x + C = (x^3 - 3x^2 + 6x - 6)e^x + C$$

13.

$$\begin{array}{lcl} & & e^x \\ x^2 - 5x & \xrightarrow{(+)} & e^x \\ 2x - 5 & \xrightarrow{(-)} & e^x \\ 2 & \xrightarrow{(+)} & e^x \\ 0 & & \end{array}$$

$$\int (x^2 - 5x)e^x\ dx = (x^2 - 5x)e^x - (2x-5)e^x + 2e^x + C = x^2 e^x - 7xe^x + 7e^x + C$$
$$= (x^2 - 7x + 7)e^x + C$$

15.

$$\begin{array}{lcl} & & e^x \\ x^5 & \xrightarrow{(+)} & e^x \\ 5x^4 & \xrightarrow{(-)} & e^x \\ 20x^3 & \xrightarrow{(+)} & e^x \\ 60x^2 & \xrightarrow{(-)} & e^x \\ 120x & \xrightarrow{(+)} & e^x \\ 120 & \xrightarrow{(-)} & e^x \\ 0 & & \end{array}$$

$$\int x^5 e^x\ dx = x^5 e^x - 5x^4 e^x + 20x^3 e^x - 60x^2 e^x + 120xe^x - 120e^x + C$$
$$= (x^5 - 5x^4 + 20x^3 - 60x^2 + 120x - 120)e^x + C$$

17.

$$\begin{array}{lcl} & & \sin 2\theta \\ \theta^2 & \xrightarrow{(+)} & -\frac{1}{2}\cos 2\theta \\ 2\theta & \xrightarrow{(-)} & -\frac{1}{4}\sin 2\theta \\ 2 & \xrightarrow{(+)} & \frac{1}{8}\cos 2\theta \end{array}$$

0 $$\int_0^{\pi/2} \theta^2 \sin 2\theta \, d\theta = \left[-\frac{\theta^2}{2}\cos 2\theta + \frac{\theta}{2}\sin 2\theta + \frac{1}{4}\cos 2\theta\right]_0^{\pi/2}$$

$$= \left[-\frac{\pi^2}{8}\cdot(-1) + \frac{\pi}{4}\cdot 0 + \frac{1}{4}\cdot(-1)\right] - \left[0 + 0 + \frac{1}{4}\cdot 1\right] = \frac{\pi^2}{8} - \frac{1}{2} = \frac{\pi^2 - 4}{8}$$

19. $u = \sec^{-1} t$, $du = \frac{dt}{t\sqrt{t^2-1}}$; $dv = t\,dt$, $v = \frac{t^2}{2}$;

$$\int_{2/\sqrt{3}}^{2} t \sec^{-1} t \, dt = \left[\frac{t^2}{2}\sec^{-1} t\right]_{2/\sqrt{3}}^{2} - \int_{2/\sqrt{3}}^{2} \left(\frac{t^2}{2}\right)\frac{dt}{t\sqrt{t^2-1}} = \left(2\cdot\frac{\pi}{3} - \frac{2}{3}\cdot\frac{\pi}{6}\right) - \int_{2/\sqrt{3}}^{2} \frac{t\,dt}{2\sqrt{t^2-1}}$$

$$= \frac{5\pi}{9} - \left[\frac{1}{2}\sqrt{t^2-1}\right]_{2/\sqrt{3}}^{2} = \frac{5\pi}{9} - \frac{1}{2}\left(\sqrt{3} - \sqrt{\frac{4}{3} - 1}\right) = \frac{5\pi}{9} - \frac{1}{2}\left(\sqrt{3} - \frac{\sqrt{3}}{3}\right) = \frac{5\pi}{9} - \frac{\sqrt{3}}{3} = \frac{5\pi - 3\sqrt{3}}{9}$$

21. $I = \int e^\theta \sin\theta \, d\theta$; $[u = \sin\theta,\ du = \cos\theta\,d\theta;\ dv = e^\theta\,d\theta,\ v = e^\theta] \Rightarrow I = e^\theta \sin\theta - \int e^\theta \cos\theta\,d\theta$;

$[u = \cos\theta,\ du = -\sin\theta\,d\theta;\ dv = e^\theta\,d\theta,\ v = e^\theta] \Rightarrow I = e^\theta \sin\theta - \left(e^\theta\cos\theta + \int e^\theta \sin\theta\,d\theta\right)$

$= e^\theta \sin\theta - e^\theta\cos\theta - I + C' \Rightarrow 2I = \left(e^\theta\sin\theta - e^\theta\cos\theta\right) + C' \Rightarrow I = \frac{1}{2}\left(e^\theta\sin\theta - e^\theta\cos\theta\right) + C$, where $C = \frac{C'}{2}$ is another arbitrary constant

23. $I = \int e^{2x}\cos 3x\,dx$; $\left[u = \cos 3x;\ du = -3\sin 3x\,dx,\ dv = e^{2x}\,dx;\ v = \frac{1}{2}e^{2x}\right]$

$\Rightarrow I = \frac{1}{2}e^{2x}\cos 3x + \frac{3}{2}\int e^{2x}\sin 3x\,dx$; $\left[u = \sin 3x,\ du = 3\cos 3x,\ dv = e^{2x}\,dx;\ v = \frac{1}{2}e^{2x}\right]$

$\Rightarrow I = \frac{1}{2}e^{2x}\cos 3x + \frac{3}{2}\left(\frac{1}{2}e^{2x}\sin 3x - \frac{3}{2}\int e^{2x}\cos 3x\,dx\right) = \frac{1}{2}e^{2x}\cos 3x + \frac{3}{4}e^{2x}\sin 3x - \frac{9}{4}I + C'$

$\Rightarrow \frac{13}{4}I = \frac{1}{2}e^{2x}\cos 3x + \frac{3}{4}e^{2x}\sin 3x + C' \Rightarrow \frac{e^{2x}}{13}(3\sin 3x + 2\cos 3x) + C$, where $C = \frac{4}{13}C'$

25. $\int e^{\sqrt{3s+9}}\,ds$; $\begin{bmatrix} 3s + 9 = x^2 \\ ds = \frac{2}{3}x\,dx \end{bmatrix} \Rightarrow \int e^x\cdot\frac{2}{3}x\,dx = \frac{2}{3}\int xe^x\,dx$; $[u = x,\ du = dx;\ dv = e^x\,dx,\ v = e^x]$;

$\frac{2}{3}\int xe^x\,dx = \frac{2}{3}\left(xe^x - \int e^x\,dx\right) = \frac{2}{3}(xe^x - e^x) + C = \frac{2}{3}\left(\sqrt{3s+9}\,e^{\sqrt{3s+9}} - e^{\sqrt{3s+9}}\right) + C$

27. $u = x$, $du = dx$; $dv = \tan^2 x\,dx$, $v = \int \tan^2 x\,dx = \int \frac{\sin^2 x}{\cos^2 x}\,dx = \int \frac{1 - \cos^2 x}{\cos^2 x}\,dx = \int \frac{dx}{\cos^2 x} - \int dx$

$= \tan x - x$; $\int_0^{\pi/3} x\tan^2 x\,dx = [x(\tan x - x)]_0^{\pi/3} - \int_0^{\pi/3}(\tan x - x)\,dx = \frac{\pi}{3}\left(\sqrt{3} - \frac{\pi}{3}\right) + \left[\ln|\cos x| + \frac{x^2}{2}\right]_0^{\pi/3}$

$= \frac{\pi}{3}\left(\sqrt{3} - \frac{\pi}{3}\right) + \ln\frac{1}{2} + \frac{\pi^2}{18} = \frac{\pi\sqrt{3}}{3} - \ln 2 - \frac{\pi^2}{18}$

29. $\int \sin(\ln x)\, dx;\ \left[\begin{array}{l} u = \ln x \\ du = \frac{1}{x}\, dx \\ dx = e^u\, du \end{array}\right] \Rightarrow \int (\sin u)\, e^u\, du$. From Exercise 21, $\int (\sin u)\, e^u\, du = e^u\left(\frac{\sin u - \cos u}{2}\right) + C$

$= \frac{1}{2}[-x \cos(\ln x) + x \sin(\ln x)] + C$

31. $y = \int x^2 e^{4x}\, dx$

Let $u = x^2$ $\quad dv = e^{4x}\, dx$

$du = 2x\, dx$ $\quad v = \frac{1}{4}e^{4x}$

$y = (x^2)\left(\frac{1}{4}e^{4x}\right) - \int \left(\frac{1}{4}e^{4x}\right)(2x\, dx)$

$= \frac{1}{4}x^2 e^{4x} - \frac{1}{2}\int x e^{4x}\, dx$

Let $u = x$ $\quad dv = e^{4x}\, dx$

$du = dx$ $\quad v = \frac{1}{4}e^{4x}$

$y = \frac{1}{4}x^2 e^{4x} - \frac{1}{2}\left[(x)\left(\frac{1}{4}e^{4x}\right) - \int \left(\frac{1}{4}e^{4x}\right) dx\right]$

$y = \frac{1}{4}x^2 e^{4x} - \frac{1}{8}xe^{4x} + \frac{1}{32}e^{4x} + C$

$y = \left(\frac{x^2}{4} - \frac{x}{8} + \frac{1}{32}\right)e^{4x} + C$

33. Let $w = \sqrt{\theta}$. Then $dw = \frac{d\theta}{2\sqrt{\theta}}$, so $d\theta = 2\sqrt{\theta}\, dw = 2w\, dw$.

$\int \sin \sqrt{\theta}\, d\theta = \int (\sin w)(2w\, dw) = 2\int w \sin w\, dw$

Let $u = w$ $\quad dv = \sin w\, dw$

$du = dw$ $\quad v = -\cos w$

$\int w \sin w\, dw = -w \cos w + \int \cos w\, dw$

$= -w \cos w + \sin w + C$

$\int \sin \sqrt{\theta}\, d\theta = 2\int w \sin w\, dw$

$= -2w \cos w + 2 \sin w + C$

$= -2\sqrt{\theta} \cos \sqrt{\theta} + 2 \sin \sqrt{\theta} + C$

35. (a) $u = x$, $du = dx$; $dv = \sin x\, dx$, $v = -\cos x$;

$S_1 = \int_0^{\pi} x \sin x\, dx = [-x \cos x]_0^{\pi} + \int_0^{\pi} \cos x\, dx = \pi + [\sin x]_0^{\pi} = \pi$

(b) $S_2 = -\int_{\pi}^{2\pi} x \sin x\, dx = \left[[-x \cos x]_{\pi}^{2\pi} + \int_{\pi}^{2\pi} \cos x\, dx\right] = -\left[-3\pi + [\sin x]_{\pi}^{2\pi}\right] = 3\pi$

(c) $S_3 = \int_{2\pi}^{3\pi} x \sin x\, dx = [-x \cos x]_{2\pi}^{3\pi} + \int_{2\pi}^{3\pi} \cos x\, dx = 5\pi + [\sin x]_{2\pi}^{3\pi} = 5\pi$

(d) $S_{n+1} = (-1)^{n+1} \int_{n\pi}^{(n+1)\pi} x \sin x\, dx = (-1)^{n+1}\left[[-x \cos x]_{n\pi}^{(n+1)\pi} + [\sin x]_{n\pi}^{(n+1)\pi}\right]$

$= (-1)^{n+1}[-(n+1)\pi(-1)^n + n\pi(-1)^{n+1}] + 0 = (2n+1)\pi$

37. $V = \int_0^{\ln 2} 2\pi(\ln 2 - x)\, e^x\, dx = 2\pi \ln 2 \int_0^{\ln 2} e^x\, dx - 2\pi \int_0^{\ln 2} xe^x\, dx$

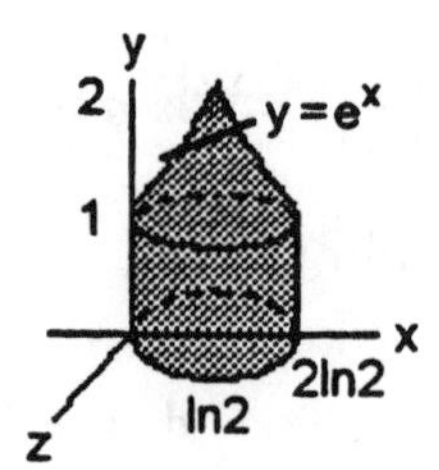

$= (2\pi \ln 2)[e^x]_0^{\ln 2} - 2\pi\left([xe^x]_0^{\ln 2} - \int_0^{\ln 2} e^x\, dx\right)$

$= 2\pi \ln 2 - 2\pi\left(2 \ln 2 + [e^x]_0^{\ln 2}\right) = -2\pi \ln 2 + 2 = 2\pi(1 - \ln 2)$

39. (a) $V = \int_0^{\pi/2} 2\pi x \cos x\, dx = 2\pi\left([x \sin x]_0^{\pi/2} - \int_0^{\pi/2} \sin x\, dx\right)$

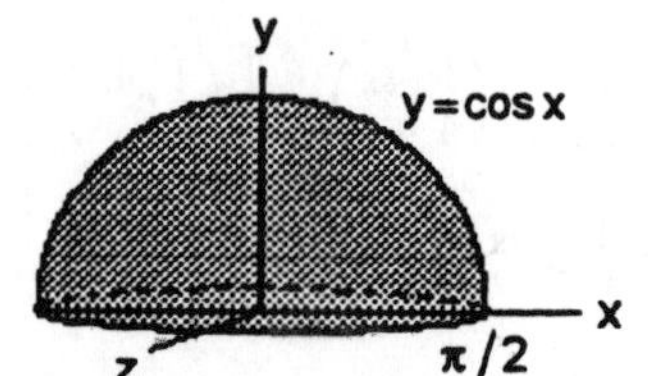

$= 2\pi\left(\frac{\pi}{2} + [\cos x]_0^{\pi/2}\right) = 2\pi\left(\frac{\pi}{2} + 0 - 1\right) = \pi(\pi - 2)$

(b) $V = \int_0^{\pi/2} 2\pi\left(\frac{\pi}{2} - x\right) \cos x\, dx$; $u = \frac{\pi}{2} - x$, $du = -dx$; $dv = \cos x\, dx$, $v = \sin x$;

$V = 2\pi\left[\left(\frac{\pi}{2} - x\right) \sin x\right]_0^{\pi/2} + 2\pi \int_0^{\pi/2} \sin x\, dx = 0 + 2\pi[-\cos x]_0^{\pi/2} = 2\pi(0+1) = 2\pi$

41. $av(y) = \frac{1}{2\pi} \int_0^{2\pi} 2e^{-t} \cos t\, dt = \frac{1}{\pi}\left[e^{-t}\left(\frac{\sin t - \cos t}{2}\right)\right]_0^{2\pi}$

(see Exercise 22) $\Rightarrow av(y) = \frac{1}{2\pi}(1 - e^{-2\pi})$

43. Let $u = x^n$ $\quad dv = \cos x\, dx$

$du = nx^{n-1}\, dx$ $\quad v = \sin x$

$\int x^n \cos x\, dx = x^n \sin x - \int (\sin x)(nx^{n-1}\, dx) = x^n \sin x - n\int x^{n-1} \sin x\, dx$

45. Let $u = x^n$ $\quad dv = e^{ax}\, dx$

$du = nx^{n-1}\, dx$ $\quad v = \frac{1}{a}e^{ax}$

$$\int x^n e^{ax}\,dx = (x^n)\left(\tfrac{1}{a}e^{ax}\right) - \int\left(\tfrac{1}{a}e^{ax}\right)(nx^{n-1}\,dx) = \frac{x^n e^{ax}}{a} - \frac{n}{a}\int x^{n-1}e^{ax}\,dx,\ a \neq 0$$

47. (a) Let $y = f^{-1}(x)$. Then $x = f(y)$, so $dx = f'(y)\,dy$.

Hence, $\int f^{-1}(x)\,dx = \int (y)[f'(y)\,dy] = \int yf'(y)\,dy$

(b) Let $u = y$ $\quad dv = f'(y)\,dy$

$du = dy$ $\quad v = f(y)$

$$\int yf'(y)\,dy = yf(y) - \int f(y)\,dy = f^{-1}(x)(x) - \int f(y)\,dy$$

Hence, $\int f^{-1}(x)\,dx = \int yf'(y)\,dy = xf^{-1}(x) - \int f(y)\,dy.$

49. (a) Using $y = f^{-1}(x) = \sin^{-1} x$ and $f(y) = \sin y$, $-\frac{\pi}{2} \le y \le \frac{\pi}{2}$, we have:

$$\int \sin^{-1} x\,dx = x\sin^{-1} x - \int \sin y\,dy = x\sin^{-1} x + \cos y + C = x\sin^{-1} x + \cos(\sin^{-1} x) + C$$

(b) $\int \sin^{-1} x\,dx = x\sin^{-1} x - \int x\left(\frac{d}{dx}\sin^{-1} x\right)dx = x\sin^{-1} x - \int x\frac{1}{\sqrt{1-x^2}}\,dx$

$u = 1 - x^2$, $du = -2x\,dx = x\sin^{-1} x + \frac{1}{2}\int u^{-1/2}\,du = x\sin^{-1} x + u^{1/2} + C = x\sin^{-1} x + \sqrt{1-x^2} + C$

(c) $\cos(\sin^{-1} x) = \sqrt{1-x^2}$

51. (a) Using $y = f^{-1}(x) = \cos^{-1} x$ and $f(y) = \cos x$, $0 \le x \le \pi$, we have:

$$\int \cos^{-1} x\,dx = x\cos^{-1} x - \int \cos y\,dy = x\cos^{-1} x - \sin y + C = x\cos^{-1} x - \sin(\cos^{-1} x) + C$$

(b) $\int \cos^{-1} x\,dx = x\cos^{-1} x - \int x\left(\frac{d}{dx}\cos^{-1} x\right)dx = x\cos^{-1} x - \int x\left(-\frac{1}{\sqrt{1-x^2}}\right)dx$

$u = 1 - x^2$, $du = -2x\,dx = x\cos^{-1} x - \frac{1}{2}\int u^{-1/2}\,du = x\cos^{-1} x - u^{1/2} + C = x\cos^{-1} x - \sqrt{1-x^2} + C$

(c) $\sin(\cos^{-1} x) = \sqrt{1-x^2}$

7.3 PARTIAL FRACTIONS

1. $\frac{5x-13}{(x-3)(x-2)} = \frac{A}{x-3} + \frac{B}{x-2} \Rightarrow 5x - 13 = A(x-2) + B(x-3) = (A+B)x - (2A+3B)$

$\Rightarrow \left.\begin{array}{c} A+B=5 \\ 2A+3B=13 \end{array}\right\} \Rightarrow -B = (10-13) \Rightarrow B = 3 \Rightarrow A = 2$; thus, $\frac{5x-13}{(x-3)(x-2)} = \frac{2}{x-3} + \frac{3}{x-2}$

3. $\frac{x+4}{(x+1)^2} = \frac{A}{x+1} + \frac{B}{(x+1)^2} \Rightarrow x + 4 = A(x+1) + B = Ax + (A+B) \Rightarrow \left.\begin{array}{c} A=1 \\ A+B=4 \end{array}\right\} \Rightarrow A = 1$ and $B = 3$;

thus, $\frac{x+4}{(x+1)^2} = \frac{1}{x+1} + \frac{3}{(x+1)^2}$

5. $\frac{z+1}{z^2(z-1)} = \frac{A}{z} + \frac{B}{z^2} + \frac{C}{z-1} \Rightarrow z+1 = Az(z-1) + B(z-1) + Cz^2 \Rightarrow z+1 = (A+C)z^2 + (-A+B)z - B$

$\Rightarrow \left.\begin{matrix} A+C=0 \\ -A+B=1 \\ -B=1 \end{matrix}\right\} \Rightarrow B=-1 \Rightarrow A=-2 \Rightarrow C=2$; thus, $\frac{z+1}{z^2(z-1)} = \frac{-2}{z} + \frac{-1}{z^2} + \frac{2}{z-1}$

7. $\frac{t^2+8}{t^2-5t+6} = 1 + \frac{5t+2}{t^2-5t+6}$ (after long division); $\frac{5t+2}{t^2-5t+6} = \frac{5t+2}{(t-3)(t-2)} = \frac{A}{t-3} + \frac{B}{t-2}$

$\Rightarrow 5t+2 = A(t-2) + B(t-3) = (A+B)t + (-2A-3B) \Rightarrow \left.\begin{matrix} A+B=5 \\ -2A-3B=2 \end{matrix}\right\} \Rightarrow -B = (10+2) = 12$

$\Rightarrow B = -12 \Rightarrow A = 17$; thus, $\frac{t^2+8}{t^2-5t+6} = 1 + \frac{17}{t-3} + \frac{-12}{t-2}$

9. $\frac{1}{1-x^2} = \frac{A}{1-x} + \frac{B}{1+x} \Rightarrow 1 = A(1+x) + B(1-x)$; $x=1 \Rightarrow A = \frac{1}{2}$; $x=-1 \Rightarrow B = \frac{1}{2}$;

$\int \frac{dx}{1-x^2} = \frac{1}{2}\int \frac{dx}{1-x} + \frac{1}{2}\int \frac{dx}{1+x} = \frac{1}{2}[\ln|1+x| - \ln|1-x|] + C$

11. $\frac{x+4}{x^2+5x-6} = \frac{A}{x+6} + \frac{B}{x-1} \Rightarrow x+4 = A(x-1) + B(x+6)$; $x=1 \Rightarrow B = \frac{5}{7}$; $x=-6 \Rightarrow A = \frac{-2}{-7} = \frac{2}{7}$;

$\int \frac{x+4}{x^2+5x-6}\,dx = \frac{2}{7}\int \frac{dx}{x+6} + \frac{5}{7}\int \frac{dx}{x-1} = \frac{2}{7}\ln|x+6| + \frac{5}{7}\ln|x-1| + C = \frac{1}{7}\ln\left|(x+6)^2(x-1)^5\right| + C$

13. $\frac{y}{y^2-2y-3} = \frac{A}{y-3} + \frac{B}{y+1} \Rightarrow y = A(y+1) + B(y-3)$; $y=-1 \Rightarrow B = \frac{-1}{-4} = \frac{1}{4}$; $y=3 \Rightarrow A = \frac{3}{4}$;

$\int_4^8 \frac{y\,dy}{y^2-2y-3} = \frac{3}{4}\int_4^8 \frac{dy}{y-3} + \frac{1}{4}\int_4^8 \frac{dy}{y+1} = \left[\frac{3}{4}\ln|y-3| + \frac{1}{4}\ln|y+1|\right]_4^8 = \left(\frac{3}{4}\ln 5 + \frac{1}{4}\ln 9\right) - \left(\frac{3}{4}\ln 1 + \frac{1}{4}\ln 5\right)$

$= \frac{1}{2}\ln 5 + \frac{1}{2}\ln 3 = \frac{\ln 15}{2}$

15. $\frac{1}{t^3+t^2-2t} = \frac{A}{t} + \frac{B}{t+2} + \frac{C}{t-1} \Rightarrow 1 = A(t+2)(t-1) + Bt(t-1) + Ct(t+2)$; $t=0 \Rightarrow A = -\frac{1}{2}$; $t=-2$

$\Rightarrow B = \frac{1}{6}$; $t=1 \Rightarrow C = \frac{1}{3}$; $\int \frac{dt}{t^3+t^2-2t} = -\frac{1}{2}\int \frac{dt}{t} + \frac{1}{6}\int \frac{dt}{t+2} + \frac{1}{3}\int \frac{dt}{t-1}$

$= -\frac{1}{2}\ln|t| + \frac{1}{6}\ln|t+2| + \frac{1}{3}\ln|t-1| + C$

17. $\frac{x^3}{x^2+2x+1} = (x-2) + \frac{3x+2}{(x+1)^2}$ (after long division); $\frac{3x+2}{(x+1)^2} = \frac{A}{x+1} + \frac{B}{(x+1)^2} \Rightarrow 3x+3 = A(x+1) + B$

$= Ax + (A+B) \Rightarrow A = 3,\ A+B = 2 \Rightarrow A = 3,\ B = -1$; $\int_0^1 \frac{x^3\,dx}{x^2+2x+1}$

$$= \int_0^1 (x-2)\,dx + 3\int_0^1 \frac{dx}{x+1} - \int_0^1 \frac{dx}{(x+1)^2} = \left[\frac{x^2}{2} - 2x + 3\ln|x+1| + \frac{1}{x+1}\right]_0^1$$

$$= \left(\frac{1}{2} - 2 + 3\ln 2 + \frac{1}{2}\right) - (1) = 3\ln 2 - 2$$

19. $\frac{1}{(x^2-1)^2} = \frac{A}{x+1} + \frac{B}{x-1} + \frac{C}{(x+1)^2} + \frac{D}{(x-1)^2} \Rightarrow 1 = A(x+1)(x-1)^2 + B(x-1)(x+1)^2 + C(x-1)^2 + D(x+1)^2;$
$x = -1 \Rightarrow C = \frac{1}{4};\ x = 1 \Rightarrow D = \frac{1}{4};$ coefficient of $x^3 = A + B \Rightarrow A + B = 0;$ constant $= A - B + C + D$
$\Rightarrow A - B + C + D = 1 \Rightarrow A - B = \frac{1}{2};$ thus, $A = \frac{1}{4} \Rightarrow B = -\frac{1}{4};\ \int \frac{dx}{(x^2-1)^2}$
$= \frac{1}{4}\int \frac{dx}{x+1} - \frac{1}{4}\int \frac{dx}{x-1} + \frac{1}{4}\int \frac{dx}{(x+1)^2} + \frac{1}{4}\int \frac{dx}{(x-1)^2} = \frac{1}{4}\ln\left|\frac{x+1}{x-1}\right| - \frac{x}{2(x^2-1)} + C$

21. $\frac{1}{(x+1)(x^2+1)} = \frac{A}{x+1} + \frac{Bx+C}{x^2+1} \Rightarrow 1 = A(x^2+1) + (Bx+C)(x+1);\ x = -1 \Rightarrow A = \frac{1}{2};$ coefficient of x^2
$= A + B \Rightarrow A + B = 0 \Rightarrow B = -\frac{1}{2};$ constant $= A + C \Rightarrow A + C = 1 \Rightarrow C = \frac{1}{2};\ \int_0^1 \frac{dx}{(x+1)(x^2+1)}$
$= \frac{1}{2}\int_0^1 \frac{dx}{x+1} + \frac{1}{2}\int_0^1 \frac{(-x+1)}{x^2+1}\,dx = \left[\frac{1}{2}\ln|x+1| - \frac{1}{4}\ln(x^2+1) + \frac{1}{2}\tan^{-1}x\right]_0^1$
$= \left(\frac{1}{2}\ln 2 - \frac{1}{4}\ln 2 + \frac{1}{2}\tan^{-1}1\right) - \left(\frac{1}{2}\ln 1 - \frac{1}{4}\ln 1 + \frac{1}{2}\tan^{-1}0\right) = \frac{1}{4}\ln 2 + \frac{1}{2}\left(\frac{\pi}{4}\right) = \frac{(\pi + 2\ln 2)}{8}$

23. $\frac{y^2+2y+1}{(y^2+1)^2} = \frac{Ay+B}{y^2+1} + \frac{Cy+D}{(y^2+1)^2} \Rightarrow y^2 + 2y + 1 = (Ay+B)(y^2+1) + Cy + D$
$= Ay^3 + By^2 + (A+C)y + (B+D) \Rightarrow A = 0,\ B = 1;\ A + C = 2 \Rightarrow C = 2;\ B + D = 1 \Rightarrow D = 0;$
$\int \frac{y^2+2y+1}{(y^2+1)^2}\,dy = \int \frac{1}{y^2+1}\,dy + 2\int \frac{y}{(y^2+1)^2}\,dy = \tan^{-1}y - \frac{1}{y^2+1} + C$

25. $\frac{2s+2}{(s^2+1)(s-1)^3} = \frac{As+B}{s^2+1} + \frac{C}{s-1} + \frac{D}{(s-1)^2} + \frac{E}{(s-1)^3} \Rightarrow 2s + 2$
$= (As+B)(s-1)^3 + C(s^2+1)(s-1)^2 + D(s^2+1)(s-1) + E(s^2+1)$
$= [As^4 + (-3A+B)s^3 + (3A-3B)s^2 + (-A+3B)s - B] + C(s^4 - 2s^3 + 2s^2 - 2s + 1) + D(s^3 - s^2 + s - 1)$
$\quad + E(s^2+1)$
$= (A+C)s^4 + (-3A+B-2C+D)s^3 + (3A-3B+2C-D+E)s^2 + (-A+3B-2C+D)s + (-B+C-D+E)$

$$\Rightarrow \left.\begin{aligned} A \quad\ + C \qquad\qquad &= 0 \\ -3A + B - 2C + D \qquad &= 0 \\ 3A - 3B + 2C - D + E &= 0 \\ -A + 3B - 2C + D \qquad &= 2 \\ -B + C - D + E &= 2 \end{aligned}\right\} \text{summing all equations} \Rightarrow 2E = 4 \Rightarrow E = 2;$$

summing eqs (2) and (3) $\Rightarrow -2B+2=0 \Rightarrow B=1$; summing eqs (3) and (4) $\Rightarrow 2A+2=2 \Rightarrow A=0$; $C=0$ from eq (1); then $-1+0-D+2=2$ from eq (5) $\Rightarrow D=-1$;

$$\int \frac{2s+2}{(s^2+1)(s-1)^3}\,ds = \int \frac{ds}{s^2+1} - \int \frac{ds}{(s-1)^2} + 2\int \frac{ds}{(s-1)^3} = -(s-1)^{-2} + (s-1)^{-1} + \tan^{-1} s + C$$

27. $\frac{2\theta^3+5\theta^2+8\theta+4}{(\theta^2+2\theta+2)^2} = \frac{A\theta+B}{\theta^2+2\theta+2} + \frac{C\theta+D}{(\theta^2+2\theta+2)^2} \Rightarrow 2\theta^3+5\theta^2+8\theta+4 = (A\theta+B)(\theta^2+2\theta+2)+C\theta+D$
$= A\theta^3+(2A+B)\theta^2+(2A+2B+C)\theta+(2B+D) \Rightarrow A=2$; $2A+B=5 \Rightarrow B=1$; $2A+2B+C=8 \Rightarrow C=2$;
$2B+D=4 \Rightarrow D=2$; $\int \frac{2\theta^3+5\theta^2+8\theta+4}{(\theta^2+2\theta+2)^2}\,d\theta = \int \frac{2\theta+1}{(\theta^2+2\theta+2)}\,d\theta + \int \frac{2\theta+2}{(\theta^2+2\theta+2)^2}\,d\theta$
$= \int \frac{2\theta+2}{\theta^2+2\theta+2}\,d\theta - \int \frac{d\theta}{\theta^2+2\theta+2} + \int \frac{d(\theta^2+2\theta+2)}{(\theta^2+2\theta+2)^2} = \int \frac{d(\theta^2+2\theta+2)}{\theta^2+2\theta+2} - \int \frac{d\theta}{(\theta+1)^2+1} - \frac{1}{\theta^2+2\theta+2}$
$= \frac{-1}{\theta^2+2\theta+2} + \ln(\theta^2+2\theta+2) - \tan^{-1}(\theta+1) + C$

29. $\frac{2x^3-2x^2+1}{x^2-x} = 2x + \frac{1}{x^2-x} = 2x + \frac{1}{x(x-1)}$; $\frac{1}{x(x-1)} = \frac{A}{x} + \frac{B}{x-1} \Rightarrow 1 = A(x-1)+Bx$; $x=0 \Rightarrow A=-1$;
$x=1 \Rightarrow B=1$; $\int \frac{2x^3-2x^2+1}{x^2-x} = \int 2x\,dx - \int \frac{dx}{x} + \int \frac{dx}{x-1} = x^2 - \ln|x| + \ln|x-1| + C = x^2 + \ln\left|\frac{x-1}{x}\right| + C$

31. $\frac{9x^3-3x+1}{x^3-x^2} = 9 + \frac{9x^2-3x+1}{x^2(x-1)}$ (after long division); $\frac{9x^2-3x+1}{x^2(x-1)} = \frac{A}{x} + \frac{B}{x^2} + \frac{C}{x-1}$
$\Rightarrow 9x^2-3x+1 = Ax(x-1)+B(x-1)+Cx^2$; $x=1 \Rightarrow C=7$; $x=0 \Rightarrow B=-1$; $A+C=9 \Rightarrow A=2$;
$\int \frac{9x^3-3x+1}{x^3-x^2}\,dx = \int 9\,dx + 2\int \frac{dx}{x} - \int \frac{dx}{x^2} + 7\int \frac{dx}{x-1} = 9x + 2\ln|x| + \frac{1}{x} + 7\ln|x-1| + C$

33. $\frac{y^4+y^2-1}{y^3+y} = y - \frac{1}{y(y^2+1)}$; $\frac{1}{y(y^2+1)} = \frac{A}{y} + \frac{By+C}{y^2+1} \Rightarrow 1 = A(y^2+1)+(By+C)y = (A+B)y^2+Cy+A$
$\Rightarrow A=1$; $A+B=0 \Rightarrow B=-1$; $C=0$; $\int \frac{y^4+y^2-1}{y^3+y}\,dy = \int y\,dy - \int \frac{dy}{y} + \int \frac{y\,dy}{y^2+1}$
$= \frac{y^2}{2} - \ln|y| + \frac{1}{2}\ln(1+y^2) + C$

35. $\int \frac{e^t\,dt}{e^{2t}+3e^t+2} = [e^t = y] \int \frac{dy}{y^2+3y+2} = \int \frac{dy}{y+1} - \int \frac{dy}{y+2} = \ln\left|\frac{y+1}{y+2}\right| + C = \ln\left|\frac{e^t+1}{e^t+2}\right| + C$

37. $\int \frac{\cos y\,dy}{\sin^2 y + \sin y - 6}$; $[\sin y = t, \cos y\,dy = dt] \Rightarrow \int \frac{dy}{t^2+t-6} = \frac{1}{5}\int \left(\frac{1}{t-2} - \frac{1}{t+3}\right) dt = \frac{1}{5}\ln\left|\frac{t-2}{t+3}\right| + C$
$= \frac{1}{5}\ln\left|\frac{\sin y - 2}{\sin y + 3}\right| + C$

39. $\int \frac{(x-2)^2 \tan^{-1}(2x) - 12x^3 - 3x}{(4x^2+1)(x-2)^2}\,dx = \int \frac{\tan^{-1}(2x)}{4x^2+1}\,dx - 3\int \frac{x}{(x-2)^2}\,dx$

$= \frac{1}{2}\int \tan^{-1}(2x)\,d\left(\tan^{-1}(2x)\right) - 3\int \frac{dx}{x-2} - 6\int \frac{dx}{(x-2)^2} = \frac{\left(\tan^{-1} 2x\right)^2}{4} - 3\ln|x-2| + \frac{6}{x-2} + C$

41. $(t^2 - 3t + 2)\frac{dx}{dt} = 1$; $x = \int \frac{dt}{t^2-3t+2} = \int \frac{dt}{t-2} - \int \frac{dt}{t-1} = \ln\left|\frac{t-2}{t-1}\right| + C$; $\frac{t-2}{t-1} = Ce^x$; $t = 3$ and $x = 0$

$\Rightarrow \frac{1}{2} = C \Rightarrow \frac{t-2}{t-1} = \frac{1}{2}e^x \Rightarrow x = \ln\left|2\left(\frac{t-2}{t-1}\right)\right| = \ln|t-2| - \ln|t-1| + \ln 2$

43. $(t^2 + 2t)\frac{dx}{dt} = 2x + 2$; $\frac{1}{2}\int \frac{dx}{x+1} = \int \frac{dt}{t^2+2t} \Rightarrow \frac{1}{2}\ln|x+1| = \frac{1}{2}\int \frac{dt}{t} - \frac{1}{2}\int \frac{dt}{t+2} \Rightarrow \ln|x+1| = \ln\left|\frac{t}{t+2}\right| + C$;

$t = 1$ and $x = 1 \Rightarrow \ln 2 = \ln\frac{1}{3} + C \Rightarrow C = \ln 2 + \ln 3 = \ln 6 \Rightarrow \ln|x+1| = \ln 6\left|\frac{t}{t+2}\right| \Rightarrow x + 1 = \frac{6t}{t+2}$

$\Rightarrow x = \frac{6t}{t+2} - 1,\ t > 0$

45. $\frac{1}{y^2-y}\,dy = e^x\,dx \Rightarrow \int \frac{1}{y(y-1)}\,dy = \int e^x\,dx = e^x + C$

$\frac{1}{y(y-1)} = \frac{A}{y} + \frac{B}{y-1} \Rightarrow 1 = A(y-1) + B(y) = (A+B)y - A$

Equating coefficients of like terms gives

$A + B = 0$ and $-A = 1$

Solving the system simultaneously yields $A = -1$, $B = 1$.

$\int \frac{1}{y(y-1)}\,dy = \int -\frac{1}{y}\,dy + \int \frac{1}{y-1}\,dy = -\ln|y| + \ln|y-1| + C_2 \Rightarrow -\ln|y| + \ln|y-1| = e^x + C$

Substitute $x = 0$, $y = 2$.

$-\ln 2 + 0 = 1 + C$ or $C = -1 - \ln 2$

The solution to the initial value problem is

$-\ln|y| + \ln|y-1| = e^x - 1 - \ln 2.$

47. $dy = \frac{dx}{x^2-3x+2}$; $x^2 - 3x + 2 = (x-2)(x-1) \Rightarrow \frac{1}{x^2-3x+2} = \frac{A}{x-2} + \frac{B}{x-1} \Rightarrow 1 = A(x-1) + B(x-2)$

$\Rightarrow 1 = (A+B)x - A - 2B$

Equating coefficients of like terms gives

$A + B = 0,\ -A - 2B = 1$

Solving the system simultaneously yields $A = 1$, $B = -1$.

$\int dy = \int \frac{dx}{x^2-3x+2} = \int \frac{dx}{x-2} - \int \frac{dx}{x-1}$

$y = \ln|x-2| - \ln|x-1| + C$

Substitute $x = 3$, $y = 0 \Rightarrow 0 = 0 - \ln 2 + C$ or $C = \ln 2$

The solution to the initial value problems is

$y = \ln|x-2| - \ln|x-1| + \ln 2$

49. $V = \pi \int_{0.5}^{2.5} y^2\,dx = \pi \int_{0.5}^{2.5} \frac{9}{3x - x^2}\,dx = 3\pi\left(\int_{0.5}^{2.5} \left(-\frac{1}{x-3} + \frac{1}{x}\right)\right)dx = \left[3\pi \ln\left|\frac{x}{x-3}\right|\right]_{0.5}^{2.5} = 3\pi \ln 25$

51. (a) $\frac{dx}{dt} = kx(N-x) \Rightarrow \int \frac{dx}{x(N-x)} = \int k\,dt \Rightarrow \frac{1}{N}\int \frac{dx}{x} + \frac{1}{N}\int \frac{dx}{N-x} = \int k\,dt \Rightarrow \frac{1}{N}\ln\left|\frac{x}{N-x}\right| = kt + C;$

$k = \frac{1}{250}$, $N = 1000$, $t = 0$ and $x = 2 \Rightarrow \frac{1}{1000}\ln\left|\frac{2}{998}\right| = C \Rightarrow \frac{1}{1000}\ln\left|\frac{x}{1000-x}\right| = \frac{t}{250} + \frac{1}{1000}\ln\left(\frac{1}{499}\right)$

$\Rightarrow \ln\left|\frac{499x}{1000-x}\right| = 4t \Rightarrow \frac{499x}{1000-x} = e^{4t} \Rightarrow 499x = e^{4t}(1000-x) \Rightarrow \left(499 + e^{4t}\right)x = 1000e^{4t} \Rightarrow x = \frac{1000e^{4t}}{499 + e^{4t}}$

(b) $x = \frac{1}{2}N = 500 \Rightarrow 500 = \frac{1000e^{4t}}{499 + e^{4t}} \Rightarrow 500 \cdot 499 + 500e^{4t} = 1000e^{4t} \Rightarrow e^{4t} = 499 \Rightarrow t = \frac{1}{4}\ln 499 \approx 1.55$ days

7.4 TRIGONOMETRIC SUBSTITUTIONS

1. $y = 3\tan\theta$, $-\frac{\pi}{2} < \theta < \frac{\pi}{2}$, $dy = \frac{3\,d\theta}{\cos^2\theta}$, $9 + y^2 = 9\left(1 + \tan^2\theta\right) = \frac{9}{\cos^2\theta} \Rightarrow \frac{1}{\sqrt{9+y^2}} = \frac{|\cos\theta|}{3} = \frac{\cos\theta}{3}$

$\left(\text{because } \cos\theta > 0 \text{ when } -\frac{\pi}{2} < \theta < \frac{\pi}{2}\right);$

$\int \frac{dy}{\sqrt{9+y^2}} = 3\int \frac{\cos\theta\,d\theta}{3\cos^2\theta} = \int \frac{d\theta}{\cos\theta} = \ln|\sec\theta + \tan\theta| + C' = \ln\left|\frac{\sqrt{9+y^2}}{3} + \frac{y}{3}\right| + C' = \ln\left|\sqrt{9+y^2} + y\right| + C$

3. $t = 5\sin\theta$, $-\frac{\pi}{2} < \theta < \frac{\pi}{2}$, $dt = 5\cos\theta\,d\theta$, $\sqrt{25 - t^2} = 5\cos\theta;$

$\int \sqrt{25-t^2}\,dt = \int (5\cos\theta)(5\cos\theta)\,d\theta = 25\int \cos^2\theta\,d\theta = 25\int \frac{1+\cos 2\theta}{2}\,d\theta = 25\left(\frac{\theta}{2} + \frac{\sin 2\theta}{4}\right) + C$

$= \frac{25}{2}(\theta + \sin\theta\cos\theta) + C = \frac{25}{2}\left[\sin^{-1}\left(\frac{t}{5}\right) + \left(\frac{t}{5}\right)\left(\frac{\sqrt{25-t^2}}{5}\right)\right] + C = \frac{25}{2}\sin^{-1}\left(\frac{t}{5}\right) + \frac{t\sqrt{25-t^2}}{2} + C$

5. $x = \frac{7}{2}\sec\theta$, $0 < \theta < \frac{\pi}{2}$, $dx = \frac{7}{2}\sec\theta\tan\theta\,d\theta$, $\sqrt{4x^2 - 49} = \sqrt{49\sec^2\theta - 49} = 7\tan\theta;$

$\int \frac{dx}{\sqrt{4x^2-49}} = \int \frac{\left(\frac{7}{2}\sec\theta\tan\theta\right)d\theta}{7\tan\theta} = \frac{1}{2}\int \sec\theta\,d\theta = \frac{1}{2}\ln|\sec\theta + \tan\theta| + C = \frac{1}{2}\ln\left|\frac{2x}{7} + \frac{\sqrt{4x^2-49}}{7}\right| + C$

7. $x = \sec\theta$, $0 < \theta < \frac{\pi}{2}$, $dx = \sec\theta\tan\theta\,d\theta$, $\sqrt{x^2-1} = \tan\theta;$

$\int \frac{dx}{x^2\sqrt{x^2-1}} = \int \frac{\sec\theta\tan\theta\,d\theta}{\sec^2\theta\tan\theta} = \int \frac{d\theta}{\sec\theta} = \sin\theta + C = \tan\theta\cos\theta + C = \frac{\sqrt{x^2-1}}{x} + C$

9. $x = 2\tan\theta,\ -\frac{\pi}{2} < \theta < \frac{\pi}{2},\ dx = \frac{2\,d\theta}{\cos^2\theta},\ \sqrt{x^2+4} = \frac{2}{\cos\theta};$

$$\int \frac{x^3\,dx}{\sqrt{x^2+4}} = \int \frac{(8\tan^3\theta)(\cos\theta)\,d\theta}{\cos^2\theta} = 8\int \frac{\sin^3\theta\,d\theta}{\cos^4\theta} = 8\int \frac{(\cos^2\theta-1)(-\sin\theta)\,d\theta}{\cos^4\theta};$$

$$[t=\cos\theta] \to 8\int \frac{t^2-1}{t^4}\,dt = 8\int\left(\frac{1}{t^2}-\frac{1}{t^4}\right)dt = 8\left(-\frac{1}{t}+\frac{1}{3t^3}\right)+C = 8\left(-\sec\theta+\frac{\sec^3\theta}{3}\right)+C$$

$$= 8\left(-\frac{\sqrt{x^2+4}}{2}+\frac{(x^2+4)^{3/2}}{8\cdot 3}\right)+C = \frac{1}{3}(x^2+4)^{3/2} - 4\sqrt{x^2+4}+C$$

11. $w = 2\sin\theta,\ -\frac{\pi}{2} < \theta < \frac{\pi}{2},\ dw = 2\cos\theta\,d\theta,\ \sqrt{4-w^2} = 2\cos\theta;$

$$\int \frac{8\,dw}{w^2\sqrt{4-w^2}} = \int \frac{8\cdot 2\cos\theta\,d\theta}{4\sin^2\theta\cdot 2\cos\theta} = 2\int \frac{d\theta}{\sin^2\theta} = -2\cot\theta + C = \frac{-2\sqrt{4-w^2}}{w}+C$$

13. $x = \sec\theta,\ 0 < \theta < \frac{\pi}{2},\ dx = \sec\theta\tan\theta\,d\theta,\ (x^2-1)^{3/2} = \tan^3\theta;$

$$\int \frac{dx}{(x^2-1)^{3/2}} = \int \frac{\sec\theta\tan\theta\,d\theta}{\tan^3\theta} = \int \frac{\cos\theta\,d\theta}{\sin^2\theta} = -\frac{1}{\sin\theta}+C = -\left(\frac{1}{\tan\theta}\right)\left(\frac{1}{\cos\theta}\right)+C$$

$$= -\left(\frac{1}{\sqrt{x^2-1}}\right)(x)+C = -\frac{x}{\sqrt{x^2-1}}+C$$

15. $x = \sin\theta,\ -\frac{\pi}{2} < \theta < \frac{\pi}{2},\ dx = \cos\theta\,d\theta,\ (1-x^2)^{3/2} = \cos^3\theta;$

$$\int \frac{(1-x^2)^{3/2}\,dx}{x^6} = \int \frac{\cos^3\theta\cdot\cos\theta\,d\theta}{\sin^6\theta} = \int \cot^4\theta\csc^2\theta\,d\theta = -\frac{\cot^5\theta}{5}+C = -\frac{1}{5}\left(\frac{\sqrt{1-x^2}}{x}\right)^5+C$$

17. $x = \frac{1}{2}\tan\theta,\ -\frac{\pi}{2} < \theta < \frac{\pi}{2},\ dx = \frac{1}{2}\sec^2\theta\,d\theta,\ (4x^2+1)^2 = \sec^4\theta;$

$$\int \frac{8\,dx}{(4x^2+1)^2} = \int \frac{8\left(\frac{1}{2}\sec^2\theta\right)d\theta}{\sec^4\theta} = 4\int \cos^2\theta\,d\theta = 2(\theta+\sin\theta\cos\theta)+C = 2(\theta+\tan\theta\cdot\cos^2\theta)+C$$

$$= 2\tan^{-1}2x + \frac{4x}{(4x^2+1)}+C$$

19. Let $e^t = 3\tan\theta,\ t = \ln(3\tan\theta),\ dt = \frac{\sec^2\theta}{\tan\theta}\,d\theta,\ \sqrt{e^{2t}+9} = \sqrt{9\tan^2\theta+9} = 3\sec\theta;$

$$\int_0^{\ln 4} \frac{e^t\,dt}{\sqrt{e^{2t}+9}} = \int_{\tan^{-1}(1/3)}^{\tan^{-1}(4/3)} \frac{3\tan\theta\cdot\sec^2\theta\,d\theta}{\tan\theta\cdot 3\sec\theta} = \int_{\tan^{-1}(1/3)}^{\tan^{-1}(4/3)} \sec\theta\,d\theta = \left[\ln|\sec\theta+\tan\theta|\right]_{\tan^{-1}(1/3)}^{\tan^{-1}(4/3)}$$

$$= \ln\left(\frac{5}{3}+\frac{4}{3}\right) - \ln\left(\frac{\sqrt{10}}{3}+\frac{1}{3}\right) = \ln 9 - \ln\left(1+\sqrt{10}\right)$$

21. $\int_{1/12}^{1/4} \frac{2\,dt}{\sqrt{t}+4t\sqrt{t}}$; $\left[u = 2\sqrt{t},\ du = \frac{1}{\sqrt{t}}\,dt\right] \Rightarrow \int_{1/\sqrt{3}}^{1} \frac{2\,du}{1+u^2}$; $u = \tan\theta$, $\frac{\pi}{6} < \theta < \frac{\pi}{4}$, $du = \sec^2\theta\,d\theta$, $1+u^2 = \sec^2\theta$;

$$\int_{1/\sqrt{3}}^{1} \frac{4\,du}{u(1+u^2)} = \int_{\pi/6}^{\pi/4} \frac{2\sec^2\theta\,d\theta}{\sec^2\theta} = [2\theta]_{\pi/6}^{\pi/4} = 2\left(\frac{\pi}{4}-\frac{\pi}{6}\right) = \frac{\pi}{6}$$

23. $x = \sec\theta$, $dx = \sec\theta\tan\theta\,d\theta$, $\sqrt{x^2-1} = \sqrt{\sec^2\theta - 1} = \tan\theta$;

$$\int \frac{dx}{x\sqrt{x^2-1}} = \int \frac{\sec\theta\tan\theta\,d\theta}{\sec\theta\tan\theta} = \theta + C = \sec^{-1}|x| + C$$

25. $x = \sec\theta$, $dx = \sec\theta\tan\theta\,d\theta$, $\sqrt{x^2-1} = \sqrt{\sec^2\theta - 1} = \tan\theta$;

$$\int \frac{x\,dx}{\sqrt{x^2-1}} = \int \frac{\sec\theta\cdot\sec\theta\tan\theta\,d\theta}{\tan\theta} = \int \sec^2\theta\,d\theta = \tan\theta + C = \sqrt{x^2-1} + C$$

27. $x\frac{dy}{dx} = \sqrt{x^2-4}$; $dy = \sqrt{x^2-4}\,\frac{dx}{x}$; $y = \int \frac{\sqrt{x^2-4}}{x}\,dx$; $\begin{bmatrix} x = 2\sec\theta,\ 0 < \theta < \frac{\pi}{2} \\ dx = 2\sec\theta\tan\theta\,d\theta \\ \sqrt{x^2-4} = 2\tan\theta \end{bmatrix}$

$$\Rightarrow y = \int \frac{(2\tan\theta)(2\sec\theta\tan\theta)\,d\theta}{2\sec\theta} = 2\int \tan^2\theta\,d\theta = 2\int (\sec^2\theta - 1)\,d\theta = 2(\tan\theta - \theta) + C$$

$$= 2\left[\frac{\sqrt{x^2-4}}{2} - \sec^{-1}\left(\frac{x}{2}\right)\right] + C;\ x = 2 \text{ and } y = 0 \Rightarrow 0 = 0 + C \Rightarrow C = 0 \Rightarrow y = 2\left[\frac{\sqrt{x^2-4}}{2} - \sec^{-1}\frac{x}{2}\right]$$

29. $(x^2+4)\frac{dy}{dx} = 3$, $dy = \frac{3\,dx}{x^2+4}$; $y = 3\int \frac{dx}{x^2+4} = \frac{3}{2}\tan^{-1}\frac{x}{2} + C$; $x = 2$ and $y = 0 \Rightarrow 0 = \frac{3}{2}\tan^{-1}1 + C$

$$\Rightarrow C = -\frac{3\pi}{8} \Rightarrow y = \frac{3}{2}\tan^{-1}\left(\frac{x}{2}\right) - \frac{3\pi}{8}$$

31. $A = \int_0^3 \frac{\sqrt{9-x^2}}{3}\,dx$; $x = 3\sin\theta$, $0 \le \theta \le \frac{\pi}{2}$, $dx = 3\cos\theta\,d\theta$, $\sqrt{9-x^2} = \sqrt{9-9\sin^2\theta} = 3\cos\theta$;

$$A = \int_0^{\pi/2} \frac{3\cos\theta\cdot 3\cos\theta\,d\theta}{3} = 3\int_0^{\pi/2} \cos^2\theta\,d\theta = \frac{3}{2}[\theta + \sin\theta\cos\theta]_0^{\pi/2} = \frac{3\pi}{4}$$

33. (a) From the figure, $\tan\frac{x}{2} = \frac{\sin x}{1+\cos x}$

(b) From part (a), $z = \frac{\sin x}{1+\cos x} \Rightarrow z(1+\cos x) = \sin x \Rightarrow z^2(1+\cos x)^2 = \sin^2 x$

$\Rightarrow z^2(1+\cos x)^2 - (1-\cos x)(1+\cos x) = 0 \Rightarrow (1+\cos x)(z^2 + z^2\cos x - 1 + \cos x) = 0$

$1 + \cos x = 0$ or $(z^2+1)\cos x = 1 - z^2$

$\cos x = -1$ $\qquad \cos x = \frac{1-z^2}{1+z^2}$

$\cos x = -1$ does not make sense in this case.

(c) From part (b), $\cos x = \frac{1-z^2}{1+z^2} \Rightarrow \sin^2 x = 1 - \frac{(1-z^2)^2}{(1+z^2)^2} = \frac{(1+z^2)^2-(1-z^2)^2}{(1+z^2)^2}$

$= \frac{1+2z^2+z^4-1+2z^2-z^4}{(1+z^2)^2} = \frac{4z^2}{(1+z^2)^2} \Rightarrow \sin x = \pm\frac{2z}{1+z^2}$

Only $\sin x = \frac{2z}{1+z^2}$ makes sense in this case.

(d) $z = \tan\frac{x}{2}$, $dz = \left(\frac{1}{2}\sec^2\frac{x}{2}\right)dx \Rightarrow dz = \frac{1}{2}\left(1+\tan^2\frac{x}{2}\right)dx \Rightarrow dz = \frac{1}{2}(1+z^2)\,dx \Rightarrow dx = \frac{2\,dz}{1+z^2}$

35. $$\int \frac{dx}{1-\cos x} = \int \frac{\frac{2\,dz}{1+z^2}}{1-\frac{1-z^2}{1+z^2}} = \int \frac{dz}{z^2} = -\frac{1}{z}+C = -\frac{1}{\tan\frac{x}{2}}+C$$

37. $$\int \frac{dt}{1+\sin t+\cos t} = \int \frac{\frac{2\,dz}{1+z^2}}{1+\frac{2z}{1+z^2}+\frac{1-z^2}{1+z^2}} = \int \frac{dz}{z+1} = \ln|z+1|+C = \ln\left|\tan\frac{t}{2}+1\right|+C$$

39. $$\int_{\pi/2}^{2\pi/3} \frac{\cos\theta\,d\theta}{\sin\theta\cos\theta+\sin\theta} = \int_1^{\sqrt{3}} \frac{\left(\frac{1-z^2}{1+z^2}\right)\left(\frac{2\,dz}{1+z^2}\right)}{\left[\frac{2z(1-z^2)}{(1+z^2)^2}+\left(\frac{2z}{1+z^2}\right)\right]} = \int_1^{\sqrt{3}} \frac{2(1-z^2)\,dz}{2z-2z^3+2z+2z^3} = \int_1^{\sqrt{3}} \frac{1-z^2}{2z}\,dz$$

$$= \left[\frac{1}{2}\ln z - \frac{z^2}{4}\right]_1^{\sqrt{3}} = \left(\frac{1}{2}\ln\sqrt{3}-\frac{3}{4}\right)-\left(0-\frac{1}{4}\right) = \frac{\ln 3}{4}-\frac{1}{2} = \frac{1}{4}(\ln 3-2) = \frac{1}{2}(\ln\sqrt{3}-1)$$

41. $$\int \frac{\cos t\,dt}{1-\cos t} = \int \frac{\left(\frac{1-z^2}{1+z^2}\right)\left(\frac{2\,dz}{1+z^2}\right)}{1-\left(\frac{1-z^2}{1+z^2}\right)} = \int \frac{2(1-z^2)\,dz}{(1+z^2)^2-(1+z^2)(1-z^2)} = \int \frac{2(1-z^2)\,dz}{(1+z^2)(1+z^2-1+z^2)}$$

$$= \int \frac{(1-z^2)\,dz}{(1+z^2)z^2} = \int \frac{dz}{z^2(1+z^2)} - \int \frac{dz}{1+z^2} = \int \frac{dz}{z^2} - 2\int \frac{dz}{z^2+1} = -\frac{1}{z}-2\tan^{-1}z+C = -\cot\left(\frac{t}{2}\right)-t+C$$

7.5 INTEGRAL TABLES, COMPUTER ALGEBRA SYSTEMS, AND MONTE CARLO INTEGRATION

1. $$\int \frac{dx}{x\sqrt{x-3}} = \frac{2}{\sqrt{3}}\tan^{-1}\sqrt{\frac{x-3}{3}}+C$$

(We used FORMULA 13(a) with $a = 1$, $b = -3$)

3. $\int x\sqrt{2x-3}\,dx = \frac{1}{2}\int (2x-3)\sqrt{2x-3}\,dx + \frac{3}{2}\int \sqrt{2x-3}\,dx = \frac{1}{2}\int \left(\sqrt{2x-3}\right)^3 dx + \frac{3}{2}\int \left(\sqrt{2x-3}\right)^1 dx$

$= \left(\frac{1}{2}\right)\left(\frac{2}{2}\right)\frac{\left(\sqrt{2x-3}\right)^5}{5} + \left(\frac{3}{2}\right)\left(\frac{2}{2}\right)\frac{\left(\sqrt{2x-3}\right)^3}{3} + C = \frac{(2x-3)^{3/2}}{2}\left[\frac{2x-3}{5}+1\right] + C = \frac{(2x-3)^{3/2}(x+1)}{5} + C$

(We used FORMULA 11 with $a = 2$, $b = -3$, $n = 3$ and $a = 2$, $b = -3$, $n = 1$)

5. $\int x\sqrt{4x-x^2}\,dx = \int x\sqrt{2\cdot 2x - x^2}\,dx = \frac{(x+2)(2x-3\cdot 2)\sqrt{2\cdot 2\cdot x - x^2}}{6} + \frac{2^3}{2}\sin^{-1}\left(\frac{x-2}{2}\right) + C$

$= \frac{(x+2)(2x-6)\sqrt{4x-x^2}}{6} + 4\sin^{-1}\left(\frac{x-2}{2}\right) + C$

(We used FORMULA 51 with $a = 2$)

7. $\int \frac{\sqrt{4-x^2}}{x}\,dx = \int \frac{\sqrt{2^2-x^2}}{x}\,dx = \sqrt{2^2-x^2} - 2\ln\left|\frac{2+\sqrt{2^2-x^2}}{x}\right| + C = \sqrt{4-x^2} - 2\ln\left|\frac{2+\sqrt{4-x^2}}{x}\right| + C$

(We used FORMULA 31 with $a = 2$)

9. $\int \frac{r^2}{\sqrt{4-r^2}}\,dr = \int \frac{r^2}{\sqrt{2^2-r^2}}\,dr = \frac{2^2}{2}\sin^{-1}\left(\frac{r}{2}\right) - \frac{1}{2}r\sqrt{2^2-r^2} + C = 2\sin^{-1}\left(\frac{r}{2}\right) - \frac{1}{2}r\sqrt{4-r^2} + C$

(We used FORMULA 33 with $a = 2$)

11. $\int e^{2t}\cos 3t\,dt = \frac{e^{2t}}{2^2+3^2}(2\cos 3t + 3\sin 3t) + C = \frac{e^{2t}}{13}(2\cos 3t + 3\sin 3t) + C$

(We used FORMULA 108 with $a = 2$, $b = 3$)

13. $\int \frac{ds}{(9-s^2)^2} = \int \frac{ds}{(3^3-s^2)^2} = \frac{s}{2\cdot 3^2\cdot(3^2-s^2)} + \frac{1}{2\cdot 3^2}\int \frac{ds}{3^2-s^2}$

(We used FORMULA 19 with $a = 3$)

$= \frac{s}{18(9-s^2)} + \frac{1}{18}\left(\frac{1}{2\cdot 3}\ln\left|\frac{s+3}{s-3}\right|\right) + C = \frac{s}{18(9-s^2)} + \frac{1}{108}\ln\left|\frac{s+3}{s-3}\right| + C$

(We used FORMULA 18 with $a = 3$)

15. $\int \frac{\sqrt{3t-4}}{t}\,dt = 2\sqrt{3t-4} + (-4)\int \frac{dt}{t\sqrt{3t-4}}$

(We used FORMULA 12 with $a = 3$, $b = -4$)

$= 2\sqrt{3t-4} - 4\left(\frac{2}{\sqrt{4}}\tan^{-1}\sqrt{\frac{3t-4}{4}}\right) + C = 2\sqrt{3t-4} - 4\tan^{-1}\sqrt{\frac{3t-4}{4}} + C$

(We used FORMULA 13(a) with $a = 3$, $b = -4$)

17. $\int \sin 3x\cos 2x\,dx = -\frac{\cos 5x}{10} - \frac{\cos x}{2} + C$

(We used FORMULA 62(a) with $a = 3$, $b = 2$)

19. $\int \cos\frac{\theta}{3}\cos\frac{\theta}{4}\,d\theta = 6\sin\left(\frac{\theta}{12}\right) + \frac{6}{7}\sin\left(\frac{7\theta}{12}\right) + C$

(We used FORMULA 62(c) with $a = \frac{1}{3}$, $b = \frac{1}{4}$)

21. $\int \frac{x^3+x+1}{(x^2+1)^2}\,dx = \int \frac{x\,dx}{x^2+1} + \int \frac{dx}{(x^2+1)^2} = \frac{1}{2}\int \frac{d(x^2+1)}{x^2+1} + \int \frac{dx}{(x^2+1)^2}$

$= \frac{1}{2}\ln|x^2+1| + \frac{x}{2(1+x^2)} + \frac{1}{2}\tan^{-1}x + C$

(For the second integral we used FORMULA 17 with $a = 1$)

23. $\int \sin^{-1}\sqrt{x}\,dx;\ \begin{bmatrix} u = \sqrt{x} \\ x = u^2 \\ dx = 2u\,du \end{bmatrix} \Rightarrow 2\int u^1 \sin^{-1}u\,du = 2\left(\frac{u^{1+1}}{1+1}\sin^{-1}u - \frac{1}{1+1}\int \frac{u^{1+1}}{\sqrt{1-u^2}}\,du\right)$

$= u^2\sin^{-1}u - \int \frac{u^2\,du}{\sqrt{1-u^2}}$

(We used FORMULA 99 with $a = 1$, $n = 1$)

$= u^2\sin^{-1}u - \left(\frac{1}{2}\sin^{-1}u - \frac{1}{2}u\sqrt{1-u^2}\right) + C = \left(u^2 - \frac{1}{2}\right)\sin^{-1}u + \frac{1}{2}u\sqrt{1-u^2} + C$

(We used FORMULA 33 with $a = 1$)

$= \left(x - \frac{1}{2}\right)\sin^{-1}\sqrt{x} + \frac{1}{2}\sqrt{x - x^2} + C$

25. $\int (\cot t)\sqrt{1-\sin^2 t}\,dt = \int \frac{\sqrt{1-\sin^2 t}\,(\cos t)\,dt}{\sin t};\ \begin{bmatrix} u = \sin t \\ du = \cos t\,dt \end{bmatrix} \Rightarrow \int \frac{\sqrt{1-u^2}\,du}{u}$

$= \sqrt{1-u^2} - \ln\left|\frac{1+\sqrt{1-u^2}}{u}\right| + C$

(We used FORMULA 31 with $a = 1$)

$= \sqrt{1-\sin^2 t} - \ln\left|\frac{1+\sqrt{1-\sin^2 t}}{\sin t}\right| + C$

27. $\int \frac{dy}{y\sqrt{3+(\ln y)^2}};\ \begin{bmatrix} u = \ln y \\ y = e^u \\ dy = e^u\,du \end{bmatrix} \Rightarrow \int \frac{e^u\,du}{e^u\sqrt{3+u^2}} = \int \frac{du}{\sqrt{3+u^2}} = \ln\left|u + \sqrt{3+u^2}\right| + C$

$= \ln\left|\ln y + \sqrt{3+(\ln y)^2}\right| + C$

(We used FORMULA 20 with $a = \sqrt{3}$)

29. $\int \frac{3\,dr}{\sqrt{9r^2-1}};\ \begin{bmatrix} u = 3r \\ du = 3\,dr \end{bmatrix} \Rightarrow \int \frac{du}{\sqrt{u^2-1}} = \ln\left|u + \sqrt{u^2-1}\right| + C = \ln\left|3r + \sqrt{9r^2-1}\right| + C$

(We used FORMULA 36 with $a = 1$)

31. $\int \cos^{-1}\sqrt{x}\,dx;\ \begin{bmatrix} t=\sqrt{x} \\ x=t^2 \\ dx=2t\,dt \end{bmatrix} \Rightarrow 2\int t\cos^{-1}t\,dt = 2\left(\frac{t^2}{2}\cos^{-1}t+\frac{1}{2}\int \frac{t^2}{\sqrt{1-t^2}}\,dt\right) = t^2\cos^{-1}t+\int \frac{t^2}{\sqrt{1-t^2}}\,dt$

(We used FORMULA 100 with $a=1$, $n=1$)

$= t^2\cos^{-1}t+\frac{1}{2}\sin^{-1}t-\frac{1}{2}t\sqrt{1-t^2}+C$

(We used FORMULA 33 with $a=1$)

$= x\cos^{-1}\sqrt{x}+\frac{1}{2}\sin^{-1}\sqrt{x}-\frac{1}{2}\sqrt{x}\sqrt{1-x}+C = x\cos^{-1}\sqrt{x}+\frac{1}{2}\sin^{-1}\sqrt{x}-\frac{1}{2}\sqrt{x-x^2}+C$

33. $\int xe^{3x}\,dx = \frac{e^{3x}}{3^2}(3x-1)+C = \frac{e^{3x}}{9}(3x-1)+C$

(We used FORMULA 104 with $a=3$)

35. $\int x^2 2^x\,dx = \frac{x^2 2^x}{\ln 2}-\frac{2}{\ln 2}\int x2^x\,dx = \frac{x^2 2^x}{\ln 2}-\frac{2}{\ln 2}\left(\frac{x2^x}{\ln 2}-\frac{1}{\ln 2}\int 2^x\,dx\right) = \frac{x^2 2^x}{\ln 2}-\frac{2}{\ln 2}\left[\frac{x2^x}{\ln 2}-\frac{2^x}{(\ln 2)^2}\right]+C$

(We used FORMULA 106 with $a=1$, $b=2$)

37. $\int \frac{1}{8}\sinh^5 3x\,dx = \frac{1}{8}\left(\frac{\sinh^4 3x\cosh 3x}{5\cdot 3}-\frac{5-1}{5}\int \sinh^3 3x\,dx\right)$

$= \frac{\sinh^4 3x\cosh 3x}{120}-\frac{1}{10}\left(\frac{\sinh^2 3x\cosh 3x}{3\cdot 3}-\frac{3-1}{3}\int \sinh 3x\,dx\right)$

(We used FORMULA 117 with $a=3$, $n=5$ and $a=1$, $n=3$)

$= \frac{\sinh^4 3x\cosh 3x}{120}-\frac{\sinh^2 3x\cosh 3x}{90}+\frac{2}{30}\left(\frac{1}{3}\cosh 3x\right)+C$

$= \frac{1}{120}\sinh^4 3x\cosh 3x-\frac{1}{90}\sinh^2 3x\cosh 3x+\frac{2}{90}\cosh 3x+C$

39. $\int x^2\cosh 3x\,dx = \frac{x^2}{3}\sinh 3x-\frac{2}{3}\int x\sinh 3x\,dx = \frac{x^2}{3}\sinh 3x-\frac{2}{3}\left(\frac{x}{3}\cosh 3x-\frac{1}{3}\int \cosh 3x\,dx\right)$

(We used FORMULA 122 with $a=3$, $n=2$ and FORMULA 121 with $a=3$, $n=1$)

$= \frac{x^2}{3}\sinh 3x-\frac{2x}{9}\cosh 3x+\frac{2}{27}\sinh 3x+C$

41. $u=ax+b \Rightarrow x=\frac{u-b}{a} \Rightarrow dx=\frac{du}{a}$;

$\int \frac{x\,dx}{(ax+b)^2} = \int \frac{(u-b)}{au^2}\frac{du}{a} = \frac{1}{a^2}\int\left(\frac{1}{u}-\frac{b}{u^2}\right)du = \frac{1}{a^2}\left[\ln|u|+\frac{b}{u}\right]+C = \frac{1}{a^2}\left[\ln|ax+b|+\frac{b}{ax+b}\right]+C$

43. $\int x^n(\ln ax)^m\,dx = \int (\ln ax)^m\,d\left(\frac{x^{n+1}}{n+1}\right) = \frac{x^{n+1}(\ln ax)^m}{n+1}-\int\left(\frac{x^{n+1}}{n+1}\right)m(\ln ax)^{m-1}\left(\frac{1}{x}\right)dx$

$= \frac{x^{n+1}(\ln ax)^m}{n+1}-\frac{m}{n+1}\int x^n(\ln ax)^{m-1}\,dx$

(We used integration by parts $\int u\,dv = uv-\int v\,du$ with $u=(\ln ax)^m$, $v=\frac{x^{n+1}}{n+1}$)

45. (a) The volume of the filled part equals the length of the tank times the area of the shaded region shown in the accompanying figure. Consider a layer of gasoline of thickness dy located at height y where $-r < y < -r + d$. The width of this layer is $2\sqrt{r^2 - y^2}$. Therefore, $A = 2\int_{-r}^{-r+d} \sqrt{r^2 - y^2}\, dy$

and $V = L \cdot A = 2L \int_{-r}^{-r+d} \sqrt{r^2 - y^2}\, dy$

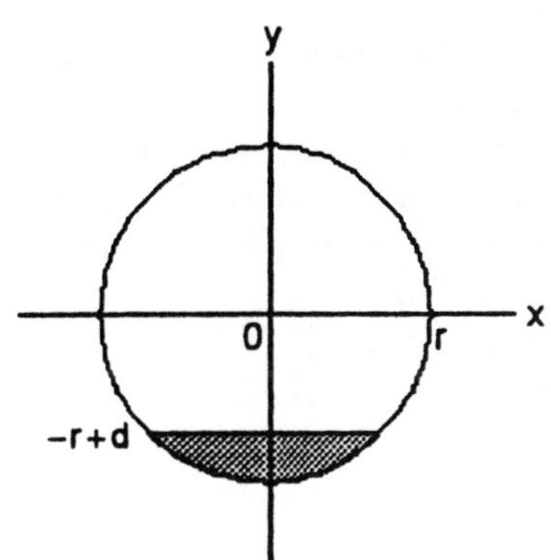

(b) $2L \int_{-r}^{-r+d} \sqrt{r^2 - y^2}\, dy = 2L\left[\frac{y\sqrt{r^2 - y^2}}{2} + \frac{r^2}{2}\sin^{-1}\frac{y}{r}\right]_{-r}^{-r+d}$

(We used FORMULA 29 with $a = r$)

$$= 2L\left[\frac{(d-r)}{2}\sqrt{2rd - d^2} + \frac{r^2}{2}\sin^{-1}\left(\frac{d-r}{r}\right) + \frac{r^2}{2}\left(\frac{\pi}{2}\right)\right] = 2L\left[\left(\frac{d-r}{2}\right)\sqrt{2rd - d^2} + \left(\frac{r^2}{2}\right)\left(\sin^{-1}\left(\frac{d-r}{r}\right) + \frac{\pi}{2}\right)\right]$$

CAS EXPLORATIONS

For MAPLE use the int(f(x),x) command, and for MATHEMATICA use the command Integrate[f(x),x], as discussed in the text.

47. (e) $\int x^n \ln x\, dx = \frac{x^{n+1}\ln x}{n+1} - \frac{1}{n+1}\int x^n\, dx,\ n \neq -1$

(We used FORMULA 110 with $a = 1$, $m = 1$)

$$= \frac{x^{n+1}\ln x}{n+1} - \frac{x^{n+1}}{(n+1)^2} + C = \frac{x^{n+1}}{n+1}\left(\ln x - \frac{1}{n+1}\right) + C$$

49. (a) Neither MAPLE nor MATHEMATICA can find this integral for arbitrary n.

(b) MAPLE and MATHEMATICA get stuck at about $n = 5$.

(c) Let $x = \frac{\pi}{2} - u \Rightarrow dx = -du$; $x = 0 \Rightarrow u = \frac{\pi}{2}$, $x = \frac{\pi}{2} \Rightarrow u = 0$;

$$I = \int_0^{\pi/2} \frac{\sin^n x\, dx}{\sin^n x + \cos^n x} = \int_{\pi/2}^{0} \frac{-\sin^n\left(\frac{\pi}{2} - u\right) du}{\sin^n\left(\frac{\pi}{2} - u\right) + \cos^n\left(\frac{\pi}{2} - u\right)} = \int_0^{\pi/2} \frac{\cos^n u\, du}{\cos^n u + \sin^n u} = \int_0^{\pi/2} \frac{\cos^n x\, dx}{\cos^n x + \sin^n x}$$

$$\Rightarrow I + I = \int_0^{\pi/2} \left(\frac{\sin^n x + \cos^n x}{\sin^n x + \cos^n x}\right) dx = \int_0^{\pi/2} dx = \frac{\pi}{2} \Rightarrow I = \frac{\pi}{4}$$

The following *Mathematica* module is used to obtain the Monte Carlo estimates of area in Problems 51 through 55.

```
monte[f_, indvar_, m_, a_, b_, n_List] :=
 Module[{g, x, xr, yr, area, lim, areaavg, y1, y2},
  g = f/. indvar -> x;
  lim = Length[n];
  area - Table[0, {k, 1, lim}];
```

```
For[k = 1, k <= lim, k++,
  For[counter = 0; i = 1, i <= n[[k]], i++,
   xr = a+ (b - a)*Random[];
   yr = m*Random[];
  If[yr <= g/. x -> xr, counter = counter + 1];];
area[[k]] = m*(b - a)*counter/n[[k]]];
areaavg = (Sum[n[[i]]*area[[i]], {i, 1, lim}]) /
  Sum[n[i]], {i, 1, lim}];
y1 = Integrate[g, {x, a, b}] // N;
y2 = Integrate[g, {x, a, b}];
Print[area];
Print[areaavg];
Print[y2];
Print y1  ;
```

The following command executes the preceding module. The arguments are the integrand function, the independent variable, an upper bound on the integrand function, the lower limit of integration, the upper limit of integration, and a list of the numbers of random points to generate in each estimation.

```
monte[z*Sqrt[1 - z], z, 0.5, 0, 1, {100, 200, 300, 400,
  500, 600, 700, 800, 900, 1000, 2000, 3000, 4000,
  5000, 6000, 8000, 10000, 15000, 20000, 30000
```

The preceding command is for Problem 51.

51.

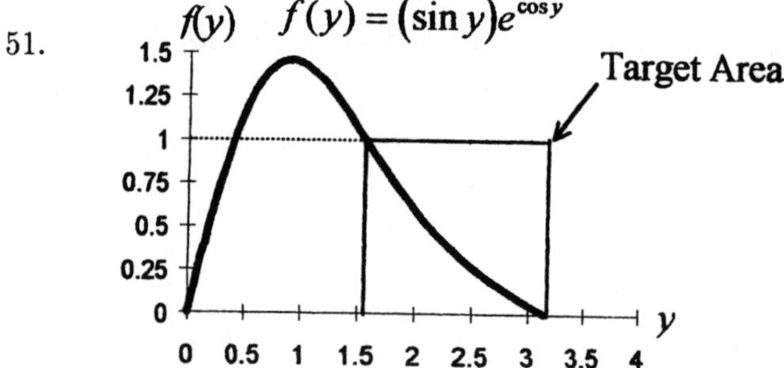

Select M = 1

The area approximations will vary depending on the random number generator and seed value that is used.

Number of Points	Approximation of Area	Number of Points	Approximation of Area
100	0.722566	2000	0.628319
200	0.628319	3000	0.646121
300	0.586431	4000	0.642456
400	0.581195	5000	0.636487
500	0.637743	6000	0.627533
600	0.560251	8000	0.643437
700	0.583439	10,000	0.62235
800	0.577268	15,000	0.625386
900	0.5621337	20,000	0.635073
1000	0.655022	30,000	0.638895

A weighted average of the areas in the table is used to estimate the integral. Therefore,

$$\int_{\pi/2}^{\pi} (\sin y)e^{\cos y}\,dy \approx \left(\sum_{i=1}^{20} n_i \cdot \text{area}(i)\right) \Big/ \left(\sum_{i=1}^{20} n(i)\right) = 0.63298 \text{ by Monte Carlo.}$$

The actual value of the integral is $1 - \frac{1}{e} \approx 0.632121$.

53.

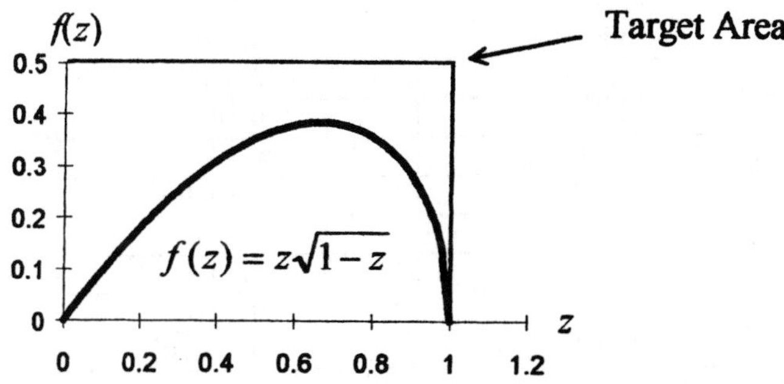

Select M = 0.5
The area approximations will vary depending on the random number generator and seed value that is used.

Number of Points	Approximation of Area	Number of Points	Approximation of Area
100	0.28	2000	0.259
200	0.265	3000	0.262167
300	0.278333	4000	0.259625
400	0.2625	5000	0.2724
500	0.261	6000	0.270583
600	0.27	8000	0.265875
700	0.254286	10,000	0.26495
800	0.270625	15,000	0.2668
900	0.277778	20,000	0.268275
1000	0.2685	30,000	0.265875

A weighted average of the areas in the table is used to estimate the integral. Therefore,

$$\int_0^1 z\sqrt{1-z}\,dz \approx \left(\sum_{i=1}^{20} n_i \cdot \text{area}(i)\right) \Big/ \left(\sum_{i=1}^{20} n(i)\right) = 0.266465 \text{ by Monte Carlo.}$$

The actual value of the integral is $\frac{4}{15} \approx 0.266667$.

55.

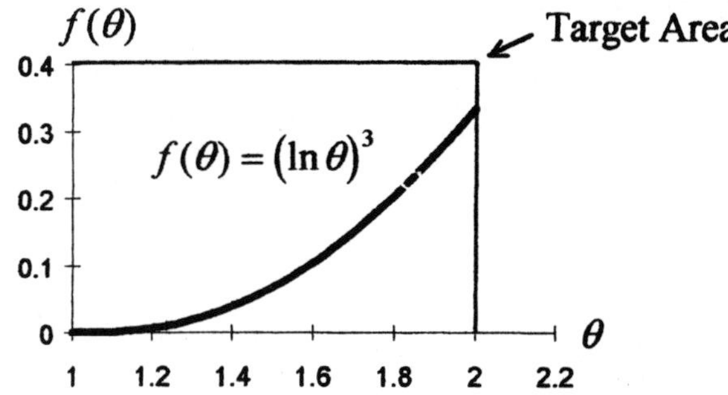

Select M = 0.4
The area approximations will vary depending on the random number generator and seed value that is used.

Number of Points	Approximation of Area	Number of Points	Approximation of Area
100	0.096	2000	0.095
200	0.104	3000	0.103467
300	0.0986667	4000	0.0999
400	0.095	5000	0.10096
500	0.0992	6000	0.1048
600	0.096	8000	0.10105
700	0.0908571	10,000	0.104
800	0.0985	15,000	0.0995733
900	0.1	20,000	0.1013
1000	0.104	30,000	0.100707

A weighted average of the areas in the table is used to estimate the integral. Therefore,

$$\int_1^2 (\ln \theta)^3 \, d\theta \approx \left(\sum_{i=1}^{20} n_i \cdot \text{area}(i) \right) \Big/ \left(\sum_{i=1}^{20} n(i) \right) = 0.101054 \text{ by Monte Carlo.}$$

The actual value of the integral is $6 + 2\left[(\ln 2)^3 - 3(\ln 2)^2 + 6 \ln 2 - 6\right] \approx 0.101097$.

7.6 L'HÔPITAL'S RULE

1. l'Hôpital: $\lim\limits_{x\to 2} \frac{x-2}{x^2-4} = \frac{1}{2x}\Big|_{x=2} = \frac{1}{4}$ or $\lim\limits_{x\to 2} \frac{x-2}{x^2-4} = \lim\limits_{x\to 2} \frac{x-2}{(x-2)(x+2)} = \lim\limits_{x\to 2} \frac{1}{x+2} = \frac{1}{4}$

3. l'Hôpital: $\lim\limits_{x\to\infty} \frac{5x^3-3x}{7x^2+1} = \lim\limits_{x\to\infty} \frac{10x-3}{14x} = \lim\limits_{x\to\infty} \frac{10}{14} = \frac{5}{7}$ or $\lim\limits_{x\to\infty} \frac{5x^2-3x}{7x^2+1} = \lim\limits_{x\to\infty} \frac{5-\frac{3}{x}}{7+\frac{1}{x}} = \frac{5}{7}$

5. l'Hôpital: $\lim\limits_{x\to 0} \frac{1-\cos x}{x^2} = \lim\limits_{x\to 0} \frac{\sin x}{2x} = \lim\limits_{x\to 0} \frac{\cos x}{2} = \frac{1}{2}$ or $\lim\limits_{x\to 0} \frac{1-\cos x}{x^2} = \lim\limits_{x\to 0} \left[\frac{1-\cos x}{x^2}\left(\frac{1+\cos x}{1+\cos x}\right)\right]$

$$= \lim_{x\to 0} \frac{\sin^2 x}{x^2(1+\cos x)} = \lim_{x\to 0} \left[\left(\frac{\sin x}{x}\right)\left(\frac{\sin x}{x}\right)\left(\frac{1}{1+\cos x}\right)\right] = \frac{1}{2}$$

7. $\lim\limits_{\theta\to 0} \frac{\sin \theta^2}{\theta} = \lim\limits_{\theta\to 0} \frac{2\theta \cos \theta^2}{1} = (2)(0) \cos (0)^2 = 0$

9. $\lim\limits_{t\to 0} \frac{\cos t - 1}{e^t - t - 1} = \lim\limits_{t\to 0} \frac{-\sin t}{e^t - 1} = \lim\limits_{t\to 0} \frac{-\cos t}{e^t} = -1$

11. $\lim\limits_{x\to\infty} \frac{\ln(x+1)}{\log_2 t} = \lim\limits_{x\to\infty} \frac{\frac{1}{x+1}}{\frac{1}{x \ln 2}} = \lim\limits_{x\to\infty} \frac{x \ln 2}{x+1} = \lim\limits_{x\to\infty} \ln 2 = \ln 2$

13. $\lim\limits_{y\to 0^+} \frac{\ln(y^2+2y)}{\ln y} = \lim\limits_{y\to 0^+} \frac{\frac{2y+2}{y^2+2y}}{\frac{1}{y}} = \lim\limits_{y\to 0^+} \frac{y(2y+2)}{y^2+2y} = \lim\limits_{y\to 0^+} \frac{2y^2+2y}{y^2+2y} = \lim\limits_{y\to 0^+} \frac{4y+2}{2y+2} = \frac{4(0)+2}{2(0)+2} = \frac{2}{2} = 1$

15. $\lim\limits_{x\to 0^+} x \ln x = \lim\limits_{x\to 0^+} \frac{\ln x}{\frac{1}{x}} = \lim\limits_{x\to 0^+} \frac{\frac{1}{x}}{-\frac{1}{x^2}} = \lim\limits_{x\to 0^+} \frac{-x^2}{x} = \lim\limits_{x\to 0^+} -x = 0$

17. $\lim\limits_{x\to 0^+} (\csc x - \cot x + \cos x) = \lim\limits_{x\to 0^+} \left(\frac{1}{\sin x} - \frac{\cos x}{\sin x} + \cos x\right) = \lim\limits_{x\to 0^+} \frac{1 - \cos x + \cos x \sin x}{\sin x}$

$= \lim\limits_{x\to 0^+} \frac{\sin x + \cos x \cos x - \sin x \sin x}{\cos x} = 1$

19. $\lim\limits_{x\to 0^+} (\ln x - \ln \sin x) = \lim\limits_{x\to 0^+} \ln \frac{x}{\sin x}$; let $f(x) = \frac{x}{\sin x} \Rightarrow \lim\limits_{x\to 0^+} \frac{x}{\sin x} = \lim\limits_{x\to 0^+} \frac{1}{\cos x} = 1$. Therefore,

$\lim\limits_{x\to 0^+} (\ln x - \ln \sin x) = \lim\limits_{x\to 0^+} \ln f(x) = \ln 1 = 0$

21. The limit leads to the indeterminate form 1^∞. Let $f(x) = (e^x + x)^{1/x} \Rightarrow \ln (e^x + x)^{1/x} = \frac{\ln (e^x + x)}{x}$

$\Rightarrow \lim\limits_{x\to 0} \frac{\ln (e^x + x)}{x} = \lim\limits_{x\to 0} \frac{\frac{e^x + 1}{e^x + x}}{1} = 2 \Rightarrow \lim\limits_{x\to 0} (e^x + x)^{1/x} \lim\limits_{x\to 0} e^{\ln f(x)} = e^2$

23. $\lim\limits_{x\to \pm\infty} \frac{3x - 5}{2x^2 - x + 2} = \lim\limits_{x\to \pm\infty} \frac{3}{4x - 1} = 0$

25. The limit leads to the indeterminate form ∞^0. Let $f(x) = (\ln x)^{1/x} \Rightarrow \ln (\ln x)^{1/x} = \frac{\ln (\ln x)}{x}$

$\Rightarrow \lim\limits_{x\to \infty} \frac{\ln (\ln x)}{x} = \lim\limits_{x\to \infty} \frac{\frac{1/x}{\ln x}}{1} = \lim\limits_{x\to \infty} \frac{1}{x \ln x} = 0 \Rightarrow \lim\limits_{x\to \infty} (\ln x)^{1/x} = \lim\limits_{x\to \infty} e^{\ln f(x)} = e^0 = 1$

27. The limit leads to the indeterminate form 0^0. Let $f(x) = (x^2 - 2x + 1)^{x-1}$

$\Rightarrow \ln (x^2 - 2x + 1)^{x-1} = (x - 1) \ln (x^2 - 2x + 1) = \frac{\ln (x^2 - 2x + 1)}{\frac{1}{x - 1}} \Rightarrow \lim\limits_{x\to 1} \frac{\ln (x^2 - 2x + 1)}{\frac{1}{x - 1}} = \lim\limits_{x\to 1} \frac{\frac{2x - 2}{x^2 - 2x + 1}}{-\frac{1}{(x - 1)^2}}$

$= \lim\limits_{x\to 1} \frac{\frac{2(x - 1)}{(x - 1)^2}}{-\frac{1}{(x - 1)^2}} = \lim\limits_{x\to 1} -2(x - 1) = 0 \Rightarrow \lim\limits_{x\to 1} (x^2 - 2x + 1)^{x-1} = \lim\limits_{x\to 1} e^{\ln f(x)} = e^0 = 1$

29. The limit leads to the indeterminate form 1^∞. Let $f(x) = (1 + x)^{1/x} \Rightarrow \ln (1 + x)^{1/x} = \frac{\ln (1 + x)}{x}$

$\Rightarrow \lim\limits_{x\to 0^+} \frac{\ln (1 + x)}{x} = \lim\limits_{x\to 0^+} \frac{\frac{1}{1 + x}}{1} = 1 \Rightarrow \lim\limits_{x\to 0^+} (1 + x)^{1/x} = \lim\limits_{x\to 0^+} e^{\ln f(x)} = e^1 = e$

31. The limit leads to the indeterminate form 0^0. Let $f(x) = (\sin x)^x \Rightarrow \ln (\sin x)^x = x \ln (\sin x) = \frac{\ln (\sin x)}{\frac{1}{x}}$

$\Rightarrow \lim\limits_{x\to 0^+} \frac{\ln (\sin x)}{\frac{1}{x}} = \lim\limits_{x\to 0^+} \frac{\frac{\cos x}{\sin x}}{-\frac{1}{x^2}} = \lim\limits_{x\to 0^+} \frac{-x^2 \cos x}{\sin x} = \lim\limits_{x\to 0^+} \frac{x^2 \sin x - 2x \cos x}{\cos x} = 0$

$\Rightarrow \lim\limits_{x\to 0^+} (\sin x)^x = \lim\limits_{x\to 0^+} e^{\ln f(x)} = e^0 = 1$

33. The limit leads to the indeterminate form $1^{-\infty}$. Let $f(x) = x^{1/(1-x)} \Rightarrow \ln x^{1/(1-x)} = \frac{\ln x}{1-x}$

$\Rightarrow \lim\limits_{x\to 1^+} \frac{\ln x}{1-x} = \lim\limits_{x\to 1^+} \frac{\frac{1}{x}}{-1} = -1 \Rightarrow \lim\limits_{x\to 1^+} x^{1/(1-x)} = \lim\limits_{x\to 1^+} e^{\ln f(x)} = e^{-1} = \frac{1}{e}$

35. $\lim\limits_{x\to\infty} \int_x^{2x} \frac{1}{t}\,dt = \lim\limits_{x\to\infty} \left[\ln|t|\right]_x^{2x} = \lim\limits_{x\to\infty} \ln\left(\frac{2x}{x}\right) = \ln 2$

37. $\lim\limits_{\theta\to 0} \frac{\cos\theta - 1}{e^\theta - \theta - 1} = \lim\limits_{\theta\to 0} \frac{-\sin\theta}{e^\theta - 1} = \lim\limits_{\theta\to 0} \frac{-\cos\theta}{e^\theta} = -1$

39. $\lim\limits_{x\to\infty} \frac{\sqrt{9x+1}}{\sqrt{x+1}} = \sqrt{\lim\limits_{x\to\infty} \frac{9x+1}{x+1}} = \sqrt{\lim\limits_{x\to\infty} \frac{9}{1}} = \sqrt{9} = 3$

41. $\lim\limits_{x\to\pi/2^-} \frac{\sec x}{\tan x} = \lim\limits_{x\to\pi/2^-} \left(\frac{1}{\cos x}\right)\left(\frac{\cos x}{\sin x}\right) = \lim\limits_{x\to\pi/2^-} \frac{1}{\sin x} = 1$

43. Part (b) is correct because part (a) is neither in the $\frac{0}{0}$ nor $\frac{\infty}{\infty}$ form and so l'Hôpital's rule may not be used.

45. If f(x) is to be continuous at $x = 0$, then $\lim\limits_{x\to 0} f(x) = f(0) \Rightarrow c = f(0) = \lim\limits_{x\to 0} \frac{9x - 3\sin 3x}{5x^3} = \lim\limits_{x\to 0} \frac{9 - 9\cos 3x}{15x^2}$

$= \lim\limits_{x\to 0} \frac{27\sin 3x}{30x} = \lim\limits_{x\to 0} \frac{81\cos 3x}{30} = \frac{27}{10}$.

47. (a) The limit leads to the indeterminate form 1^∞. Let $f(k) = \left(1 + \frac{r}{k}\right)^{kt} \Rightarrow \ln f(k) = kt\ln\left(1 + \frac{r}{k}\right) = \frac{t\ln\left(1 + \frac{r}{k}\right)}{\frac{1}{k}}$

$\Rightarrow \lim\limits_{k\to\infty} \frac{t\ln\left(1+\frac{r}{k}\right)}{\frac{1}{k}} = \lim\limits_{k\to\infty} \frac{t\left(-\frac{r}{k^2}\right)\left(1+\frac{r}{k}\right)^{-1}}{-\frac{1}{k^2}} = \lim\limits_{k\to\infty} \frac{rt}{1+\frac{r}{k}} = \frac{rt}{1} = rt$

$\Rightarrow \lim\limits_{k\to\infty} A_0\left(1+\frac{r}{k}\right)^{kt} = A_0 \lim\limits_{k\to\infty} \left(1+\frac{r}{k}\right)^{kt} = A_0 \lim\limits_{k\to\infty} e^{\ln f(k)} = A_0 e^{rt}$

(b) Part (a) shows that as the number of compoundings per year increases toward infinity, the limit of interest compounded k times per year is interest compounded continuously.

49. (a) The graph indicates a limit near -0.225. The limit leads to the indeterminate form $\frac{0}{0}$: $\lim\limits_{x\to 1} \frac{(x-1)^2}{x\ln x - x - \cos(\pi x)}$

$= \lim\limits_{x\to 1} \frac{2(x-1)}{\ln x + 1 - 1 + \pi\sin(\pi x)} = \lim\limits_{x\to 1} \frac{2}{\frac{1}{x} + \pi^2\cos(\pi x)}$

$= \frac{2}{1 + \pi^2(-1)} = \frac{2}{1-\pi^2}$

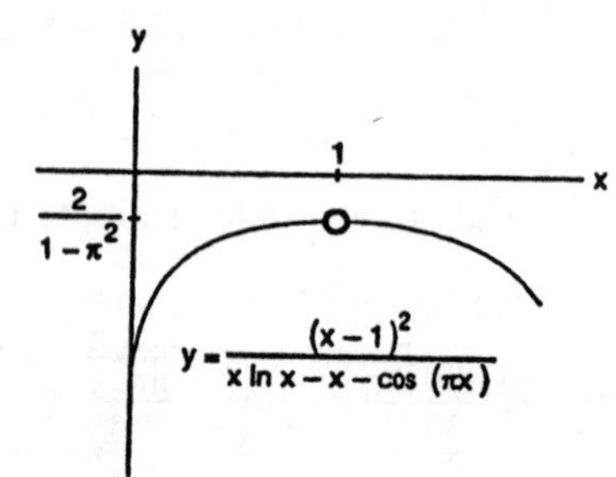

(b) The graph of $y = \dfrac{(x-1)^2}{x \ln x - x - \cos(\pi x)}$ has a vertical asymptote near $x = 2.552$.

51. (a) Because the difference in the numerator is so small compared to the values being subtracted, any calculator or computer with limited precision will give the incorrect result that $1 - \cos x^6$ is 0 for even moderately small values of x. For example, at $x = 0.1$, $\cos x^6 \approx 0.9999999999995$ (13 places), so on a 10-place calculator, $\cos x^6 = 1$ and $1 - \cos x^6 = 0$.

(b) Same reason as in part (a) applies.

(c) $\lim\limits_{x\to 0} \dfrac{1-\cos x^6}{x^{12}} = \lim\limits_{x\to 0} \dfrac{6x^5 \sin x^6}{12x^{11}} = \lim\limits_{x\to 0} \dfrac{\sin x^6}{2x^6} = \lim\limits_{x\to 0} \dfrac{6x^5 \cos x^6}{12x^5} = \lim\limits_{x\to 0} \dfrac{\cos x^6}{2} = \dfrac{1}{2}$

(d) The graph and/or table on a grapher shows the value of the function to be 0 for x-values moderately close to 0, but the limit is 1/2. The calculator is giving unreliable information because there is significant round-off error in computing values of this function on a limited precision device.

53. (a) $f(x) = e^{x \ln(1+1/x)}$

$1 + \frac{1}{x} > 0$ when $x < -1$ or $x > 0$

Domain: $(-\infty, -1) \cup (0, \infty)$

(b) The form is 0^{-1}, so $\lim\limits_{x\to -1} f(x) = \infty$

(c) $\lim\limits_{x\to -\infty} x \ln\left(1+\frac{1}{x}\right) = \lim\limits_{x\to -\infty} \dfrac{\ln\left(1+\frac{1}{x}\right)}{\frac{1}{x}} = \lim\limits_{x\to -\infty} \dfrac{\left(-\frac{1}{x^2}\right)\left(1+\frac{1}{x}\right)^{-1}}{-\frac{1}{x^2}} = \lim\limits_{x\to -\infty} \dfrac{1}{1+\frac{1}{x}} = 1$

$\Rightarrow \lim\limits_{x\to -\infty} f(x) = \lim\limits_{x\to -\infty} e^{x \ln(1+1/x)} = e$

55. (a)

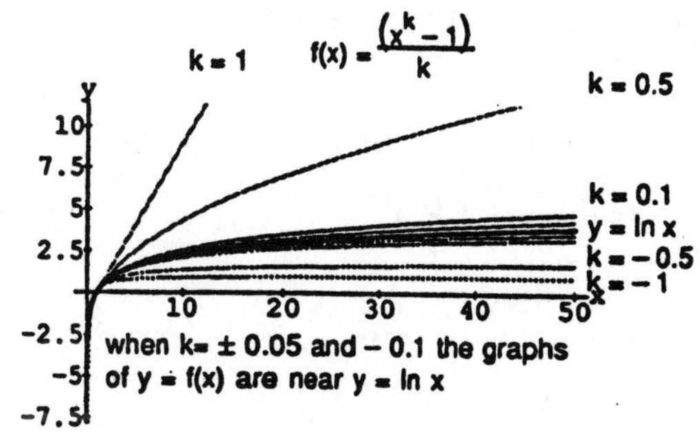

(b) $\lim\limits_{k\to 0} \frac{x^k - 1}{k} = \lim\limits_{k\to 0} \frac{x^k \ln x}{1} = \ln x$

7.7 IMPROPER INTEGRALS

1. (a) The integral is improper because of an infinite limit of integration.

(b) $\int_0^\infty \frac{dx}{x^2+1} = \lim\limits_{b\to\infty} \int_0^b \frac{dx}{x^2+1} = \lim\limits_{b\to\infty} \left[\tan^{-1} x\right]_0^b = \lim\limits_{b\to\infty} (\tan^{-1} b - 0) = \frac{\pi}{2}$

The integral converges.

(c) $\frac{\pi}{2}$

3. (a) The integral involves improper integrals because the integrand has an infinite discontinuity at $x = 0$.

(b) $\int_{-8}^{1} \frac{dx}{x^{1/3}} = \int_{-8}^{0} \frac{dx}{x^{1/3}} + \int_{0}^{1} \frac{dx}{x^{1/3}}$

$\int_{-8}^{0} \frac{dx}{x^{1/3}} = \lim\limits_{b\to 0^-} \int_{-8}^{b} \frac{dx}{x^{1/3}} = \lim\limits_{b\to 0^-} \left[\frac{3}{2}x^{2/3}\right]_{-8}^{b} = \lim\limits_{b\to 0^-} \left(\frac{3}{2}b^{2/3} - 6\right) = -6$

$\int_{0}^{1} \frac{dx}{x^{1/3}} = \lim\limits_{b\to 0^+} \int_{b}^{1} \frac{dx}{x^{1/3}} = \lim\limits_{b\to 0^+} \left[\frac{3}{2}x^{2/3}\right]_{b}^{1} = \lim\limits_{b\to 0^+} \left(\frac{3}{2} - \frac{3}{2}b^{2/3}\right) = \frac{3}{2}$

$\int_{-8}^{1} \frac{dx}{x^{1/3}} = -6 + \frac{3}{2} = -\frac{9}{2}$

The integral converges.

(c) $-\frac{9}{2}$

5. (a) The integral is improper because the integrand has an infinite discontinuity at 0.

(b) $\int_0^{\ln 2} x^{-2} e^{1/x}\,dx = \lim\limits_{b\to 0^+} \int_b^{\ln 2} x^{-2} e^{1/x}\,dx = \lim\limits_{b\to 0^+} \left[-e^{1/x}\right]_b^{\ln 2} = \lim\limits_{b\to 0^+} \left[-e^{1/\ln 2} + e^{1/b}\right] = \infty$

The integral diverges.

(c) No value

7. $\int_1^\infty \frac{dx}{x^{1.001}} = \lim\limits_{b\to\infty} \int_1^b \frac{dx}{x^{1.001}} = \lim\limits_{b\to\infty} \left[-1000x^{-0.001}\right]_1^b = \lim\limits_{b\to\infty} \left(\frac{-1000}{b^{0.001}} + 1000\right) = 1000$

9. $\int_0^4 \frac{dr}{\sqrt{4-r}} = \lim\limits_{b\to 4^-} \left[-2\sqrt{4-r}\right]_0^b = \lim\limits_{b\to 4^-} \left[-2\sqrt{4-b} - (-2\sqrt{4})\right] = 0 + 4 = 4$

11. $\int_0^1 \frac{dx}{\sqrt{1-x^2}} = \lim_{b\to 1^-} \left[\sin^{-1} x\right]_0^b = \lim_{b\to 1^-} \left(\sin^{-1} b - \sin^{-1} 0\right) = \frac{\pi}{2} - 0 = \frac{\pi}{2}$

13. $\int_{-\infty}^{-2} \frac{2\,dx}{x^2+1} = \int_{-\infty}^{-2} \frac{dx}{x-1} - \int_{-\infty}^{-2} \frac{dx}{x+1} = \lim_{b\to -\infty} \left[\ln|x-1|\right]_b^{-2} - \lim_{b\to -\infty} \left[\ln|x+1|\right]_b^{-2} = \lim_{b\to -\infty} \left[\ln\left|\frac{x-1}{x+1}\right|\right]_b^{-2}$

$= \lim_{b\to -\infty} \left(\ln\left|\frac{-3}{-1}\right| - \ln\left|\frac{b-1}{b+1}\right|\right) = \ln 3 - \ln\left(\lim_{b\to -\infty} \frac{b-1}{b+1}\right) = \ln 3 - \ln 1 = \ln 3$

15. $\int_0^1 \frac{\theta+1}{\sqrt{\theta^2+2\theta}}\,d\theta; \left[\begin{array}{c} u = \theta^2 + 2\theta \\ du = 2(\theta+1)\,d\theta \end{array}\right] \Rightarrow \int_0^3 \frac{du}{2\sqrt{u}} = \lim_{b\to 0^+} \int_b^3 \frac{du}{2\sqrt{u}} = \lim_{b\to 0^+} \left[\sqrt{u}\right]_b^3 = \lim_{b\to 0^+} \left(\sqrt{3} - \sqrt{b}\right)$

$= \sqrt{3} - 0 = \sqrt{3}$

17. $\int_0^\infty \frac{dx}{(1+x)\sqrt{x}}; \left[\begin{array}{c} u = \sqrt{x} \\ du = \frac{dx}{2\sqrt{x}} \end{array}\right] \Rightarrow \int_0^\infty \frac{2\,du}{u^2+1} = \lim_{b\to\infty} \int_0^b \frac{2\,du}{u^2+1} = \lim_{b\to\infty} \left[2\tan^{-1} u\right]_0^b$

$= \lim_{b\to\infty} \left(2\tan^{-1} b - 1\tan^{-1} 0\right) = 2\left(\frac{\pi}{2}\right) - 2(0) = \pi$

19. $\int_1^2 \frac{ds}{s\sqrt{s^2-1}} = \lim_{b\to 1^+} \left[\sec^{-1} s\right]_b^2 = \sec^{-1} 2 - \lim_{b\to 1^+} \sec^{-1} b = \frac{\pi}{3} - 0 = \frac{\pi}{3}$

21. $\int_2^\infty \frac{2\,dv}{v^2-v} = \lim_{b\to\infty} \left[2\ln\left|\frac{v-1}{v}\right|\right]_2^b = \lim_{b\to\infty} \left(2\ln\left|\frac{b-1}{b}\right| - 2\ln\left|\frac{2-1}{2}\right|\right) = 2\ln(1) - 2\ln\left(\frac{1}{2}\right) = 0 + 2\ln 2 = \ln 4$

23. $\int_0^2 \frac{ds}{\sqrt{4-s^2}} = \lim_{b\to 2^-} \left[\sin^{-1} \frac{s}{2}\right]_0^b = \lim_{b\to 2^-} \left(\sin^{-1} \frac{b}{2}\right) - \sin^{-1} 0 = \frac{\pi}{2} - 0 = \frac{\pi}{2}$

25. $\int_0^\infty \frac{dv}{(1+v^2)(1+\tan^{-1} v)} = \lim_{b\to\infty} \left[\ln|1+\tan^{-1} v|\right]_0^b = \lim_{b\to\infty} \left[\ln|1+\tan^{-1} b|\right] - \ln|1+\tan^{-1} 0|$

$= \ln\left(1+\frac{\pi}{2}\right) - \ln(1+0) = \ln\left(1+\frac{\pi}{2}\right)$

27. $\int_{-1}^4 \frac{dx}{\sqrt{|x|}} = \lim_{b\to 0^-} \int_{-1}^b \frac{dx}{\sqrt{-x}} + \lim_{c\to 0^+} \int_c^4 \frac{dx}{\sqrt{x}} = \lim_{b\to 0^-} \left[-2\sqrt{-x}\right]_{-1}^b + \lim_{c\to 0^+} \left[2\sqrt{x}\right]_c^4$

$= \lim_{b\to 0^-} \left(-2\sqrt{-b}\right) - \left(-2\sqrt{-(-1)}\right) + 2\sqrt{4} - 6 \lim_{c\to 0^+} 2\sqrt{c} = 0 + 2 + 2\cdot 2 - 0 = 6$

29. $\int_{-\infty}^{0} \theta e^{\theta}\, d\theta = \lim_{b\to-\infty} \left[\theta e^{\theta} - e^{\theta}\right]_b^0 = \left(0 \cdot e^0 - e^0\right) - \lim_{b\to-\infty} \left[be^b - e^b\right] = -1 - \lim_{b\to-\infty} \left(\frac{b-1}{e^{-b}}\right)$

$= -1 - \lim_{b\to-\infty} \left(\frac{1}{-e^{-b}}\right)$ (l'Hôpital's rule for $\frac{\infty}{\infty}$ form)

$= -1 - 0 = -1$

31. $\int_{-\infty}^{0} e^{-|x|}\, dx = \int_{-\infty}^{0} e^x\, dx = \lim_{b\to-\infty} \left[e^x\right]_b^0 = \lim_{b\to-\infty} \left(1 - e^b\right) = (1-0) = 1$

33. $\int_0^1 x \ln x\, dx = \lim_{b\to0^+} \left[\frac{x^2}{2} \ln x - \frac{x^2}{4}\right]_b^1 = \left(\frac{1}{2} \ln 1 - \frac{1}{4}\right) - \lim_{b\to0^+} \left(\frac{b^2}{2} \ln b - \frac{b^2}{4}\right) = -\frac{1}{4} - \lim_{b\to0^+} \frac{\ln b}{\left(\frac{2}{b^2}\right)} + 0$

$= -\frac{1}{4} - \lim_{b\to0^+} \frac{\left(\frac{1}{b}\right)}{\left(-\frac{4}{b^3}\right)} = -\frac{1}{4} + \lim_{b\to0^+} \left(\frac{b^2}{4}\right) = -\frac{1}{4} + 0 = -\frac{1}{4}$

35. $\int_0^{\pi/2} \tan \theta\, d\theta = \lim_{b\to\frac{\pi}{2}^-} \left[-\ln|\cos \theta|\right]_0^b = \lim_{b\to\frac{\pi}{2}^-} \left[-\ln|\cos b|\right] + \ln 1 = \lim_{b\to\frac{\pi}{2}^-} \left[-\ln|\cos b|\right] = -\infty,$

the integral diverges

37. $\int_0^{\pi} \frac{\sin \theta\, d\theta}{\sqrt{\pi-\theta}}$; $[\pi - \theta = x] \Rightarrow -\int_{\pi}^{0} \frac{\sin x\, dx}{\sqrt{x}} = \int_0^{\pi} \frac{\sin x\, dx}{\sqrt{x}}$. Since $0 \le \frac{\sin x}{\sqrt{x}} \le \frac{1}{\sqrt{x}}$ for all $0 \le x \le \pi$ and $\int_0^{\pi} \frac{dx}{\sqrt{x}}$

converges, then $\int_0^{\pi} \frac{\sin x}{\sqrt{x}}\, dx$ converges by the Direct Comparison Test.

39. $\int_0^{\ln 2} x^{-2} e^{-1/x}\, dx$; $\left[\frac{1}{x} = y\right] \Rightarrow \int_{\infty}^{1/\ln 2} \frac{y^2 e^{-y}\, dy}{-y^2} = \int_{1/\ln 2}^{\infty} e^{-y}\, dy = \lim_{b\to\infty} \left[-e^{-y}\right]_{1/\ln 2}^{b} = \lim_{b\to\infty} \left[-e^{-b}\right] - \left[-e^{-1/\ln 2}\right]$

$= 0 + e^{-1/\ln 2} = e^{-1/\ln 2}$, so the integral converges.

41. $\int_0^{\pi} \frac{dt}{\sqrt{t} + \sin t}$. Since for $0 \le t \le \pi$, $0 \le \frac{1}{\sqrt{t} + \sin t} \le \frac{1}{\sqrt{t}}$ and $\int_0^{\pi} \frac{dt}{\sqrt{t}}$ converges, then the original integral

converges as well by the Direct Comparison Test.

43. $\int_0^2 \frac{dx}{1-x^2} = \int_0^1 \frac{dx}{1-x^2} + \int_1^2 \frac{dx}{1-x^2}$ and $\int_0^1 \frac{dx}{1-x^2} = \lim_{b\to1^-} \left[\frac{1}{2} \ln\left|\frac{1+x}{1-x}\right|\right]_0^b = \lim_{b\to1^-} \left[\frac{1}{2} \ln\left|\frac{1+b}{1-b}\right|\right] - 0 = \infty$, which

diverges $\Rightarrow \int_0^2 \frac{dx}{1-x^2}$ diverges as well.

45. $\int_{-1}^{1} \ln|x|\,dx = \int_{-1}^{0} \ln(-x)\,dx + \int_{0}^{1} \ln x\,dx$; $\int_{0}^{1} \ln x\,dx = \lim_{b\to 0^+} [x\ln x - x]_b^1 = [1\cdot 0 - 1] - \lim_{b\to 0^+} [b\ln b - b]$

$= -1 - 0 = -1$; $\int_{-1}^{0} \ln(-x)\,dx = -1 \Rightarrow \int_{-1}^{1} \ln|x|\,dx = -2$ converges.

47. $\int_{1}^{\infty} \frac{dx}{1+x^3}$; $0 \le \frac{1}{x^3+1} \le \frac{1}{x^3}$ for $1 \le x < \infty$ and $\int_{1}^{\infty} \frac{dx}{x^3}$ converges $\Rightarrow \int_{1}^{\infty} \frac{dx}{1+x^3}$ converges by the Direct Comparison Test.

49. $\int_{2}^{\infty} \frac{dv}{\sqrt{v-1}}$; $\lim_{v\to\infty} \frac{\left(\frac{1}{\sqrt{v-1}}\right)}{\left(\frac{1}{\sqrt{v}}\right)} = \lim_{v\to\infty} \frac{\sqrt{v}}{\sqrt{v-1}} = \lim_{v\to\infty} \frac{1}{\sqrt{1-\frac{1}{v}}} = \frac{1}{\sqrt{1-0}} = 1$ and $\int_{2}^{\infty} \frac{dv}{\sqrt{v}} = \lim_{b\to\infty} \left[2\sqrt{v}\right]_2^b = \infty$,

which diverges $\Rightarrow \int_{2}^{\infty} \frac{dv}{\sqrt{v-1}}$ diverges by the Limit Comparison Test.

51. $\int_{0}^{\infty} \frac{dx}{\sqrt{x^6+1}} = \int_{0}^{1} \frac{dx}{\sqrt{x^6+1}} + \int_{1}^{\infty} \frac{dx}{\sqrt{x^6+1}} < \int_{0}^{1} \frac{dx}{\sqrt{x^6+1}} + \int_{1}^{\infty} \frac{dx}{x^3}$ and $\int_{1}^{\infty} \frac{dx}{x^3} = \lim_{b\to\infty} \left[-\frac{1}{2x^2}\right]_1^b$

$= \lim_{b\to\infty} \left(-\frac{1}{2b^2} + \frac{1}{2}\right) = \frac{1}{2} \Rightarrow \int_{0}^{\infty} \frac{dx}{\sqrt{x^6+1}}$ converges by the Direct Comparison Test.

53. $\int_{1}^{\infty} \frac{\sqrt{x+1}}{x^2}\,dx$; $\lim_{x\to\infty} \frac{\left(\frac{\sqrt{x}}{x^2}\right)}{\left(\frac{\sqrt{x+1}}{x^2}\right)} = \lim_{x\to\infty} \frac{\sqrt{x}}{\sqrt{x+1}} = \lim_{x\to\infty} \frac{1}{\sqrt{1+\frac{1}{x}}} = 1$; $\int_{1}^{\infty} \frac{\sqrt{x}}{x^2}\,dx = \int_{1}^{\infty} \frac{dx}{x^{3/2}}$

$= \lim_{b\to\infty} \left[-2x^{-1/2}\right]_1^b = \lim_{b\to\infty} \left(\frac{-2}{\sqrt{x}} + 2\right) = 2 \Rightarrow \int_{1}^{\infty} \frac{\sqrt{x+1}}{x^2}\,dx$ converges by the Limit Comparison Test.

55. $\int_{\pi}^{\infty} \frac{2+\cos x}{x}\,dx$; $0 < \frac{1}{x} \le \frac{2+\cos x}{x}$ for $x \ge \pi$ and $\int_{\pi}^{\infty} \frac{dx}{x} = \lim_{b\to\infty} [\ln x]_\pi^b = \infty$, which diverges

$\Rightarrow \int_{\pi}^{\infty} \frac{2+\cos x}{x}\,dx$ diverges by the Direct Comparison Test.

57. $\int_{0}^{\infty} \frac{d\theta}{1+e^\theta}$; $0 \le \frac{1}{1+e^\theta} \le \frac{1}{e^\theta}$ for $0 \le \theta < \infty$ and $\int_{0}^{\infty} \frac{d\theta}{e^\theta} = \lim_{b\to 0} \left[-e^{-\theta}\right]_0^b = \lim_{b\to\infty} \left(-e^{-b}+1\right) = 1$

$\Rightarrow \int_0^{\infty} \frac{d\theta}{e^{\theta}}$ converges $\Rightarrow \int_0^{\infty} \frac{d\theta}{1+e^{\theta}}$ converges by the Direct Comparison Test.

59. $\int_1^{\infty} \frac{e^x}{x}\,dx$; $0 < \frac{1}{x} < \frac{e^x}{x}$ for $x > 1$ and $\int_1^{\infty} \frac{dx}{x}$ diverges $\Rightarrow \int_1^{\infty} \frac{e^x\,dx}{x}$ diverges by the Direct Comparison Test.

61. $\int_1^{\infty} \frac{dx}{\sqrt{e^x - x}}$; $\lim\limits_{x\to\infty} \frac{\left(\frac{1}{\sqrt{e^x-x}}\right)}{\left(\frac{1}{\sqrt{e^x}}\right)} = \lim\limits_{x\to\infty} \frac{\sqrt{e^x}}{\sqrt{e^x-x}} = \lim\limits_{x\to\infty} \frac{1}{\sqrt{1-\frac{x}{e^x}}} = \frac{1}{\sqrt{1-0}} = 1$; $\int_1^{\infty} \frac{dx}{\sqrt{e^x}} = \int_1^{\infty} e^{-x/2}\,dx$

$= \lim\limits_{b\to\infty} \left[-2e^{-x/2}\right]_1^b = \lim\limits_{b\to\infty} \left(-2e^{-b/2} + 2e^{-1/2}\right) = \frac{2}{\sqrt{e}} \Rightarrow \int_1^{\infty} e^{-x/2}\,dx$ converges $\Rightarrow \int_1^{\infty} \frac{dx}{\sqrt{e^x - x}}$ converges

by the Limit Comparison Test.

63. $\int_{-\infty}^{\infty} \frac{dx}{\sqrt{x^4+1}} = 2\int_0^{\infty} \frac{dx}{\sqrt{x^4+1}}$; $\lim\limits_{x\to\infty} \frac{x^2}{\sqrt{x^4+1}} = 1$; $\int_0^{\infty} \frac{dx}{\sqrt{x^4+1}} = \int_0^1 \frac{dx}{\sqrt{x^4+1}} + \int_1^{\infty} \frac{dx}{\sqrt{x^4+1}}$

$< \int_0^1 \frac{dx}{\sqrt{x^4+1}} + \int_1^{\infty} \frac{dx}{x^2}$ and $\int_1^{\infty} \frac{dx}{x^2} = \lim\limits_{b\to\infty} \left[-\frac{1}{x}\right]_1^b = \lim\limits_{b\to\infty} \left(-\frac{1}{b}+1\right) = 1 \Rightarrow \int_{-\infty}^{\infty} \frac{dx}{\sqrt{x^4+1}}$ converges by the

Direct Comparison Test.

65. (a) $\int_1^2 \frac{dx}{x(\ln x)^p}$; $[t = \ln x] \Rightarrow \int_0^{\ln 2} \frac{dt}{t^p} = \lim\limits_{b\to 0^+} \left[\frac{1}{-p+1} t^{1-p}\right]_b^{\ln 2} = \lim\limits_{b\to 0^+} \frac{b^{1-p}}{p-1} + \frac{1}{1-p}(\ln 2)^{1-p}$

$\Rightarrow$ the integral converges for $p < 1$ and diverges for $p \geq 1$

(b) $\int_2^{\infty} \frac{dx}{x(\ln x)^p}$; $[t = \ln x] \Rightarrow \int_{\ln 2}^{\infty} \frac{dt}{t^p}$ and this integral is essentially the same as in Exercise 67(a): it converges

for $p > 1$ and diverges for $p \leq 1$

67. $A = \int_0^{\infty} e^{-x}\,dx = \lim\limits_{b\to\infty} \left[-e^{-x}\right]_0^b = \lim\limits_{b\to\infty} \left(-e^{-b}\right) - \left(-e^{-0}\right)$

$= 0 + 1 = 1$

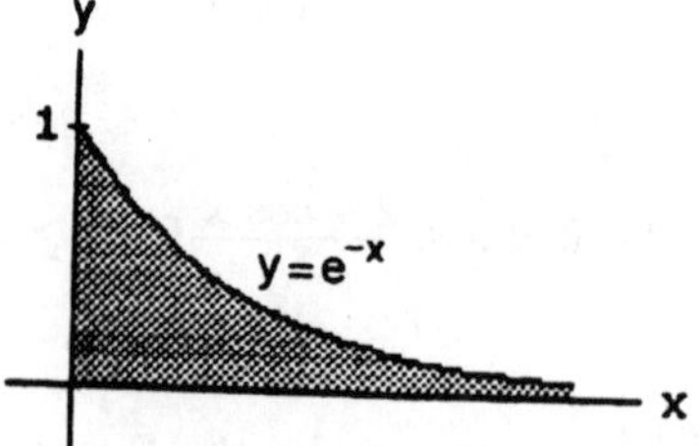

69. $V = \int_0^{\infty} \pi\left(e^{-x}\right)^2 dx = \pi \int_0^{\infty} e^{-2x}\,dx = \pi \lim\limits_{b\to\infty} \left[-\frac{1}{2}e^{-2x}\right]_0^b = \pi \lim\limits_{b\to\infty} \left(-\frac{1}{2}e^{-2b} + \frac{1}{2}\right) = \frac{\pi}{2}$

71. $\int_3^{\infty} \left(\frac{1}{x-2} - \frac{1}{x}\right) dx \neq \int_3^{\infty} \frac{dx}{x-2} - \int_3^{\infty} \frac{dx}{x}$, since the left hand integral converges but both of the right hand

integrals diverge.

73. (a) $\int_1^{\infty} e^{-3x}\,dx = \lim_{b\to\infty}\left[-\frac{1}{3}e^{-3x}\right]_3^b = \lim_{b\to\infty}\left(-\frac{1}{3}e^{-3b}\right)-\left(-\frac{1}{3}e^{-3\cdot 3}\right) = 0+\frac{1}{3}\cdot e^{-9} = \frac{1}{3}e^{-9}$

$\approx 0.0000411 < 0.000042$. Since $e^{-x^2} \le e^{-3x}$ for $x > 3$, then $\int_3^{\infty} e^{-x^2}\,dx < 0.000042$ and therefore

$\int_0^{\infty} e^{-x^2}\,dx$ can be replaced by $\int_0^3 e^{-x^2}\,dx$ without introducing an error greater than 0.000042.

(b) $\int_0^3 e^{-x^2}\,dx \cong 0.88621$

75. (a)

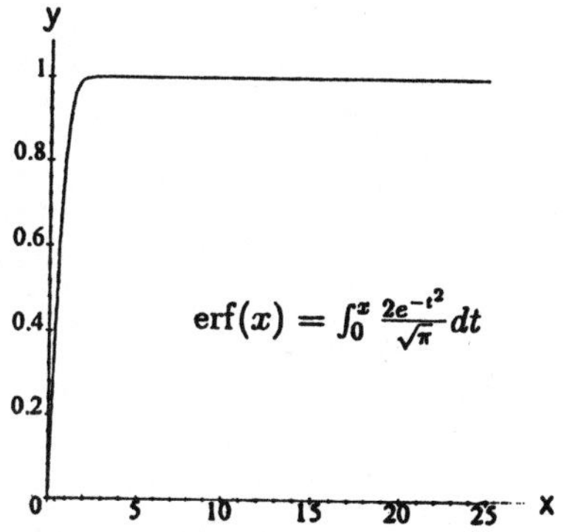

$\mathrm{erf}(x) = \int_0^x \frac{2e^{-t^2}}{\sqrt{\pi}}\,dt$

(b) Maple commands:

```
> f:= 2*exp(-t^2)/sqrt(Pi);
> int(f, t=0..infinity); (answer is 1)
```

77-79. Use the MAPLE or MATHEMATICA integration commands, as discussed in the text.

CHAPTER 7 PRACTICE EXERCISES

1. $\int x\sqrt{4x^2-9}\,dx;\ \begin{bmatrix} u = 4x^2-9 \\ du = 8x\,dx \end{bmatrix} \Rightarrow \frac{1}{8}\int \sqrt{u}\,du = \frac{1}{8}\cdot\frac{2}{3}u^{3/2}+C = \frac{1}{12}\left(4x^2-9\right)^{3/2}+C$

3. $\int \frac{x\,dx}{\sqrt{8x^2+1}};\ \begin{bmatrix} u = 8x^2+1 \\ du = 16x\,dx \end{bmatrix} \Rightarrow \frac{1}{16}\int \frac{du}{\sqrt{u}} = \frac{1}{16}\cdot 2u^{1/2}+C = \frac{\sqrt{8x^2+1}}{8}+C$

5. $\int \frac{t^3\,dt}{\sqrt{9-4t^4}};\ \begin{bmatrix} u = 9-4t^4 \\ du = -16t^3\,dt \end{bmatrix} \Rightarrow -\frac{1}{16}\int \frac{du}{\sqrt{u}} = -\frac{1}{16}\cdot 2u^{1/2}+C = -\frac{\sqrt{9-4t^4}}{8}+C$

7. $\int \frac{\sin 2\theta\,d\theta}{(1-\cos 2\theta)^2};\ \begin{bmatrix} u = 1-\cos 2\theta \\ du = 2\sin 2\theta\,d\theta \end{bmatrix} \Rightarrow \frac{1}{2}\int \frac{du}{u^2} = -\frac{1}{2u}+C = -\frac{1}{2(1-\cos 2\theta)}+C$

9. $\int (\sin 2x)\, e^{\cos 2x}\, dx;\ \begin{bmatrix} u = \cos 2x \\ du = -2 \sin 2x\, dx \end{bmatrix} \Rightarrow -\frac{1}{2}\int e^u\, du = -\frac{1}{2}e^u + C = -\frac{1}{2}e^{\cos 2x} + C$

11. $\int 2^{x-1}\, dx = \frac{2^{x-1}}{\ln 2} + C$

13. $\int \frac{dx}{(x^2+1)(2+\tan^{-1}x)};\ \begin{bmatrix} u = 2 + \tan^{-1} x \\ du = \frac{dx}{x^2+1} \end{bmatrix} \Rightarrow \int \frac{du}{u} = \ln|u| + C = \ln\left|2 + \tan^{-1} x\right| + C$

15. $\int \frac{dt}{\sqrt{16-9t^2}} = \frac{1}{4}\int \frac{dt}{\sqrt{1-\left(\frac{3t}{4}\right)^2}};\ \begin{bmatrix} u = \frac{3}{4}t \\ du = \frac{3}{4}\, dt \end{bmatrix} \Rightarrow \frac{1}{3}\int \frac{du}{\sqrt{1-u^2}} = \frac{1}{3}\sin^{-1} u + C = \frac{1}{3}\sin^{-1}\left(\frac{3t}{4}\right) + C$

17. $\int \frac{4\, dx}{5x\sqrt{25x^2-16}} = \frac{4}{25}\int \frac{dx}{x\sqrt{x^2-\frac{16}{25}}} = \frac{1}{5}\sec^{-1}\left|\frac{5x}{4}\right| + C$

19. $\int \frac{dy}{y^2-4y+8} = \int \frac{d(y-2)}{(y-2)^2+4} = \frac{1}{2}\tan^{-1}\left(\frac{y-2}{2}\right) + C$

21. $\int \cos^2 3x\, dx = \int \frac{1+\cos 6x}{2}\, dx = \frac{x}{2} + \frac{\sin 6x}{12} + C$

23. $\int \tan^3 2t\, dt = \int (\tan 2t)\left(\sec^2 2t - 1\right) dt = \int \tan 2t \sec^2 2t\, dt - \int \tan 2t\, dt;\ \begin{bmatrix} u = 2t \\ du = 2\, dt \end{bmatrix}$

$\Rightarrow \frac{1}{2}\int \tan u \sec^2 u\, du - \frac{1}{2}\int \tan u\, du = \frac{1}{4}\tan^2 u + \frac{1}{2}\ln|\cos u| + C = \frac{1}{4}\tan^2 2t + \frac{1}{2}\ln|\cos 2t| + C$

$= \frac{1}{4}\tan^2 2t - \frac{1}{2}\ln|\sec 2t| + C$

25. $\int \frac{2\, dx}{\cos^2 x - \sin^2 x} = \int \frac{2\, dx}{\cos 2x};\ \begin{bmatrix} u = 2x \\ du = 2\, dx \end{bmatrix} \Rightarrow \int \frac{du}{\cos u} = \int \sec u\, du = \ln|\sec u + \tan u| + C$

$= \ln|\sec 2x + \tan 2x| + C$

27. $\int_{\pi/4}^{3\pi/4} \sqrt{\cot^2 t + 1}\, dt = \int_{\pi/4}^{3\pi/4} \csc t\, dt = \left[-\ln|\csc t + \cot t|\right]_{\pi/4}^{3\pi/4} = -\ln\left|\csc\frac{3\pi}{4} + \cot\frac{3\pi}{4}\right| + \ln\left|\csc\frac{\pi}{4} + \cot\frac{\pi}{4}\right|$

$= -\ln\left|\sqrt{2}-1\right| + \ln\left|\sqrt{2}+1\right| = \ln\left|\frac{\sqrt{2}+1}{\sqrt{2}-1}\right| = \ln\left|\frac{(\sqrt{2}+1)(\sqrt{2}+1)}{2-1}\right| = \ln(3+2\sqrt{2})$

29. $\int_{-\pi/2}^{\pi/2} \sqrt{1-\cos 2t}\, dt = \sqrt{2} \int_{-\pi/2}^{\pi/2} |\sin t|\, dt = 2\sqrt{2} \int_0^{\pi/2} \sin t\, dt = \left[-2\sqrt{2} \cos t\right]_0^{\pi/2} = 2\sqrt{2}\,[0-(-1)] = 2\sqrt{2}$

31. $\int \frac{x^2\, dx}{x^2+4} = x - \int \frac{4\, dx}{x^2+4} = x - 2 \tan^{-1}\left(\frac{x}{2}\right) + C$

33. $\int \frac{2y-1}{y^2+4}\, dy = \int \frac{2y\, dy}{y^2+4} - \int \frac{dy}{y^2+4} = \ln\left(y^2+4\right) - \frac{1}{2} \tan^{-1}\left(\frac{y}{2}\right) + C$

35. $\int \frac{t+2}{\sqrt{4-t^2}}\, dt = \int \frac{t\, dt}{\sqrt{4-t^2}} + 2 \int \frac{dt}{\sqrt{4-t^2}} = -\sqrt{4-t^2} + 2 \sin^{-1}\left(\frac{t}{2}\right) + C$

37. $\int \frac{\tan x\, dx}{\tan x + \sec x} = \int \frac{\sin x\, dx}{\sin x + 1} = \int \frac{(\sin x)(1-\sin x)}{1-\sin^2 x}\, dx = \int \frac{\sin x - 1 + \cos^2 x}{\cos^2 x}\, dx$

$= -\int \frac{d(\cos x)}{\cos^2 x} - \int \frac{dx}{\cos^2 x} + \int dx = \frac{1}{\cos x} - \tan x + x + C = x - \tan x + \sec x + C$

39. $\int \cot\left(\frac{x}{4}\right) dx = 4 \int \cot\left(\frac{x}{4}\right) d\left(\frac{x}{4}\right) = 4 \ln\left|\sin\left(\frac{x}{4}\right)\right| + C$

41. $\int \left(16+z^2\right)^{-3/2} dz; \begin{bmatrix} z = 4 \tan \theta \\ dz = 4 \sec^2 \theta\, d\theta \end{bmatrix} \Rightarrow \int \frac{4 \sec^2 \theta\, d\theta}{64 \sec^3 \theta\, d\theta} = \frac{1}{16} \int \cos \theta\, d\theta = \frac{1}{16} \sin \theta + C = \frac{z}{16\sqrt{16+z^2}} + C$

$= \frac{z}{16\left(16+z^2\right)^{1/2}} + C$

43. $\int \frac{dx}{x^2\sqrt{1-x^2}}; \begin{bmatrix} x = \sin \theta \\ dx = \cos \theta\, d\theta \end{bmatrix} \Rightarrow \int \frac{\cos \theta\, d\theta}{\sin^2 \theta \cos \theta} = \int \csc^2 \theta\, d\theta = -\cot \theta + C = -\frac{\cos \theta}{\sin \theta} + C = \frac{-\sqrt{1-x^2}}{x} + C$

45. $\int \frac{dx}{\sqrt{x^2-9}}; \begin{bmatrix} x = 3 \sec \theta \\ dx = 3 \sec \theta \tan \theta\, d\theta \end{bmatrix} \Rightarrow \int \frac{3 \sec \theta \tan \theta\, d\theta}{\sqrt{9 \sec^2 \theta - 9}} = \int \frac{3 \sec \theta \tan \theta\, d\theta}{3 \tan \theta} = \int \sec \theta\, d\theta$

$= \ln|\sec \theta + \tan \theta| + C_1 = \ln\left|\frac{x}{3} + \sqrt{\left(\frac{x}{3}\right)^2 - 1}\right| + C_1 = \ln\left|\frac{x+\sqrt{x^2-9}}{3}\right| + C_1 = \ln\left|x+\sqrt{x^2-9}\right| + C$

47. $u = \ln(x+1),\ du = \frac{dx}{x+1};\ dv = dx,\ v = x;$

$\int \ln(x+1)\, dx = x \ln(x+1) - \int \frac{x}{x+1}\, dx = x \ln(x+1) - \int dx + \int \frac{dx}{x+1} = x \ln(x+1) - x + \ln(x+1) + C_1$

$= (x+1) \ln(x+1) - x + C_1 = (x+1) \ln(x+1) - (x+1) + C$, where $C = C_1 + 1$

49. $u = \tan^{-1} 3x,\ du = \frac{3\, dx}{1+9x^2};\ dv = dx,\ v = x;$

$\int \tan^{-1} 3x\, dx = x \tan^{-1} 3x - \int \frac{3x\, dx}{1+9x^2}; \begin{bmatrix} y = 1+9x^2 \\ dy = 18x\, dx \end{bmatrix} \Rightarrow x \tan^{-1} 3x - \frac{1}{6} \int \frac{dy}{y}$

$= x \tan^{-1}(3x) - \frac{1}{6} \ln(1+9x^2) + C$

51.

		e^x
$(x+1)^2$	$\xrightarrow{(+)}$	e^x
$2(x+1)$	$\xrightarrow{(-)}$	e^x
2	$\xrightarrow{(+)}$	e^x
0		

$\Rightarrow \int (x+1)^2 e^x\, dx = [(x+1)^2 - 2(x+1) + 2]e^x + C$

53. $u = \cos 2x,\ du = -2 \sin 2x\, dx;\ dv = e^x\, dx,\ v = e^x;$

$I = \int e^x \cos 2x\, dx = e^x \cos 2x + 2 \int e^x \sin 2x\, dx;$

$u = \sin 2x,\ du = 2 \cos 2x\, dx;\ dv = e^x\, dx,\ v = e^x;$

$I = e^x \cos 2x + 2\left[e^x \sin 2x - 2 \int e^x \cos 2x\, dx\right] = e^x \cos 2x + 2e^x \sin 2x - 4I \Rightarrow I = \frac{e^x \cos 2x}{5} + \frac{2e^x \sin 2x}{5} + C$

55. $\int \frac{x\, dx}{x^2 - 3x + 2} = \int \frac{2\, dx}{x-2} - \int \frac{dx}{x-1} = 2 \ln|x-2| - \ln|x-1| + C$

57. $\int \frac{\sin\theta\, d\theta}{\cos^2\theta + \cos\theta - 2}; [\cos\theta = y] \Rightarrow -\int \frac{dy}{y^2+y-2} = -\frac{1}{3}\int \frac{dy}{y-1} + \frac{1}{3}\int \frac{dy}{y+2} = \frac{1}{3} \ln\left|\frac{y+2}{y-1}\right| + C$

$= \frac{1}{3} \ln\left|\frac{\cos\theta + 2}{\cos\theta - 1}\right| + C = -\frac{1}{3} \ln\left|\frac{\cos\theta - 1}{\cos\theta + 2}\right| + C$

59. $\int \frac{(v+3)\, dv}{2v^3 - 8v} = \frac{1}{2} \int \left(-\frac{3}{4v} + \frac{5}{8(v-2)} + \frac{1}{8(v+2)}\right) dv = -\frac{3}{8} \ln|v| + \frac{5}{16} \ln|v-2| + \frac{1}{16} \ln|v+2| + C$

$= \frac{1}{16} \ln\left|\frac{(v-2)^5(v+2)}{v^6}\right| + C$

61. $\int \frac{x^3+x^2}{x^2+x-2}\, dx = \int \left(x + \frac{2x}{x^2+x-2}\right) dx = \int x\, dx + \frac{2}{3}\int \frac{dx}{x-1} + \frac{4}{3}\int \frac{dx}{x+2}$

$= \frac{x^2}{2} + \frac{4}{3} \ln|x+2| + \frac{2}{3} \ln|x-1| + C$

63. $\int \frac{2x^3+x^2-21x+24}{x^2+2x-8}\, dx = \int \left[(2x-3) + \frac{x}{x^2+2x-8}\right] dx = \int (2x-3)\, dx + \frac{1}{3}\int \frac{dx}{x-2} + \frac{2}{3}\int \frac{dx}{x+4}$

$= x^2 - 3x + \frac{2}{3} \ln|x+4| + \frac{1}{3} \ln|x-2| + C$

65. $\int \frac{ds}{e^s-1}; \begin{bmatrix} u=e^s-1 \\ du=e^s\,ds \\ ds=\frac{du}{u+1} \end{bmatrix} \Rightarrow \int \frac{du}{u(u-1)} = \int \frac{du}{u-1} - \int \frac{du}{u} = \ln\left|\frac{u-1}{u}\right| + C = \ln\left|\frac{e^s-1}{e^s}\right| + C = \ln|1-e^{-s}| + C$

67. (a) $\int \frac{y\,dy}{\sqrt{16-y^2}} = -\frac{1}{2}\int \frac{d(16-y^2)}{\sqrt{16-y^2}} = -\sqrt{16-y^2} + C$

(b) $\int \frac{y\,dy}{\sqrt{16-y^2}}; [y = 4\sin x] \Rightarrow 4\int \frac{\sin x \cos x\,dx}{\cos x} = -4\cos x + C = -\frac{4\sqrt{16-y^2}}{4} + C = -\sqrt{16-y^2} + C$

69. (a) $\int \frac{x\,dx}{4-x^2} = -\frac{1}{2}\int \frac{d(4-x^2)}{4-x^2} = -\frac{1}{2}\ln|4-x^2| + C$

(b) $\int \frac{x\,dx}{4-x^2}; [x = 2\sin\theta] \Rightarrow \int \frac{2\sin\theta \cdot 2\cos\theta\,d\theta}{4\cos^2\theta} = \int \tan\theta\,d\theta = -\ln|\cos\theta| + C = -\ln\sqrt{4-x^2} + C$

$= -\frac{1}{2}\ln|4-x^2| + C$

71. $\int \frac{x\,dx}{9-x^2}; \begin{bmatrix} u=9-x^2 \\ du=-2x\,dx \end{bmatrix} \Rightarrow -\frac{1}{2}\int \frac{du}{u} = -\frac{1}{2}\ln|u| + C = \ln\frac{1}{\sqrt{u}} + C = \ln\frac{1}{\sqrt{9-x^2}} + C$

73. $\int \frac{dx}{9-x^2} = \frac{1}{6}\int \frac{dx}{3-x} + \frac{1}{6}\int \frac{dx}{3+x} = -\frac{1}{6}\ln|3-x| + \frac{1}{6}\ln|3+x| + C = \frac{1}{6}\ln\left|\frac{x+3}{x-3}\right| + C$

75. $\int \frac{x\,dx}{1+\sqrt{x}}; \begin{bmatrix} u=\sqrt{x} \\ du=\frac{dx}{2\sqrt{x}} \end{bmatrix} \Rightarrow \int \frac{u^2 \cdot 2u\,du}{1+u} = \int \left(2u^2 - 2u + 2 - \frac{2}{1+u}\right) du = \frac{2}{3}u^3 - u^2 + 2u - 2\ln|1+u| + C$

$= \frac{2x^{3/2}}{3} - x + 2\sqrt{x} - 2\ln(1+\sqrt{x}) + C$

77. $\int \frac{\cos\sqrt{x}}{\sqrt{x}}\,dx; \begin{bmatrix} u=\sqrt{x} \\ du=\frac{dx}{2\sqrt{x}} \end{bmatrix} \Rightarrow \int \frac{\cos u \cdot 2u\,du}{u} = 2\int \cos u\,du = 2\sin u + C = 2\sin\sqrt{x} + C$

79. $\int \frac{du}{\sqrt{1+u^2}}; [u = \tan\theta] \Rightarrow \int \frac{\sec^2\theta\,d\theta}{\sec\theta} = \ln|\sec\theta + \tan\theta| + C = \ln\left|\sqrt{1+u^2} + u\right| + C$

81. $\int \frac{9\,dv}{81-v^4} = \frac{1}{2}\int \frac{dv}{v^2+9} + \frac{1}{12}\int \frac{dv}{3-v} + \frac{1}{12}\int \frac{dv}{3+v} = \frac{1}{12}\ln\left|\frac{3+v}{3-v}\right| + \frac{1}{6}\tan^{-1}\frac{v}{3} + C$

83. $\int \frac{x^3\,dx}{x^2-2x+1} = \int \left(x + 2 + \frac{3x+2}{x^2-2x+1}\right) dx = \int (x+2)\,dx + 3\int \frac{dx}{x-1} + \int \frac{dx}{(x-1)^2}$

$= \frac{x^2}{2} + 2x + 3\ln|x-1| - \frac{1}{x-1} + C$

85. $\int \frac{2 \sin \sqrt{x}\, dx}{\sqrt{x} \sec \sqrt{x}}; \begin{bmatrix} y = \sqrt{x} \\ dy = \frac{dx}{2\sqrt{x}} \end{bmatrix} \Rightarrow \int \frac{2 \sin y \cdot 2y\, dy}{y \sec y} = \int 2 \sin 2y\, dy = -\cos(2y) + C = -\cos(2\sqrt{x}) + C$

87. $\int \frac{d\theta}{\theta^2 - 2\theta + 4} = \int \frac{d\theta}{(\theta - 1)^2 + 3} = \frac{\sqrt{3}}{3} \tan^{-1}\left(\frac{\theta - 1}{\sqrt{3}}\right) + C$

89. $\int \frac{\sin 2\theta\, d\theta}{(1 + \cos 2\theta)^2} = -\frac{1}{2} \int \frac{d(1 + \cos 2\theta)}{(1 + \cos 2\theta)^2} = \frac{1}{2(1 + \cos 2\theta)} + C = \frac{1}{4} \sec^2 \theta + C$

91. $\int \frac{x\, dx}{\sqrt{2 - x}}; \begin{bmatrix} y = 2 - x \\ dy = -dx \end{bmatrix} \Rightarrow -\int \frac{(2 - y)\, dy}{\sqrt{y}} = \frac{2}{3} y^{3/2} - 4y^{1/2} + C = \frac{2}{3}(2 - x)^{3/2} - 4(2 - x)^{1/2} + C$

$= 2\left[\frac{\left(\sqrt{2 - x}\right)^3}{3} - 2\sqrt{2 - x}\right] + C$

93. $\int \ln \sqrt{x - 1}\, dx; \begin{bmatrix} y = \sqrt{x - 1} \\ dy = \frac{dx}{2\sqrt{x - 1}} \end{bmatrix} \Rightarrow \int \ln y \cdot 2y\, dy;\ u = \ln y,\ du = \frac{dy}{y};\ dv = 2y\, dy,\ v = y^2$

$\Rightarrow \int 2y \ln y\, dy = y^2 \ln y - \int y\, dy = y^2 \ln y - \frac{1}{2} y^2 + C = (x - 1) \ln \sqrt{x - 1} - \frac{1}{2}(x - 1) + C_1$

$= \frac{1}{2}[(x - 1) \ln |x - 1| - x] + \left(C_1 + \frac{1}{2}\right) = \frac{1}{2}[x \ln |x - 1| - x - \ln |x - 1|] + C$

95. $\int \frac{z + 1}{z^2(z^2 + 4)}\, dz = \frac{1}{4} \int \left(\frac{1}{z} + \frac{1}{z^2} - \frac{z + 1}{z^2 + 4}\right) dz = \frac{1}{4} \ln |z| - \frac{1}{4z} - \frac{1}{8} \ln(z^2 + 4) - \frac{1}{8} \tan^{-1} \frac{z}{2} + C$

97. $u = \tan^{-1} x,\ du = \frac{dx}{1 + x^2};\ dv = \frac{dx}{x^2},\ v = -\frac{1}{x};$

$\int \frac{\tan^{-1} x\, dx}{x^2} = -\frac{1}{x} \tan^{-1} x + \int \frac{dx}{x(1 + x^2)} = -\frac{1}{x} \tan^{-1} x + \int \frac{dx}{x} - \int \frac{x\, dx}{1 + x^2}$

$= -\frac{1}{x} \tan^{-1} x + \ln |x| - \frac{1}{2} \ln(1 + x^2) + C = -\frac{\tan^{-1} x}{x} + \ln |x| - \ln \sqrt{1 + x^2} + C$

99. $\int \frac{1 - \cos 2x}{1 + \cos 2x}\, dx = \int \tan^2 x\, dx = \int (\sec^2 x - 1)\, dx = \tan x - x + C$

101. $\int \frac{\cos x\, dx}{\sin^3 x - \sin x} = -\int \frac{\cos x\, dx}{(\sin x)(1 - \sin^2 x)} = -\int \frac{\cos x\, dx}{(\sin x)(\cos^2 x)} = -\int \frac{2\, dx}{\sin 2x} = -2 \int \csc 2x\, dx$

$= \ln \left|\csc(2x) + \cot(2x)\right| + C$

103. $\int_1^\infty \frac{\ln y\, dy}{y^3}$; $\begin{bmatrix} x = \ln y \\ dx = \frac{dy}{y} \\ dy = e^x\, dx \end{bmatrix} \Rightarrow \int_0^\infty \frac{x \cdot e^x}{e^{3x}}\, dx = \int_0^\infty xe^{-2x}\, dx = \lim_{b\to\infty} \left[-\frac{x}{2}e^{-2x} - \frac{1}{4}e^{-2x}\right]_0^b$

$= \lim_{b\to\infty} \left(\frac{-b}{2e^{2b}} - \frac{1}{4e^{2b}}\right) - \left(0 - \frac{1}{4}\right) = \frac{1}{4}$

105. $\int \frac{dx}{(2x-1)\sqrt{x^2-x}} = \int \frac{2\, dx}{(2x-1)\sqrt{4x^2-4x}} = \int \frac{2\, dx}{(2x-1)\sqrt{(2x-1)^2-1}}$; $\begin{bmatrix} u = 2x-1 \\ du = 2\, dx \end{bmatrix} \Rightarrow \int \frac{du}{u\sqrt{u^2-1}}$

$= \sec^{-1}|u| + C = \sec^{-1}|2x-1| + C$

107. $\int e^\theta \sqrt{3+4e^\theta}\, d\theta$; $\begin{bmatrix} u = 4e^\theta \\ du = 4e^\theta\, d\theta \end{bmatrix} \Rightarrow \frac{1}{4}\int \sqrt{3+u}\, du = \frac{1}{4}\cdot\frac{2}{3}(3+u)^{3/2} + C = \frac{1}{6}\left(3+4e^\theta\right)^{3/2} + C$

109. $\int (27)^{3\theta+1}\, d\theta = \frac{1}{3}\int (27)^{3\theta+1}\, d(3\theta+1) = \frac{1}{3\ln 27}(27)^{3\theta+1} + C = \frac{1}{3}\left(\frac{27^{3\theta+1}}{\ln 27}\right) + C$

111. $\int \frac{dr}{1+\sqrt{r}}$; $\begin{bmatrix} u = \sqrt{r} \\ du = \frac{dr}{2\sqrt{r}} \end{bmatrix} \Rightarrow \int \frac{2u\, du}{1+u} = \int \left(2 - \frac{2}{1+u}\right) du = 2u - 2\ln|1+u| + C = 2\sqrt{r} - 2\ln\left(1+\sqrt{r}\right) + C$

113. $\int \frac{8\, dm}{m\sqrt{49m^2-4}} = \frac{8}{7}\int \frac{dm}{m\sqrt{m^2-\left(\frac{2}{7}\right)^2}} = 4\sec^{-1}\left(\frac{7m}{2}\right) + C$

115. $\lim_{t\to 0} \frac{t - \ln(1+2t)}{t^2} = \lim_{t\to 0} \frac{1 - \frac{2}{1+2t}}{2t} = \infty$ for $t \to 0^-$ and $-\infty$ for $t \to 0^+$

The limit does not exist.

117. $\lim_{x\to 0} \frac{x \sin x}{1-\cos x} = \lim_{x\to 0} \frac{x\cos x + \sin x}{\sin x} = \lim_{x\to 0} \frac{-x\sin x + \cos x + \cos x}{\cos x} = 2$

119. The limit leads to the indeterminate form ∞^0. $f(x) = x^{1/x} \Rightarrow \ln f(x) = \frac{\ln x}{x} \Rightarrow \lim_{x\to\infty} \frac{\ln x}{x} = \lim_{x\to\infty} \frac{1/x}{1} = 0$

$\Rightarrow \lim_{x\to\infty} x^{1/x} = \lim_{x\to\infty} e^{\ln f(x)} = e^0 = 1$

121. $\lim_{r\to\infty} \frac{\cos r}{\ln r} = 0$ since $|\cos r| \le 1$ and $\ln r \to \infty$ as $r \to \infty$.

123. $\lim_{x\to 1} \left(\frac{1}{x-1} - \frac{1}{\ln x}\right) = \lim_{x\to 1} \left[\frac{\ln x - x + 1}{(x-1)\ln x}\right] = \lim_{x\to 1} \frac{\frac{1}{x} - 1}{\frac{x-1}{x} + \ln x} = \lim_{x\to 1} \frac{1-x}{x-1+x\ln x} = \lim_{x\to 1} \frac{-1}{1+x/x+\ln x} = -\frac{1}{2}$

125. The limit leads to the indeterminate form 0^0. $f(\theta) = (\tan\theta)^\theta \Rightarrow \ln f(\theta) = \theta\ln(\tan\theta) = \frac{\ln(\tan\theta)}{1/\theta}$

$$\Rightarrow \lim_{x\to 0^+} \frac{\ln(\tan\theta)}{1/\theta} = \lim_{x\to 0^+} \frac{\frac{\sec^2\theta}{\tan\theta}}{-\frac{1}{\theta^2}} = \lim_{x\to 0^+} -\frac{\theta^2}{\sin\theta\cos\theta} = \lim_{x\to 0^+} \frac{-2\theta}{-\sin^2\theta + \cos^2\theta} = 0$$

$$\Rightarrow \lim_{x\to 0^+} (\tan\theta)^\theta = \lim_{x\to 0^+} e^{\ln f(\theta)} = e^0 = 1$$

127. $\lim\limits_{x\to\infty} \frac{x^3 - 3x^2 + 1}{2x^2 + x - 3} = \lim\limits_{x\to\infty} \frac{3x^2 - 6x}{4x + 1} = \lim\limits_{x\to\infty} \frac{6x - 6}{4} = \infty$

129. $\int_0^3 \frac{dx}{\sqrt{9 - x^2}} = \lim\limits_{b\to 3^-} \int_0^b \frac{dx}{\sqrt{9 - x^2}} = \lim\limits_{b\to 3^-} \left[\sin^{-1}\left(\frac{x}{3}\right)\right]_0^b = \lim\limits_{b\to 3^-} \sin^{-1}\left(\frac{b}{3}\right) - \sin^{-1}\left(\frac{0}{3}\right) = \frac{\pi}{2} - 0 = \frac{\pi}{2}$

131. $\int_{-1}^1 \frac{dy}{y^{2/3}} = \int_{-1}^0 \frac{dy}{y^{2/3}} + \int_0^1 \frac{dy}{y^{2/3}} = 2\int_0^1 \frac{dy}{y^{2/3}} = 2\cdot 3 \lim\limits_{b\to 0^+} \left[y^{1/3}\right]_b^1 = 6\left(1 - \lim\limits_{b\to 0^+} b^{1/3}\right) = 6$

133. $\int_3^\infty \frac{2\,du}{u^2 - 2u} = \int_3^\infty \frac{du}{u - 2} - \int_3^\infty \frac{du}{u} = \lim\limits_{b\to\infty} \left[\ln\left|\frac{u - 2}{u}\right|\right]_3^b = \lim\limits_{b\to\infty} \left[\ln\left|\frac{b - 2}{b}\right|\right] - \ln\left|\frac{3 - 2}{3}\right| = 0 - \ln\left(\frac{1}{3}\right) = \ln 3$

135. $\int_0^\infty x^2 e^{-x}\,dx = \lim\limits_{b\to\infty} \left[-x^2e^{-x} - 2xe^{-x} - 2e^{-x}\right]_0^b = \lim\limits_{b\to\infty} \left(-b^2e^{-b} - 2be^{-b} - 2e^{-b}\right) - (-2) = 0 + 2 = 2$

137. $\int_{-\infty}^\infty \frac{dx}{4x^2 + 9} = 2\int_0^\infty \frac{dx}{4x^2 + 9} = \frac{1}{2}\int_0^\infty \frac{dx}{x^2 + \frac{9}{4}} = \frac{1}{2} \lim\limits_{b\to\infty} \left[\frac{2}{3}\tan^{-1}\left(\frac{2x}{3}\right)\right]_0^b = \frac{1}{2} \lim\limits_{b\to\infty} \left[\frac{2}{3}\tan^{-1}\left(\frac{2b}{3}\right)\right] - \frac{1}{3}\tan^{-1}(0)$

$= \frac{1}{2}\left(\frac{2}{3}\cdot\frac{\pi}{2}\right) - 0 = \frac{\pi}{6}$

139. $\lim\limits_{\theta\to\infty} \frac{\theta}{\sqrt{\theta^2 + 1}} = 1$ and $\int_6^\infty \frac{d\theta}{\theta}$ diverges $\Rightarrow \int_6^\infty \frac{d\theta}{\sqrt{\theta^2 + 1}}$ diverges

141. $\int_1^\infty \frac{\ln z}{z}\,dz = \int_1^e \frac{\ln z}{z}\,dz + \int_e^\infty \frac{\ln z}{z}\,dz = \left[(\ln z)^2\right]_1^e + \lim\limits_{b\to\infty} \left[(\ln z)^2\right]_e^b = \left(1^2 - 0\right) + \lim\limits_{b\to\infty} \left[(\ln b)^2 - 1\right]$

$= \infty \Rightarrow$ diverges

143. $0 < \frac{e^{-x}}{3 + e^{-2x}} = \frac{1}{3e^x + e^{-x}} < \frac{1}{e^x + e^{-x}}$ and $\int_{-\infty}^\infty \frac{dx}{e^x + e^{-x}} = 2\int_0^\infty \frac{dx}{e^x + e^{-x}} < \int_0^\infty \frac{2\,dx}{e^x}$ converges

$\Rightarrow \int_{-\infty}^\infty \frac{e^{-x}}{3 + e^{-2x}}\,dx$ converges

145. $\frac{1}{y^2-y}\,dy = e^x\,dx \Rightarrow \int \frac{1}{y(y-1)}\,dy = \int e^x\,dx = e^x + C;\ \frac{1}{y(y-1)} = \frac{A}{y} + \frac{B}{y-1} \Rightarrow 1 = A(y-1) + B(y)$
$= (A+B)y - A$

Equating coefficients of like terms gives $A + B = 0$ and $-A = 1$. Solving the system simultaneously yields $A = -1$, $B = 1$.

$\int \frac{1}{y(y-1)}\,dy = \int -\frac{1}{y}\,dy + \int \frac{1}{y-1}\,dy = -\ln|y| + \ln|y-1| + C_2 \Rightarrow -\ln|y| + \ln|y-1| = e^x + C$

Substitute $x = 0$, $y = 2 \Rightarrow -\ln 2 + 0 = 1 + C$ or $C = -1 - \ln 2$.

The solution to the initial value problem is $-\ln|y| + \ln|y-1| = e^x - 1 - \ln 2$.

147. $dy = \frac{dx}{x^2-3x+2};\ x^2 - 3x + 2 = (x-2)(x-1) \Rightarrow \frac{1}{x^2-3x+2} = \frac{A}{x-2} + \frac{B}{x-1} \Rightarrow 1 = A(x-1) + B(x-2)$
$\Rightarrow 1 = (A+B)x - A - 2B$

Equating coefficients of like terms gives $A + B = 0$, $-A - 2B = 1$. Solving the system simultaneously yields $A = 1$, $B = -1$.

$\int dy = \int \frac{dx}{x^2-3x+2} = \int \frac{dx}{x-2} - \int \frac{dx}{x-1} \Rightarrow y = \ln|x-2| - \ln|x-1| + C$

Substitute $x = 3$, $y = 0 \Rightarrow 0 = 0 - \ln 2 + C$ or $C = \ln 2$.

The solution to the initial value problem is $y = \ln|x-2| - \ln|x-1| + \ln 2$.

CHAPTER 7 ADDITIONAL EXERCISES–THEORY, EXAMPLES, APPLICATIONS

1. $u = \left(\sin^{-1} x\right)^2$, $du = \frac{2\sin^{-1} x\,dx}{\sqrt{1-x^2}}$; $dv = dx$, $v = x$;

$\int \left(\sin^{-1} x\right)^2 dx = x\left(\sin^{-1} x\right)^2 - \int \frac{2x\sin^{-1} x\,dx}{\sqrt{1-x^2}}$;

$u = \sin^{-1} x$, $du = \frac{dx}{\sqrt{1-x^2}}$; $dv = -\frac{2x\,dx}{\sqrt{1-x^2}}$, $v = 2\sqrt{1-x^2}$;

$\int \frac{2x\sin^{-1} x\,dx}{\sqrt{1-x^2}} = 2\left(\sin^{-1} x\right)\sqrt{1-x^2} - \int 2\,dx = 2\left(\sin^{-1} x\right)\sqrt{1-x^2} - 2x + C$; therefore

$\int \left(\sin^{-1} x\right)^2 dx = x\left(\sin^{-1} x\right)^2 + 2\left(\sin^{-1} x\right)\sqrt{1-x^2} - 2x + C$

3. $u = \sin^{-1} x$, $du = \frac{dx}{\sqrt{1-x^2}}$; $dv = x\,dx$, $v = \frac{x^2}{2}$;

$\int x\sin^{-1} x\,dx = \frac{x^2}{2}\sin^{-1} x - \int \frac{x^2\,dx}{2\sqrt{1-x^2}};\ \begin{bmatrix} x = \sin\theta \\ dx = \cos\theta\,d\theta \end{bmatrix} \Rightarrow \int x\sin^{-1} x\,dx = \frac{x^2}{2}\sin^{-1} x - \int \frac{\sin^2\theta\cos\theta\,d\theta}{2\cos\theta}$

$= \frac{x^2}{2} \sin^{-1} x - \frac{1}{2} \int \sin^2 \theta \, d\theta = \frac{x^2}{2} \sin^{-1} x - \frac{1}{2}\left(\frac{\theta}{2} - \frac{\sin 2\theta}{4}\right) + C = \frac{x^2}{2} \sin^{-1} x + \frac{\sin \theta \cos \theta - \theta}{4} + C$

$= \frac{x^2}{2} \sin^{-1} x + \frac{x\sqrt{1-x^2} - \sin^{-1} x}{4} + C$

5. $\int \frac{d\theta}{1 - \tan^2 \theta} = \int \frac{\cos^2 \theta}{\cos^2 \theta - \sin^2 \theta} \, d\theta = \int \frac{1 + \cos 2\theta}{2 \cos 2\theta} \, d\theta = \frac{1}{2} \int (\sec 2\theta + 1) \, d\theta = \frac{\ln|\sec 2\theta + \tan 2\theta| + 2\theta}{4} + C$

7. $\int \frac{dt}{t - \sqrt{1-t^2}}; \begin{bmatrix} t = \sin \theta \\ dt = \cos \theta \, d\theta \end{bmatrix} \Rightarrow \int \frac{\cos \theta \, d\theta}{\sin \theta - \cos \theta} = \int \frac{d\theta}{\tan \theta - 1}; \begin{bmatrix} u = \tan \theta \\ du = \sec^2 \theta \, d\theta \\ d\theta = \frac{du}{u^2+1} \end{bmatrix} \Rightarrow \int \frac{du}{(u-1)(u^2+1)}$

$= \frac{1}{2} \int \frac{du}{u-1} - \frac{1}{2} \int \frac{du}{u^2+1} - \frac{1}{2} \int \frac{u \, du}{u^2+1} = \frac{1}{2} \ln\left|\frac{u-1}{\sqrt{u^2+1}}\right| - \frac{1}{2} \tan^{-1} u + C = \frac{1}{2} \ln\left|\frac{\tan \theta - 1}{\sec \theta}\right| - \frac{1}{2}\theta + C$

$= \frac{1}{2} \ln\left(t - \sqrt{1-t^2}\right) - \frac{1}{2} \sin^{-1} t + C$

9. $\int \frac{1}{x^4+4} \, dx = \int \frac{1}{(x^2+2)^2 - 4x^2} \, dx = \int \frac{1}{(x^2+2x+2)(x^2-2x+2)} \, dx$

$= \frac{1}{16} \int \left[\frac{2x+2}{x^2+2x+2} + \frac{2}{(x+1)^2+1} - \frac{2x-2}{x^2-2x+2} + \frac{2}{(x-1)^2+1}\right] dx$

$= \frac{1}{16} \ln\left|\frac{x^2+2x+2}{x^2-2x+2}\right| + \frac{1}{8}\left[\tan^{-1}(x+1) + \tan^{-1}(x-1)\right] + C$

11. $\lim_{b \to 1^-} \int_0^b \frac{1}{\sqrt{1-x^2}} \, dx = \lim_{b \to 1^-} \left[\sin^{-1} x\right]_0^b = \lim_{b \to 1^-} \left(\sin^{-1} b - \sin^{-1} 0\right) = \lim_{b \to 1^-} \left(\sin^{-1} b - 0\right) = \lim_{b \to 1^-} \sin^{-1} b = \frac{\pi}{2}$

13. $y = \left(\cos \sqrt{x}\right)^{1/x} \Rightarrow \ln y = \frac{1}{x} \ln\left(\cos \sqrt{x}\right)$ and $\lim_{x \to 0^+} \frac{\ln\left(\cos \sqrt{x}\right)}{x} = \lim_{x \to 0^+} \frac{-\sin \sqrt{x}}{2\sqrt{x} \cos \sqrt{x}} = \frac{-1}{2} \lim_{x \to 0^+} \frac{\tan \sqrt{x}}{\sqrt{x}}$

$= -\frac{1}{2} \lim_{x \to 0^+} \frac{\frac{1}{2} x^{-1/2} \sec^2 \sqrt{x}}{\frac{1}{2} x^{-1/2}} = -\frac{1}{2} \Rightarrow \lim_{x \to 0^+} \left(\cos \sqrt{x}\right)^{1/x} = e^{-1/2} = \frac{1}{\sqrt{e}}$

15. $\lim_{x \to \infty} \int_{-x}^{x} \sin t \, dt = \lim_{x \to \infty} \left[-\cos t\right]_{-x}^{x} = \lim_{x \to \infty} \left[-\cos x + \cos(-x)\right] = \lim_{x \to \infty} (-\cos x + \cos x) = \lim_{x \to \infty} 0 = 0$

17. $\frac{dy}{dx} = \sqrt{\cos 2x} \Rightarrow 1 + \left(\frac{dy}{dx}\right)^2 = 1 + \cos 2x = 2 \cos^2 x; \; L = \int_0^{\pi/4} \sqrt{1 + \left(\sqrt{\cos 2t}\right)^2} \, dt = \sqrt{2} \int_0^{\pi/4} \sqrt{\cos^2 t} \, dt$

$= \sqrt{2}\,[\sin t]_0^{\pi/4} = 1$

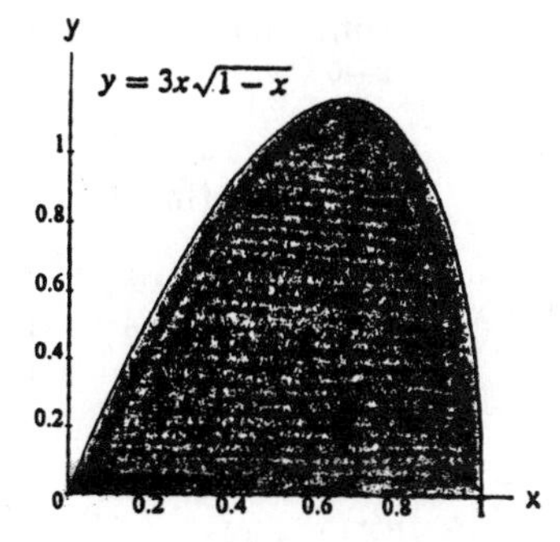

19. $V = \int_a^b 2\pi\left(\begin{matrix}\text{shell}\\\text{radius}\end{matrix}\right)\left(\begin{matrix}\text{shell}\\\text{height}\end{matrix}\right) dx = \int_0^1 2\pi xy\, dx$

$= 6\pi \int_0^1 x^2\sqrt{1-x}\, dx; \begin{bmatrix} u = 1-x \\ du = -\,dx \\ x^2 = (1-u)^2 \end{bmatrix}$

$\Rightarrow -6\pi \int_1^0 (1-u)^2\sqrt{u}\, du = -6\pi \int_1^0 \left(u^{1/2} - 2u^{3/2} + u^{5/2}\right) du$

$= -6\pi\left[\frac{2}{3}u^{3/2} - \frac{4}{5}u^{5/2} + \frac{2}{7}u^{7/2}\right]_1^0 = 6\pi\left(\frac{2}{3} - \frac{4}{5} + \frac{2}{7}\right) = 6\pi\left(\frac{70-84+30}{105}\right) = 6\pi\left(\frac{16}{105}\right) = \frac{32\pi}{35}$

21. $V = \int_a^b 2\pi\left(\begin{matrix}\text{shell}\\\text{radius}\end{matrix}\right)\left(\begin{matrix}\text{shell}\\\text{height}\end{matrix}\right) dx = \int_0^1 2\pi x e^x\, dx$

$= 2\pi\left[xe^x - e^x\right]_0^1 = 2\pi$

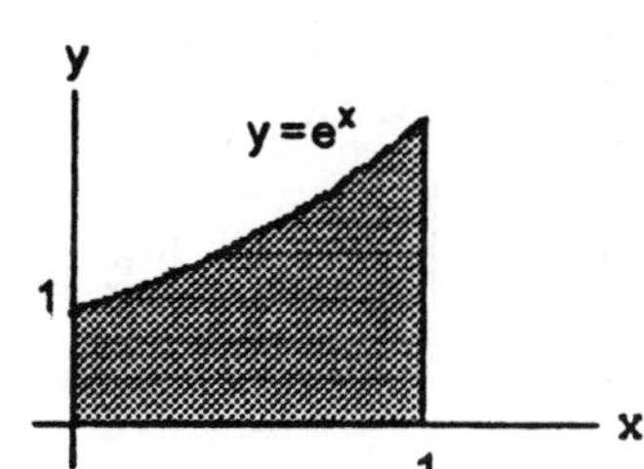

23. (a) $V = \int_1^e \pi\left[1 - (\ln x)^2\right] dx$

$= \pi\left[x - x(\ln x)^2\right]_1^e - 2\pi \int_1^e \ln x\, dx$ (FORMULA 110)

$= \pi\left[x - x(\ln x)^2 + 2(x\ln x - x)\right]_1^e$

$= \pi\left[-x - x(\ln x)^2 + 2x\ln x\right]_1^e = \pi\left[-e - e + 2e - (-1)\right] = \pi$

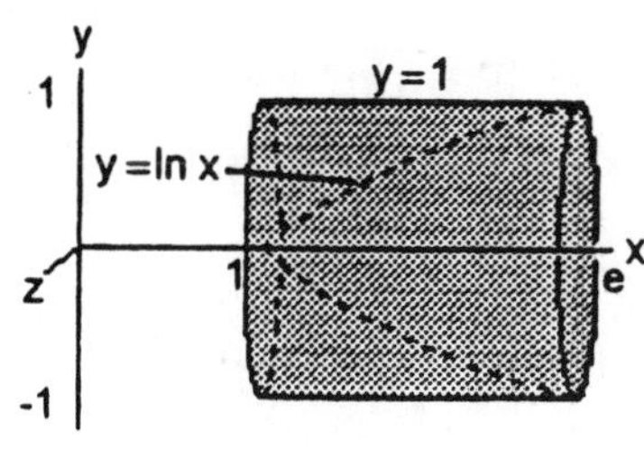

(b) $V = \int_1^e \pi(1 - \ln x)^2\, dx = \pi \int_1^e \left[1 - 2\ln x + (\ln x)^2\right] dx$

$= \pi\left[x - 2(x\ln x - x) + x(\ln x)^2\right]_1^e - 2\pi \int_1^e \ln x\, dx$

$= \pi\left[x - 2(x\ln x - x) + x(\ln x)^2 - 2(x\ln x - x)\right]_1^e$

$= \pi\left[5x - 4x\ln x + x(\ln x)^2\right]_1^e = \pi\left[(5e - 4e + e) - (5)\right]$

$= \pi(2e - 5)$

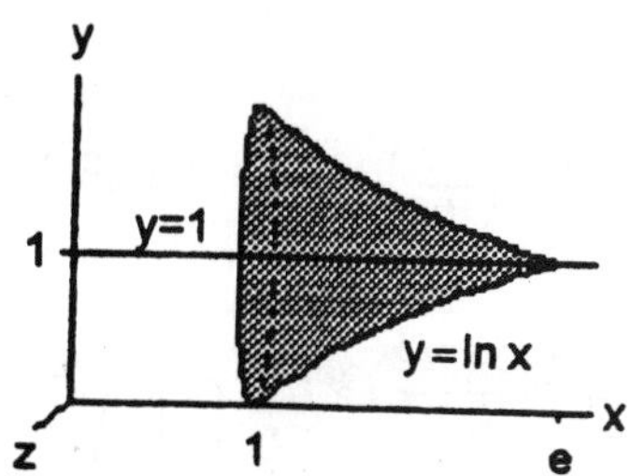

25. (a) $\lim_{x\to0^+} x \ln x = 0 \Rightarrow \lim_{x\to0^+} f(x) = 0 = f(0) \Rightarrow$ f is continuous

(b) $V = \int_0^2 \pi x^2(\ln x)^2\,dx;\ \left[\begin{array}{c} u = (\ln x)^2 \\ du = (2\ln x)\frac{dx}{x} \\ dv = x^2 \\ v = \frac{x^3}{3}\end{array}\right] \Rightarrow \pi\left(\lim_{b\to0^+}\left[\frac{x^3}{3}(\ln x)^2\right]_b^2 - \int_0^2\left(\frac{x^3}{3}\right)(2\ln x)\frac{dx}{x}\right)$

$= \pi\left[\left(\frac{8}{3}\right)(\ln 2)^2 - \left(\frac{2}{3}\right)\lim_{b\to0^+}\left[\frac{x^3}{3}\ln x - \frac{x^3}{9}\right]_b^2\right] = \pi\left[\frac{8(\ln 2)^2}{3} - \frac{16(\ln 2)}{9} + \frac{16}{27}\right]$

27. $u = \frac{1}{1+y}$, $du = -\frac{dy}{(1+y)^2}$; $dv = ny^{n-1}\,dy$, $v = y^n$;

$\lim_{n\to\infty}\int_0^1 \frac{ny^{n-1}}{1+y}\,dy = \lim_{n\to\infty}\left(\left[\frac{y^n}{1+y}\right]_0^1 + \int_0^1 \frac{y^n}{1+y^2}\,dy\right) = \frac{1}{2} + \lim_{n\to\infty}\int_0^1 \frac{y^n}{1+y^2}\,dy$. Now, $0 \le \frac{y^n}{1+y^2} \le y^n$

$\Rightarrow 0 \le \lim_{n\to\infty}\int_0^1 \frac{y^n}{1+y^2}\,dy \le \lim_{n\to\infty}\int_0^1 y^n\,dy = \lim_{n\to\infty}\left[\frac{y^{n+1}}{n+1}\right]_0^1 = \lim_{n\to\infty}\frac{1}{n+1} = 0 \Rightarrow \lim_{n\to\infty}\int_0^1 \frac{ny^{n-1}}{1+y}\,dy$

$= \frac{1}{2} + 0 = \frac{1}{2}$

29. $\frac{\pi}{6} = \sin^{-1}\frac{1}{2} = \left[\sin^{-1}\frac{x}{2}\right]_0^1 = \int_0^1 \frac{dx}{\sqrt{4-x^2}} < \int_0^1 \frac{dx}{\sqrt{4-x^2-x^3}} < \int_0^1 \frac{dx}{\sqrt{4-2x^2}} = \frac{1}{\sqrt{2}}\int_0^{\sqrt{2}} \frac{du}{\sqrt{4-u^2}}$

$= \frac{1}{\sqrt{2}}\left[\sin^{-1}\frac{u}{2}\right]_0^{\sqrt{2}} = \frac{1}{\sqrt{2}}\sin^{-1}\frac{\sqrt{2}}{2} = \frac{1}{\sqrt{2}}\left(\frac{\pi}{4}\right) = \frac{\pi\sqrt{2}}{8}$

31. Let $u = f(x) \Rightarrow du = f'(x)\,dx$ and $dv = dx \Rightarrow v = x$;

$\int_{\pi/2}^{3\pi/2} f(x)\,dx = [x\,f(x)]_{\pi/2}^{3\pi/2} - \int_{\pi/2}^{3\pi/2} xf'(x)\,dx = \left[\frac{3\pi}{2}f\left(\frac{3\pi}{2}\right) - \frac{\pi}{2}f\left(\frac{\pi}{2}\right)\right] - \int_{\pi/2}^{3\pi/2}\cos x\,dx$

$= \frac{3\pi}{2}b - \frac{\pi}{2}a - [\sin x]_{\pi/2}^{3\pi/2} = \frac{\pi}{2}(3b-a) - [-1-1] = \frac{\pi}{2}(3b-a) + 2$

33. $L = 4\int_0^1 \sqrt{1+\left(\frac{dy}{dx}\right)^2}\,dy$; $x^{2/3} + y^{2/3} = 1 \Rightarrow y = \left(1-x^{2/3}\right)^{3/2} \Rightarrow \frac{dy}{dx} = -\frac{3}{2}\left(1-x^{2/3}\right)^{1/2}\left(x^{-1/3}\right)\left(\frac{2}{3}\right)$

$\Rightarrow \left(\frac{dy}{dx}\right)^2 = \frac{1-x^{2/3}}{x^{2/3}} \Rightarrow L = 4\int_0^1 \sqrt{1+\left(\frac{1-x^{2/3}}{x^{2/3}}\right)}\,dx = 4\int_0^1 \frac{dx}{x^{1/3}} = 6\left[x^{2/3}\right]_0^1 = 6$

35. $P(x) = ax^2 + bx + c$, $P(0) = c = 1$ and $P'(0) = 0 \Rightarrow b = 0 \Rightarrow P(x) = ax^2 + 1$. Next,

$\frac{ax^2+1}{x^3(x-1)^2} = \frac{A}{x} + \frac{B}{x^2} + \frac{C}{x^3} + \frac{D}{x-1} + \frac{E}{(x-1)^2}$; for the integral to be a rational function, we must have $A = 0$ and $D = 0$. Thus, $ax^2 + 1 = Bx(x-1)^2 + C(x-1)^2 + Ex^3 = (B+E)x^3 + (C-2B)x^2 + (B-2C)x + C$

$\Rightarrow C = 1$; B-2C $= 0 \Rightarrow B = 2$; $C - 2B = a \Rightarrow a = -3$; therefore, $P(x) = -3x^2 + 1$

37. $A = \int_1^\infty \frac{dx}{x^p}$ converges if $p > 1$ and diverges if $p \le 1$ (Exercise 67 in Section 7.6). Thus, $p \le 1$ for infinite area.

The volume of the solid of revolution about the x-axis is $V = \int_1^\infty \pi\left(\frac{1}{x^p}\right)^2 dx = \pi \int_1^\infty \frac{dx}{x^{2p}}$ which converges if $2p > 1$ and diverges if $2p \le 1$. Thus we want $p > \frac{1}{2}$ for finite volume. In conclusion, the curve $y = x^{-p}$ gives infinite area and finite volume for values of p satisfying $\frac{1}{2} < p \le 1$.

39. (a)

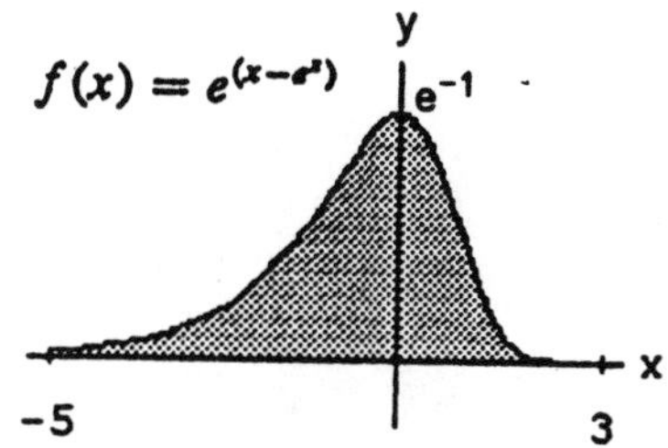

(b) $\int_{-\infty}^{\infty} e^{(x-e^x)}\,dx = \int_{-\infty}^{\infty} e^{(-e^x)}e^x\,dx$

$= \lim_{a\to-\infty} \int_a^0 e^{(-e^x)}e^x\,dx + \lim_{b\to+\infty} \int_0^b e^{(-e^x)}e^x\,dx$;

$\begin{bmatrix} u = e^x \\ du = e^x\,dx \end{bmatrix} \to \lim_{a\to-\infty} \int_{e^a}^{1} e^{-u}\,du + \lim_{b\to+\infty} \int_1^{e^b} e^{-u}\,du$

$= \lim_{a\to-\infty} \left[-e^{-u}\right]_{e^a}^{1} + \lim_{b\to-\infty} \left[-e^{-u}\right]_1^{e^b} = \lim_{a\to-\infty} \left[-\frac{1}{e} + e^{-(e^a)}\right] + \lim_{b\to+\infty} \left[-e^{-(e^b)} + \frac{1}{e}\right] = \left(-\frac{1}{e} + e^0\right) + \left(0 + \frac{1}{e}\right) = 1$

41. e^{2x} (+) $\cos 3x$

$2e^{2x}$ (−) $\frac{1}{3}\sin 3x$

$4e^{2x}$ $\underline{\ (+)\ }$ $-\frac{1}{9}\cos 3x$

$I = \frac{e^{2x}}{3}\sin 3x + \frac{2e^{2x}}{9}\cos 3x - \frac{4}{9}I \Rightarrow \frac{13}{9}I = \frac{e^{2x}}{9}(3\sin 3x + 2\cos 3x) \Rightarrow I = \frac{e^{2x}}{13}(3\sin 3x + 2\cos 3x) + C$

43. $\sin 3x \quad (+) \quad \sin x$

$3\cos 3x \quad (-) \quad -\cos x$

$-9\sin 3x \xrightarrow{(+)} -\sin x$

$I = -\sin 3x\cos x + 3\cos 3x\sin x + 9I \Rightarrow -8I = -\sin 3x\cos x + 3\cos 3x\sin x$

$\Rightarrow I = \frac{\sin 3x\cos x - 3\cos 3x\sin x}{8} + C$

45. $e^{ax} \quad (+) \quad \sin bx$

$ae^{ax} \quad (-) \quad -\frac{1}{b}\cos bx$

$a^2e^{ax} \xrightarrow{(+)} -\frac{1}{b^2}\sin bx$

$I = -\frac{e^{ax}}{b}\cos bx + \frac{ae^{ax}}{b^2}\sin bx - \frac{a^2}{b^2}I \Rightarrow \left(\frac{a^2+b^2}{b^2}\right)I = \frac{e^{ax}}{b^2}(a\sin bx - b\cos bx)$

$\Rightarrow I = \frac{e^{ax}}{a^2+b^2}(a\sin bx - b\cos bx) + C$

47. $\ln(ax) \quad (+) \quad 1$

$\frac{1}{x} \xrightarrow{(-)} x$

$I = x\ln(ax) - \int\left(\frac{1}{x}\right)x\,dx = x\ln(ax) - x + C$

49. (a) $\Gamma(1) = \int_0^\infty e^{-t}\,dt = \lim_{b\to\infty}\int_0^b e^{-t}\,dt = \lim_{b\to\infty}\left[-e^{-t}\right]_0^b = \lim_{b\to\infty}\left[-\frac{1}{e^b} - (-1)\right] = 0 + 1 = 1$

(b) $u = t^x$, $du = xt^{x-1}\,dt$; $dv = e^{-t}\,dt$, $v = -e^{-t}$; x = fixed positive real

$\Rightarrow \Gamma(x+1) = \int_0^\infty t^xe^{-t}\,dt = \lim_{b\to\infty}\left[-t^xe^{-t}\right]_0^b + x\int_0^\infty t^{x-1}e^{-t}\,dt = \lim_{b\to\infty}\left(-\frac{b^x}{e^b} + 0^xe^0\right) + x\Gamma(x) = x\Gamma(x)$

(c) $\Gamma(n+1) = n\Gamma(n) = n!$:

$n = 0$: $\Gamma(0+1) = \Gamma(1) = 0!$;

$n = k$: Assume $\Gamma(k+1) = k!$ for some $k > 0$;

$n = k+1$: $\Gamma(k+1+1) = (k+1)\Gamma(k+1)$ from part (b)

$= (k+1)k!$ induction hypothesis

$= (k+1)!$ definition of factorial

Thus, $\Gamma(n+1) = n\Gamma(n) = n!$ for every positive integer n.

CHAPTER 8 INFINITE SERIES

8.1 LIMITS OF SEQUENCES OF NUMBERS

1. $a_1 = \frac{1-1}{1^2} = 0$, $a_2 = \frac{1-2}{2^2} = -\frac{1}{4}$, $a_3 = \frac{1-3}{3^2} = -\frac{2}{9}$, $a_4 = \frac{1-4}{4^2} = -\frac{3}{16}$

3. $a_1 = \frac{(-1)^2}{2-1} = 1$, $a_2 = \frac{(-1)^3}{4-1} = -\frac{1}{3}$, $a_3 = \frac{(-1)^4}{6-1} = \frac{1}{5}$, $a_4 = \frac{(-1)^5}{8-1} = -\frac{1}{7}$

5. $a_n = (-1)^{n+1}$, $n = 1, 2, \ldots$

7. $a_n = n^2 - 1$, $n = 1, 2, \ldots$

9. $a_n = 4n - 3$, $n = 1, 2, \ldots$

11. $a_n = \frac{1 + (-1)^{n+1}}{2}$, $n = 1, 2, \ldots$

13. $\lim_{n\to\infty} 2 + (0.1)^n = 2 \Rightarrow$ converges (Table 8.1, #4)

15. $\lim_{n\to\infty} \frac{1-2n}{1+2n} = \lim_{n\to\infty} \frac{\left(\frac{1}{n}\right) - 2}{\left(\frac{1}{n}\right) + 2} = \lim_{n\to\infty} \frac{-2}{2} = -1 \Rightarrow$ converges

17. $\lim_{n\to\infty} \frac{n^2 - 2n + 1}{n-1} = \lim_{n\to\infty} \frac{(n-1)(n-1)}{n-1} = \lim_{n\to\infty} (n-1) = \infty \Rightarrow$ diverges

19. $\lim_{n\to\infty} (1 + (-1)^n)$ does not exist $\Rightarrow$ diverges

21. $\lim_{n\to\infty} \left(\frac{n+1}{2n}\right)\left(1 - \frac{1}{n}\right) = \lim_{n\to\infty} \left(\frac{1}{2} + \frac{1}{2n}\right)\left(1 - \frac{1}{n}\right) = \frac{1}{2} \Rightarrow$ converges

23. $\lim_{n\to\infty} \sqrt{\frac{2n}{n+1}} = \sqrt{\lim_{n\to\infty} \frac{2n}{n+1}} = \sqrt{\lim_{n\to\infty} \left(\frac{2}{1 + \frac{1}{n}}\right)} = \sqrt{2} \Rightarrow$ converges

25. $\lim_{n\to\infty} \frac{\sin n}{n} = 0$ because $-\frac{1}{n} \le \frac{\sin n}{n} \le \frac{1}{n} \Rightarrow$ converges by the Sandwich Theorem for sequences

27. $\lim_{n\to\infty} \frac{n}{2^n} = \lim_{n\to\infty} \frac{1}{2^n \ln 2} = 0 \Rightarrow$ converges (using l'Hôpital's rule)

29. $\lim_{n\to\infty} \frac{\ln n}{n^{1/n}} = \frac{\lim_{n\to\infty} \ln n}{\lim_{n\to\infty} n^{1/n}} = \frac{\infty}{1} = \infty \Rightarrow$ diverges (Table 8.1, #2)

31. $\lim_{n\to\infty} \left(1+\frac{7}{n}\right)^n = e^7 \Rightarrow$ converges (Table 8.1, #5)

33. $\lim_{n\to\infty} \sqrt[n]{10n} = \lim_{n\to\infty} 10^{1/n} \cdot n^{1/n} = 1 \cdot 1 = 1 \Rightarrow$ converges (Table 8.1, #3 and #2)

35. $\lim_{n\to\infty} \left(\frac{3}{n}\right)^{1/n} = \frac{\lim_{n\to\infty} 3^{1/n}}{\lim_{n\to\infty} n^{1/n}} = \frac{1}{1} = 1 \Rightarrow$ converges (Table 8.1, #3 and #2)

37. $\lim_{n\to\infty} \sqrt[n]{4^n n} = \lim_{n\to\infty} 4\sqrt[n]{n} = 4 \cdot 1 = 4 \Rightarrow$ converges (Table 8.1, #2)

39. $\lim_{n\to\infty} \frac{n!}{n^n} = \lim_{n\to\infty} \frac{1 \cdot 2 \cdot 3 \cdots (n-1)(n)}{n \cdot n \cdot n \cdots n \cdot n} \le \lim_{n\to\infty} \left(\frac{1}{n}\right) = 0$ and $\frac{n!}{n^n} \ge 0 \Rightarrow \lim_{n\to\infty} \frac{n!}{n^n} = 0 \Rightarrow$ converges

41. $\lim_{n\to\infty} \frac{n!}{10^{6n}} = \lim_{n\to\infty} \frac{1}{\left(\frac{(10^6)^n}{n!}\right)} = \infty \Rightarrow$ diverges (Table 8.1, #6)

43. $\lim_{n\to\infty} \left(\frac{1}{n}\right)^{1/(\ln n)} = \lim_{n\to\infty} \exp\left(\frac{1}{\ln n} \ln\left(\frac{1}{n}\right)\right) = \lim_{n\to\infty} \exp\left(\frac{\ln 1 - \ln n}{\ln n}\right) = e^{-1} \Rightarrow$ converges

45. $\lim_{n\to\infty} \left(\frac{3n+1}{3n-1}\right)^n = \lim_{n\to\infty} \exp\left(n \ln\left(\frac{3n+1}{3n-1}\right)\right) = \lim_{n\to\infty} \exp\left(\frac{\ln(3n+1) - \ln(3n-1)}{\frac{1}{n}}\right)$

$= \lim_{n\to\infty} \exp\left(\frac{\frac{3}{3n+1} - \frac{3}{3n-1}}{\left(-\frac{1}{n^2}\right)}\right) = \lim_{n\to\infty} \exp\left(\frac{6n^2}{(3n+1)(3n-1)}\right) = \exp\left(\frac{6}{9}\right) = e^{2/3} \Rightarrow$ converges

47. $\lim_{n\to\infty} \left(\frac{x^n}{2n+1}\right)^{1/n} = \lim_{n\to\infty} x\left(\frac{1}{2n+1}\right)^{1/n} = x \lim_{n\to\infty} \exp\left(\frac{1}{n} \ln\left(\frac{1}{2n+1}\right)\right) = x \lim_{n\to\infty} \exp\left(\frac{-\ln(2n+1)}{n}\right)$

$= x \lim_{n\to\infty} \exp\left(\frac{-2}{2n+1}\right) = xe^0 = x,\ x > 0 \Rightarrow$ converges

49. $\lim_{n\to\infty} \frac{3^n \cdot 6^n}{2^{-n} \cdot n!} = \lim_{n\to\infty} \frac{36^n}{n!} = 0 \Rightarrow$ converges (Table 8.1, #6)

51. $\lim_{n\to\infty} \tan^{-1} n = \frac{\pi}{2} \Rightarrow$ converges

53. $\lim_{n\to\infty} \left(\frac{1}{3}\right)^n + \frac{1}{\sqrt{2^n}} = \lim_{n\to\infty} \left(\left(\frac{1}{3}\right)^n + \left(\frac{1}{\sqrt{2}}\right)^n\right) = 0 \Rightarrow$ converges (Table 8.1, #4)

55. $\lim_{n\to\infty} \frac{(\ln n)^5}{\sqrt{n}} = \lim_{n\to\infty} \left[\frac{\left(\frac{5(\ln n)^4}{n}\right)}{\left(\frac{1}{2\sqrt{n}}\right)}\right] = \lim_{n\to\infty} \frac{10(\ln n)^4}{\sqrt{n}} = \lim_{n\to\infty} \frac{80(\ln n)^3}{\sqrt{n}} = \ldots = \lim_{n\to\infty} \frac{3840}{\sqrt{n}} = 0 \Rightarrow$ converges

57. $\left|\sqrt[n]{0.5}-1\right| < 10^{-3} \Rightarrow -\frac{1}{1000} < \left(\frac{1}{2}\right)^{1/n} - 1 < \frac{1}{1000} \Rightarrow \left(\frac{999}{1000}\right)^n < \frac{1}{2} < \left(\frac{1001}{1000}\right)^n \Rightarrow n > \frac{\ln\left(\frac{1}{2}\right)}{\ln\left(\frac{999}{1000}\right)} \Rightarrow n > 692.8$

$\Rightarrow N = 692;\ a_n = \left(\frac{1}{2}\right)^{1/n}$ and $\lim_{n\to\infty} a_n = 1$

59. $(0.9)^n < 10^{-3} \Rightarrow n \ln(0.9) < -3 \ln 10 \Rightarrow n > \frac{-3 \ln 10}{\ln(0.9)} \approx 65.54 \Rightarrow N = 65;\ a_n = \left(\frac{9}{10}\right)^n$ and $\lim_{n\to\infty} a_n = 0$

61. (a) $1^2 - 2(1)^2 = -1,\ 3^2 - 2(2)^2 = 1$; let $f(a,b) = (a+2b)^2 - 2(a+b)^2 = a^2 + 4ab + 4b^2 - 2a^2 - 4ab - 2b^2$
$= 2b^2 - a^2;\ a^2 - 2b^2 = -1 \Rightarrow f(a,b) = 2b^2 - a^2 = 1;\ a^2 - 2b^2 = 1 \Rightarrow f(a,b) = 2b^2 - a^2 = -1$

(b) $r_n^2 - 2 = \left(\frac{a+2b}{a+b}\right)^2 - 2 = \frac{a^2 + 4ab + 4b^2 - 2a^2 - 4ab - 2b^2}{(a+b)^2} = \frac{-\left(a^2 - 2b^2\right)}{(a+b)^2} = \frac{\pm 1}{y_n^2} \Rightarrow r_n = \sqrt{2 \pm \left(\frac{1}{y_n}\right)^2}$

In the first and second fractions, $y_n \ge n$. Let $\frac{a}{b}$ represent the $(n-1)$th fraction where $\frac{a}{b} \ge 1$ and $b \ge n-1$ for n a positive integer ≥ 3. Now the nth fraction is $\frac{a+2b}{a+b}$ and $a + b \ge 2b \ge 2n - 2 \ge n \Rightarrow y_n \ge n$. Thus, $\lim_{n\to\infty} r_n = \sqrt{2}$.

63. (a) If $a = 2n+1$, then $b = \lfloor \frac{a^2}{2} \rfloor = \lfloor \frac{4n^2 + 4n + 1}{2} \rfloor = \lfloor 2n^2 + 2n + \frac{1}{2} \rfloor = 2n^2 + 2n$, $c = \lceil \frac{a^2}{2} \rceil = \lceil 2n^2 + 2n + \frac{1}{2} \rceil$
$= 2n^2 + 2n + 1$ and $a^2 + b^2 = (2n+1)^2 + \left(2n^2 + 2n\right)^2 = 4n^2 + 4n + 1 + 4n^4 + 8n^3 + 4n^2$
$= 4n^4 + 8n^3 + 8n^2 + 4n + 1 = \left(2n^2 + 2n + 1\right)^2 = c^2$.

(b) $\lim_{a\to\infty} \frac{\lfloor \frac{a^2}{2} \rfloor}{\lceil \frac{a^2}{2} \rceil} = \lim_{a\to\infty} \frac{2n^2 + 2n}{2n^2 + 2n + 1} = 1$ or $\lim_{a\to\infty} \frac{\lfloor \frac{a^2}{2} \rfloor}{\lceil \frac{a^2}{2} \rceil} = \lim_{a\to\infty} \sin\theta = \lim_{\theta\to\pi/2} \sin\theta = 1$

65. (a) $\lim_{n\to\infty} \frac{\ln n}{n^c} = \lim_{n\to\infty} \frac{\left(\frac{1}{n}\right)}{cn^{c-1}} = \lim_{n\to\infty} \frac{1}{cn^c} = 0$

(b) For all $\epsilon > 0$, there exists an $N = e^{-(\ln \epsilon)/c}$ such that $n > e^{-(\ln \epsilon)/c} \Rightarrow \ln n > -\frac{\ln \epsilon}{c} \Rightarrow \ln n^c > \ln\left(\frac{1}{\epsilon}\right)$
$\Rightarrow n^c > \frac{1}{\epsilon} \Rightarrow \frac{1}{n^c} < \epsilon \Rightarrow \left|\frac{1}{n^c} - 0\right| < \epsilon \Rightarrow \lim_{n\to\infty} \frac{1}{n^c} = 0$

67. $\lim_{n\to\infty} n^{1/n} = \lim_{n\to\infty} \exp\left(\frac{1}{n} \ln n\right) = \lim_{n\to\infty} \exp\left(\frac{1}{n}\right) = e^0 = 1$

69. Assume the hypotheses of the theorem and let ϵ be a positive number. For all ϵ there exists a N_1 such that when $n > N_1$ then $|a_n - L| < \epsilon \Rightarrow -\epsilon < a_n - L < \epsilon \Rightarrow L - \epsilon < a_n$, and there exists a N_2 such that when $n > N_2$ then $|c_n - L| < \epsilon \Rightarrow -\epsilon < c_n - L < \epsilon \Rightarrow c_n < L + \epsilon$. If $n > \max\{N_1, N_2\}$, then $L - \epsilon < a_n \le b_n \le c_n < L + \epsilon \Rightarrow |b_n - L| < \epsilon \Rightarrow \lim_{n\to\infty} b_n = L$.

71. Let L be the limit of the convergent sequence $\{a_n\}$. Then by definition of convergence, for $\frac{\epsilon}{2}$ there corresponds an N such that for all m and n, $m > N \Rightarrow |a_m - L| < \frac{\epsilon}{2}$ and $n > N \Rightarrow |a_n - L| < \frac{\epsilon}{2}$. Now

$|a_m - a_n| = |a_m - L + L - a_n| \le |a_m - L| + |L - a_n| < \frac{\epsilon}{2} + \frac{\epsilon}{2} = \epsilon$ whenever $m > N$ and $n > N$.

73. Assume $a_n \to 0$. This implies that given an $\epsilon > 0$ there corresponds an N such that $n > N \Rightarrow |a_n - 0| < \epsilon$ $\Rightarrow |a_n| < \epsilon \Rightarrow ||a_n|| < \epsilon \Rightarrow ||a_n| - 0| < \epsilon \Rightarrow |a_n| \to 0$. On the other hand, assume $|a_n| \to 0$. This implies that given an $\epsilon > 0$ there corresponds an N such that for $n > N$, $||a_n| - 0| < \epsilon \Rightarrow ||a_n|| < \epsilon \Rightarrow |a_n| < \epsilon$ $\Rightarrow |a_n - 0| < \epsilon \Rightarrow a_n \to 0$.

8.2 SUBSEQUENCES, BOUNDED SEQUENCES, AND PICARD'S METHOD

1. $a_1 = 1$, $a_2 = 1 + \frac{1}{2} = \frac{3}{2}$, $a_3 = \frac{3}{2} + \frac{1}{2^2} = \frac{7}{4}$, $a_4 = \frac{7}{4} + \frac{1}{2^3} = \frac{15}{8}$, $a_5 = \frac{15}{8} + \frac{1}{2^4} = \frac{31}{16}$, $a_6 = \frac{63}{32}$, $a_7 = \frac{127}{64}$, $a_8 = \frac{255}{128}$, $a_9 = \frac{511}{256}$, $a_{10} = \frac{1023}{512}$

3. $a_1 = 2$, $a_2 = \frac{(-1)^2(2)}{2} = 1$, $a_3 = \frac{(-1)^3(1)}{2} = -\frac{1}{2}$, $a_4 = \frac{(-1)^4\left(-\frac{1}{2}\right)}{2} = -\frac{1}{4}$, $a_5 = \frac{(-1)^5\left(-\frac{1}{4}\right)}{2} = \frac{1}{8}$, $a_6 = \frac{1}{16}$, $a_7 = -\frac{1}{32}$, $a_8 = -\frac{1}{64}$, $a_9 = \frac{1}{128}$, $a_{10} = \frac{1}{256}$

5. $a_1 = 1$, $a_2 = 1$, $a_3 = 1 + 1 = 2$, $a_4 = 2 + 1 = 3$, $a_5 = 3 + 2 = 5$, $a_6 = 8$, $a_7 = 13$, $a_8 = 21$, $a_9 = 34$, $a_{10} = 55$

7. (a) $f(x) = x^2 - a \Rightarrow f'(x) = 2x \Rightarrow x_{n+1} = x_n - \frac{x_n^2 - a}{2x_n} \Rightarrow x_{n+1} = \frac{2x_n^2 - \left(x_n^2 - a\right)}{2x_n} = \frac{x_n^2 + a}{2x_n} = \frac{\left(x_n + \frac{a}{x_n}\right)}{2}$

   (b) $x_1 = 2$, $x_2 = 1.75$, $x_3 = 1.732142857$, $x_4 = 1.73205081$, $x_5 = 1.732050808$; we are finding the positive number where $x^2 - 3 = 0$; that is, where $x^2 = 3$, $x > 0$, or where $x = \sqrt{3}$.

9. (a) $f(x) = x^2 - 2$; the sequence converges to $1.414213562 \approx \sqrt{2}$

   (b) $f(x) = \tan(x) - 1$; the sequence converges to $0.7853981635 \approx \frac{\pi}{4}$

   (c) $f(x) = e^x$; the sequence $1, 0, -1, -2, -3, -4, -5, \ldots$ diverges

11. $a_{n+1} \ge a_n \Rightarrow \frac{3(n+1)+1}{(n+1)+1} > \frac{3n+1}{n+1} \Rightarrow \frac{3n+4}{n+2} > \frac{3n+1}{n+1} \Rightarrow 3n^2 + 3n + 4n + 4 > 3n^2 + 6n + n + 2$ $\Rightarrow 4 > 2$; the steps are reversible so the sequence is nondecreasing; $\frac{3n+1}{n+1} < 3 \Rightarrow 3n + 1 < 3n + 3$ $\Rightarrow 1 < 3$; the steps are reversible so the sequence is bounded above by 3

13. $a_{n+1} \le a_n \Rightarrow \frac{2^{n+1}3^{n+1}}{(n+1)!} \le \frac{2^n 3^n}{n!} \Rightarrow \frac{2^{n+1}3^{n+1}}{2^n 3^n} \le \frac{(n+1)!}{n!} \Rightarrow 2 \cdot 3 \le n + 1$ which is true for $n \ge 5$; the steps are reversible so the sequence is decreasing after a_5, but it is not nondecreasing for all its terms; $a_1 = 6$, $a_2 = 18$, $a_3 = 36$, $a_4 = 54$, $a_5 = \frac{324}{5} = 64.8 \Rightarrow$ the sequence is bounded from above by 64.8

15. $a_n = 1 - \frac{1}{n}$ converges because $\frac{1}{n} \to 0$ by Example 6 in Section 8.1; also it is a nondecreasing sequence bounded above by 1

17. $a_n = \frac{2^n - 1}{2^n} = 1 - \frac{1}{2^n}$ and $0 < \frac{1}{2^n} < \frac{1}{n}$; since $\frac{1}{n} \to 0$ (by Example 6 in Section 8.1) $\Rightarrow \frac{1}{2^n} \to 0$, the sequence converges; also it is a nondecreasing sequence bounded above by 1

19. $a_n = ((-1)^n + 1)\left(\frac{n+1}{n}\right)$ diverges because $a_n = 0$ for n odd, while for n even $a_n = 2\left(1 + \frac{1}{n}\right)$ converges to 2; it diverges by definition of divergence

21. $a_n \ge a_{n+1} \Leftrightarrow \frac{n+1}{n} \ge \frac{(n+1)+1}{n+1} \Leftrightarrow n^2 + 2n + 1 \ge n^2 + 2n \Leftrightarrow 1 \ge 0$ and $\frac{n+1}{n} \ge 1$; thus the sequence is nonincreasing and bounded below by 1 $\Rightarrow$ it converges

23. $a_n \ge a_{n+1} \Leftrightarrow \frac{1-4^n}{2^n} \ge \frac{1-4^{n+1}}{2^{n+1}} \Leftrightarrow 2^{n+1} - 2^{n+1}4^n \ge 2^n - 2^n 4^{n+1} \Leftrightarrow 2^{n+1} - 2^n \ge 2^{n+1}4^n - 2^n 4^{n+1}$

$\Leftrightarrow 2 - 1 \ge 2 \cdot 4^n - 4^{n+1} \Leftrightarrow 1 \ge 4^n(2-4) \Leftrightarrow 1 \ge (-2) \cdot 4^n$; thus the sequence is nonincreasing. However, $a_n = \frac{1}{2^n} - \frac{4^n}{2^n} = \frac{1}{2^n} - 2^n$ which is not bounded below so the sequence diverges

25. Let k(n) and i(n) be two order-preserving functions whose domains are the set of positive integers and whose ranges are a subset of the positive integers. Consider the two subsequences $a_{k(n)}$ and $a_{i(n)}$, where $a_{k(n)} \to L_1$, $a_{i(n)} \to L_2$ and $L_1 \ne L_2$. Given an $\epsilon > 0$ there corresponds an N_1 such that for $k(n) > N_1$, $\left|a_{k(n)} - L_1\right| < \epsilon$, and an N_2 such that for $i(n) > N_2$, $\left|a_{i(n)} - L_2\right| < \epsilon$. Let $N = \max\{N_1, N_2\}$. Then for $n > N$, we have that $\left|a_n - L_1\right| < \epsilon$ and $\left|a_n - L_2\right| < \epsilon$. This implies $a_n \to L_1$ and $a_n \to L_2$ where $L_1 \ne L_2$. Since the limit of a sequence is unique (by Exercise 72, Section 8.1), a_n does not converge and hence diverges.

27. $g(x) = \sqrt{x}$; $2 \to 1.00000132$ in 20 iterations; $.1 \to 0.9999956$ in 20 iterations; a root is 1

29. $g(x) = -\cos x$; $x_0 = .1 \to x \approx -0.739085$

31. $g(x) = 0.1 + \sin x$; $x_0 = -2 \to x \approx 0.853750$

33. $x_0 =$ initial guess $> 0 \Rightarrow x_1 = \sqrt{x_0} = (x_0)^{1/2} \Rightarrow x_2 = \sqrt{{x_0}^{1/2}} = {x_0}^{1/4}, \ldots \Rightarrow x_n = {x_0}^{1/(2n)} \Rightarrow x_n \to 1$ as $n \to \infty$

8.3 INFINITE SERIES

1. $s_n = \frac{a(1-r^n)}{(1-r)} = \frac{2\left(1-\left(\frac{1}{3}\right)^n\right)}{1-\left(\frac{1}{3}\right)} \Rightarrow \lim_{n\to\infty} s_n = \frac{2}{1-\left(\frac{1}{3}\right)} = 3$

3. $s_n = \frac{a(1-r^n)}{(1-r)} = \frac{1-\left(-\frac{1}{2}\right)^n}{1-\left(-\frac{1}{2}\right)} \Rightarrow \lim_{n\to\infty} s_n = \frac{1}{\left(\frac{3}{2}\right)} = \frac{2}{3}$

5. $\frac{1}{(n+1)(n+2)} = \frac{1}{n+1} - \frac{1}{n+2} \Rightarrow s_n = \left(\frac{1}{2} - \frac{1}{3}\right) + \left(\frac{1}{3} - \frac{1}{4}\right) + \ldots + \left(\frac{1}{n+1} - \frac{1}{n+2}\right) = \frac{1}{2} - \frac{1}{n+2} \Rightarrow \lim_{n\to\infty} s_n = \frac{1}{2}$

7. $1-\frac{1}{4}+\frac{1}{16}-\frac{1}{64}+\ldots$, the sum of this geometric series is $\frac{1}{1-\left(-\frac{1}{4}\right)}=\frac{1}{1+\left(\frac{1}{4}\right)}=\frac{4}{5}$

9. $(5+1)+\left(\frac{5}{2}+\frac{1}{3}\right)+\left(\frac{5}{4}+\frac{1}{9}\right)+\left(\frac{5}{8}+\frac{1}{27}\right)+\ldots$, is the sum of two geometric series; the sum is $\frac{5}{1-\left(\frac{1}{2}\right)}+\frac{1}{1-\left(\frac{1}{3}\right)}=10+\frac{3}{2}=\frac{23}{2}$

11. $(1+1)+\left(\frac{1}{2}-\frac{1}{5}\right)+\left(\frac{1}{4}+\frac{1}{25}\right)+\left(\frac{1}{8}-\frac{1}{125}\right)+\ldots$, is the sum of two geometric series; the sum is $\frac{1}{1-\left(\frac{1}{2}\right)}+\frac{1}{1+\left(\frac{1}{5}\right)}=2+\frac{5}{6}=\frac{17}{6}$

13. $\frac{4}{(4n-3)(4n+1)}=\frac{1}{4n-3}-\frac{1}{4n+1} \Rightarrow s_n=\left(1-\frac{1}{5}\right)+\left(\frac{1}{5}-\frac{1}{9}\right)+\left(\frac{1}{9}-\frac{1}{13}\right)+\ldots+\left(\frac{1}{4n-7}-\frac{1}{4n-3}\right)$
$+\left(\frac{1}{4n-3}-\frac{1}{4n+1}\right)=1-\frac{1}{4n+1} \Rightarrow \lim\limits_{n\to\infty} s_n=\lim\limits_{n\to\infty}\left(1-\frac{1}{4n+1}\right)=1$

15. $\frac{40n}{(2n-1)^2(2n+1)^2}=\frac{A}{(2n-1)}+\frac{B}{(2n-1)^2}+\frac{C}{(2n+1)}+\frac{D}{(2n+1)^2}$

$$=\frac{A(2n-1)(2n+1)^2+B(2n+1)^2+C(2n+1)(2n-1)^2+D(2n-1)^2}{(2n-1)^2(2n+1)^2}$$

$\Rightarrow A(2n-1)(2n+1)^2+B(2n+1)^2+C(2n+1)(2n-1)^2+D(2n-1)^2=40n$

$\Rightarrow A\left(8n^3+4n^2-2n-1\right)+B\left(4n^2+4n+1\right)+C\left(8n^3-4n^2-2n+1\right)=D\left(4n^2-4n+1\right)=40n$

$\Rightarrow (8A+8C)n^3+(4A+4B-4C+4D)n^2+(-2A+4B-2C-4D)n+(-A+B+C+D)=40n$

$$\Rightarrow \begin{cases} 8A+8C=0 \\ 4A+4B-4C+4D=0 \\ -2A+4B-2C-4D=40 \\ -A+B+C+D=0 \end{cases} \Rightarrow \begin{cases} 8A+8C=0 \\ A+B-C+D=0 \\ -A+2B-C-2D=20 \\ -A+B+C+D=0 \end{cases} \Rightarrow \begin{cases} B+D=0 \\ 2B-2D=20 \end{cases} \Rightarrow 4B=20 \Rightarrow B=5 \text{ and}$$

$D=-5 \Rightarrow \begin{cases} A+C=0 \\ -A+5+C-5=0 \end{cases} \Rightarrow C=0$ and $A=0$. Hence, $\sum\limits_{n=1}^{k}\left[\frac{40n}{(2n-1)^2(2n+1)^2}\right]$

$$=5\sum_{n=1}^{k}\left[\frac{1}{(2n-1)^2}-\frac{1}{(2n+1)^2}\right]=5\left(\frac{1}{1}-\frac{1}{9}+\frac{1}{9}-\frac{1}{25}+\frac{1}{25}-\ldots-\frac{1}{(2(k-1)+1)^2}+\frac{1}{(2k-1)^2}-\frac{1}{(2k+1)^2}\right)$$

$=5\left(1-\frac{1}{(2k+1)^2}\right) \Rightarrow$ the sum is $\lim\limits_{n\to\infty} 5\left(1-\frac{1}{(2k+1)^2}\right)=5$

17. $s_n=\left(1-\frac{1}{\sqrt{2}}\right)+\left(\frac{1}{\sqrt{2}}-\frac{1}{\sqrt{3}}\right)+\left(\frac{1}{\sqrt{3}}-\frac{1}{\sqrt{4}}\right)+\ldots+\left(\frac{1}{\sqrt{n-1}}+\frac{1}{\sqrt{n}}\right)+\left(\frac{1}{\sqrt{n}}-\frac{1}{\sqrt{n+1}}\right)=1-\frac{1}{\sqrt{n+1}}$
$\Rightarrow \lim\limits_{n\to\infty} s_n=\lim\limits_{n\to\infty}\left(1-\frac{1}{\sqrt{n+1}}\right)=1$

19. convergent geometric series with sum $\frac{1}{1-\left(\frac{1}{\sqrt{2}}\right)}=\frac{\sqrt{2}}{\sqrt{2}-1}=2+\sqrt{2}$

21. convergent geometric series with sum $\frac{\left(\frac{3}{2}\right)}{1-\left(-\frac{1}{2}\right)} = 1$

23. convergent geometric series with sum $\frac{1}{1-\left(\frac{1}{e^2}\right)} = \frac{e^2}{e^2-1}$

25. convergent geometric series with sum $\frac{1}{1-\left(\frac{1}{x}\right)} = \frac{x}{x-1}$

27. $\lim_{n\to\infty}\left(1-\frac{1}{n}\right)^n = \lim_{n\to\infty}\left(1+\frac{-1}{n}\right)^n = e^{-1} \neq 0 \Rightarrow$ diverges

29. $\sum_{n=1}^{\infty} \ln\left(\frac{n}{n+1}\right) = \sum_{n=1}^{\infty} [\ln(n) - \ln(n+1)] \Rightarrow s_n = [\ln(1) - \ln(2)] + [\ln(2) - \ln(3)] + [\ln(3) - \ln(4)] + \ldots$
$+ [\ln(n-1) - \ln(n)] + [\ln(n) - \ln(n+1)] = \ln(1) - \ln(n+1) = -\ln(n+1) \Rightarrow \lim_{n\to\infty} s_n = -\infty, \Rightarrow$ diverges

31. $\lim_{n\to\infty} \frac{n!}{1000^n} = \infty \neq 0 \Rightarrow$ diverges

33. $\sum_{n=0}^{\infty} (-1)^n x^n = \sum_{n=0}^{\infty} (-x)^n$; $a = 1$, $r = -x$; converges to $\frac{1}{1-(-x)} = \frac{1}{1+x}$ for $|x| < 1$

35. $a = 3$, $r = \frac{x-1}{2}$; converges to $\frac{3}{1-\left(\frac{x-1}{2}\right)} = \frac{6}{3-x}$ for $-1 < \frac{x-1}{2} < 1$ or $-1 < x < 3$

37. $a = 1$, $r = 2x$; converges to $\frac{1}{1-2x}$ for $|2x| < 1$ or $|x| < \frac{1}{2}$

39. $a = 1$, $r = \frac{3-x}{2}$; converges to $\frac{1}{1-\left(\frac{3-x}{2}\right)} = \frac{2}{x-1}$ for $\left|\frac{3-x}{2}\right| < 1$ or $1 < x < 5$

41. $0.\overline{23} = \sum_{n=0}^{\infty} \frac{23}{100}\left(\frac{1}{10^2}\right)^n = \frac{\left(\frac{23}{100}\right)}{1-\left(\frac{1}{100}\right)} = \frac{23}{99}$

43. $0.\overline{7} = \sum_{n=0}^{\infty} \frac{7}{10}\left(\frac{1}{10}\right)^n = \frac{\left(\frac{7}{10}\right)}{1-\left(\frac{1}{10}\right)} = \frac{7}{9}$

45. $1.24\overline{123} = \frac{124}{100} + \sum_{n=0}^{\infty} \frac{123}{10^5}\left(\frac{1}{10^3}\right)^n = \frac{124}{100} + \frac{\left(\frac{123}{10^5}\right)}{1-\left(\frac{1}{10^3}\right)} = \frac{124}{100} + \frac{123}{10^5 - 10^2} = \frac{124}{100} + \frac{123}{99{,}900} = \frac{123{,}999}{99{,}900} = \frac{41{,}333}{33{,}300}$

47. distance $= 4 + 2\left[(4)\left(\frac{3}{4}\right) + (4)\left(\frac{3}{4}\right)^2 + \ldots\right] = 4 + 2\left(\frac{3}{1-\left(\frac{3}{4}\right)}\right) = 28$ m

49. area $= 2^2 + \left(\sqrt{2}\right)^2 + (1)^2 + \left(\frac{1}{\sqrt{2}}\right)^2 + \ldots = 4 + 2 + 1 + \frac{1}{2} + \ldots = \frac{4}{1-\frac{1}{2}} = 8\ \text{m}^2$

51. (a) $L_1 = 3, L_2 = 3\left(\frac{4}{3}\right), L_3 = 3\left(\frac{4}{3}\right)^2, \ldots, L_n = 3\left(\frac{4}{3}\right)^{n-1} \Rightarrow \lim_{n\to\infty} L_n = \lim_{n\to\infty} 3\left(\frac{4}{3}\right)^{n-1} = \infty$

(b) $A_1 = \frac{1}{2}(1)\left(\frac{\sqrt{3}}{2}\right) = \frac{\sqrt{3}}{4}, A_2 = A_1 + 3\left(\frac{1}{2}\right)\left(\frac{1}{3}\right)\left(\frac{\sqrt{3}}{6}\right) = \frac{\sqrt{3}}{4} + \frac{\sqrt{3}}{12}, A_3 = A_2 + 12\left(\frac{1}{2}\right)\left(\frac{1}{9}\right)\left(\frac{\sqrt{3}}{18}\right)$

$= \frac{\sqrt{3}}{4} + \frac{\sqrt{3}}{12} + \frac{\sqrt{3}}{27}, A_4 = A_3 + 48\left(\frac{1}{2}\right)\left(\frac{1}{27}\right)\left(\frac{\sqrt{3}}{54}\right), \ldots, A_n = \frac{\sqrt{3}}{4} + \frac{27\sqrt{3}}{64}\left(\frac{4}{9}\right)^2 + \frac{27\sqrt{3}}{64}\left(\frac{4}{9}\right)^3 + \ldots$

$= \frac{\sqrt{3}}{4} + \sum_{n=2}^{\infty} \frac{27\sqrt{3}}{64}\left(\frac{4}{9}\right)^n = \frac{\sqrt{3}}{4} + \frac{\left(\frac{27\sqrt{3}}{64}\right)\left(\frac{4}{9}\right)^2}{1-\left(\frac{4}{9}\right)} = \frac{\sqrt{3}}{4} + \frac{\left(\frac{27\sqrt{3}}{64}\right)\left(\frac{16}{9}\right)}{9-4} = \frac{\sqrt{3}}{4} + \frac{3\sqrt{3}}{4\cdot 5} = \frac{5\sqrt{3}+3\sqrt{3}}{20} = \frac{2\sqrt{3}}{5}$

53. (a) $\sum_{n=-2}^{\infty} \frac{1}{(n+4)(n+5)}$ (b) $\sum_{n=0}^{\infty} \frac{1}{(n+2)(n+3)}$ (c) $\sum_{n=5}^{\infty} \frac{1}{(n-3)(n-2)}$

55. $1 + e^b + e^{2b} + \ldots = \frac{1}{1-e^b} = 9 \Rightarrow \frac{1}{9} = 1 - e^b \Rightarrow e^b = \frac{8}{9} \Rightarrow b = \ln\left(\frac{8}{9}\right)$

57. $L - s_n = \frac{a}{1-r} - \frac{a\left(1-r^n\right)}{1-r} = \frac{ar^n}{1-r}$

59. Let $a_n = \left(\frac{1}{4}\right)^n$ and $b_n = \left(\frac{1}{2}\right)^n$. Then $A = \sum_{n=1}^{\infty} a_n = \frac{1}{3}$, $B = \sum_{n=1}^{\infty} b_n = 1$ and $\sum_{n=1}^{\infty} \left(\frac{a_n}{b_n}\right) = \sum_{n=1}^{\infty} \left(\frac{1}{2}\right)^n = 1 \neq \frac{A}{B}$.

61. Yes: $\sum \left(\frac{1}{a_n}\right)$ diverges. The reasoning: $\sum a_n$ converges $\Rightarrow a_n \to 0 \Rightarrow \frac{1}{a_n} \to \infty \Rightarrow \sum \left(\frac{1}{a_n}\right)$ diverges by the nth-Term Test.

63. Let $A_n = a_1 + a_2 + \ldots + a_n$ and $\lim_{n\to\infty} A_n = A$. Assume $\sum (a_n + b_n)$ converges to S. Let $S_n = (a_1 + b_1) + (a_2 + b_2) + \ldots + (a_n + b_n) \Rightarrow S_n = (a_1 + a_2 + \ldots + a_n) + (b_1 + b_2 + \ldots + b_n)$ $\Rightarrow b_1 + b_2 + \ldots + b_n = S_n - A_n \Rightarrow \lim_{n\to\infty} (b_1 + b_2 + \ldots + b_n) = S - A \Rightarrow \sum b_n$ converges. This contradicts the assumption that $\sum b_n$ diverges; therefore, $\sum (a_n + b_n)$ diverges.

8.4 SERIES OF NONNEGATIVE TERMS

1. diverges by the Integral Test; $\int_1^n \frac{5}{x+1}\,dx = \ln(n+1) - \ln 2 \Rightarrow \int_1^{\infty} \frac{5}{x+1}\,dx \to \infty$

3. diverges by the Integral Test: $\int_2^n \frac{\ln x}{x}\,dx = \frac{1}{2}(\ln^2 n - \ln 2) \Rightarrow \int_2^{\infty} \frac{\ln x}{x}\,dx \to \infty$

5. converges by the Integral Test: $\int_1^{\infty} \frac{e^x}{1+e^{2x}}\,dx; \left[\begin{array}{c} u = e^x \\ du = e^x\,dx \end{array}\right] \to \int_e^{\infty} \frac{1}{1+u^2}\,du = \lim_{b\to\infty} \left[\tan^{-1} u\right]_e^b$

$= \lim_{b \to \infty} \left(\tan^{-1} b - \tan^{-1} e\right) = \frac{\pi}{2} - \tan^{-1} e \approx 0.35$

7. converges by the Integral Test: $\int_3^{\infty} \frac{\left(\frac{1}{x}\right)}{(\ln x)\sqrt{(\ln x)^2 - 1}}\,dx;\ \begin{bmatrix} u = \ln x \\ du = \frac{1}{x}\,dx \end{bmatrix} \to \int_{\ln 3}^{\infty} \frac{1}{u\sqrt{u^2 - 1}}\,du$

$= \lim_{b \to \infty} \left[\sec^{-1} |u|\right]_{\ln 3}^{b} = \lim_{b \to \infty} \left[\sec^{-1} b - \sec^{-1}(\ln 3)\right] = \lim_{b \to \infty} \left[\cos^{-1}\left(\frac{1}{b}\right) - \sec^{-1}(\ln 3)\right]$

$= \cos^{-1}(0) - \sec^{-1}(\ln 3) = \frac{\pi}{2} - \sec^{-1}(\ln 3) \approx 1.1439$

9. diverges by the Limit Comparison Test (part 1) when compared with $\sum_{n=1}^{\infty} \frac{1}{\sqrt{n}}$, a divergent p-series:

$$\lim_{n \to \infty} \frac{\left(\frac{1}{2\sqrt{n} + \sqrt[3]{n}}\right)}{\left(\frac{1}{\sqrt{n}}\right)} = \lim_{n \to \infty} \frac{\sqrt{n}}{2\sqrt{n} + \sqrt[3]{n}} = \lim_{n \to \infty} \left(\frac{1}{2 + n^{-1/6}}\right) = \frac{1}{2}$$

11. converges by the Direct Comparison Test; $\frac{\sin^2 n}{2^n} \le \frac{1}{2^n}$, which is the nth term of a convergent geometric series

13. converges by the Direct Comparison Test; $\left(\frac{n}{3n+1}\right)^n < \left(\frac{n}{3n}\right)^n < \left(\frac{1}{3}\right)^n$, the nth term of a convergent geometric series

15. diverges by the Limit Comparison Test (part 3) when compared with $\sum_{n=2}^{\infty} \frac{1}{n}$, a divergent p-series:

$$\lim_{n \to \infty} \frac{\left(\frac{1}{(\ln n)^2}\right)}{\left(\frac{1}{n}\right)} = \lim_{n \to \infty} \frac{n}{(\ln n)^2} = \lim_{n \to \infty} \frac{1}{2(\ln n)\left(\frac{1}{n}\right)} = \frac{1}{2} \lim_{n \to \infty} \frac{n}{\ln n} = \frac{1}{2} \lim_{n \to \infty} \frac{1}{\left(\frac{1}{n}\right)} = \frac{1}{2} \lim_{n \to \infty} n = \infty$$

17. converges by the Limit Comparison Test (part 2) when compared with $\sum_{n=1}^{\infty} \frac{1}{n^2}$, a convergent p-series:

$$\lim_{n \to \infty} \frac{\left[\frac{(\ln n)^3}{n^3}\right]}{\left(\frac{1}{n^2}\right)} = \lim_{n \to \infty} \frac{(\ln n)^3}{n} = \lim_{n \to \infty} \frac{3(\ln n)^2\left(\frac{1}{n}\right)}{1} = 3 \lim_{n \to \infty} \frac{(\ln n)^2}{n} = 3 \lim_{n \to \infty} \frac{2(\ln n)\left(\frac{1}{n}\right)}{1} = 6 \lim_{n \to \infty} \frac{\ln n}{n}$$

$= 6 \cdot 0 = 0$ (Table 8.1)

19. converges by the Limit Comparison Test (part 2) with $\frac{1}{n^{5/4}}$, the nth term of a convergent p-series:

$$\lim_{n \to \infty} \frac{\left[\frac{(\ln n)^2}{n^{3/2}}\right]}{\left(\frac{1}{n^{5/4}}\right)} = \lim_{n \to \infty} \frac{(\ln n)^2}{n^{1/4}} = \lim_{n \to \infty} \frac{\left(\frac{2 \ln n}{n}\right)}{\left(\frac{1}{4n^{3/4}}\right)} = 8 \lim_{n \to \infty} \frac{\ln n}{n^{1/4}} = 8 \lim_{n \to \infty} \frac{\left(\frac{1}{n}\right)}{\left(\frac{1}{4n^{3/4}}\right)} = 32 \lim_{n \to \infty} \frac{1}{n^{1/4}} = 32 \cdot 0 = 0$$

21. converges by the Ratio Test: $\lim_{n\to\infty} \frac{a_{n+1}}{a_n} = \lim_{n\to\infty} \frac{\left[\frac{(n+1)^{\sqrt{2}}}{2^{n+1}}\right]}{\left[\frac{n^{\sqrt{2}}}{2^n}\right]} = \lim_{n\to\infty} \frac{(n+1)^{\sqrt{2}}}{2^{n+1}} \cdot \frac{2^n}{n^{\sqrt{2}}}$

$= \lim_{n\to\infty} \left(1+\frac{1}{n}\right)^{\sqrt{2}} \left(\frac{1}{2}\right) = \frac{1}{2} < 1$

23. diverges by the Ratio Test: $\lim_{n\to\infty} \frac{a_{n+1}}{a_n} = \lim_{n\to\infty} \frac{\left(\frac{(n+1)!}{e^{n+1}}\right)}{\left(\frac{n!}{e^n}\right)} = \lim_{n\to\infty} \frac{(n+1)!}{e^{n+1}} \cdot \frac{e^n}{n!} = \lim_{n\to\infty} \frac{n+1}{e} = \infty$

25. converges by the Ratio Test: $\lim_{n\to\infty} \frac{a_{n+1}}{a_n} = \lim_{n\to\infty} \frac{\left(\frac{(n+1)^{10}}{10^{n+1}}\right)}{\left(\frac{n^{10}}{10^n}\right)} = \lim_{n\to\infty} \frac{(n+1)^{10}}{10^{n+1}} \cdot \frac{10^n}{n^{10}} = \lim_{n\to\infty} \left(1+\frac{1}{n}\right)^{10} \left(\frac{1}{10}\right)$

$= \frac{1}{10} < 1$

27. converges by the Ratio Test: $\lim_{n\to\infty} \frac{a_{n+1}}{a_n} = \lim_{n\to\infty} \frac{(n+2)(n+3)}{(n+1)!} \cdot \frac{n!}{(n+1)(n+2)} = 0 < 1$

29. converges by the nth-Root Test: $\lim_{n\to\infty} \sqrt[n]{a_n} = \lim_{n\to\infty} \sqrt[n]{\frac{(\ln n)^n}{n^n}} = \lim_{n\to\infty} \frac{((\ln n)^n)^{1/n}}{(n^n)^{1/n}} = \lim_{n\to\infty} \frac{\ln n}{n}$

$= \lim_{n\to\infty} \frac{\left(\frac{1}{n}\right)}{1} = 0 < 1$

31. converges by the Root Test: $\lim_{n\to\infty} \sqrt[n]{a_n} = \lim_{n\to\infty} \sqrt[n]{\frac{n}{(\ln n)^n}} = \lim_{n\to\infty} \frac{\sqrt[n]{n}}{\ln n} = \lim_{n\to\infty} \frac{1}{\ln n} = 0 < 1$

33. diverges by the Root Test: $\lim_{n\to\infty} \sqrt[n]{a_n} \equiv \lim_{n\to\infty} \sqrt[n]{\frac{(n!)^n}{(n^n)^2}} = \lim_{n\to\infty} \frac{n!}{n^2} = \infty > 1$

35. converges; a geometric series with $r = \frac{1}{e} < 1$

37. diverges; $\sum_{n=1}^{\infty} \frac{3}{\sqrt{n}} = 3 \sum_{n=1}^{\infty} \frac{1}{\sqrt{n}}$, which is a divergent p-series

39. diverges by the Limit Comparison Test (part 3) with $\frac{1}{n}$, the nth term of the divergent harmonic series:

$\lim_{n\to\infty} \frac{\left(\frac{1}{(1+\ln n)^2}\right)}{\left(\frac{1}{n}\right)} = \lim_{n\to\infty} \frac{n}{(1+\ln n)^2} = \lim_{n\to\infty} \frac{1}{\left[\frac{2(1+\ln n)}{n}\right]}$ (by L'Hôpital's Rule) $= \lim_{n\to\infty} \frac{n}{2(1+\ln n)}$

$= \lim_{n\to\infty} \frac{1}{\left(\frac{2}{n}\right)}$ (by L'Hôpital's Rule) $= \lim_{n\to\infty} \frac{n}{2} = \infty$

41. converges by the Direct Comparison Test with $\frac{1}{n^{3/2}}$, the nth term of a convergent p-series: $n^2 - 1 > n$ for $n \geq 2 \Rightarrow n^2(n^2-1) > n^3 \Rightarrow n\sqrt{n^2-1} > n^{3/2} \Rightarrow \frac{1}{n^{3/2}} > \frac{1}{n\sqrt{n^2-1}}$

43. converges by the Ratio Test: $\lim_{n\to\infty} \frac{a_{n+1}}{a_n} = \lim_{n\to\infty} \frac{(n+4)!}{3!\,(n+1)!\,3^{n+1}} \cdot \frac{3!\,n!\,3^n}{(n+3)!} = \lim_{n\to\infty} \frac{n+4}{3(n+1)} = \frac{1}{3} < 1$

45. converges by the Ratio Test: $\lim_{n\to\infty} \frac{a_{n+1}}{a_n} = \lim_{n\to\infty} \frac{(n+1)!}{(2n+3)!} \cdot \frac{(2n+1)!}{n!} = \lim_{n\to\infty} \frac{n+1}{(2n+3)(2n+2)} = 0 < 1$

47. converges by the Integral Test: $\int_1^\infty \frac{8\tan^{-1}x}{1+x^2}\,dx;\ \begin{bmatrix} u = \tan^{-1}x \\ du = \frac{dx}{1+x^2} \end{bmatrix} \to \int_{\pi/4}^{\pi/2} 8u\,du = \left[4u^2\right]_{\pi/4}^{\pi/2} = 4\left(\frac{\pi^2}{4} - \frac{\pi^2}{16}\right) = \frac{3\pi^2}{4}$

49. converges by the Integral Test: $\int_1^\infty \operatorname{sech} x\,dx = 2\lim_{b\to\infty} \int_1^b \frac{e^x}{1+(e^x)^2}\,dx = 2\lim_{b\to\infty} \left[\tan^{-1} e^x\right]_1^b$

$= 2\lim_{b\to\infty} \left(\tan^{-1} e^b - \tan^{-1} e\right) = \pi - 2\tan^{-1} e$

51. converges by the Direct Comparison Test: $\frac{2+(-1)^n}{(1.25)^n} = \left(\frac{4}{5}\right)^n \left[2+(-1)^n\right] \leq \left(\frac{4}{5}\right)^n (3)$

53. converges by the Direct Comparison Test: $\frac{\ln n}{n^3} < \frac{n}{n^3} = \frac{1}{n^2}$ for $n \geq 2$

55. converges by the Limit Comparison Test (part 1) with $\frac{1}{n^2}$, the nth term of a convergent p-series:

$$\lim_{n\to\infty} \frac{\left(\frac{10n+1}{n(n+1)(n+2)}\right)}{\left(\frac{1}{n^2}\right)} = \lim_{n\to\infty} \frac{10n^2+n}{n^2+3n+2} = \lim_{n\to\infty} \frac{20n+1}{2n+3} = \lim_{n\to\infty} \frac{20}{2} = 10$$

57. converges by the Direct Comparison Test: $\frac{\tan^{-1} n}{n^{1.1}} < \frac{\frac{\pi}{2}}{n^{1.1}}$ and $\sum_{n=1}^{\infty} \frac{\frac{\pi}{2}}{n^{1.1}} = \frac{\pi}{2} \sum_{n=1}^{\infty} \frac{1}{n^{1.1}}$ is the product of a convergent p-series and a nonzero constant

59. diverges by the nth-Term Test for divergence; $\lim_{n\to\infty} n \sin\left(\frac{1}{n}\right) = \lim_{n\to\infty} \frac{\sin\left(\frac{1}{n}\right)}{\left(\frac{1}{n}\right)} = \lim_{x\to 0} \frac{\sin x}{x} = 1 \neq 0$

61. converges by the Ratio Test: $\lim_{n\to\infty} \frac{a_{n+1}}{a_n} = \lim_{n\to\infty} \frac{\left(\frac{1+\sin n}{n}\right)a_n}{a_n} = 0 < 1$

63. diverges by the Ratio Test: $\lim_{n\to\infty} \frac{a_{n+1}}{a_n} = \lim_{n\to\infty} \frac{\left(\frac{3n-1}{2n+1}\right)a_n}{a_n} = \lim_{n\to\infty} \frac{3n-1}{2n+1} = \frac{3}{2} > 1$

65. diverges by the nth-Term Test: $a_1 = \frac{1}{3}$, $a_2 = \sqrt[2]{\frac{1}{3}}$, $a_3 = \sqrt[3]{\sqrt[2]{\frac{1}{3}}} = \sqrt[6]{\frac{1}{3}}$, $a_4 = \sqrt[4]{\sqrt[3]{\sqrt[2]{\frac{1}{3}}}} = \sqrt[4!]{\frac{1}{3}}, \ldots,$

$a_n = \sqrt[n!]{\frac{1}{3}} \Rightarrow \lim_{n\to\infty} a_n = 1$ because $\left\{\sqrt[n!]{\frac{1}{3}}\right\}$ is a subsequence of $\left\{\sqrt[n]{\frac{1}{3}}\right\}$ whose limit is 1 by Table 8.1

67. (a) If $\lim_{n\to\infty} \frac{a_n}{b_n} = 0$, then there exists an integer N such that for all $n > N$, $\left|\frac{a_n}{b_n} - 0\right| < 1 \Rightarrow -1 < \frac{a_n}{b_n} < 1$ $\Rightarrow a_n < b_n$. Thus, if $\sum b_n$ converges, then $\sum a_n$ converges by the Direct Comparison Test.

(b) If $\lim_{n\to\infty} \frac{a_n}{b_n} = \infty$, then there exists an integer N such that for all $n > N$, $\frac{a_n}{b_n} > 1 \Rightarrow a_n > b_n$. Thus, if $\sum b_n$ diverges, then $\sum a_n$ diverges by the Direct Comparison Test.

69. $\lim_{n\to\infty} \frac{a_n}{b_n} = \infty \Rightarrow$ there exists an integer N such that for all $n > N$, $\frac{a_n}{b_n} > 1 \Rightarrow a_n > b_n$. If $\sum a_n$ converges, then $\sum b_n$ converges by the Direct Comparison Test

71. $\int_1^\infty \left(\frac{a}{x+2} - \frac{1}{x+4}\right) dx = \lim_{b\to\infty} \left[a \ln|x+2| - \ln|x+4|\right]_1^b = \lim_{b\to\infty} \ln \frac{(b+2)^a}{b+4} - \ln\left(\frac{3^a}{5}\right)$;

$\lim_{b\to\infty} \frac{(b+2)^a}{b+4} = a \lim_{b\to\infty} (b+2)^{a-1} = \begin{cases} \infty, & a > 1 \\ 1, & a = 1 \end{cases} \Rightarrow$ the series converges to $\ln\left(\frac{5}{3}\right)$ if $a = 1$ and diverges to ∞ if $a > 1$. If $a < 1$, the terms of the series eventually become negative and the Integral Test does not apply. From that point on, however, the series behaves like a negative multiple of the harmonic series, and so it diverges.

73. Let $A_n = \sum_{k=1}^{n} a_k$ and $B_n = \sum_{k=1}^{n} 2^k a_{(2^k)}$, where $\{a_k\}$ is a nonincreasing sequence of positive terms converging to 0. Note that $\{A_n\}$ and $\{B_n\}$ are nondecreasing sequences of positive terms. Now,

$B_n = 2a_2 + 4a_4 + 8a_8 + \ldots + 2^n a_{(2^n)} = 2a_2 + (2a_4 + 2a_4) + (2a_8 + 2a_8 + 2a_8 + 2a_8) + \ldots$

$+ \underbrace{\left(2a_{(2^n)} + 2a_{(2^n)} + \ldots + 2a_{(2^n)}\right)}_{2^{n-1} \text{ terms}} \le 2a_1 + 2a_2 + (2a_3 + 2a_4) + (2a_5 + 2a_6 + 2a_7 + 2a_8) + \ldots$

$+ \left(2a_{(2^{n-1})} + 2a_{(2^{n-1}+1)} + \ldots + 2a_{(2^n)}\right) = 2A_{(2^n)} \le 2\sum_{k=1}^{\infty} a_k$. Therefore if $\sum a_k$ converges, then $\{B_n\}$ is bounded above $\Rightarrow \sum 2^k a_{(2^k)}$ converges. Conversely,

$A_n = a_1 + (a_2 + a_3) + (a_4 + a_5 + a_6 + a_7) + \ldots + a_n < a_1 + 2a_2 + 4a_4 + \ldots + 2^n a_{(2^n)} = a_1 + B_n < a_1 + \sum_{k=1}^{\infty} 2^k a_{(2^k)}$.

Therefore, if $\sum_{k=1}^{\infty} 2^k a_{(2^k)}$ converges, then $\{A_n\}$ is bounded above and hence converges.

75. (a) $\int_2^\infty \frac{dx}{x(\ln x)^p}$; $\left[\begin{array}{l} u = \ln x \\ du = \frac{dx}{x} \end{array}\right] \to \int_{\ln 2}^{\infty} u^{-p}\, du = \lim_{b\to\infty} \left[\frac{u^{-p+1}}{-p+1}\right]_{\ln 2}^{b} = \lim_{b\to\infty} \left(\frac{1}{1-p}\right)\left[b^{-p+1} - (\ln 2)^{-p+1}\right]$

$= \begin{cases} \frac{1}{p-1}(\ln 2)^{-p+1}, & p > 1 \\ \infty, & p < 1 \end{cases}$ $\Rightarrow$ the improper integral converges if $p > 1$ and diverges

if $p < 1$. For $p = 1$: $\int_2^{\infty} \frac{dx}{x \ln x} = \lim_{b\to\infty} [\ln(\ln x)]_2^b = \lim_{b\to\infty} [\ln(\ln b) - \ln(\ln 2)] = \infty$, so the improper

integral diverges if $p = 1$.

(b) Since the series and the integral converge or diverge together, $\sum_{n=2}^{\infty} \frac{1}{n(\ln n)^p}$ converges if and only if $p > 1$.

77. Ratio: $\lim_{n\to\infty} \frac{a_{n+1}}{a_n} = \lim_{n\to\infty} \frac{1}{(\ln(n+1))^p} \cdot \frac{(\ln n)^p}{1} = \left[\lim_{n\to\infty} \frac{\ln n}{\ln(n+1)}\right]^p = \left[\lim_{n\to\infty} \frac{\left(\frac{1}{n}\right)}{\left(\frac{1}{n+1}\right)}\right]^p = \left(\lim_{n\to\infty} \frac{n+1}{n}\right)^p$

$= (1)^p = 1 \Rightarrow$ no conclusion

Root: $\lim_{n\to\infty} \sqrt[n]{a_n} = \lim_{n\to\infty} \sqrt[n]{\frac{1}{(\ln n)^p}} = \frac{1}{\left(\lim_{n\to\infty} (\ln n)^{1/n}\right)^p}$; let $f(n) = (\ln n)^{1/n}$, then $\ln f(n) = \frac{\ln(\ln n)}{n}$

$\Rightarrow \lim_{n\to\infty} \ln f(n) = \lim_{n\to\infty} \frac{\ln(\ln n)}{n} = \lim_{n\to\infty} \frac{\left(\frac{1}{n \ln n}\right)}{1} = \lim_{n\to\infty} \frac{1}{n \ln n} = 0 \Rightarrow \lim_{n\to\infty} (\ln n)^{1/n}$

$= \lim_{n\to\infty} e^{\ln f(n)} = e^0 = 1$; therefore $\lim_{n\to\infty} \sqrt[n]{a_n} = \frac{1}{\left(\lim_{n\to\infty} (\ln n)^{1/n}\right)^p} = \frac{1}{(1)^p} = 1 \Rightarrow$ no conclusion

79. Ratio: $\lim_{n\to\infty} \frac{a_{n+1}}{a_n} = \lim_{n\to\infty} \frac{1}{(n+1)^p} \cdot \frac{n^p}{1} = \lim_{n\to\infty} \left(\frac{n}{n+1}\right)^p = 1^p = 1 \Rightarrow$ no conclusion

Root: $\lim_{n\to\infty} \sqrt[n]{a_n} = \lim_{n\to\infty} \sqrt[n]{\frac{1}{n^p}} = \lim_{n\to\infty} \frac{1}{\left(\sqrt[n]{n}\right)^p} = \frac{1}{(1)^p} = 1 \Rightarrow$ no conclusion

8.5 ALTERNATING SERIES, ABSOLUTE AND CONDITIONAL CONVERGENCE

1. converges absolutely $\Rightarrow$ converges by the Absolute Convergence Test since $\sum_{n=1}^{\infty} |a_n| = \sum_{n=1}^{\infty} \frac{1}{n^2}$ which is a convergent p-series

3. diverges by the nth-Term Test since for $n > 10 \Rightarrow \frac{n}{10} > 1 \Rightarrow \lim_{n\to\infty} \left(\frac{n}{10}\right)^n \neq 0 \Rightarrow \sum_{n=1}^{\infty} (-1)^{n+1}\left(\frac{n}{10}\right)^n$ diverges

5. converges by the Alternating Series Test because $f(x) = \ln x$ is an increasing function of $x \Rightarrow \frac{1}{\ln x}$ is decreasing

$\Rightarrow u_n \geq u_{n+1}$ for $n \geq 1$; also $u_n \geq 0$ for $n \geq 1$ and $\lim_{n\to\infty} \frac{1}{\ln n} = 0$

7. diverges by the nth-Term Test since $\lim_{n\to\infty} \frac{\ln n}{\ln n^2} = \lim_{n\to\infty} \frac{\ln n}{2 \ln n} = \lim_{n\to\infty} \frac{1}{2} = \frac{1}{2} \neq 0$

9. converges by the Alternating Series Test since $f(x) = \frac{\sqrt{x}+1}{x+1} \Rightarrow f'(x) = \frac{1-x-2\sqrt{x}}{2\sqrt{x}(x+1)^2} < 0 \Rightarrow f(x)$ is decreasing $\Rightarrow u_n \ge u_{n+1}$; also $u_n \ge 0$ for $n \ge 1$ and $\lim_{n\to\infty} u_n = \lim_{n\to\infty} \frac{\sqrt{n}+1}{n+1} = 0$

11. converges absolutely since $\sum_{n=1}^{\infty} |a_n| = \sum_{n=1}^{\infty} \left(\frac{1}{10}\right)^n$ a convergent geometric series

13. The series $\sum_{n=1}^{\infty} (-1)^n \frac{1}{\sqrt{n+1}}$ converges by the Alternating Series Test since $\left(\frac{1}{\sqrt{n+1}}\right) > \left(\frac{1}{\sqrt{n+2}}\right)$ and $\left(\frac{1}{\sqrt{n+1}}\right) \to 0$. The series diverges absolutely by the Integral Test: $\int_1^{\infty} \frac{1}{\sqrt{x+1}}\,dx = \lim_{b\to\infty} 2\sqrt{x+1}\Big|_1^b$ $= \lim_{b\to\infty} \left[2\sqrt{b+1} - 2\sqrt{2}\right] = \infty.$

15. converges absolutely since $\sum_{n=1}^{\infty} |a_n| = \sum_{n=1}^{\infty} \frac{n}{n^3+1}$ and $\frac{n}{n^3+1} < \frac{1}{n^2}$ which is the nth-term of a converging p-series

17. converges conditionally since $\frac{1}{n+3} > \frac{1}{(n+1)+3} > 0$ and $\lim_{n\to\infty} \frac{1}{n+3} = 0 \Rightarrow$ convergence; but $\sum_{n=1}^{\infty} |a_n|$ $= \sum_{n=1}^{\infty} \frac{1}{n+3}$ diverges because $\frac{1}{n+3} \ge \frac{1}{4n}$ and $\sum_{n=1}^{\infty} \frac{1}{n}$ is a divergent series

19. diverges by the nth-Term Test since $\lim_{n\to\infty} \frac{3+n}{5+n} = 1 \ne 0$

21. converges conditionally since $f(x) = \frac{1}{x^2} + \frac{1}{x} \Rightarrow f'(x) = -\left(\frac{2}{x^3} + \frac{1}{x^2}\right) < 0 \Rightarrow f(x)$ is decreasing and hence $u_n > u_{n+1} > 0$ for $n \ge 1$ and $\lim_{n\to\infty} \left(\frac{1}{n^2} + \frac{1}{n}\right) = 0 \Rightarrow$ convergence; but $\sum_{n=1}^{\infty} |a_n| = \sum_{n=1}^{\infty} \frac{1+n}{n^2}$ $= \sum_{n=1}^{\infty} \frac{1}{n^2} + \sum_{n=1}^{\infty} \frac{1}{n}$ is the sum of a convergent and divergent series, and hence diverges

23. converges absolutely by the Ratio Test: $\lim_{n\to\infty} \left(\frac{u_{n+1}}{u_n}\right) = \lim_{n\to\infty} \left[\frac{(n+1)^2\left(\frac{2}{3}\right)^{n+1}}{n^2\left(\frac{2}{3}\right)^n}\right] = \frac{2}{3} < 1$

25. converges absolutely by the Integral Test since $\int_1^{\infty} (\tan^{-1} x)\left(\frac{1}{1+x^2}\right) dx = \lim_{b\to\infty} \left[\frac{(\tan^{-1} x)^2}{2}\right]_1^b$ $= \lim_{b\to\infty} \left[(\tan^{-1} b)^2 - (\tan^{-1} 1)^2\right] = \frac{1}{2}\left[\left(\frac{\pi}{2}\right)^2 - \left(\frac{\pi}{4}\right)^2\right] = \frac{3\pi^2}{32}$

27. diverges by the nth-Term Test since $\lim_{n\to\infty} \frac{n}{n+1} = 1 \ne 0$

29. converges absolutely by the Ratio Test: $\lim_{n\to\infty}\left(\frac{u_{n+1}}{u_n}\right) = \lim_{n\to\infty}\frac{(100)^{n+1}}{(n+1)!}\cdot\frac{n!}{(100)^n} = \lim_{n\to\infty}\frac{100}{n+1} = 0 < 1$

31. converges absolutely by the Direct Comparison Test since $\sum_{n=1}^{\infty}|a_n| = \sum_{n=1}^{\infty}\frac{1}{n^2+2n+1}$ and

$\frac{1}{n^2+2n+1} < \frac{1}{n^2}$ which is the nth-term of a convergent p-series

33. converges absolutely since $\sum_{n=1}^{\infty}|a_n| = \sum_{n=1}^{\infty}\left|\frac{(-1)^n}{n\sqrt{n}}\right| = \sum_{n=1}^{\infty}\frac{1}{n^{3/2}}$ is a convergent p-series

35. converges absolutely by the Root Test: $\lim_{n\to\infty}\sqrt[n]{|a_n|} = \lim_{n\to\infty}\left(\frac{(n+1)^n}{(2n)^n}\right)^{1/n} = \lim_{n\to\infty}\frac{n+1}{2n} = \frac{1}{2}$

37. diverges by the nth-Term Test since $\lim_{n\to\infty}|a_n| = \lim_{n\to\infty}\frac{(2n)!}{2^n n!\,n} = \lim_{n\to\infty}\frac{(n+1)(n+2)\cdots(2n)}{2^n n}$

$= \lim_{n\to\infty}\frac{(n+1)(n+2)\cdots(n+(n-1))}{2^{n-1}} > \lim_{n\to\infty}\left(\frac{n+1}{2}\right)^{n-1} = \infty \ne 0$

39. converges conditionally since $\frac{\sqrt{n+1}-\sqrt{n}}{1}\cdot\frac{\sqrt{n+1}+\sqrt{n}}{\sqrt{n+1}+\sqrt{n}} = \frac{1}{\sqrt{n+1}+\sqrt{n}}$ and $\left\{\frac{1}{\sqrt{n+1}+\sqrt{n}}\right\}$ is a

decreasing sequence of positive terms which converges to $0 \Rightarrow \sum_{n=1}^{\infty}\frac{(-1)^n}{\sqrt{n+1}+\sqrt{n}}$ converges; but $n > \frac{1}{3} \Rightarrow 3n > 1$

$\Rightarrow 4n > n+1 \Rightarrow 2\sqrt{n} > \sqrt{n+1} \Rightarrow 3\sqrt{n} > \sqrt{n+1}+\sqrt{n} \Rightarrow \frac{1}{3\sqrt{n}} < \frac{1}{\sqrt{n+1}+\sqrt{n}} \Rightarrow \sum_{n=1}^{\infty}\frac{1}{\sqrt{n+1}+\sqrt{n}}$

diverges by the Direct Comparison Test

41. diverges by the nth-Term Test since $\lim_{n\to\infty}\left(\sqrt{n+\sqrt{n}}-\sqrt{n}\right) = \lim_{n\to\infty}\left[\left(\sqrt{n+\sqrt{n}}-\sqrt{n}\right)\left(\frac{\sqrt{n+\sqrt{n}}+\sqrt{n}}{\sqrt{n+\sqrt{n}}+\sqrt{n}}\right)\right]$

$= \lim_{n\to\infty}\frac{\sqrt{n}}{\sqrt{n+\sqrt{n}}+\sqrt{n}} = \lim_{n\to\infty}\frac{1}{\sqrt{1+\frac{1}{\sqrt{n}}}+1} = \frac{1}{2} \ne 0$

43. converges absolutely by the Direct Comparison Test since $\operatorname{sech}(n) = \frac{2}{e^n+e^{-n}} = \frac{2e^n}{e^{2n}+1} < \frac{2e^n}{e^{2n}} = \frac{2}{e^n}$ which is the nth term of a convergent geometric series

45. $|\text{error}| < \left|(-1)^6\left(\frac{1}{5}\right)\right| = 0.2$

47. $|\text{error}| < \left|(-1)^6\frac{(0.01)^5}{5}\right| = 2\times 10^{-11}$

49. $\frac{1}{(2n)!} < \frac{5}{10^6} \Rightarrow (2n)! > \frac{10^6}{5} = 200{,}000 \Rightarrow n \ge 5 \Rightarrow 1 - \frac{1}{2!} + \frac{1}{4!} - \frac{1}{6!} + \frac{1}{8!} \approx 0.54030$

51. (a) $a_n \ge a_{n+1}$ fails since $\frac{1}{3} < \frac{1}{2}$

(b) Since $\sum_{n=1}^{\infty}|a_n| = \sum_{n=1}^{\infty}\left[\left(\frac{1}{3}\right)^n + \left(\frac{1}{2}\right)^n\right] = \sum_{n=1}^{\infty}\left(\frac{1}{3}\right)^n + \sum_{n=1}^{\infty}\left(\frac{1}{2}\right)^n$ is the sum of two absolutely convergent series, we can rearrange the terms of the original series to find its sum:

$$\left(\tfrac{1}{3}+\tfrac{1}{9}+\tfrac{1}{27}+\ldots\right)-\left(\tfrac{1}{2}+\tfrac{1}{4}+\tfrac{1}{8}+\ldots\right)=\frac{\left(\frac{1}{3}\right)}{1-\left(\frac{1}{3}\right)}-\frac{\left(\frac{1}{2}\right)}{1-\left(\frac{1}{2}\right)}=\tfrac{1}{2}-1=-\tfrac{1}{2}$$

53. The unused terms are $\sum\limits_{j=n+1}^{\infty}(-1)^{j+1}a_j=(-1)^{n+1}(a_{n+1}-a_{n+2})+(-1)^{n+3}(a_{n+3}-a_{n+4})+\ldots$ $=(-1)^{n+1}[(a_{n+1}-a_{n+2})+(a_{n+3}-a_{n+4})+\ldots]$. Each grouped term is positive, so the remainder has the same sign as $(-1)^{n+1}$, which is the sign of the first unused term.

55. Using the Direct Comparison Test, since $|a_n|\ge a_n$ and $\sum\limits_{n=1}^{\infty}a_n$ diverges we must have that $\sum\limits_{n=1}^{\infty}|a_n|$ diverges.

57. (a) $\sum\limits_{n=1}^{\infty}|a_n+b_n|$ converges by the Direct Comparison Test since $|a_n+b_n|\le|a_n|+|b_n|$ and hence $\sum\limits_{n=1}^{\infty}(a_n+b_n)$ converges absolutely

(b) $\sum\limits_{n=1}^{\infty}|b_n|$ converges $\Rightarrow \sum\limits_{n=1}^{\infty}-b_n$ converges absolutely; since $\sum\limits_{n=1}^{\infty}a_n$ converges absolutely and $\sum\limits_{n=1}^{\infty}-b_n$ converges absolutely, we have $\sum\limits_{n=1}^{\infty}[a_n+(-b_n)]=\sum\limits_{n=1}^{\infty}(a_n-b_n)$ converges absolutely by part (a)

(c) $\sum\limits_{n=1}^{\infty}|a_n|$ converges $\Rightarrow |k|\sum\limits_{n=1}^{\infty}|a_n|=\sum\limits_{n=1}^{\infty}|ka_n|$ converges $\Rightarrow \sum\limits_{n=1}^{\infty}ka_n$ converges absolutely

59. $s_1=-\frac{1}{2}$, $s_2=-\frac{1}{2}+1=\frac{1}{2}$,

$s_3=-\frac{1}{2}+1-\frac{1}{4}-\frac{1}{6}-\frac{1}{8}-\frac{1}{10}-\frac{1}{12}-\frac{1}{14}-\frac{1}{16}-\frac{1}{18}-\frac{1}{20}-\frac{1}{22}\approx-0.5099$,

$s_4=s_3+\frac{1}{3}\approx-0.1766$,

$s_5=s_4-\frac{1}{24}-\frac{1}{26}-\frac{1}{28}-\frac{1}{30}-\frac{1}{32}-\frac{1}{34}-\frac{1}{36}-\frac{1}{38}-\frac{1}{40}-\frac{1}{42}-\frac{1}{44}\approx-0.512$,

$s_6=s_5+\frac{1}{5}\approx-0.312$,

$s_7=s_6-\frac{1}{46}-\frac{1}{48}-\frac{1}{50}-\frac{1}{52}-\frac{1}{54}-\frac{1}{56}-\frac{1}{58}-\frac{1}{60}-\frac{1}{62}-\frac{1}{64}-\frac{1}{66}\approx-0.51106$

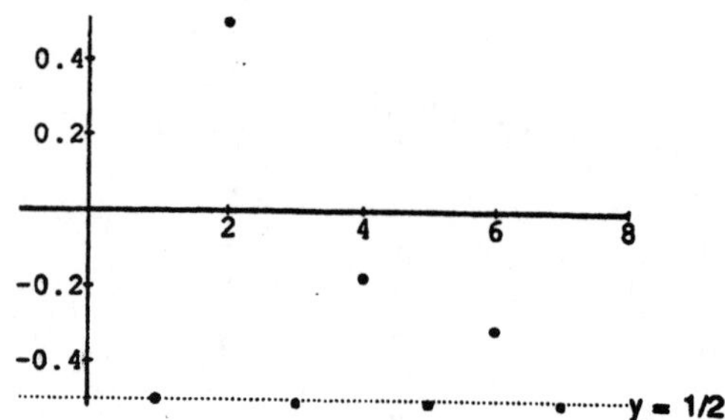

61. (a) If $\sum_{n=1}^{\infty} |a_n|$ converges, then $\sum_{n=1}^{\infty} a_n$ converges and $\frac{1}{2}\sum_{n=1}^{\infty} a_n + \frac{1}{2}\sum_{n=1}^{\infty} |a_n| = \sum_{n=1}^{\infty} \frac{a_n + |a_n|}{2}$

converges where $b_n = \frac{a_n + |a_n|}{2} = \begin{cases} a_n, & \text{if } a_n \geq 0 \\ 0, & \text{if } a_n < 0 \end{cases}$.

(b) If $\sum_{n=1}^{\infty} |a_n|$ converges, then $\sum_{n=1}^{\infty} a_n$ converges and $\frac{1}{2}\sum_{n=1}^{\infty} a_n - \frac{1}{2}\sum_{n=1}^{\infty} |a_n| = \sum_{n=1}^{\infty} \frac{a_n - |a_n|}{2}$

converges where $c_n = \frac{a_n - |a_n|}{2} = \begin{cases} 0, & \text{if } a_n \geq 0 \\ a_n, & \text{if } a_n < 0 \end{cases}$.

63. Here is an example figure when N = 5. Notice that $u_3 > u_2 > u_1$ and $u_3 > u_5 > u_4$, but $u_n \geq u_{n+1}$ for $n \geq 5$.

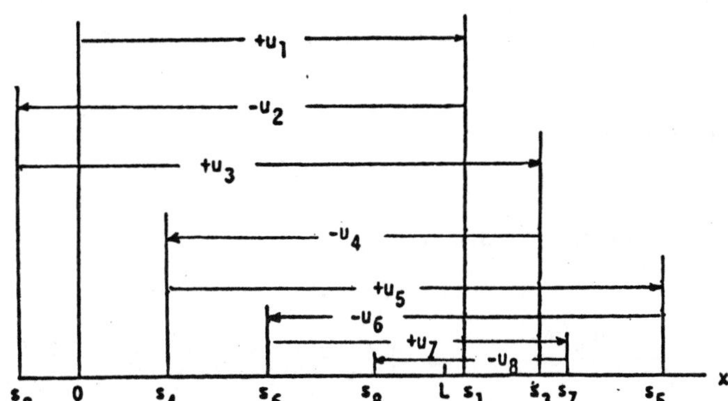

8.6 POWER SERIES

1. $\lim_{n\to\infty} \left|\frac{u_{n+1}}{u_n}\right| < 1 \Rightarrow \lim_{n\to\infty} \left|\frac{x^{n+1}}{x^n}\right| < 1 \Rightarrow |x| < 1 \Rightarrow -1 < x < 1$; when $x = -1$ we have $\sum_{n=1}^{\infty} (-1)^n$, a divergent series; when $x = 1$ we have $\sum_{n=1}^{\infty} 1$, a divergent series

(a) the radius is 1; the interval of convergence is $-1 < x < 1$

(b) the interval of absolute convergence is $-1 < x < 1$

(c) there are no values for which the series converges conditionally

3. $\lim_{n\to\infty} \left|\frac{u_{n+1}}{u_n}\right| < 1 \Rightarrow \lim_{n\to\infty} \left|\frac{(4x+1)^{n+1}}{(4x+1)^n}\right| < 1 \Rightarrow |4x+1| < 1 \Rightarrow -1 < 4x+1 < 1 \Rightarrow -\frac{1}{2} < x < 0$; when $x = -\frac{1}{2}$ we have $\sum_{n=1}^{\infty} (-1)^n(-1)^n = \sum_{n=1}^{\infty} (-1)^{2n} = \sum_{n=1}^{\infty} 1^n$, a divergent series; when $x = 0$ we have $\sum_{n=1}^{\infty} (-1)^n(1)^n$ $= \sum_{n=1}^{\infty} (-1)^n$, a divergent series

(a) the radius is $\frac{1}{4}$; the interval of convergence is $-\frac{1}{2} < x < 0$

(b) the interval of absolute convergence is $-\frac{1}{2} < x < 0$

(c) there are no values for which the series converges conditionally

5. $\lim_{n\to\infty} \left|\frac{u_{n+1}}{u_n}\right| < 1 \Rightarrow \lim_{n\to\infty} \left|\frac{(x-2)^{n+1}}{10^{n+1}} \cdot \frac{10^n}{(x-2)^n}\right| < 1 \Rightarrow \frac{|x-2|}{10} < 1 \Rightarrow |x-2| < 10 \Rightarrow -10 < x-2 < 10$

$\Rightarrow -8 < x < 12$; when $x = -8$ we have $\sum_{n=1}^{\infty} (-1)^n$, a divergent series; when $x = 12$ we have $\sum_{n=1}^{\infty} 1$, a divergent series

(a) the radius is 10; the interval of convergence is $-8 < x < 12$

(b) the interval of absolute convergence is $-8 < x < 12$

(c) there are no values for which the series converges conditionally

7. $\lim_{n\to\infty} \left|\frac{u_{n+1}}{u_n}\right| < 1 \Rightarrow \lim_{n\to\infty} \left|\frac{(n+1)x^{n+1}}{(n+3)} \cdot \frac{(n+2)}{nx^n}\right| < 1 \Rightarrow |x| \lim_{n\to\infty} \frac{(n+1)(n+2)}{(n+3)(n)} < 1 \Rightarrow |x| < 1$

$\Rightarrow -1 < x < 1$; when $x = -1$ we have $\sum_{n=1}^{\infty} (-1)^n \frac{n}{n+2}$, a divergent series by the nth-term Test; when $x = 1$ we have $\sum_{n=1}^{\infty} \frac{n}{n+2}$, a divergent series

(a) the radius is 1; the interval of convergence is $-1 < x < 1$

(b) the interval of absolute convergence is $-1 < x < 1$

(c) there are no values for which the series converges conditionally

9. $\lim_{n\to\infty} \left|\frac{u_{n+1}}{u_n}\right| < 1 \Rightarrow \lim_{n\to\infty} \left|\frac{x^{n+1}}{(n+1)\sqrt{n+1}\,3^{n+1}} \cdot \frac{n\sqrt{n}\,3^n}{x^n}\right| < 1 \Rightarrow \frac{|x|}{3}\left(\lim_{n\to\infty} \frac{n}{n+1}\right)\left(\sqrt{\lim_{n\to\infty} \frac{n}{n+1}}\right) < 1$

$\Rightarrow \frac{|x|}{3}(1)(1) < 1 \Rightarrow |x| < 3 \Rightarrow -3 < x < 3$; when $x = -3$ we have $\sum_{n=1}^{\infty} \frac{(-1)^n}{n^{3/2}}$, an absolutely convergent series; when $x = 3$ we have $\sum_{n=1}^{\infty} \frac{1}{n^{3/2}}$, a convergent p-series

(a) the radius is 3; the interval of convergence is $-3 \le x \le 3$

(b) the interval of absolute convergence is $-3 \le x \le 3$

(c) there are no values for which the series converges conditionally

11. $\lim_{n\to\infty} \left|\frac{u_{n+1}}{u_n}\right| < 1 \Rightarrow \lim_{n\to\infty} \left|\frac{x^{n+1}}{(n+1)!} \cdot \frac{n!}{x^n}\right| < 1 \Rightarrow |x| \lim_{n\to\infty} \left(\frac{1}{n+1}\right) < 1$ for all x

(a) the radius is ∞; the series converges for all x

(b) the series converges absolutely for all x

(c) there are no values for which the series converges conditionally

13. $\lim_{n\to\infty} \left|\frac{u_{n+1}}{u_n}\right| < 1 \Rightarrow \lim_{n\to\infty} \left|\frac{x^{2n+3}}{(n+1)!} \cdot \frac{n!}{x^{2n+1}}\right| < 1 \Rightarrow x^2 \lim_{n\to\infty} \left(\frac{1}{n+1}\right) < 1$ for all x

(a) the radius is ∞; the series converges for all x

(b) the series converges absolutely for all x

(c) there are no values for which the series converges conditionally

15. $\lim_{n\to\infty} \left|\frac{u_{n+1}}{u_n}\right| < 1 \Rightarrow \lim_{n\to\infty} \left|\frac{x^{n+1}}{\sqrt{(n+1)^2+3}} \cdot \frac{\sqrt{n^2+3}}{x^n}\right| < 1 \Rightarrow |x| \sqrt{\lim_{n\to\infty} \frac{n^2+3}{n^2+2n+4}} < 1 \Rightarrow |x| < 1$

$\Rightarrow -1 < x < 1$; when $x = -1$ we have $\sum_{n=1}^{\infty} \frac{(-1)^n}{\sqrt{n^2+3}}$, a conditionally convergent series; when $x = 1$ we have $\sum_{n=1}^{\infty} \frac{1}{n^2+3}$, a divergent series

(a) the radius is 1; the interval of convergence is $-1 \le x < 1$

(b) the interval of absolute convergence is $-1 < x < 1$

(c) the series converges conditionally at $x = -1$

17. $\lim_{n\to\infty} \left|\frac{u_{n+1}}{u_n}\right| < 1 \Rightarrow \lim_{n\to\infty} \left|\frac{(n+1)(x+3)^{n+1}}{5^{n+1}} \cdot \frac{5^n}{n(x+3)^n}\right| < 1 \Rightarrow \frac{|x+3|}{5} \lim_{n\to\infty} \left(\frac{n+1}{n}\right) < 1 \Rightarrow \frac{|x+3|}{5} < 1$

$\Rightarrow |x+3| < 5 \Rightarrow -5 < x+3 < 5 \Rightarrow -8 < x < 2$; when $x = -8$ we have $\sum_{n=1}^{\infty} \frac{n(-5)^n}{5^n} = \sum_{n=1}^{\infty} (-1)^n n$, a divergent series; when $x = 2$ we have $\sum_{n=1}^{\infty} \frac{n5^n}{5^n} = \sum_{n=1}^{\infty} n$, a divergent series

(a) the radius is 5; the interval of convergence is $-8 < x < 2$

(b) the interval of absolute convergence is $-8 < x < 2$

(c) there are no values for which the series converges conditionally

19. $\lim_{n\to\infty} \left|\frac{u_{n+1}}{u_n}\right| < 1 \Rightarrow \lim_{n\to\infty} \left|\frac{\sqrt{n+1}\,x^{n+1}}{3^{n+1}} \cdot \frac{3^n}{\sqrt{n}\,x^n}\right| < 1 \Rightarrow \frac{|x|}{3} \sqrt{\lim_{n\to\infty} \left(\frac{n+1}{n}\right)} < 1 \Rightarrow \frac{|x|}{3} < 1 \Rightarrow |x| < 3$

$\Rightarrow -3 < x < 3$; when $x = -3$ we have $\sum_{n=1}^{\infty} (-1)^n \sqrt{n}$, a divergent series; when $x = 3$ we have $\sum_{n=1}^{\infty} \sqrt{n}$, a divergent series

(a) the radius is 3; the interval of convergence is $-3 < x < 3$

(b) the interval of absolute convergence is $-3 < x < 3$

(c) there are no values for which the series converges conditionally

21. $\lim_{n\to\infty} \left|\frac{u_{n+1}}{u_n}\right| < 1 \Rightarrow \lim_{n\to\infty} \left|\frac{\left(1+\frac{1}{n+1}\right)^{n+1} x^{n+1}}{\left(1+\frac{1}{n}\right)^n x^n}\right| < 1 \Rightarrow |x| \left(\frac{\lim_{t\to\infty} \left(1+\frac{1}{t}\right)^t}{\lim_{n\to\infty} \left(1+\frac{1}{n}\right)^n}\right) < 1 \Rightarrow |x|\left(\frac{e}{e}\right) < 1 \Rightarrow |x| < 1$

$\Rightarrow -1 < x < 1$; when $x = -1$ we have $\sum_{n=1}^{\infty} (-1)^n \left(1+\frac{1}{n}\right)^n$, a divergent series by the nth-Term Test since $\lim_{n\to\infty} \left(1+\frac{1}{n}\right)^n = e \ne 0$; when $x = 1$ we have $\sum_{n=1}^{\infty} \left(1+\frac{1}{n}\right)^n$, a divergent series

(a) the radius is 1; the interval of convergence is $-1 < x < 1$

(b) the interval of absolute convergence is $-1 < x < 1$

(c) there are no values for which the series converges conditionally

23. $\lim_{n\to\infty} \left|\frac{u_{n+1}}{u_n}\right| < 1 \Rightarrow \lim_{n\to\infty} \left|\frac{(n+1)^{n+1}x^{n+1}}{n^n x^n}\right| < 1 \Rightarrow |x|\left(\lim_{n\to\infty}\left(1+\frac{1}{n}\right)^n\right)\left(\lim_{n\to\infty}(n+1)\right) < 1$

$\Rightarrow e|x| \lim_{n\to\infty}(n+1) < 1 \Rightarrow$ only $x = 0$ satisfies this inequality

(a) the radius is 0; the series converges only for $x = 0$

(b) the series converges absolutely only for $x = 0$

(c) there are no values for which the series converges conditionally

25. $\lim_{n\to\infty} \left|\frac{u_{n+1}}{u_n}\right| < 1 \Rightarrow \lim_{n\to\infty} \left|\frac{(x+2)^{n+1}}{(n+1)\,2^{n+1}} \cdot \frac{n2^n}{(x+2)^n}\right| < 1 \Rightarrow \frac{|x+2|}{2} \lim_{n\to\infty}\left(\frac{n}{n+1}\right) < 1 \Rightarrow \frac{|x+2|}{2} < 1 \Rightarrow |x+2| < 2$

$\Rightarrow -2 < x+2 < 2 \Rightarrow -4 < x < 0$; when $x = -4$ we have $\sum_{n=1}^{\infty} \frac{-1}{n}$, a divergent series; when $x = 0$ we have $\sum_{n=1}^{\infty} \frac{(-1)^{n+1}}{n}$, the alternating harmonic series which converges conditionally

(a) the radius is 2; the interval of convergence is $-4 < x \le 0$

(b) the interval of absolute convergence is $-4 < x < 0$

(c) the series converges conditionally at $x = 0$

27. $\lim_{n\to\infty} \left|\frac{u_{n+1}}{u_n}\right| < 1 \Rightarrow \lim_{n\to\infty} \left|\frac{x^{n+1}}{(n+1)(\ln(n+1))^2} \cdot \frac{n(\ln n)^2}{x^n}\right| < 1 \Rightarrow |x| \left(\lim_{n\to\infty} \frac{n}{n+1}\right)\left(\lim_{n\to\infty} \frac{\ln n}{\ln(n+1)}\right)^2 < 1$

$\Rightarrow |x|(1)\left(\lim_{n\to\infty} \frac{\left(\frac{1}{n}\right)}{\left(\frac{1}{n+1}\right)}\right)^2 < 1 \Rightarrow |x|\left(\lim_{n\to\infty} \frac{n+1}{n}\right)^2 < 1 \Rightarrow |x| < 1 \Rightarrow -1 < x < 1$; when $x = -1$ we have

$\sum_{n=1}^{\infty} \frac{(-1)^n}{n(\ln n)^2}$ which converges absolutely; when $x = 1$ we have $\sum_{n=1}^{\infty} \frac{1}{n(\ln n)^2}$ which converges

(a) the radius is 1; the interval of convergence is $-1 \le 1 \le 1$

(b) the interval of absolute convergence is $-1 \le x \le 1$

(c) there are no values for which the series converges conditionally

29. $\lim_{n\to\infty} \left|\frac{u_{n+1}}{u_n}\right| < 1 \Rightarrow \lim_{n\to\infty} \left|\frac{(4x-5)^{2n+3}}{(n+1)^{3/2}} \cdot \frac{n^{3/2}}{(4x-5)^{2n+1}}\right| < 1 \Rightarrow (4x-5)^2\left(\lim_{n\to\infty} \frac{n}{n+1}\right)^{3/2} < 1 \Rightarrow (4x-5)^2 < 1$

$\Rightarrow |4x-5| < 1 \Rightarrow -1 < 4x-5 < 1 \Rightarrow 1 < x < \frac{3}{2}$; when $x = 1$ we have $\sum_{n=1}^{\infty} \frac{(-1)^{2n+1}}{n^{3/2}} = \sum_{n=1}^{\infty} \frac{-1}{n^{3/2}}$ which is absolutely convergent; when $x = \frac{3}{2}$ we have $\sum_{n=1}^{\infty} \frac{(1)^{2n+1}}{n^{3/2}}$, a convergent p-series

(a) the radius is $\frac{1}{4}$; the interval of convergence is $1 \le x \le \frac{3}{2}$

(b) the interval of absolute convergence is $1 \le x \le \frac{3}{2}$

(c) there are no values for which the series converges conditionally

31. $\lim_{n\to\infty}\left|\frac{u_{n+1}}{u_n}\right| < 1 \Rightarrow \lim_{n\to\infty}\left|\frac{(x+\pi)^{n+1}}{\sqrt{n+1}}\cdot\frac{\sqrt{n}}{(x+\pi)^n}\right| < 1 \Rightarrow |x+\pi|\lim_{n\to\infty}\left|\sqrt{\frac{n}{n+1}}\right| < 1$

$\Rightarrow |x+\pi|\sqrt{\lim_{n\to\infty}\left(\frac{n}{n+1}\right)} < 1 \Rightarrow |x+\pi| < 1 \Rightarrow -1 < x+\pi < 1 \Rightarrow -1-\pi < x < 1-\pi$;

when $x = -1-\pi$ we have $\sum_{n=1}^{\infty}\frac{(-1)^n}{\sqrt{n}} = \sum_{n=1}^{\infty}\frac{(-1)^n}{n^{1/2}}$, a conditionally convergent series; when $x = 1-\pi$ we have

$\sum_{n=1}^{\infty}\frac{1^n}{\sqrt{n}} = \sum_{n=1}^{\infty}\frac{1}{n^{1/2}}$, a divergent p-series

(a) the radius is 1; the interval of convergence is $(-1-\pi) \le x < (1-\pi)$

(b) the interval of absolute convergence is $-1-\pi < x < 1-\pi$

(c) the series converges conditionally at $x = -1-\pi$

33. $\lim_{n\to\infty}\left|\frac{u_{n+1}}{u_n}\right| < 1 \Rightarrow \lim_{n\to\infty}\left|\frac{(x-1)^{2n+2}}{4^{n+1}}\cdot\frac{4^n}{(x-1)^{2n}}\right| < 1 \Rightarrow \frac{(x-1)^2}{4}\lim_{n\to\infty}|1| < 1 \Rightarrow (x-1)^2 < 4 \Rightarrow |x-1| < 2$

$\Rightarrow -2 < x-1 < 2 \Rightarrow -1 < x < 3$; at $x = -1$ we have $\sum_{n=0}^{\infty}\frac{(-2)^{2n}}{4^n} = \sum_{n=0}^{\infty}\frac{4^n}{4^n} = \sum_{n=0}^{\infty} 1$, which diverges; at $x = 3$

we have $\sum_{n=0}^{\infty}\frac{2^{2n}}{4^n} = \sum_{n=0}^{\infty}\frac{4^n}{4^n} = \sum_{n=0}^{\infty} 1$, a divergent series; the interval of convergence is $-1 < x < 3$; the series

$\sum_{n=0}^{\infty}\frac{(x-1)^{2n}}{4^n} = \sum_{n=0}^{\infty}\left(\left(\frac{x-1}{2}\right)^2\right)^n$ is a convergent geometric series when $-1 < x < 3$ and the sum is

$$\frac{1}{1-\left(\frac{x-1}{2}\right)^2} = \frac{1}{\left[\frac{4-(x-1)^2}{4}\right]} = \frac{4}{4-x^2+2x-1} = \frac{4}{3+2x-x^2}$$

35. $\lim_{n\to\infty}\left|\frac{u_{n+1}}{u_n}\right| < 1 \Rightarrow \lim_{n\to\infty}\left|\frac{\left(\sqrt{x}-2\right)^{n+1}}{2^{n+1}}\cdot\frac{2^n}{\left(\sqrt{x}-2\right)^n}\right| < 1 \Rightarrow \left|\sqrt{x}-2\right| < 2 \Rightarrow -2 < \sqrt{x}-2 < 2 \Rightarrow 0 < \sqrt{x} < 4$

$\Rightarrow 0 < x < 16$; when $x = 0$ we have $\sum_{n=0}^{\infty}(-1)^n$, a divergent series; when $x = 16$ we have $\sum_{n=0}^{\infty}(1)^n$, a divergent

series; the interval of convergence is $0 < x < 16$; the series $\sum_{n=0}^{\infty}\left(\frac{\sqrt{x}-2}{2}\right)^n$ is a convergent geometric series when

$0 < x < 16$ and its sum is $\frac{1}{1-\left(\frac{\sqrt{x}-2}{2}\right)} = \frac{1}{\left(\frac{2-\sqrt{x}+2}{2}\right)} = \frac{2}{4-\sqrt{x}}$

37. $\lim_{n\to\infty}\left|\frac{u_{n+1}}{u_n}\right| < 1 \Rightarrow \lim_{n\to\infty}\left|\left(\frac{x^2+1}{3}\right)^{n+1}\cdot\left(\frac{3}{x^2+1}\right)^n\right| < 1 \Rightarrow \frac{(x^2+1)}{3}\lim_{n\to\infty}|1| < 1 \Rightarrow \frac{x^2+1}{3} < 1 \Rightarrow x^2 < 2$

$\Rightarrow |x| < \sqrt{2} \Rightarrow -\sqrt{2} < x < \sqrt{2}$; at $x = \pm\sqrt{2}$ we have $\sum_{n=0}^{\infty}(1)^n$ which diverges; the interval of convergence is

$-\sqrt{2} < x < \sqrt{2}$; the series $\sum_{n=0}^{\infty}\left(\frac{x^2+1}{3}\right)^n$ is a convergent geometric series when $-\sqrt{2} < x < \sqrt{2}$ and its sum is

$$\frac{1}{1-\left(\frac{x^2+1}{3}\right)} = \frac{1}{\left(\frac{3-x^2-1}{3}\right)} = \frac{3}{2-x^2}$$

39. $\lim_{n\to\infty}\left|\frac{(x-3)^{n+1}}{2^{n+1}}\cdot\frac{2^n}{(x-3)^n}\right|<1 \Rightarrow |x-3|<2 \Rightarrow 1<x<5$; when $x=1$ we have $\sum_{n=1}^{\infty}(1)^n$ which diverges; when $x=5$ we have $\sum_{n=1}^{\infty}(-1)^n$ which also diverges; the interval of convergence is $1<x<5$; the sum of this convergent geometric series is $\frac{1}{1+\left(\frac{x-3}{2}\right)}=\frac{2}{x-1}$. If $f(x)=1-\frac{1}{2}(x-3)+\frac{1}{4}(x-3)^2+\ldots+\left(-\frac{1}{2}\right)^n(x-3)^n+\ldots$ $=\frac{2}{x-1}$ then $f'(x)=-\frac{1}{2}+\frac{1}{2}(x-3)+\ldots+\left(-\frac{1}{2}\right)^n n(x-3)^{n-1}+\ldots$ is convergent when $1<x<5$, and diverges when $x=1$ or 5. The sum for $f'(x)$ is $\frac{-2}{(x-1)^2}$, the derivative of $\frac{2}{x-1}$.

41. (a) Differentiate the series for sin x to get $\cos x = 1-\frac{3x^2}{3!}+\frac{5x^4}{5!}-\frac{7x^6}{7!}+\frac{9x^8}{9!}-\frac{11x^{10}}{11!}+\ldots$ $=1-\frac{x^2}{2!}+\frac{x^4}{4!}-\frac{x^6}{6!}+\frac{x^8}{8!}-\frac{x^{10}}{10!}+\ldots$. The series converges for all values of x since

$$\lim_{n\to\infty}\left|\frac{x^{n+1}}{(n+1)!}\cdot\frac{n!}{x^n}\right|=|x|\lim_{n\to\infty}\left(\frac{1}{n+1}\right)=0<1 \text{ for all } x$$

(b) $\sin 2x = 2x-\frac{2^3x^3}{3!}+\frac{2^5x^5}{5!}-\frac{2^7x^7}{7!}+\frac{2^9x^9}{9!}-\frac{2^{11}x^{11}}{11!}+\ldots = 2x-\frac{8x^3}{3!}+\frac{32x^5}{5!}-\frac{128x^7}{7!}+\frac{512x^9}{9!}-\frac{2048x^{11}}{11!}+\ldots$

(c) $2\sin x\cos x = 2\left[(0\cdot 1)+(0\cdot 0+1\cdot 1)x+\left(0\cdot\frac{-1}{2}+1\cdot 0+0\cdot 1\right)x^2+\left(0\cdot 0-1\cdot\frac{1}{2}+0\cdot 0-1\cdot\frac{1}{3!}\right)x^3\right.$
$+\left(0\cdot\frac{1}{4!}+1\cdot 0-0\cdot\frac{1}{2}-0\cdot\frac{1}{3!}+0\cdot 1\right)x^4+\left(0\cdot 0+1\cdot\frac{1}{4!}+0\cdot 0+\frac{1}{2}\cdot\frac{1}{3!}+0\cdot 0+1\cdot\frac{1}{5!}\right)x^5$
$\left.+\left(0\cdot\frac{1}{6!}+1\cdot 0+0\cdot\frac{1}{4!}+0\cdot\frac{1}{3!}+0\cdot\frac{1}{2}+0\cdot\frac{1}{5!}+0\cdot 1\right)x^6+\ldots\right]=2\left[x-\frac{4x^3}{3!}+\frac{16x^5}{5!}-\ldots\right]$
$=2x-\frac{2^3x^3}{3!}+\frac{2^5x^5}{5!}-\frac{2^7x^7}{7!}+\frac{2^9x^9}{9!}-\frac{2^{11}x^{11}}{11!}+\ldots$

43. (a) $\ln|\sec x|+C=\int\tan x\,dx=\int\left(x+\frac{x^3}{3}+\frac{2x^5}{15}+\frac{17x^7}{315}+\frac{62x^9}{2835}+\ldots\right)dx$
$=\frac{x^2}{2}+\frac{x^4}{12}+\frac{x^6}{45}+\frac{17x^8}{2520}+\frac{31x^{10}}{14{,}175}+\ldots+C$; $x=0\Rightarrow C=0\Rightarrow \ln|\sec x|=\frac{x^2}{2}+\frac{x^4}{12}+\frac{x^6}{45}+\frac{17x^8}{2520}+\frac{31x^{10}}{14{,}175}+\ldots,$ converges when $-\frac{\pi}{2}<x<\frac{\pi}{2}$

(b) $\sec^2 x=\frac{d(\tan x)}{dx}=\frac{d}{dx}\left(x+\frac{x^3}{3}+\frac{2x^5}{15}+\frac{17x^7}{315}+\frac{62x^9}{2835}+\ldots\right)=1+x^2+\frac{2x^4}{3}+\frac{17x^6}{45}+\frac{62x^8}{315}+\ldots,$ converges when $-\frac{\pi}{2}<x<\frac{\pi}{2}$

(c) $\sec^2 x=(\sec x)(\sec x)=\left(1+\frac{x^2}{2}+\frac{5x^4}{24}+\frac{61x^6}{720}+\ldots\right)\left(1+\frac{x^2}{2}+\frac{5x^4}{24}+\frac{61x^6}{720}+\ldots\right)$
$=1+\left(\frac{1}{2}+\frac{1}{2}\right)x^2+\left(\frac{5}{24}+\frac{1}{4}+\frac{5}{24}\right)x^4+\left(\frac{61}{720}+\frac{5}{48}+\frac{5}{48}+\frac{61}{720}\right)x^6+\ldots$
$=1+x^2+\frac{2x^4}{3}+\frac{17x^6}{45}+\frac{62x^8}{315}+\ldots,\ -\frac{\pi}{2}<x<\frac{\pi}{2}$

45. (a) If $f(x) = \sum_{n=0}^{\infty} a_n x^n$, then $f^{(k)}(x) = \sum_{n=k}^{\infty} n(n-1)(n-2)\cdots(n-(k-1))\, a_n x^{n-k}$ and $f^{(k)}(0) = k!a_k$

$\Rightarrow a_k = \frac{f^{(k)}(0)}{k!}$; likewise if $f(x) = \sum_{n=0}^{\infty} b_n x^n$, then $b_k = \frac{f^{(k)}(0)}{k!} \Rightarrow a_k = b_k$ for every nonnegative integer k

(b) If $f(x) = \sum_{n=0}^{\infty} a_n x^n = 0$ for all x, then $f^{(k)}(x) = 0$ for all x $\Rightarrow$ from part (a) that $a_k = 0$ for every nonnegative integer k

47. The series $\sum_{n=1}^{\infty} \frac{x^n}{n}$ converges conditionally at the left-hand endpoint of its interval of convergence $[-1,1]$; the series $\sum_{n=1}^{\infty} \frac{x^n}{(n^2)}$ converges absolutely at the left-hand endpoint of its interval of convergence $[-1,1]$

8.7 TAYLOR AND MACLAURIN SERIES

1. $f(x) = \ln x$, $f'(x) = \frac{1}{x}$, $f''(x) = -\frac{1}{x^2}$, $f'''(x) = \frac{2}{x^3}$; $f(1) = \ln 1 = 0$, $f'(1) = 1$, $f''(1) = -1$, $f'''(1) = 2 \Rightarrow P_0(x) = 0$, $P_1(x) = (x-1)$, $P_2(x) = (x-1) - \frac{1}{2}(x-1)^2$, $P_3(x) = (x-1) - \frac{1}{2}(x-1)^2 + \frac{1}{3}(x-1)^3$

3. $f(x) = (x+2)^{-1}$, $f'(x) = -(x+2)^{-2}$, $f''(x) = 2(x+2)^{-3}$, $f'''(x) = -6(x+2)^{-4}$; $f(0) = (2)^{-1} = \frac{1}{2}$, $f'(0) = -(2)^{-2} = -\frac{1}{4}$, $f''(0) = 2(2)^{-3} = \frac{1}{4}$, $f'''(0) = -6(2)^{-4} = -\frac{3}{8} \Rightarrow P_0(x) = \frac{1}{2}$, $P_1(x) = \frac{1}{2} - \frac{x}{4}$, $P_2(x) = \frac{1}{2} - \frac{x}{4} + \frac{x^2}{8}$, $P_3(x) = \frac{1}{2} - \frac{x}{4} + \frac{x^2}{8} - \frac{x^3}{16}$

5. $f(x) = \cos x$, $f'(x) = -\sin x$, $f''(x) = -\cos x$, $f'''(x) = \sin x$; $f\left(\frac{\pi}{4}\right) = \cos\frac{\pi}{4} = \frac{1}{\sqrt{2}}$, $f'\left(\frac{\pi}{4}\right) = -\sin\frac{\pi}{4} = -\frac{1}{\sqrt{2}}$, $f''\left(\frac{\pi}{4}\right) = -\cos\frac{\pi}{4} = -\frac{1}{\sqrt{2}}$, $f'''\left(\frac{\pi}{4}\right) = \sin\frac{\pi}{4} = \frac{1}{\sqrt{2}} \Rightarrow P_0(x) = \frac{1}{\sqrt{2}}$, $P_1(x) = \frac{1}{\sqrt{2}} - \frac{1}{\sqrt{2}}\left(x - \frac{\pi}{4}\right)$, $P_2(x) = \frac{1}{\sqrt{2}} - \frac{1}{\sqrt{2}}\left(x - \frac{\pi}{4}\right) - \frac{1}{2\sqrt{2}}\left(x - \frac{\pi}{4}\right)^2$, $P_3(x) = \frac{1}{\sqrt{2}} - \frac{1}{\sqrt{2}}\left(x - \frac{\pi}{4}\right) - \frac{1}{2\sqrt{2}}\left(x - \frac{\pi}{4}\right)^2 + \frac{1}{6\sqrt{2}}\left(x - \frac{\pi}{4}\right)^3$

7. $e^x = \sum_{n=0}^{\infty} \frac{x^n}{n!} \Rightarrow e^{-x} = \sum_{n=0}^{\infty} \frac{(-x)^n}{n!} = 1 - x + \frac{x^2}{2!} - \frac{x^3}{3!} + \frac{x^4}{4!} - \ldots$

9. $\sin x = \sum_{n=0}^{\infty} \frac{(-1)^n x^{2n+1}}{(2n+1)!} \Rightarrow \sin 3x = \sum_{n=0}^{\infty} \frac{(-1)^n (3x)^{2n+1}}{(2n+1)!} = \sum_{n=0}^{\infty} \frac{(-1)^n 3^{2n+1} x^{2n+1}}{(2n+1)!} = 3x - \frac{3^3 x^3}{3!} + \frac{3^5 x^5}{5!} - \ldots$

11. $\cosh x = \frac{e^x + e^{-x}}{2} = \frac{1}{2}\left[\left(1 + x^2 + \frac{x^2}{2!} + \frac{x^3}{3!} + \frac{x^4}{4!} + \ldots\right) + \left(1 - x + \frac{x^2}{2!} - \frac{x^3}{3!} + \frac{x^4}{4!} - \ldots\right)\right] = 1 + \frac{x^2}{2!} + \frac{x^4}{4!} + \frac{x^6}{6!} + \ldots = \sum_{n=0}^{\infty} \frac{x^{2n}}{(2n)!}$

13. $f(x) = x^4 - 2x^3 - 5x + 4 \Rightarrow f'(x) = 4x^3 - 6x^2 - 5$, $f''(x) = 12x^2 - 12x$, $f'''(x) = 24x - 12$, $f^{(4)}(x) = 24$
$\Rightarrow f^{(n)}(x) = 0$ if $n \geq 5$; $f(0) = 4$, $f'(0) = -5$, $f''(0) = 0$, $f'''(0) = -12$, $f^{(4)}(0) = 24$, $f^{(n)}(0) = 0$ if $n \geq 5$
$\Rightarrow x^4 - 2x^3 - 5x + 4 = 4 - 5x - \frac{12}{3!}x^3 + \frac{24}{4!}x^4 = x^4 - 2x^3 - 5x + 4$ itself

15. $f(x) = x^3 - 2x + 4 \Rightarrow f'(x) = 3x^2 - 2$, $f''(x) = 6x$, $f'''(x) = 6 \Rightarrow f^{(n)}(x) = 0$ if $n \geq 4$; $f(2) = 8$, $f'(2) = 10$,
$f''(2) = 12$, $f'''(2) = 6$, $f^{(n)}(2) = 0$ if $n \geq 4 \Rightarrow x^3 - 2x + 4 = 8 + 10(x-2) + \frac{12}{2!}(x-2)^2 + \frac{6}{3!}(x-2)^3$
$= 8 + 10(x-2) + 6(x-2)^2 + (x-2)^3$

17. $f(x) = x^{-2} \Rightarrow f'(x) = -2x^{-3}$, $f''(x) = 3!\,x^{-4}$, $f'''(x) = -4!\,x^{-5} \Rightarrow f^{(n)}(x) = (-1)^n(n+1)!\,x^{-n-2}$;
$f(1) = 1$, $f'(1) = -2$, $f''(1) = 3!$, $f'''(1) = -4!$, $f^{(n)}(1) = (-1)^n(n+1)! \Rightarrow \frac{1}{x^2}$
$= 1 - 2(x-1) + 3(x-1)^2 - 4(x-1)^3 + \ldots = \sum_{n=0}^{\infty} (-1)^n(n+1)(x-1)^n$

19. $f(x) = e^x \Rightarrow f'(x) = e^x$, $f''(x) = e^x \Rightarrow f^{(n)}(x) = e^x$; $f(2) = e^2$, $f'(2) = e^2$, $\ldots$ $f^{(n)}(2) = e^2$
$\Rightarrow e^x = e^2 + e^2(x-2) + \frac{e^2}{2}(x-2)^2 + \frac{e^3}{3!}(x-2)^3 + \ldots = \sum_{n=0}^{\infty} \frac{e^2}{n!}(x-2)^n$

21. $e^x = 1 + x + \frac{x^2}{2!} + \ldots = \sum_{n=0}^{\infty} \frac{x^n}{n!} \Rightarrow e^{-5x} = 1 + (-5x) + \frac{(-5x)^2}{2!} + \ldots = 1 - 5x + \frac{5^2x^2}{2!} - \frac{5^3x^3}{3!} + \ldots = \sum_{n=0}^{\infty} \frac{(-1)^n 5^n x^n}{n!}$

23. $\sin x = x - \frac{x^3}{3!} + \frac{x^5}{5!} - \ldots = \sum_{n=0}^{\infty} \frac{(-1)^n x^{2n+1}}{(2n+1)!} \Rightarrow \sin\frac{\pi x}{2} = \frac{\pi x}{2} - \frac{\left(\frac{\pi x}{2}\right)^3}{3!} + \frac{\left(\frac{\pi x}{2}\right)^5}{5!} - \frac{\left(\frac{\pi x}{2}\right)^7}{7!} + \ldots$
$= \sum_{n=0}^{\infty} \frac{(-1)^n \pi^{2n+1} x^{2n+1}}{2^{2n+1}(2n+1)!}$

25. $e^x = \sum_{n=0}^{\infty} \frac{x^n}{n!} \Rightarrow xe^x = x\left(\sum_{n=0}^{\infty} \frac{x^n}{n!}\right) = \sum_{n=0}^{\infty} \frac{x^{n+1}}{n!} = x + x^2 + \frac{x^3}{2!} + \frac{x^4}{3!} + \frac{x^5}{4!} + \ldots$

27. $\cos x = \sum_{n=0}^{\infty} \frac{(-1)^n x^{2n}}{(2n)!} \Rightarrow \frac{x^2}{2} - 1 + \cos x = \frac{x^2}{2} - 1 + \sum_{n=0}^{\infty} \frac{(-1)^n x^{2n}}{(2n)!} = \frac{x^2}{2} - 1 + 1 - \frac{x^2}{2} + \frac{x^4}{4!} - \frac{x^6}{6!} + \frac{x^8}{8!} - \frac{x^{10}}{10!} + \ldots$
$= \frac{x^4}{4!} - \frac{x^6}{6!} + \frac{x^8}{8!} - \frac{x^{10}}{10!} + \ldots = \sum_{n=2}^{\infty} \frac{(-1)^n x^{2n}}{(2n)!}$

29. $\cos x = \sum_{n=0}^{\infty} \frac{(-1)^n x^{2n}}{(2n)!} \Rightarrow x \cos \pi x = x \sum_{n=0}^{\infty} \frac{(-1)^n (\pi x)^{2n}}{(2n)!} = \sum_{n=0}^{\infty} \frac{(-1)^n \pi^{2n} x^{2n+1}}{(2n)!} = x - \frac{\pi^2 x^3}{2!} + \frac{\pi^4 x^5}{4!} - \frac{\pi^6 x^7}{6!} + \ldots$

31. $\sin^2 x = \left(\frac{1 - \cos 2x}{2}\right) = \frac{1}{2} - \frac{1}{2}\cos 2x = \frac{1}{2} - \frac{1}{2}\left(1 - \frac{(2x)^2}{2!} + \frac{(2x)^4}{4!} - \frac{(2x)^6}{6!} + \ldots\right) = \frac{(2x)^2}{2 \cdot 2!} - \frac{(2x)^4}{2 \cdot 4!} + \frac{(2x)^6}{2 \cdot 6!} - \ldots$
$= \sum_{n=1}^{\infty} \frac{(-1)^{n+1}(2x)^{2n}}{2 \cdot (2n)!}$

33. $x \ln(1 + 2x) = x \sum_{n=1}^{\infty} \frac{(-1)^{n-1}(2x)^n}{n} = \sum_{n=1}^{\infty} \frac{(-1)^{n-1} 2^n x^{n+1}}{n} = 2x^2 - \frac{2^2x^3}{2} + \frac{2^3x^4}{3} - \frac{2^4x^5}{4} + \ldots$

35. By the Alternating Series Estimation Theorem, the error is less than $\frac{|x|^5}{5!} \Rightarrow |x|^5 < (5!)(5 \times 10^{-4})$

$\Rightarrow |x|^5 < 600 \times 10^{-4} \Rightarrow |x| < \sqrt[5]{6 \times 10^{-2}} \approx 0.56968$

37. If $\sin x = x$ and $|x| < 10^{-3}$, then the $|\text{error}| = |R_2(x)| = \left|\frac{-\cos c}{3!}x^3\right| < \frac{(10^{-3})^3}{3!} \approx 1.67 \times 10^{-10}$, where c is between 0 and x. The Alternating Series Estimation Theorem says $R_2(x)$ has the same sign as $-\frac{x^3}{3!}$. Moreover, $x < \sin x \Rightarrow 0 < \sin x - x = R_2(x) \Rightarrow x < 0 \Rightarrow -10^{-3} < x < 0$.

39. (a) $|R_2(x)| = \left|\frac{e^c x^3}{3!}\right| < \frac{3^{(0.1)}(0.1)^3}{3!} < 1.87 \times 10^{-4}$, where c is between 0 and x

(b) $|R_2(x)| = \left|\frac{e^c x^3}{3!}\right| < \frac{(0.1)^3}{3!} = 1.67 \times 10^{-4}$, where c is between 0 and x

41. If we approximate e^h with $1 + h$ and $0 \le h \le 0.01$, then $|\text{error}| < \left|\frac{e^c h^2}{2}\right| \le \frac{e^{0.01} h \cdot h}{2} = \left(\frac{e^{0.01}(0.01)}{2}\right)h$ $= 0.005005h < 0.006h = (0.6\%)h$, where c is between 0 and h.

43. $\tan^{-1} x = x - \frac{x^3}{3} + \frac{x^5}{5} - \frac{x^7}{7} + \ldots \Rightarrow \frac{\pi}{4} = \tan^{-1} 1 = 1 - \frac{1}{3} + \frac{1}{5} - \frac{1}{7} + \ldots$; $|\text{error}| < \frac{1}{2n+1} < .01$ $\Rightarrow 2n + 1 > 100 \Rightarrow n > 49$

45. $f(x) = \ln(\cos x) \Rightarrow f'(x) = -\tan x$ and $f''(x) = -\sec^2 x$; $f(0) = 0$, $f'(0) = 0$, $f''(0) = -1$

$\Rightarrow L(x) = 0$ and $Q(x) = -\frac{x^2}{2}$

47. $f(x) = (1 - x^2)^{-1/2} \Rightarrow f'(x) = x(1 - x^2)^{-3/2}$ and $f''(x) = (1 - x^2)^{-3/2} + 3x^2(1 - x^2)^{-5/2}$; $f(0) = 1$, $f'(0) = 0$, $f''(0) = 1 \Rightarrow L(x) = 1$ and $Q(x) = 1 = \frac{x^2}{2}$

49. A special case of Taylor's Formula is $f(x) = f(a) + f'(c)(x - a)$. Let $x = b$ and this becomes $f(b) - f(a) = f'(c)(b - a)$, the Mean Value Theorem

51. (a) $f'' \le 0$, $f'(a) = 0$ and $x = a$ interior to the interval I $\Rightarrow f(x) - f(a) = \frac{f''(c_2)}{2}(x - a)^2 \le 0$ throughout I $\Rightarrow f(x) \le f(a)$ throughout I $\Rightarrow$ f has a local maximum at $x = a$

(b) similar reasoning gives $f(x) - f(a) = \frac{f''(c_2)}{2}(x - a)^2 \ge 0$ throughout I $\Rightarrow f(x) \ge f(a)$ throughout I $\Rightarrow$ f has a local minimum at $x = a$

53. Let $P = x + \pi \Rightarrow |x| = |P - \pi| < .5 \times 10^{-n}$ since P approximates π accurate to n decimals. Then, $P + \sin P = (\pi + x) + \sin(\pi + x) = (\pi + x) - \sin x = \pi + (x - \sin x) \Rightarrow |(P + \sin P) - \pi|$

$= |\sin x - x| \le \frac{|x|^3}{3!} < \frac{0.125}{3!} \times 10^{-3n} < .5 \times 10^{-3n} \Rightarrow P + \sin P$ gives an approximation to π correct to 3n decimals.

55. Note: f even $\Rightarrow f(-x) = f(x) \Rightarrow -f'(-x) = f'(x) \Rightarrow f'(-x) = -f'(x) \Rightarrow f'$ odd;

f odd $\Rightarrow f(-x) = -f(x) \Rightarrow -f'(-x) = -f'(x) \Rightarrow f'(-x) = f'(x) \Rightarrow f'$ even;

also, f odd $\Rightarrow f(-0) = f(0) \Rightarrow 2f(0) = 0 \Rightarrow f(0) = 0$

(a) If f(x) is even, then any odd-order derivative is odd and equal to 0 at $x = 0$. Therefore, $a_1 = a_3 = a_5 = \ldots = 0$; that is, the Maclaurin series for f contains only even powers.

(b) If f(x) is odd, then any even-order derivative is odd and equal to 0 at $x = 0$. Therefore, $a_0 = a_2 = a_4 = \ldots = 0$; that is, the Maclaurin series for f contains only odd powers.

57. (a)

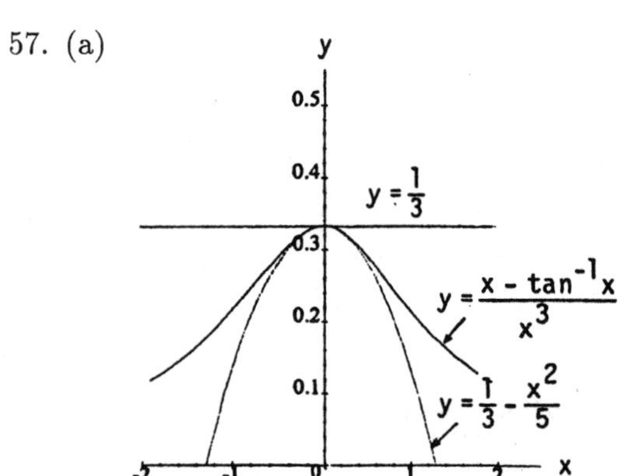

(b) $\tan^{-1} x = x - \frac{x^3}{3} + \frac{x^5}{5} - \ldots \Rightarrow \frac{x - \tan^{-1} x}{x^3} = \frac{1}{3} - \frac{x^2}{5} + \ldots$; from the Alternating Series Estimation Theorem, $\frac{x - \tan^{-1} x}{x^3} - \frac{1}{3} < 0$ $\Rightarrow \frac{x - \tan^{-1} x}{x^3} - \left(\frac{1}{3} - \frac{x^2}{5}\right) > 0 \Rightarrow \frac{1}{3} < \frac{x - \tan^{-1} x}{x^3}$ $< \frac{1}{3} - \frac{x^2}{5}$; therefore, the $\lim_{x \to 0} \frac{x - \tan^{-1} x}{x^3} = \frac{1}{3}$

8.8 APPLICATIONS OF POWER SERIES

1. $(1+x)^{1/2} = 1 + \frac{1}{2}x + \frac{\left(\frac{1}{2}\right)\left(-\frac{1}{2}\right)x^2}{2!} + \frac{\left(\frac{1}{2}\right)\left(-\frac{1}{2}\right)\left(-\frac{3}{2}\right)x^3}{3!} + \ldots = 1 + \frac{1}{2}x - \frac{1}{8}x^2 + \frac{1}{16}x^3 - \ldots$

3. $(1-x)^{-1/2} = 1 - \frac{1}{2}(-x) + \frac{\left(-\frac{1}{2}\right)\left(-\frac{3}{2}\right)(-x)^2}{2!} + \frac{\left(-\frac{1}{2}\right)\left(-\frac{3}{2}\right)\left(-\frac{5}{2}\right)(-x)^3}{3!} + \ldots = 1 + \frac{1}{2}x + \frac{3}{8}x^2 + \frac{5}{16}x^3 + \ldots$

5. $\left(1+\frac{x}{2}\right)^{-2} = 1 - 2\left(\frac{x}{2}\right) + \frac{(-2)(-3)\left(\frac{x}{2}\right)^2}{2!} + \frac{(-2)(-3)(-4)\left(\frac{x}{2}\right)^3}{3!} + \ldots = 1 - x + \frac{3}{4}x^2 - \frac{1}{2}x^3$

7. $(1+x^3)^{-1/2} = 1 - \frac{1}{2}x^3 + \frac{\left(-\frac{1}{2}\right)\left(-\frac{3}{2}\right)(x^3)^2}{2!} + \frac{\left(-\frac{1}{2}\right)\left(-\frac{3}{2}\right)\left(-\frac{5}{2}\right)(x^3)^3}{3!} + \ldots = 1 - \frac{1}{2}x^3 + \frac{3}{8}x^6 - \frac{5}{16}x^9 + \ldots$

9. $\left(1+\frac{1}{x}\right)^{1/2} = 1 + \frac{1}{2}\left(\frac{1}{x}\right) + \frac{\left(\frac{1}{2}\right)\left(-\frac{1}{2}\right)\left(\frac{1}{x}\right)^2}{2!} + \frac{\left(\frac{1}{2}\right)\left(-\frac{1}{2}\right)\left(-\frac{3}{2}\right)\left(\frac{1}{x}\right)^3}{3!} + \ldots = 1 + \frac{1}{2x} - \frac{1}{8x^2} + \frac{1}{16x^3}$

11. $(1+x)^4 = 1 + 4x + \frac{(4)(3)x^2}{2!} + \frac{(4)(3)(2)x^3}{3!} + \frac{(4)(3)(2)x^4}{4!} = 1 + 4x + 6x^2 + 4x^3 + x^4$

13. $(1-2x)^3 = 1 + 3(-2x) + \frac{(3)(2)(-2x)^2}{2!} + \frac{(3)(2)(1)(-2x)^3}{3!} = 1 - 6x + 12x^2 - 8x^3$

15. Assume the solution has the form $y = a_0 + a_1x + a_2x^2 + \ldots + a_{n-1}x^{n-1} + a_nx^n + \ldots$

$\Rightarrow \frac{dy}{dx} = a_1 + 2a_2x + \ldots + na_nx^{n-1} + \ldots$

$\Rightarrow \frac{dy}{dx} + y = (a_1 + a_0) + (2a_2 + a_1)x + (3a_3 + a_2)x^2 + \ldots + (na_n + a_{n-1})x^{n-1} + \ldots = 0$

$\Rightarrow a_1 + a_0 = 0,\ 2a_2 + a_1 = 0,\ 3a_3 + a_2 = 0$ and in general $na_n + a_{n-1} = 0$. Since $y = 1$ when $x = 0$ we have $a_0 = 1$. Therefore $a_1 = -1,\ a_2 = \frac{-a_1}{2\cdot 1} = \frac{1}{2},\ a_3 = \frac{-a_2}{3} = -\frac{1}{3\cdot 2}, \ldots,\ a_n = \frac{-a_{n-1}}{n} = \frac{(-1)^n}{n!}$

$\Rightarrow y = 1 - x + \frac{1}{2}x^2 - \frac{1}{3!}x^3 + \ldots + \frac{(-1)^n}{n!}x^n + \ldots = \sum_{n=0}^{\infty} \frac{(-1)^n x^n}{n!} = e^{-x}$

17. Assume the solution has the form $y = a_0 + a_1x + a_2x^2 + \ldots + a_{n-1}x^{n-1} + a_nx^n + \ldots$

$\Rightarrow \frac{dy}{dx} = a_1 + 2a_2x + \ldots + na_nx^{n-1} + \ldots$

$\Rightarrow \frac{dy}{dx} - y = (a_1 - a_0) + (2a_2 - a_1)x + (3a_3 - a_2)x^2 + \ldots + (na_n - a_{n-1})x^{n-1} + \ldots = 1$

$\Rightarrow a_1 - a_0 = 1,\ 2a_2 - a_1 = 0,\ 3a_3 - a_2 = 0$ and in general $na_n - a_{n-1} = 0$. Since $y = 0$ when $x = 0$ we have $a_0 = 0$. Therefore $a_1 = 1,\ a_2 = \frac{a_1}{2} = \frac{1}{2},\ a_3 = \frac{a_2}{3} = \frac{1}{3\cdot 2},\ a_4 = \frac{a_3}{4} = \frac{1}{4\cdot 3\cdot 2}, \ldots,\ a_n = \frac{a_{n-1}}{n} = \frac{1}{n!}$

$\Rightarrow y = 0 + 1x + \frac{1}{2}x^2 + \frac{1}{3\cdot 2}x^3 + \frac{1}{4\cdot 3\cdot 2}x^4 + \ldots + \frac{1}{n!}x^n + \ldots$

$= \left(1 + 1x + \frac{1}{2}x^2 + \frac{1}{3\cdot 2}x^3 + \frac{1}{4\cdot 3\cdot 2}x^4 + \ldots + \frac{1}{n!}x^n + \ldots\right) - 1 = \sum_{n=0}^{\infty} \frac{x^n}{n!} - 1 = e^x - 1$

19. Assume the solution has the form $y = a_0 + a_1x + a_2x^2 + \ldots + a_{n-1}x^{n-1} + a_nx^n + \ldots$

$\Rightarrow \frac{dy}{dx} = a_1 + 2a_2x + \ldots + na_nx^{n-1} + \ldots$

$\Rightarrow \frac{dy}{dx} - y = (a_1 - a_0) + (2a_2 - a_1)x + (3a_3 - a_2)x^2 + \ldots + (na_n - a_{n-1})x^{n-1} + \ldots = x$

$\Rightarrow a_1 - a_0 = 0,\ 2a_2 - a_1 = 1,\ 3a_3 - a_2 = 0$ and in general $na_n - a_{n-1} = 0$. Since $y = 0$ when $x = 0$ we have $a_0 = 0$. Therefore $a_1 = 0,\ a_2 = \frac{1 + a_1}{2} = \frac{1}{2},\ a_3 = \frac{a_2}{3} = \frac{1}{3\cdot 2},\ a_4 = \frac{a_3}{4} = \frac{1}{4\cdot 3\cdot 2}, \ldots,\ a_n = \frac{a_{n-1}}{n} = \frac{1}{n!}$

$\Rightarrow y = 0 + 0x + \frac{1}{2}x^2 + \frac{1}{3\cdot 2}x^3 + \frac{1}{4\cdot 3\cdot 2}x^4 + + \ldots + \frac{1}{n!}x^n + \ldots$

$= \left(1 + 1x + \frac{1}{2}x^2 + \frac{1}{3\cdot 2}x^3 + \frac{1}{4\cdot 3\cdot 2}x^4 + \ldots + \frac{1}{n!}x^n + \ldots\right) - 1 - x = \sum_{n=0}^{\infty} \frac{x^n}{n!} - 1 - x = e^x - x - 1$

21. $y' - xy = a_1 + (2a_2 - a_0)x + (3a_3 - a_1)x + \ldots + (na_n - a_{n-2})x^{n-1} + \ldots = 0 \Rightarrow a_1 = 0,\ 2a_2 - a_0 = 0,\ 3a_3 - a_1 = 0,$ $4a_4 - a_2 = 0$ and in general $na_n - a_{n-2} = 0$. Since $y = 1$ when $x = 0$, we have $a_0 = 1$. Therefore $a_2 = \frac{a_0}{2} = \frac{1}{2},$ $a_3 = \frac{a_1}{3} = 0,\ a_4 = \frac{a_2}{4} = \frac{1}{2\cdot 4},\ a_5 = \frac{a_3}{5} = 0, \ldots,\ a_{2n} = \frac{1}{2\cdot 4\cdot 6\cdots 2n}$ and $a_{2n+1} = 0$

$\Rightarrow y = 1 + \frac{1}{2}x^2 + \frac{1}{2\cdot 4}x^4 + \frac{1}{2\cdot 4\cdot 6}x^6 + \ldots + \frac{1}{2\cdot 4\cdot 6\cdots 2n}x^{2n} + \ldots = \sum_{n=0}^{\infty} \frac{x^{2n}}{2^n n!} = \sum_{n=0}^{\infty} \frac{\left(\frac{x^2}{2}\right)^n}{n!} = e^{x^2/2}$

23. $(1 - x)y' - y = (a_1 - a_0) + (2a_2 - a_1 - a_1)x + (3a_3 - 2a_2 - a_2)x^2 + (4a_4 - 3a_3 - a_3)x^3 + \ldots$ $+ (na_n - (n-1)a_{n-1} - a_{n-1})x^{n-1} + \ldots = 0 \Rightarrow a_1 - a_0 = 0,\ 2a_2 - 2a_1 = 0,\ 3a_3 - 3a_2 = 0$ and in

general $(na_n - na_{n-1}) = 0$. Since $y = 2$ when $x = 0$, we have $a_0 = 2$. Therefore $a_1 = 2,\ a_2 = 2,\ \ldots,\ a_n = 2 \Rightarrow y = 2 + 2x + 2x^2 + \ldots = \sum_{n=0}^{\infty} 2x^n = \frac{2}{1-x}$

25. $y = a_0 + a_1x + a_2x^2 + \ldots + a_nx^n + \ldots \Rightarrow y'' = 2a_2 + 3 \cdot 2a_3x + \ldots + n(n-1)a_nx^{n-2} + \ldots \Rightarrow y'' - y$ $= (2a_2 - a_0) + (3 \cdot 2a_3 - a_1)x + (4 \cdot 3a_4 - a_2)x^2 + \ldots + (n(n-1)a_n - a_{n-2})x^{n-2} + \ldots = 0 \Rightarrow 2a_2 - a_0 = 0,$ $3 \cdot 2a_3 - a_1 = 0,\ 4 \cdot 3a_4 - a_2 = 0$ and in general $n(n-1)a_n - a_{n-2} = 0$. Since $y' = 1$ and $y = 0$ when $x = 0$, we have $a_0 = 0$ and $a_1 = 1$. Therefore $a_2 = 0,\ a_3 = \frac{1}{3 \cdot 2},\ a_4 = 0,\ a_5 = \frac{1}{5 \cdot 4 \cdot 3 \cdot 2},\ \ldots,\ a_{2n+1} = \frac{1}{(2n+1)!}$ and $a_{2n} = 0 \Rightarrow y = x + \frac{1}{3!}x^3 + \frac{1}{5!}x^5 + \ldots = \sum_{n=0}^{\infty} \frac{x^{2n+1}}{(2n+1)!} = \sinh x$

27. $y = a_0 + a_1x + a_2x^2 + \ldots + a_nx^n + \ldots \Rightarrow y'' = 2a_2 + 3 \cdot 2a_3x + \ldots + n(n-1)a_nx^{n-2} + \ldots \Rightarrow y'' + y$ $= (2a_2 + a_0) + (3 \cdot 2a_3 + a_1)x + (4 \cdot 3a_4 + a_2)x^2 + \ldots + (n(n-1)a_n + a_{n-2})x^{n-2} + \ldots = x \Rightarrow 2a_2 + a_0 = 0,$ $3 \cdot 2a_3 + a_1 = 1,\ 4 \cdot 3a_4 + a_2 = 0$ and in general $n(n-1)a_n + a_{n-2} = 0$. Since $y' = 1$ and $y = 2$ when $x = 0$, we have $a_0 = 2$ and $a_1 = 1$. Therefore $a_2 = -1,\ a_3 = 0,\ a_4 = \frac{1}{4 \cdot 3},\ a_5 = 0,\ \ldots,\ a_{2n} = -2 \cdot \frac{(-1)^{n+1}}{(2n)!}$ and $a_{2n+1} = 0 \Rightarrow y = 2 + x - x^2 + 2 \cdot \frac{x^4}{4!} + \ldots = 2 + x - 2\sum_{n=1}^{\infty} \frac{(-1)^{n+1}x^{2n}}{(2n)!}$

29. $y = a_0 + a_1(x-2) + a_2(x-2)^2 + \ldots + a_n(x-2)^n + \ldots$

$\Rightarrow y'' = 2a_2 + 3 \cdot 2a_3(x-2) + \ldots + n(n-1)a_n(x-2)^{n-2} + \ldots \Rightarrow y'' - y$

$= (2a_2 - a_0) + (3 \cdot 2a_3 - a_1)(x-2) + (4 \cdot 3a_4 - a_2)(x-2)^2 + \ldots + (n(n-1)a_n - a_{n-2})(x-2)^{n-2} + \ldots$

$= -2 - (x-2) \Rightarrow 2a_2 - a_0 = -2,\ 3 \cdot 2a_3 - a_1 = -1,\ 4 \cdot 3a_4 - a_2$ and in general $n(n-1)a_n - a_{n-2} = 0$. Since $y' = -2$ and $y = 0$ when $x = 2$, we have $a_0 = 0$ and $a_1 = -2$. Therefore $a_2 = \frac{-2}{2} = -1,$

$a_3 = \frac{-2-1}{3 \cdot 2} = -\frac{3}{3 \cdot 2},\ a_4 = -\frac{2}{4 \cdot 3 \cdot 2},\ a_5 = -\frac{3}{5 \cdot 4 \cdot 3 \cdot 2},\ \ldots,\ a_{2n} = -\frac{2}{(2n)!},\ a_{2n+1} = -\frac{3}{(2n+1)!}$

$\Rightarrow y = -2(x-2) - \frac{2}{2!}(x-2)^2 - \frac{3}{3!}(x-2)^3 - \frac{2}{4!}(x-2)^4 - \frac{3}{5!}(x-2)^5 - \ \ldots$

$= -2(x-2) - \sum_{n=1}^{\infty}\left[\frac{2(x-2)^{2n}}{(2n)!} + \frac{3(x-2)^{2n+1}}{(2n+1)!}\right]$

31. $y'' + x^2y = 2a_2 + 6a_3x + (4 \cdot 3a_4 + a_0)x^2 + \ldots + (n(n-1)a_n + a_{n-4})x^{n-2} + \ldots = x \Rightarrow 2a_2 = 0,\ 6a_3 = 1,$ $4 \cdot 3a_4 + a_0 = 0,\ 5 \cdot 4a_5 + a_1 = 0$, and in general $n(n-1)a_n + a_{n-4} = 0$. Since $y' = b$ and $y = a$ when $x = 0$, we have $a_0 = a$ and $a_1 = b$. Therefore $a_2 = 0,\ a_3 = \frac{1}{2 \cdot 3},\ a_4 = -\frac{a}{3 \cdot 4},\ a_5 = -\frac{b}{4 \cdot 5},\ a_6 = 0,\ a_7 = \frac{1}{2 \cdot 3 \cdot 6 \cdot 7}$

$\Rightarrow y = a + bx + \frac{1}{2 \cdot 3}x^3 - \frac{a}{3 \cdot 4}x^4 - \frac{b}{4 \cdot 5}x^5 - \frac{1}{2 \cdot 3 \cdot 6 \cdot 7}x^7 + \frac{ax^8}{3 \cdot 4 \cdot 7 \cdot 8} + \frac{bx^9}{4 \cdot 5 \cdot 8 \cdot 9} + \ldots$

33. $F(x) = \int_0^x \left(t^2 - \frac{t^6}{3!} + \frac{t^{10}}{5!} - \frac{t^{14}}{7!} + \ldots\right) dt = \left[\frac{t^3}{3} - \frac{t^7}{7 \cdot 3!} + \frac{t^{11}}{11 \cdot 5!} - \frac{t^{15}}{15 \cdot 7!} + \ldots\right]_0^x \approx \frac{x^3}{3} - \frac{x^7}{7 \cdot 3!}$

$\Rightarrow |\text{error}| < \frac{1}{11 \cdot 5!} \approx 0.0008$

35. (a) $F(x) = \int_0^x \left(t - \frac{t^3}{3} + \frac{t^5}{5} - \frac{t^7}{7} + \ldots\right) dt = \left[\frac{t^2}{2} - \frac{t^4}{12} + \frac{t^6}{30} - \ldots\right]_0^x \approx \frac{x^2}{2} - \frac{x^4}{12} \Rightarrow |\text{error}| < \frac{(0.5)^6}{30} \approx .00052$

(b) $|\text{error}| < \frac{1}{33 \cdot 34} \approx .00089$ so $F(x) \approx \frac{x^2}{2} - \frac{x^4}{3 \cdot 4} + \frac{x^6}{5 \cdot 6} - \frac{x^8}{7 \cdot 8} + \ldots + (-1)^{15} \frac{x^{32}}{31 \cdot 32}$

37. $\frac{1}{x^2}(e^x - (1+x)) = \frac{1}{x^2}\left(\left(1 + x + \frac{x^2}{2} + \frac{x^3}{3!} + \ldots\right) - 1 - x\right) = \frac{1}{2} + \frac{x}{3!} + \frac{x^2}{4!} + \ldots \Rightarrow \lim_{x \to 0} \frac{e^x - (1+x)}{x^2}$

$= \lim_{x \to 0} \left(\frac{1}{2} + \frac{x}{3!} + \frac{x^2}{4!} + \ldots\right) = \frac{1}{2}$

39. $x^2\left(-1 + e^{-1/x^2}\right) = x^2\left(-1 + 1 - \frac{1}{x^2} + \frac{1}{2x^4} - \frac{1}{6x^6} + \ldots\right) = -1 + \frac{1}{2x^2} - \frac{1}{6x^4} + \ldots \Rightarrow \lim_{x \to \infty} x^2\left(e^{-1/x^2} - 1\right)$

$= \lim_{x \to \infty} \left(-1 + \frac{1}{2x^2} - \frac{1}{6x^4} + \ldots\right) = -1$

41. $\frac{\ln(1+x^2)}{1 - \cos x} = \frac{\left(x^2 - \frac{x^4}{2} + \frac{x^6}{3} - \ldots\right)}{1 - \left(1 - \frac{x^2}{2!} + \frac{x^4}{4!} - \ldots\right)} = \frac{\left(1 - \frac{x^2}{2} + \frac{x^4}{3} - \ldots\right)}{\left(\frac{1}{2!} - \frac{x^2}{4!} + \ldots\right)} \Rightarrow \lim_{x \to 0} \frac{\ln(1+x^2)}{1 - \cos x} = \lim_{x \to 0} \frac{\left(1 - \frac{x^2}{2} + \frac{x^4}{3} - \ldots\right)}{\left(\frac{1}{2!} - \frac{x^2}{4!} + \ldots\right)} = 2!$

$= 2$

43. $\ln\left(\frac{1+x}{1-x}\right) = \ln(1+x) - \ln(1-x) = \left(x - \frac{x^2}{2} + \frac{x^3}{3} - \frac{x^4}{4} + \ldots\right) - \left(-x - \frac{x^2}{2} - \frac{x^3}{3} - \frac{x^4}{4} - \ldots\right) = 2\left(x + \frac{x^3}{3} + \frac{x^5}{5} + \ldots\right)$

45. $\tan^{-1} x = x - \frac{x^3}{3} + \frac{x^5}{5} - \frac{x^7}{7} + \frac{x^9}{9} - \ldots + \frac{(-1)^{n-1} x^{2n-1}}{2n-1} + \ldots \Rightarrow |\text{error}| = \left|\frac{(-1)^{n-1} x^{2n-1}}{2n-1}\right| = \frac{1}{2n-1}$ when $x = 1$;

$\frac{1}{2n-1} < \frac{1}{10^3} \Rightarrow n > \frac{1001}{2} = 500.5 \Rightarrow$ the first term not used is the 501st $\Rightarrow$ we must use 500 terms

47. (a) $(1 - x^2)^{-1/2} \approx 1 + \frac{x^2}{2} + \frac{3x^4}{8} + \frac{5x^6}{16} \Rightarrow \sin^{-1} x \approx x + \frac{x^3}{6} + \frac{3x^5}{40} + \frac{5x^7}{112}$;

$$\lim_{n \to \infty} \left|\frac{1 \cdot 3 \cdot 5 \cdots (2n-1)(2n+1)x^{2n+3}}{2 \cdot 4 \cdot 6 \cdots (2n)(2n+2)(2n+3)} \cdot \frac{2 \cdot 4 \cdot 6 \cdots (2n)(2n+1)}{1 \cdot 3 \cdot 5 \cdots (2n-1)x^{2n+1}}\right| < 1 \Rightarrow x^2 \lim_{n \to \infty} \left|\frac{(2n+1)(2n+1)}{(2n+2)(2n+3)}\right| < 1$$

$\Rightarrow |x| < 1 \Rightarrow$ the radius of convergence is 1

(b) $\frac{d}{dx}(\cos^{-1} x) = -(1 - x^2)^{-1/2} \Rightarrow \cos^{-1} x = \frac{\pi}{2} - \sin^{-1} x \approx \frac{\pi}{2} - \left(x + \frac{x^3}{6} + \frac{3x^5}{40} + \frac{5x^7}{112}\right) \approx \frac{\pi}{2} - x - \frac{x^3}{6} - \frac{3x^5}{40} - \frac{5x^7}{112}$

49. $\left[\tan^{-1} t\right]_x^{\infty} = \frac{\pi}{2} - \tan^{-1} x = \int_x^{\infty} \frac{dt}{1+t^2} = \int_x^{\infty} \left[\frac{\left(\frac{1}{t^2}\right)}{1 + \left(\frac{1}{t^2}\right)}\right] dt = \int_x^{\infty} \frac{1}{t^2}\left(1 - \frac{1}{t^2} + \frac{1}{t^4} - \frac{1}{t^6} + \ldots\right) dt$

$= \int_x^{\infty} \left(\frac{1}{t^2} - \frac{1}{t^4} + \frac{1}{t^6} - \frac{1}{t^8} + \ldots\right) dt = \lim_{b \to \infty} \left[-\frac{1}{t} + \frac{1}{3t^3} - \frac{1}{5t^5} + \frac{1}{7t^7} - \ldots\right]_x^b = \frac{1}{x} - \frac{1}{3x^3} + \frac{1}{5x^5} - \frac{1}{7x^7} + \ldots$

$$\Rightarrow \tan^{-1} x = \frac{\pi}{2} - \frac{1}{x} + \frac{1}{3x^3} - \frac{1}{5x^5} + \ldots, \; x > 1; \; \left[\tan^{-1} t\right]_{-\infty}^{x} = \tan^{-1} x + \frac{\pi}{2} = \int_{-\infty}^{x} \frac{dt}{1+t^2}$$

$$= \lim_{b \to -\infty} \left[-\frac{1}{t} + \frac{1}{3t^3} - \frac{1}{5t^5} + \frac{1}{7t^7} - \ldots\right]_b^x = -\frac{1}{x} + \frac{1}{3x^3} - \frac{1}{5x^5} + \frac{1}{7x^7} - \ldots \Rightarrow \tan^{-1} x = -\frac{\pi}{2} - \frac{1}{x} + \frac{1}{3x^3} - \frac{1}{5x^5} + \ldots,$$

$x < -1$

8.9 FOURIER SERIES

1. $$a_0 = \frac{1}{\pi} \int_{-\pi}^{\pi} 1 \, dx = \left(\frac{1}{\pi}\right) x \Big]_{-\pi}^{\pi} = 2$$

$$a_n = \frac{1}{\pi} \int_{-\pi}^{\pi} \cos nx \, dx = \frac{1}{\pi n} \sin nx \Big]_{-\pi}^{\pi} = 0$$

$$b_n = \frac{1}{\pi} \int_{-\pi}^{\pi} \sin nx \, dx = -\frac{1}{n\pi} \cos nx \Big]_{-\pi}^{\pi} = \frac{1}{nx}\left[\cos(-n\pi) - \cos(n\pi)\right] = 0$$

Therefore,

$$f(x) = \frac{a_0}{2} = 1.$$

3. $$a_0 = \frac{1}{\pi} \int_{-\pi}^{\pi} x \, dx = \frac{1}{2\pi} x^2 \Big]_{-\pi}^{\pi} = 0. \qquad \text{(Note: x is an odd function)}$$

$$a_n = \frac{1}{\pi} \int_{-\pi}^{\pi} x \cos nx \, dx = 0. \qquad \text{(because x cos nx is an odd function)}$$

$$b_n = \frac{1}{\pi} \int_{-\pi}^{\pi} x \sin nx \, dx = \frac{2}{\pi} \int_{0}^{\pi} x \sin nx \, dx \qquad \text{(because x sin nx is even)}$$

$$= \frac{2}{\pi}\left(-\frac{x}{n} \cos nx + \frac{1}{n^2} \sin nx \Big]_0^{\pi}\right) = -\frac{2}{n} \cos nx = \frac{2}{n}(-1)^{n+1}$$

Therefore,

$$f(x) = \sum_{n=1}^{\infty} \frac{2(-1)^{n+1}}{n} \sin nx.$$

5. $$a_0 = \frac{1}{4\pi} \int_{-\pi}^{\pi} x^2 \, dx = \frac{1}{2\pi} \int_{0}^{\pi} x^2 \, dx = \frac{\pi^2}{6}$$

$$a_n = \frac{1}{4\pi} \int_{-\pi}^{\pi} x^2 \cos nx \, dx = \frac{1}{2\pi} \int_{0}^{\pi} x^2 \cos nx \, dx \qquad \text{(even function)}$$

$$= \frac{1}{2\pi}\left[\frac{x^2}{n} \sin nx + \frac{2x}{n^2} \cos nx - \frac{2}{n^3} \sin nx\right]_0^{\pi} = \frac{1}{n^2} \cos n\pi = \frac{(-1)^n}{n^2}$$

$$b^n = \frac{1}{4\pi}\int_{-\pi}^{\pi} x^2 \sin nx\, dx = 0 \qquad \text{(odd function)}$$

Therefore,

$$f(x) = \frac{\pi^2}{12} + \sum_{n=1}^{\infty} \frac{(-1)^n}{n^2} \cos nx.$$

7. $$a_0 = \frac{1}{\pi}\int_{-\pi}^{\pi} e^x\, dx = \frac{1}{\pi}(e^{\pi} - e^{-\pi}) = \frac{2}{\pi}\sinh \pi$$

$$a_n = \frac{1}{\pi}\int_{-\pi}^{\pi} e^x \cos nx\, dx = \frac{1}{\pi}\left[\frac{n^2 e^x}{1+n^2}\left(\frac{1}{n}\sin nx + \frac{1}{n^2}\cos nx\right)\right]_{-\pi}^{\pi} = \frac{1}{\pi}\frac{2\cos n\pi}{(1+n^2)}\left(\frac{e^{\pi} - e^{-\pi}}{2}\right) = \frac{2(-1)^n}{\pi(n^2+1)}\sinh \pi$$

$$b_n = \frac{1}{\pi}\int_{-\pi}^{\pi} e^x \sin nx\, dx = \frac{1}{\pi}\left[\frac{n^2 e^x}{1+n^2}\left(-\frac{1}{n}\cos nx + \frac{1}{n^2}\sin nx\right)\right]_{-\pi}^{\pi} = \frac{1}{\pi}\frac{2n\cos n\pi}{(1+n^2)}\left(\frac{e^{-\pi} - e^{\pi}}{2}\right) = \frac{2n(-1)^{n+1}}{\pi(n^2+1)}\sinh \pi$$

Therefore,

$$f(x) = \frac{1}{\pi}\sinh \pi + \sum_{n=1}^{\infty}\frac{2(-1)^n}{\pi(n^2+1)}\sinh \pi \cos nx + \sum_{n=1}^{\infty}\frac{2n(-1)^{n+1}}{\pi(n^2+1)}\sinh \pi \sin nx$$

$$= \frac{\sinh \pi}{\pi}\left[1 + \sum_{n=1}^{\infty}\frac{2(-1)^n}{n^2+1}\cos nx + \sum_{n=1}^{\infty}\frac{2n(-1)^{n+1}}{n^2+1}\sin nx\right]$$

$$= \frac{2\sinh \pi}{\pi}\left[\frac{1}{2} + \sum_{n=1}^{\infty}\frac{(-1)^n}{n^2+1}(\cos nx - n \sin nx)\right].$$

9. $$a_0 = \frac{1}{\pi}\int_0^{\pi} \cos x\, dx = \frac{1}{\pi}\sin x\Big|_0^{\pi} = 0$$

$$a_n = \frac{1}{\pi}\int_0^{\pi} \cos x \cos nx\, dx = \begin{cases} 0, & n \neq 1 \\ \frac{1}{2}, & n = 1 \end{cases}$$

$$b_n = \frac{1}{\pi}\int_0^{\pi} \cos x \sin nx\, dx = \begin{cases} \frac{1}{2\pi}\sin^2 x\Big|_0^{\pi} = 0, & n = 1 \\ \left(-\frac{\cos(n-1)x}{2\pi(n-1)} - \frac{\cos(n+1)x}{2\pi(n+1)}\right)\Big|_0^{\pi}, & n \neq 1 \end{cases} = \begin{cases} 0, & n = 1 \\ (1+(-1)^n)\frac{n}{\pi(n^2-1)}, & n \neq 1 \end{cases}$$

Therefore,

$$f(x) = \frac{1}{2}\cos x + \frac{1}{\pi}\sum_{n=2}^{\infty}\frac{n(1+(-1)^n)}{n^2-1}\sin nx.$$

11. $$a_0 = \frac{1}{\pi}\int_{-\pi/2}^{\pi/2} dx = 1$$

$$a_n = \frac{1}{\pi}\int_{-\pi/2}^{\pi/2} \cos nx \, dx = \frac{1}{n\pi}\sin nx\Big|_{-\pi/2}^{\pi/2} = \frac{2}{n\pi}\sin\frac{n\pi}{2}$$

$$b_n = \frac{1}{\pi}\int_{-\pi/2}^{\pi/2} \sin nx \, dx = 0$$

Therefore,

$$f(x) = \frac{1}{2}\sum_{n=1}^{\infty} \frac{2}{n\pi}\sin\frac{n\pi}{2}\cos nx.$$

Note: $\sin\frac{n\pi}{2} = \begin{cases} 0, & n = 2k \quad \text{(even)} \\ (-1)^k, & n = 2k+1 \quad \text{(odd)} \end{cases}$

Thus we can write f(x) in the form:

$$f(x) = \frac{1}{2} + \frac{2}{\pi}\sum_{k=0}^{\infty} \frac{(-1)^k}{2k+1}\cos(2k+1)x.$$

13. $a_0 = \int_{-1}^{1/2} -(2x-1)\,dx + \int_{1/2}^{1} (2x-1)\,dx = \left(x - x^2\right]_{-1}^{1/2} + \left(x^2 - x\right]_{1/2}^{1} = \frac{5}{2}$

$$a_n = \int_{-1}^{1/2} -(2x-1)\cos(n\pi x)\,dx + \int_{1/2}^{1} (2x-1)\cos(n\pi x)\,dx$$

$$= \left[\frac{(1-2x)}{n\pi}\sin(n\pi x) - \frac{2}{n^2\pi^2}\cos(n\pi x)\right]_{-1}^{1/2} + \left[\frac{(2x-1)}{n\pi}\sin(n\pi x) + \frac{2}{n^2\pi^2}\cos(n\pi x)\right]_{1/2}^{1}$$

$$= \frac{2}{n^2\pi^2}\left[(-1)^n - \cos\frac{n\pi}{2}\right] + \frac{2}{n^2\pi^2}\left[(-1)^n - \cos\frac{n\pi}{2}\right] = \frac{4}{n^2\pi^2}\left[(-1)^n - \cos\frac{n\pi}{2}\right]$$

$$b_n = \int_{-1}^{1/2} -(2x-1)\sin(n\pi x)\,dx + \int_{1/2}^{1} (2x-1)\sin(n\pi x)\,dx$$

$$= \left[-\frac{(1-2x)}{n\pi}\cos(n\pi x) - \frac{2}{n^2\pi^2}\sin(n\pi x)\right]_{-1}^{1/2} + \left[\frac{(2x-1)}{n\pi}\cos(n\pi x) + \frac{2}{n^2\pi^2}\sin(n\pi x)\right]_{1/2}^{1}$$

$$= \left[\frac{3}{n\pi}\cos n\pi - \frac{2}{n^2\pi^2}\sin\frac{n\pi}{2}\right] + \left[-\frac{1}{n\pi}\cos n\pi - \frac{2}{n^2\pi^2}\sin\frac{n\pi}{2}\right]$$

$$= \frac{2}{n\pi}(-1)^n - \frac{4}{n^2\pi^2}\sin\frac{n\pi}{2} = \frac{2}{n\pi}\left[(-1)^n - \frac{2}{n\pi}\sin\frac{n\pi}{2}\right]$$

Therefore,

$$f(x) = \frac{5}{4} + \frac{4}{\pi^2}\sum_{n=1}^{\infty}\frac{1}{n^2}\left[(-1)^n - \cos\frac{n\pi}{2}\right]\cos(n\pi x) + \frac{2}{\pi}\sum_{n=1}^{\infty}\frac{1}{n}\left[(-1)^n - \frac{2}{n\pi}\sin\frac{n\pi}{2}\right]\sin(n\pi x).$$

15. From exercise #5,

$$\frac{x^2}{4} = \frac{\pi^2}{12} + \sum_{n=1}^{\infty} \frac{(-1)^n}{n^2} \cos nx.$$

Setting $x = \pi$,

$$\frac{\pi^2}{4} = \frac{\pi^2}{12} + \sum_{n=1}^{\infty} \frac{(-1)^n}{n^2} \cos n\pi$$

$$\frac{3\pi^2}{12} - \frac{\pi^2}{12} = \sum_{n=1}^{\infty} \frac{(-1)^n(-1)^n}{n^2},$$

or

$$\frac{\pi^2}{6} = \sum_{n=1}^{\infty} \frac{1}{n^2} = 1 + \frac{1}{4} + \frac{1}{9} + \frac{1}{16} + \ldots + \frac{1}{n^2} + \ldots.$$

17. $\int_{-L}^{L} \cos \frac{m\pi x}{L}\, dx = \frac{L}{m\pi} \sin \frac{m\pi x}{L}\Big|_{-L}^{L} = \frac{L}{m\pi}[\sin m\pi - \sin(-m\pi)] = \frac{L}{m\pi}(0 - 0) = 0.$

19. $\cos A \cos B = \frac{1}{2}[\cos(A+B) + \cos(A-B)]$

If $m \neq n$,

$$\int_{-L}^{L} \cos \frac{n\pi x}{L} \cos \frac{m\pi x}{L}\, dx = \frac{1}{2}\int_{-L}^{L} \left[\cos \frac{(m+n)\pi x}{L} + \cos \frac{(n-m)\pi x}{L}\right] dx = 0, \text{ by exercise 17}$$

If $m = n$,

$$\int_{-L}^{L} \cos \frac{n\pi x}{L} \cos \frac{m\pi x}{L}\, dx = \frac{1}{2}\int_{-L}^{L} \left(\cos \frac{2m\pi x}{L} + 1\right) dx = \frac{1}{2}\int_{-L}^{L} \cos \frac{2m\pi x}{L}\, dx + \frac{1}{2}\int_{-L}^{L} dx$$

$$= 0 + L, \quad \text{by exercise 17}$$

$$= L.$$

21. $\sin A \cos B = \frac{1}{2}[\sin(A+B) + \sin(A-B)]$

If $m \neq n$,

$$\int_{-L}^{L} \sin \frac{n\pi x}{L} \cos \frac{m\pi x}{L}\, dx = \frac{1}{2}\int_{-L}^{L} \left[\sin \frac{(n+m)\pi x}{L} + \sin \frac{(n-m)\pi x}{L}\right] dx = 0, \text{ by exercise 18}$$

If $m = n$,

$$\int_{-L}^{L} \sin \frac{n\pi x}{L} \cos \frac{m\pi x}{L}\, dx = \frac{1}{2}\int_{-L}^{L} \left(\sin \frac{2m\pi x}{L} + 0\right) dx = 0, \text{ by exercise 18}$$

23. (a) Since the function $f(x) = x$ and its derivative $f'(x) = 1$ are continuous on $-\pi < x < \pi$, the function f satisfies conditions of Theorem 18, and $f(x) = x = \sum_{n=1}^{\infty} (-1)^{n+1} \frac{2}{n} \sin(nx)$.

(b) $\frac{d}{dx}\sum_{n=1}^{\infty}(-1)^{n+1}\frac{2}{n}\sin(nx) = \sum_{n=1}^{\infty}(-1)^{n+1}\frac{2}{n}\frac{d}{dx}(\sin(nx)) = \sum_{n=1}^{\infty}(-1)^{n+1}2\cos(nx)$

This series diverges by the nth term test because $\lim_{x\to\infty}\left((-1)^{n+1}2\cos(nx)\right) \neq 0$.

(c) We cannot be assured that term-by-term differentiation of the Fourier series of a piecewise continuous function gives a Fourier series that converges on the derivative of the function and, in fact, the series might not converge at all.

8.10 FOURIER COSINE AND SINE SERIES

1. $a_0 = \frac{2}{\pi}\int_0^{\pi} x\,dx = \frac{1}{\pi}x^2\Big|_0^{\pi} = \pi$

$$a_n = \frac{2}{\pi}\int_0^{\pi} x\cos nx\,dx = \frac{2}{\pi}\left[\frac{x}{n}\sin nx + \frac{1}{n^2}\cos nx\right]_0^{\pi}$$

$$= \frac{2}{\pi}\left[\frac{1}{n^2}\cos n\pi - \frac{1}{n^2}\right] = \frac{2}{\pi n^2}[(-1)^n - 1]$$

$b_n = 0$

Therefore,

$$f(x) = \frac{\pi}{2} + \frac{2}{\pi}\sum_{n=1}^{\infty}\frac{[(-1)^n - 1]}{n^2}\cos nx.$$

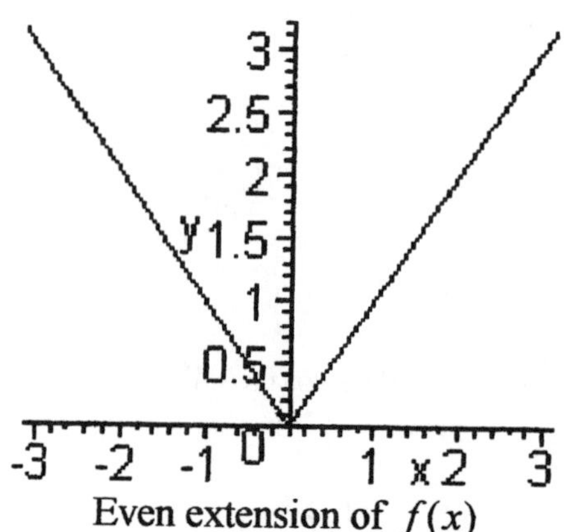

Even extension of $f(x)$

3. $a_0 = 2\int_0^1 e^x\,dx = 2e^x\Big|_0^1 = 2(e-1)$

$$a_n = 2\int_0^1 e^x\cos n\pi x\,dx$$

$$= 2\left(\frac{n^2\pi^2}{1+n^2\pi^2}\right)\left[e^x\left(\frac{1}{n\pi}\sin n\pi x + \frac{1}{n^2\pi^2}\cos n\pi x\right)\right]_0^1$$

$$= \frac{2}{1+n^2\pi^2}(e\cos n\pi - 1) = \frac{2\left[e(-1)^n - 1\right]}{1+n^2\pi^2}$$

$b_n = 0$

Therefore

$$f(x) = (e-1) + 2\sum_{n=1}^{\infty}\frac{\left[e(-1)^n - 1\right]}{1+n^2\pi^2}\cos n\pi x.$$

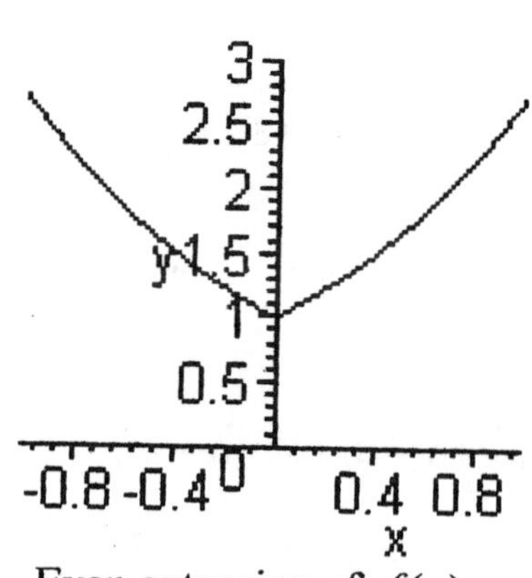

Even extension of $f(x)$

5. $a_0 = \frac{2}{2}\int_0^1 dx + \frac{2}{2}\int_1^2 -x\,dx = 1 + \left(-\frac{1}{2}x^2\Big|_1^2\right) = -\frac{1}{2}$

$$a_n = \int_0^1 \cos\frac{n\pi x}{2}\,dx + \int_1^2 -x\cos\frac{n\pi x}{2}\,dx$$

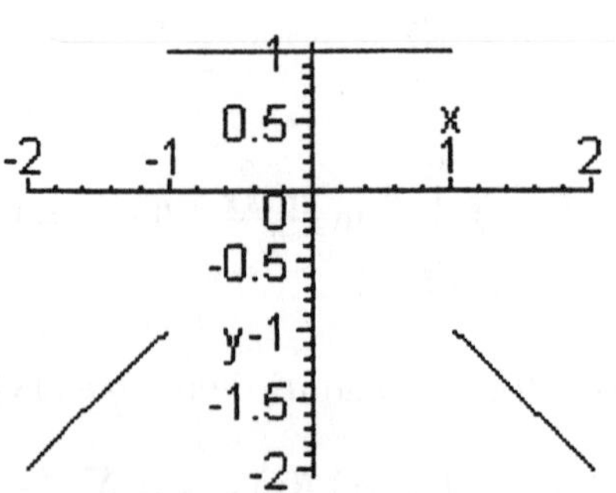

Even extension of $f(x)$

$$= \frac{2}{n\pi} \sin \frac{n\pi x}{2}\Big|_0^1 - \left[\frac{2x}{n\pi} \sin \frac{n\pi x}{2} + \frac{4}{n^2\pi^2} \cos \frac{n\pi x}{2}\right]_1^2$$

$$= \frac{2}{n\pi} \sin \frac{n\pi}{2} - \frac{4}{n\pi} \sin n\pi - \frac{4}{n^2\pi^2} \cos n\pi + \frac{2}{n\pi} \sin \frac{n\pi}{2} + \frac{4}{n^2\pi^2} \cos \frac{n\pi}{2}$$

$$= \frac{4}{n\pi} \sin \frac{n\pi}{2} + \frac{4}{n^2\pi^2}\left[(-1)^{n+1} + \cos \frac{n\pi}{2}\right]$$

Therefore,

$$f(x) = -\frac{1}{4} + \frac{4}{\pi} \sum_{n=1}^{\infty} \left[\frac{1}{n} \sin \frac{n\pi}{2} + \frac{1}{\pi n^2}\left((-1)^{n+1} + \cos \frac{n\pi}{2}\right)\right] \cos \frac{n\pi x}{2}.$$

7. $a_0 = 2 \int_0^{1/2} -(2x-1)\, dx + 2 \int_{1/2}^{1} (2x-1)\, dx = 1$

$$a_n = 2 \int_0^{1/2} -(2x-1) \cos n\pi x\, dx + 2 \int_{1/2}^{1} (2x-1) \cos n\pi x\, dx$$

$$= 2\left[\frac{(1-2x)}{n\pi} \sin n\pi x - \frac{2}{n^2\pi^2} \cos n\pi x\right]_0^{1/2}$$

$$+ 2\left[\frac{(2x-1)}{n\pi} \sin n\pi x + \frac{2}{n^2\pi^2} \cos n\pi x\right]_{1/2}^{1}$$

$$= 2\left[0 - \frac{2}{n^2\pi^2} \cos \frac{n\pi}{2} - 0 + \frac{2}{n^2\pi^2}\right]$$

$$+ 2\left[0 + \frac{2}{n^2\pi^2} \cos n\pi - 0 - \frac{2}{n^2\pi^2} \cos \frac{n\pi}{2}\right]$$

$$= \frac{4}{n^2\pi^2}\left[1 - 2 \cos \frac{n\pi}{2} + (-1)^n\right]$$

Therefore,

$$f(x) = \frac{1}{2} + \sum_{n=1}^{\infty} \frac{4}{n^2\pi^2}\left[1 + (-1)^n - 2 \cos \frac{n\pi}{2}\right] \cos n\pi x.$$

Even extension of $f(x)$

9. $b_n = 2 \int_0^1 -x \sin n\pi x\, dx = 2\left[\frac{x}{n\pi} \cos n\pi x\right.$

$$\left. - \frac{1}{n^2\pi^2} \sin n\pi x\right]_0^1 = \frac{2}{n\pi} \cos n\pi = \frac{2(-1)^n}{n\pi}$$

Therefore,

$$f(x) = 2 \sum_{n=1}^{\infty} \frac{(-1)^n}{n\pi} \sin n\pi x.$$

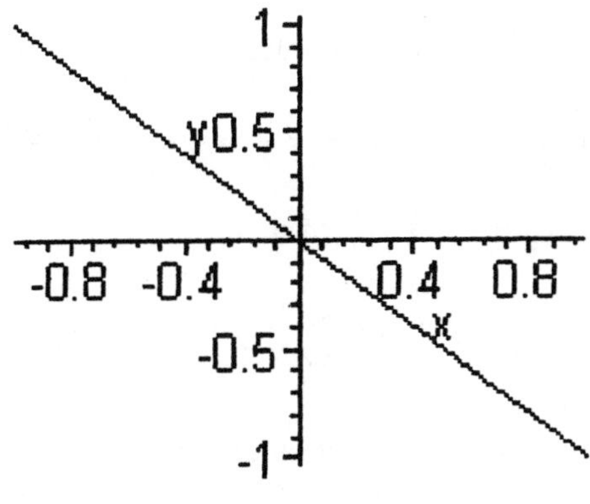

Even extension of $f(x)$

11. $b_n = \frac{2}{\pi} \int_0^{\pi} \cos x \sin nx\, dx$

$$= \frac{2}{\pi}\left[\frac{n^2}{n^2-1}\left(-\frac{1}{n} \cos x \cos nx - \frac{1}{n^2} \sin x \sin nx\right)\right]_0^{\pi}$$

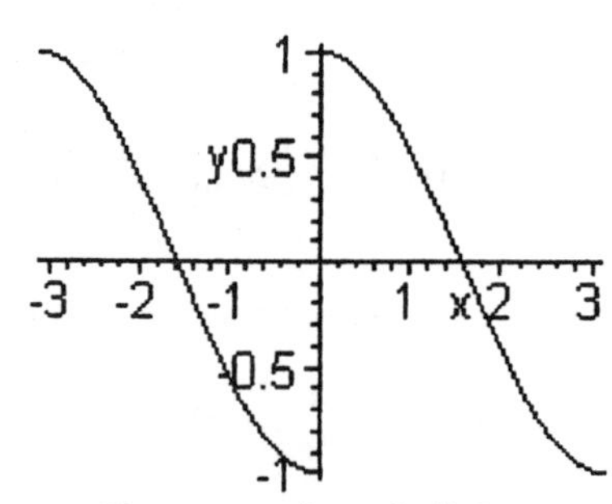

Even extension of $f(x)$

$$= \frac{2}{\pi}\left[\frac{n}{n^2-1}(-\cos\pi\cos n\pi + 1)\right] = \frac{2n}{\pi(n^2-1)}[(-1)^n + 1]$$

Therefore,

$$f(x) = \frac{2}{\pi}\sum_{n=1}^{\infty}\frac{n[(-1)^n+1]}{n^2-1}\sin nx = \frac{8}{\pi}\sum_{k=1}^{\infty}\frac{k}{4k^2-1}\sin 2kx.$$

13. $b_n = \frac{2}{\pi}\int_0^{\pi}\sin x \sin nx\,dx$

$= 0$, if $n \neq 1$

$$b_1 = \frac{2}{\pi}\int_0^{\pi}\sin^2 x\,dx = \frac{2}{\pi}\left[\frac{x}{2} - \frac{1}{4}\sin 2x\right]_0^{\pi} = \frac{2}{\pi}\left(\frac{\pi}{2}\right) = 1.$$

Therefore, $f(x) = b_1 \sin x = \sin x$.

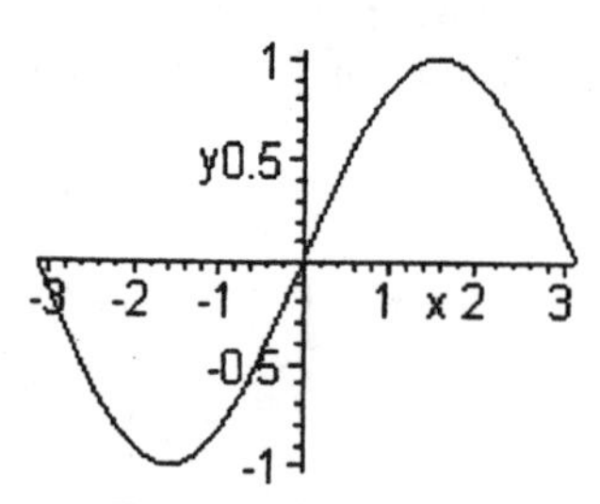

Even extension of $f(x)$

15. $b_n = \int_0^1 (1-x)\sin\frac{n\pi x}{2}\,dx$

$$= \left[-\frac{2(1-x)}{n\pi}\cos\frac{n\pi x}{2} - \frac{4}{n^2\pi^2}\sin\frac{n\pi x}{2}\right]_0^1 = \frac{-4}{n^2\pi^2}\sin\frac{n\pi}{2} + \frac{2}{n\pi}.$$

Therefore,

$$f(x) = \frac{2}{\pi}\sum_{n=1}^{\infty}\left[\frac{1}{n} - \frac{2}{n^2\pi}\sin\frac{n\pi}{2}\right]\sin\frac{n\pi x}{2}.$$

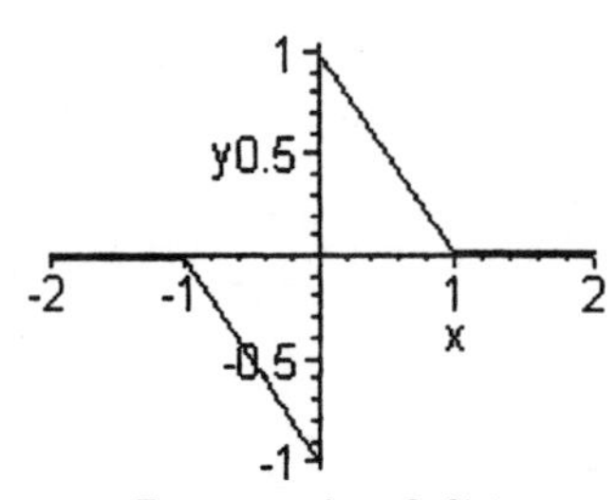

Even extension of $f(x)$

17. (a) $b_n = \frac{2}{\pi}\int_0^{\pi}\sin nx\,dx = \frac{2}{n\pi}[-\cos nx]_0^{\pi} = \frac{2}{n\pi}(-\cos n\pi + 1) = \frac{2}{n\pi}[1-(-1)^n] \Rightarrow f(x) = \sum_{n=1}^{\infty}\frac{2(1-(-1)^n)}{n\pi}\sin nx$

$$= \sum_{n=1}^{\infty}\frac{4}{(2n-1)\pi}\sin[(2n-1)x] = \frac{4}{\pi}\sin x + \frac{4}{3\pi}\sin 3x + \frac{4}{5\pi}\sin 5x + \ldots$$

$$\Rightarrow f(x) = \frac{4}{\pi}\left[\sin x + \frac{\sin 3x}{3} + \frac{\sin 5x}{5} + \frac{\sin 7x}{7} + \ldots\right] \Rightarrow \frac{\pi}{4}f(x) = \sin x + \frac{\sin 3x}{3} + \frac{\sin 5x}{5} + \frac{\sin 7x}{7} + \ldots.$$

(b) Evaluate $f(x)$ at $x = \frac{\pi}{2} \Rightarrow \frac{\pi}{4}\cdot 1 = \sin\left(\frac{\pi}{2}\right) + \frac{1}{3}\sin\left(\frac{3\pi}{2}\right) + \frac{1}{5}\sin\left(\frac{5\pi}{2}\right) + \frac{1}{7}\sin\left(\frac{7\pi}{2}\right) + \ldots$

$$\Rightarrow \frac{\pi}{4} = 1 - \frac{1}{3} + \frac{1}{5} - \frac{1}{7} + \ldots$$

19. $f(x) = \sin x = \frac{2}{\pi} + \frac{2}{\pi}\sum_{n=1}^{\infty}\frac{[(-1)^n+1]}{1-n^2}\cos nx = \frac{2}{\pi} - \frac{4}{\pi}\sum_{n=1}^{\infty}\frac{1}{4n^2-1}\cos(2nx)$ for $0 < x < \pi$

Evaluate the function and its series representation at $x = \frac{\pi}{2}$.

$$1 = \frac{2}{\pi} - \frac{4}{\pi}\sum_{n=1}^{\infty}\frac{1}{4n^2-1}\cos(n\pi) = \frac{2}{\pi} - \frac{4}{\pi}\sum_{n=1}^{\infty}\frac{(-1)^n}{4n^2-1} \Rightarrow \sum_{n=1}^{\infty}\frac{(-1)^n}{4n^2-1} = \frac{\pi}{4}\left(\frac{2}{\pi} - 1\right) = \frac{1}{2} - \frac{\pi}{4}$$

CHAPTER 8 PRACTICE EXERCISES

1. converges to 1, since $\lim_{n\to\infty} a_n = \lim_{n\to\infty} \left(1 + \frac{(-1)^n}{n}\right) = 1$

3. converges to -1, since $\lim_{n\to\infty} a_n = \lim_{n\to\infty} \left(\frac{1-2^n}{2^n}\right) = \lim_{n\to\infty} \left(\frac{1}{2^n} - 1\right) = -1$

5. diverges, since $\left\{\sin \frac{n\pi}{2}\right\} = \{0, 1, 0, -1, 0, 1, \ldots\}$

7. converges to 0, since $\lim_{n\to\infty} a_n = \lim_{n\to\infty} \frac{\ln n^2}{n} = 2 \lim_{n\to\infty} \frac{\left(\frac{1}{n}\right)}{1} = 0$

9. converges to 1, since $\lim_{n\to\infty} a_n = \lim_{n\to\infty} \left(\frac{n + \ln n}{n}\right) = \lim_{n\to\infty} \frac{1 + \left(\frac{1}{n}\right)}{1} = 1$

11. converges to e^{-5}, since $\lim_{n\to\infty} a_n = \lim_{n\to\infty} \left(\frac{n-5}{n}\right)^n = \lim_{n\to\infty} \left(1 + \frac{(-5)}{n}\right)^n = e^{-5}$ by Table 8.1

13. converges to 3, since $\lim_{n\to\infty} a_n = \lim_{n\to\infty} \left(\frac{3^n}{n}\right)^{1/n} = \lim_{n\to\infty} \frac{3}{n^{1/n}} = \frac{3}{1} = 3$ by Table 8.1

15. converges to ln 2, since $\lim_{n\to\infty} a_n = \lim_{n\to\infty} n\left(2^{1/n} - 1\right) = \lim_{n\to\infty} \frac{2^{1/n} - 1}{\left(\frac{1}{n}\right)} = \lim_{n\to\infty} \frac{\left[\frac{\left(-2^{1/n} \ln 2\right)}{n^2}\right]}{\left(\frac{-1}{n^2}\right)} = \lim_{n\to\infty} 2^{1/n} \ln 2$

$= 2^0 \cdot \ln 2 = \ln 2$

17. diverges, since $\lim_{n\to\infty} a_n = \lim_{n\to\infty} \frac{(n+1)!}{n!} = \lim_{n\to\infty} (n+1) = \infty$

19. $\frac{1}{(2n-3)(2n-1)} = \frac{\left(\frac{1}{2}\right)}{2n-3} - \frac{\left(\frac{1}{2}\right)}{2n-1} \Rightarrow s_n = \left[\frac{\left(\frac{1}{2}\right)}{3} - \frac{\left(\frac{1}{2}\right)}{5}\right] + \left[\frac{\left(\frac{1}{2}\right)}{5} - \frac{\left(\frac{1}{2}\right)}{7}\right] + \ldots + \left[\frac{\left(\frac{1}{2}\right)}{2n-3} - \frac{\left(\frac{1}{2}\right)}{2n-1}\right] = \frac{\left(\frac{1}{2}\right)}{3} - \frac{\left(\frac{1}{2}\right)}{2n-1}$

$\Rightarrow \lim_{n\to\infty} s_n = \lim_{n\to\infty} \left[\frac{1}{6} - \frac{\left(\frac{1}{2}\right)}{2n-1}\right] = \frac{1}{6}$

21. $\frac{9}{(3n-1)(3n+2)} = \frac{3}{3n-1} - \frac{3}{3n+2} \Rightarrow s_n = \left(\frac{3}{2} - \frac{3}{5}\right) + \left(\frac{3}{5} - \frac{3}{8}\right) + \left(\frac{3}{8} - \frac{3}{11}\right) + \ldots + \left(\frac{3}{3n-1} - \frac{3}{3n+2}\right)$

$= \frac{3}{2} - \frac{3}{3n+2} \Rightarrow \lim_{n\to\infty} s_n = \lim_{n\to\infty} \left(\frac{3}{2} - \frac{3}{3n+2}\right) = \frac{3}{2}$

23. $\sum_{n=0}^{\infty} e^{-n} = \sum_{n=0}^{\infty} \frac{1}{e^n}$, a convergent geometric series with $r = \frac{1}{e}$ and $a = 1 \Rightarrow$ the sum is $\frac{1}{1 - \left(\frac{1}{e}\right)} = \frac{e}{e-1}$

25. diverges, a p-series with $p = \frac{1}{2}$

27. Since $f(x) = \frac{1}{x^{1/2}} \Rightarrow f'(x) = -\frac{1}{2x^{3/2}} < 0 \Rightarrow f(x)$ is decreasing $\Rightarrow a_{n+1} < a_n$, and $\lim_{n\to\infty} a_n = \lim_{n\to\infty} \frac{(-1)^n}{\sqrt{n}} = 0$, the series $\sum_{n=1}^{\infty} \frac{(-1)^n}{\sqrt{n}}$ converges by the Alternating Series Test. Since $\sum_{n=1}^{\infty} \frac{1}{\sqrt{n}}$ diverges, the given series converges conditionally.

29. The given series does not converge absolutely by the Direct Comparison Test since $\frac{1}{\ln(n+1)} > \frac{1}{n+1}$, which is the nth term of a divergent series. Since $f(x) = \frac{1}{\ln(x+1)} \Rightarrow f'(x) = -\frac{1}{(\ln(x+1))^2(x+1)} < 0 \Rightarrow f(x)$ is decreasing $\Rightarrow a_{n+1} < a_n$, and $\lim_{n\to\infty} a_n = \lim_{n\to\infty} \frac{1}{\ln(n+1)} = 0$, the given series converges conditionally by the Alternating Series Test.

31. converges absolutely by the Direct Comparison Test since $\frac{\ln n}{n^3} < \frac{n}{n^3} = \frac{1}{n^2}$, the nth term of a convergent p-series

33. $\lim_{n\to\infty} \frac{\left(\frac{1}{n\sqrt{n^2+1}}\right)}{\left(\frac{1}{n^2}\right)} = \sqrt{\lim_{n\to\infty} \frac{n^2}{n^2+1}} = \sqrt{1} = 1 \Rightarrow$ converges absolutely by the Limit Comparison Test

35. converges absolutely by the Ratio Test since $\lim_{n\to\infty} \left[\frac{n+2}{(n+1)!} \cdot \frac{n!}{n+1}\right] = \lim_{n\to\infty} \frac{n+2}{(n+1)^2} = 0 < 1$

37. converges absolutely by the Ratio Test since $\lim_{n\to\infty} \left[\frac{3^{n+1}}{(n+1)!} \cdot \frac{n!}{3^n}\right] = \lim_{n\to\infty} \frac{3}{n+1} = 0 < 1$

39. converges absolutely by the Limit Comparison Test since $\lim_{n\to\infty} \frac{\left(\frac{1}{n^{3/2}}\right)}{\left(\frac{1}{\sqrt{n(n+1)(n+2)}}\right)}$
$= \sqrt{\lim_{n\to\infty} \frac{n(n+1)(n+2)}{n^3}} = 1$

41. $\lim_{n\to\infty} \left|\frac{u_{n+1}}{u_n}\right| < 1 \Rightarrow \lim_{n\to\infty} \left|\frac{(x+4)^{n+1}}{(n+1)3^{n+1}} \cdot \frac{n3^n}{(x+4)^n}\right| < 1 \Rightarrow \frac{|x+4|}{3} \lim_{n\to\infty} \left(\frac{n}{n+1}\right) < 1 \Rightarrow \frac{|x+4|}{3} < 1$
$\Rightarrow |x+4| < 3 \Rightarrow -3 < x+4 < 3 \Rightarrow -7 < x < -1$; at $x = -7$ we have $\sum_{n=1}^{\infty} \frac{(-1)^n 3^n}{n3^n} = \sum_{n=1}^{\infty} \frac{(-1)^n}{n}$, the alternating harmonic series, which converges conditionally; at $x = -1$ we have $\sum_{n=1}^{\infty} \frac{3^n}{n3^n} = \sum_{n=1}^{\infty} \frac{1}{n}$, the divergent harmonic series

(a) the radius is 3; the interval of convergence is $-7 \le x < -1$

(b) the interval of absolute convergence is $-7 < x < -1$

(c) the series converges conditionally at $x = -7$

43. $\lim_{n\to\infty} \left|\frac{u_{n+1}}{u_n}\right| < 1 \Rightarrow \lim_{n\to\infty} \left|\frac{(3x-1)^{n+1}}{(n+1)^2} \cdot \frac{n^2}{(3x-1)^n}\right| < 1 \Rightarrow |3x-1| \lim_{n\to\infty} \frac{n^2}{(n+1)^2} < 1 \Rightarrow |3x-1| < 1$

$\Rightarrow -1 < 3x - 1 < 1 \Rightarrow 0 < 3x < 2 \Rightarrow 0 < x < \frac{2}{3}$; at $x = 0$ we have $\sum_{n=1}^{\infty} \frac{(-1)^{n-1}(-1)^n}{n^2} = \sum_{n=1}^{\infty} \frac{(-1)^{2n-1}}{n^2}$

$= -\sum_{n=1}^{\infty} \frac{1}{n^2}$, a nonzero constant multiple of a convergent p-series, which is absolutely convergent; at $x = \frac{2}{3}$ we have $\sum_{n=1}^{\infty} \frac{(-1)^{n-1}(1)^n}{n^2} = \sum_{n=1}^{\infty} \frac{(-1)^{n-1}}{n^2}$, which converges absolutely

(a) the radius is $\frac{1}{3}$; the interval of convergence is $0 \le x \le \frac{2}{3}$

(b) the interval of absolute convergence is $0 \le x \le \frac{2}{3}$

(c) there are no values for which the series converges conditionally

45. $\lim_{n\to\infty} \left|\frac{u_{n+1}}{u_n}\right| < 1 \Rightarrow \lim_{n\to\infty} \left|\frac{x^{n+1}}{(n+1)^{n+1}} \cdot \frac{n^n}{x^n}\right| < 1 \Rightarrow |x| \lim_{n\to\infty} \left|\left(\frac{n}{n+1}\right)^n \left(\frac{1}{n+1}\right)\right| < 1 \Rightarrow \frac{|x|}{e} \lim_{n\to\infty} \left(\frac{1}{n+1}\right) < 1$

$\Rightarrow \frac{|x|}{e} \cdot 0 < 1$, which holds for all x

(a) the radius is ∞; the series converges for all x

(b) the series converges absolutely for all x

(c) there are no values for which the series converges conditionally

47. $\lim_{n\to\infty} \left|\frac{u_{n+1}}{u_n}\right| < 1 \Rightarrow \lim_{n\to\infty} \left|\frac{(n+2)x^{2n+1}}{3^{n+1}} \cdot \frac{3^n}{(n+1)x^{2n-1}}\right| < 1 \Rightarrow \frac{x^2}{3} \lim_{n\to\infty} \left(\frac{n+2}{n+1}\right) < 1 \Rightarrow -\sqrt{3} < x < \sqrt{3}$;

the series $\sum_{n=1}^{\infty} -\frac{n+1}{\sqrt{3}}$ and $\sum_{n=1}^{\infty} \frac{n+1}{\sqrt{3}}$, obtained with $x = \pm\sqrt{3}$, both diverge

(a) the radius is $\sqrt{3}$; the interval of convergence is $-\sqrt{3} < x < \sqrt{3}$

(b) the interval of absolute convergence is $-\sqrt{3} < x < \sqrt{3}$

(c) there are no values for which the series converges conditionally

49. $\lim_{n\to\infty} \left|\frac{u_{n+1}}{u_n}\right| < 1 \Rightarrow \lim_{n\to\infty} \left|\frac{\operatorname{csch}(n+1)x^{n+1}}{\operatorname{csch}(n)x^n}\right| < 1 \Rightarrow |x| \lim_{n\to\infty} \left|\frac{\left(\frac{2}{e^{n+1} - e^{-n-1}}\right)}{\left(\frac{2}{e^n - e^{-n}}\right)}\right| < 1$

$\Rightarrow |x| \lim_{n\to\infty} \left|\frac{e^{-1} - e^{-2n-1}}{1 - e^{-2n-2}}\right| < 1 \Rightarrow \frac{|x|}{e} < 1 \Rightarrow -e < x < e$; the series $\sum_{n=1}^{\infty} (\pm e)^n \operatorname{csch} n$, obtained with $x = \pm e$,

both diverge since $\lim_{n\to\infty} (\pm e)^n \operatorname{csch} n \ne 0$

(a) the radius is e; the interval of convergence is $-e < x < e$

(b) the interval of absolute convergence is $-e < x < e$

(c) there are no values for which the series converges conditionally

51. The given series has the form $1 - x + x^2 - x^3 + \ldots + (-x)^n + \ldots = \frac{1}{1+x}$, where $x = \frac{1}{4}$; the sum is $\frac{1}{1+\left(\frac{1}{4}\right)} = \frac{4}{5}$

53. The given series has the form $x - \frac{x^3}{3!} + \frac{x^5}{5!} - \ldots + (-1)^n \frac{x^{2n+1}}{(2n+1)!} + \ldots = \sin x$, where $x = \pi$; the sum is $\sin \pi = 0$

55. The given series has the form $1 + x + \frac{x^2}{2!} + \frac{x^2}{3!} + \ldots + \frac{x^n}{n!} + \ldots = e^x$, where $x = \ln 2$; the sum is $e^{\ln(2)} = 2$

57. Consider $\frac{1}{1-2x}$ as the sum of a convergent geometric series with $a = 1$ and $r = 2x \Rightarrow \frac{1}{1-2x}$ $= 1 + (2x) + (2x)^2 + (2x)^3 + \ldots = \sum_{n=0}^{\infty} (2x)^n = \sum_{n=0}^{\infty} 2^n x^n$ where $|2x| < 1 \Rightarrow |x| < \frac{1}{2}$

59. $\sin x = \sum_{n=0}^{\infty} \frac{(-1)^n x^{2n+1}}{(2n+1)!} \Rightarrow \sin \pi x = \sum_{n=0}^{\infty} \frac{(-1)^n (\pi x)^{2n+1}}{(2n+1)!} = \sum_{n=0}^{\infty} \frac{(-1)^n \pi^{2n+1} x^{2n+1}}{(2n+1)!}$

61. $\cos x = \sum_{n=0}^{\infty} \frac{(-1)^n x^{2n}}{(2n)!} \Rightarrow \cos\left(x^{5/2}\right) = \sum_{n=0}^{\infty} \frac{(-1)^n \left(x^{5/2}\right)^{2n}}{(2n)!} = \sum_{n=0}^{\infty} \frac{(-1)^n x^{5n}}{(2n)!}$

63. $e^x = \sum_{n=0}^{\infty} \frac{x^n}{n!} \Rightarrow e^{(\pi x/2)} = \sum_{n=0}^{\infty} \frac{\left(\frac{\pi x}{2}\right)^n}{n!} = \sum_{n=0}^{\infty} \frac{\pi^n x^n}{2^n n!}$

65. $f(x) = \sqrt{3+x^2} = \left(3+x^2\right)^{1/2} \Rightarrow f'(x) = x\left(3+x^2\right)^{-1/2} \Rightarrow f''(x) = -x^2\left(3+x^2\right)^{-3/2} + \left(3+x^2\right)^{-1/2}$ $\Rightarrow f'''(x) = 3x^3\left(3+x^2\right)^{-5/2} - 3x\left(3+x^2\right)^{-3/2}$; $f(-1) = 2$, $f'(-1) = -\frac{1}{2}$, $f''(-1) = -\frac{1}{8} + \frac{1}{2} = \frac{3}{8}$, $f'''(-1) = -\frac{3}{32} + \frac{3}{8} = \frac{9}{32} \Rightarrow \sqrt{3+x^2} = 2 - \frac{(x+1)}{2 \cdot 1!} + \frac{3(x+1)^2}{2^3 \cdot 2!} + \frac{9(x+1)^3}{2^5 \cdot 3!} + \ldots$

67. $f(x) = \frac{1}{x+1} = (x+1)^{-1} \Rightarrow f'(x) = -(x+1)^{-2} \Rightarrow f''(x) = 2(x+1)^{-3} \Rightarrow f'''(x) = -6(x+1)^{-4}$; $f(3) = \frac{1}{4}$, $f'(3) = -\frac{1}{4^2}$, $f''(3) = \frac{2}{4^3}$, $f'''(2) = \frac{-6}{4^4} \Rightarrow \frac{1}{x+1} = \frac{1}{4} - \frac{1}{4^2}(x-3) + \frac{1}{4^3}(x-3)^2 - \frac{1}{4^4}(x-3)^3 + \ldots$

69. Assume the solution has the form $y = a_0 + a_1 x + a_2 x^2 + \ldots + a_{n-1} x^{n-1} + a_n x^n + \ldots$ $\Rightarrow \frac{dy}{dx} = a_1 + 2a_2 x + \ldots + n a_n x^{n-1} + \ldots \Rightarrow \frac{dy}{dx} + y$ $= (a_1 + a_0) + (2a_2 + a_1)x + (3a_3 + a_2)x^2 + \ldots + (na_n + a_{n-1})x^{n-1} + \ldots = 0 \Rightarrow a_1 + a_0 = 0$, $2a_2 + a_1 = 0$, $3a_3 + a_2 = 0$ and in general $na_n + a_{n-1} = 0$. Since $y = -1$ when $x = 0$ we have $a_0 = -1$. Therefore $a_1 = 1$, $a_2 = \frac{-a_1}{2 \cdot 1} = -\frac{1}{2}$, $a_3 = \frac{-a_2}{3} = \frac{1}{3 \cdot 2}$, $a_4 = \frac{-a_3}{4} = -\frac{1}{4 \cdot 3 \cdot 2}$, …, $a_n = \frac{-a_{n-1}}{n} = \frac{-1}{n} \frac{(-1)^n}{(n-1)!} = \frac{(-1)^{n+1}}{n!}$ $\Rightarrow y = -1 + x - \frac{1}{2}x^2 + \frac{1}{3 \cdot 2}x^3 - \ldots + \frac{(-1)^{n+1}}{n!}x^n + \ldots = -\sum_{n=0}^{\infty} \frac{(-1)^n x^n}{n!} = -e^{-x}$

71. Assume the solution has the form $y = a_0 + a_1 x + a_2 x^2 + \ldots + a_{n-1} x^{n-1} + a_n x^n + \ldots$ $\Rightarrow \frac{dy}{dx} = a_1 + 2a_2 x + \ldots + n a_n x^{n-1} + \ldots \Rightarrow \frac{dy}{dx} + 2y$

$= (a_1 + 2a_0) + (2a_2 + 2a_1)x + (3a_3 + 2a_2)x^2 + \ldots + (na_n + 2a_{n-1})x^{n-1} + \ldots = 0$. Since $y = 3$ when $x = 0$ we have $a_0 = 3$. Therefore $a_1 = -2a_0 = -2(3) = -3(2)$, $a_2 = -\frac{2}{2}a_1 = -\frac{2}{2}(-2\cdot 3) = 3\left(\frac{2^2}{2}\right)$, $a_3 = -\frac{2}{3}a_2$

$$= -\frac{2}{3}\left[3\left(\frac{2^2}{2}\right)\right] = -3\left(\frac{2^3}{3\cdot 2}\right), \ldots, a_n = \left(-\frac{2}{n}\right)a_{n-1} = \left(-\frac{2}{n}\right)\left(3\left(\frac{(-1)^{n-1}2^{n-1}}{(n-1)!}\right)\right) = 3\left(\frac{(-1)^n 2^n}{n!}\right)$$

$$\Rightarrow y = 3 - 3(2x) + 3\frac{(2)^2}{2}x^2 - 3\frac{(2)^3}{3\cdot 2}x^3 + \ldots + 3\frac{(-1)^n 2^n}{n!}x^n + \ldots$$

$$= 3\left[1 - (2x) + \frac{(2x)^2}{2!} - \frac{(2x)^3}{3!} + \ldots + \frac{(-1)^n(2x)^n}{n!} + \ldots\right] = 3\sum_{n=0}^{\infty}\frac{(-1)^n(2x)^n}{n!} = 3e^{-2x}$$

73. Assume the solution has the form $y = a_0 + a_1x + a_2x^2 + \ldots + a_{n-1}x^{n-1} + a_nx^n + \ldots$

$\Rightarrow \frac{dy}{dx} = a_1 + 2a_2x + \ldots + na_nx^{n-1} + \ldots \Rightarrow \frac{dy}{dx} - y$

$= (a_1 - a_0) + (2a_2 - a_1)x + (3a_3 - a_2)x^2 + \ldots + (na_n - a_{n-1})x^{n-1} + \ldots = 3x \Rightarrow a_1 - a_0 = 0$, $2a_2 - a_1 = 3$, $3a_3 - a_2 = 0$ and in general $na_n - a_{n-1} = 0$ for $n > 2$. Since $y = -1$ when $x = 0$ we have $a_0 = -1$. Therefore

$$a_1 = -1,\ a_2 = \frac{3 + a_1}{2} = \frac{2}{2},\ a_3 = \frac{a_2}{3} = \frac{2}{3\cdot 2},\ a_4 = \frac{a_3}{4} = \frac{2}{4\cdot 3\cdot 2}, \ldots, a_n = \frac{a_{n-1}}{n} = \frac{2}{n!}$$

$$\Rightarrow y = -1 - x + \left(\frac{2}{2}\right)x^2 + \frac{3}{3\cdot 2}x^3 + \frac{2}{4\cdot 3\cdot 2}x^4 + \ldots + \frac{2}{n!}x^n + \ldots$$

$$= 2\left(1 + x + \frac{1}{2}x^2 + \frac{1}{3\cdot 2}x^3 + \frac{1}{4\cdot 3\cdot 2}x^4 + \ldots + \frac{1}{n!}x^n + \ldots\right) - 3 - 3x = 2\sum_{n=0}^{\infty}\frac{x^n}{n!} - 3 - 3x = 2e^x - 3x - 3$$

75. Assume the solution has the form $y = a_0 + a_1x + a_2x^2 + \ldots + a_{n-1}x^{n-1} + a_nx^n + \ldots$

$\Rightarrow \frac{dy}{dx} = a_1 + 2a_2x + \ldots + na_nx^{n-1} + \ldots \Rightarrow \frac{dy}{dx} - y$

$= (a_1 - a_0) + (2a_2 - a_1)x + (3a_3 - a_2)x^2 + \ldots + (na_n - a_{n-1})x^{n-1} + \ldots = x \Rightarrow a_1 - a_0 = 0$, $2a_2 - a_1 = 1$, $3a_3 - a_2 = 0$ and in general $na_n - a_{n-1} = 0$ for $n > 2$. Since $y = 1$ when $x = 0$ we have $a_0 = 1$. Therefore

$$a_1 = 1,\ a_2 = \frac{1 + a_1}{2} = \frac{2}{2},\ a_3 = \frac{a_2}{3} = \frac{2}{3\cdot 2},\ a_4 = \frac{a_3}{4} = \frac{2}{4\cdot 3\cdot 2}, \ldots, a_n = \frac{a_{n-1}}{n} = \frac{2}{n!}$$

$$\Rightarrow y = 1 + x + \left(\frac{2}{2}\right)x^2 + \frac{2}{3\cdot 2}x^3 + \frac{2}{4\cdot 2\cdot 2}x^4 + \ldots + \frac{2}{n!}x^n + \ldots$$

$$= 2\left(1 + x + \frac{1}{2}x^2 + \frac{1}{3\cdot 2}x^3 + \frac{1}{4\cdot 3\cdot 2}x^4 + \ldots + \frac{1}{n!}x^n + \ldots\right) - 1 - x = 2\sum_{n=0}^{\infty}\frac{x^n}{n!} - 1 - x = 2e^x - x - 1$$

77. $$\lim_{x\to 0}\frac{7\sin x}{e^{2x} - 1} = \lim_{x\to 0}\frac{7\left(x - \frac{x^3}{3!} + \frac{x^5}{5!} - \ldots\right)}{\left(2x + \frac{2^2x^2}{2!} + \frac{2^3x^3}{3!} + \ldots\right)} = \lim_{x\to 0}\frac{7\left(1 - \frac{x^2}{3!} + \frac{x^4}{5!} - \ldots\right)}{\left(2 + \frac{2^2x}{2!} + \frac{2^3x^2}{3!} + \ldots\right)} = \frac{7}{2}$$

79. $$\lim_{t\to 0}\left(\frac{1}{2 - 2\cos t} - \frac{1}{t^2}\right) = \lim_{t\to 0}\frac{t^2 - 2 + 2\cos t}{2t^2(1 - \cos t)} = \lim_{t\to 0}\frac{t^2 - 2 + 2\left(1 - \frac{t^2}{2} + \frac{t^4}{4!} - \ldots\right)}{2t^2\left(1 - 1 + \frac{t^2}{2} - \frac{t^4}{4!} + \ldots\right)} = \lim_{t\to 0}\frac{2\left(\frac{t^4}{4!} - \frac{t^6}{6!} + \ldots\right)}{\left(t^4 - \frac{2t^6}{4!} + \ldots\right)}$$

$$= \lim_{t\to 0} \frac{2\left(\frac{1}{4!} - \frac{t^2}{6!} + \ldots\right)}{\left(1 - \frac{2t^2}{4!} + \ldots\right)} = \frac{1}{12}$$

81. $$\lim_{z\to 0} \frac{1-\cos^2 z}{\ln(1-z)+\sin z} = \lim_{z\to 0} \frac{1-\left(1-z^2+\frac{z^4}{3}-\ldots\right)}{\left(-z-\frac{z^2}{2}-\frac{z^3}{3}-\ldots\right)+\left(z-\frac{z^3}{3!}+\frac{z^5}{5!}-\ldots\right)} = \lim_{z\to 0} \frac{\left(z^2-\frac{z^4}{3}+\ldots\right)}{\left(-\frac{z^2}{2}-\frac{2z^3}{3}-\frac{z^4}{4}-\ldots\right)}$$

$$= \lim_{z\to 0} \frac{\left(1-\frac{z^2}{3}+\ldots\right)}{\left(-\frac{1}{2}-\frac{2z}{3}-\frac{z^2}{4}-\ldots\right)} = -2$$

83. $$\lim_{x\to 0}\left(\frac{\sin 3x}{x^3}+\frac{r}{x^2}+s\right) = \lim_{x\to 0}\left[\frac{\left(3x-\frac{(3x)^3}{6}+\frac{(3x)^5}{120}-\ldots\right)}{x^3}+\frac{r}{x^2}+s\right] = \lim_{x\to 0}\left(\frac{3}{x^2}-\frac{9}{2}+\frac{81x^2}{40}+\ldots+\frac{r}{x^2}+s\right) = 0$$

$$\Rightarrow \frac{r}{x^2}+\frac{3}{x^2} = 0 \text{ and } s-\frac{9}{2} = 0 \Rightarrow r = -3 \text{ and } s = \frac{9}{2}$$

85. $$a_0 = \frac{1}{\pi}\int_{-\pi}^{\pi} f(x)\,dx = 1$$

$$a_n = \frac{1}{\pi}\int_{-\pi}^{\pi} f(x)\cos nx\,dx = \frac{1}{\pi}\int_{-\pi}^{0} -\cos nx\,dx + \frac{1}{\pi}\int_{0}^{\pi} 2\cos nx\,dx = -\frac{1}{n\pi}\sin nx\Big|_{-\pi}^{0} + \frac{2}{n\pi}\sin nx\Big|_{0}^{\pi} = 0$$

$$b_n = \frac{1}{\pi}\int_{-\pi}^{\pi} f(x)\sin nx\,dx = \frac{1}{\pi}\int_{-\pi}^{0} -\sin nx\,dx + \frac{1}{\pi}\int_{0}^{\pi} 2\sin nx\,dx = \frac{1}{n\pi}\cos nx\Big|_{-\pi}^{0} - \frac{2}{n\pi}\cos nx\Big|_{0}^{\pi}$$

$$= \frac{1}{n\pi}(1-\cos n\pi) - \frac{2}{n\pi}(\cos n\pi - 1) = \frac{3}{n\pi}(1-\cos n\pi) = \frac{3}{n\pi}(1-(-1)^n)$$

Therefore, $f(x) = \frac{1}{2} + \sum_{n=1}^{\infty} \frac{3}{n\pi}(1-(-1)^n)\sin nx = \frac{1}{2} + \frac{6}{\pi}\sum_{n=1}^{\infty} \frac{1}{2n-1}\sin[(2n-1)x]$

87. $$a_0 = \frac{1}{\pi}\int_{-\pi}^{\pi} f(x)\,dx = \frac{1}{\pi}\int_{-\pi}^{\pi}(x+\pi)\,dx = \frac{1}{\pi}(2\pi^2) = 2\pi$$

$a_n = 0$ because $f(x) - \pi$ is an odd function

$$b_n = \frac{1}{\pi}\int_{-\pi}^{\pi}(x+\pi)\sin nx\,dx = \frac{1}{\pi}\left[\frac{\sin nx}{n^2} - \frac{(x+\pi)\cos nx}{n}\right]_{-\pi}^{\pi} = -\frac{2}{n}\cos n\pi = -\frac{2(-1)^n}{n}$$

Therefore, $f(x) = \pi - 2\sum_{n=1}^{\infty} \frac{(-1)^n}{n}\sin nx$

89. $$a_0 = \frac{1}{2}\int_{-2}^{2} f(x)\,dx = \frac{1}{2}[2+4] = 3$$

$$a_n = \frac{1}{2}\int_{-2}^{2} f(x)\cos\left(\frac{n\pi x}{2}\right)dx = \frac{1}{2}\left[\int_{-2}^{0}\cos\left(\frac{n\pi x}{2}\right)dx + \int_{0}^{2}(1+x)\cos\left(\frac{n\pi x}{2}\right)dx\right]$$

$$= \frac{1}{n\pi}\sin\left(\frac{n\pi x}{2}\right)\Big|_{-2}^{0} + \frac{1}{2}\left[\frac{4}{n^2\pi^2}\cos\left(\frac{n\pi x}{2}\right) + \frac{2(1+x)}{n\pi}\sin\left(\frac{n\pi x}{2}\right)\right]\Big|_{0}^{2} = \frac{2}{n^2\pi^2}[\cos n\pi - 1] = \frac{2((-1)^n - 1)}{n^2\pi^2}$$

$$b_n = \frac{1}{2}\int_{-2}^{2} f(x)\sin\left(\frac{n\pi x}{2}\right)dx = \frac{1}{2}\left[\int_{-2}^{0}\sin\left(\frac{n\pi x}{2}\right)dx + \int_{0}^{2}(1+x)\sin\left(\frac{n\pi x}{2}\right)dx\right]$$

$$= -\frac{1}{n\pi}\cos\left(\frac{n\pi x}{2}\right)\Big|_{-2}^{0} + \left[-\frac{(1+x)}{n\pi}\cos\left(\frac{n\pi x}{2}\right) + \frac{2}{n^2\pi^2}\sin\left(\frac{n\pi x}{2}\right)\right]\Big|_{0}^{2} = \frac{1}{n\pi}(\cos n\pi - 1) + \frac{1}{n\pi}(1 - 3\cos n\pi)$$

$$= \frac{(-1)^n - 1}{n\pi} + \frac{1 - 3(-1)^n}{n\pi} = -\frac{2(-1)^n}{n\pi}$$

Therefore, $f(x) = \frac{3}{2} + \frac{2}{\pi^2}\sum_{n=1}^{\infty}\frac{(-1)^n - 1}{n^2}\cos\left(\frac{n\pi x}{2}\right) - \frac{2}{\pi}\sum_{n=1}^{\infty}\frac{(-1)^n}{n}\sin\left(\frac{n\pi x}{2}\right)$

91. (a) $a_0 = \frac{2}{1}\int_{0}^{1} f(x)\,dx = 2\int_{0}^{1/2} dx = 1$; $a_n = \frac{2}{1}\int_{0}^{1} f(x)\cos n\pi x\,dx = 2\int_{0}^{1/2}\cos n\pi x\,dx$

$= \frac{2}{n\pi}\sin n\pi x\Big|_{0}^{1/2} = \frac{2}{n\pi}\sin\left(\frac{n\pi}{2}\right)$. Therefore, $f(x) = \frac{1}{2} + \frac{2}{\pi}\sum_{n=1}^{\infty}\frac{\sin\left(\frac{n\pi}{2}\right)}{n}\cos n\pi x$

(b) $b_n = \frac{2}{1}\int_{0}^{1} f(x)\sin n\pi x\,dx = 2\int_{0}^{1/2}\sin n\pi x\,dx = -\frac{2}{n\pi}\cos n\pi x\Big|_{0}^{1/2} = \frac{2}{n\pi}\left(1 - \cos\left(\frac{n\pi}{2}\right)\right)$

Therefore, $f(x) = \frac{2}{\pi}\sum_{n=1}^{\infty}\frac{1}{n}\left(1 - \cos\left(\frac{n\pi}{2}\right)\right)\sin n\pi x$

93. (a) $a_0 = \frac{2}{1}\int_{0}^{1} f(x)\,dx = 2\int_{0}^{1}\sin \pi x\,dx = -\frac{2}{\pi}\cos \pi x\Big|_{0}^{1} = \frac{4}{\pi}$; $a_1 = \frac{2}{1}\int_{0}^{1} f(x)\cos \pi x\,dx = 2\int_{0}^{1}\sin \pi x\cos \pi x\,dx$

$= -\frac{1}{2\pi}\cos^2 \pi x\Big|_{0}^{1} = 0$. For $n \geq 2$, $a_n = 2\int_{0}^{1} f(x)\cos n\pi x\,dx = 2\int_{0}^{1}\sin \pi x\cos n\pi x\,dx$

$$= \left[\frac{\cos[(n-1)\pi x]}{n-1} - \frac{\cos[(n+1)\pi x]}{n+1}\right]\Big|_{0}^{1} = \frac{1}{\pi}\left[\frac{1}{n-1} + \frac{1}{n+1} + \frac{\cos[(n-1)\pi]}{n-1} - \frac{\cos[(n+1)\pi]}{n+1}\right]$$

Therefore, $f(x) = \frac{2}{\pi} + \frac{1}{\pi}\sum_{n=1}^{\infty}\left[\frac{1}{n+1} - \frac{1}{n-1} + \frac{\cos[(n-1)\pi]}{n-1} - \frac{\cos[(n+1)\pi]}{n+1}\right]\cos n\pi x$

(b) $b_1 = \frac{2}{1}\int_{0}^{1} f(x)\sin \pi x\,dx = 2\int_{0}^{1}\sin^2 \pi x\,dx = \left[x - \frac{\sin 2\pi x}{2\pi}\right]\Big|_{0}^{1} = 1$; $b_n = 0$ for $n \geq 2$.

Therefore, $f(x) = \sin \pi x$, as expected.

95. (a) $a_0 = \frac{2}{3}\int_0^3 f(x)\,dx = \frac{2}{3}\int_0^3 (2x + x^2)dx = \frac{2}{3}\left(x^2 + \frac{x^3}{3}\right)\Big|_0^3 = \frac{2}{3}(18) = 12$

$a_n = \frac{2}{3}\int_0^3 f(x)\cos\left(\frac{n\pi x}{3}\right)dx = \frac{2}{3}\int_0^3 (2x + x^2)\cos\left(\frac{n\pi x}{3}\right)dx =$ (using CAS)

$= \frac{2}{n^3\pi^3}\left[6n\pi(1+x)\cos\left(\frac{n\pi x}{3}\right) + \left(n^2\pi^2 x(x+2) - 18\right)\sin\left(\frac{n\pi x}{3}\right)\right]\Big|_0^3 = \frac{12}{n^2\pi^2}[4(-1)^n - 1]$

Therefore, $f(x) = 6 + \frac{12}{\pi^2}\sum_{n=1}^{\infty}\frac{4(-1)^n - 1}{n^2}\cos\left(\frac{n\pi x}{3}\right)$

(b) $b_n = \frac{2}{3}\int_0^3 f(x)\sin\left(\frac{n\pi x}{3}\right)dx = \frac{2}{3}\int_0^3 (2x + x^2)\sin\left(\frac{n\pi x}{3}\right)dx =$ (using CAS)

$= \frac{2}{n^3\pi^3}\left[-18n\pi(1+x)\sin\left(\frac{n\pi x}{3}\right) - \left(n^2\pi^2 x(x+2) - 18\right)\cos\left(\frac{n\pi x}{3}\right)\right]\Big|_0^3 = \frac{2\left[(18 - 15n^2\pi^2)(-1)^n - 18\right]}{n^3\pi^3}$

Therefore, $f(x) = \frac{6}{\pi^3}\sum_{n=1}^{\infty}\frac{(6 - 5n^2\pi^2)(-1)^n - 6}{n^3}\sin\left(\frac{n\pi x}{3}\right)$

97. (a) $\sum_{n=1}^{\infty}\left(\sin\frac{1}{2n} - \sin\frac{1}{2n+1}\right) = \left(\sin\frac{1}{2} - \sin\frac{1}{3}\right) + \left(\sin\frac{1}{4} - \sin\frac{1}{5}\right) + \left(\sin\frac{1}{6} - \sin\frac{1}{7}\right) + \ldots + \left(\sin\frac{1}{2n} - \sin\frac{1}{2n+1}\right)$

$+ \ldots = \sum_{n=2}^{\infty}(-1)^n\sin\frac{1}{n}$; $f(x) = \sin\frac{1}{x} \Rightarrow f'(x) = \frac{-\cos\left(\frac{1}{x}\right)}{x^2} < 0$ if $x \ge 2 \Rightarrow \sin\frac{1}{n+1} < \sin\frac{1}{n}$, and

$\lim_{n\to\infty}\sin\frac{1}{n} = 0 \Rightarrow \sum_{n=2}^{\infty}(-1)^n\sin\frac{1}{n}$ converges by the Alternating Series Test

(b) $|\text{error}| < \left|\sin\frac{1}{42}\right| \approx 0.02381$ and the sum is an underestimate because the remainder is positive

99. $\lim_{n\to\infty}\left|\frac{2\cdot 5\cdot 8\cdots(3n-1)(3n+2)x^{n+1}}{2\cdot 4\cdot 6\cdots(2n)(2n+2)}\cdot\frac{2\cdot 4\cdot 6\cdots(2n)}{2\cdot 5\cdot 8\cdots(3n-1)x^n}\right| < 1 \Rightarrow |x|\lim_{n\to\infty}\left|\frac{3n+2}{2n+2}\right| < 1 \Rightarrow |x| < \frac{2}{3}$

$\Rightarrow$ the radius of convergence is $\frac{2}{3}$

101. $\sum_{k=2}^{n}\ln\left(1 - \frac{1}{k^2}\right) = \sum_{k=2}^{n}\left[\ln\left(1 + \frac{1}{k}\right) + \ln\left(1 - \frac{1}{k}\right)\right] = \sum_{k=2}^{n}\left[\ln(k+1) - \ln k + \ln(k-1) - \ln k\right]$

$= [\ln 3 - \ln 2 + \ln 1 - \ln 2] + [\ln 4 - \ln 3 + \ln 2 - \ln 3] + [\ln 5 - \ln 4 + \ln 3 - \ln 4] + [\ln 6 - \ln 5 + \ln 4 - \ln 5]$
$+ \ldots + [\ln(n+1) - \ln n + \ln(n-1) - \ln n] = [\ln 1 - \ln 2] + [\ln(n+1) - \ln n]$ after cancellation

$\Rightarrow \sum_{k=2}^{n}\ln\left(1 - \frac{1}{k^2}\right) = \ln\left(\frac{n+1}{2n}\right) \Rightarrow \sum_{k=2}^{\infty}\ln\left(1 - \frac{1}{k^2}\right) = \lim_{n\to\infty}\ln\left(\frac{n+1}{2n}\right) = \ln\frac{1}{2}$ is the sum

103. (a) $\lim_{n\to\infty}\left|\frac{1\cdot 4\cdot 7\cdots(3n-2)(3n+1)x^{3n+3}}{(3n+3)!}\cdot\frac{(3n)!}{1\cdot 4\cdot 7\cdots(3n-2)x^{3n}}\right| < 1 \Rightarrow |x^3|\lim_{n\to\infty}\frac{(3n+1)}{(3n+1)(3n+2)(3n+3)}$

$= |x^3|\cdot 0 < 1 \Rightarrow$ the radius of convergence is ∞

(b) $y = 1 + \sum_{n=1}^{\infty} \frac{1\cdot 4\cdot 7\cdots(3n-2)}{(3n)!}x^{3n} \Rightarrow \frac{dy}{dx} = \sum_{n=1}^{\infty} \frac{1\cdot 4\cdot 7\cdots(3n-2)}{(3n-1)!}x^{3n-1}$

$\Rightarrow \frac{d^2y}{dx^2} = \sum_{n=1}^{\infty} \frac{1\cdot 4\cdot 7\cdots(3n-2)}{(3n-2)!}x^{3n-2} = x + \sum_{n=2}^{\infty} \frac{1\cdot 4\cdot 7\cdots(3n-5)}{(3n-3)!}x^{3n-2}$

$= x\left(1 + \sum_{n=1}^{\infty} \frac{1\cdot 4\cdot 7\cdots(3n-2)}{(3n)!}x^{3n}\right) = xy + 0 \Rightarrow a = 1$ and $b = 0$

105. Yes, the series $\sum_{n=1}^{\infty} a_n b_n$ converges as we now show. Since $\sum_{n=1}^{\infty} a_n$ converges it follows that $a_n \to 0 \Rightarrow a_n < 1$ for $n >$ some index $N \Rightarrow a_n b_n < b_n$ for $n > N \Rightarrow \sum_{n=1}^{\infty} a_n b_n$ converges by the Direct Comparison Test with $\sum_{n=1}^{\infty} b_n$

107. $\sum_{n=1}^{\infty} (x_{n+1} - x_n) = \lim_{n\to\infty} \sum_{k=1}^{\infty} (x_{k+1} - x_k) = \lim_{n\to\infty} (x_{n+1} - x_1) = \lim_{n\to\infty} (x_{n+1}) - x_1 \Rightarrow$ both the series and sequence must either converge or diverge.

109. (a) $\sum_{n=1}^{\infty} \frac{a_n}{n} = a_1 + \frac{a_2}{2} + \frac{a_3}{3} + \frac{a_4}{4} + \ldots \geq a_1 + \left(\frac{1}{2}\right)a_2 + \left(\frac{1}{3} + \frac{1}{4}\right)a_4 + \left(\frac{1}{5} + \frac{1}{6} + \frac{1}{7} + \frac{1}{8}\right)a_8$

$+\left(\frac{1}{9} + \frac{1}{10} + \frac{1}{11} + \ldots + \frac{1}{16}\right)a_{16} + \ldots \geq \frac{1}{2}(a_2 + a_4 + a_8 + a_{16} + \ldots)$ which is a divergent series

(b) $a_n = \frac{1}{\ln n}$ for $n \geq 2 \Rightarrow a_2 \geq a_3 \geq a_4 \geq \ldots$, and $\frac{1}{\ln 2} + \frac{1}{\ln 4} + \frac{1}{\ln 8} + \ldots = \frac{1}{\ln 2} + \frac{1}{2\ln 2} + \frac{1}{3\ln 2} + \ldots$

$= \frac{1}{\ln 2}\left(1 + \frac{1}{2} + \frac{1}{3} + \ldots\right)$ which diverges so that $1 + \sum_{n=2}^{\infty} \frac{1}{n \ln n}$ diverges by part (a)

111. (a) $\int_0^x \frac{1}{1+t^2}dt = \int_0^x \left(1 - t^2 + t^4 - t^6 + \ldots + (-1)^n t^{2n} + \frac{(-1)^{n+1}t^{2n+2}}{1+t^2}\right)dt$

$\tan^{-1} x = x - \frac{x^3}{3} + \frac{x^5}{5} - \frac{x^7}{7} + \ldots + \frac{(-1)^n}{2n+1}x^{2n+1} + \int_0^x \frac{(-1)^{n+1}t^{2n+2}}{1+t^2}dt$

(b) By definition,

$R_n(x) = f(x) - P_n(x) = \tan^{-1} x - \left(x - \frac{x^3}{3} + \frac{x^5}{5} - \frac{x^7}{7} + \ldots + \frac{(-1)^n}{2n+1}x^{2n+1}\right) = \int_0^x \frac{(-1)^{n+1}t^{2n+2}}{1+t^2}dt$

If the integrand goes to zero in the limit, then so will the value of the integral.

$|x| < 1 \Rightarrow |t| < 1 \Rightarrow \lim_{n\to\infty} \frac{(-1)^{n+1}t^{2n+2}}{1+t^2} = \frac{1}{1+t^2} \lim_{n\to\infty} (-1)^{n+1}t^{2n+2} = 0$. If $|x| = 1$, then the value of the integrand will approach 0 for all values of t between 0 and x, while at $t = x$, it will oscillate between $\pm\frac{1}{1+t^2}$. However, the integral of a function will converge provided the function is piecewise continuous in the interval $0 < t < x$. Therefore, we would expect that convergence of $R_n(x)$ to zero would not be affected by the value of the integrand at the single value $t = x$ provided it is finite, which it is. Therefore, $\lim_{n\to\infty} R_n(x) = 0$ for $|x| \leq 1$.

(c) For $|x| \le 1$, $\tan^{-1} x = \sum_{n=0}^{\infty} \frac{(-1)^n}{2n+1} x^{2n+1}$.

(d) $\tan^{-1} 1 = \frac{\pi}{4} = \sum_{n=0}^{\infty} \frac{(-1)^n}{2n+1}(1)^{2n+1} = \sum_{n=0}^{\infty} 5\,\frac{(-1)^n}{2n+1} = 1 - \frac{1}{3} + \frac{1}{5} - \frac{1}{7} + \frac{1}{9} - \ldots + \frac{(-1)^n}{2n+1} + \ldots$

113. (a) $g(x) = 2x + 3 \Rightarrow g^{-1}(x) = \frac{x-3}{2}$ and when the iterative method is applied to $g^{-1}(x)$ we have $x_0 = 2$ $\Rightarrow -2.99999881$ in 23 iterations $\Rightarrow -3$ is the fixed point

(b) $g(x) = 1 - 4x \Rightarrow g^{-1}(x) = \frac{1-x}{4}$ and when the iterative method is applied to $g^{-1}(x)$ we have $x_0 = 2$ $\Rightarrow 0.199999571$ in 12 iterations $\Rightarrow 0.2$ is the fixed point

CHAPTER 8 ADDITIONAL EXERCISES–THEORY, EXAMPLES, APPLICATIONS

1. converges since $\frac{1}{(3n-2)^{(2n+1)/2}} < \frac{1}{(3n-2)^{3/2}}$ and $\sum_{n=1}^{\infty} \frac{1}{(3n-2)^{3/2}}$ converges by the Limit Comparison Test:

$$\lim_{n\to\infty} \frac{\left(\frac{1}{n^{3/2}}\right)}{\left(\frac{1}{(3n-2)^{3/2}}\right)} = \lim_{n\to\infty} \left(\frac{3n-2}{n}\right)^{3/2} = 3^{3/2}$$

3. diverges by the nth-Term Test since $\lim_{n\to\infty} a_n = \lim_{n\to\infty} (-1)^n \tanh n = \lim_{b\to\infty} (-1)^n \left(\frac{1-e^{-2n}}{1+e^{-2n}}\right) = \lim_{n\to\infty} (-1)^n$ does not exist

5. converges by the Direct Comparison Test: $a_1 = 1 = \frac{12}{(1)(3)(2)^2}$, $a_2 = \frac{1\cdot 2}{3\cdot 4} = \frac{12}{(2)(4)(3)^2}$, $a_3 = \left(\frac{2\cdot 3}{4\cdot 5}\right)\left(\frac{1\cdot 2}{3\cdot 4}\right) = \frac{12}{(3)(5)(4)^2}$, $a_4 = \left(\frac{3\cdot 4}{5\cdot 6}\right)\left(\frac{2\cdot 3}{4\cdot 5}\right)\left(\frac{1\cdot 2}{3\cdot 4}\right) = \frac{12}{(4)(6)(5)^2}$, $\ldots \Rightarrow 1 + \sum_{n=1}^{\infty} \frac{12}{(n+1)(n+3)(n+2)^2}$ represents the given series and $\frac{12}{(n+1)(n+3)(n+2)^2} < \frac{12}{n^4}$, which is the nth-term of a convergent p-series

7. diverges by the nth-Term Test since if $a_n \to L$ as $n \to \infty$, then $L = \frac{1}{1+L} \Rightarrow L^2 + L - 1 = 0 \Rightarrow L = \frac{-1 \pm \sqrt{5}}{2} \ne 0$

9. $f(x) = \cos x$ with $a = \frac{\pi}{3} \Rightarrow f\left(\frac{\pi}{3}\right) = 0.5$, $f'\left(\frac{\pi}{3}\right) = -\frac{\sqrt{3}}{2}$, $f''\left(\frac{\pi}{3}\right) = -0.5$, $f'''\left(\frac{\pi}{3}\right) = \frac{\sqrt{3}}{2}$, $f^{(4)}\left(\frac{\pi}{3}\right) = 0.5$;

$$\cos x = \frac{1}{2} - \frac{\sqrt{3}}{2}\left(x - \frac{\pi}{3}\right) - \frac{1}{4}\left(x - \frac{\pi}{3}\right)^2 + \frac{\sqrt{3}}{12}\left(x - \frac{\pi}{3}\right)^3 + \ldots$$

11. $e^x = 1 + x + \frac{x^2}{2!} + \frac{x^3}{2!} + \ldots$ with $a = 0$

13. $f(x) = \cos x$ with $a = 22\pi \Rightarrow f(22\pi) = 1$, $f'(22\pi) = 0$, $f''(22\pi) = -1$, $f'''(22\pi) = 0$, $f^{(4)}(22\pi) = 1$, $f^{(5)}(22\pi) = 0$, $f^{(6)}(22\pi) = -1$; $\cos x = 1 - \frac{1}{2}(x - 22\pi)^2 + \frac{1}{4!}(x - 22\pi)^4 - \frac{1}{6!}(x - 22\pi)^6 + \ldots$

15. Yes, the sequence converges: $c_n = \left(a^n + b^n\right)^{1/n} \Rightarrow c_n = b\left(\left(\frac{a}{b}\right)^n + 1\right)^{1/n} \Rightarrow \lim\limits_{n\to\infty} c_n = \lim\limits_{n\to\infty} b\left(\left(\frac{a}{b}\right)^n + 1\right)^{1/n} = b$ since $0 < a < b$

17. $s_n = \sum\limits_{k=0}^{n-1} \int\limits_k^{k+1} \frac{dx}{1+x^2} \Rightarrow s_n = \int\limits_0^1 \frac{dx}{1+x^2} + \int\limits_1^2 \frac{dx}{1+x^2} + \ldots + \int\limits_{n-1}^n \frac{dx}{1+x^2} \Rightarrow s_n = \int\limits_0^n \frac{dx}{1+x^2}$

$\Rightarrow \lim\limits_{n\to\infty} s_n = \lim\limits_{n\to\infty} \left(\tan^{-1} n - \tan^{-1} 0\right) = \frac{\pi}{2}$

19. (a) From Fig. 8.13 in the text with $f(x) = \frac{1}{x}$ and $a_k = \frac{1}{k}$, we have $\int\limits_1^{n+1} \frac{1}{x}\,dx \le 1 + \frac{1}{2} + \frac{1}{3} + \ldots + \frac{1}{n}$

$\le 1 + \int\limits_1^n f(x)\,dx \Rightarrow \ln(n+1) \le 1 + \frac{1}{2} + \frac{1}{3} + \ldots + \frac{1}{n} \le 1 + \ln n \Rightarrow 0 \le \ln(n+1) - \ln n$

$\le \left(1 + \frac{1}{2} + \frac{1}{3} + \ldots + \frac{1}{n}\right) - \ln n \le 1$. Therefore the sequence $\left\{\left(1 + \frac{1}{2} + \frac{1}{3} + \ldots + \frac{1}{n}\right) - \ln n\right\}$ is bounded above by 1 and below by 0.

(b) From the graph in Fig. 8.13(a) with $f(x) = \frac{1}{x}$, $\frac{1}{n+1} < \int\limits_n^{n+1} \frac{1}{x}\,dx = \ln(n+1) - \ln n$

$\Rightarrow 0 > \frac{1}{n+1} - [\ln(n+1) - \ln n] = \left(1 + \frac{1}{2} + \frac{1}{3} + \ldots + \frac{1}{n+1} - \ln(n+1)\right) - \left(1 + \frac{1}{2} + \frac{1}{3} + \ldots + \frac{1}{n} - \ln n\right)$.

If we define $a_n = 1 + \frac{1}{2} = \frac{1}{3} + \frac{1}{n} - \ln n$, then $0 > a_{n+1} - a_n \Rightarrow a_{n+1} < a_n \Rightarrow \{a_n\}$ is a decreasing sequence of nonnegative terms.

21. The number of triangles removed at stage n is 3^{n-1}; the side length at stage n is $\frac{b}{2^{n-1}}$; the area of a triangle at stage n is $\frac{\sqrt{3}}{4}\left(\frac{b}{2^{n-1}}\right)^2$.

(a) $\frac{\sqrt{3}}{4}b^2 + 3\frac{\sqrt{3}}{4}\left(\frac{b^2}{2^2}\right) + 3^2\frac{\sqrt{3}}{4}\left(\frac{b^2}{2^4}\right) + 3^3\frac{\sqrt{3}}{4}\left(\frac{b^2}{2^6}\right) + \ldots = \frac{\sqrt{3}}{4}b^2 \sum\limits_{n=0}^{\infty} \frac{3^n}{2^{2n}} = \frac{\sqrt{3}}{4}b^2 \sum\limits_{n=0}^{\infty} \left(\frac{3}{4}\right)^n$

(b) a geometric series with sum $\frac{\left(\frac{\sqrt{3}}{4}b^2\right)}{1 - \left(\frac{3}{4}\right)} = \sqrt{3}b^2$

(c) No; for instance, the three vertices of the original triangle are not removed. However the total area removed is $\sqrt{3}b^2$ which equals the area of the original triangle. Thus the set of points not removed has area 0.

23. (a) No, the limit does not appear to depend on the value of the constant a.

(b) Yes, the limit depends on the value of b. The answer to part (c) shows how the limit depends on the value of b.

(c) $s=\left(1-\frac{\cos\left(\frac{a}{n}\right)}{n}\right)^n \Rightarrow \log s = \frac{\log\left(1-\frac{\cos\left(\frac{a}{n}\right)}{n}\right)}{\left(\frac{1}{n}\right)} \Rightarrow \lim\limits_{n\to\infty} \log s = \frac{\left(\frac{1}{1-\frac{\cos\left(\frac{a}{n}\right)}{n}}\right)\left(\frac{-\frac{a}{n}\sin\left(\frac{a}{n}\right)+\cos\left(\frac{a}{n}\right)}{n^2}\right)}{\left(-\frac{1}{n^2}\right)}$

$= \lim\limits_{n\to\infty} \frac{\frac{a}{n}\sin\left(\frac{a}{n}\right)-\cos\left(\frac{a}{n}\right)}{1-\frac{\cos\left(\frac{a}{n}\right)}{n}} = \frac{0-1}{1-0} = -1 \Rightarrow \lim\limits_{n\to\infty} s = e^{-1} \approx 0.3678794412$; similarly,

$\lim\limits_{n\to\infty}\left(1-\frac{\cos\left(\frac{a}{n}\right)}{bn}\right)^n = e^{-1/b}$

25. $\lim\limits_{n\to\infty}\left|\frac{u_{n+1}}{u_n}\right| < 1 \Rightarrow \lim\limits_{n\to\infty}\left|\frac{b^{n+1}x^{n+1}}{\ln(n+1)}\cdot\frac{\ln n}{b^n x^n}\right| < 1 \Rightarrow |bx| < 1 \Rightarrow -\frac{1}{b} < x < \frac{1}{b} = 5 \Rightarrow b = \pm\frac{1}{5}$

27. (a) $\frac{u_n}{u_{n+1}} = \frac{(n+1)^2}{n^2} = 1+\frac{2}{n}+\frac{1}{n^2} \Rightarrow C = 2 > 1$ and $\sum\limits_{n=1}^{\infty}\frac{1}{n^2}$ converges

(b) $\frac{u_n}{u_{n+1}} = \frac{n+1}{n} = 1+\frac{1}{n}+\frac{0}{n^2} \Rightarrow C = 1 \le 1$ and $\sum\limits_{n=1}^{\infty}\frac{1}{n}$ diverges

29. (a) $\sum\limits_{n=1}^{\infty} a_n = L \Rightarrow a_n^2 \le a_n \sum\limits_{n=1}^{\infty} a_n = a_n L \Rightarrow \sum\limits_{n=1}^{\infty} a_n^2$ converges by the Direct Comparison Test

(b) converges by the Limit Comparison Test: $\lim\limits_{n\to\infty}\frac{\left(\frac{a_n}{1-a_n}\right)}{a_n} = \lim\limits_{n\to\infty}\frac{1}{1-a_n} = 1$ since $\sum\limits_{n=1}^{\infty} a_n$ converges and therefore $\lim\limits_{x\to\infty} a_n = 0$

31. $(1-x)^{-1} = 1+\sum\limits_{n=1}^{\infty} x^n$ where $|x|<1 \Rightarrow \frac{1}{(1-x)^2} = \frac{d}{dx}(1-x)^{-1} = \sum\limits_{n=1}^{\infty} nx^{n-1}$ and when $x=\frac{1}{2}$ we have $4 = 1+2\left(\frac{1}{2}\right)+3\left(\frac{1}{2}\right)^2+4\left(\frac{1}{2}\right)^3+\ldots+n\left(\frac{1}{2}\right)^{n-1}+\ldots$

33. $e^{-x^2} \le e^{-x}$ for $x \ge 1$, and $\int\limits_1^{\infty} e^{-x}\,dx = \lim\limits_{b\to\infty}\left[-e^{-x}\right]_1^b = \lim\limits_{b\to\infty}\left(-e^{-b}+e^{-1}\right) = e^{-1} \Rightarrow \int\limits_1^{\infty} e^{-x^2}\,dx$ converges by the Comparison Test for improper integrals $\Rightarrow \sum\limits_{n=0}^{\infty} e^{-n^2} = 1+\sum\limits_{n=1}^{\infty} e^{-n^2}$ converges by the Integral Test.

35. (a) $\frac{1}{(1-x)^2} = \frac{d}{dx}\left(\frac{1}{1-x}\right) = \frac{d}{dx}\left(1+x+x^2+x^3+\ldots\right) = 1+2x+3x^2+4x^3+\ldots = \sum\limits_{n=1}^{\infty} nx^{n-1}$

(b) from part (a) we have $\sum\limits_{n=1}^{\infty} n\left(\frac{5}{6}\right)^{n-1}\left(\frac{1}{6}\right) = \left(\frac{1}{6}\right)\left[\frac{1}{1-\left(\frac{5}{6}\right)}\right] = 6$

(c) from part (a) we have $\sum_{n=1}^{\infty} np^{n-1}q = \frac{q}{(1-p)^2} = \frac{q}{q^2} = \frac{1}{q}$

37. (a) $R_n = C_0e^{-kt_0} + C_0e^{-2kt_0} + \ldots + C_0e^{-nkt_0} = \frac{C_0e^{-kt_0}\left(1-e^{-nkt_0}\right)}{1-e^{-kt_0}} \Rightarrow R = \lim_{n\to\infty} R_n = \frac{C_0e^{-kt_0}}{1-e^{-kt_0}} = \frac{C_0}{e^{kt_0}-1}$

(b) $R_n = \frac{e^{-1}\left(1-e^{-n}\right)}{1-e^{-1}} \Rightarrow R_1 = e^{-1} \approx 0.36787944$ and $R_{10} = \frac{e^{-1}\left(1-e^{-10}\right)}{1-e^{-1}} \approx 0.58195028$;

$R = \frac{1}{e-1} \approx 0.58197671$; $R - R_{10} \approx 0.00002643 \Rightarrow \frac{R-R_{10}}{R} < 0.0001$

(c) $R_n = \frac{e^{-.1}\left(1-e^{-.1n}\right)}{1-e^{-.1}}$, $\frac{R}{2} = \frac{1}{2}\left(\frac{1}{e^{.1}-1}\right) \approx 4.7541659$; $R_n > \frac{R}{2} \Rightarrow \frac{1-e^{-.1n}}{e^{.1}-1} > \left(\frac{1}{2}\right)\left(\frac{1}{e^{.1}-1}\right)$

$\Rightarrow 1 - e^{-n/10} > \frac{1}{2} \Rightarrow e^{-n/10} < \frac{1}{2} \Rightarrow -\frac{n}{10} < \ln\left(\frac{1}{2}\right) \Rightarrow \frac{n}{10} > -\ln\left(\frac{1}{2}\right) \Rightarrow n > 6.93 \Rightarrow n = 7$

39. The convergence of $\sum_{n=1}^{\infty} |a_n|$ implies that $\lim_{n\to\infty} |a_n| = 0$. Let $N > 0$ be such that $|a_n| < \frac{1}{2} \Rightarrow 1 - |a_n| > \frac{1}{2}$

$\Rightarrow \frac{|a_n|}{1-|a_n|} < 2|a_n|$ for all $n > N$. Now $\left|\ln(1+a_n)\right| = \left|a_n - \frac{a_n^2}{2} + \frac{a_n^3}{3} - \frac{a_n^4}{4} + \ldots\right| \le |a_n| + \left|\frac{a_n^2}{2}\right| + \left|\frac{a_n^3}{3}\right| + \left|\frac{a_n^4}{4}\right| + \ldots$

$< |a_n| + |a_n|^2 + |a_n|^3 + |a_n|^4 + \ldots = \frac{|a_n|}{1-|a_n|} < 2|a_n|$. Therefore $\sum_{n=1}^{\infty} \ln(1+a_n)$ converges by the Direct Comparison Test since $\sum_{n=1}^{\infty} |a_n|$ converges.

NOTES:

CHAPTER 9 VECTORS IN THE PLANE AND POLAR FUNCTIONS

9.1 VECTORS IN THE PLANE

1. (a) $\langle 3(3), 3(-2)\rangle = \langle 9, -6\rangle$

   (b) $\sqrt{9^2 + (-6)^2} = \sqrt{117} = 3\sqrt{13}$

3. (a) $\langle 3 + (-2), -2 + 5\rangle = \langle 1, 3\rangle$

   (b) $\sqrt{1^2 + 3^2} = \sqrt{10}$

5. (a) $2\mathbf{u} = \langle 2(3), 2(-2)\rangle = \langle 6, -4\rangle$
   $3\mathbf{v} = \langle 3(-2), 3(5)\rangle = \langle -6, 15\rangle$
   $2\mathbf{u} - 3\mathbf{v} = \langle 6 - (-6), -4 - 15\rangle = \langle 12, -19\rangle$

   (b) $\sqrt{12^2 + (-19)^2} = \sqrt{505}$

7. (a) $\frac{3}{5}\mathbf{u} = \left\langle \frac{3}{5}(3), \frac{3}{5}(-2)\right\rangle = \left\langle \frac{9}{5}, -\frac{6}{5}\right\rangle$
   $\frac{4}{5}\mathbf{v} = \left\langle \frac{4}{5}(-2), \frac{4}{5}(5)\right\rangle = \left\langle -\frac{8}{5}, 4\right\rangle$
   $\frac{3}{5}\mathbf{u} + \frac{4}{5}\mathbf{v} = \left\langle \frac{9}{5} + \left(-\frac{8}{5}\right), -\frac{6}{5} + 4\right\rangle = \left\langle \frac{1}{5}, \frac{14}{5}\right\rangle$

   (b) $\sqrt{\left(\frac{1}{5}\right)^2 + \left(\frac{14}{5}\right)^2} = \frac{\sqrt{197}}{5}$

9. $\langle 2 - 1, -1 - 3\rangle = \langle 1, -4\rangle$

11. $\langle 0 - 2, 0 - 3\rangle = \langle -2, -3\rangle$

13. $\left\langle \cos\frac{2\pi}{3}, \sin\frac{2\pi}{3}\right\rangle = \left\langle -\frac{1}{2}, \frac{\sqrt{3}}{2}\right\rangle$

15. This is the unit vector which makes an angle of $120° + 90° = 210°$ with the positive x-axis;

    $\langle \cos 210°, \sin 210°\rangle = \left\langle -\frac{\sqrt{3}}{2}, -\frac{1}{2}\right\rangle$

17. The vector $\mathbf{v}$ is horizontal and 1 in. long. The vectors $\mathbf{u}$ and $\mathbf{w}$ are $\frac{11}{16}$ in. long. $\mathbf{w}$ is vertical and $\mathbf{u}$ makes a 45° angle with the horizontal. All vectors must be drawn to scale.

    (a) 

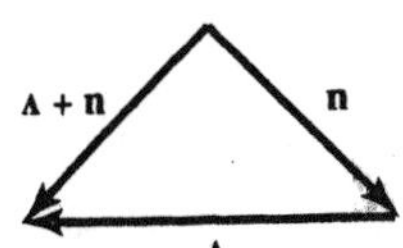

(b) 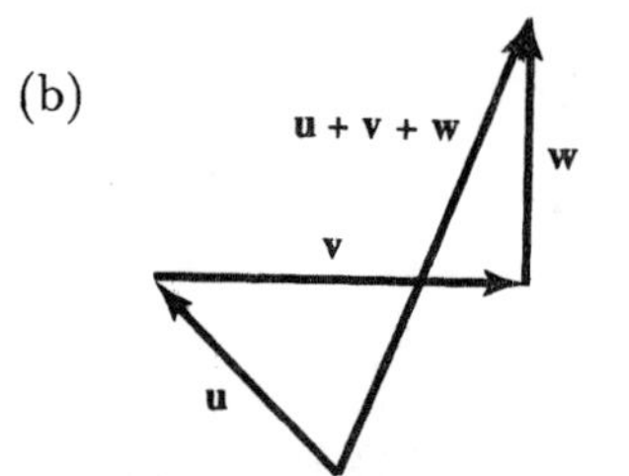

(c) 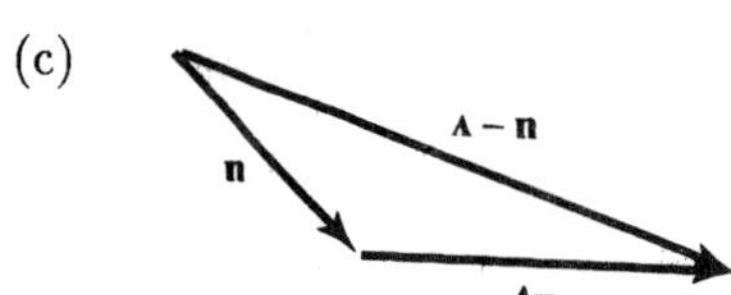

(d) 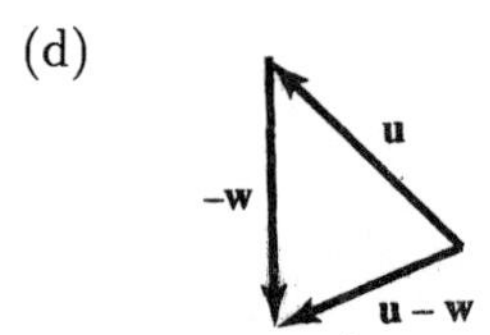

19. $\overrightarrow{P_1P_2} = (2-5)\mathbf{i} + (9-7)\mathbf{j} = -3\mathbf{i} + 2\mathbf{j}$

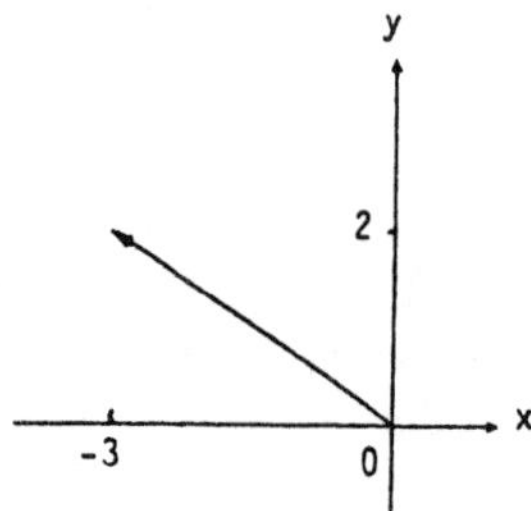

21. $\overrightarrow{AB} = (-10-(-5))\mathbf{i} + (8-3)\mathbf{j} = -5\mathbf{i} + 5\mathbf{j}$

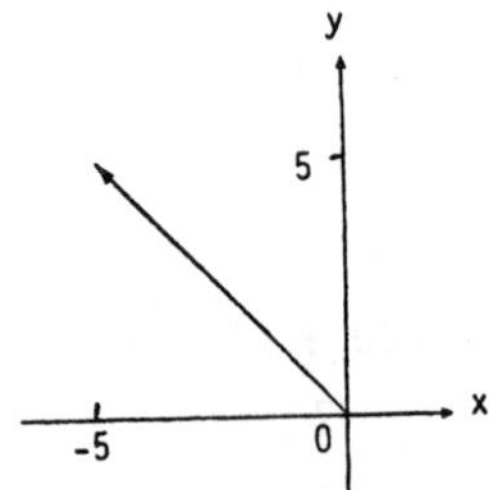

23. $\overrightarrow{P_1P_2} = (2-1)\mathbf{i} + (-1-3)\mathbf{j} = \mathbf{i} - 4\mathbf{j}$

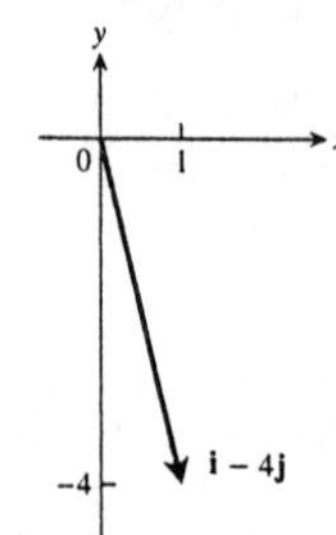

25. $\mathbf{u} = \left(\cos\frac{\pi}{6}\right)\mathbf{i} + \left(\sin\frac{\pi}{6}\right)\mathbf{j} = \frac{\sqrt{3}}{2}\mathbf{i} + \frac{1}{2}\mathbf{j}$;

$\mathbf{u} = \left(\cos\frac{2\pi}{3}\right)\mathbf{i} + \left(\sin\frac{2\pi}{3}\right)\mathbf{j} = -\frac{1}{2}\mathbf{i} + \frac{\sqrt{3}}{2}\mathbf{j}$

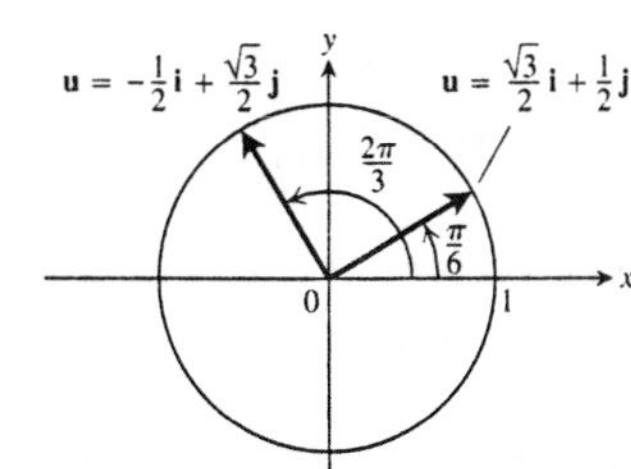

27. $\mathbf{u} = \left(\cos\left(\frac{\pi}{2} + \frac{3\pi}{4}\right)\right)\mathbf{i} + \left(\sin\left(\frac{\pi}{2} + \frac{3\pi}{4}\right)\right)\mathbf{j}$

$= \left(\cos\left(\frac{5\pi}{4}\right)\right)\mathbf{i} + \left(\sin\left(\frac{5\pi}{4}\right)\right)\mathbf{j}$

$= -\frac{\sqrt{2}}{2}\mathbf{i} - \frac{\sqrt{2}}{2}\mathbf{j}$

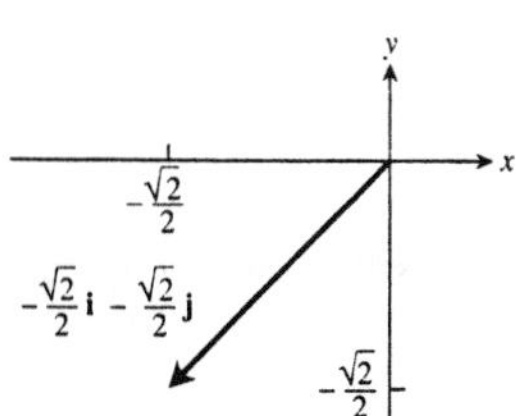

29. $\sqrt{3^2 + 4^2} = 5$; $\frac{1}{5}\langle 3, 4\rangle = \left\langle \frac{3}{5}, \frac{4}{5} \right\rangle$

31. $\sqrt{(-15)^2 + 8^2} = 17$; $\frac{1}{17}\langle -15, 8\rangle = \left\langle -\frac{15}{17}, \frac{8}{17} \right\rangle$

33. $|6\mathbf{i} - 8\mathbf{j}| = \sqrt{36 + 64} = 10 \Rightarrow \frac{\mathbf{v}}{|\mathbf{v}|} = \frac{6}{10}\mathbf{i} - \frac{8}{10}\mathbf{j} = \frac{3}{5}\mathbf{i} - \frac{4}{5}\mathbf{j}$

35. $\mathbf{v} = 5\mathbf{i} + 12\mathbf{j} \Rightarrow |\mathbf{v}| = \sqrt{25 + 144} = 13 \Rightarrow \mathbf{v} = |\mathbf{v}|\left(\frac{\mathbf{v}}{|\mathbf{v}|}\right) = 13\left(\frac{5}{13}\mathbf{i} + \frac{12}{13}\mathbf{j}\right)$

37. $\mathbf{v} = 3\mathbf{i} - 4\mathbf{j} \Rightarrow |\mathbf{v}| = \sqrt{9 + 16} = 5 \Rightarrow \mathbf{u} = \pm\left(\frac{\mathbf{v}}{|\mathbf{v}|}\right) = \pm\frac{1}{5}(3\mathbf{i} - 4\mathbf{j})$

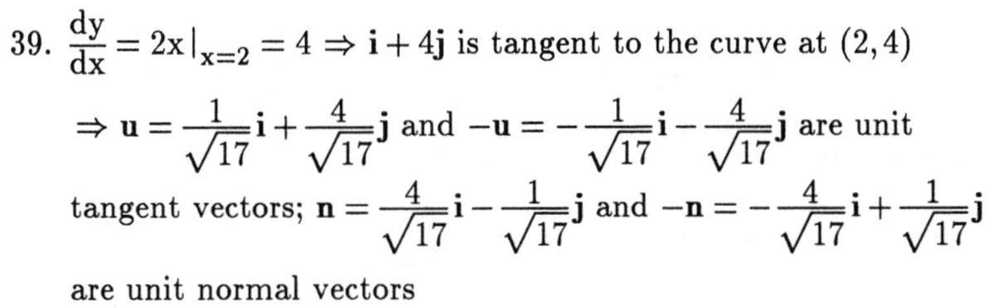

39. $\frac{dy}{dx} = 2x\,|_{x=2} = 4 \Rightarrow \mathbf{i} + 4\mathbf{j}$ is tangent to the curve at $(2, 4)$

$\Rightarrow \mathbf{u} = \frac{1}{\sqrt{17}}\mathbf{i} + \frac{4}{\sqrt{17}}\mathbf{j}$ and $-\mathbf{u} = -\frac{1}{\sqrt{17}}\mathbf{i} - \frac{4}{\sqrt{17}}\mathbf{j}$ are unit tangent vectors; $\mathbf{n} = \frac{4}{\sqrt{17}}\mathbf{i} - \frac{1}{\sqrt{17}}\mathbf{j}$ and $-\mathbf{n} = -\frac{4}{\sqrt{17}}\mathbf{i} + \frac{1}{\sqrt{17}}\mathbf{j}$ are unit normal vectors

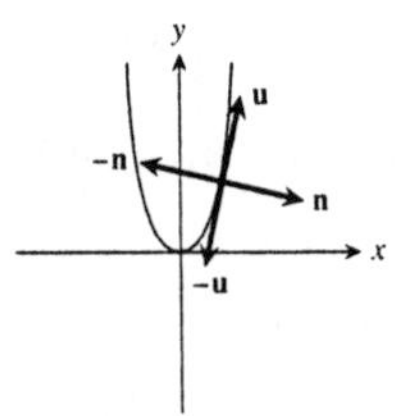

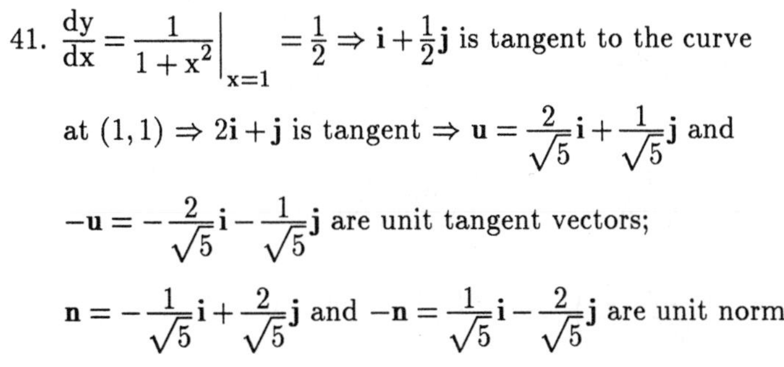

41. $\frac{dy}{dx} = \frac{1}{1 + x^2}\Big|_{x=1} = \frac{1}{2} \Rightarrow \mathbf{i} + \frac{1}{2}\mathbf{j}$ is tangent to the curve at $(1, 1) \Rightarrow 2\mathbf{i} + \mathbf{j}$ is tangent $\Rightarrow \mathbf{u} = \frac{2}{\sqrt{5}}\mathbf{i} + \frac{1}{\sqrt{5}}\mathbf{j}$ and

$-\mathbf{u} = -\frac{2}{\sqrt{5}}\mathbf{i} - \frac{1}{\sqrt{5}}\mathbf{j}$ are unit tangent vectors;

$\mathbf{n} = -\frac{1}{\sqrt{5}}\mathbf{i} + \frac{2}{\sqrt{5}}\mathbf{j}$ and $-\mathbf{n} = \frac{1}{\sqrt{5}}\mathbf{i} - \frac{2}{\sqrt{5}}\mathbf{j}$ are unit normal vectors

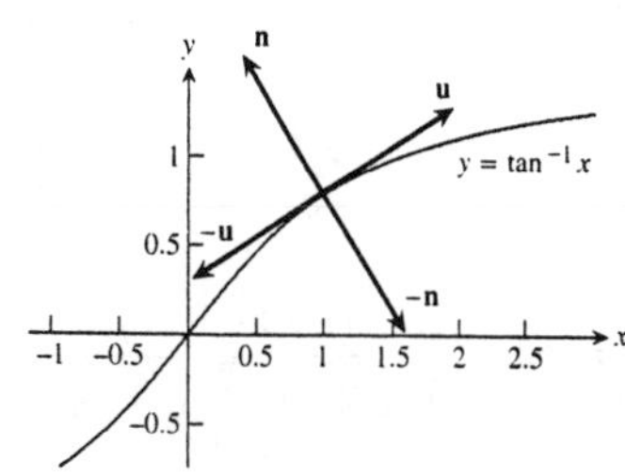

43. $6x+8y+8x\frac{dy}{dx}+4y\frac{dy}{dx}=0 \Rightarrow \frac{dy}{dx}=-\frac{3x+4y}{4x+2y}\Big|_{(1,0)}=-\frac{3}{4} \Rightarrow 4\mathbf{i}-3\mathbf{j}$ is tangent to the curve at $(1,0)$

$\Rightarrow \mathbf{u}=\pm\frac{1}{5}(4\mathbf{i}-3\mathbf{j})$ are unit tangent vectors and $\mathbf{v}=\pm\frac{1}{5}(3\mathbf{i}+4\mathbf{j})$ are unit normal vectors

45. $\frac{dy}{dx}=\sqrt{3+x^4}\Big|_{(0,0)}=\sqrt{3} \Rightarrow \mathbf{i}+\sqrt{3}\mathbf{j}$ is tangent to the curve at $(0,0) \Rightarrow \mathbf{u}=\pm\frac{1}{2}(\mathbf{i}+\sqrt{3}\mathbf{j})$ are unit tangent vectors and $\mathbf{v}=\pm\frac{1}{2}(-\sqrt{3}\mathbf{i}+\mathbf{j})$ are unit normal vectors

47. $2\mathbf{i}+\mathbf{j}=a(\mathbf{i}+\mathbf{j})+b(\mathbf{i}-\mathbf{j})=(a+b)\mathbf{i}+(a-b)\mathbf{j} \Rightarrow a+b=2$ and $a-b=1 \Rightarrow 2a=3 \Rightarrow a=\frac{3}{2}$ and $b=a-1=\frac{1}{2}$

49. If $|\mathbf{x}|$ is the magnitude of the x-component, then $\cos 30^\circ=\frac{|\mathbf{x}|}{|\mathbf{F}|} \Rightarrow |\mathbf{x}|=|\mathbf{F}|\cos 30^\circ=(10)\left(\frac{\sqrt{3}}{2}\right)=5\sqrt{3}$ lb $\Rightarrow \mathbf{F}_x=5\sqrt{3}\,\mathbf{i}$;
if $|\mathbf{y}|$ is the magnitude of the y-component, then $\sin 30^\circ=\frac{|\mathbf{y}|}{|\mathbf{F}|} \Rightarrow |\mathbf{y}|=|\mathbf{F}|\sin 30^\circ=(10)\left(\frac{1}{2}\right)=5$ lb $\Rightarrow \mathbf{F}_y=5\mathbf{j}$.

51. 25° west of north is $90^\circ+25^\circ=115^\circ$ north of east.
$800\langle\cos 115^\circ, \sin 115^\circ\rangle \approx \langle -338.095, 725.045\rangle$

53. (a) The tree is located at the tip of the vector $\overrightarrow{OP}=(5\cos 60^\circ)\mathbf{i}+(5\sin 60^\circ)\mathbf{j}=\frac{5}{2}\mathbf{i}+\frac{5\sqrt{3}}{2}\mathbf{j} \Rightarrow P=\left(\frac{5}{2},\frac{5\sqrt{3}}{2}\right)$
(b) The telephone pole is located at the point Q, which is the tip of the vector $\overrightarrow{OP}+\overrightarrow{PQ}$
$=\left(\frac{5}{2}\mathbf{i}+\frac{5\sqrt{3}}{2}\mathbf{j}\right)+(10\cos 315^\circ)\mathbf{i}+(10\sin 315^\circ)\mathbf{j}=\left(\frac{5}{2}+\frac{\sqrt{2}}{2}\right)\mathbf{i}+\left(\frac{5\sqrt{3}}{2}-\frac{10\sqrt{2}}{2}\right)\mathbf{j}$
$\Rightarrow Q=\left(\frac{5+\sqrt{2}}{2},\frac{5\sqrt{3}-10\sqrt{2}}{2}\right)$

9.2 DOT PRODUCTS

NOTE: In Exercises 1-5 below we calculate $\text{proj}_{\mathbf{v}}\,\mathbf{u}$ as the vector $\left(\frac{|\mathbf{u}|\cos\theta}{|\mathbf{v}|}\right)\mathbf{v}$, so the scalar multiplier of $\mathbf{v}$ is the number in column 5 divided by the number in column 2.

	$\mathbf{v}\cdot\mathbf{u}$	$\lvert\mathbf{v}\rvert$	$\lvert\mathbf{u}\rvert$	$\cos\theta$	$\lvert\mathbf{u}\rvert\cos\theta$	$\text{proj}_{\mathbf{v}}\,\mathbf{u}$
1.	-12	$2\sqrt{5}$	$2\sqrt{5}$	$-\frac{3}{5}$	$-\frac{6\sqrt{5}}{5}$	$-\frac{6}{5}\mathbf{i}+\frac{12}{5}\mathbf{j}$
3.	$\sqrt{3}-\sqrt{2}$	$\sqrt{2}$	$\sqrt{5}$	$\frac{\sqrt{30}-\sqrt{20}}{10}$	$\frac{\sqrt{6}-2}{2}$	$\frac{\sqrt{3}-\sqrt{2}}{2}(-\mathbf{i}+\mathbf{j})$
5.	$\frac{1}{6}$	$\frac{\sqrt{30}}{6}$	$\frac{\sqrt{30}}{6}$	$\frac{1}{5}$	$\frac{1}{\sqrt{30}}$	$\frac{1}{5}\left\langle\frac{1}{\sqrt{2}},\frac{1}{\sqrt{3}}\right\rangle$

7. $\theta=\cos^{-1}\left(\frac{\mathbf{v}\cdot\mathbf{u}}{|\mathbf{v}||\mathbf{u}|}\right)=\cos^{-1}\left(\frac{(2)(1)+(1)(2)}{\sqrt{2^2+1^2}\,\sqrt{1^2+2^2}}\right)=\cos^{-1}\left(\frac{4}{\sqrt{5}\,\sqrt{5}}\right)=\cos^{-1}\left(\frac{4}{5}\right)\approx 0.64$ rad

9. $\mathbf{v} = \sqrt{3}\mathbf{i} - 7\mathbf{j}$, $\mathbf{u} = \sqrt{3}\mathbf{i} + \mathbf{j} \Rightarrow |\mathbf{v}| = \sqrt{(\sqrt{3})^2 + (-7)^2} = 2\sqrt{13}$, $|\mathbf{u}| = \sqrt{(\sqrt{3})^2 + 1^2} = 2$, and

$\mathbf{v} \cdot \mathbf{u} = (\sqrt{3})(\sqrt{3}) + (-7)(1) = -4 \Rightarrow \mathbf{v} \cdot \mathbf{u} = |\mathbf{v}||\mathbf{u}| \cos\theta$ gives $-4 = (2\sqrt{13})(2) \cos\theta$

$$\Rightarrow \theta = \cos^{-1}\left(-\frac{\sqrt{13}}{13}\right) \approx 1.85$$

11. $\overrightarrow{AB} = \langle 3, 1\rangle$, $\overrightarrow{BC} = \langle -1, -3\rangle$, and $\overrightarrow{AC} = \langle 2, -2\rangle$. $\overrightarrow{BA} = \langle -3, -1\rangle$, $\overrightarrow{CB} = \langle 1, 3\rangle$, and $\overrightarrow{CA} = \langle -2, 2\rangle$.

$|\overrightarrow{AB}| = |\overrightarrow{BA}| = \sqrt{10}$, $|\overrightarrow{BC}| = |\overrightarrow{CB}| = \sqrt{10}$, and $|\overrightarrow{AC}| = |\overrightarrow{CA}| = 2\sqrt{2}$.

$$\text{Angle at A} = \cos^{-1}\left(\frac{\overrightarrow{AB} \cdot \overrightarrow{AC}}{|\overrightarrow{AB}||\overrightarrow{AC}|}\right) = \cos^{-1}\left(\frac{3(2) + 1(-2)}{(\sqrt{10})(2\sqrt{2})}\right) = \cos^{-1}\left(\frac{1}{\sqrt{5}}\right) \approx 63.435^\circ,$$

$$\text{Angle at B} = \cos^{-1}\left(\frac{\overrightarrow{BC} \cdot \overrightarrow{BA}}{|\overrightarrow{BC}||\overrightarrow{BA}|}\right) = \cos^{-1}\left(\frac{(-1)(-3) + (-3)(-1)}{(\sqrt{10})(\sqrt{10})}\right) = \cos^{-1}\left(\frac{3}{5}\right) \approx 53.130^\circ, \text{ and}$$

$$\text{Angle at C} = \cos^{-1}\left(\frac{\overrightarrow{CB} \cdot \overrightarrow{CA}}{|\overrightarrow{CB}||\overrightarrow{CA}|}\right) = \cos^{-1}\left(\frac{1(-2) + 3(2)}{(\sqrt{10})(2\sqrt{2})}\right) = \cos^{-1}\left(\frac{1}{\sqrt{5}}\right) \approx 63.435^\circ,$$

13. The sum of two vectors of equal length is *always* orthogonal to their difference, as we can see from the equation $(\mathbf{v}_1 + \mathbf{v}_2) \cdot (\mathbf{v}_1 - \mathbf{v}_2) = \mathbf{v}_1 \cdot \mathbf{v}_1 + \mathbf{v}_2 \cdot \mathbf{v}_1 - \mathbf{v}_1 \cdot \mathbf{v}_2 - \mathbf{v}_2 \cdot \mathbf{v}_2 = |\mathbf{v}_1|^2 - |\mathbf{v}_2|^2 = 0$

15. Let $\mathbf{u}$ and $\mathbf{v}$ be the sides of a rhombus $\Rightarrow$ the diagonals are $\mathbf{d}_1 = \mathbf{u} + \mathbf{v}$ and $\mathbf{d}_2 = -\mathbf{u} + \mathbf{v}$

$\Rightarrow \mathbf{d}_1 \cdot \mathbf{d}_2 = (\mathbf{u} + \mathbf{v}) \cdot (-\mathbf{u} + \mathbf{v}) = -\mathbf{u} \cdot \mathbf{u} + \mathbf{u} \cdot \mathbf{v} - \mathbf{v} \cdot \mathbf{u} + \mathbf{v} \cdot \mathbf{v} = |\mathbf{v}|^2 - |\mathbf{u}|^2 = 0$ because $|\mathbf{u}| = |\mathbf{v}|$, since a rhombus has equal sides.

17. Clearly the diagonals of a rectangle are equal in length. What is not as obvious is the statement that equal diagonals happen only in a rectangle. We show this is true by letting the opposite sides of a parallelogram be the vectors $(v_1\mathbf{i} + v_2\mathbf{j})$ and $(u_1\mathbf{i} + u_2\mathbf{j})$. The equal diagonals of the parallelogram are

$d_1 = (v_1\mathbf{i} + v_2\mathbf{j}) + (u_1\mathbf{i} + u_2\mathbf{j})$ and $d_2 = (v_1\mathbf{i} + v_2\mathbf{j}) - (u_1\mathbf{i} + u_2\mathbf{j})$. Hence $|d_1| = |d_2| = |(v_1\mathbf{i} + v_2\mathbf{j}) + (u_1\mathbf{i} + u_2\mathbf{j})|$
$= |(v_1\mathbf{i} + v_2\mathbf{j}) - (u_1\mathbf{i} + u_2\mathbf{j})| \Rightarrow |(v_1 + u_1)\mathbf{i} + (v_2 + u_2)\mathbf{j}| = |(v_1 - u_1)\mathbf{i} + (v_2 - u_2)\mathbf{j}|$

$\Rightarrow \sqrt{(v_1 + u_1)^2 + (v_2 + u_2)^2} = \sqrt{(v_1 - u_1)^2 + (v_2 - u_2)^2} \Rightarrow v_1^2 + 2v_1u_1 + u_1^2 + v_2^2 + 2v_2u_2 + u_2^2$
$= v_1^2 - 2v_1u_1 + u_1^2 + v_2^2 - 2v_2u_2 + u_2^2 \Rightarrow 2(v_1u_1 + v_2u_2) = -2(v_1u_1 + v_2u_2) \Rightarrow v_1u_1 + v_2u_2 = 0$

$\Rightarrow (v_1\mathbf{i} + v_2\mathbf{j}) \cdot (u_1\mathbf{i} + u_2\mathbf{j}) = 0 \Rightarrow$ the vectors $(v_1\mathbf{i} + v_2\mathbf{j})$ and $(u_1\mathbf{i} + u_2\mathbf{j})$ are perpendicular and the parallelogram must be a rectangle.

19. horizontal component: $1200 \cos(8^\circ) \approx 1188$ ft/s; vertical component: $1200 \sin(8^\circ) \approx 167$ ft/s

21. (a) Since $|\cos\theta| \le 1$, we have $|\mathbf{u} \cdot \mathbf{v}| = |\mathbf{u}||\mathbf{v}||\cos\theta| \le |\mathbf{u}||\mathbf{v}|(1) = |\mathbf{u}||\mathbf{v}|$.
(b) We have equality precisely when $|\cos\theta| = 1$ or when one or both of $\mathbf{u}$ and $\mathbf{v}$ is $\mathbf{0}$. In the case of nonzero vectors, we have equality when $\theta = 0$ or π, i.e., when the vectors are parallel.

23. $\mathbf{v} \cdot \mathbf{u}_1 = (a\mathbf{u}_1 + b\mathbf{u}_2) \cdot \mathbf{u}_1 = a\mathbf{u}_1 \cdot \mathbf{u}_1 + b\mathbf{u}_2 \cdot \mathbf{u}_1 = a|\mathbf{u}_1|^2 + b(\mathbf{u}_2 \cdot \mathbf{u}_1) = a(1)^2 + b(0) = a$

25. $P(x_1, y_1) = P\left(x_1, \frac{c}{b} - \frac{a}{b}x_1\right)$ and $Q(x_2, y_2) = Q\left(x_2, \frac{c}{b} - \frac{a}{b}x_2\right)$ are any two points P and Q on the line with $b \neq 0$ $\Rightarrow \overrightarrow{PQ} = (x_2 - x_1)\mathbf{i} + \frac{a}{b}(x_2 - x_1)\mathbf{j} \Rightarrow \overrightarrow{PQ} \cdot \mathbf{v} = \left[(x_2 - x_1)\mathbf{i} + \frac{a}{b}(x_2 - x_1)\mathbf{j}\right] \cdot (a\mathbf{i} + b\mathbf{j}) = a(x_2 - x_1) + b\left(\frac{a}{b}\right)(x_2 - x_1)$ $= 0 \Rightarrow \mathbf{v}$ is perpendicular to $\overrightarrow{PQ}$ for $b \neq 0$. If $b = 0$, then $\mathbf{v} = a\mathbf{i}$ is perpendicular to the vertical line $ax = c$.

Alternatively, the slope of $\mathbf{v}$ is $\frac{b}{a}$ and the slope of the line $ax + by = c$ is $-\frac{a}{b}$, so the slopes are negative reciprocals $\Rightarrow$ the vector $\mathbf{v}$ and the line are perpendicular.

27. $\mathbf{v} = \mathbf{i} + 2\mathbf{j}$ is perpendicular to the line $x + 2y = c$;
P(2, 1) on the line $\Rightarrow 2 + 2 = c \Rightarrow x + 2y = 4$

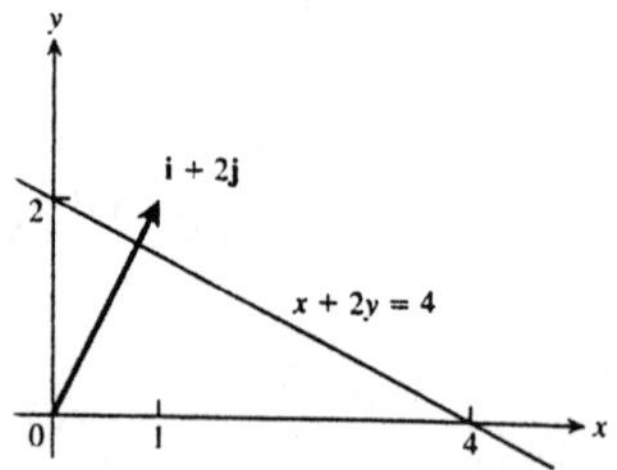

29. $\mathbf{v} = -2\mathbf{i} + \mathbf{j}$ is perpendicular to the line $-2x + y = c$;
P(−2, −7) on the line $\Rightarrow (-2)(-2) - 7 = c \Rightarrow -2x + y = -3$

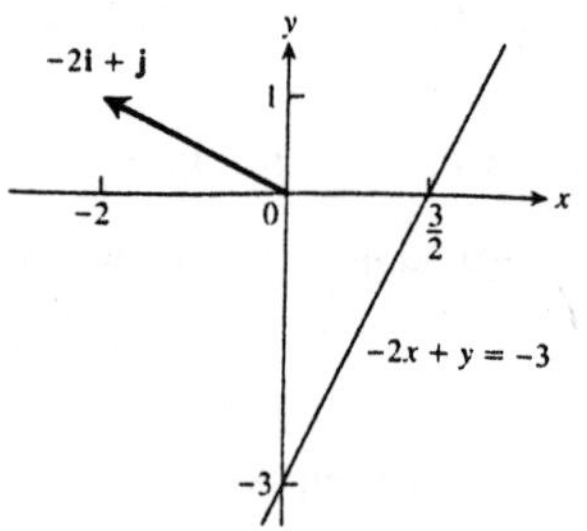

31. $\mathbf{v} = \mathbf{i} - \mathbf{j}$ is parallel to the line $x + y = c$;
P(−2, 1) on the line $\Rightarrow -2 + 1 = c \Rightarrow x + y = -1$

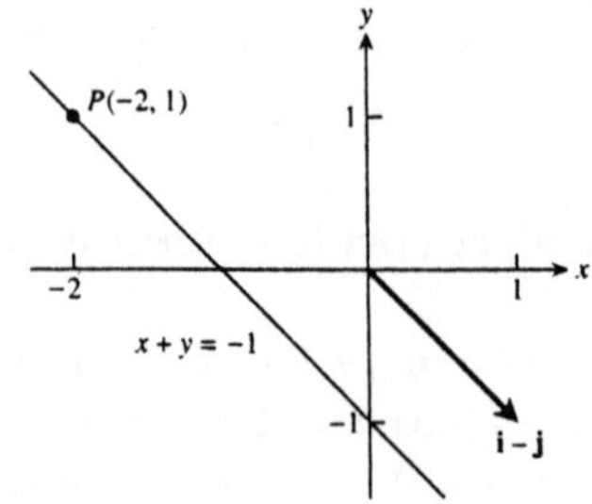

33. $\mathbf{v} = -\mathbf{i} - 2\mathbf{j}$ is parallel to the line $2x - y = c$;

$P(1,2)$ on the line $\Rightarrow (2)(1) - 2 = c \Rightarrow 2x - y = 0$

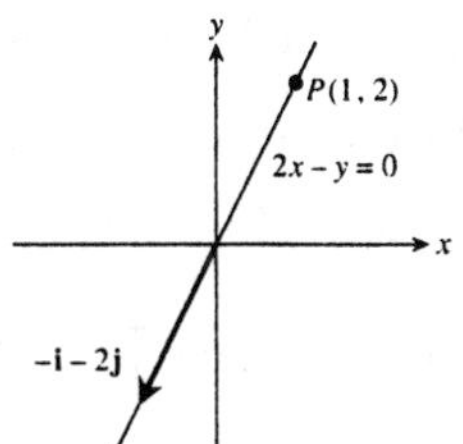

35. $P(0,0)$, $Q(1,1)$ and $\mathbf{F} = 5\mathbf{j} \Rightarrow \overrightarrow{PQ} = \mathbf{i} + \mathbf{j}$ and $\mathbf{W} = \mathbf{F} \cdot \overrightarrow{PQ} = (5\mathbf{j}) \cdot (\mathbf{i} + \mathbf{j}) = 5\ \mathrm{N \cdot m} = 5\ \mathrm{J}$

37. $\mathbf{W} = |\mathbf{F}|\left|\overrightarrow{PQ}\right| \cos\theta = (200)(20)(\cos 30^\circ) = 2000\sqrt{3} = 3464.10\ \mathrm{N \cdot m} = 3464.10\ \mathrm{J}$

In Exercises 39-43 we use the fact that $\mathbf{n} = a\mathbf{i} + b\mathbf{j}$ is normal to the line $ax + by = c$.

39. $\mathbf{n}_1 = 3\mathbf{i} + \mathbf{j}$ and $\mathbf{n}_2 = 2\mathbf{i} - \mathbf{j} \Rightarrow \theta = \cos^{-1}\left(\frac{\mathbf{n}_1 \cdot \mathbf{n}_2}{|\mathbf{n}_1||\mathbf{n}_2|}\right) = \cos^{-1}\left(\frac{6-1}{\sqrt{10}\,\sqrt{5}}\right) = \cos^{-1}\left(\frac{1}{\sqrt{2}}\right) = \frac{\pi}{4}$

41. $\mathbf{n}_1 = \sqrt{3}\mathbf{i} - \mathbf{j}$ and $\mathbf{n}_2 = \mathbf{i} - \sqrt{3}\mathbf{j} \Rightarrow \theta = \cos^{-1}\left(\frac{\mathbf{n}_1 \cdot \mathbf{n}_2}{|\mathbf{n}_1||\mathbf{n}_2|}\right) = \cos^{-1}\left(\frac{\sqrt{3}+\sqrt{3}}{\sqrt{4}\,\sqrt{4}}\right) = \cos^{-1}\left(\frac{\sqrt{3}}{2}\right) = \frac{\pi}{6}$

43. $\mathbf{n}_1 = 3\mathbf{i} - 4\mathbf{j}$ and $\mathbf{n}_2 = \mathbf{i} - \mathbf{j} \Rightarrow \theta = \cos^{-1}\left(\frac{\mathbf{n}_1 \cdot \mathbf{n}_2}{|\mathbf{n}_1||\mathbf{n}_2|}\right) = \cos^{-1}\left(\frac{3+4}{\sqrt{25}\,\sqrt{2}}\right) = \cos^{-1}\left(\frac{7}{5\sqrt{2}}\right) \approx 0.14$ rad

45. The angle between the corresponding normals is equal to the angle between the corresponding tangents. The points of intersection are $\left(-\frac{\sqrt{3}}{2}, \frac{3}{4}\right)$ and $\left(\frac{\sqrt{3}}{2}, \frac{3}{4}\right)$. At $\left(-\frac{\sqrt{3}}{2}, \frac{3}{4}\right)$ the tangent line for $f(x) = x^2$ is $y - \frac{3}{4} = f'\left(-\frac{\sqrt{3}}{2}\right)\left(x - \left(-\frac{\sqrt{3}}{2}\right)\right) \Rightarrow y = -\sqrt{3}\left(x + \frac{\sqrt{3}}{2}\right) + \frac{3}{4} \Rightarrow y = -\sqrt{3}x - \frac{3}{4}$, and the tangent line for $f(x) = \left(\frac{3}{2}\right) - x^2$ is $y - \frac{3}{4} = f'\left(-\frac{\sqrt{3}}{2}\right)\left(x - \left(-\frac{\sqrt{3}}{2}\right)\right) \Rightarrow y = \sqrt{3}\left(x + \frac{\sqrt{3}}{2}\right) + \frac{3}{4} = \sqrt{3}x + \frac{9}{4}$. The corresponding normals are $\mathbf{n}_1 = \sqrt{3}\mathbf{i} + \mathbf{j}$ and $\mathbf{n}_2 = -\sqrt{3}\mathbf{i} + \mathbf{j}$. The angle at $\left(-\frac{\sqrt{3}}{2}, \frac{3}{4}\right)$ is $\theta = \cos^{-1}\left(\frac{\mathbf{n}_1 \cdot \mathbf{n}_2}{|\mathbf{n}_1||\mathbf{n}_2|}\right)$ $= \cos^{-1}\left(\frac{-3+1}{\sqrt{4}\,\sqrt{4}}\right) = \cos^{-1}\left(-\frac{1}{2}\right) = \frac{2\pi}{3}$; the angles are $\frac{\pi}{3}$ and $\frac{2\pi}{3}$. At $\left(\frac{\sqrt{3}}{2}, \frac{3}{4}\right)$ the tangent line for $f(x) = x^2$ is $y = \sqrt{3}\left(x + \frac{\sqrt{3}}{2}\right) + \frac{3}{4} = \sqrt{3}x + \frac{9}{4}$ and the tangent line for $f(x) = \frac{3}{2} - x^2$ is $y = -\sqrt{3}\left(x + \frac{\sqrt{3}}{2}\right) + \frac{3}{4}$ $= -\sqrt{3}x - \frac{3}{4}$. The corresponding normals are $\mathbf{n}_1 = -\sqrt{3}\mathbf{i} + \mathbf{j}$ and $\mathbf{n}_2 = \sqrt{3}\mathbf{i} + \mathbf{j}$. The angle at $\left(\frac{\sqrt{3}}{2}, \frac{3}{4}\right)$ is

$\theta = \cos^{-1}\left(\frac{\mathbf{n}_1 \cdot \mathbf{n}_2}{|\mathbf{n}_1||\mathbf{n}_2|}\right) = \cos^{-1}\left(\frac{-3+1}{\sqrt{4}\sqrt{4}}\right) = \cos^{-1}\left(-\frac{1}{2}\right) = \frac{2\pi}{3}$; the angles are $\frac{\pi}{3}$ and $\frac{2\pi}{3}$.

47. The curves intersect when $y = x^3 = (y^2)^3 = y^6 \Rightarrow y = 0$ or $y = 1$. The points of intersection are $(0,0)$ and $(1,1)$. Note that $y \geq 0$ since $y = y^6$. At $(0,0)$ the tangent line for $y = x^3$ is $y = 0$ and the tangent line for $y = \sqrt{x}$ is $x = 0$. Therefore, the angle of intersection at $(0,0)$ is $\frac{\pi}{2}$. At $(1,1)$ the tangent line for $y = x^3$ is $y = 3x - 2$ and the tangent line for $y = \sqrt{x}$ is $y = \frac{1}{2}x + \frac{1}{2}$. The corresponding normal vectors are $\mathbf{n}_1 = -3\mathbf{i} + \mathbf{j}$ and $\mathbf{n}_2 = -\frac{1}{2}\mathbf{i} + \mathbf{j} \Rightarrow \theta = \cos^{-1}\left(\frac{\mathbf{n}_1 \cdot \mathbf{n}_2}{|\mathbf{n}_1||\mathbf{n}_2|}\right) = \cos^{-1}\left(\frac{1}{\sqrt{2}}\right) = \frac{\pi}{4}$; the angles are $\frac{\pi}{4}$ and $\frac{3\pi}{4}$.

9.3 VECTOR-VALUED FUNCTIONS

1. (a)

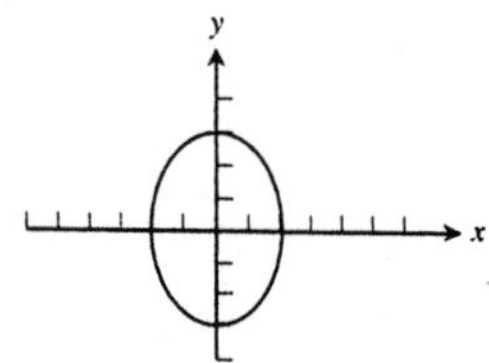

(b) $\mathbf{v}(t) = \frac{d}{dt}(2\cos t)\mathbf{i} + \frac{d}{dt}(3\sin t)\mathbf{j}$
$= (-2\sin t)\mathbf{i} + (3\cos t)\mathbf{j}$
$\mathbf{a}(t) = \frac{d}{dt}(-2\sin t)\mathbf{i} + \frac{d}{dt}(3\cos t)\mathbf{j}$
$= (-2\cos t)\mathbf{i} - (3\sin t)\mathbf{j}$

(c) $\mathbf{v}\left(\frac{\pi}{2}\right) = \langle -2, 0\rangle$; speed $= \sqrt{(-2)^2 + 0^2} = 2$,
direction $= \frac{1}{2}\langle -2, 0\rangle = \langle -1, 0\rangle$

(d) Velocity $= 2\langle -1, 0\rangle$

3. (a)

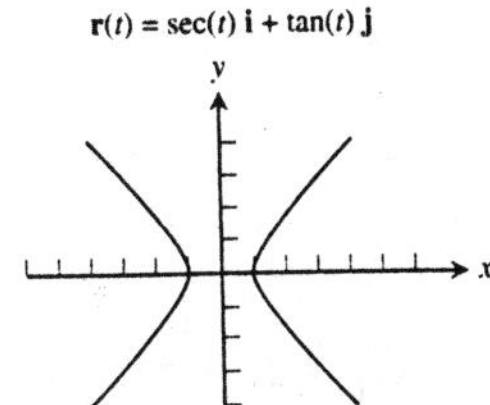

(b) $\mathbf{v}(t) = \frac{d}{dt}(\sec t)\mathbf{i} + \frac{d}{dt}(\tan t)\mathbf{i} = (\sec t \tan t)\mathbf{i} + (\sec^2 t)\mathbf{j}$

$\mathbf{a}(t) = \frac{d}{dt}(\sec t \tan t)\mathbf{i} + \frac{d}{dt}(\sec^2 t)\mathbf{j}$

$= (\sec t \tan^2 t + \sec^3 t)\mathbf{i} + (2 \sec^2 t \tan t)\mathbf{j}$

(c) $\mathbf{v}\left(\frac{\pi}{6}\right) = \left\langle \frac{2}{3}, \frac{4}{3} \right\rangle$; speed $= \sqrt{\left(\frac{2}{3}\right)^2 + \left(\frac{4}{3}\right)^2} = \frac{2\sqrt{5}}{3}$, direction $= \frac{3}{2\sqrt{5}}\left\langle \frac{2}{3}, \frac{4}{3} \right\rangle = \left\langle \frac{1}{\sqrt{5}}, \frac{2}{\sqrt{5}} \right\rangle$

(d) Velocity $= \frac{2\sqrt{5}}{3}\left\langle \frac{1}{\sqrt{5}}, \frac{2}{\sqrt{5}} \right\rangle$

5. $\mathbf{v}(t) = (1 - \cos t)\mathbf{i} + (\sin t)\mathbf{j}$ and $\mathbf{a}(t) = (\sin t)\mathbf{i} + (\cos t)\mathbf{j}$. Solve $\mathbf{v} \cdot \mathbf{a} = 0$: $(\sin t - \sin t \cos t) + (\sin t \cos t) = 0$ implies $\sin t = 0$, which is true for $t = 0, \pi$, or 2π.

7. $\mathbf{v}(t) = (-3 \sin t)\mathbf{i} + (4 \cos t)\mathbf{j}$, and $\mathbf{a}(t) = (-3 \cos t)\mathbf{i} + (-4 \sin t)\mathbf{j}$. Solve $\mathbf{v} \cdot \mathbf{a} = 0$: $(9 \sin t \cos t) - (16 \sin t \cos t) = 0$, which is true when $\sin t = 0$ or $\cos t = 0$, i.e., for $t = \frac{k\pi}{2}$, k any nonnegative integer.

9. $\mathbf{v}(t) = (-2 \sin t)\mathbf{i} + (\cos t)\mathbf{j}$, and $\mathbf{a}(t) = (-2 \cos t)\mathbf{i} + (-\sin t)\mathbf{j}$. So $\mathbf{v}\left(\frac{\pi}{4}\right) = (-\sqrt{2})\mathbf{i} + \left(\frac{1}{\sqrt{2}}\right)\mathbf{j}$, and $\mathbf{a}\left(\frac{\pi}{4}\right) = (-\sqrt{2})\mathbf{i} + \left(-\frac{1}{\sqrt{2}}\right)\mathbf{j}$. Then $|\mathbf{v}| = |\mathbf{a}| = \sqrt{\frac{5}{2}}$. $\mathbf{v} \cdot \mathbf{a} = \frac{3}{2}$ and $\theta = \cos^{-1}\left(\frac{\mathbf{v} \cdot \mathbf{a}}{|\mathbf{v}||\mathbf{a}|}\right) = \cos^{-1}\left(\frac{3}{5}\right) \approx 53.130°$.

11. (a) Both components are continuous at $t = 3$, so the limit is $3\mathbf{i} + \left(\frac{3^2 - 9}{3^2 + 3(3)}\right)\mathbf{j} = 3\mathbf{i}$.

(b) Continuous so long as $t^2 + 3t \neq 0$, i.e., $t \neq 0, -3$

(c) Discontinuous when $t^2 + 3t = 0$, i.e., $t = 0$ or -3

13. $\mathbf{v}(t) = (\cos t)\mathbf{i} + (2t + \sin t)\mathbf{j}$, $\mathbf{r}(0) = -\mathbf{j}$ and $\mathbf{v}(0) = \mathbf{i}$. So the slope is zero (the velocity vector is horizontal).
(a) The horizontal line through $(0, -1)$: $y = -1$.
(b) The vertical line through $(0, -1)$: $x = 0$.

15. $\left(\int_1^2 (6 - 6t)\, dt\right)\mathbf{i} + \left(\int_1^2 3\sqrt{t}\, dt\right)\mathbf{j} = \left[6t - 3t^2\right]_1^2 \mathbf{i} + \left[2t^{3/2}\right]_1^2 \mathbf{j} = -3\mathbf{i} + (4\sqrt{2} - 2)\mathbf{j}$

17. $\left(\int \sec t \tan t\ dt\right)\mathbf{i}+\left(\int \tan t\ dt\right)\mathbf{j} = (\sec t + C_1)\mathbf{i}+(\ln|\sec t|+C_2)\mathbf{j} = (\sec t)\mathbf{i}+(\ln|\sec t|)\mathbf{j}+\mathbf{C}$

19. $\mathbf{r}(t) = (t+1)^{3/2}\mathbf{i} - e^{-t}\mathbf{j}+\mathbf{C}$, and $\mathbf{r}(0) = \mathbf{i}-\mathbf{j}+\mathbf{C} = \mathbf{0}$, so $\mathbf{C} = -(\mathbf{i}-\mathbf{j}) = -\mathbf{i}+\mathbf{j}$
$\mathbf{r}(t) = \left((t+1)^{3/2}-1\right)\mathbf{i} - (e^{-t}-1)\mathbf{j}$

21. $\frac{d\mathbf{r}}{dt} = (-32t)\mathbf{j}+\mathbf{C}_1$ and $\mathbf{r}(t) = (-16t^2)\mathbf{j}+\mathbf{C}_1 t+\mathbf{C}_2$. $\mathbf{r}(0) = \mathbf{C}_2 = 100\mathbf{i}$ and $\left.\frac{d\mathbf{r}}{dt}\right|_{t=0} = \mathbf{C}_1 = 8\mathbf{i}+8\mathbf{j}$. So
$\mathbf{r}(t) = (-16t^2)\mathbf{j}+(8\mathbf{i}+8\mathbf{j})t+100\mathbf{i} = (8t+100)\mathbf{i}+(-16t^2+8t)\mathbf{j}$.

23. $\mathbf{v}(t) = (\sin t)\mathbf{i}+(1-\cos t)\mathbf{j}$; i.e., $\frac{dx}{dt} = \sin t$, and $\frac{dy}{dt} = 1-\cos t$

$$\text{Distance} = \int_0^{2\pi/3} \sqrt{(\sin t)^2+(1-\cos t)^2}\ dt = \int_0^{2\pi/3} \sqrt{2-2\cos t}\ dt = \int_0^{2\pi/3} 2\sin\left(\frac{t}{2}\right) dt = \left[-4\cos\left(\frac{t}{2}\right)\right]_0^{2\pi/3} = 2$$

25. (a) $\mathbf{v}(t) = (\cos t)\mathbf{i} - (2\sin 2t)\mathbf{j}$

(b) $\mathbf{v}(t) = \mathbf{0}$ when both $\cos t = 0$ and $\sin 2t = 0$. $\cos t = 0$ at $t = \frac{\pi}{2}$ and $\frac{3\pi}{2}$; $\sin 2t = 0$ at $t = 0, \frac{\pi}{2}, \pi, \frac{3\pi}{2}$, and 2π. So $\mathbf{v}(t) = \mathbf{0}$ at $t = \frac{\pi}{2}, \frac{3\pi}{2}$.

(c) $x = \sin t$, $y = \cos 2t$. Relate the two using the identity $\cos 2u = 1-2\sin^2 u$: $y = 1-2x^2$, where as x ranges over all possible values, $-1 \le x \le 1$. When t increases from 0 to 2π, the particle starts at $(0,1)$, goes to $(1,-1)$, then goes to $(-1,-1)$, and then goes to $(0,1)$, tracing the curve twice.

27. $\mathbf{a}(t) = 3\mathbf{i}-\mathbf{j}$, so $\mathbf{v}(t) = (3t)\mathbf{i}-t\mathbf{j}+\mathbf{C}_1$ and $\mathbf{r}(t) = \left(\frac{3}{2}t^2\right)\mathbf{i}-\left(\frac{1}{2}t^2\right)\mathbf{j}+\mathbf{C}_1 t+\mathbf{C}_2$. $\mathbf{r}(0) = \mathbf{C}_2 = \mathbf{i}+2\mathbf{j}$, and since $\mathbf{v}(0)$ must point directly from $(1,2)$ toward $(4,1)$ with magnitude 2,

$$\mathbf{v}(0) = \mathbf{C}_1 = 2\left(\frac{(4-1)\mathbf{i}+(1-2)\mathbf{j}}{\sqrt{(4-1)^2+(1-2)^2}}\right) = \frac{6}{\sqrt{10}}\mathbf{i}-\frac{2}{\sqrt{10}}\mathbf{j} = \frac{3\sqrt{10}}{5}\mathbf{i}-\frac{\sqrt{10}}{5}\mathbf{j}$$

So $\mathbf{r}(t) = \left(\frac{3}{2}t^2+\frac{3\sqrt{10}}{5}t+1\right)\mathbf{i}+\left(-\frac{1}{2}t^2-\frac{\sqrt{10}}{5}t+2\right)\mathbf{j}$.

29. (a) $\mathbf{v}(t) = -(\sin t)\mathbf{i}+(\cos t)\mathbf{j} \Rightarrow \mathbf{a}(t) = -(\cos t)\mathbf{i}-(\sin t)\mathbf{j}$;

(i) $|\mathbf{v}(t)| = \sqrt{(-\sin t)^2+(\cos t)^2} = 1 \Rightarrow$ constant speed;

(ii) $\mathbf{v}\cdot\mathbf{a} = (\sin t)(\cos t)-(\cos t)(\sin t) = 0 \Rightarrow$ yes, orthogonal;

(iii) counterclockwise movement;

(iv) yes, $\mathbf{r}(0) = \mathbf{i}+0\mathbf{j}$

(b) $\mathbf{v}(t) = -(2\sin 2t)\mathbf{i}+(2\cos 2t)\mathbf{j} \Rightarrow \mathbf{a}(t) = -(4\cos 2t)\mathbf{i}-(4\sin 2t)\mathbf{j}$;

(i) $|\mathbf{v}(t)| = \sqrt{4\sin^2 2t+4\cos^2 2t} = 2 \Rightarrow$ constant speed;

(ii) $\mathbf{v}\cdot\mathbf{a} = 8\sin 2t\cos 2t - 8\cos 2t\sin 2t = 0 \Rightarrow$ yes, orthogonal;

(iii) counterclockwise movement;

(iv) yes, $\mathbf{r}(0) = \mathbf{i}+0\mathbf{j}$

(c) $\mathbf{v}(t) = -\sin\left(t-\frac{\pi}{2}\right)\mathbf{i}+\cos\left(t-\frac{\pi}{2}\right)\mathbf{j} \Rightarrow \mathbf{a}(t) = -\cos\left(t-\frac{\pi}{2}\right)\mathbf{i}-\sin\left(t-\frac{\pi}{2}\right)\mathbf{j}$;

(i) $|\mathbf{v}(t)| = \sqrt{\sin^2\left(t-\frac{\pi}{2}\right)+\cos^2\left(t-\frac{\pi}{2}\right)} = 1 \Rightarrow$ constant speed;

(ii) $\mathbf{v}\cdot\mathbf{a} = \sin\left(t-\frac{\pi}{2}\right)\cos\left(t-\frac{\pi}{2}\right)-\cos\left(t-\frac{\pi}{2}\right)\sin\left(t-\frac{\pi}{2}\right) = 0 \Rightarrow$ yes, orthogonal;

(iii) counterclockwise movement;

(iv) no, $\mathbf{r}(0) = 0\mathbf{i}-\mathbf{j}$ instead of $\mathbf{i}+0\mathbf{j}$

(d) $\mathbf{v}(t) = -(\sin t)\mathbf{i}-(\cos t)\mathbf{j} \Rightarrow \mathbf{a}(t) = -(\cos t)\mathbf{i}+(\sin t)\mathbf{j}$;

(i) $|\mathbf{v}(t)| = \sqrt{(-\sin t)^2+(-\cos t)^2} = 1 \Rightarrow$ constant speed;

(ii) $\mathbf{v}\cdot\mathbf{a} = (\sin t)(\cos t)-(\cos t)(\sin t) = 0 \Rightarrow$ yes, orthogonal;

(iii) clockwise movement;

(iv) yes, $\mathbf{r}(0) = \mathbf{i}-0\mathbf{j}$

(e) $\mathbf{v}(t) = -2t\sin(t^2)\mathbf{i}+2t\cos(t^2)\mathbf{j} \Rightarrow \mathbf{a}(t) = -\left(4t^2\cos(t^2)+2\sin(t^2)\right)\mathbf{i}+\left(2\cos(t^2)-4t^2\sin(t^2)\right)\mathbf{j}$;

(i) $|\mathbf{v}(t)| = \sqrt{\left(-2t\sin(t^2)\right)^2+\left(2t\cos(t^2)\right)^2} = 2t \Rightarrow$ variable speed

(ii) $\mathbf{v}\cdot\mathbf{a} = 2t\cos(t^2)\left(2\cos(t^2)-4t^2\sin(t^2)\right)+2t\sin(t^2)\left(2\sin(t^2)+4t^2\cos(t^2)\right)$

$= 4t\left(\left(\sin(t^2)\right)^2+\left(\cos(t^2)\right)^2\right) = 4t \Rightarrow$ orthogonal only at $t = 0$

(iii) counterclockwise movement;

(iv) yes, $\mathbf{r}(0) = 1\mathbf{i}+0\mathbf{j}$

31. (a) The $\mathbf{j}$-component is zero at $t = 0$ and $t = 160$: 160 seconds.

(b) $-\frac{3}{64}(40)(40-160) = 225$ m

(c) $\frac{d}{dt}\left[-\frac{3}{64}t(t-160)\right] = -\frac{3}{32}t+\frac{15}{2}$, which for $t = 40$ equals $\frac{15}{4}$ meters per second.

(d) $\mathbf{v}(t) = -\frac{3}{32}t+\frac{15}{2}$ equals 0 at $t = 80$ seconds (and is negative after that time).

33. (a) Referring to the figure, look at the circular arc from the point where $t = 0$ to the point "m." On one hand, this arc has length given by $r_0\theta$, but it also has length given by vt. Setting those two quantities equal gives the result.

(b) $\mathbf{v}(t) = \left(-v\sin\frac{vt}{r_0}\right)\mathbf{i}+\left(v\cos\frac{vt}{r_0}\right)\mathbf{j}$, and $\mathbf{a}(t) = \left(-\frac{v^2}{r_0}\cos\frac{vt}{r_0}\right)\mathbf{i}+\left(-\frac{v^2}{r_0}\sin\frac{vt}{r_0}\right)\mathbf{j} = -\frac{v^2}{r_0}\left[\left(\cos\frac{vt}{r_0}\right)\mathbf{i}+\left(\sin\frac{vt}{r_0}\right)\mathbf{j}\right]$

(c) From part (b) above, $\mathbf{a}(t) = -\left(\frac{v}{r_0}\right)^2\mathbf{r}(t)$. So, by Newton's second law, $\mathbf{F} = -m\left(\frac{v}{r_0}\right)^2\mathbf{r}$. Substituting for $\mathbf{F}$ in the law of gravitation gives the result.

(d) Set $\frac{vT}{r_0} = 2\pi$ and solve for vT.

(e) Substitute $\frac{2\pi r_0}{T}$ for v in $v^2 = \frac{GM}{r_0}$ and solve for T^2.

$$\left(\frac{2\pi r_0}{T}\right)^2 = \frac{GM}{r_0} \Rightarrow \frac{4\pi^2 r_0^2}{T^2} = \frac{GM}{r_0} \Rightarrow \frac{1}{T^2} = \frac{GM}{4\pi^2 r_0^3} \Rightarrow T^2 = \frac{4\pi^2}{GM}r_0^3$$

35. (a) Apply Corollary 2 to each component separately. If the components all differ by scalar constants, the difference vector is a constant vector.
(b) Follows immediately from (a) since any two anti-derivatives of $\mathbf{r}(t)$ must have identical derivatives, namely $\mathbf{r}(t)$.

37. Let $\mathbf{u} = \mathbf{C} = \langle C_1, C_2 \rangle$. $\frac{d\mathbf{u}}{dt} = \frac{d\mathbf{C}}{dt} = \left\langle \frac{dC_1}{dt}, \frac{dC_2}{dt} \right\rangle = \langle 0, 0 \rangle$.

39. $\mathbf{u} = \langle u_1, u_2 \rangle$, $\mathbf{v} = \langle v_1, v_2 \rangle$

(a) $\frac{d}{dt}(\mathbf{u}+\mathbf{v}) = \frac{d}{dt}(\langle u_1 + v_1, u_2 + v_2 \rangle) = \left\langle \frac{d}{dt}(u_1 + v_1), \frac{d}{dt}(u_2 + v_2) \right\rangle = \langle u_1' + v_1', u_2' + v_2' \rangle$

$= \langle u_1', u_2' \rangle + \langle v_1', v_2' \rangle = \frac{d\mathbf{u}}{dt} + \frac{d\mathbf{v}}{dt}$

(b) $\frac{d}{dt}(\mathbf{u}-\mathbf{v}) = \frac{d}{dt}(\langle u_1 - v_1, u_2 - v_2 \rangle) = \left\langle \frac{d}{dt}(u_1 - v_1), \frac{d}{dt}(u_2 - v_2) \right\rangle = \langle u_1' - v_1', u_2' - v_2' \rangle$

$= \langle u_1', u_2' \rangle - \langle v_1', v_2' \rangle = \frac{d\mathbf{u}}{dt} - \frac{d\mathbf{v}}{dt}$

41. $f(t)$ and $g(t)$ differentiable at $c \Rightarrow f(t)$ and $g(t)$ continuous at $c \Rightarrow \mathbf{r}(t) = f(t)\mathbf{i} + g(t)\mathbf{j}$ is continuous at c.

43. (a) Let $\mathbf{r}(t) = f(t)\mathbf{i} + g(t)\mathbf{j}$. Then

$$\frac{d}{dt}\int_a^t \mathbf{r}(q)\,dq = \frac{d}{dt}\int_a^t [f(q)\mathbf{i} + g(q)\mathbf{j}]\,dq = \frac{d}{dt}\left[\left(\int_a^t f(q)\,dq\right)\mathbf{i} + \left(\int_a^t g(q)\,dq\right)\mathbf{j}\right]$$

$$= \left(\frac{d}{dt}\int_a^t f(1)\,dq\right)\mathbf{i} + \left(\frac{d}{dt}\int_a^t g(q)\,dq\right)\mathbf{j} = f(t)\mathbf{i} + g(t)\mathbf{j} = \mathbf{r}(t).$$

(b) Let $S(t) = \int_a^t \mathbf{r}(q)\,dq$. Then part (a) shows that $\mathbf{S}(t)$ is an antiderivative of $\mathbf{r}(t)$. Let $\mathbf{R}(t)$ be any antiderivative of $\mathbf{r}(t)$. Then according to 35(b), $\mathbf{S}(t) = \mathbf{R}(t) + \mathbf{C}$. Letting $t = a$, we have $\mathbf{0} = \mathbf{S}(a) = \mathbf{R}(a) + \mathbf{C}$. Therefore, $\mathbf{C} = -\mathbf{R}(a)$ and $\mathbf{S}(t) = \mathbf{R}(t) - \mathbf{R}(a)$. The result follows by letting $t = b$.

9.4 MODELING PROJECTILE MOTION

1. $x = (v_0 \cos\alpha)t \Rightarrow (21 \text{ km})\left(\frac{1000 \text{ m}}{1 \text{ km}}\right) = (840 \text{ m/s})(\cos 60°)t \Rightarrow t = \frac{21{,}000 \text{ m}}{(840 \text{ m/s})(\cos 60°)} = 50$ seconds

3. (a) $t = \frac{2v_0 \sin\alpha}{g} = \frac{2(500 \text{ m/s})(\sin 45°)}{9.8 \text{ m/s}^2} = 72.2$ seconds; $R = \frac{v_0^2}{g}\sin 2\alpha = \frac{(500 \text{ m/s})^2}{9.8 \text{ m/s}^2}(\sin 90°) = 25{,}510.2$ m

(b) $x = (v_0 \cos\alpha)t \Rightarrow 5000 \text{ m} = (500 \text{ m/s})(\cos 45°)t \Rightarrow t = \frac{5000 \text{ m}}{(500 \text{ m/s})(\cos 45°)} \approx 14.14$ s; thus,

$y = (v_0 \sin\alpha)t - \frac{1}{2}gt^2 \Rightarrow y \approx (500 \text{ m/s})(\sin 45°)(14.14 \text{ s}) - \frac{1}{2}(9.8 \text{ m/s}^2)(14.14 \text{ s})^2 \approx 4020$ m

(c) $y_{max} = \frac{(v_0 \sin\alpha)^2}{2g} = \frac{((500 \text{ m/s})(\sin 45°))^2}{2(9.8 \text{ m/s}^2)} = 6378$ m

5. $x = x_0 + (v_0 \cos\alpha)t = 0 + (44\cos 45°)t = 22\sqrt{2}t$ and $y = y_0 + (v_0 \sin\alpha)t - \frac{1}{2}gt^2 = 6.5 + (44\sin 45°)t - 16t^2$

$= 6.5 + 22\sqrt{2}t - 16t^2$; the shot lands when $y = 0 \Rightarrow t = \frac{22\sqrt{2} \pm \sqrt{968+416}}{32} \approx 2.135$ sec since $t > 0$; thus $x = 22\sqrt{2}t \approx (22\sqrt{2})(2.134839) \approx 66.42$ ft

7. (a) $R = \frac{v_0^2}{g}\sin 2\alpha \Rightarrow 10\text{ m} = \left(\frac{v_0^2}{9.8\text{ m/s}^2}\right)(\sin 90°) \Rightarrow v_0^2 = 98\text{ m}^2\text{s}^2 \Rightarrow v_0 \approx 9.9$ m/s;

(b) $6\text{m} \approx \frac{(9.9\text{ m/s})^2}{9.8\text{ m/s}^2}(\sin 2\alpha) \Rightarrow \sin 2\alpha \approx 0.59999 \Rightarrow 2\alpha \approx 36.87°$ or $143.12° \Rightarrow \alpha \approx 18.4°$ or $71.6°$

9. $R = \frac{v_0^2}{g}\sin 2\alpha \Rightarrow 3(248.8)\text{ ft} = \left(\frac{v_0^2}{32\text{ ft/sec}^2}\right)(\sin 18°) \Rightarrow v_0^2 \approx 77{,}292.84\text{ ft}^2/\text{sec}^2 \Rightarrow v_0 \approx 278.01\text{ ft/sec} \approx 190$ mph

11. $x = (v_0\cos\alpha)t \Rightarrow 135\text{ ft} = (90\text{ ft/sec})(\cos 30°)t \Rightarrow t \approx 1.732$ sec; $y = (v_0 \sin\alpha)t - \frac{1}{2}gt^2$

$\Rightarrow y \approx (90\text{ ft/sec})(\sin 30°)(1.732\text{ sec}) - \frac{1}{2}(32\text{ ft/sec}^2)(1.732\text{ sec})^2 \Rightarrow y \approx 29.94\text{ ft} \Rightarrow$ the golf ball will clip the leaves at the top

13. $x = (v_0\cos\alpha)t \Rightarrow 315\text{ ft} = (v_0\cos 20°)t \Rightarrow v_0 = \frac{315}{t\cos 20°}$; also $y = (v_0\sin\alpha)t - \frac{1}{2}gt^2$

$\Rightarrow 34\text{ ft} = \left(\frac{315}{t\cos 20°}\right)(t\sin 20°) - \frac{1}{2}(32)t^2 \Rightarrow 34 = 315\tan 20° - 16t^2 \Rightarrow t^2 \approx 5.04\text{ sec}^2 \Rightarrow t \approx 2.25$ sec

$\Rightarrow v_0 = \frac{315}{(2.25)(\cos 20°)} \approx 149$ ft/sec

15. $R = \frac{v_0^2}{g}\sin 2\alpha \Rightarrow 16{,}000\text{ m} = \frac{(400\text{ m/s})^2}{9.8\text{ m/s}^2}\sin 2\alpha \Rightarrow \sin 2\alpha = 0.98 \Rightarrow 2\alpha \approx 78.5°$ or $2\alpha \approx 101.5° \Rightarrow \alpha \approx 39.3°$ or $50.7°$

17. $x = x_0 + (v_0\cos\alpha)t = 0 + (v_0\cos 40°)t \approx 0.766\,v_0t$ and $y = y_0 + (v_0\sin\alpha)t - \frac{1}{2}gt^2 = 6.5 + (v_0\sin 40°)t - 16t^2$

$\approx 6.5 + 0.643\,v_0t - 16t^2$; now the shot went 73.833 ft $\Rightarrow 73.833 = 0.766\,v_0t \Rightarrow t \approx \frac{96.383}{v_0}$ sec; the shot lands

when $y = 0 \Rightarrow 0 = 6.5 + (0.643)(96.383) - 16\left(\frac{96.383}{v_0}\right)^2 \Rightarrow 0 \approx 68.474 - \frac{148{,}634}{v_0^2} \Rightarrow v_0 \approx \sqrt{\frac{148{,}634}{68.474}}$

≈ 46.6 ft/sec, the shot's initial speed

19. $\frac{d\mathbf{r}}{dt} = \int(-g\mathbf{j})\,dt = -gt\mathbf{j} + \mathbf{C}_1$ and $\frac{d\mathbf{r}}{dt}(0) = (v_0\cos\alpha)\mathbf{i} + (v_0\sin\alpha)\mathbf{j} \Rightarrow -g(0)\mathbf{j} + \mathbf{C}_1 = (v_0\cos\alpha)\mathbf{i} + (v_0\sin\alpha)\mathbf{j}$

$\Rightarrow \mathbf{C}_1 = (v_0\cos\alpha)\mathbf{i} + (v_0\sin\alpha)\mathbf{j} \Rightarrow \frac{d\mathbf{r}}{dt} = (v_0\cos\alpha)\mathbf{i} + (v_0\sin\alpha - gt)\mathbf{j}$; $\mathbf{r} = \int[(v_0\cos\alpha)\mathbf{i} + (v_0\sin\alpha - gt)\mathbf{j}]\,dt$

$= (v_0t\cos\alpha)\mathbf{i} + \left(v_0t\sin\alpha - \frac{1}{2}gt^2\right)\mathbf{j} + \mathbf{C}_2$ and $\mathbf{r}(0) = x_0\mathbf{i} + y_0\mathbf{j} \Rightarrow [v_0(0)\cos\alpha]\mathbf{i} + \left[v_0(0)\sin\alpha - \frac{1}{2}g(0)^2\right]\mathbf{j} + \mathbf{C}_2$

$= x_0\mathbf{i} + y_0\mathbf{j} \Rightarrow \mathbf{C}_2 = x_0\mathbf{i} + y_0\mathbf{j} \Rightarrow \mathbf{r} = (x_0 + v_0t\cos\alpha)\mathbf{i} + \left(y_0 + v_0t\sin\alpha - \frac{1}{2}gt^2\right)\mathbf{j} \Rightarrow x = x_0 + v_0t\cos\alpha$ and

$y = y_0 + v_0t\sin\alpha - \frac{1}{2}gt^2$

21. The horizontal distance from Rebollo to the center of the cauldron is 90 ft $\Rightarrow$ the horizontal distance to the nearest rim is $x = 90 - \frac{1}{2}(12) = 84 \Rightarrow 84 = x_0 + (v_0 \cos \alpha)t \approx 0 + \left(\frac{90g}{v_0 \sin \alpha}\right)t \Rightarrow 84 = \frac{(90)(32)}{\sqrt{(68)(64)}}t$ $\Rightarrow t = 1.92$ sec. The vertical distance at this time is $y = y_0 + (v_0 \sin \alpha)t - \frac{1}{2}gt^2$ $\approx 6 + \sqrt{(68)(64)}\,(1.92) - 16(1.92)^2 \approx 73.7$ ft $\Rightarrow$ the arrow clears the rim by 3.7 ft

23. Flight time = 1 sec and the measure of the angle of elevation is about 64° (using a protractor) so that $t = \frac{2v_0 \sin \alpha}{g} \Rightarrow 1 = \frac{2v_0 \sin 64^\circ}{32} \Rightarrow v_0 \approx 17.80$ ft/sec. Then $y_{max} = \frac{(17.80 \sin 64^\circ)^2}{2(32)} \approx 4.00$ ft and $R = \frac{v_0^2}{g} \sin 2\alpha \Rightarrow R = \frac{(17.80)^2}{32} \sin 128^\circ \approx 7.80$ ft $\Rightarrow$ the engine traveled about 7.80 ft in 1 sec $\Rightarrow$ the engine velocity was about 7.80 ft/sec

25. (a) At the time t when the projectile hits the line OR we have $\tan \beta = \frac{y}{x}$; $x = [v_0 \cos(\alpha - \beta)]t$ and $y = [v_0 \sin(\alpha - \beta)]t - \frac{1}{2}gt^2 < 0$ since R is below level ground. Therefore let $|y| = \frac{1}{2}gt^2 - [v_0 \sin(\alpha - \beta)]t > 0$

so that $\tan \beta = \frac{\left[\frac{1}{2}gt^2 - (v_0 \sin(\alpha - \beta))t\right]}{[v_0 \cos(\alpha - \beta)]t} = \frac{\left[\frac{1}{2}gt - v_0 \sin(\alpha - \beta)\right]}{v_0 \cos(\alpha - \beta)}$

$\Rightarrow v_0 \cos(\alpha - \beta) \tan \beta = \frac{1}{2}gt - v_0 \sin(\alpha - \beta)$

$\Rightarrow t = \frac{2v_0 \sin(\alpha - \beta) + 2v_0 \cos(\alpha - \beta) \tan \beta}{g}$, which is the time when the projectile hits the downhill slope. Therefore,

$x = [v_0 \cos(\alpha - \beta)]\left[\frac{2v_0 \sin(\alpha - \beta) + 2v_0 \cos(\alpha - \beta) \tan \beta}{g}\right]$

$= \frac{2v_0^2}{g}\left[\cos^2(\alpha - \beta) \tan \beta + \sin(\alpha - \beta) \cos(\alpha - \beta)\right]$. If x is maximized, then OR is maximized:

$\frac{dx}{d\alpha} = \frac{2v_0^2}{g}[-\sin 2(\alpha - \beta) \tan \beta + \cos 2(\alpha - \beta)] = 0 \Rightarrow -\sin 2(\alpha - \beta) \tan \beta + \cos 2(\alpha - \beta) = 0$

$\Rightarrow \tan \beta = \cot 2(\alpha - \beta) \Rightarrow 2(\alpha - \beta) = 90^\circ - \beta \Rightarrow \alpha - \beta = \frac{1}{2}(90^\circ - \beta) \Rightarrow \alpha = \frac{1}{2}(90^\circ + \beta) = \frac{1}{2}$ of $\angle AOR$.

(b) At the time t when the projectile hits OR we have $\tan \beta = \frac{y}{x}$;

$x = [v_0 \cos(\alpha + \beta)]t$ and $y = [v_0 \sin(\alpha + \beta)]t - \frac{1}{2}gt^2$

$\Rightarrow \tan \beta = \frac{[v_0 \sin(\alpha + \beta)]t - \frac{1}{2}gt^2}{[v_0 \cos(\alpha + \beta)]t} = \frac{\left[v_0 \sin(\alpha + \beta) - \frac{1}{2}gt\right]}{v_0 \cos(\alpha + \beta)}$

$\Rightarrow v_0 \cos(\alpha + \beta) \tan \beta = v_0 \sin(\alpha + \beta) - \frac{1}{2}gt$

$\Rightarrow t = \frac{2v_0 \sin(\alpha + \beta) - 2v_0 \cos(\alpha + \beta) \tan \beta}{g}$, which is the time when the projectile hits the uphill slope. Therefore,

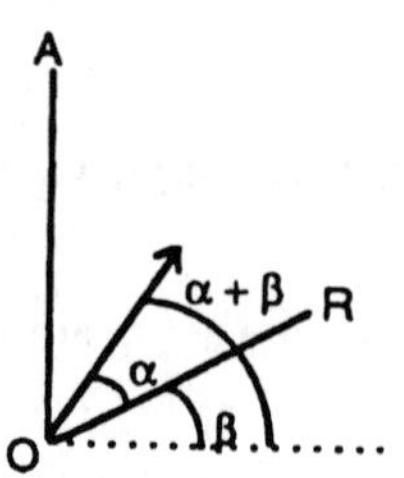

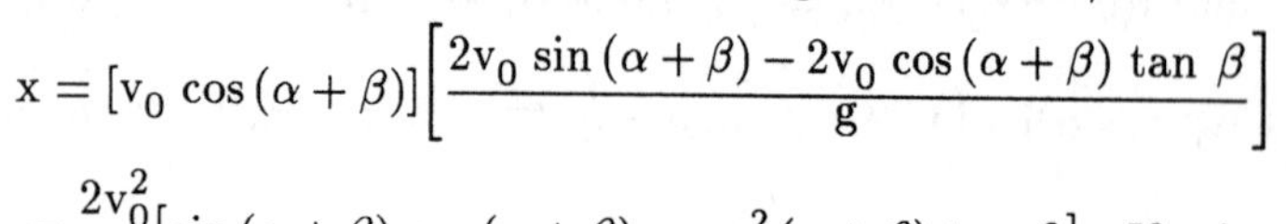

$x = [v_0 \cos(\alpha + \beta)]\left[\frac{2v_0 \sin(\alpha + \beta) - 2v_0 \cos(\alpha + \beta) \tan \beta}{g}\right]$

$= \frac{2v_0^2}{g}\left[\sin(\alpha + \beta) \cos(\alpha + \beta) - \cos^2(\alpha + \beta) \tan \beta\right]$. If x is maximized, then OR is maximized:

$\frac{dx}{d\alpha} = \frac{2v_0^2}{g}[\cos 2(\alpha+\beta) + \sin 2(\alpha+\beta)\tan\beta] = 0 \Rightarrow \cos 2(\alpha+\beta) + \sin 2(\alpha+\beta)\tan\beta = 0$

$\Rightarrow \cot 2(\alpha+\beta) + \tan\beta = 0 \Rightarrow \cot 2(\alpha+\beta) = -\tan\beta = \tan(-\beta) \Rightarrow 2(\alpha+\beta) = 90^\circ - (-\beta)$

$= 90^\circ + \beta \Rightarrow \alpha = \frac{1}{2}(90^\circ - \beta) = \frac{1}{2}$ of $\angle AOR$. Therefore v_0 would bisect $\angle AOR$ for maximum range uphill.

27. (a) (Assuming that "x" is zero at the point of impact.)
$\mathbf{r}(t) = (x(t))\mathbf{i} + (y(t))\mathbf{j}$, where $x(t) = (35\cos 27^\circ)t$ and $y(t) = 4 + (35\sin 27^\circ)t - 16t^2$.

(b) $y_{max} = \frac{(v_0\sin\alpha)^2}{2g} + 4 = \frac{(35\sin 27^\circ)^2}{64} + 4 \approx 7.945$ feet, which is reached at $t = \frac{v_0\sin\alpha}{g}$

$= \frac{35\sin 27^\circ}{32} \approx 0.497$ seconds.

(c) For the time, solve $y = 4 + (35\sin 27^\circ)t - 16t^2 = 0$ for t, using the quadratic formula

$t = \frac{35\sin 27^\circ + \sqrt{(-35\sin 27^\circ)^2 + 256}}{32} \approx 1.201$ seconds. Then the range is about

$x(1.201) = (35\cos 27^\circ)(1.201) \approx 37.453$ feet.

(d) For the time, solve $y = 4 + (35\sin 27^\circ)t - 16t^2 = 7$ for t, using the quadratic formula

$t = \frac{35\sin 27^\circ \pm \sqrt{(-35\sin 27^\circ)^2 - 192}}{32} \approx 0.254$ and 0.740 seconds. At those times the ball is about $x(0.254) = (35\cos 27^\circ)(0.254) \approx 7.906$ feet and $x(0.740) = (35\cos 27^\circ)(0.740) \approx 23.064$ feet from the impact point, or about $37.460 - 7.906 \approx 29.554$ feet and $37.460 - 23.064 \approx 14.396$ feet from the landing spot.

(e) Yes. It changes things because the ball won't clear the net ($y_{max} \approx 7.945$ ft).

29. $\frac{d^2\mathbf{r}}{dt^2} + k\frac{d\mathbf{r}}{dt} = -g\mathbf{j} \Rightarrow P(t) = k$ and $\mathbf{Q}(t) = -g\mathbf{j} \Rightarrow \int P(t)\,dt = kt \Rightarrow v(t) = e^{\int P(t)\,dt} = e^{kt} \Rightarrow \frac{d\mathbf{r}}{dt}$

$= \frac{1}{v(t)}\int v(t)\mathbf{Q}(t)\,dt = -ge^{-kt}\int \mathbf{j}e^{kt}\,dt = -ge^{-kt}\left[\frac{e^{kt}}{k}\mathbf{j} + \mathbf{C}_1\right] = -\frac{g}{k}\mathbf{j} + \mathbf{C}e^{-kt}$, where $\mathbf{C} = -g\mathbf{C}_1$; apply the

initial condition: $\left.\frac{d\mathbf{r}}{dt}\right|_{t=0} = (v_0\cos\alpha)\mathbf{i} + (v_0\sin\alpha)\mathbf{j} = -\frac{g}{k}\mathbf{j} + \mathbf{C} \Rightarrow \mathbf{C} = (v_0\cos\alpha)\mathbf{i} + \left(\frac{g}{k} + v_0\sin\alpha\right)\mathbf{j}$

$\Rightarrow \frac{d\mathbf{r}}{dt} = \left(v_0e^{-kt}\cos\alpha\right)\mathbf{i} + \left[-\frac{g}{k} + e^{-kt}\left(\frac{g}{k} + v_0\sin\alpha\right)\right]\mathbf{j}$, $\mathbf{r} = \int \left\{\left(v_0e^{-kt}\cos\alpha\right)\mathbf{i} + \left[-\frac{g}{k} + e^{-kt}\left(\frac{g}{k} + v_0\sin\alpha\right)\right]\mathbf{j}\right\}dt$

$= \left(-\frac{v_0}{k}e^{-kt}\cos\alpha\right)\mathbf{i} + \left[-\frac{gt}{k} - \frac{e^{-kt}}{k}\left(\frac{g}{k} + v_0\sin\alpha\right)\right]\mathbf{j} + \mathbf{C}_2$; apply the initial condition:

$\mathbf{r}(0) = \mathbf{0} = \left(-\frac{v_0}{k}\cos\alpha\right)\mathbf{i} + \left(-\frac{g}{k^2} - \frac{v_0}{k}\sin\alpha\right)\mathbf{j} + \mathbf{C}_2 \Rightarrow \mathbf{C}_2 = \left(\frac{v_0}{k}\cos\alpha\right)\mathbf{i} + \left(\frac{g}{k^2} + \frac{v_0}{k}\sin\alpha\right)\mathbf{j}$

$\Rightarrow \mathbf{r}(t) = \left(\frac{v_0}{k}(1 - e^{-kt})\cos\alpha\right)\mathbf{i} + \left[\frac{v_0}{k}(1 - e^{-kt})\sin\alpha + \frac{g}{k^2}(1 - kt - e^{-kt})\right]\mathbf{j}$

31. (a) $\mathbf{r}(t) = (x(t))\mathbf{i} + (y(t))\mathbf{j}$, where

$x(t) = \left(\frac{1}{0.08}\right)(1 - e^{-0.08t})(152\cos 20^\circ - 17.6)$ and

$y(t) = 3 + \left(\frac{152}{0.08}\right)(1 - e^{-0.08t})(\sin 20^\circ)$

$$+\left(\frac{32}{0.08^2}\right)(1-0.08t-e^{-0.08t})$$

(b) Solve graphically using a calculator or CAS: At $t \approx 1.527$ seconds the ball reaches a maximum height of about 41.893 feet.

(c) Use a graphing calculator or CAS to find that $y = 0$ when the ball has traveled for ≈ 3.181 seconds. The range is about $x(3.181)$ $= \left(\frac{1}{0.08}\right)(1-e^{-0.08(3.181)})(152 \cos 20° - 17.6)$ ≈ 351.734 feet.

(d) Use a graphing calculator or CAS to find that $y = 35$ for $t \approx 0.877$ and 2.190 seconds, at which times the ball is about $x(0.877)$ ≈ 106.028 feet and $x(2.190) \approx 251.530$ feet from home plate.

(e) No; the range is less than 380 feet. To find the wind needed for a home run, first use the method of part (d) to find that $y = 20$ at $t \approx 0.376$ and 2.716 seconds. Then define

$$x(w) = \left(\frac{1}{0.08}\right)(1-e^{-0.08(2.716)})(152 \cos 20° + w),$$

and solve $x(w) = 380$ to find $w \approx 12.846$ ft/sec. This is the speed of a wind gust needed in the direction of the hit for the ball to clear the fence for a home run.

9.5 POLAR COORDINATES AND GRAPHS

For exercise 1, two pairs of polar coordinates label the same point if the r-coordinates are the same and the θ coordinates differ by an even multiple of π, or if the r-coordinates are opposites and the θ-coordinates differ by an odd multiple of π.

1. (a) and (e) are the same.
   (b) and (g) are the same.
   (c) and (h) are the same.
   (d) and (f) are the same.

3.

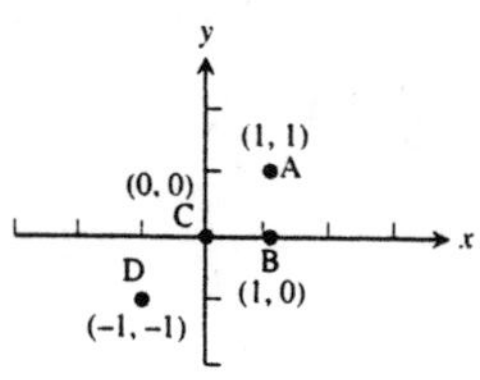

(a) $\left(\sqrt{2}\cos\frac{\pi}{4}, \sqrt{2}\sin\frac{\pi}{4}\right) = (1,1)$

(b) $(1\cos 0, 1\sin 0) = (1,0)$

(c) $\left(0\cos\frac{\pi}{2}, 0\sin\frac{\pi}{2}\right) = (0,0)$

(d) $\left(-\sqrt{2}\cos\frac{\pi}{4}, -\sqrt{2}\sin\frac{\pi}{4}\right) = (-1,-1)$

5.

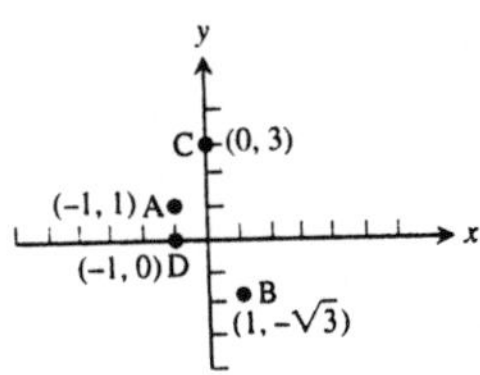

(a) $r = \sqrt{(-1)^2 + 1^2} = \sqrt{2}$, $\tan\theta = \frac{1}{-1} = -1$ with θ in quadrant II. The coordinates are $\left(\sqrt{2}, \frac{3\pi}{4}\right)$. $\left(\sqrt{2}, -\frac{5\pi}{4}\right)$ also works, since r is the same and θ differs by 2π.

(b) $r = \sqrt{1^2 + (-\sqrt{3})^2} = 2$, $\tan\theta = -\frac{\sqrt{3}}{1} = -\sqrt{3}$ with θ in quadrant IV. The coordinates are $\left(2, -\frac{\pi}{3}\right)$. $\left(-2, \frac{2\pi}{3}\right)$ also works, since r has the opposite sign and θ differs by π.

(c) $r = \sqrt{0^2 + 3^2} = 3$, $\tan\theta = \frac{3}{0}$ is undefined with θ on the positive y-axis. The coordinates are $\left(3, \frac{\pi}{2}\right)$. $\left(3, \frac{5\pi}{2}\right)$ also works, since r is the same and θ differs by 2π.

(d) $r = \sqrt{(-1)^2 + 0^2} = 1$, $\tan\theta = \frac{0}{-1} = 0$ with θ on the negative x-axis. The coordinates are $(1, \pi)$. $(-1, 0)$ also works, since r has the opposite sign and θ differs by π.

7.

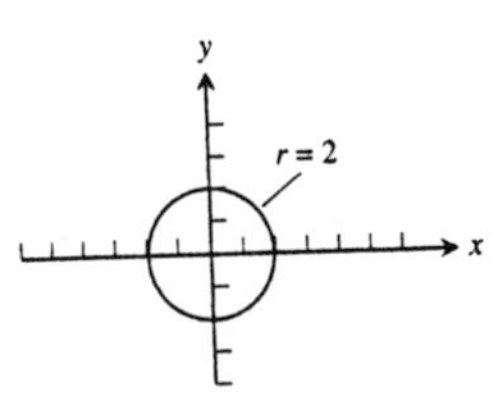

9.

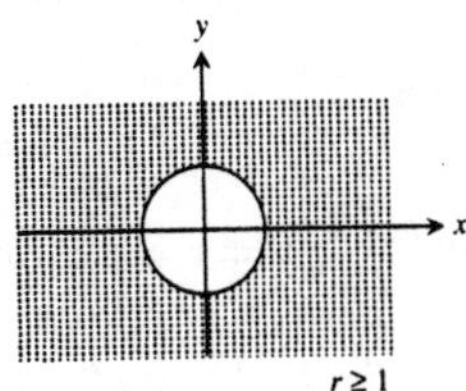

11.

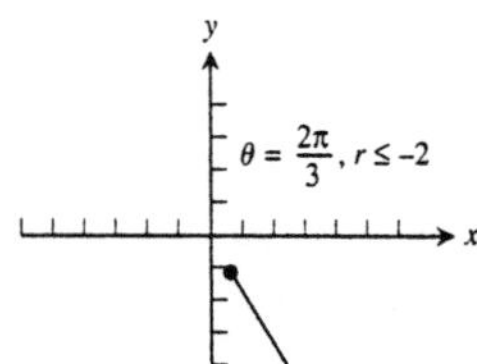

13.

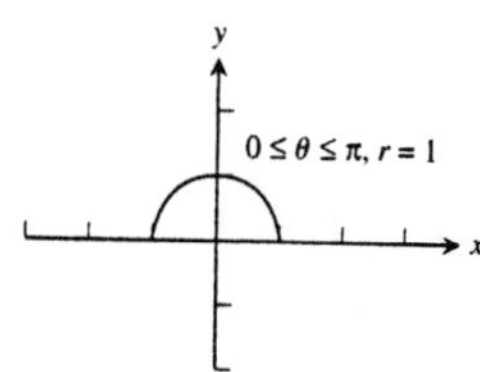

15.

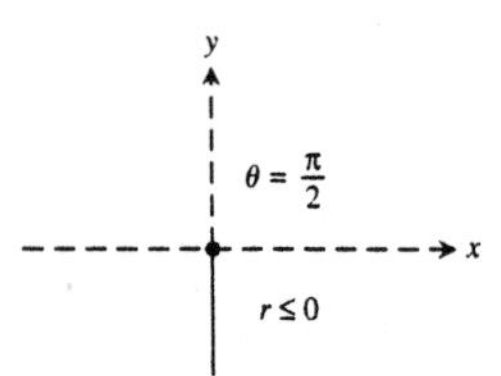

17.

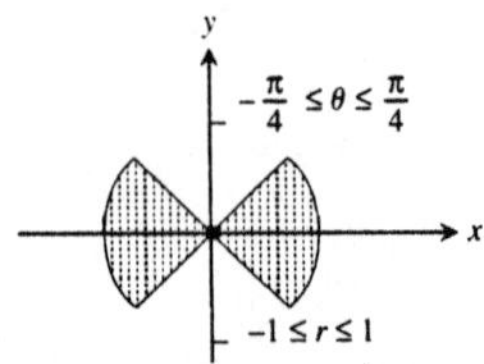

19. $y = r \sin \theta$, so the equation is $y = 0$, which is the x-axis.

21. $r = 4 \csc \theta$
$r \sin \theta = 4$
$y = r \sin \theta$, so the equation is $y = 4$,
a horizontal line.

23. $x = r \cos \theta$ and $y = r \sin \theta$, so the equation is $x + y = 1$, a line (slope $= -1$, y-intercept $= 1$).

25. $x^2 + y^2 = r^2$ and $y = r \sin \theta$, so the equation is $x^2 + y^2 = 4y \Rightarrow x^2 + (y-2)^2 = 4$, a circle (center $= (0, 2)$, radius $= 2$).

27. $r^2 \sin 2\theta = 2 \Rightarrow 2r^2 \sin \theta \cos \theta = 2 \Rightarrow (r \sin \theta)(r \cos \theta) = 1$; $x = r \cos \theta$ and $y = r \sin \theta$, so the equation is $xy = 1 \left(\text{or, } y = \frac{1}{x}\right)$, a hyperbola.

29. $r = \csc \theta \, e^{r \cos \theta} \Rightarrow r \sin \theta = e^{r \cos \theta}$; $x = r \cos \theta$ and $y = r \sin \theta$, so the equation is $y = e^x$, the exponential curve.

31. $r \sin \theta = \ln r + \ln \cos \theta \Rightarrow r \sin \theta = \ln (r \cos \theta) \Rightarrow y = \ln x$, the logarithmic curve.

33. $r^2 = -4r \sin \theta \Rightarrow x^2 + y^2 = -4x \Rightarrow x^2 + (y-4)^2 = 16$, a circle (center $= (0, 4)$, radius $= 4$)

35. $r = 2 \cos \theta + 2 \sin \theta \Rightarrow r^2 = 2r \cos \theta + 2r \sin \theta \Rightarrow x^2 + y^2 = 2x + 2y \Rightarrow (x-1)^2 + (y-1)^2 = 2$, a circle (center $= (1, 1)$, radius $= \sqrt{2}$)

37. $x = 7 \Rightarrow r \cos \theta = 7$; The graph is a vertical line.

39. $x = y \Rightarrow r \cos\theta = r \sin\theta \Rightarrow \tan\theta = 1 \Rightarrow \theta = \frac{\pi}{4}$. More generally, $\theta = \frac{\pi}{4} + 2k\pi$ for any integer k.

The graph is a slanted line.

41. $x^2 + y^2 = 4 \Rightarrow r^2 = 4$ or $r = 2$ (or $r = -2$)

43. $\frac{x^2}{9} + \frac{y^2}{4} = 1 \Rightarrow \frac{r^2 \cos^2\theta}{9} + \frac{r^2 \sin^2\theta}{4} = 1 \Rightarrow r^2(4\cos^2\theta + 9\sin^2\theta) = 36$

45. $y^2 = 4x \Rightarrow r^2 \sin^2\theta = 4r\cos\theta \Rightarrow r\sin^2\theta = 4\cos\theta$

47. $x^2 + (y-2)^2 = 4 \Rightarrow r^2\cos^2\theta + (r\sin\theta - 2)^2 = 4 \Rightarrow r^2\cos^2\theta + r^2\sin^2\theta - 4r\sin\theta + 4 = 4$
$\Rightarrow r^2 - 4r\sin\theta = 0 \Rightarrow r = 4\sin\theta$. The graph is a circle centered at $(0, 2)$ with radius 2.

In Exercises 49-57, find the minimum θ-interval by trying different intervals on a graphing calculator.

49. (a)

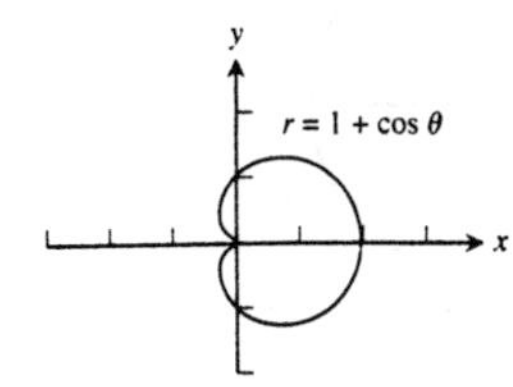

(b) Length of interval = 2π

51. (a)

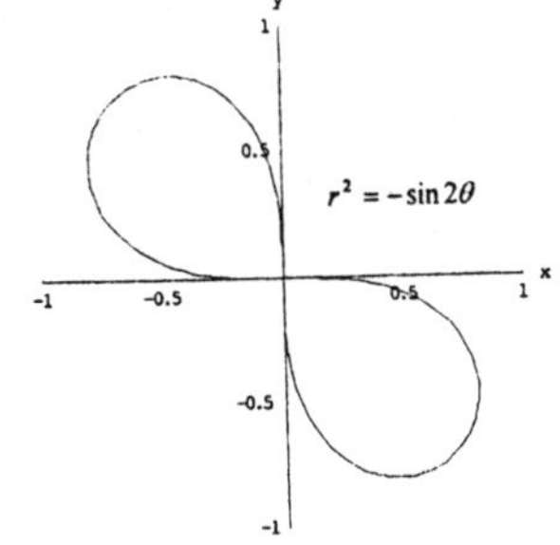

(b) Length of interval = $\frac{\pi}{2}$

53. (a)

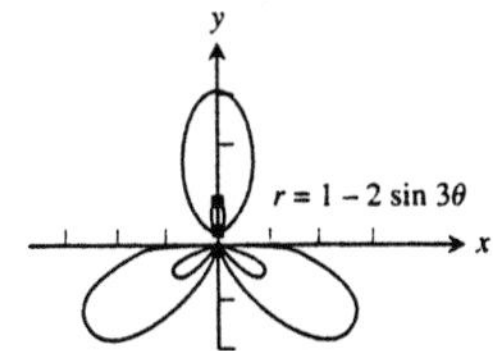

(b) Length of interval = 2π

55. (a)

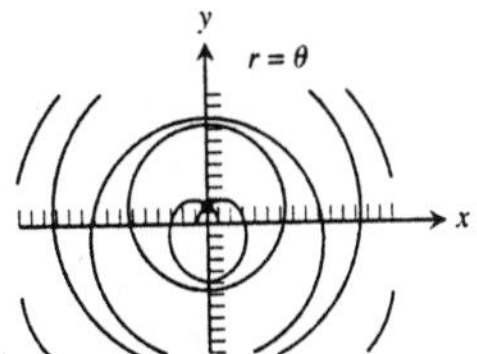

(b) Required interval = $(-\infty, -\infty)$

57. (a)

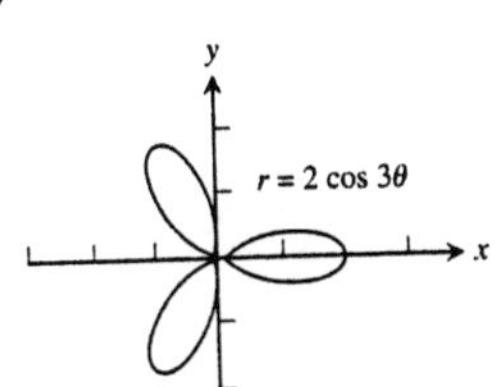

(b) Length of interval $= \pi$

59. If (r,θ) is a solution, so is $(-r,\theta)$. Therefore, the curve is symmetric about the origin. And if (r,θ) is a solution, so is $(r,-\theta)$. Therefore, the curve is symmetric about the x-axis. And since any curve with x-axis and origin symmetry also has y-axis symmetry, the curve is symmetric about the y-axis.

61. If (r,θ) is a solution, so is $(r,\pi-\theta)$. Therefore, the curve is symmetric about the y-axis. The curve does not have x-axis or origin symmetry.

63. (a) Because $r = a \sec \theta$ is equivalent to $r \cos \theta = a$, which is equivalent to the Cartesian equation $x = a$.
(b) $r = a \csc \theta$ is equivalent to $y = a$.

65. $\left(2,\frac{3\pi}{4}\right)$ is the same point as $\left(-2,-\frac{\pi}{4}\right)$; $r = 2 \sin 2\left(-\frac{\pi}{4}\right) = 2 \sin\left(-\frac{\pi}{2}\right) = -2 \Rightarrow \left(-2,-\frac{\pi}{4}\right)$ is on the graph $\Rightarrow \left(2,\frac{3\pi}{4}\right)$ is on the graph

67. $1+\cos\theta = 1-\cos\theta \Rightarrow \cos\theta = 0 \Rightarrow \theta = \frac{\pi}{2}, \frac{3\pi}{2} \Rightarrow r = 1$; points of intersection are $\left(1,\frac{\pi}{2}\right)$ and $\left(1,\frac{3\pi}{2}\right)$. The point of intersection $(0,0)$ is found by graphing.

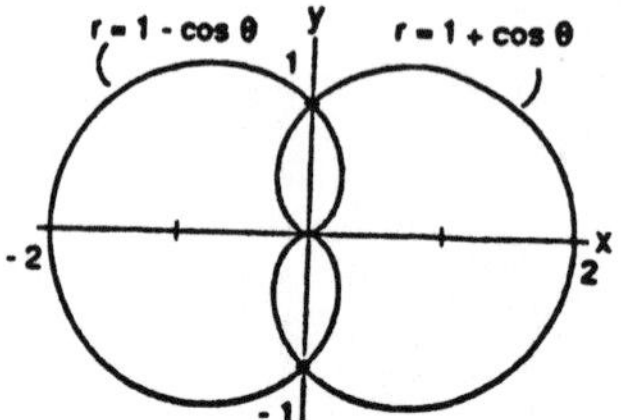

69. $\cos\theta = 1-\cos\theta \Rightarrow 2\cos\theta = 1 \Rightarrow \cos\theta = \frac{1}{2}$ $\Rightarrow \theta = \frac{\pi}{3}, -\frac{\pi}{3} \Rightarrow r = \frac{1}{2}$; points of intersection are $\left(\frac{1}{2},\frac{\pi}{3}\right)$ and $\left(\frac{1}{2},-\frac{\pi}{3}\right)$. The point $(0,0)$ is found by graphing.

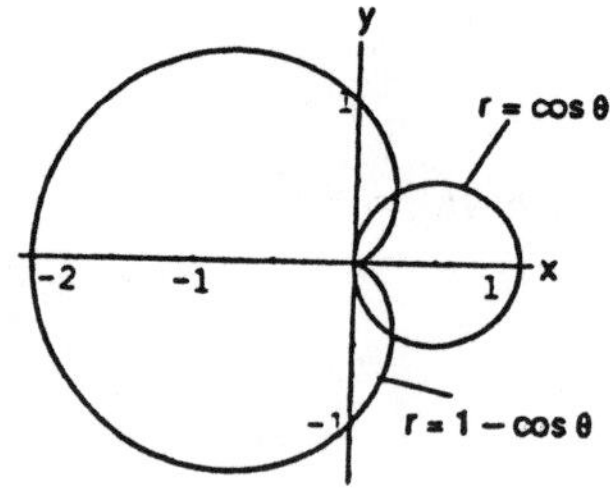

71. $r^2 = \sin 2\theta$ and $r^2 = \cos 2\theta$ are generated completely for $0 \le \theta \le \frac{\pi}{2}$. Then $\sin 2\theta = \cos 2\theta \Rightarrow 2\theta = \frac{\pi}{4}$ is the only solution on that interval $\Rightarrow \theta = \frac{\pi}{8} \Rightarrow r^2 = \sin 2\left(\frac{\pi}{8}\right) = \frac{1}{\sqrt{2}}$ $\Rightarrow r = \pm\frac{1}{\sqrt[4]{2}}$; points of intersection are $\left(\pm\frac{1}{\sqrt[4]{2}},\frac{\pi}{8}\right)$. The point of intersection $(0,0)$ is found by graphing.

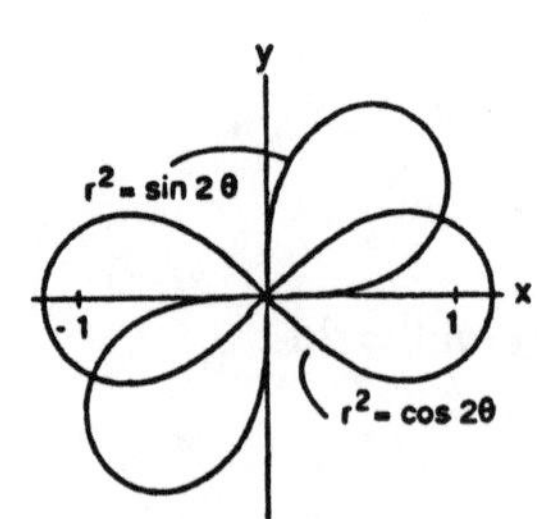

73. $1 = 2\sin 2\theta \Rightarrow \sin 2\theta = \frac{1}{2} \Rightarrow 2\theta = \frac{\pi}{6}, \frac{5\pi}{6}, \frac{13\pi}{6}, \frac{17\pi}{6}$

$\Rightarrow \theta = \frac{\pi}{12}, \frac{5\pi}{12}, \frac{13\pi}{12}, \frac{17\pi}{12}$; points of intersection are $\left(1, \frac{\pi}{12}\right), \left(1, \frac{5\pi}{12}\right), \left(1, \frac{13\pi}{12}\right)$, and $\left(1, \frac{17\pi}{12}\right)$. The points of intersection $\left(1, \frac{7\pi}{12}\right), \left(1, \frac{11\pi}{12}\right), \left(1, \frac{19\pi}{12}\right)$ and $\left(1, \frac{23\pi}{12}\right)$ are found by graphing and symmetry.

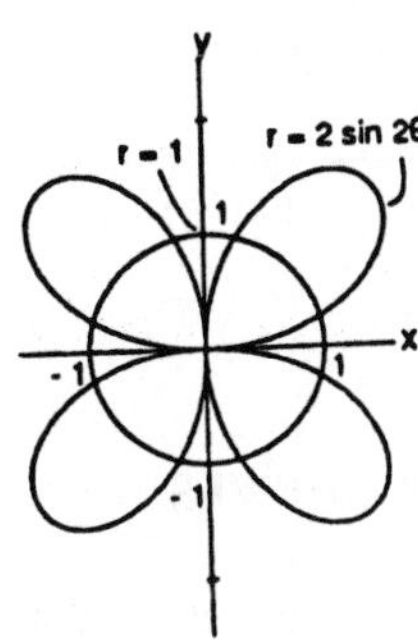

75. Note that (r, θ) and $(-r, \theta + \pi)$ describe the same point in the plane. Then $r = 1 - \cos\theta \Leftrightarrow -1 - \cos(\theta + \pi)$ $= -1 - (\cos\theta\cos\pi - \sin\theta\sin\pi) = -1 + \cos\theta = -(1 - \cos\theta) = -r$; therefore (r, θ) is on the graph of $r = 1 - \cos\theta \Leftrightarrow (-r, \theta + \pi)$ is on the graph of $r = -1 - \cos\theta \Rightarrow$ the answer is (a).

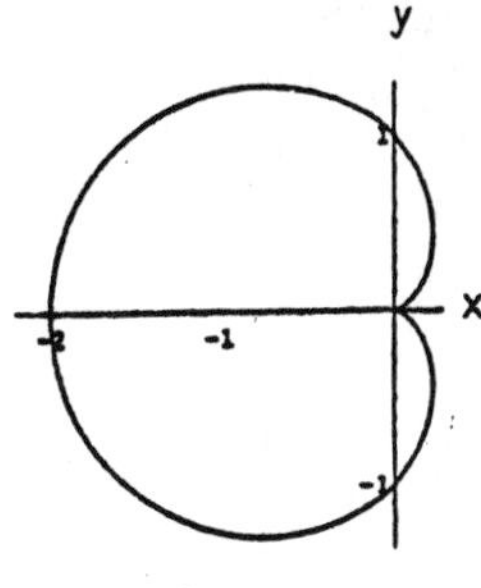

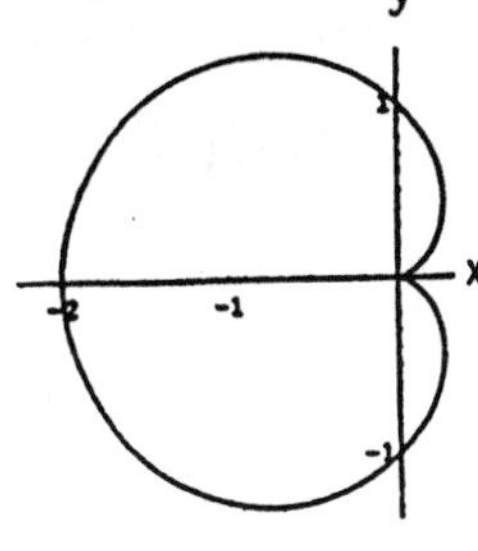

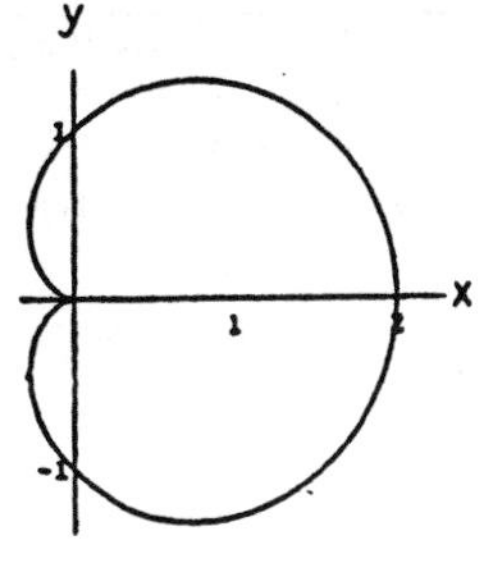

77.

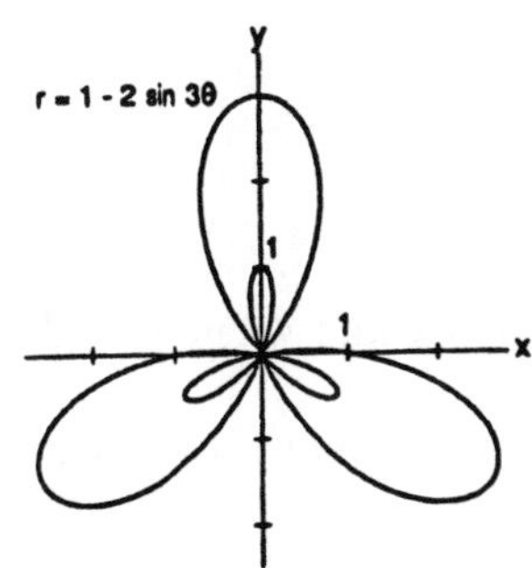

79. (a) $r^2 = -4\cos\theta \Rightarrow \cos\theta = -\frac{r^2}{4}$; $r = 1 - \cos\theta \Rightarrow r = 1 - \left(-\frac{r^2}{4}\right) \Rightarrow 0 = r^2 - 4r + 4 \Rightarrow (r-2)^2 = 0$

$\Rightarrow r = 2$; therefore $\cos\theta = -\frac{2^2}{4} = -1 \Rightarrow \theta = \pi \Rightarrow (2, \pi)$ is a point of intersection

(b) $r = 0 \Rightarrow 0^2 = 4\cos\theta \Rightarrow \cos\theta = 0 \Rightarrow \theta = \frac{\pi}{2}, \frac{3\pi}{2} \Rightarrow \left(0, \frac{\pi}{2}\right)$ or $\left(0, \frac{3\pi}{2}\right)$ is on the graph; $r = 0 \Rightarrow 0 = 1 - \cos\theta$ $\Rightarrow \cos\theta = 1 \Rightarrow \theta = 0 \Rightarrow (0,0)$ is on the graph. Since $(0,0) = \left(0, \frac{\pi}{2}\right)$ for polar coordinates, the graphs intersect at the origin.

81. $d = \sqrt{(x_2 - x_1)^2 + (y_1 - y_2)^2}$

$= \left[(r_2\cos\theta_2 - r_1\cos\theta_1)^2 + (r_2\sin\theta_2 - r_1\sin\theta_1)^2\right]^{1/2}$

$$= \left[r_2^2 \cos^2\theta_2 - 2r_2r_1 \cos\theta_2 \cos\theta_1 + r_1^2 \cos^2\theta_1 + r_2^2 \sin^2\theta_2 - 2r_2r_1 \sin\theta_2 \sin\theta_1 + r_1^2 \sin^2\theta_1\right]^{1/2}$$

$$= \sqrt{r_1^2 + r_2^2 - 2r_1r_2 \cos(\theta_1 - \theta_2)}$$

9.6 CALCULUS OF POLAR CURVES

1. $\frac{dy}{dx} = \frac{f'(\theta)\sin\theta + f(\theta)\cos\theta}{f'(\theta)\cos\theta - f(\theta)\sin\theta} = \frac{\cos\theta\sin\theta + (-1+\sin\theta)\cos\theta}{\cos\theta\cos\theta - (-1+\sin\theta)\sin\theta} = \frac{2\sin\theta\cos\theta - \cos\theta}{\cos^2\theta - \sin^2\theta + \sin\theta}$

$\left.\frac{dy}{dx}\right|_{\theta=0} = -\frac{1}{1} = -1$, $\left.\frac{dy}{dx}\right|_{\theta=\pi} = \frac{1}{1} = 1$

3. $\frac{dy}{dx} = \frac{f'(\theta)\sin\theta + f(\theta)\cos\theta}{f'(\theta)\cos\theta - f(\theta)\sin\theta} = \frac{-3\cos\theta\sin\theta + (2-3\sin\theta)\cos\theta}{-3\cos\theta\cos\theta - (2-3\sin\theta)\sin\theta} = \frac{2\cos\theta - 6\sin\theta\cos\theta}{-2\sin\theta - 3(\cos^2\theta - \sin^2\theta)}$

$\left.\frac{dy}{dx}\right|_{(2,0)} = \left.\frac{dy}{dx}\right|_{\theta=0} = \frac{2}{-3} = -\frac{2}{3}$, $\left.\frac{dy}{dx}\right|_{(-1,\pi/2)} = \left.\frac{dy}{dx}\right|_{\theta=\pi/2} = \frac{0}{-1} = 0$, $\left.\frac{dy}{dx}\right|_{(2,\pi)} = \left.\frac{dy}{dx}\right|_{\theta=\pi} = \frac{2}{3}$,

and $\left.\frac{dy}{dx}\right|_{(5,3\pi/2)} = \left.\frac{dy}{dx}\right|_{\theta=3\pi/2} = \frac{0}{-5} = 0.$

5.

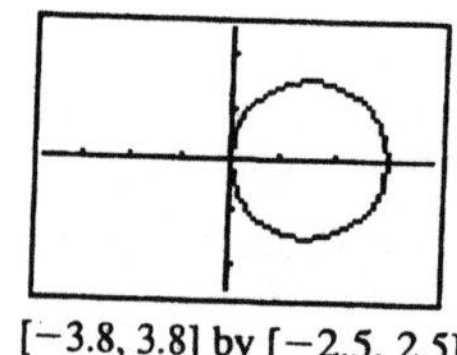

[−3.8, 3.8] by [−2.5, 2.5]

The graph passes through the pole when $r = 3\cos\theta = 0$, which occurs when $\theta = \frac{\pi}{2}$ and when $\theta = \frac{3\pi}{2}$. Since the θ-interval $0 \le \theta \le \pi$ produce the entire graph, we need only consider $\theta = \frac{\pi}{2}$. At this point, there appears to be a vertical tangent line with equation $\theta = \frac{\pi}{2}$ (or $x = 0$). Confirm analytically:

$x = (3\cos\theta)\cos\theta = 3\cos^2\theta$ and $y = (3\cos\theta)\sin\theta$;

$\frac{dy}{d\theta} = (-3\sin\theta)\sin\theta + (3\cos\theta)\cos\theta = 3(\cos^2\theta - \sin^2\theta)$ and $\frac{dx}{d\theta} = 6\cos\theta(-\sin\theta)$. At $\left(0, \frac{\pi}{2}\right)$, $\left.\frac{dx}{d\theta}\right|_{\theta=\pi/2} = 0$, and $\left.\frac{dy}{d\theta}\right|_{\theta=\pi/2} = 3(0^2 - 1^2) = -3$. So at $\left(0, \frac{\pi}{2}\right)$, $\frac{dx}{d\theta} = 0$ and $\frac{dy}{d\theta} \ne 0$, so $\frac{dy}{dx}$ is undefined and the tangent line is vertical.

7.

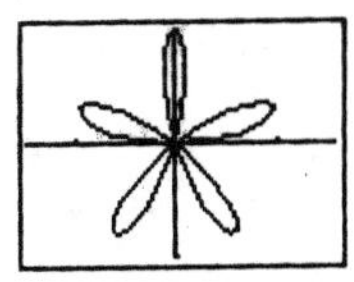

[−1.5, 1.5] by [−1, 1]

The polar solutions are $\left(0, \frac{k\pi}{5}\right)$ for k = 0, 1, 2, 3, 4, and for a given k, the line $\theta = \frac{k\pi}{5}$ appears to be tangent to the curve at $\left(0, \frac{k\pi}{5}\right)$. This can be confirmed analytically by noting that the slope of the curve, $\frac{dy}{dx}$, equals the slope of the line, tan $\frac{k\pi}{5}$. so the tangent lines are $\theta = 0$ [y = 0], $\theta = \frac{\pi}{5}\left[y = \left(\tan \frac{\pi}{5}\right)x\right]$, $\theta = \frac{2\pi}{5}\left[y = \left(\tan \frac{2\pi}{5}\right)x\right]$, $\theta = \frac{3\pi}{5}\left[y = \left(\tan \frac{3\pi}{5}\right)x\right]$, and $\theta = \frac{4\pi}{5}\left[y = \left(\tan \frac{4\pi}{5}\right)x\right]$.

9. $\frac{dy}{d\theta} = \cos\theta \sin\theta + (-1 + \sin\theta)\cos\theta = \cos\theta(2\sin\theta - 1) = \sin 2\theta - \cos\theta$

$\frac{dx}{d\theta} = \cos^2\theta - (-1 + \sin\theta)\sin\theta = \cos^2\theta + \sin\theta - \sin^2\theta = -2\sin^2\theta + \sin\theta + 1$

$\frac{dy}{d\theta} = 0$ when $\theta = \frac{\pi}{2}, \frac{3\pi}{2}$ ($\cos\theta = 0$) or when $\theta = \frac{\pi}{6}, \frac{5\pi}{6}$ ($2\sin\theta - 1 = 0$). $\frac{dx}{d\theta} = 0$ when $\sin\theta = \frac{-1 \pm \sqrt{9}}{-4}$ $= -\frac{1}{2}$ or 1, i.e., when $\theta = \frac{7\pi}{6}, \frac{11\pi}{6}$, or $\frac{\pi}{2}$. So there is a horizontal tangent line for $\theta = \frac{3\pi}{2}$, r = −2 [the line $y = -2\sin\frac{3\pi}{2} = 2$], for $\theta = \frac{\pi}{6}$, $r = -\frac{1}{2}$ [the line $y = -\frac{1}{2}\sin\frac{\pi}{6} = -\frac{1}{4}$] and for $\theta = \frac{5\pi}{6}$, $r = -\frac{1}{2}$ [again, the line $y = -\frac{1}{2}\sin\frac{5\pi}{6} = -\frac{1}{4}$]. There is a vertical tangent line for $\theta = \frac{7\pi}{6}$, $r = -\frac{3}{2}$ [the line $x = -\frac{3}{2}\cos\frac{7\pi}{6} = \frac{3\sqrt{3}}{4}$] and for $\theta = \frac{11\pi}{6}$, $r = -\frac{3}{2}$ [the line $x = -\frac{3}{2}\cos\frac{11\pi}{6} = -\frac{3\sqrt{3}}{4}$]. For $\theta = \frac{\pi}{2}$, $\frac{dy}{d\theta} = \frac{dx}{d\theta} = 0$, but $\frac{d}{d\theta}\left(\frac{dy}{d\theta}\right) = 2\cos 2\theta + \sin\theta = -1$ for $\theta = \frac{\pi}{2}$ and $\frac{d}{d\theta}\left(\frac{dx}{d\theta}\right) = -4\sin\theta\cos\theta + \cos\theta = 0$ for $\theta = \frac{\pi}{2}$, so by L'Hôpital's rule $\frac{dy}{dx}$ is undefined and the tangent line is vertical at $\theta = \frac{\pi}{2}$, r = 0 [the line x = 0]. This information can be summarized as follows.

Horizontal at: $\left(-\frac{1}{2}, \frac{\pi}{6}\right)$ $[y = -\frac{1}{4}]$, $\left(-\frac{1}{2}, \frac{5\pi}{6}\right)$ $[y = -\frac{1}{4}]$, $\left(-3, \frac{3\pi}{2}\right)$ [y = 2]

Vertical at: $\left(0, \frac{\pi}{2}\right)$ [x = 0], $\left(-\frac{3}{2}, \frac{7\pi}{6}\right)$ $[x = \frac{3\sqrt{3}}{4}]$, $\left(-\frac{3}{5}, \frac{11\pi}{6}\right)$ $[x = -\frac{3\sqrt{3}}{4}]$

11. $y = 2\sin^2\theta \Rightarrow \frac{dy}{d\theta} = 4\sin\theta\cos\theta = 2\sin 2\theta$

$x = 2\sin\theta\cos\theta = \sin 2\theta \Rightarrow \frac{dx}{d\theta} = 2\cos 2\theta$

$\frac{dy}{d\theta} = 0$ when $\theta = 0, \frac{\pi}{2}, \pi$, and $\frac{dx}{d\theta} = 0$ when $\theta = \frac{\pi}{4}, \frac{3\pi}{4}$. They are never both zero. For $\theta = 0, \frac{\pi}{2}, \pi$ the curve has horizontal asymptotes at (0,0) [y = 0 sin 0 = 0], $\left(2, \frac{\pi}{2}\right)$ $[y = 2\sin\frac{\pi}{2} = 2]$, and $(0, \pi)$ $[y = 0\sin\pi = 0]$. For

$\theta = \frac{\pi}{4}, \frac{3\pi}{4}$ the curve has vertical asymptotes at $\left(\sqrt{2}, \frac{\pi}{4}\right)$ $[x = \sqrt{2} \cos \frac{\pi}{4} = 1]$ and $\left(\sqrt{2}, \frac{3\pi}{4}\right)$ $[x = \sqrt{2} \cos \frac{3\pi}{4} = -1]$. This information can be summarized as follows.

Horizontal at: $(0,0)$ $[y = 0]$, $\left(2, \frac{\pi}{2}\right)$ $[y = 2]$, $(0, \pi)$ $[y = 0]$

Vertical at: $\left(\sqrt{2}, \frac{\pi}{4}\right)$ $[x = 1]$, $\left(\sqrt{2}, \frac{3\pi}{4}\right)$ $[x = -1]$

13. $A = \int_0^{2\pi} \frac{1}{2}(4 + 2 \cos \theta)^2 \, d\theta = \int_0^{2\pi} \frac{1}{2}(16 + 16 \cos \theta + 4 \cos^2 \theta) \, d\theta = \int_0^{2\pi} \left[8 + 8 \cos \theta + 2\left(\frac{1 + \cos 2\theta}{2}\right)\right] d\theta$

$= \int_0^{2\pi} (9 + 8 \cos \theta + \cos 2\theta) \, d\theta = \left[9\theta + 8 \sin \theta + \frac{1}{2} \sin 2\theta\right]_0^{2\pi} = 18\pi$

15. $A = 2 \int_0^{\pi/4} \frac{1}{2} \cos^2 2\theta \, d\theta = \int_0^{\pi/4} \frac{1 + \cos 4\theta}{2} \, d\theta = \frac{1}{2}\left[\theta + \frac{\sin 4\theta}{4}\right]_0^{\pi/4} = \frac{\pi}{8}$

17. $A = 2 \int_0^{\pi/2} \frac{1}{2}(4 \sin 2\theta) \, d\theta = \int_0^{\pi/2} 2 \sin 2\theta \, d\theta = [-\cos 2\theta]_0^{\pi/2} = 2$

19. $r = 2 \cos \theta$ and $r = 2 \sin \theta \Rightarrow 2 \cos \theta = 2 \sin \theta$

$\Rightarrow \cos \theta = \sin \theta \Rightarrow \theta = \frac{\pi}{4}$; therefore

$A = 2 \int_0^{\pi/4} \frac{1}{2}(2 \sin \theta)^2 \, d\theta = \int_0^{\pi/4} 4 \sin^2 \theta \, d\theta$

$= \int_0^{\pi/4} 4\left(\frac{1 - \cos 2\theta}{2}\right) d\theta = \int_0^{\pi/4} (2 - 2 \cos 2\theta) \, d\theta$

$= [2\theta - \sin 2\theta]_0^{\pi/4} = \frac{\pi}{2} - 1$

21. $r = 2$ and $r = 2(1 - \cos \theta) \Rightarrow 2 = 2(1 - \cos \theta) \Rightarrow \cos \theta = 0$

$\Rightarrow \theta = \pm\frac{\pi}{2}$; therefore $A = 2 \int_0^{\pi/2} \frac{1}{2}[2(1 - \cos \theta)]^2 \, d\theta$

$+ \frac{1}{2}$ area of the circle $= \int_0^{\pi/2} 4(1 - 2 \cos \theta + \cos^2 \theta) \, d\theta + \left(\frac{1}{2}\pi\right)(2)^2$

$= \int_0^{\pi/2} 4\left(1 - 2 \cos \theta + \frac{1 + \cos 2\theta}{2}\right) d\theta + 2\pi$

$= \int_0^{\pi/2} (4 - 8 \cos \theta + 2 + 2 \cos 2\theta) \, d\theta + 2\pi$

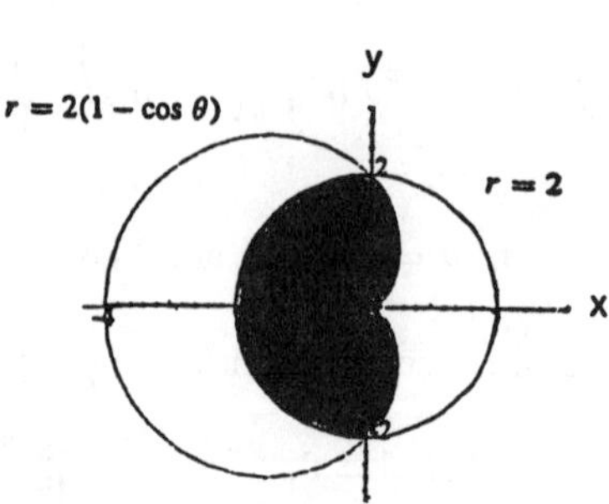

$= \left[6\theta - 8 \sin \theta + \sin 2\theta\right]_0^{\pi/2} + 2\pi = 5\pi - 8$

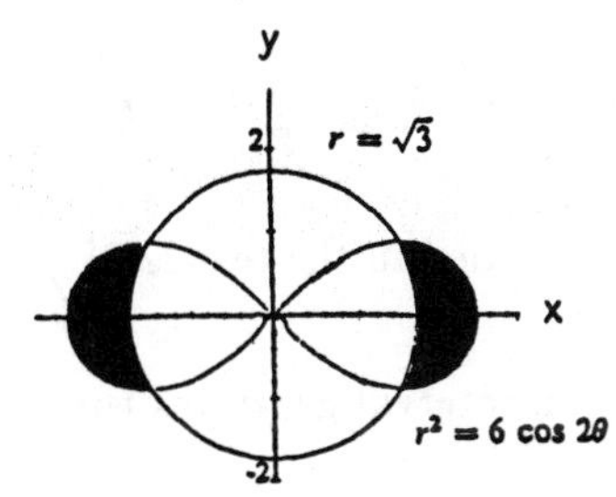

23. $r = \sqrt{3}$ and $r^2 = 6 \cos 2\theta \Rightarrow 3 = 6 \cos 2\theta \Rightarrow \cos 2\theta = \frac{1}{2}$ $\Rightarrow \theta = \frac{\pi}{6}$ (in the 1st quadrant); we use symmetry of the graph to find the area, so $A = 4 \int_0^{\pi/6} \left[\frac{1}{2}(6 \cos 2\theta) - \frac{1}{2}\left(\sqrt{3}\right)^2\right] d\theta$

$= 2 \int_0^{\pi/6} (6 \cos 2\theta - 3)\, d\theta = 2[3 \sin 2\theta - 3\theta]_0^{\pi/6} = 3\sqrt{3} - \pi$

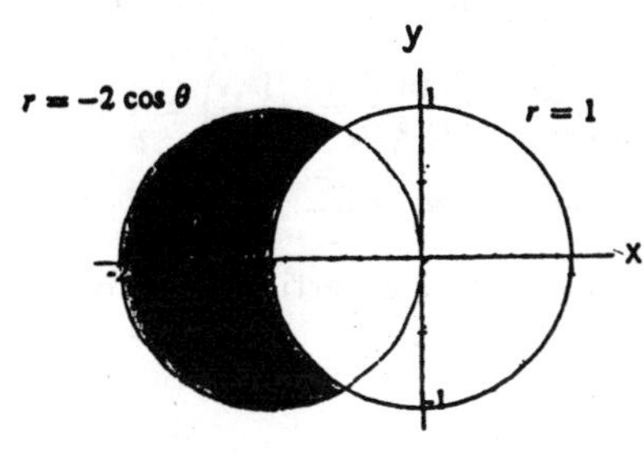

25. $r = 1$ and $r = -2 \cos \theta \Rightarrow 1 = -2 \cos \theta \Rightarrow \cos \theta = -\frac{1}{2}$ $\Rightarrow \theta = \frac{2\pi}{3}$ in quadrant II; therefore

$A = 2 \int_{2\pi/3}^{\pi} \frac{1}{2}[(-2 \cos \theta)^2 - 1^2]\, d\theta = \int_{2\pi/3}^{\pi} \left(4 \cos^2 \theta - 1\right) d\theta$

$= \int_{2\pi/3}^{\pi} [2(1 + \cos 2\theta) - 1]\, d\theta = \int_{2\pi/3}^{\pi} (1 + 2 \cos 2\theta)\, d\theta$

$= \left[\theta + \sin 2\theta\right]_{2\pi/3}^{\pi} = \frac{\pi}{3} + \frac{\sqrt{3}}{2}$

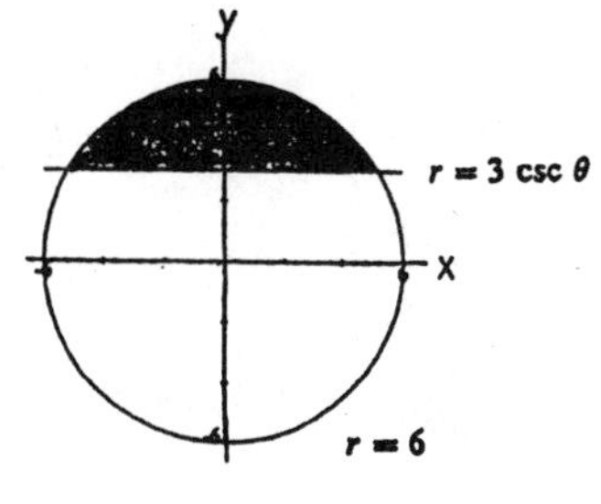

27. $r = 6$ and $r = 3 \csc \theta \Rightarrow 6 \sin \theta = 3 \Rightarrow \sin \theta = \frac{1}{2} \Rightarrow \theta = \frac{\pi}{6}$ or $\frac{5\pi}{6}$; therefore $A = \int_{\pi/6}^{5\pi/6} \frac{1}{2}\left(6^2 - 9 \csc^2 \theta\right) d\theta$

$= \int_{\pi/6}^{5\pi/6} \left(18 - \frac{9}{2} \csc^2 \theta\right) d\theta = \left[18\theta + \frac{9}{2} \cot \theta\right]_{\pi/6}^{5\pi/6}$

$= \left(15\pi - \frac{9}{2}\sqrt{3}\right) - \left(3\pi + \frac{9}{2}\sqrt{3}\right) = 12\pi - 9\sqrt{3}$

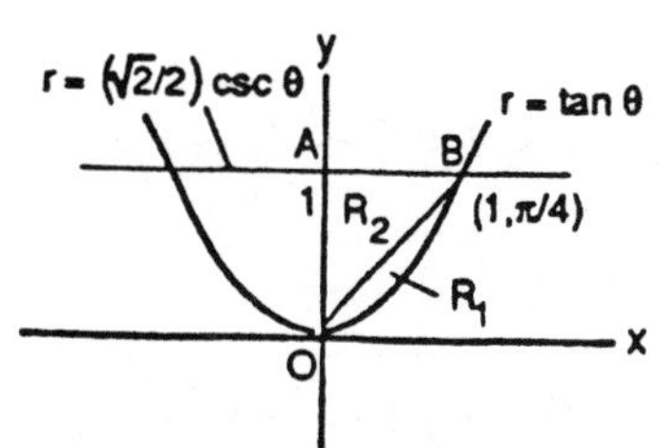

29. (a) $r = \tan \theta$ and $r = \left(\frac{\sqrt{2}}{2}\right) \csc \theta \Rightarrow \tan \theta = \left(\frac{\sqrt{2}}{2}\right) \csc \theta$ $\Rightarrow \sin^2 \theta = \left(\frac{\sqrt{2}}{2}\right) \cos \theta \Rightarrow 1 - \cos^2 \theta = \left(\frac{\sqrt{2}}{2}\right) \cos \theta$ $\Rightarrow \cos^2 \theta + \left(\frac{\sqrt{2}}{2}\right) \cos \theta - 1 = 0 \Rightarrow \cos \theta = -\sqrt{2}$ or $\frac{\sqrt{2}}{2}$ (use the quadratic formula) $\Rightarrow \theta = \frac{\pi}{4}$ (the solution in the first quadrant); therefore the area of R_1 is $A_1 = \int_0^{\pi/4} \frac{1}{2} \tan^2 \theta\, d\theta = \frac{1}{2} \int_0^{\pi/4} \left(\sec^2 \theta - 1\right) d\theta$

$= \frac{1}{2}[\tan \theta - \theta]_0^{\pi/4} = \frac{1}{2}\left(\tan \frac{\pi}{4} - \frac{\pi}{4}\right) = \frac{1}{2} - \frac{\pi}{8}$; $AO = \left(\frac{\sqrt{2}}{2}\right) \csc \frac{\pi}{2} = \frac{\sqrt{2}}{2}$ and $OB = \left(\frac{\sqrt{2}}{2}\right) \csc \frac{\pi}{4} = 1$

$\Rightarrow AB = \sqrt{1^2 - \left(\frac{\sqrt{2}}{2}\right)^2} = \frac{\sqrt{2}}{2} \Rightarrow$ the area of R_2 is $A_2 = \frac{1}{2}\left(\frac{\sqrt{2}}{2}\right)\left(\frac{\sqrt{2}}{2}\right) = \frac{1}{4}$; therefore the area of the region shaded in the text is $2\left(\frac{1}{2} - \frac{\pi}{8} + \frac{1}{4}\right) = \frac{3}{2} - \frac{\pi}{4}$. Note: The area must be found this way since no common interval generates the region. For example, the interval $0 \le \theta \le \frac{\pi}{4}$ generates the arc OB of $r = \tan \theta$ but does not generate the segment AB of the line $r = \frac{\sqrt{2}}{2} \csc \theta$. Instead the interval generates the half-line from B to $+\infty$ on the line $r = \frac{\sqrt{2}}{2} \csc \theta$.

(b) $\lim\limits_{\theta \to \pi/2^-} \tan \theta = \infty$ and the line $x = 1$ is $r = \sec \theta$ in polar coordinates; then $\lim\limits_{\theta \to \pi/2^-} (\tan \theta - \sec \theta)$ $= \lim\limits_{\theta \to \pi/2^-} \left(\frac{\sin \theta}{\cos \theta} - \frac{1}{\cos \theta}\right) = \lim\limits_{\theta \to \pi/2^-} \left(\frac{\sin \theta - 1}{\cos \theta}\right) = \lim\limits_{\theta \to \pi/2^-} \left(\frac{\cos \theta}{-\sin \theta}\right) = 0 \Rightarrow r = \tan \theta$ approaches $r = \sec \theta$ as $\theta \to \frac{\pi}{2}^- \Rightarrow r = \sec \theta$ (or $x = 1$) is a vertical asymptote of $r = \tan \theta$. Similarly, $r = -\sec \theta$ (or $x = -1$) is a vertical asymptote of $r = \tan \theta$.

31. $r = \theta^2$, $0 \le \theta \le \sqrt{5} \Rightarrow \frac{dr}{d\theta} = 2\theta$; therefore Length $= \int_0^{\sqrt{5}} \sqrt{(\theta^2)^2 + (2\theta)^2}\, d\theta = \int_0^{\sqrt{5}} \sqrt{\theta^4 + 4\theta^2}\, d\theta$

$= \int_0^{\sqrt{5}} |\theta| \sqrt{\theta^2 + 4}\, d\theta =$ (since $\theta \ge 0$) $\int_0^{\sqrt{5}} \theta\sqrt{\theta^2 + 4}\, d\theta$; $\left[u = \theta^2 + 4 \Rightarrow \frac{1}{2}\, du = \theta\, d\theta;\ \theta = 0 \Rightarrow u = 4,\right.$

$\left.\theta = \sqrt{5} \Rightarrow u = 9\right] \to \int_4^9 \frac{1}{2}\sqrt{u}\, du = \frac{1}{2}\left[\frac{2}{3}u^{3/2}\right]_4^9 = \frac{19}{3}$

33. $r = 1 + \cos \theta \Rightarrow \frac{dr}{d\theta} = -\sin \theta$; therefore Length $= \int_0^{2\pi} \sqrt{(1 + \cos \theta)^2 + (-\sin \theta)^2}\, d\theta$

$= 2\int_0^{\pi} \sqrt{2 + 2\cos \theta}\, d\theta = 2\int_0^{\pi} \sqrt{\frac{4(1 + \cos \theta)}{2}}\, d\theta = 4\int_0^{\pi} \sqrt{\frac{1 + \cos \theta}{2}}\, d\theta = 4\int_0^{\pi} \cos\left(\frac{\theta}{2}\right) d\theta = 4\left[2 \sin \frac{\theta}{2}\right]_0^{\pi} = 8$

35. $r = \frac{6}{1 + \cos \theta}$, $0 \le \theta \le \frac{\pi}{2} \Rightarrow \frac{dr}{d\theta} = \frac{6 \sin \theta}{(1 + \cos \theta)^2}$; therefore Length $= \int_0^{\pi/2} \sqrt{\left(\frac{6}{1 + \cos \theta}\right)^2 + \left(\frac{6 \sin \theta}{(1 + \cos \theta)^2}\right)^2}\, d\theta$

$= \int_0^{\pi/2} \sqrt{\frac{36}{(1 + \cos \theta)^2} + \frac{36 \sin^2 \theta}{(1 + \cos^2 \theta)^4}}\, d\theta = 6\int_0^{\pi/2} \left|\frac{1}{1 + \cos \theta}\right| \sqrt{1 + \frac{\sin^2 \theta}{(1 + \cos \theta)^2}}\, d\theta$

$= \left(\text{since } \frac{1}{1 + \cos \theta} > 0 \text{ on } 0 \le \theta \le \frac{\pi}{2}\right)\ 6\int_0^{\pi/2} \left(\frac{1}{1 + \cos \theta}\right) \sqrt{\frac{1 + 2\cos \theta + \cos^2 \theta + \sin^2 \theta}{(1 + \cos \theta)^2}}\, d\theta$

$$= 6\int_0^{\pi/2} \left(\frac{1}{1+\cos\theta}\right)\sqrt{\frac{2+2\cos\theta}{(1+\cos\theta)^2}}\,d\theta = 6\sqrt{2}\int_0^{\pi/2} \frac{d\theta}{(1+\cos\theta)^{3/2}} = 6\sqrt{2}\int_0^{\pi/2} \frac{d\theta}{\left(2\cos^2\frac{\theta}{2}\right)^{3/2}} = 6\int_0^{\pi/2} \left|\sec^3\frac{\theta}{2}\right| d\theta$$

$$= 6\int_0^{\pi/2} \sec^3\frac{\theta}{2}\,d\theta = 12\int_0^{\pi/4} \sec^3 u\,du = (\text{use tables})\ 6\left(\left[\frac{\sec u\tan u}{2}\right]_0^{\pi/4} + \frac{1}{2}\int_0^{\pi/4} \sec u\,du\right)$$

$$= 6\left(\frac{1}{\sqrt{2}} + \left[\frac{1}{2}\ln|\sec u + \tan u|\right]_0^{\pi/4}\right) = 3\left[\sqrt{2} + \ln\left(1+\sqrt{2}\right)\right]$$

37. $r = \cos^3\frac{\theta}{3} \Rightarrow \frac{dr}{d\theta} = -\sin\frac{\theta}{3}\cos^2\frac{\theta}{3}$; therefore Length $= \int_0^{\pi/4} \sqrt{\left(\cos^3\frac{\theta}{3}\right)^2 + \left(-\sin\frac{\theta}{3}\cos^2\frac{\theta}{3}\right)^2}\,d\theta$

$$= \int_0^{\pi/4} \sqrt{\cos^6\left(\frac{\theta}{3}\right) + \sin^2\left(\frac{\theta}{3}\right)\cos^4\left(\frac{\theta}{3}\right)}\,d\theta = \int_0^{\pi/4} \left(\cos^2\frac{\theta}{3}\right)\sqrt{\cos^2\left(\frac{\theta}{3}\right) + \sin^2\left(\frac{\theta}{3}\right)}\,d\theta = \int_0^{\pi/4} \cos^2\left(\frac{\theta}{3}\right) d\theta$$

$$= \int_0^{\pi/4} \frac{1+\cos\left(\frac{2\theta}{3}\right)}{2}\,d\theta = \frac{1}{2}\left[\theta + \frac{3}{2}\sin\frac{2\theta}{3}\right]_0^{\pi/4} = \frac{\pi}{8} + \frac{3}{8}$$

39. $r = \sqrt{1+\cos 2\theta} \Rightarrow \frac{dr}{d\theta} = \frac{1}{2}(1+\cos 2\theta)^{-1/2}(-2\sin 2\theta)$; therefore Length $= \int_0^{\pi\sqrt{2}} \sqrt{(1+\cos 2\theta) + \frac{\sin^2 2\theta}{(1+\cos 2\theta)}}\,d\theta$

$$= \int_0^{\pi\sqrt{2}} \sqrt{\frac{1+2\cos 2\theta + \cos^2 2\theta + \sin^2 2\theta}{1+\cos 2\theta}}\,d\theta = \int_0^{\pi\sqrt{2}} \sqrt{\frac{2+2\cos 2\theta}{1+\cos 2\theta}}\,d\theta = \int_0^{\pi\sqrt{2}} \sqrt{2}\,d\theta = \left[\sqrt{2}\,\theta\right]_0^{\pi\sqrt{2}} = 2\pi$$

41. Let $r = f(\theta)$. Then $x = f(\theta)\cos\theta \Rightarrow \frac{dx}{d\theta} = f'(\theta)\cos\theta - f(\theta)\sin\theta \Rightarrow \left(\frac{dx}{d\theta}\right)^2 = \left[f'(\theta)\cos\theta - f(\theta)\sin\theta\right]^2$
$= \left[f'(\theta)\right]^2\cos^2\theta - 2f'(\theta)\,f(\theta)\sin\theta\cos\theta + [f(\theta)]^2\sin^2\theta$; $y = f(\theta)\sin\theta \Rightarrow \frac{dy}{d\theta} = f'(\theta)\sin\theta + f(\theta)\cos\theta$
$\Rightarrow \left(\frac{dy}{d\theta}\right)^2 = \left[f'(\theta)\sin\theta + f(\theta)\cos\theta\right]^2 = \left[f'(\theta)\right]^2\sin^2\theta + 2f'(\theta)f(\theta)\sin\theta\cos\theta + [f(\theta)]^2\cos^2\theta$. Therefore
$\left(\frac{dx}{d\theta}\right)^2 + \left(\frac{dy}{d\theta}\right)^2 = \left[f'(\theta)\right]^2\left(\cos^2\theta + \sin^2\theta\right) + [f(\theta)]^2\left(\cos^2\theta + \sin^2\theta\right) = \left[f'(\theta)\right]^2 + [f(\theta)]^2 = r^2 + \left(\frac{dr}{d\theta}\right)^2$.
Thus, $L = \int_\alpha^\beta \sqrt{\left(\frac{dx}{d\theta}\right)^2 + \left(\frac{dy}{d\theta}\right)^2}\,d\theta = \int_\alpha^\beta \sqrt{r^2 + \left(\frac{dr}{d\theta}\right)^2}\,d\theta.$

43. $r = 2f(\theta),\ \alpha \le \theta \le \beta \Rightarrow \frac{dr}{d\theta} = 2f'(\theta) \Rightarrow r^2 + \left(\frac{dr}{d\theta}\right)^2 = [2f(\theta)]^2 + \left[2f'(\theta)\right]^2 \Rightarrow$ Length $= \int_\alpha^\beta \sqrt{4[f(\theta)]^2 + 4\left[f'(\theta)\right]^2}\,d\theta$

$= 2\int_\alpha^\beta \sqrt{[f(\theta)]^2 + \left[f'(\theta)\right]^2}\,d\theta$ which is twice the length of the curve $r = f(\theta)$ for $\alpha \le \theta \le \beta$.

45. (a) Use the approximation, L_a, from Exercise #45(e). If the reel has made n complete turns, then the angle is $2\pi n$. So from the integral, $L_a = \pi bn^2 + 2\pi r_0 n$. Solving for n gives $n = \left(\frac{r_0}{b}\right)\left(\sqrt{\frac{bL}{r_0^2\pi} + 1} - 1\right)$.

(b) The take up reel slows down as time progresses.

(c) Since L is proportional to time, the formula in part (a) shows that n will grow roughly as the square root of time.

CHAPTER 9 PRACTICE EXERCISES

1. (a) $3\langle -3,4\rangle - 4\langle 2,-5\rangle = \langle -9-8, 12+20\rangle = \langle -17, 32\rangle$

(b) $\sqrt{17^2 + 32^2} = \sqrt{1313}$

3. (a) $\langle -2(-3), -2(4)\rangle = \langle 6, -8\rangle$

(b) $\sqrt{6^2 + 8^2} = 10$

5. $\frac{\pi}{6}$ radians below the negative x-axis: $\left\langle -\frac{\sqrt{3}}{2}, -\frac{1}{2}\right\rangle$ [assuming counterclockwise].

7. $2\left(\frac{1}{\sqrt{4^2+1^2}}\right)(4\mathbf{i} - \mathbf{j}) = \left(\frac{8}{\sqrt{17}}\mathbf{i} - \frac{2}{\sqrt{17}}\mathbf{j}\right)$

9. $\text{length} = \left|\sqrt{2}\mathbf{i} + \sqrt{2}\mathbf{j}\right| = \sqrt{2+2} = 2$, $\sqrt{2}\mathbf{i} + \sqrt{2}\mathbf{j} = 2\left(\frac{1}{\sqrt{2}}\mathbf{i} + \frac{1}{\sqrt{2}}\mathbf{j}\right) \Rightarrow$ the direction is $\frac{1}{\sqrt{2}}\mathbf{i} + \frac{1}{\sqrt{2}}\mathbf{j}$

11. $\frac{d\mathbf{r}}{dt} = (-2\sin t)\mathbf{i} + (2\cos t)\mathbf{j}$; at the point $(0,2)$, $t = \frac{\pi}{2} \Rightarrow \left.\frac{d\mathbf{r}}{dt}\right|_{t=\pi/2} = -2\mathbf{i}$; length $= |-2\mathbf{i}| = 2$;

direction $= -\mathbf{i} \Rightarrow \left.\frac{d\mathbf{r}}{dt}\right|_{t=\pi/2} = 2(-\mathbf{i})$

13. $y = \tan x \Rightarrow [y']_{\pi/4} = [\sec^2 x]_{\pi/4} = 2 = \frac{2}{1} \Rightarrow \mathbf{T} = \mathbf{i} + 2\mathbf{j} \Rightarrow$ the unit tangents are $\pm\left(\frac{1}{\sqrt{5}}\mathbf{i} + \frac{2}{\sqrt{5}}\mathbf{j}\right)$ and the unit normals are $\pm\left(-\frac{2}{\sqrt{5}}\mathbf{i} + \frac{1}{\sqrt{5}}\mathbf{j}\right)$

15.

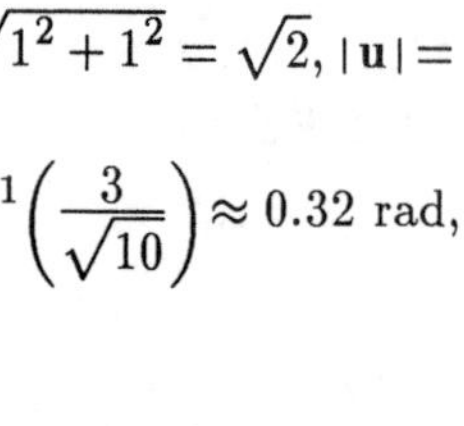

17. $|\mathbf{v}| = \sqrt{1^2+1^2} = \sqrt{2}$, $|\mathbf{u}| = \sqrt{2^2+1^2} = \sqrt{5}$, $\mathbf{u}\cdot\mathbf{v} = \mathbf{v}\cdot\mathbf{u} = 1(2) + 1(1) = 3$, $\theta = \cos^{-1}\left(\frac{\mathbf{u}\cdot\mathbf{v}}{|\mathbf{u}||\mathbf{v}|}\right)$

$= \cos^{-1}\left(\frac{3}{\sqrt{10}}\right) \approx 0.32$ rad, $|\mathbf{u}|\cos\theta = \sqrt{5}\left(\frac{3}{\sqrt{10}}\right) = \frac{3\sqrt{2}}{2}$, $\text{proj}_{\mathbf{v}}\,\mathbf{u} = (|\mathbf{u}|\cos\theta)\left(\frac{\mathbf{v}}{|\mathbf{v}|}\right)$

$$= \left(\frac{3\sqrt{2}}{2}\right)\left(\frac{\mathbf{i}+\mathbf{j}}{\sqrt{2}}\right) = \frac{3}{2}(\mathbf{i}+\mathbf{j})$$

19. Vector component of **u** parallel to **v**: $\text{proj}_{\mathbf{v}}\mathbf{u} = \left(\frac{\mathbf{u}\cdot\mathbf{v}}{|\mathbf{v}|^2}\right)\mathbf{v} = \frac{(1)(2)+(-1)(1)}{2^2+1^2}(2\mathbf{i}-\mathbf{j}) = \frac{2}{5}\mathbf{i} - \frac{1}{5}\mathbf{j}$

Vector component of **u** orthogonal to **v**: $\mathbf{u} - \text{proj}_{\mathbf{v}}\mathbf{u} = (\mathbf{i}+\mathbf{j}) - \left(\frac{2}{5}\mathbf{i} - \frac{1}{5}\mathbf{j}\right) = \frac{3}{5}\mathbf{i} + \frac{6}{5}\mathbf{j}$

Therefore, $\mathbf{u} = \text{proj}_{\mathbf{v}}\mathbf{u} + (\mathbf{u} - \text{proj}_{\mathbf{v}}\mathbf{u}) = \left(\frac{2}{5}\mathbf{i} - \frac{1}{5}\mathbf{j}\right) + \left(\frac{3}{5}\mathbf{i} + \frac{6}{5}\mathbf{j}\right)$.

21. (a) $\mathbf{v}(t) = \frac{d}{dt}\left[(4\cos t)\mathbf{i} + \left(\sqrt{2}\sin t\right)\mathbf{j}\right]$

$$= (-4\sin t)\mathbf{i} + \left(\sqrt{2}\cos t\right)\mathbf{j}$$

$$\mathbf{a}(t) = \frac{d}{dt}\left[(-4\sin t)\mathbf{i} + \left(\sqrt{2}\cos t\right)\mathbf{j}\right]$$

$$= (-4\cos t)\mathbf{i} + \left(-\sqrt{2}\sin t\right)\mathbf{j}$$

(b) $\left|\mathbf{v}\left(\frac{\pi}{4}\right)\right| = \sqrt{\left(-4\sin\frac{\pi}{4}\right)^2 + \left(\sqrt{2}\cos\frac{\pi}{4}\right)^2} = \sqrt{8+1} = 3$

(c) At $t = \frac{\pi}{4}$, $\mathbf{v} = -2\sqrt{2}\mathbf{i} + \mathbf{j}$, $\mathbf{a} = -2\sqrt{2}\mathbf{i} - \mathbf{j}$, and

$$\theta = \cos^{-1}\frac{\mathbf{v}\cdot\mathbf{a}}{|\mathbf{v}||\mathbf{a}|} = \cos^{-1}\frac{8-1}{(3)(3)} = \cos^{-1}\frac{7}{9} \approx 38.94°.$$

23. $\mathbf{v}(t) = -\frac{t}{(1+t^2)^{3/2}}\mathbf{i} + \frac{1}{(1+t^2)^{3/2}}\mathbf{j}$

$\left|\frac{d\mathbf{r}}{dt}\right| = |\mathbf{v}(t)| = \sqrt{\left(-\frac{t}{(1+t^2)^{3/2}}\right)^2 + \left(\frac{1}{(1+t^2)^{3/2}}\right)^2} = \frac{1}{1+t^2}$ which is at a maximum of 1 when $t = 0$.

25. $\left(\int_0^t (3+6t)\,dt\right)\mathbf{i} + \left(\int_0^1 6\pi\cos\pi t\,dt\right)\mathbf{j}$

$$= \left[3t + 3t^2\right]_0^1\mathbf{i} + [6\sin\pi t]_0^1\mathbf{j} = 6\mathbf{i}$$

27. $\mathbf{r}(t) = \int \frac{d\mathbf{r}}{dt}\,dt = (\cos t)\mathbf{i} + (\sin t)\mathbf{j} + \mathbf{C}$

$\mathbf{r}(0) = \mathbf{i} + \mathbf{C} = \mathbf{j}$, so $\mathbf{C} = -\mathbf{i} + \mathbf{j}$, and

$\mathbf{r}(t) = (\cos t - 1)\mathbf{i} + (\sin t + 1)\mathbf{j}$

29. $\frac{d\mathbf{r}}{dt} = \int \frac{d^2\mathbf{r}}{dt^2}\,dt = 2t\mathbf{j} + \mathbf{C}_1$, $\mathbf{r}(t) = \int \frac{d\mathbf{r}}{dt}\,dt = t^2\mathbf{j} + \mathbf{C}_1 t + \mathbf{C}_2$

$\left.\frac{d\mathbf{r}}{dt}\right|_{t=0} = \mathbf{C}_1 = \mathbf{0}$, so $\mathbf{r}(t) = t^2\mathbf{j} + \mathbf{C}_2$. And $\mathbf{r}(0) = \mathbf{C}_2 = \mathbf{i}$, so $\mathbf{r}(t) = \mathbf{i} + t^2\mathbf{j}$

31.

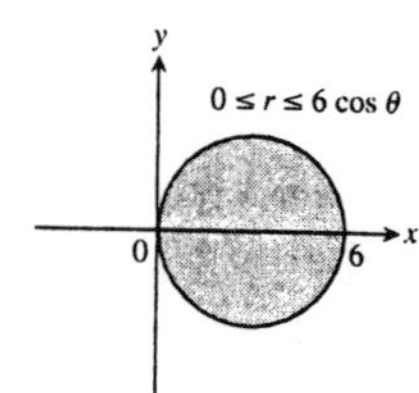

33. d

35. l

37. k

39. i

41. (a)

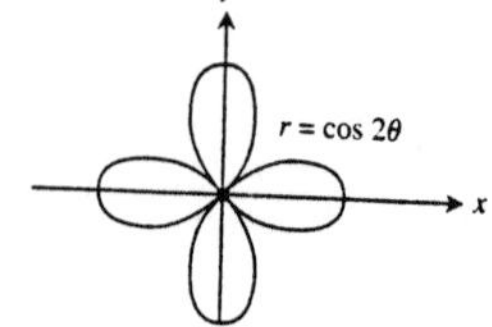

(b) 2π

43. (a)

(b) $\frac{\pi}{2}$

45. $\frac{dy}{dx} = \frac{f'(\theta)\sin\theta + f(\theta)\cos\theta}{f'(\theta)\cos\theta - f(\theta)\sin\theta} = \frac{-2\sin 2\theta\sin\theta + \cos 2\theta\cos\theta}{-2\sin 2\theta\cos\theta - \cos 2\theta\sin\theta}$

$\left(0, \frac{\pi}{4}\right)$, $\left(0, \frac{3\pi}{4}\right)$, $\left(0, \frac{5\pi}{4}\right)$ and $\left(0, \frac{7\pi}{4}\right)$ are polar solutions.

$\left.\frac{dy}{dx}\right|_{\theta=\pi/4} = \frac{-2/\sqrt{2}}{-2\sqrt{2}} = 1$, $\left.\frac{dy}{dx}\right|_{\theta=3\pi/4} = \frac{2/\sqrt{2}}{-2\sqrt{2}} = -1$, $\left.\frac{dy}{dx}\right|_{\theta=5\pi/4} = \frac{2/\sqrt{2}}{2\sqrt{2}} = 1$, $\left.\frac{dy}{dx}\right|_{\theta=7\pi/4} = \frac{-2/\sqrt{2}}{2\sqrt{2}} = -1$.

The Cartesian equations are $y = \pm x$.

47. $\frac{dy}{d\theta} = \frac{d}{d\theta}\left[\left(1 - \cos\left(\frac{\theta}{2}\right)\right)\sin\theta\right] = \frac{1}{2}\sin\left(\frac{\theta}{2}\right)\sin\theta + \cos\theta - \cos\left(\frac{\theta}{2}\right)\cos\theta$

$\frac{dx}{d\theta} = \frac{d}{d\theta}\left[\left(1 - \cos\left(\frac{\theta}{2}\right)\right)\cos\theta\right] = \frac{1}{2}\sin\left(\frac{\theta}{2}\right)\cos\theta - \sin\theta + \cos\left(\frac{\theta}{2}\right)\sin\theta$

Solve $\frac{dy}{d\theta} = 0$ for θ with a graphing calculator: the solutions are 0, ≈ 2.243, ≈ 4.892, ≈ 7.675, ≈ 10.323, and 4π. Using the middle four solutions to find $y = r\sin\theta$ reveals horizontal tangent lines at $y \approx \pm 0.443$ and $y \approx \pm 1.739$. Solve $\frac{dx}{d\theta} = 0$ for θ with a graphing calculator: the solutions are 0, ≈ 1.070, ≈ 3.531, 2π, ≈ 9.035, ≈ 11.497, and 4π. Using the middle five solutions to find $x = r\cos\theta$ reveals vertical tangent lines at $x = 2$, $x \approx 0.067$, and $x \approx -1.104$. Where $\frac{dy}{dt}$ and $\frac{dx}{dt}$ both equal zero ($\theta = 0, 4\pi$), close inspection of the plot shows that the tangent lines are horizontal, with equation $y = 0$. (This can be confirmed using L'Hôpital's rule.)

49.

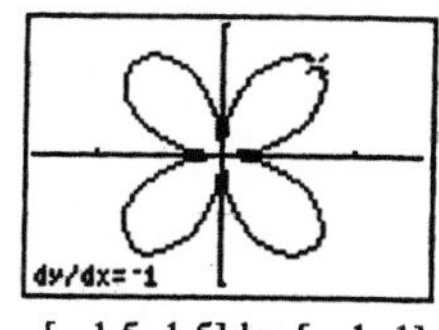

[−1.5, 1.5] by [−1, 1]

The tips have Cartesian coordinates $\left(\frac{1}{\sqrt{2}},\frac{1}{\sqrt{2}}\right)$, $\left(-\frac{1}{\sqrt{2}},\frac{1}{\sqrt{2}}\right)$, $\left(-\frac{1}{\sqrt{2}},-\frac{1}{\sqrt{2}}\right)$, and $\left(\frac{1}{\sqrt{2}},-\frac{1}{\sqrt{2}}\right)$. From the curve's symmetries, it is evident that the tangent lines at those points have slopes of −1, 1, −1, and 1, respectively. So the equations of the tangent lines are

$y-\frac{1}{\sqrt{2}}=-\left(x-\frac{1}{\sqrt{2}}\right)$ or $y=-x+\sqrt{2}$, $y-\frac{1}{\sqrt{2}}=x+\frac{1}{\sqrt{2}}$ or $y=x+\sqrt{2}$,

$y+\frac{1}{\sqrt{2}}=-\left(x+\frac{1}{\sqrt{2}}\right)$ or $y=-x-\sqrt{2}$, and $y+\frac{1}{\sqrt{2}}=x-\frac{1}{\sqrt{2}}$ or $y=x-\sqrt{2}$.

51. $r\cos\theta=r\sin\theta \Rightarrow x=y$, a line

53. $r=4\tan\theta\sec\theta \Rightarrow r\cos\theta=4\,\frac{r\sin\theta}{r\cos\theta} \Rightarrow x=4\frac{y}{x}$ or $x^2=4y$, a parabola

55. $r=2\sec\theta \Rightarrow r=\frac{2}{\cos\theta} \Rightarrow r\cos\theta=2 \Rightarrow x=2$

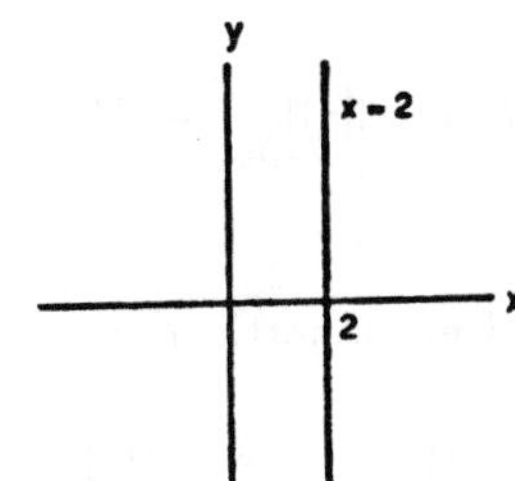

57. $x^2+y^2+5y=0$
$r^2+5r\sin\theta=0$
$r=-5\sin\theta$

59. $x^2+4y^2=16$
$(r\cos\theta)^2+4(r\sin\theta)^2=16$
$r^2\cos^2\theta+4r^2\sin^2\theta=16$, or $r^2=\frac{16}{\cos^2\theta+4\sin^2\theta}$

61. $A=2\int_0^{\pi}\frac{1}{2}r^2\,d\theta=\int_0^{\pi}(2-\cos\theta)^2\,d\theta=\int_0^{\pi}\left(4-2\cos\theta+\cos^2\theta\right)d\theta=\int_0^{\pi}\left(4-2\cos\theta+\frac{1+\cos 2\theta}{2}\right)d\theta$

$$= \int_0^{\pi} \left(\frac{9}{2} - 2\cos\theta + \frac{\cos 2\theta}{2}\right) d\theta = \left[\frac{9}{2}\theta - 2\sin\theta + \frac{\sin 2\theta}{4}\right]_0^{\pi} = \frac{9}{2}\pi$$

63. $r = 1 + \cos 2\theta$ and $r = 1 \Rightarrow 1 = 1 + \cos 2\theta \Rightarrow 0 = \cos 2\theta \Rightarrow 2\theta = \frac{\pi}{2} \Rightarrow \theta = \frac{\pi}{4}$; therefore

$$A = 4\int_0^{\pi/4} \frac{1}{2}[(1+\cos 2\theta)^2 - 1^2]\, d\theta = 2\int_0^{\pi/4} (1 + 2\cos 2\theta + \cos^2 2\theta - 1)\, d\theta$$

$$= 2\int_0^{\pi/4} \left(2\cos 2\theta + \frac{1}{2} + \frac{\cos 4\theta}{2}\right) d\theta = 2\left[\sin 2\theta + \frac{1}{2}\theta + \frac{\sin 4\theta}{8}\right]_0^{\pi/4} = 2\left(1 + \frac{\pi}{8} + 0\right) = 2 + \frac{\pi}{4}$$

65. $r = -1 + \cos\theta \Rightarrow \frac{dr}{d\theta} = -\sin\theta$; Length $= \int_0^{2\pi} \sqrt{(-1+\cos\theta)^2 + (-\sin\theta)^2}\, d\theta = \int_0^{2\pi} \sqrt{2 - 2\cos\theta}\, d\theta$

$$= \int_0^{2\pi} \sqrt{\frac{4(1-\cos\theta)}{2}}\, d\theta = \int_0^{2\pi} 2\sin\frac{\theta}{2}\, d\theta = \left[-4\cos\frac{\theta}{2}\right]_0^{2\pi} = (-4)(-1) - (-4)(1) = 8$$

67. $r = 8\sin^3\left(\frac{\theta}{3}\right),\ 0 \le \theta \le \frac{\pi}{4} \Rightarrow \frac{dr}{d\theta} = 8\sin^2\left(\frac{\theta}{3}\right)\cos\left(\frac{\theta}{3}\right)$; $r^2 + \left(\frac{dr}{d\theta}\right)^2 = \left[8\sin^3\left(\frac{\theta}{3}\right)\right]^2 + \left[8\sin^2\left(\frac{\theta}{3}\right)\cos\left(\frac{\theta}{3}\right)\right]^2$

$$= 64\sin^4\left(\frac{\theta}{3}\right) \Rightarrow L = \int_0^{\pi/4} \sqrt{64\sin^4\left(\frac{\theta}{3}\right)}\, d\theta = \int_0^{\pi/4} 8\sin^2\left(\frac{\theta}{3}\right) d\theta = \int_0^{\pi/4} 8\left[\frac{1-\cos\left(\frac{2\theta}{3}\right)}{2}\right] d\theta$$

$$= \int_0^{\pi/4} \left[4 - 4\cos\left(\frac{2\theta}{3}\right)\right] d\theta = \left[4\theta - 6\sin\left(\frac{2\theta}{3}\right)\right]_0^{\pi/4} = 4\left(\frac{\pi}{4}\right) - 6\sin\left(\frac{\pi}{6}\right) - 0 = \pi - 3$$

69. x degrees east of north is $(90 - x)$ degrees north of east.
Add the vectors:
$\langle 540\cos 10^\circ, 540\sin 10^\circ\rangle + \langle 55\cos(-10^\circ), 55\sin(-10^\circ)\rangle = \langle 595\cos 10^\circ, 485\sin 10^\circ\rangle \approx \langle 585.961, 84.219\rangle$.
Speed $\approx \sqrt{585.961^2 + 84.219^2} \approx 591.982$ mph.
Direction $\approx \tan^{-1}\left(\frac{585.961}{84.219}\right) \approx 81.821^\circ$ east of north

71. Taking the launch point as the origin, $y = (44\sin 45^\circ)t - 16t^2$ equals -6.5 when $t \approx 2.135$ sec (as can be determined graphically or using the quadratic formula). Then $x \approx (44\cos 45^\circ)(2.135) \approx 66.421$ horizontal feet from where it left the thrower's hand. Assuming it doesn't bounce or roll, it will still be there 3 seconds after it was thrown.

73. (a)

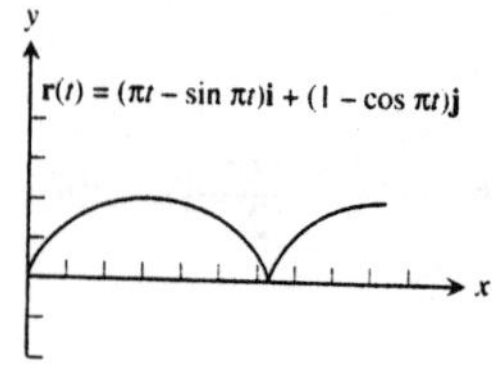

(b) $\mathbf{v}(t) = \left\langle \frac{dx}{dt}, \frac{dy}{dt} \right\rangle = \langle \pi - \pi \cos \pi t, \pi \sin \pi t \rangle$ and $\mathbf{a}(t) = \left\langle \frac{d^2x}{dt^2}, \frac{d^2y}{dt^2} \right\rangle = \langle \pi^2 \sin \pi t, \pi^2 \cos \pi t \rangle$

$\mathbf{v}(0) = \langle 0, 0 \rangle$ $\mathbf{v}(1) = \langle 2\pi, 0 \rangle$ $\mathbf{v}(2) = \langle 0, 0 \rangle$ $\mathbf{v}(3) = \langle 2\pi, 0 \rangle$

$\mathbf{a}(0) = \langle 0, \pi^2 \rangle$ $\mathbf{a}(1) = \langle 0, -\pi^2 \rangle$ $\mathbf{a}(2) = \langle 0, \pi^2 \rangle$ $\mathbf{a}(3) = \langle 0, -\pi^2 \rangle$

(c) Topmost point: 2π ft/sec; center of wheel: π ft/sec
Reasons: Since the wheel rolls half a circumference, or π feet every second, the center of the wheel will move π feet every second. Since the rim of the wheel is turning at a rate of π ft/sec about the center, the velocity of the topmost point relative to the center is π ft/sec, giving it a total velocity of 2π ft/sec.

75. (a) $v_0 = \sqrt{\frac{Rg}{\sin 2\alpha}} = \sqrt{(109.5)(32)} \approx 59.195$ ft/sec

(b) The cork lands at $y = -4$, $x = 177.75$.

Solve $y = -\left(\frac{g}{2v_0^2 \cos^2 \alpha}\right)x^2 + (\tan \alpha)x$ for v_0, with $\alpha = 45°$; $v_0 = \sqrt{-\frac{gx^2}{y - x}} \approx 74.584$ ft/sec

77. We have $x = (v_0 t) \cos \alpha$ and $y + \frac{gt^2}{2} = (v_0 t) \sin \alpha$. Squaring and adding gives

$$x^2 + \left(y + \frac{gt^2}{2}\right)^2 = (v_0 t)^2(\cos^2 \alpha + \sin^2 \alpha) = v_0^2 t^2.$$

79. (a) $\mathbf{r}(t) = \left[(155 \cos 18° - 11.7)\frac{1}{0.09}(1 - e^{-0.09t})\right]\mathbf{i}$

$$+\left[4 + \left(\frac{155 \sin 18°}{0.09}\right)(1 - e^{-0.09t}) + \frac{32}{0.09^2}(1 - 0.09t - e^{-0.09t}\right]\mathbf{j}$$

$x(t) = (155 \cos 18° - 11.7)\frac{1}{0.09}(1 - e^{-0.09t})$

$y(t) = 4 + \left(\frac{155 \sin 18°}{0.09}\right)(1 - e^{-0.09t}) + \frac{32}{0.09^2}(1 - 0.09t - e^{-0.09t})$

(b) Plot $y(t)$ and use the maximum function fo find $y \approx 36.921$ feet at $t \approx 1.404$ seconds.

(c) Plot $y(t)$ and find that $y(t) = 0$ at $t \approx 2.959$ sec, then plug this into the expression for $x(t)$ to find $x(2.959) \approx 352.520$ ft.

(d) Plot $y(t)$ and find that $y(t) = 30$ at $t \approx 0.753$ and 2.068 seconds. At those times, $x \approx 98.799$ and 256.138 feet (from home plate).

(e) No, the batter has not hit a home run. If the drag coefficient k is less than ≈ 0.011, the hit will be a home run. (This result can be found by trying different k-values until the parametrically plotted curve has $y \geq 10$ for $x = 380$.)

81. The widths between the successive turns are constant and are given by $2\pi a$.

CHAPTER 9 ADDITIONAL EXERCISES–THEORY, EXAMPLES, APPLICATIONS

1. (a) Let $a\mathbf{i} + b\mathbf{j}$ be the velocity of the boat. The velocity of the boat relative to an observer on the bank of the river is $\mathbf{v} = a\mathbf{i} + \left[b - \frac{3x(20 - x)}{100}\right]\mathbf{j}$. The distance x of the boat as it crosses the river is related to time by

$x = at \Rightarrow \mathbf{v} = a\mathbf{i} + \left[b - \frac{3at(20-at)}{100}\right]\mathbf{j} = a\mathbf{i} + \left(b + \frac{3a^2t^2 - 60at}{100}\right)\mathbf{j} \Rightarrow \mathbf{r}(t) = at\mathbf{i} + \left(bt + \frac{a^2t^3}{100} - \frac{30at^2}{100}\right)\mathbf{j} + \mathbf{C};$

$\mathbf{r}(0) = 0\mathbf{i} + 0\mathbf{j} \Rightarrow \mathbf{C} = 0 \Rightarrow \mathbf{r}(t) = at\mathbf{i} + \left(bt + \frac{a^2t^3 - 30at^2}{100}\right)\mathbf{j}$. The boat reaches the shore when $x = 20$

$\Rightarrow 20 = at \Rightarrow t = \frac{20}{a}$ and $y = 0 \Rightarrow 0 = b\left(\frac{20}{a}\right) + \frac{a^2\left(\frac{20}{a}\right)^3 - 30a\left(\frac{20}{a}\right)^2}{100} = \frac{20b}{a} + \frac{(20)^3 - 30(20)^2}{100a}$

$= \frac{2000b + 8000 - 12{,}000}{100a} \Rightarrow b = 2$; the speed of the boat is $\sqrt{20} = |\mathbf{v}| = \sqrt{a^2 + b^2} = \sqrt{a^2 + 4} \Rightarrow a^2 = 16$

$\Rightarrow a = 4$; thus, $\mathbf{v} = 4\mathbf{i} + 2\mathbf{j}$ is the velocity of the boat

(b) $\mathbf{r}(t) = at\mathbf{i} + \left(bt + \frac{a^2t^3 - 30at^2}{100}\right)\mathbf{j} = 4t\mathbf{i} + \left(2t + \frac{16t^3}{100} - \frac{120t^2}{100}\right)\mathbf{j}$ by part (a), where $0 \le t \le 5$

(c) $x = 4t$ and $y = 2t + \frac{16t^3}{100} - \frac{120t^2}{100}$

$= \frac{4}{25}t^3 - \frac{6}{5}t^2 + 2t = \frac{2}{25}t(2t^2 - 15t + 25)$

$= \frac{2}{25}t(2t-5)(t-5)$, which is the graph of the cubic displayed here

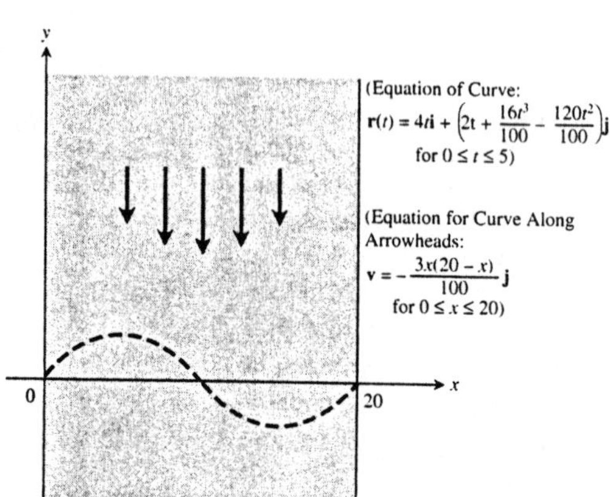

3. $\mathbf{r} = (e^t \cos t)\mathbf{i} + (e^t \sin t)\mathbf{j} \Rightarrow \mathbf{v} = (e^t \cos t - e^t \sin t)\mathbf{i} + (e^t \sin t + e^t \cos t)\mathbf{j}$

$\Rightarrow \mathbf{a} = (e^t \cos t - e^t \sin t - e^t \sin t - e^t \cos t)\mathbf{i} + (e^t \sin t + e^t \cos t + e^t \cos t - e^t \sin t)\mathbf{j}$

$= (-2e^t \sin t)\mathbf{i} + (2e^t \cos t)\mathbf{j}$. Let θ be the angle between $\mathbf{r}$ and $\mathbf{a}$. Then $\theta = \cos^{-1}\left(\frac{\mathbf{r}\cdot\mathbf{a}}{|\mathbf{r}||\mathbf{a}|}\right)$

$= \cos^{-1}\left(\frac{-2e^{2t}\sin t \cos t + 2e^{2t}\sin t \cos t}{\sqrt{(e^t \cos t)^2 + (e^t \sin t)^2}\sqrt{(-2e^t \sin t)^2 + (2e^t \cos t)^2}}\right) = \cos^{-1}\left(\frac{0}{2e^{2t}}\right) = \cos^{-1} 0 = \frac{\pi}{2}$ for all t

5. $9y = x^3 \Rightarrow 9\frac{dy}{dt} = 3x^2\frac{dx}{dt} \Rightarrow \frac{dy}{dt} = \frac{1}{3}x^2\frac{dx}{dt}$. If $\mathbf{r} = x\mathbf{i} + y\mathbf{j}$, where x and y are differentiable functions of t, then $\mathbf{v} = \frac{dx}{dt}\mathbf{i} + \frac{dy}{dt}\mathbf{j}$. Hence $\mathbf{v}\cdot\mathbf{i} = 4 \Rightarrow \frac{dx}{dt} = 4$ and $\mathbf{v}\cdot\mathbf{j} = \frac{dy}{dt} = \frac{1}{3}x^2\frac{dx}{dt} = \frac{1}{3}(3)^2(4) = 12$ at $(3,3)$. Also, $\mathbf{a} = \frac{d^2x}{dt^2}\mathbf{i} + \frac{d^2y}{dt^2}\mathbf{j}$ and $\frac{d^2y}{dt^2} = \left(\frac{2}{3}x\right)\left(\frac{dx}{dt}\right)^2 + \left(\frac{1}{3}x^2\right)\frac{d^2x}{dt^2}$. Hence $\mathbf{a}\cdot\mathbf{i} = -2 \Rightarrow \frac{d^2x}{dt^2} = -2$ and $\mathbf{a}\cdot\mathbf{j} = \frac{d^2y}{dt^2} = \frac{2}{3}(3)(4)^2 + \frac{1}{3}(3)^2(-2) = 26$ at the point $(x,y) = (3,3)$.

7. (a) $x = e^{2t}\cos t$ and $y = e^{2t}\sin t \Rightarrow x^2 + y^2 = e^{4t}\cos^2 t + e^{4t}\sin^2 t = e^{4t}$. Also $\frac{y}{x} = \frac{e^{2t}\sin t}{e^{2t}\cos t} = \tan t$

$\Rightarrow t = \tan^{-1}\left(\frac{y}{x}\right) \Rightarrow x^2 + y^2 = e^{4\tan^{-1}(y/x)}$ is the Cartesian equation. Since $r^2 = x^2 + y^2$ and

$\theta = \tan^{-1}\left(\frac{y}{x}\right)$, the polar equation is $r^2 = e^{4\theta}$ or $r = e^{2\theta}$ for $r > 0$

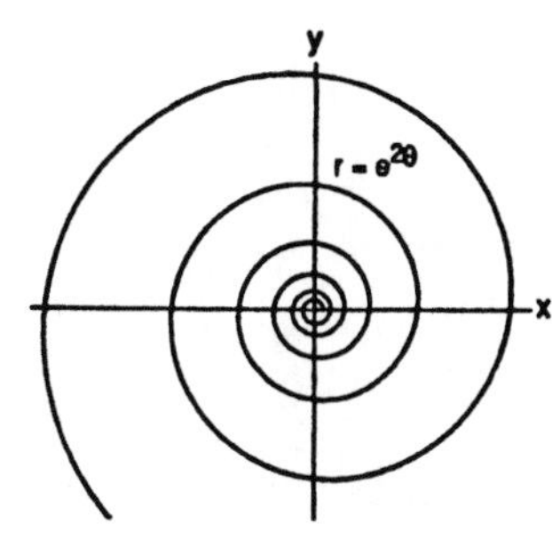

(b) $ds^2 = r^2\, d\theta^2 + dr^2$; $r = e^{2\theta} \Rightarrow dr = 2e^{2\theta}\, d\theta$

$$\Rightarrow ds^2 = r^2\, d\theta^2 + \left(2e^{2\theta}\, d\theta\right)^2 = \left(e^{2\theta}\right)^2 d\theta^2 + 4e^{4\theta}\, d\theta^2$$

$$= 5e^{4\theta}\, d\theta^2 \Rightarrow ds = \sqrt{5}\, e^{2\theta}\, d\theta \Rightarrow L = \int_0^{2\pi} \sqrt{5}\, e^{2\theta}\, d\theta$$

$$= \left[\frac{\sqrt{5}\, e^{2\theta}}{2}\right]_0^{2\pi} = \frac{\sqrt{5}}{2}\left(e^{4\pi} - 1\right)$$

9. The region in question is the figure eight in the middle. The arc of $r = 2a\sin^2\left(\frac{\theta}{2}\right)$ in the first quadrant gives $\frac{1}{4}$ of that region. Therefore the area is $A = 4\int_0^{\pi/2} \frac{1}{2} r^2\, d\theta$

$$= 4\int_0^{\pi/2} \frac{1}{2}\left[2a\sin^2\left(\frac{\theta}{2}\right)\right]^2 d\theta = 8a^2 \int_0^{\pi/2} \sin^4\left(\frac{\theta}{2}\right) d\theta$$

$$= 8a^2 \int_0^{\pi/2} \sin^2\left(\frac{\theta}{2}\right)\left[1 - \cos^2\left(\frac{\theta}{2}\right)\right] d\theta = 8a^2 \int_0^{\pi/2} \left[\sin^2\left(\frac{\theta}{2}\right) - \sin^2\left(\frac{\theta}{2}\right)\cos^2\left(\frac{\theta}{2}\right)\right] d\theta = 8a^2 \int_0^{\pi/2} \left(\frac{1-\cos\theta}{2} - \frac{\sin^2\theta}{4}\right) d\theta$$

$$= 2a^2 \int_0^{\pi/2} \left(2 - 2\cos\theta - \frac{1-\cos 2\theta}{2}\right) d\theta = a^2 \int_0^{\pi/2} (3 - 4\cos\theta + \cos 2\theta)\, d\theta = a^2\left[3\theta - 4\sin\theta + \frac{1}{2}\sin 2\theta\right]_0^{\pi/2}$$

$$= a^2\left(\frac{3\pi}{2} - 4\right)$$

NOTES: